Cultural Studies of Science Education

Volume 14

The series is unique in focusing on the publication of scholarly works that employ social and cultural perspectives as foundations for research and other scholarly activities in the three fields implied in its title: science education, education, and social studies of science.

The aim of the series is to establish bridges to related fields, such as those concerned with the social studies of science, public understanding of science, science/technology and human values, or science and literacy. Cultural Studies of Science Education, the book series explicitly aims at establishing such bridges and at building new communities at the interface of currently distinct discourses. In this way, the current almost exclusive focus on science education on school learning would be expanded becoming instead a focus on science education as a cultural, cross-age, cross-class, and cross-disciplinary phenomenon.

The book series is conceived as a parallel to the journal Cultural Studies of Science Education, opening up avenues for publishing works that do not fit into the limited amount of space and topics that can be covered within the same text. Book proposals for this series may be submitted to the Publishing Editor: Bernadette Ohmer E-mail: Bernadette.Ohmer@springer.com.

More information about this series at http://www.springer.com/series/8286

Larry Bencze
Editor

Science and Technology Education Promoting Wellbeing for Individuals, Societies and Environments

STEPWISE

Springer

Editor
Larry Bencze
Ontario Institute for Studies in Education
(OISE)
University of Toronto
Toronto, ON, Canada

ISSN 1879-7229 ISSN 1879-7237 (electronic)
Cultural Studies of Science Education
ISBN 978-3-319-55503-4 ISBN 978-3-319-55505-8 (eBook)
DOI 10.1007/978-3-319-55505-8

Library of Congress Control Number: 2017943331

Printed on acid-free paper

This Springer imprint is published by Springer Nature
The registered company is Springer International Publishing AG
The registered company address is: Gewerbestrasse 11, 6330 Cham, Switzerland

Acknowledgements

Welcome to the STEPWISE edited collection. This book represents a decade of curriculum development, implementation and research surrounding the *Science and Technology Education Promoting Wellbeing for Individuals, Societies and Environments* programme. There are many people and groups to thank for inspiring and helping to realize this collection. It is difficult to know where to start, given that tacit inspiration seemed to begin many years ago – likely at the time of completion of my bachelor's degree in biology in 1974, a story which is reviewed here in the introductory chapter (2). However, in terms of particular people to thank, I want to begin with Derek Hodson – author of the foreword here (1). He was my Ph.D. doctoral supervisor and early career mentor. He still is inspirational to me, but much impetus for this book came from his 2003 publication – 'Time for Action: Science Education for an Alternative Future'. He also brought tones of critique of power and caring (often for the disadvantaged) that moved me greatly. Much of that inspiration informed an earlier book I helped edit with my colleague, Steve Alsop (York University, Toronto), *Activist Science and Technology Education* (ASTE). Steve is a brilliant and eloquent writer, also committed to social justice and environmental sustainability – and, while he was unable to participate in this book, has contributed to it in important ways. There are, of course, many others to thank who have been part of this programme. I have had the privilege of working with many fine and committed graduate students and teachers, many of whom have contributed to chapters in this book. For their experiences, expertise and dedication to causes of STEPWISE, I am extremely grateful. Much of this project would not, as well, have been possible without collegial support and friendship of the 'academic' authors who have contributed to this book – many of whom have worked with me on various other publications and research grant applications. It is a joy to be part of our critical and activist collective. In terms of collegiality, I must also acknowledge support from editors-in-chief of *Cultural Studies of Science Education* (CSSE) – a journal that aligns very well with ideals of STEPWISE. The annual CSSE forum they sponsor, along with Springer, the publisher of this book, and ASTE, has been a great inspiration to this work, and, related to that, I am extremely grateful to have this book published as part of its CSSE book series with Springer. In this regard, I would like to express particular gratitude to Michael Mueller, the CSSE editor-in-chief who helped facilitate the development and review of this book. About that, I am also grateful to the thoughtful and thorough feedback provided by two anonymous reviewers of this book. At Springer, meanwhile, I want to also express my gratitude to Bernadette Ohmer (senior publishing editor), who has been very supportive of my work and that of many of my closest colleagues and students. Finally, preparation of this book would not have been possible without the loving support of my lifelong partner, Eva, who continues to amaze me with her thoughtful and caring ways.

Dr. John Lawrence Bencze B.Sc., M.Sc., B.Ed., Ph.D.

Contents

Part III Commentaries

Contributors

Jesse Bazzul Faculty of Education, University of Regina, Regina, SK, Canada

Larry Bencze Ontario Institute for Studies in Education (OISE), University of Toronto, Toronto, ON, Canada

Lyn Carter Faculty of Education and Arts, Australian Catholic University, Melbourne, VIC, Australia

Laura Colucci-Gray School of Education, University of Aberdeen, Aberdeen, UK

Interdisciplinary Research Institute on Sustainability, Turin, Italy

Alexander DiGiacomo Ontario Institute for Studies in Education (OISE), University of Toronto, Toronto, ON, Canada

Sarah El Halwany Ontario Institute for Studies in Education (OISE), University of Toronto, Toronto, ON, Canada

Audrey Groleau Département des sciences de l'éducation, Université du Québec à Trois-Rivières, Trois-Rivières, QC, Canada

Angeliki Grundy Toronto District School Board, Toronto, ON, Canada

Derek Hodson University of Auckland, Auckland, New Zealand

Darren Hoeg Department of Secondary Education, University of Alberta, Edmonton, AB, Canada

Mellita Jones Faculty of Education and Arts, Australian Catholic University, Ballarat, VIC, Australia

Shakhnoza Kayumova Department of STEM Education & Teacher Development, University of Massachusetts-Dartmouth, Dartmouth, MA, USA

Mirjan Krstovic Peel District School Board, Mississauga, ON, Canada

Ralph Levinson University College London, Institute of Education, London, UK

Jenny Martin Faculty of Education and Arts, Australian Catholic University, Melbourne, VIC, Australia

Isabel Martins Universidade Federal do Rio de Janeiro, Rio de Janeiro, Brasil

Minja Milanovic Ontario Institute for Studies in Education (OISE), University of Toronto, Toronto, ON, Canada

Tomo Nishizawa Faculty of Education, University of Victoria, Victoria, BC, Canada

Varsha Patel Toronto District School Board, Toronto, ON, Canada

Christina Phillips-MacNeil Ontario Institute for Studies in Education (OISE), University of Toronto, Toronto, ON, Canada

Clayton Pierce Fairhaven College of Interdisciplinary Studies, Western Washington University, Bellingham, WA, USA

Chantal Pouliot Faculté des sciences de l'éducation, Université Laval, Québec City, QC, Canada

Cassie F. Quigley College of Education, Clemson University, Clemson, SC, USA

Neil T. Ramjewan Ontario Institute for Studies in Education (OISE), University of Toronto, Toronto, ON, Canada

Wolff-Michael Roth Faculty of Education, University of Victoria, Victoria, BC, Canada

Ajay Sharma College of Education, University of Georgia, Athens, GA, USA

Jean Simonneaux UMR EFTS, ENSFEA, University of Toulouse, Toulouse, France

Laurence Simonneaux UMR EFTS, ENSFEA, University of Toulouse, Toulouse, France

Erin Sperling Ontario Institute for Studies in Education (OISE), University of Toronto, Toronto, ON, Canada

Matthew Weinstein School of Education, University of Washington, Tacoma, WA, USA

Tanya Williamson Halton District School Board, Burlington, ON, Canada

Brandon Zoras Toronto District School Board, Toronto, ON, Canada

Majd Zouda Ontario Institute for Studies in Education (OISE), University of Toronto, Toronto, ON, Canada

Part I
Preamble

Overview

Welcome to the STEPWISE edited book! It has four main Parts, with this 'Preamble' providing readers with a general introduction to STEPWISE—a curricular and pedagogical framework that encourages students/citizens to 'spend' some of their cultural and social capital, especially in terms of fields of science and technology, on promoting 'wellbeing' for individuals, societies and environments. This Part begins with a 'Foreword' Chap. 1 by Derek Hodson, who was my Ph.D. Supervisor and has been an inspiration to me for years. More particularly, much of STEPWISE was conceived as a response to Derek's 2003 article, entitled *Time for Action: Science Education for an Alternative Future*. He is, therefore, well-positioned to introduce perspectives and practices relating to STEPWISE, including inspiration and guidance regarding needs for socio-political actions to address many of our personal, social and environmental problems relating to fields of science and technology (often with other fields). His Foreword is then followed by my introductory Chap. 2, in which I provide some history of development of the framework(s) and general suggestions for its uses. Prior to writing their chapters, all other authors were provided with an early draft of my introductory chapter—possibly contributing to some consistencies in discussions throughout the book.

The Chapters

Chapter 1
Foreword: The Significance of STEPWISE for Fostering Life-Long Sociopolitical Activism

Derek Hodson

1.1 The Need for Radical Overhaul of Science Education

There is an old saying that optimists believe they live in the best of all possible worlds, while pessimists are afraid they might be right. It is true that we are living in an increasingly complex, rapidly changing, uncertain and challenging world, with extraordinary possibilities and opportunities brought about by scientific discovery and technological innovation. But it is also true that we are faced with rapid growth in social, economic, political and environmental problems at the local, regional and global levels. And it is true that there is massive and growing disparity between societies and within societies in terms of income, access to proper housing, food and water security, educational opportunity, health care, freedom, justice and safety. While many of these disparities are a consequence of geographical location, climatic conditions and levels of political and economic stability, many are also closely linked to the practices of the global industrial complex and the consumerism it promotes, and to restriction of opportunity, persecution and outright violence related to racial, ethno-cultural and religious differences, sexual orientation, values and political leanings. In too many parts of the world, too many people are quick to exploit others and to engage in corrupt practices, unethical conduct of all kinds and both overt and covert distrust and intolerance of those different from themselves. As teachers, teacher educators and educational researchers, we need to ask, as a matter of some urgency, whether our schools, colleges and universities are responding appropriately to these new realities. We need to know whether existing educational priorities are any longer relevant for the new world in which we live. We need to consider whether our educational goals and practices are still relevant. In particular, we need to ask fundamental questions about the overall purpose of education and,

D. Hodson (✉)
University of Auckland, Auckland, New Zealand
e-mail: d.hodson@auckland.ac.nz

L. Bencze (ed.), *Science and Technology Education Promoting Wellbeing for Individuals, Societies and Environments*, Cultural Studies of Science Education 14,
DOI 10.1007/978-3-319-55505-8_1

in the context of this book, about the fundamental purpose of science education. One could argue that there are at least five key purposes for science education at the school level.

- *Economic* purposes – ensuring a steady supply of people with strong backgrounds in science and technology to create and maintain future prosperity for all.
- *Utilitarian* purposes – ensuring that all members of society have sufficient knowledge of science to operate effectively and critically in spheres and activities where science can make a contribution to their personal well-being and quality of life.
- *Personal development* purposes – ensuring that all members of society benefit from the contribution that the analytical skills, investigative strategies and values of science can make to their ability to learn, deal successfully with day-to-day issues and problems and gain rewarding employment.
- *Cultural* purposes – ensuring that all members of society develop a robust understanding of the history, development and contemporary scope of science and scientific practice.
- *Democratic* purposes – ensuring that all students develop sufficient scientific knowledge and skills, familiarity with scientific language and argumentation, capability and confidence in appraising scientific reports and media literacy to be active participants in debate and decision-making about scientific and technological issues.

So how do our schools measure up? Are we doing enough to meet the challenges of the contemporary world in terms of these five purposes? In most cases, the answer is a resounding "No". In most cases, priority is afforded to purposes higher in this list, with insufficient, little or no attention given to those lower in the list. In most cases, there is way too much emphasis on preparing students for later study of science or subsequent employment in science-based careers and little or nothing to prepare them for responsible and active citizenship. Over the past two decades, there has been major emphasis on competition between students rather than cooperation and collaboration among them, way too much emphasis on pre-specified and highly detailed (but often educationally trivial) learning outcomes, way too much emphasis on rigorous and systematic testing for so-called educational standards and way too much emphasis on teacher-centred pedagogy. In consequence, many students have been led to distrust and devalue their own knowledge, skills, values and experiences; they look to experts as the source of all views, solutions to problems and decisions on key issues. It is a matter of some urgency that we shift our priorities. As John Dewey is reputed to have said: "If we teach today as we taught yesterday, we rob our children of tomorrow".

If we are to meet the cultural and democratic goals above, we need a school science curriculum that: (i) ensures all students acquire a robust understanding of scientific knowledge, scientific practice and the language, norms of behaviour and values that guide scientists in their work; (ii) equips all students with the knowledge, skills and confidence to make judgements and reach decisions on the complex socioscientific issues (SSI) that confront them; (iii) develops their capacity to deal with change, uncertainty and unpredictability; (iv) cultivates their ability to ascertain what is desirable/undesirable and what is possible in both the long and short terms; (v) pays much more attention than has been usual in the past to values issues in the deployment of scientific developments and technological innovations and to the active promotion of democracy and social justice; and (vi) prepares students for taking direct and indirect action in pursuit of changes they consider desirable, both individually and collectively. Such is the curriculum advocated in this important book.

1.2 How STEPWISE Can Play a Key Role

Advocating a radical shift of curriculum emphasis is the easy part; developing a rigorous framework that enables teachers to build and implement such a curriculum is a much more demanding task. This is the task that Larry Bencze and his colleagues and co-workers undertook in establishing STEPWISE, the acronym for *Science and Technology Education Promoting Wellbeing for Individuals, Societies and Environments*. In this timely and ground-breaking book, reflecting several years of careful research and development, a rigorous framework for a curriculum that prioritizes the development of what some have called *civic scientific literacy* in pursuit of a better world is clearly articulated and extensively theorized (principally, through actor-network theory and notions of cultural capital). Case studies and personal experiences are outlined, appropriate teaching and learning strategies and resources are discussed, successes are celebrated and shortcomings are acknowledged. The book serves two major purposes. First, it makes a powerful case for the adoption of an SSI-based curriculum culminating in preparation for, and experience of, sociopolitical action. Second, and more importantly, it provides some sound, research-validated advice for how such a curriculum can be organized and implemented.

When teachers are presented with a significantly different way of thinking about science education, they need time to address mismatches between old and new practice, resolve conflicts, establish new priorities and build personal and practical schemes of action. This is no easy task. Teachers need time to acquire new knowledge and skills to be incorporated into new classroom practices, and they need encouragement and support in doing so. They need time to resolve conflicts, clarify uncertainties, deal with anxieties and cope with the increased stress, pressure and feelings of vulnerability that inevitably ensue. To respond appropriately and effectively to the kind of radical change advocated in this book, teachers will even need to reconsider what it means to be a science teacher and seek to establish a new identity within the community of practitioners. At the very least, they will need to shift their view of a teacher's prime responsibility: from a dispenser of information and knowledge to a facilitator of learning. They will need to put much more emphasis on development of reasoning skills than on knowledge acquisition, shift from closed and authoritative classroom discourse to open and dialogic interactions, and change from classroom, laboratory and field activities that are selected and designed by the teacher to activities that are owned, planned and directed by the students. Teachers' views of learning are crucial to effecting these changes. Those who subscribe to a transmission of knowledge view of learning are inclined to regard problem solving as a matter of applying rules and algorithms, they prefer to lower the level of cognitive demand, spoon-feed students with correct answers, utilize standard problems that are amenable to routine and carefully focused approaches, while avoiding problems that require students to adopt an independent approach, largely because they believe students will find them difficult, frustrating or confusing, or because they are reluctant to cede any measure of classroom control to students. In contrast, those teachers who view learning as knowledge construction make greater cognitive demands on students and encourage more independent thinking. Put bluntly, making the curriculum changes advocated in this book may require some teachers to radically shift their views about learning and radically change their classroom practice. Doing so is likely to be extraordinarily stressful and support from others engaged in this same educational reform is essential. It is here that this book and the insight it provides through critical accounts of a large number of case studies can play a key role.

It is highly unlikely that a set of guidelines issued by the Ministry/Department of Education or School Board, or a one-off teacher workshop organized by an in-service education provider, will bring about the level of professional knowledge, expertise, confidence and emotional support needed. These attributes are best acquired and consolidated through experience, critical reflection and the interest, cooperation, advice and support of other teachers with whom ideas, feelings, successes and failures can be shared. It is here that the real strengths of the STEPWISE project become evident. First, the project recognizes that if we are to politicize students we need to politicize teachers, too. Because teachers hold a pivotal position between the state, parental influence, media power and the dictates of institutional norms, they have enormous opportunities to foster development of democratic values and influence attitudes of students. Using those opportunities wisely, effectively and responsibly requires a substantial measure of political awareness. Teachers need to turn critical spotlights on schooling, curriculum, teaching and learning, assessment and evaluation strategies, on knowledge, beliefs, values and aspirations that underpin them, and on ways in which key curriculum decisions are made – in particular, by whom they are made and in whose interests they are made. If critical thinking, creativity and skilful problem solving are to be developed by students, it is essential that those who are responsible for that education also possess and practise these attributes. If we want students to address SSI in a critical and independent way, their teachers must be afforded opportunities to do so, too. If we want students to experience sociopolitical actions in pursuit of ideas and practices they consider important, and against those that they consider undesirable, then teachers need to have had these experiences. It is absurd to expect teachers to create the necessary experiences for students to develop these abilities if they, themselves, have not had similar experiences. It is unrealistic to expect students to have confidence in their own knowledge, skills and judgement (i.e., to be intellectually independent) if their teachers have been socialized into blind acceptance of the views and decisions of others. It is unrealistic to expect students to recognize the value of collaborative learning and collaborative action if their teachers are denied opportunities to work collaboratively with their teaching colleagues. STEPWISE provides such opportunities for teachers. It creates opportunities to build the professional learning communities that can sustain and support teacher development and curriculum development over time. It is in this kind of forum that teachers can share 'war stories' and reflect on their triumphs, difficulties, anxieties and failures. Sharing stories about classroom experiences, curriculum building issues and engagement in sociopolitical action are important sources of learning for other teachers. Through narrative, we begin to understand the actor's reasons for the action, and are thereby encouraged to make sense of these actions through the eyes of the actor and to see how they might relate to situations in which we find ourselves. This understanding constitutes an enormous contribution to learning about and getting better at teaching.

The basic approach of STEPWISE can be briefly summarized as follows: (i) encourage students to express and discuss their current understanding of important SSI, either chosen by the teacher or by the students; (ii) discuss and enrich the science content knowledge that informs a better understanding of the issue, including whatever contemporary research information is available; (iii) consider the social, political, economic and ethical considerations underpinning the issue; (iv) provide opportunities for students to conduct their own research into the issue; (v) consider the kinds of interventions, both direct and indirect, that are relevant and possible; (vi) engage in action at the individual and/or group level; (vii) evaluate the effectiveness and appropriateness of the action(s) taken, and communicate the findings and critique to others. The editor (Larry Bencze) and the numerous cooperating authors and researchers never claim that following the STEPWISE framework is the only way of approaching an SSI-based and action-oriented curriculum. Indeed, there is substantial variation across the chapters in Part II of the book in terms of the underlying rationale articulated by the teacher/researcher, the teaching and learning strategies deployed, the research projects followed by the students and interventions they propose. Further, in Part III of the book, Ralph Levinson (Chap. 22) describes the work of a consortium of science teacher educators in universities across Europe involved with development and implementation of PARRISE (Promoting Attainment of Responsible Research and Innovation in Science Education), a project directed towards inquiry-based learning focused on SSI, while Laurence and Jean Simonneaux (Chap. 27) address an initiative in French schools to involve students in critical scrutiny of *Questions Socialement Vives* (Socially Acute Questions). Both projects are concerned with relationships among the sciences, societal issues, politics and business; both projects have some elements in common with STEPWISE; both projects can inform and enrich other SSI-oriented approaches. Thus, the book is not so much a 'How to do it' book as an opportunity to consider the STEPWISE approach, critique it, use it, modify it to personal circumstances or reject it. Any teacher with an interest in an SSI-rich and action-oriented curriculum would gain a great deal from reading this book: possibilities for curriculum organization, teaching and learning strategies, evidence of successes in terms of impact on students, recognition of teething troubles and encouragement from reading the personal stories of others who have used the approach.

Many of the chapters in the book illustrate the strength of apprenticeship models of learning: first, teachers model the desired behaviour, strategy or skills; second, the learners engage in carefully structured activities with the encouragement, advice and support of the teacher; third, the learners are afforded the opportunity to design, implement and evaluate their own learning activity, investigation or action project. There is ample evidence in this book of the effectiveness of this approach in both preservice and inservice education of teachers and in school curricula that these teachers provide for their students. There is ample evidence that apprenticeship models of learning stimulate the transition from a top-down style of curriculum design and an authoritative style of teaching to a teacher/student-owned curriculum and a participatory, cooperative and democratic style of learning. Returning to an earlier point, a top-down approach to development of an SSI-oriented curriculum, that would likely entail the use of centrally-generated curriculum materials or even the use of an approved textbook (a common practice in North America), would reinforce notions that SSI can be clarified and decisions about appropriate responses and actions reached simply by applying a set of guidelines located in the text. Not only would this approach invite students to be passive consumers of someone else's knowledge, it would tell teachers that the solution to the educational problems of teaching about SSI is located in those materials and in the expertise of others, thus fostering conformity and passivity in both students and teachers. Adopting a top-down approach to an SSI-oriented education that aims to inform students about and introduce them to appropriate sociopolitical action would be the ultimate irony. If we advocate a form of education that aims to enhance critical thinking and decision-making, and with encouraging personal action among science students, it would be absurd to deny science teachers opportunities to utilize these same critical thinking and decision-making skills to develop their own teaching practices. It is not unreasonable to assert that a curriculum that aims to achieve a critical scientific and technological literacy should be based on a model of curriculum development that seeks to encourage and support teachers in becoming critically literate about their own educational practice. By ceding a substantial measure of control of the curriculum to teachers, the STEPWISE project creates the sense of ownership essential to effective, long-lasting and radical change. In short, the approach is predicated on the notion that teachers can (and must) become significant curriculum makers rather than mere implementers of curricula designed by others.

1.3 Addressing Anxieties and Overcoming Problems

It is important to note that although the prospect of implementing an SSI-based, action-oriented curriculum may be daunting for some teachers, there is ample evidence that we all get better at it by doing it and by reflecting on our experiences. STEPWISE, and this book in particular, has done teachers a great service by compiling detailed accounts of experiences, successes and failures of teachers seeking to engage students in action-oriented and/or community-based projects. Similarly, students who are apprehensive about engaging in direct or indirect sociopolitical action, whether individually or in groups, will be reassured, motivated and inspired when they hear or read about activist work by other students. We can all learn a great deal from these accounts of actual practice – that is, learning through the eyes and ears of others, reading about hopes, fears and anxieties of real people, learning of difficulties, surprises and often messy details of intervention projects, being made aware of the likelihood of opposition and obstruction, and of strategies for dealing with them. Those who engage in this radical new form of science education share a common obligation to contribute to this oral history. Teachers, researchers, teacher educators and students can compile this history by asking questions such as: How did the project come about? What motivated and inspired you to undertake this project? Who were the people involved? What were your goals, hopes and expectations? What barriers and problems were encountered? How were they addressed? What successes have there been? What failures? What have you learned? What surprised, delighted or disappointed you? What would you do differently if you were starting again? Would you do it again? Common sense tells us that not all projects will be successful in promoting, developing and sustaining an activist stance among all students. We need to recognize that some failures are inevitable and we need to develop the evaluation skills to ascertain possible causes of failure, seek to modify our approach in the light of critical feedback and have the courage and resilience to try again.

On a related matter, there is an ever-present danger that actions can reflect the teacher's agenda rather than the interests and concerns of the students, and a danger that students merely go through the motions of engaging in action, without any real commitment or sense of empowerment, in order to satisfy the course requirements or meet the expectations of the teacher. At the extreme, teachers may be led to compile a list of approved, scripted and politically safe actions in which to engage successive groups of students without ever engaging them in the critical debate that should precede and determine action. Students may be directed towards uncontroversial issues, guided away from conflict, dissuaded from political debate and censored when their proposals seem likely to challenge school practices, local government policies or the interests of local businesses. In short, both teachers and the students they teach may be anxious not to 'rock the boat' or 'go against the grain' of what they perceive to be public opinion. Indeed, much of what is advocated for the science curriculum in this book may be disturbing to some science teachers and Ministry or Department of Education officials, and so is likely to meet some stern resistance from those who favour a more traditional approach. There are people (including some teachers) who will argue that politicization is not a legitimate goal of science and technology education, or of any school-based education, for that matter, and that sociopolitical action has no place in school. They may be concerned that students will express views and values at variance with those prevailing in the local community. Being critical of forestry clear-cutting, production and promotion of gas-guzzling SUVs and environmentally destructive mining practices in communities where many of the residents are employed in the forestry, automobile-building and mining industries is likely to stir up considerable opposition. There are many teachers, educational administrators and members of the wider community who will perceive the capacity for effecting social change located in a body of students who are scientifically literate, environmentally aware, socially critical, and perhaps most controversially of all, politically literate, as a threat rather than a benefit – a threat to the established order of power and control. Indeed, the very success of such an education is likely to draw opposition. Avoiding controversial issues, especially those with significant political dimensions, and avoiding engagement in soiopolitical action may be regarded by many teachers as taking a neutral view. In reality, it is not neutral. Because it fails to confront and challenge the underlying sociopolitical causes of environmental problems and social inequities, for example, it implicitly supports current social practices, current institutions and current values. Thus, it has to be regarded as education for social reproduction. There is no such thing as political non-involvement. Non-involvement is, in itself, a form of involvement by default and constitutes implicit support for the dominant ideology. Avoiding political matters is, in effect, leaving it for others to decide. STEPWISE has the power to enable all students to reach this understanding early enough to ensure lifelong sociopolitical awareness.

Even teachers who express an interest in teaching about SSI may not implement such a curriculum, citing lack of time to plan lessons and prepare materials capable of integrating coverage of content with social concerns, economic considerations and moral-ethical dilemmas. Others may cite difficulties associated with design of assessment and evaluation strategies. Conventional assessment methods do not cope well when there is no clearly defined outcome, no certain and unambiguous solution; when the curriculum is extended to include sociopolitical action, evaluation is as much about what the community learns from the activity, or how it is changed, as it is about what the students learn. Clearly, much work will be needed to develop appropriate assessment and evaluation strategies if an issues-based and action-oriented curriculum is to become a reality. Some teachers will be concerned that a shift from the supposed certainties of science to the uncertainties of SSI-oriented teaching will constitute a threat to their classroom authority and to their role as 'gatekeepers of scientific knowledge'. Some will claim that they lack resources for addressing SSI, although this is patently untrue: newspapers, television reports and Internet websites abound with suitable material. What these teachers are really claiming is that they do not have access to carefully constructed instructional materials that meet the specifications of the official curriculum. Given the track record of such materials in seeking to promote particular political and economic ideologies or inculcate attitudes and codes of approved behaviour derived from particular social, political, cultural and religious beliefs and the dictates of the globalized consumerist society, it is perhaps no bad thing that teachers lack ready-made official resources. Indeed, we all need to be constantly aware of dangers of large corporations and government agencies responding to teachers' concerns about lack of materials by producing glossy and user-friendly materials that use sophisticated communications techniques to promote particular positions or points of view that are difficult to counter with relatively unsophisticated curriculum materials generated by teachers themselves. STEPWISE has done teachers a major service by showing how they can produce their own resources and design their own classroom activities.

1.4 Defending Charges of Bias and Indoctrination

It is inevitable that some teachers will lack confidence and expertise in handling unstructured, open-ended discussions and it is unsurprising that teachers unfamiliar with such an approach commonly express a concern, bordering on anxiety, that they will be accused of bias, and may possibly lay themselves open to charges of indoctrination or preaching about activism rather than teaching about it, as Mellita Jones comments in Chap. 23. I would make two points in response. First, when teachers present and/or support students in gathering multiple perspectives on an issue or argument, assist them in identifying, clarifying and challenging the assumptions of *all* positions (including the teacher's), acknowledge influences of sociocultural contexts, religious beliefs, emotions and feelings, address issues of rationality, equity and social justice, and encourage critical reflection, they are free of such charges. We are guilty of indoctrination when, and only when, we intend students to believe a proposition in the absence of evidence or despite/regardless of evidence to the contrary, or when we deliberately distort the evidence. We are guilty of preaching when we fail to take action ourselves and fail to create opportunities for our students to do so. What is promoted via STEPWISE is best described as *adopting a critical perspective*. Second, the views of students often indicate the exact opposite, with many of the students with whom I have worked expressing the view that confronting SSI in this critical and collaborative way "opened my eyes to other perspectives", "helped me to sort out my own views" and "enabled me to think more clearly and more carefully about such matters".

Some teachers may be uncertain about what constitutes appropriate, acceptable and worthwhile action. Many questions spring to mind. Who decides what is acceptable action and responsible action? What are relevant criteria? What are balances to be drawn between socially acceptable actions that may be politically ineffective and effective actions that may be socially unacceptable – at least to some? Will teachers be prepared to support student actions that provoke disapproval from parents, school administrators, local politicians or local businesses? Will they be prepared for a situation in which students who are well-coached in action skills choose to direct those skills against aspects of the institution in which they study and/or the community in which they live? We need courage and the encouragement and support of others if we are to tread this path. The war stories in this book provide a great deal in the way of encouragement and support.

The more radical and critical stance towards science, scientists and scientific and technological practice embodied in the STEPWISE approach is in direct conflict not only with the traditional school model of science but also with the image that universities and the science professions have tended to promote. Thus, there may be opposition to its implementation from scientists and from universities. There may be opposition from parents, some of whom may regard this approach as a soft option to 'proper science' (i.e., abstract, theoretical science taught by traditional teacher-centred methods and assessed by conventional means) and/or consider that an SSI-oriented approach takes too much time and diverts attention away from content aspects of the curriculum, thus diluting the science curriculum, reducing science to the level of social science and alienating those with a passion for science. There may also be assertions that the social, economic, historical, political and moral-ethical dimensions of SSI are likely to be poorly addressed because science teachers lack expertise in these areas. Some may argue that this will lead to lower levels of job satisfaction for science teachers, loss of experienced teachers from the profession and a decline in science teacher recruitment. Change may even be resisted by students, especially by the more academically successful ones. They, too, have expectations of science lessons and vested interests in maintaining classroom practices that have served them well in the past. As noted above, navigating these multiple resistances to change will require considerable courage and determination, and high levels of support and encouragement. Teachers may find encouragement in accounts of teachers engaged in similar efforts to overthrow the stultifying shackles of convention, as in the stories told in the chapters of this book. The real breakthrough comes when individual teachers are able to find and work with like-minded colleagues to form pressure groups that can begin to influence key decision-making bodies. Success in deconstructing the hierarchy of subject-disciplines and the tyranny of centrally-mandated curriculum depends on the extent to which the goals and values of the reformers become broadly accepted within the teaching profession and supported by the community. Perhaps teachers, teacher educators, researchers and committed students need to develop educational equivalents of public forums (consensus panels, citizen juries, focus groups, and the like) that have been used by scientists, governments and NGOs to directly engage the public.

1.5 Turning the Critical Spotlight on Ourselves

In keeping with principles underpinning STEPWISE, there are a few chapters in the book that are critical of the approach, though couched in language and terms that seek to help the project extend its influence and enhance its effectiveness. For example, Ajay Sharma notes in Chap. 31 that actions and interventions advocated or engaged in by the students involved in the STEPWISE project are generally individual actions (or small classroom group actions) conducted at the local level, whereas most important SSI exist on both the local and the regional, national and international levels and are only likely to be significantly impacted by collective actions. He also points out that many of the environmental problems we face are largely social problems rather than scientific ones, but the STEPWISE framework is focused exclusively on science education. It would be advisable, he argues, for the STEPWISE framework to integrate school science more closely with social studies. In Chap. 28, Matthew Weinstein asks to what extent STEPWISE can hope to achieve its goals when it is an integral part of the powerful technoscientific structure it seeks to critique and aims to change. The problem is, Weinstein argues, that STEPWISE is, in many of its innovative elements (including social, political and moral-ethical considerations within the science curriculum, rigorous questioning of authority, promoting sociopolitical action, and so on) a resistance to traditional school science education. Yet, in other ways, it is located within the formal structure of school-based education (subjects, assessment schemes, specific groupings of students, etc.). Perhaps, he implies, STEPWISE would be better located within the informal education movement. Wolff-Michael Roth (Chap. 32) also argues for a deinstitutionalization of science education because, he argues, in life we do not engage in science, technology, history or mathematics separate from the concerns arising from life, unless we are working in a profession that is explicitly associated with one of these domains (e.g., as mathematician, physicist, or historian). Real life issues, he reminds us, are messy and science and technology can never provide the sole or correct answer.

I feel that a brief response to these remarks might be appropriate here. It is evident in most chapters of the book that as soon as students begin to address real world problems they encounter layers of increasing complexity and uncertainty that cannot be contained within a particular disciplinary framework. They quickly recognize that problems related to SSI are inextricably linked with considerations in economics, politics, cultural location, aesthetics and moral philosophy. They recognize how dangerously misleading it is to suggest that science or technology can solve problems by simple technical means. They quickly recognize that environmental problems, for example, are pre-eminently social problems – problems of people, their lifestyles and their relations with the natural world. They cannot be solved by a simple scientific or technological fix. Indiscriminate clearance of tropical rainforest for non-forest use has brought about local problems of erosion, floods and fuel wood shortage, and global problems related to global warming, climate change and loss of biodiversity. While science provides an understanding of values of forests and raises concerns about problems, it does not contribute much to solving them. Solutions will be found, if at all, by dealing with issues relating to poverty (at individual and regional levels), patterns of land ownership, exploitation of the poor and powerless by the rich and powerful, the often unjust terms of international trade deals and the stresses of burgeoning populations. A crucial step in environmental education is helping students to recognize that environments are socially constructed: first, in the sense that we act upon and change natural environments and, so, construct and reconstruct them through our social actions; second, in the sense that we perceive environments in ways that are dependent on prevailing socio-cultural frameworks. In both senses, environments could be constructed differently. If environment is a social construct, environmental problems are not natural and inevitable; rather, they are social problems caused by societal practices and structures (how we construct/reconstruct environments) and justified by societies' current values (how we perceive the environment). It follows that solving environmental problems means addressing and changing the social conditions that give rise to them and the values that sustain them. Thus, science education for sociopolitical action is inescapably an exercise in values clarification and values change. Environmental problems will not just go away, nor will they be solved by a quick technical fix while we blithely maintain our profligate lifestyle. We have to change the way we live; the planet can no longer sustain our present (Western) way of life. Changing our way of life entails changing our values. Acid rain, global climate change, toxic waste, ozone depletion, loss of biodiversity, increasing deforestation and desertification are all located in our impoverished values. It is now a well-worn cliché to say that we live in a global village, and that what we do in our own backyard can impact quite significantly on people living elsewhere in the world. We all need to be cognizant of this reality and to act in a way that minimizes adverse impact on others. What is also true is that our actions now impact on the lives of future citizens. It is not too much of an exaggeration to say that the degree to which young citizens incorporate sustainable practices in their professional and personal lives will determine the quality of life for future generations. Students engaged in STEPWISE-supported activities are well-positioned to develop these insights.

1.6 Taking Action: Direct or Indirect; Individual or Group?

With regard to student action, we need to distinguish between direct and indirect action. In the context of addressing environmental issues, the former includes such things as recycling, cleaning up a stream or a beach, building a compost heap and using a bicycle rather than a car; the latter includes compiling petitions, distributing leaflets, writing to newspapers, composing blogs and making submissions to the local council. While direct action can be enormously important and can have some significant impact, it can also divert attention from root causes of problems in our social, political and economic activities. It depoliticizes environmental problems, for example, and shifts the burden of responsibility onto individuals and families and away from governments, corporations, the policies that might have long-term and significant impact, and the political negotiations that might lead to change. Cleaning up a beach will have immediate beneficial impact, but without an investigation of the causes and appropriate intervention aimed at those causes, there will be no long-lasting solution. Although reducing our personal use of cars is a small step in reducing air pollution and conserving energy, it fails to tackle the ways in which sociopolitical and economic decisions about modern urban developments have made the car a virtually indispensable means of transport. Setting up a recycling programme may prolong the active life of one or two landfill sites but it doesn't address (and it certainly doesn't change) the unsustainable economy of resource use, production and consumption. Does this mean that indirect action is the better option? Possibly, if it is *authentic* action: not just a classroom exercise in which a letter to an imaginary newspaper editor is composed, but a real letter to a real newspaper editor to express real concerns or to make a series of real debating points or policy recommendations, or the preparation of a report for submission to a local government body or provision of material assistance for an individual or group involved in a local dispute. The ideal involvement of students comprises a mix of direct and indirect actions, but it is both appropriate and necessary for these actions to be pitched at the local level. There are many local manifestations of global problems; local actions are likely to have substantially more impact; feedback from local action is likely to be more immediate; a group of school age students (few of whom will be of voting age) have insufficient power to have significant impact at the global level, but what they learn and the values they develop through action at the local level will prepare them well for later action on the wider stage. The slogan *Think Globally, Act Locally* gained currency because it is an excellent rule of thumb.

The STEPWISE rationale has an inbuilt recognition that, while action at the individual or small group level is a good beginning, being able to motivate groups or populations to exert pressure on governments, businesses and the wider public to dismantle barriers to change and create alternative, more equitable and ethically and environmentally responsible policies and practices will have greater impact, as Angeliki Grundy points out (in Chap. 5) in her insightful discussion of the importance of sound techniques of persuasion. Changes at a fundamental level will only result when three key elements of persuasion are in place: *legitimacy* – convincing people that the action is desirable and morally right; *urgency* – illustrating the need for the issue to be addressed quickly and decisively; and *power* – establishing the capacity to force another to do something counter to their current practice, using financial means, voting power, etc. Persuasion involves careful preparation and consideration of arguments, presentation of vivid supporting evidence and establishment of a strong emotional match with the target audience. Understanding multiple perspectives enhances the ability to persuade others to action. Hence the value of the case studies included in Part II of the book. But it is group action that provides the final element and is the only route to fundamental change in society. As well as teaching students the need to be sufficiently resilient and determined to try again, experiences of failure of individual actions may also impress upon them the need to mobilize others and to engage in collective action. Collective actions are almost always more effective than individual actions and, in some circumstances, may be the only means of bringing about change. Thus, a key part of preparation for activism is helping students to recognize, mobilize and coordinate the knowledge and skills that are distributed across communities.

1.7 Engaging Students

It follows that we should make strenuous efforts to involve students in public hearings and town hall meetings, consensus conferences, study circles, focus groups, citizen juries/panels, negotiated rule-making forums, public/citizen advisory committees and the like. It is through community-based activities that young people gain autonomy, a sense of worth, a sense of personal and civic identity, respect for other people's views, negotiation skills and so on. When engaged with real problems and issues, students encounter real barriers and obstacles; working with community members to overcome these barriers cultivates students' competency and sense of competency. When people work together, there are opportunities for doing things that individuals would not even contemplate doing alone. By working on a sub-task within a group effort, individuals acquire a level of expertise that would not be achieved alone, at least not so quickly and so painlessly. They also come into contact with perspectives on issues and problems that differ from their own. Sharing experiences, action strategies and success stories, as well as building friendships, can be inspirational and highly motivating, and can lead to lifelong sociopolitical activism. These experiences are immensely valuable because they run counter to the trend of growing social isolation of individuals and individual families, and counter to the values that underpin the pervasive competition and conspicuous consumption of contemporary society. By focusing on the community and the issues and problems that residents confront in their everyday lives, students come to recognize their own experiences as shared, social and political. It is through direct experience of confronting social and environmental problems in the immediate community that public issues acquire personal meaning for young people. For example, working in shelters for the homeless, participating in breakfast programmes, doing volunteer work in hospitals, drug rehabilitation centres, HIV-AIDS support groups and homes for the elderly, involvement in environmental clean-up projects, renovating dilapidated homes, replanting degraded areas, building and maintaining community gardens, creating parks and conservation areas, organizing community festivals and information fairs, producing a local newsletter or community blog, and so on. STEPWISE has made and will continue to make an effective contribution to this kind of involvement. To enhance it further we need to build a mutually supportive relationship between school and surrounding community. Traditional barriers between school and community need to be dissolved or rendered permeable, with community members present and active in the school, and students and teachers active and involved in the community – in part, contributing to the deinstitutionalization advocated by Wolff-Michael Roth (Chap. 32). Of course, the difficulty of building such an atmosphere of interest, trust and shared responsibility and commitment should not be under-estimated. It requires strenuous effort on the part of teachers, students and wider community members.

As part of those efforts, students should be encouraged and enabled to use aspects of youth culture, particularly music, chat rooms and other communications media to spread a youth-oriented message concerning civic and environmental responsibility.

We should be encouraging students to use their interest and skills in contemporary communications technology, especially with social media such as *Facebook* and *Twitter*, to establish networks, express concerns, share thoughts and spread messages about the need for action. New forms of ICT enable forms of participation that were not previously possible and may engage significant numbers of people who would previously have been uninvolved. They have the potential to facilitate the building of a more inclusive, participatory, socially just and politically engaged community. First, *raising awareness and advocacy* by providing access to independent and alternative sources of information that may be ignored or suppressed by mainstream media; second, developing *community-oriented sites* to spread awareness, share experiences and ideas and build networks within communities; third, creating *action groups* to raise public support for actions related to specific issues at the local, regional, national and international levels. There are numerous examples of splendid work in this sphere embedded in the STEPWISE research reports. Music can play a powerful role in identity construction and reinforcement, enabling a better understanding of one's own experiences and the experiences of others, raising political awareness, and building the solidarity and sense of community that can lead to activism. For many urban youth the rap music of hip-hop culture can be a particularly powerful vehicle, enabling them to put their feelings, emotions, needs, aspirations, hopes, joys, fears, disappointments and anger into a form that is respectful of their immediate cultural experiences and will be readily understood by their peers. Its potential for the kind of education envisaged and promoted by STEPWISE is considerable.

Though my inclination would be to give over the entire school curriculum to an issues-based action-oriented curriculum embracing all subjects, along the lines advocated by Sharma (Chap. 31), there is no possibility that this will happen in the foreseeable future. A first step in this direction would be to turn over the whole science curriculum to this kind of approach, but I am not so naive as to think that is likely to happen any time soon, either. However, it is possible to implement the kind of issues-based approach advocated in this book alongside a more conventional subject-oriented curriculum, provided that neither students nor teachers see it as a mere add-on or motivational adornment. Confrontation of issues, consideration of underlying values, and taking action need to be fully integrated into the curriculum. Engaging students in confrontation of SSI as part of informal education, as advocated by Weinstein (Chap. 28) and Roth (Chap. 32), is entirely possible and is to be valued and encouraged, but alongside rather than instead of school-based provision. Involvement in STEPWISE lays a sound foundation and preparation for such initiatives. I am confident that students exposed to such an SSI-rich and action-oriented curriculum, and the approach advocated in this exemplary book, will be much more likely to give serious consideration to the social, political, environmental and moral-ethical aspects of SSI in their daily lives outside and beyond school. I am confident that such experiences will assist them in moving firmly and positively in the direction of Oxfam's conception of a *global citizen*: someone who is aware of the wider world and has a sense of his or her own role as a world citizen; respects and values diversity; has an understanding of how the world works economically, politically, socially, culturally, technologically and environmentally; is outraged by social injustice; participates in and contributes to the community at a range of levels from the local to the global; is willing to act to make the world a more equitable and sustainable place; and takes responsibility for their actions.

Chapter 2
STEPWISE: A Framework Prioritizing Altruistic Actions to Address Socioscientific Issues

Larry Bencze

2.1 Introduction

'STEPWISE,' which is the acronym for *Science and Technology Education Promoting Wellbeing for Individuals, Societies and Environments*, is a framework for curriculum and instruction prioritizing students' uses of at least some of their literacy in fields of science and technology (and from other sources) to try to bring about what they perceive to be a better world. In its theoretical form, STEPWISE takes the shape—as shown in Fig. 2.1—of a *tetrahedron*, with four learning domains at its peripheral points and 'STSE Actions' in the 3-dimensional centre. Although there are reciprocal relationships between all domains, as indicated by the 2-way arrows, this tetrahedral configuration is meant to indicate that students can use their 'literacy'—at least in terms of *STSE Education* (e.g., conceptions of relationships among fields of science & technology and societies & environments), *SkillsEducation* (e.g., abilities to use a microscope and design/evaluate an experiment), *Products Education* (e.g., understanding of cell metabolism, nuclear radiation, genetic engineering, etc.) and *Students' Research* (e.g., findings from studies of people's plastic water bottle uses)—to inform decisions about actions they could take to address *STSE issues* (e.g., questions about government regulation of bottled water companies). This particular arrangement of the five domains is intended, in other words, to encourage and enable students to take a more 'altruistic' view of their education and the world; that is, to feel that they can—and perhaps *should*—'spend' some of their literacy, not just on themselves, but also on efforts to improve wellbeing of other individuals, societies and environments (WISE).

After developing the STEPWISE tetrahedral framework in the summer of 2006, I have facilitated action research to understand its efficacy in various educational

L. Bencze (✉)
Ontario Institute for Studies in Education (OISE), University of Toronto, Toronto, ON, Canada
e-mail: larry.bencze@utoronto.ca

L. Bencze (ed.), *Science and Technology Education Promoting Wellbeing for Individuals, Societies and Environments*, Cultural Studies of Science Education 14,
DOI 10.1007/978-3-319-55505-8_2

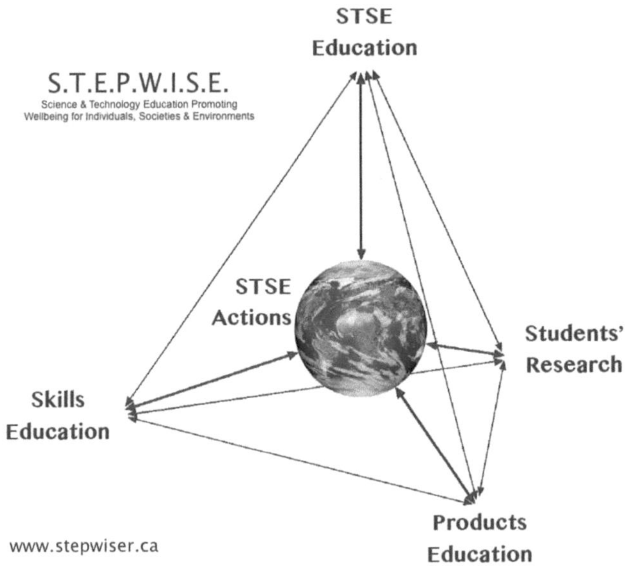

Fig. 2.1 The STEPWISE theoretical framework

contexts—including regarding my university-based science teacher education courses and in science education in school and non-school settings. This book provides documentaries of implementation of STEPWISE in such contexts, along with chapters emphasizing theoretical analyses and evaluations of the framework by invited scholars. In this introductory chapter (following Derek Hodson's *Foreword*), I provide theoretical justification for the framework—in the context of descriptions of its historical development—and brief overviews of its uses in teaching for research purposes. Summaries of sections of the book and chapters in them also are provided here.

2.2 Etiology of STEPWISE

Although constructed in 2006, the STEPWISE framework represents somewhat of a culmination of events in my education career dating back to at least 1974—the year I graduated from Queen's University in Kingston, Ontario, with a B.Sc. in biology. After unsuccessful efforts to find work using my degree, a visit with professors of biology (Drs. Klaus Brasch & Bradley White) from whom I had taken undergraduate courses changed my life and career. They offered me a position in graduate studies leading to an M.Sc. degree that involved working on their research team in molecular genetics, with me trying to locate a rare, but very active, gene on a fruit fly chromosome. Although I did not locate the gene, I did generate some knowledge about fruit fly chromosomes and earn my degree. More importantly, I feel that I became 'addicted' to *primary* research—to generating knowledge through approaches in which I had a significant role in decision-making. At the same time, however, because I thoroughly enjoyed working as a teaching assistant for university biology courses, I enrolled in a programme leading to a B.Ed. degree qualifying me to become a school science teacher. In that context, although learning to teach topics like cell metabolism, electric circuit function and types of chemical change, I also was introduced to 'inquiry science'—focusing on teaching and learning of discrete science 'processes' (e.g., variable control), which was a current movement (e.g., AAAS, 1967). When required to write a major curriculum unit, I chose to create one focused on teaching 'methods of science' and, gratefully, my instructor (Bert Horwood) commented that my unit was 'superb!' This positive response spurred me on to emphasize such student-led science investigations (mostly experiments) in my teaching career. At first, this worked well—apparently because I began my teaching career working in small schools, in which I was the only teacher of my courses and, therefore, free to design them as I pleased. In those early years, I really came to appreciate youth's creativity and the extent to which such projects enabled them, often including those who did not achieve high grades in traditional science education components (e.g., explaining photosynthesis), to become deeply engaged in problem solving. As part of this work, I also organized school science fairs—events at which students displayed and explained their research projects and were judged for their uses of science processes (e.g., control of variables, duplication of tests and graph construction) and clarity and logic of presentations to judges and others. These events brought praise to me for my students' enthusiastic and creative work and they further reinforced my interests in promoting student-led primary research. However, when my teaching career led me to work in large high schools, in which I was just one of perhaps 2–3 other teachers teaching the same courses, I experienced opposition to promotion of primary research. My colleagues seemed to be embedded in a system that prioritized celebratory teaching of large amounts of knowledge ('Products Education' in Fig. 2.1) generated by 'science and technology' (in societal and environmental contexts). Describing the nature of school science much later, Randy Bell (2006) seemed to concisely express approaches many of my colleagues were using at the time:

> In the typical classroom, instruction has focused almost exclusively on the well-established products of science ['Products' Education' in Fig. 2.1] and cookbook approaches to laboratory exercises, using authoritarian teaching modes (p. 430).

In terms of Roger Lock's (1990) learning control schema depicted in Fig. 2.2, it appears there is a preference—particularly in secondary schools—for teacher-directed, closed-ended, approaches to ensure students develop particular conceptions about products of science and technology. Such a priority and associated didactic teaching approaches seemed to leave little time or motivation for encouraging and enabling students to self-determine methods of investigation and conclusions about their findings. For two school years, I acquiesced—minimizing my promotion of student-led investigative work and focusing on instruction and assessment/evaluation of 'products' of science and technology. This was very disappointing

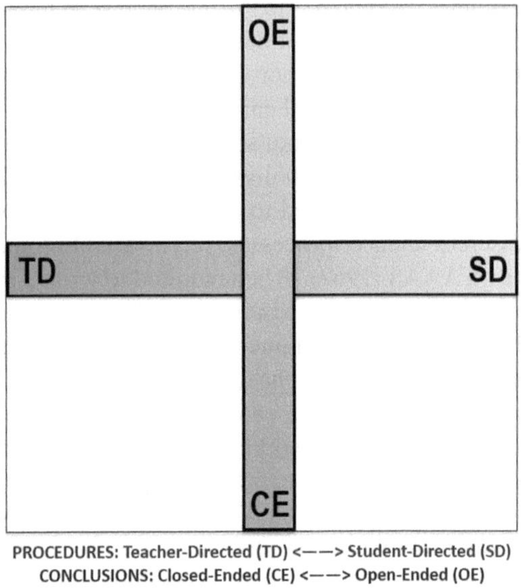

Fig. 2.2 Variations in control of learning (Lock,1990)

for me. I felt defeated; that a fundamental feature of my professional character had 'died' (although, admittedly, I had chosen to acquiesce). However, 'flames' for student-led knowledge generation seemed to be soon re-kindled—due to my fortuitous appointment to the position of science consultant for my school district. Because I had considerable flexibility in this role, I was happy to resume promotion of student-led primary research—along with assisting teachers with teaching and learning in other domains (mainly 'Products Education' in Fig. 2.1). Yet another turning point occurred around this time, however. It soon became apparent to me that, although teachers of science in elementary schools appreciated my efforts to promote student-led research, secondary school science teachers generally resisted this possibility. Teachers of science in elementary schools tend not to have significant university education in fields of science and/or engineering and, consequently, tend to lack self-efficacy for teaching science (Harlen & Holroyd, 1997). Perhaps, as Jennifer Helms (1998) suggested, secondary school teachers of science with at least undergraduate science degrees tend to identify with their area(s) of subject specialization (e.g., biology, chemistry and/or physics) and, moreover, mainly prioritize instruction in 'achievements' (i.e., 'Products Education' in Fig. 2.1) of them. Elementary teachers of science, by contrast, may lack such affinities and, accordingly, may be amenable—with support—to allowing students to self-determine claims about the world.

After my career as a science consultant, I enrolled in a graduate programme leading to a Ph.D. in education. Drawing from my experiences as a consultant, I focused my thesis project on facilitating action research by a group of secondary school teachers of science who agreed to try to find ways to promote student-led primary research in their teaching. This decision was supported by my thesis supervisor, Dr.

Derek Hodson, who had published articles in this domain (e.g., Hodson, 1993, 1996). This tack seemed to be fruitful, leading to several relevant publications (e.g., Bencze, 1996, 2000, 2001a; Bencze & Hodson, 1998, 1999). Nevertheless, in various contexts, particularly regarding secondary school science teaching (e.g., Bencze, et al., 2003), it seemed that some sort of 'invisible hand' was inhibiting most teachers from supported student-led primary research. This was a source of great frustration and puzzlement for me. I soon concluded, however, that the 'invisible hand' may be *capitalism*, particularly in its *neoliberal* form. Briefly, although its meaning is debatable, it seems that neoliberalism is a more comprehensive and strategic form of economic liberalism that, to a great extent, governs people (and, perhaps, other living things) in subliminal ways through global networks of influence (Foucault, 2008; Springer, Birch, & MacLeavy, 2016). I had been particularly affected by John McMurtry's (1999) book, *The Cancer Stage of Capitalism*, in which he suggested that capitalists often were like cancer cells, resembling other human body cells, but alien and destructive to them. At about the same time, I also became enamoured with John Ziman's (2000) book, *Real Science*, in which he discussed positive and negative aspects of government-sanctioned business partnerships with fields of science and technology. When engineers and scientists enter into contractual agreements with financiers and corporations, it seems they risk compromising the integrity of topic choices, methods and the nature and extent of dissemination practices (Mirowski, 2011)—often because of corporations' rights to minimize costs, at least partly by *externalizing* them; that is, by arranging for others to bear costs of, for example, infrastructure (e.g., roads and electrical/electronic networks), labour (e.g., low minimum wages), materials (e.g., lower quality metals) and medical care (e.g., diseases, like cancer, caused by products, like cigarettes) (Bakan, 2004). As a consequence of such 'business-science partnerships' (Krimsky, 2003), there appear to be various potential problems for wellbeing of individuals, societies and environments (WISE)—including health problems associated with genetically-modified foods, etc. (Kleinman, 2003), household cleaning and hygiene products (Leonard, 2010), pesticides (Hileman, 1998), tobacco (Barnes, Hammond, & Glantz, 2006) and pharmaceuticals (Angell, 2004), climate change due to excessive fossil fuel uses (Klein, 2014) and psycho-social harms due, for example, to gender and cultural stereotyping in girls' toys (Steinberg, 2010). In light of such potential problems, I presented several papers at refereed conferences that culminated in publication of an article summarizing ways in which government-sponsored science-business partnerships may compromise WISE and implications for science education (Bencze, 2008).

Apparently, congruent with various harms to WISE associated with capitalists' influences on fields of science and technology, it also seemed that school science was being influenced in ways that contributed to furtherance of such harms. My reviews of literature in education and other fields led me to conclude, broadly, that school science appeared to be doing so by, on the one hand, selecting future knowledge producers (e.g., scientists & engineers) while, on the other hand, leading many or most students to become knowledge consumers—in at least two senses; that is, as obedient followers of labour instructions from knowledge producers and as enthusiastic and unquestioning purchasers of for-profit products and services (Bencze, 2001b).

In light of experiences and perspectives about science and science education described above, I set about developing a theoretical framework for science education that may counter problematic neoliberal influences. While neoliberalism seemed extremely powerful, developing an oppositional science education schema seemed somewhat feasible in the jurisdiction of my work (Ontario). Among its three overall goals ('domains') for its curricula, including 'Products Education' and 'Skills Education,' it also made provision for 'STSE Education' (MoET, 1999)—a domain that, in principle, might involve education about problematic business-science partnerships. In thinking about helping teachers implement these three goals, however, I made four decisions, each for a different reason: (i) because there are reciprocal relationships among these domains, there should be 2-way arrows between each pair; (ii) 'STSE Actions' needs to be presented separately from 'STSE Education' (which the curriculum combined) to highlight needs to actively address many serious potential problems for WISE; (iii) 'Students' Research' needed to be presented separately from 'Skills Education' (which the curriculum combined) to highlight needs for students to self-direct open-ended investigations (in addition, for example, to teacher-directed and closed-ended activities; refer to Fig. 2.2); and, (iv) 'STSE Actions' should be in the centre of the resulting tetrahedron to emphasize the (perhaps more) *altruistic* nature of the framework. The latter decision was greatly influenced by an article by Hodson (2003) and book by Wolff-Michael Roth and Jacques Désautels (2002) calling for science education reform prioritizing sociopolitical actions to address STSE issues. At this point, then, I had a framework (Fig. 2.1)—but not a name for it. Concurrently, however, popular media reported efforts in Canada to define 'success' beyond traditional economic indicators, such as 'GDP' (Gross Domestic Product)—thinking more broadly in terms of 'wellbeing,' which considers indicators like: levels of education, political engagement, attention to environment, etc. (uwaterloo.ca/canadian-index-wellbeing). By moving relevant words around, I developed the phrase, 'Science & Technology Education Promoting Wellbeing for Individuals, Societies & Environments,' which happened to spell a word that could symbolize a gradual ('stepwise') process of implementing a new framework. And, so, 'STEPWISE' was born.

There appear to be many reasons to place 'STSE Actions' at the centre of the STEPWISE tetrahedron. On the one hand, researchers, educators and others have—for at least the last 45 years—been exploring and promoting STSE education (Pedretti & Nazir, 2011). In Canada, a non-binding national standards document placed it first among four 'foundation statements' for scientific literacy (CMEC, 1997). Ontario, where I have mostly worked, soon also included it in its curriculum—listing it third among three literacy goals (MoET, 1999). About a decade later, however, it was listed <u>first</u> among curricular goals (MoE, 2008)—a change that appeared to motivate many teachers to increase its emphasis in their teaching. With such a mandate, students may, for instance, learn that private members of societies have compromised ingredients in many manufactured foods (technologies, based, to some extent, on science) in ways that are associated with health problems (e.g., diabetes and heart disease) and, consequently, develop and implement plans of action to try to improve human health. However, it is apparent that much 'STSE' education (acknowledging different movements around the world with similar goals) emphasizes students' development of personal, well-argued, positions on *controversies* in such relationships—often called *socioscientific issues* (SSIs) (Sadler, 2011). Students may, for example, investigate and debate merits of various controversial technologies, such as: weapons; manufactured foods and beverages; cosmetic surgery; mass surveillance and data-mining; household hygiene and cleaning products; electronics with parts made in conflict zones; etcetera. Evidence suggests that engaging in such deliberations has, indeed, led to some significant learning gains for students—including, for instance, development of *socioscientific reasoning skills* (Sadler, Barab, & Scott, 2007) and learning of products of science (e.g., laws & theories) (Venville & Dawson, 2010).

Although logical choice approaches have led to some admirable personal learning outcomes, there appear to be some 'imbalances' with such deliberations. Ralph Levinson (2010) concluded that many such approaches—especially those he called *Deficit* experiences but, also, *Deliberative* ones—continue to place students/citizens in roles of dependency on power of political leaders, experts and others. Such personal choice roles may be appropriate for citizenship in *representative* democracies—in which citizens exercise democratic choices largely through periodic election of political representatives, often informed by experts like scientists, engineers, lawyers and other professionals (Wood, 1998). However, as discussed above, some people are very concerned about influences powerful members of societies sometimes have on fields of science and technology that frequently seem to result in various personal, social and environmental harms—with climate change associated with fossil fuel industries, lobbyists, etc. arguably our most pressing problem (Klein, 2014). Although such harms linked to fields of science and technology appear to be quite serious, it seems difficult for citizens to deeply learn about them. It is apparent, for instance, that members of the private sector *promote* controversies and associated confusion about possible harms from their products and services by paying reputable scientists, physicians, journalists and others to cast doubt on evidence from fields of science that—if the public knew about and respected it—may dissuade them from purchasing products like weapons, cigarettes, pesticides, and fossil fuel powered vehicles (Oreskes & Conway, 2010). Some scholars and others suggest, therefore, that science education needs to help students to develop, expertise, confidence and motivation for engaging in more critical and active citizenship in *participatory* forms of democracy (e.g., Hodson, 2011; Santos, 2009). Such critical and activist societies could experience improvements to wellbeing of individuals, societies and environments. Rationale for such more engaged citizenry also seems philosophically and ethically sound. Based on social epistemology (e.g., Fuller, 2002), in which it is claimed that individual achievements invariably involve communities, past and present, we each may have the responsibility to share (and can benefit from sharing) our wealth and wellbeing with our communities (e.g., Alperovitz & Daly, 2008). Such a noble position about citizenship was well-expressed by the famous physicist, Albert Einstein:

> The aim (of education) must be the training of independently acting and thinking individuals who, however, can see in the service to the community their highest life achievement (Calaprice, 2000).

Although such a stance on life and citizenship may seem highly *altruistic*, actions we take to improve others' wellbeing may be returned to us—because we are connected to many other people in our society (and, likely, worldwide) (Batson, 1994).

2.3 STEPWISE in Theory and Practice

Soon after developing the STEPWISE framework in Fig. 2.1, I began working with graduate students to conduct research to understand its effectiveness in various formal and informal science education contexts and learning about factors apparently influencing related outcomes. Most of these contexts were in urban or suburban schools in and around Toronto, Canada, as well as in my pre-service science teacher education courses. We also, however, facilitated one teacher's implementation of STEPWISE in an international school in Venezuela—which appeared to lead to some interesting insights relating to socio-political context (see Zouda, Nishizawa & Bencze, Chap. 15, this volume).

Particularly when visiting teachers in their schools and discussing approaches they might take, it soon became apparent that they found the framework in Fig. 2.1 to be *impractical* (Bencze & Carter, 2011). It seemed particularly difficult to imagine lessons and activities that, as the model suggests, would engage different students in multiple learning domains (e.g., 'Products Education' and 'Skills Education') at the same time, leading in different directions within the framework for each student—a somewhat *rhizomatic* education that Gilles Deleuze and Félix Guattari (1987) suggested may be more democratic than more hierarchical ('arborescent') approaches commonly-found in schools. Through a series of negotiations with teachers, such as those described by Darren Hoeg, Tanya Williamson and Larry Bencze in Chap. 3 of this volume, more hierarchical schemas like that in Fig. 2.3 were acceptable to them. They seemed to prefer, in other words, a relatively-structured, *deficit,* approach—which assumes students need teachers to provide missing expertise and confidence by guiding them through a series of lessons and activities for, eventually, enabling and motivating them to self-direct research-informed and negotiated action projects to address SSIs of their interest/concern.

There are various ways elements of the tetrahedral version of STEPWISE (Fig. 2.1) could have been re-arranged to be more practical for teachers' uses. As argued above, a priority was to prepare students for engaging in socio-political activism to address socioscientific issues/problems of concern to them. A key consideration in that, however, had to be student *motivation.* We asked, 'What would incline students to spend some of their cultural (and, likely, social) capital (Bourdieu, 1986) on acts for others?' To answer this question, we drew on Etienne Wenger's (1998) *knowledge duality theory*—which posits that learning engagement is

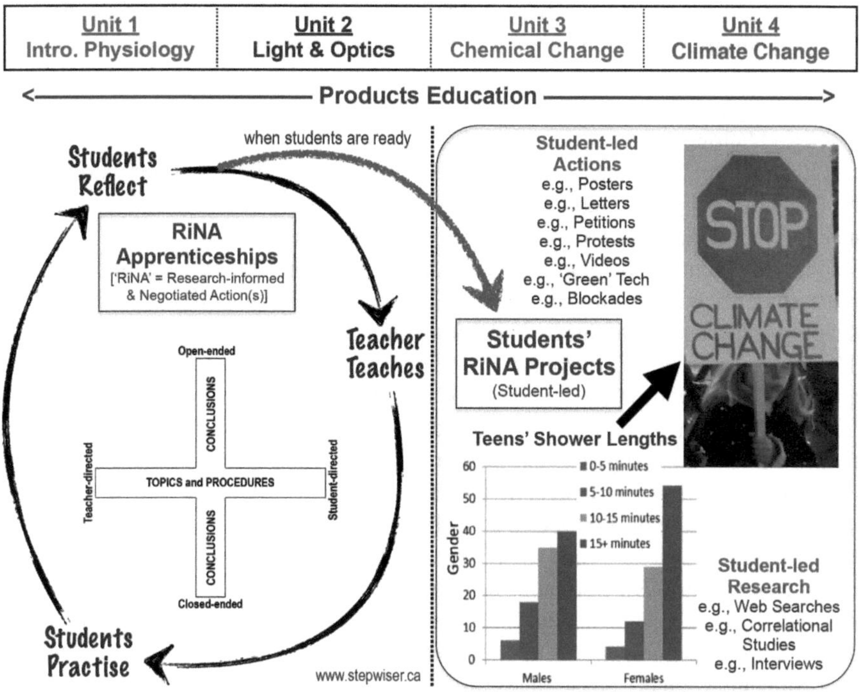

Fig. 2.3 STEPWISE pedagogical approach

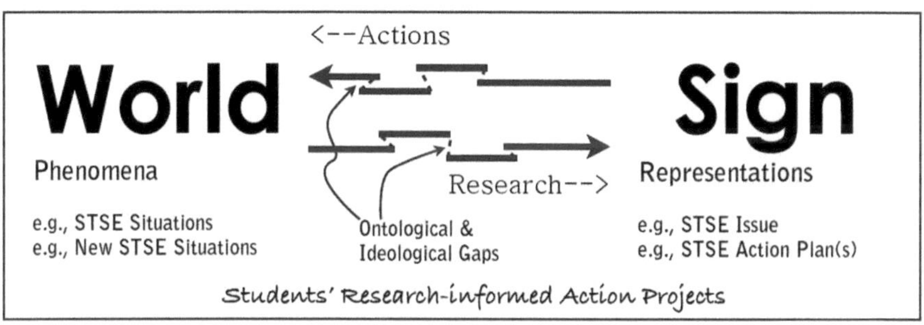

Fig. 2.4 A model for research-informed actions on STSE issues

enhanced the more students control decisions about reciprocal translations between phenomena and representations of them. This concept also relates to a schema provided by Wolff-Michael Roth (2001), which depicts reciprocal relationships between conduct of 'science' and 'technology' (or engineering). As shown in Fig. 2.4, we adapted this schema to explain research-informed and negotiated actions (RiNA) (Bencze & Carter, 2015). Students should, in other words, become highly committed to RiNA projects if they have maximum control over 'research' (World → Sign translations) and, using findings from their research, negotiations of actions (Sign → World translations). (Note: ontological and ideological gaps are discussed below.)

Given the relative paucity of STSE education approaches, apparently dominated by those that emphasize empirical-rational controversies (Pedretti & Nazir, 2011), teachers' tendencies to control students' inquiry decisions, and relative dearth of student engagement in socio-political actions (Hodson, 2011), it seems clear that most students will require preparatory lessons and activities that may help them develop expertise, confidence and motivation for self-directing RiNA projects. These are shown in the schema in Fig. 2.3 as RiNA *apprenticeships*. These lessons and activities are structured in three phases, based on fundamental constructivist learning theory (e.g., Osborne & Wittrock, 1985) (acknowledging social constructivist forms). These suggest that learners can benefit from first reflecting on and expressing their existing conceptions ("Students Reflect"), about which they often are not consciously-aware, before a teacher provides them with alternative conceptions ("Teacher Teaches") and then encourages them to evaluate competing conceptions through various personally-meaningful implementation activities ("Students Practise"). As indicated in Fig. 2.3, such apprenticeship activities should vary in terms of the extent to which teachers or students control decisions (Lock, 1990). In the "Students Reflect" phase, it seems that, while teachers need to provide stimuli (e.g., looking at a picture) for reflections, most interactions should be student-directed and, if students are to relatively-freely express *their* conceptions (without feeling pressure to provide ideas planned by the teacher, for instance), such activities should be open-ended; that is, not pre-determined by the teacher and dependent on each student's prior experiences and perspectives. Afterwards, however, because students in many classes will vary in their cultural and social capital (Bourdieu, 1986), the teacher should use relatively teacher-directed and closed-ended approaches (in the "Teacher Teaches" phase) to ensure all students have full access to essential attitudes, skills and knowledge ('ASK') that may benefit them. In the last phase of the apprenticeship ("Students Practise"), students are given considerable control over decisions, to the point of being nearly, if not fully, student-directed and open-ended. Teachers can, however, provide students with support (more teacher-directed) when students request it. If, moreover, the teacher feels it is necessary, students can be provided with a second apprenticeship cycle before being asked to self-direct projects. Once teachers feel students are ready, they would 'leave' the apprenticeship stage and be asked to self-direct RiNA projects—and, as 'situational experts,' ultimately define 'expertise' in ways matching contexts of their projects (Bencze, 2000).

2.4 Some Research-Informed Practical Suggestions for STEPWISE Uses

In working with teachers implementing the schemas in Figs. 2.3 and 2.4 (mainly) and in using them as bases for my instruction in science teacher education, it seems that, while many insights are still likely to arise, we have learned much about STEPWISE. Some of this knowledge has been published elsewhere (e.g., Bencze & Alsop, 2014), with more perspectives and practices provided through the 'Documentaries' section of this book. The sub-sections below, however, provide readers with some highlights from our research and development findings relating to these documentaries:

2.4.1 Students Reflect

In getting students to reflect on and 'express' their existing conceptions about STSE relationships and RiNA projects, educators may consider various—largely interrelated—factors, including: (i) Learning control: Regarding Fig. 2.2, encouraging students to reflect on/express their existing notions suggests needs for student-directed and open-ended experiences. However, experience demonstrates that students typically benefit from some stimuli from teachers for reflection—such as, as described next, provision of possible reflection topics; (ii) Topics of reflection: Broadly, if our purpose is to encourage students to express some of their existing conceptions of STSE relationships and RiNA projects, while teachers may keep in mind models of STSE like that in Fig. 2.5 and RiNA like that in Fig. 2.4, they also need to remember to minimize their directivity—erring on the side of encouraging students to

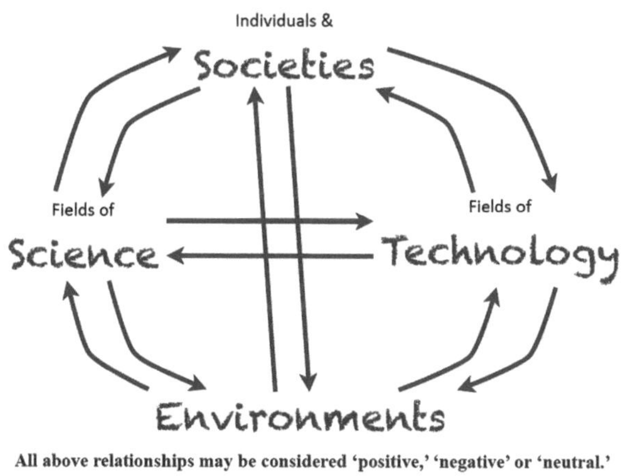

Fig. 2.5 A model for STSE relationships

freely-express their ASK, which typically implies more divergent than convergent directions and questions (examples of which are provided below). After later apprenticeship lessons and activities, however, especially in the "Teacher Teaches" phase, students may be asked to represent (in the "Students Reflect" phase) their existing conceptions in ways the teacher might desire, such as by constructing actor-network maps about them; (iii) Stimuli for reflection: Keeping in mind that teachers should attempt to minimize their influence on students' expressions, stimuli provided by them should be as divergent as possible, at least at first (Hudson, 1967). In doing so, there appears to be merit in thinking about 'stimuli' teachers can provide in terms of phenomena and teacher instructions and questions. Teachers might, for instance, begin by asking students to simply note in a journal observations and conclusions they make over a few days of, for example, being at home, travelling to and from school and being at school that they believe to relate to 'science.' Perhaps in a somewhat more convergent vein, teachers could present students with photographs, videos, etc. and/or physical examples of various 'products' of science and technology (e.g., commodities, like cell phones) and ask them to describe what they think is 'good' and 'bad' about them, which people and groups may support such claims, what solutions they imagine for problems they identify and 'preparation' (e.g., research) they may need to conduct prior to acting on their solutions. Additionally or instead, teachers might ask students to evaluate various suggestions for actions to address STSE problems; such as, 'The federal government should encourage companies to extract oil and gas from Canada's arctic' and 'It's OK to smoke cigarettes because the pleasure outweighs health risks.' In this volume, Mirjan Krstovic (Chap. 6) and Varsha Patel (Chap. 4) provide some concrete examples of similar kinds of reflection/expression activities; (iv) Forms of expression: In providing instructions and posing questions, teachers can ask students to represent their pre-instructional conceptions in many different forms. These include, but are not limited to: speech; written text; drawings; graphs; models; actor-network maps (refer below); sketches; and, poems. Again, decisions about these will depend on students' ages and stages of development in abilities and understandings of STSE, research and actions and ways of expression of them—much of which will depend on the extent to which such ASK have been previously taught and learned; and, (v) Expected outcomes: Finally, aligned with the open-ended (Fig. 2.2) nature of such reflections, it should be clear to teachers that natural variations in student responses are to be expected and, indeed, celebrated. Students should, generally, only be evaluated in terms of effort—not regarding their specific responses.

2.4.2 *Teacher Teaches*

Although it is important to honour students' pre-conceived notions about STSE relationships and actions possibly informed by research (and other resources), given the relative dearth of relevant education (e.g., Pedretti & Nazir, 2011), it seems inevitable that they will need access to attitudes, skills and knowledge (ASK) that

Fig. 2.6 A Gestalt image

may benefit their lives and that of other individuals, societies and environments. There is considerable debate, however, how students might gain access to such ASK. In science education, it is common for educational scholars, administrators and others to engage students in so-called 'authentic' inquiry-based learning experiences—such as experiments scientists might conduct—and, although prompts may be given, expect students to derive relevant laws and theories of science. Reneé Schwartz, Norman Lederman and Barbara Crawford (2004), for instance, who have done considerable work in this area in the USA, have said that such experiences often can be characterized as follows:

> Within a classroom, scientific inquiry involves student-centered projects, with students actively engaged in inquiry processes and meaning construction, *with teacher guidance*, to achieve meaningful understanding of scientifically accepted ideas targeted by the curriculum (p. 612).

Such a conception of teaching, where particular, pre-determined, learning is expected, seems flawed (Bencze & Alsop, 2009). One problem is that not all students will 'discover' or interpret phenomena in the same way, since all of them have had different prior experiences and, consequently, have different mental structures in their brains that may or may not match stimuli from the world. Regarding the black and white image in Fig. 2.6, for instance, some people may 'see' geographic shapes, others may 'see' the side of a cow and others may 'see' something else (Hodson, 1986). Therefore, a particularly problematic aspect of expecting students to discover certain, pre-determined, ASK is that, due to differences in cultural and social capital (Bourdieu, 1986), disadvantaged (e.g., poorer) students are less likely to discover intended ASK and, therefore, social stratification may be reinforced (and, perhaps, magnified) through such approaches. Accordingly, it seems best for teachers to use more teacher-directed (and, perhaps, closed-ended) approaches to ensure all students have access to ASK (in this case, about STSE relationships and RiNA projects) that may be useful to them. Having made this claim, as elaborated below ("Students Practise"), direct teacher instruction is likely to be insufficient for deep, meaningful, learning. It is apparent that such instruction must be later combined with opportunities to *practise* ASK that have been taught. Logic for this comes from *knowledge duality theory* (Wenger, 1998), which posits that deep, meaningful, learning occurs best when students have maximum control over both inductive (World ➔ Sign) and deductive (Sign ➔ World) aspects of learning (see Fig. 2.4). Teachers should, accordingly, often attempt to balance teacher instruction (phase 2) with student practice (phase 3) in contexts having meaning for students.

Although there are many teacher-led instructional approaches for phase 2 of the apprenticeship in Fig. 2.3, including lectures and/or Socratic lessons, with demonstrations, photographs, drawings, videos, etc., to teach students about particular STSE relationships and RiNA projects, a common approach, which often combines deductive and inductive thinking, is use of *case methods*; that is, documentaries ('cases') summarizing phenomena, with instructions and questions ('methods') to engage students in learning from and through the cases (Bencze, Hewitt, & Pedretti, 2001). A very effective kind of case/documentary is one (or more) depiction(s) of STSE-RiNA projects conducted by students of a similar age as those in the class. A case method could be created around the image in Fig. 2.7, which shows students presenting a report of their study of content (e.g., minerals) of different types of bottled water and tap water, along with their action banner (not shown here was this group's action video comparing bottled to tap water, which they posted to YouTube™). Other such examples could be taken from the special issue of the *Journal for Activist Science & Technology Education* (JASTE) containing several reports of student-led RiNA projects (goo.gl/N00b3s). Another excellent source of cases is provided by *The Story of Stuff* project (storyofstuff.org), which focuses on problematic aspects of 'the materials economy,' involving processes from extraction, through production, distribution, consumption and disposal (Leonard, 2010). It includes videos about such controversial commodities as bottled water, cosmetics and electronic devices (e.g., cell phones and digital music players). Teachers could develop such cases into *case methods* by asking students to review the reports and answer a range of questions about them that would help students to understand aspects of STSE relationships and research-informed and negotiated action projects that could be conducted. One common schema for developing instructions and questions for students' interactions with cases is Bloom's Taxonomy of the Cognitive Domain (Anderson et al., 2001)—which suggests that students can be engaged in a range of types of thought, perhaps balanced among memorization and explanation, analyses and evaluations.

In using case methods (and other approaches) to teach students about STSE relationships and RiNA projects, there are several specific perspectives and practices that likely should be shared with them, depending on students' ages and stages of development, using certain teaching/learning strategies. Many of these are shared in the Documentaries set of chapters provided in this volume. Perhaps most fundamentally, students should understand cases like those above in terms of the STSE model in Fig. 2.5. Discussions with students about positions (e.g., 'positive' or 'negative') of various 'stakeholders' regarding possible problems for wellbeing of individuals, societies and environments (WISE) might include claims, for example,

Sample	A	B	C	D	E	F	G	H	I	J	K
Chlorine	0 ppm	0 ppm	0 ppm	0 ppm	0 ppm	0 ppm	0 ppm	0 ppm	0 ppm	0 ppm	1 ppm
Copper	0 ppm	0 ppm	0 ppm	0 ppm	0 ppm	0 ppm	0 ppm	0 ppm	0 ppm	0 ppm	0 ppm
Dissolved Oxygen	4 ppm	4 ppm	4 ppm	4 ppm	4 ppm	4 ppm	4 ppm	4 ppm	4 ppm	4 ppm	4 ppm
Hardness	1 tablet	7 tablets	2 tablets	1 tablet	1 tablet	3 tablets	8 tablets	7 tablets	9 tablets	1 tablet	3 tablets
Iron	0 ppm	0 ppm	0 ppm	0 ppm	0 ppm	0 ppm	0 ppm	0 ppm	0 ppm	0 ppm	0 ppm
pH	6	8	7	6	6	8	8	8	8	6	7

Fig. 2.7 Students' RiNA project presentation

of those supporting and opposing government-sanctioned private sector funding of professional science and/or engineering (e.g., Mirowski, 2011). Useful related discussions with students might include, as noted above, those about private sector influences on public perceptions about risks and doubt regarding commercial products and services (Oreskes & Conway, 2010 [also see: www.merchantsofdoubt.org]). Teachers should, then, share with students various actions people have taken to address perceived problems—perhaps using a graphic like that in Fig. 2.8 as a focus of discussion; asking students, for instance, about the extent to which they would be willing and interested in engaging in each type of action. Angeliki Grundy's chapter (5) in this volume, in which she describes and discusses merits of persuasive writing as a form of action, also should be useful in this regard. In

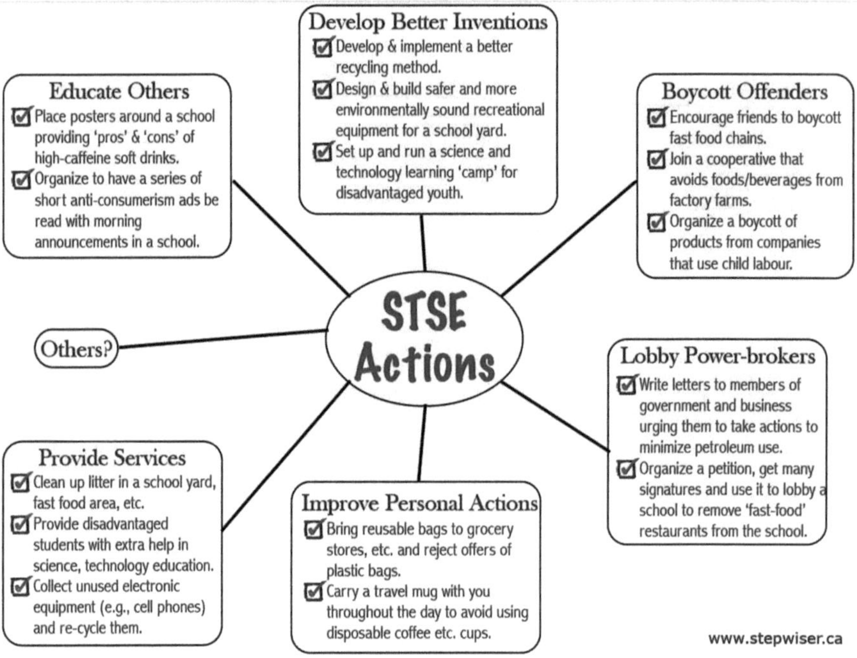

Fig. 2.8 Some STSE action types

discussing such actions, it also is important for teachers to advise students of importance of conducting effective research (with more suggestions below) as bases for action decisions. An effective technique for this is to show students a video depicting citizens' support for banning 'DHMO' (e.g., youtube.com/watch?v=A58XI4pzGU0), helping them realize that people can often agree to actions based on minimal evidence or, perhaps, in terms of social cues like friends' status. In association with such discussions, teachers should show students examples of secondary and primary research methods and results (e.g., Figs. 2.3 and 2.7) people have used as (at least partial) bases for their action decisions.

Students' initiations into STSE relationships and RiNA projects often may be relatively 'basic,' particularly for younger students. As they become familiar with fundamental principles and practices like those noted above, however, teachers can introduce them to increasingly complex relationships, issues, problems, actions and research. As recommended by Clayton Pierce (2013), students can gain perhaps much more enlightened knowledge and perspectives about the nature of STSE relationships (and research, negotiation and actions) if they are taught about aspects of *actor-network theory* (ANT) and how to develop actor-network maps illustrating reciprocal relationships among living, nonliving and symbolic (semiotic) entities ('actants'). He suggests, also, that such enlightenment can be *democratizing*, exposing students to potentially problematic actants (e.g., corporations, government [de-]regulations and transnational think tanks) often omitted from public and educational discussions about for-profit commodities. We have, accordingly, explored, in the last few years, uses of such mapping in various contexts and for various purposes. Implications of uses of ANT did, indeed, seemed to be so significant that ten of the seventeen chapters in the Documentaries section of the book emphasized it. As described by Larry Bencze and Mirjan Krstovic (Chap. 9) and Mirjan Krstovic (Chap. 6), introduction of ANT appeared to help students to accommodate enlightening critical perspectives on commodities commonly consumed by youth, consciousness that they then used to educate peers about relevant controversies and potential problems. Particularly useful in this regard seemed to be the Trojan horse metaphor and the concept of *punctualization* (Callon, 1991); that is, that commodities can be 'black-boxed,' made to appear as idealized isolated entities, distracting consumers from attending to possibly-problematic actants. This line of reasoning was, then, extended to encourage students to critically analyze commodities in ways that seemed to lead them to design inventions that would be functional and be sympathetic to wellbeing of individuals, societies and environments they found to be networked with commodities (Bencze & Krstovic, Chap. 10, this volume). Critiques of technologies, meanwhile, also involved merits of uses of online communication tools (wiki pages) and, through complex sets of reasoning, actor-network theory as an ontological tool (Ramjewan, Zoras & Bencze, Chap. 12, this volume). Such self-critical arguments seem reminiscent of Jacques Derrida's (1998) concept of *deconstruction*, acknowledging complexities and uncertainties in dialectical relationships like those in actor-networks. At the same time, an extension of ANT in the form of Foucault's (2008) concept of *dispositif* (i.e., assemblage of mutually-supporting actants for common purposes) seemed useful in analyzing an ongoing socioscientific controversy between citizen activists and city and corporate interests—a documentary perhaps worth sharing with students as an illustration of competition among power structures (Bencze & Pouliot, Chap. 17, this volume).

Finally, although we might like to think that our teaching about STSE and RiNA projects fairly and accurately represents them, it seems that educators may be—in ways perhaps not unlike that of neoliberal capitalists—subject to ontological and ideological gaps (Fig. 2.4). Those producing educational materials (e.g., textbook publishers, school district personnel & teachers) may, to some extent, misrepresent the 'World' with their 'Signs' (e.g., PowerPoint slide series, videos, written cases, etc.) that, in turn, may be translated (Sign ➔ World) into misinterpretations of the 'World' (e.g., of STSE & RiNA) by students. There are, perhaps, many opportunities for such mistranslations. According to Lilian Pozzer and Wolff-Michael Roth (2003), for instance, there may be a series of steps in World ➔ Sign translations (Fig. 2.4) that involve progressive loss of detail (and representativeness); such as: flower ➔ picture ➔ drawing ➔ graph (e.g., rate of photosynthesis vs. light intensity) ➔ chemical equation (e.g., photosynthesis). Accordingly, teachers need to be mindful of potential for misrepresentation, depending on the nature of objects (e.g., photographs vs. drawings) they use. Students may, for instance, receive printed copies of cells or they may view them on a screen containing a projected image of a cell. In such translations, one can imagine ontological and ideological gaps. The former are, likely, unavoidable. The latter, however, may depend on teachers' purposes. For example, if the cell image already is idealized, perhaps the teacher may want to point this out to students. There are, however, situations in which science teachers may misrepresent STSE relationships through planned or unconscious *omission* of certain problematic details. Indeed, this often appears to be the case, for instance, regarding the extent to which problematic aspects of business-science partnerships are avoided in school science (Carter, 2005).

2.4.3 Products Education

Clearly, students' understanding of STSE relationships (including controversies and problems) and (refer to the next sub-section) abilities to design and conduct research (secondary and primary) and personal and social actions to address perceived problems may, *at least*, depend on all other elements of the STEPWISE tetrahedron. Among these, an element that often will influence teachers' and students' engagement in STSE education and RiNA projects is 'Products Education' (Fig. 2.1)—which tends to dominate science education. Given that educators often find that teaching about STSE (*socioscientific*) issues can motivate students to learn about 'Products' (Venville & Dawson, 2010) and, to satisfy teachers' frequent interest in teaching/learning in this domain, it appears to make sense for teachers using the STEPWISE pedagogical framework (Fig. 2.3) to teach various laws, theories and technologies *as* they teach about STSE relationships and relevant RiNA projects. For example, when teaching students about potential health problems (e.g., heart disease) from 'trans-fats' (solid lipids produced from liquid oil through chemical addition of hydrogen), students also can be taught, for example, about chemistry and biology of lipids (fats and oils).

2.4.4 Students Practise

In light of problems of authenticity of representation of STSE relationships and RiNA projects, despite merits of purposely teaching students about some of them and teachers' attempts to accurately represent them, it seems that students need opportunities to engage more directly in them. Their learning about STSE-RiNA should not be—essentially—*vicarious*. Indeed, given Wenger's (1998) claim that deep, committed, learning occurs when learners have considerable control over both directions of the World ← → Sign dialectic (Fig. 2.4), the more students should have opportunities to influence decisions over representing (researching) the World of phenomena and over using (acting with) their representations (Signs) to attempt to affect changes in the World. Accordingly, students need to be engaged (perhaps in groups) in practice RiNA projects to address problematic STSE relationships of their interest ("Students Practise" in Fig. 2.3). Typically, this would involve picking an STSE issue to investigate, both in terms of secondary and primary research, and using findings to inform their development and implementation of actions they believe necessary to rectify STSE problems of their concern. Again, depending on students' ages and levels of expertise, confidence and motivation (e.g., with more or less experience with such projects), teachers may have to provide varying levels of support.

Frequently, despite having been exposed to various RiNA projects in phase 2 of one or more apprenticeship(s) (Fig. 2.3), students often request suggestions for project topics. In doing so, teachers can provide students with ideas at various levels of specificity. Some students may only need basic questions, such as: 'Should genetically engineered organisms be introduced into natural ecological systems?,' 'To what extent should governments take steps to promote public transportation and self-propelled modes of transport (e.g., bicycles) at the expense of private, motorized vehicular transportation?' and 'To what extent should people in the richest countries of the world attempt to address diseases (e.g., malaria) affecting people in poorer countries?' At a slightly higher level of specificity, teachers can provide brief issue descriptions, such as:

> People in various places throughout the world often mediate their social relations through various forms of electronic technologies, including cell phones, personal music players and organizers, video games, television, movies, internet, etc. Among various concerns about these and other technologies is that they may carry with them particular 'messages' (often in the form of instructions for or limitations of use) that may surreptitiously govern people's lives. This often is debated through the concepts of techno- and social-determinism.

More specificity, such as in terms of a greater range of actants, can be provided through 'commodity actant arrays,' like that in Fig. 2.9. Such resources, of course, give students results of some initial secondary research—while leaving room for them to conduct more secondary and primary (e.g., a correlational study of age, gender, etc. vs. drone (non-)use) research that students can then use for informing development and implementation of action(s) they deem appropriate. More information-rich and multimedia versions of such arrays also have been prepared by us using Prezi™ presentation software; such as one dealing with fast and manufactured foods at: goo.gl/zuBb7w. Our preliminary studies of such multimedia arrays suggest, for instance, that teachers often may prefer to adapt them in ways suiting needs and interests of students in their classes (Hoeg et al., Chap. 16, this volume).

In examining various stimuli for topics like those described above, students (perhaps in small groups) should then be required to develop STSE cases. This will involve secondary research (often via the Internet) to determine positive and negative effects on wellbeing of individuals, societies and/or environments due to influences of powerful people/groups on fields of science and technology. In doing so, students should be asked to determine likely 'stakeholders' (e.g., members of government, companies and activist groups) concerned and not concerned about apparent 'problems') and, perhaps, to construct actor-network maps to describe problem(s) in STSE relationships they would like to address.

To supplement what they have learned about the STSE relationship of their interest, students should then be asked to design and conduct one or more *empirical* inquiries—in which investigators base conclusions, at least in part, on experiences

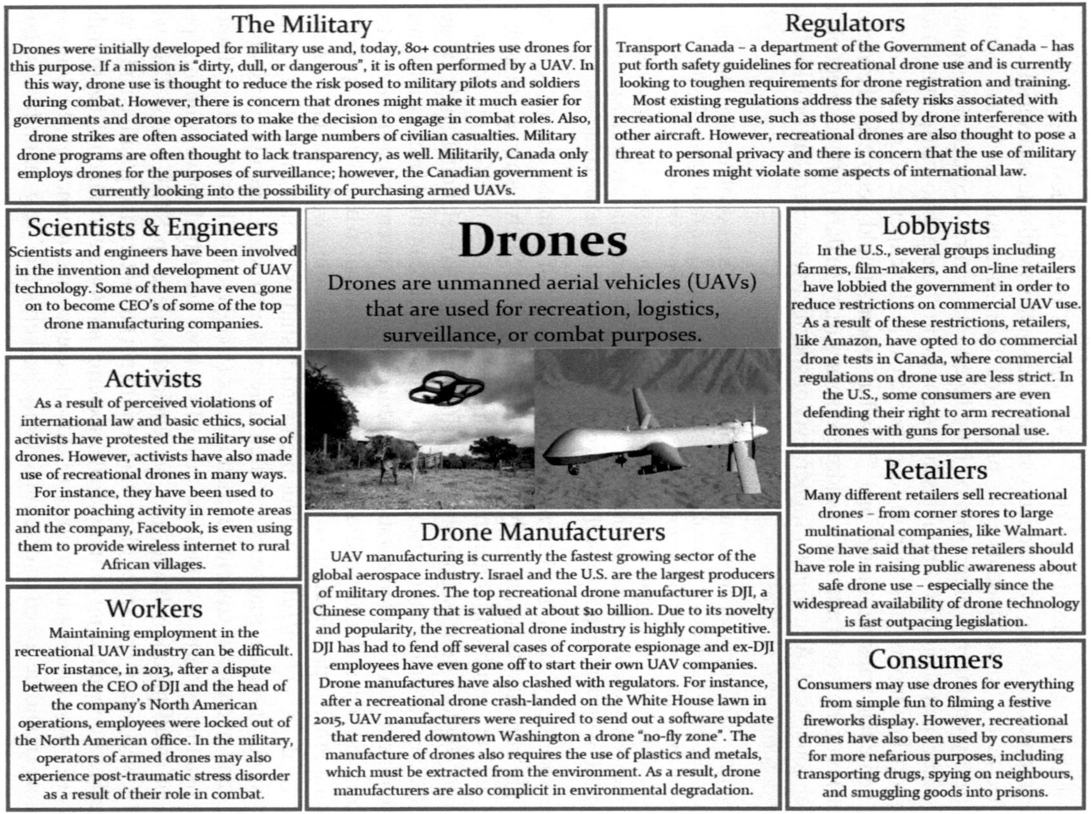

Fig. 2.9 A commodity actant array for 'Drones'

with physical phenomena (including energy). There is considerable disagreement, however, how investigators carry out such inquiries. Although experimentation is commonly associated with science inquiry, it seems appropriate to suggest that investigations into possible problems associated with STSE relationships be carried out as studies, particularly *correlational* studies—in which investigators monitor natural changes in possible independent and dependent variables, such as amount of nicotine (in cigarettes) consumed and cancer rates (Bencze, 1996). Our research suggests that correlational studies work well for teenagers when data and actions are local and social (Bencze & Krstovic, Chap. 7, this volume).

Using findings from their secondary and primary research, students should then be encouraged to negotiate conclusions and plans of action (e.g., from Fig. 2.8) to address problems they identify. In our research, most students tend to choose *educational* forms of action, such as posters, PowerPoint™ presentations, and pamphlets—examples of which have been reported elsewhere (e.g., Bencze, Sperling, & Carter, 2012; Sperling & Bencze, 2010) and in several chapters in this volume. In Fig. 2.7, an example is provided of students' RiNA project results—in this case, dealing with relative merits of bottled vs. tap water, illustrating their poster and data table used in combination with a class presentation and activist video they posted to YouTube™. It also was popular for students to post educational information through social media (Zoras & Bencze, 2014). Much less frequently, students chose to lobby power-brokers—sometimes aimed at companies (e.g., Ramjewan, Zoras & Bencze, Chaps. 12 and 14, this volume), but mostly focusing on local individuals and groups, such as school principals and/or service staff (e.g., Krstovic, 2014). Finally, apart from 'boycotting offenders' and 'providing services' (Fig. 2.8), perhaps the least common action in school science programmes we have studied was for students to design technologies that took into consideration wellbeing of individuals, societies and environments. We have reported some successes with this in an elementary school 'science and technology' course (Wilkinson & Bencze, 2015), and some successes enabling students to *imagine* (rather than build and test) technologies in a senior high school chemistry course (Bencze & Krstovic, Chap. 10, this volume)—although several students did appear to develop critical awareness of and ideas for reform of for-profit commodities, particularly with uses of actor-network analyses.

Despite being shown examples of research in the "Teacher Teaches" phase of the apprenticeship (Fig. 2.3), because students, perhaps ironically, rarely have chances to self-direct science research in science classes, which would help them develop expertise, confidence and motivation for such research, teachers often have to provide students with extra lessons and activities, as part of what may be called a 'skills' (acknowledging their complexity) apprenticeship, to help them develop said expertise, etcetera (Bencze, 2000). To assist with this, a set of sample skills development lessons and activities is provided at: goo.gl/tPILNi. Since these were developed in the early 1990s, teachers now could/should adapt them to include more STSE contexts for the activities. Such apprenticeship lessons and activities can be patterned after the 3-phase apprenticeship in Fig. 2.3. In such lessons and activities, again, students could be shown samples of reports of projects conducted by students (e.g., from JASTE, as noted above).

After various apprenticeship activities like those described above based on the schema in Fig. 2.3, teachers are urged to decide if students are ready to self-direct RiNA projects to address SSIs of their choice or whether they may benefit from another set of apprenticeship lessons and activities. We have found that this depends on ages and educational stages of students, along with the teacher's comfort-level with STEPWISE-informed practices. A group of students enrolled in a grade nine course leading to an 'IB' (International Baccalaureate) programme, for example, only needed one set of apprenticeship activities before they were able to mentor younger students in their RiNA projects and, also benefiting from that mentoring work, effectively design and conduct RiNA projects on topics of their interest/concern (Ramjewan, Zoras & Bencze, Chap. 14, this volume). For many students, however, a second apprenticeship seemed necessary. To supplement the second round of activities, however, we have found that engaging students in activities to reflect on and discuss the nature of RiNA projects (Bencze & Krstovic, Chap. 8, this volume), along with public reporting (in an 'STSE Action Fair') (Krstovic, Chap. 6, this volume), seems to help expedite students' progress towards self-directing RiNA projects. Based on discussions above, for example, it may be particularly important for teachers to share with students the idea that effective actions should be *networked*; that is, single actions, like a poster, can be, for example, 'mobilized' by sharing it via social media, such as Twitter™, Instagram™, YouTube™ and Facebook™. Such a lesson about importance of networked actions is elaborated in Chap. 17 (Bencze & Pouliot) in this volume.

2.4.5 Student-Led RiNA Projects

Finally, after the teacher feels that most—if not all—students are adequately prepared, s/he may choose to give students an assignment—with a range of deadlines for parts of the project—asking them to self-direct RiNA projects to address power-related problems of interest to them in STSE relationships. Examples of students' *self-directed* research-informed and negotiated action projects to address socioscientific issues of interest/concern to them are provided in several chapters in this book, including: Bencze and Krstovic (Chaps. 7, 8, 9, 10, and 11), Krstovic (Chap. 6), Ramjewan, Zoras and Bencze (Chaps. 12 and 14) and Zouda, Nishizawa and Bencze (Chap. 15).

2.4.6 STEPWISE in Science Teacher Education

In parallel with my work with science (and technology) teachers since 2006 based on STEPWISE, I have worked to apply it to my teaching in science teacher education—which is described in Chap. 18 in this volume—and in after-school contexts with youth. In my university-based science teacher education teaching, I had some moderate successes implementing it in the context of mainstream, mandatory, science teaching methods courses. However, for a range of reasons, I found it was necessary to withdraw from that teaching and focus on using STEPWISE as a basis for an *optional* course in science teacher education. That course has shown potential for impacting school systems, particularly as—ironically—government financial austerity measures increased average class sizes for my course, allowing me to 'reach' more student-teachers. Meanwhile, especially in the last 4–5 years, we have attempted to implement STEPWISE in non-school settings, with hopes that lack of school system structural pressures would allow for increased freedom required by our approaches. In Chap. 19 (Sperling & Bencze) in this volume, however, we describe that, while youth involved in after-school programmes can make considerable gains—in terms, for example, of active citizenship orientations—such contexts can be limiting, influenced by different situational factors.

2.5 Summary and Futures

The STEPWISE framework as originally conceived—in its *tetrahedral* form—seems to be based on some sound theories. For it to be successful in science education and science teacher education contexts, however, it appeared to need *simplification*. As a relatively structured linear approach, which conceives of many students as needing a series of 'apprenticeship' lessons and activities to help them to develop expertise, confidence and motivation for eventually self-directing research-informed and negotiated action projects to address socioscientific issues/problems of their choice, it seemed to be somewhat successful. Research reported here and elsewhere into uses of this schema suggest that, under certain conditions, the following kinds of outcomes for students are common: (i) knowledge and understanding of aspects of relationships among fields of science and technology and members of societies (including corporations) and environments that they deem 'positive' and 'negative,' including in light of awareness of often-hidden problematic actants, such as private sector agents; (ii) facility with and motivation for primary research, including as social correlational studies, to learn more about socioscientific issues; and, (iii) facility with and motivation for a range—although mostly educational—of personal and social actions that may help improve wellbeing of individuals, societies and/or environments.

Many educational approaches engaging students in consideration of socioscientific (STSE) issues tend to view citizens (including students) as receivers of expert knowledge who, while sometimes providing experts with feedback, mainly are thought of as exercising influence over experts only indirectly—through periodic elections, for example (Levinson, 2010). STEPWISE-informed approaches discussed in this book, by contrast, appear to view citizens/students as capable of engaging in 'praxis' (e.g., primary research) and 'dissent and conflict' (e.g., critical views of STSE and socio-political actions) (Levinson, 2010). Moreover, such outcomes did not appear to be limited to advantaged students. An important finding, reported in this book in Chap. 13 (Phillips-MacNeil, Krstovic & Bencze) and previously (Krstovic 2014), was that students who tended to not experience successes in their previous science education, which often emphasized more teacher-directed activities aimed at supporting relatively decontextualized, pre-specified, knowledge claims from fields of science and technology, appeared to enjoy and succeed with STEPWISE-influenced pedagogies.

Although reports in the Documentaries section of this book may suggest that STEPWISE-informed perspectives and practices can help students to achieve a range of relevant outcomes that, together, may contribute to improvements in well-being of individuals, societies and environments, contexts for such successes seem relatively rare. Again, as described above, there seems to be some sort of 'invisible hand' largely-confining science education to instruction in achievements of and practices in fields of science and technology. Perhaps, however, through insights gained from perspectives and practices provided by scholars who have provided chapters in the Commentaries section of this book, effective approaches enabling broader uses of STEPWISE—and/or modifications of it—may be achieved.

References

Alperovitz, G., & Daly, L. (2008). *Unjust deserts: How the rich are taking our common inheritance*. New York: New Press.

American Association for the Advancement of Science [AAAS]. (1967). *Science—A process approach*. Washington, DC: Ginn & Co..

Anderson, L. W., Krathwohl, D. R., Airasian, P. W., Cruikshank, K. A., Mayer, R. E., Pintrich, P. R., et al. (2001). *A taxonomy for learning, teaching, and assessing: A revision of Bloom's taxonomy of educational objectives*. New York: Pearson, Allyn & Bacon.

Angell, M. (2004). *The truth about the drug companies: How they deceive us and what to do about it*. New York: Random House.

Bakan, J. (2004). *The corporation: The pathological pursuit of profit and power*. Toronto, ON: Viking.

Barnes, R. L., Hammond, S. K., & Glantz, S. A. (2006). The tobacco industry's role in the 16 Cities Study of Secondhand Tobacco Smoke: Do the data support the stated conclusions? *Environmental Health Perspectives, 114*(12), 1890–1897.

Batson, C. D. (1994). Why act for the public good? Four answers. *Personality and Social Psychology Bulletin, 20*(5), 603–610.

Bell, R. L. (2006). Perusing Pandora's Box: Exploring the what, when, and how of nature of science instruction. In L. B. Flick & N. G. Lederman (Eds.), *Scientific inquiry and nature of science: Implications for teaching, learning, and teacher education* (pp. 427–446). Dordrecht, The Netherlands: Springer.

Bencze, J. L. (1996). Correlational studies in school science: Breaking the science-experiment-certainty connection. *School Science Review, 78*(282), 95–101.

Bencze, J. L. (2000). Procedural apprenticeship in school science: Constructivist enabling of connoisseurship. *Science Education, 84*(6), 727–739.

Bencze, J. L. (2001a). 'Technoscience' education: Empowering citizens against the tyranny of school science. *International Journal of Technology and Design Education, 11*(3), 273–298.

Bencze, J. L. (2001b). Subverting corporatism in school science. *Canadian Journal of Science, Mathematics, and Technology Education, 1*(3), 349–355.

Bencze, J. L. (2008). Private profit, science and science education: Critical problems and possibilities for action. *Canadian Journal of Science, Mathematics, and Technology Education, 8*(4), 297–312.

Bencze, J. L., & Alsop, S. (2009). A critical and creative inquiry into school science inquiry. In W.-M. Roth & K. Tobin (Eds.), *The world of science education: North America* (pp. 27–47). Rotterdam, The Netherlands: Sense.

Bencze, J. L., & Alsop, S. (Eds.). (2014). *Activist science & technology education*. Dordrecht, The Netherlands: Springer.

Bencze, J. L., & Carter, L. (2015). Capitalists' profitable virtual worlds: Roles for science & technology education. In P. P. Trifonas (Ed.), *International handbook of semiotics* (Vol. 1 & 2, pp. 1197–1212). Dordrecht, The Netherlands: Springer.

Bencze, J. L., & Sperling, E. R. (2012). Student-teachers as advocates for student-led research-informed socioscientific activism. *Canadian Journal of Science, Mathematics & Technology Education, 12*(1), 62–85.

Bencze, L., Hewitt, J., & Pedretti, E. (2001). Multi-media case methods in pre-service science education: Enabling an apprenticeship for praxis. *Research in Science Education, 31*(2), 191–209.

Bencze, L., Di Giuseppe, M., Hodson, D., Pedretti, E., Serebrin, L., & Decoito, I. (2003). Paradigmic road blocks in elementary school science 'reform': Reconsidering nature-of-science teaching within a rational-realist milieu. *Systemic Practice and Action Research, 16*(5), 285–308.

Bencze, L., & Hodson, D. (1998). Coping with uncertainty in elementary school science: A case study in collaborative action research. *Teachers and Teaching: Theory and Practice, 4*(1), 77–94.

Bencze, L., & Hodson, D. (1999). Changing practice by changing practice: Toward more authentic science and science curriculum development. *Journal of Research in Science Teaching, 36*(5), 521–539.

Bencze, L., Sperling, E., & Carter, L. (2012). Students' research-informed socioscientific activism: Re/Visions for a sustainable future. *Research in Science Education, 42*(1), 129–148.

Bourdieu, P. (1986). The forms of capital. In J. G. Richardson (Ed.), *The handbook of theory: Research for the sociology of education* (pp. 241–258). New York: Greenwood Press.

Calaprice, A. (2000). *The expanded quotable Einstein*. Princeton, NJ: Princeton University Press.

Callon, M. (1991). Techno-economic networks and irreversibility. In J. Law (Ed.), *A sociology of monsters: Essays on power, technology and domination* (pp. 132–161). London: Routledge.

Carter, L. (2005). Globalisation and science education: Rethinking science education reforms. *Journal of Research in Science Teaching, 42*(5), 561–580.

Council of Ministers of Education, Canada [CMEC]. (1997). *Common framework of science learning outcomes K-12*. Ottawa, ON: CMEC.

Deleuze, G., & Guattari, F. (1987). *A thousand plateaus: Capitalism and schizophrenia* (B. Massumi, Trans.). Minneapolis, MN: University of Minnesota Press.

Derrida, J. (1998). *Of grammatology*. Baltimore: John Hopkins University Press.

dos Santos, W. L. P. (2009). Scientific literacy: A Freirean perspective as a radical view of humanistic science education. *Science Education, 93*(2), 361–382.

Foucault, M. (2008). In M. Senellart (Ed.), *The birth of biopolitics: Lectures at the Collége de France, 1978–1979*. New York: Palgrave MacMillan.

Fuller, S. (2002). *Social epistemology* (2nd ed.). Bloomington, IN: Indiana University Press.

Harlen, W., & Holroyd, C. (1997). Primary teachers' understanding of concepts of science: Impact on confidence and teaching. *International Journal of Science Education, 19*(1), 93–105.

Helms, J. V. (1998). Science—and me: Subject matter and identity in secondary school science teachers. *Journal of Research in Science Teaching, 35*(7), 811–834.

Hileman, B. (1998, August 17). Industry's privacy rights: Is science shortchanged? *Chemical & Engineering News, 76*, p. 36.

Hodson, D. (1986). The nature of scientific observation. *School Science Review, 68*, 17–29.
Hodson, D. (1993). Re-thinking old ways: Towards a more critical approach to practical work in school science. *Studies in Science Education, 22*, 85–142.
Hodson, D. (1996). Laboratory work as scientific method: Three decades of confusion and distortion. *Journal of Curriculum Studies, 28*(2), 115–135.
Hodson, D. (2003). Time for action: Science education for an alternative future. *International Journal of Science Education, 25*(6), 645–670.
Hodson, D. (2011). *Looking to the future: Building a curriculum for social activism*. Rotterdam, The Netherlands: Sense.
Hudson, L. (1967). *Contrary imaginations: A psychological study of the English Schoolboy*. Harmondsworth, UK: Penguin.
Klein, N. (2014). *This changes everything: Capitalism vs. the climate*. Toronto, ON: Simon & Schuster.
Kleinman, D. L. (2003). *Impure cultures: University biology and the world of commerce*. Madison, WI: University of Wisconsin Press.
Krimsky, S. (2003). *Science in the private interest: Has the lure of profits corrupted biomedical research?* Lanham, MD: Rowman & Littlefield.
Krstovic, M. (2014). Preparing students for self-directed research-informed actions on socio-scientific issues. In L. Bencze & S. Alsop (Eds.), *Activist science and technology education* (pp. 399–417). Dordrecht, The Netherlands: Springer.
Leonard, A. (2010). *The Story of Stuff: How our obsession with stuff is trashing the planet, our communities, and our health – and a vision for change*. New York: Free Press.
Levinson, R. (2010). Science education and democratic participation: An uneasy congruence? *Studies in Science Education, 46*(1), 69–119.
Lock, R. (1990). Open-ended, problem-solving investigations—What do we mean and how can we use them? *School Science Review, 71*(256), 63–72.
McMurtry, J. (1999). *The cancer stage of capitalism*. London: Pluto.
Ministry of Education and Training [MoET]. (1999). *The Ontario Curriculum, Grades 9 and 10: Science*. Toronto, ON: Queen's Printer for Ontario.
Ministry of Education [MoE]. (2008). *The Ontario curriculum, grades 9 and 10: Science*. Toronto, ON: Queen's Printer for Ontario.
Mirowski, P. (2011). *Science-mart: Privatizing American science*. Cambridge, MA: Harvard University Press.
Oreskes, N., & Conway, E. (2010). *Merchants of doubt*. London: Bloomsbury Press.
Osborne, R., & Wittrock, M. (1985). The Generative Learning Model and its implications for science education. *Studies in Science Education, 12*, 59–87.
Pedretti, E., & Nazir, J. (2011). Currents in STSE education: Mapping a complex field, 40 years on. *Science Education, 95*(4), 601–626.
Pierce, C. (2013). *Education in the age of biocapitalism: Optimizing educational life for a flat world*. New York: Palgrave MacMillan.
Pozzer, L. L., & Roth, W.-M. (2003). Prevalence, function, and structure of photographs in high school biology textbooks. *Journal of Research in Science Teaching, 40*(10), 1089–1114.
Roth, W.-M. (2001). Learning science through technological design. *Journal of Research in Science Teaching, 38*(7), 768–790.
Roth, W.-M., & Désautels, J. (Eds.). (2002). *Science education as/for sociopolitical action*. New York: Peter Lang.
Sadler, T. (Ed.). (2011). *Socio-scientific issues in the classroom: Teaching, learning and trends*. Dordrecht, The Netherlands: Springer.
Sadler, T. D., Barab, S. A., & Scott, B. (2007). What do students gain by engaging in socioscientific inquiry? *Research in Science Education, 37*(4), 371–391.

Schwartz, R. S., Lederman, N. G., & Crawford, B. A. (2004). Developing views of nature of science in an authentic context: An explicit approach to bridging the gap between nature of science and scientific inquiry. *Science Education, 88*(4), 610–645.

Sperling, E., & Bencze, J. L. (2010). 'More than particle theory': Citizenship through school science. *Canadian Journal of Science, Mathematics, and Technology Education, 10*(3), 255–266.

Springer, S., Birch, K., & MacLeavy, J. (Eds.). (2016). *The handbook of neoliberalism*. New York: Routledge.

Steinberg, S. R. (2010). Barbie: The bitch can buy anything. In J. A. Sandlin & P. McLaren (Eds.), *Critical pedagogies of consumption: Living and learning in the shadow of the "Shopocalypse"* (pp. 148–156). New York: Routledge.

Venville, G. J., & Dawson, V. M. (2010). The impact of an argumentation intervention on Grade 10 students' conceptual understanding of genetics. *Journal of Research in Science Teaching, 48*(8), 952–977.

Wenger, E. (1998). *Communities of practice: Learning, meaning, and identity*. Cambridge, UK: Cambridge University Press.

Wilkinson, T., & Bencze, L. (2015). With head, hand, and heart: Children address ethical issues of design in technology education. In K. Stables & S. Keirl (Eds.), *Environment, ethics and cultures: Design and technology education's contribution to sustainable global futures* (pp. 231–244). Rotterdam, The Netherlands: Sense.

Wood, G. H. (1998). Democracy and the curriculum. In L. E. Beyer & M. W. Apple (Eds.), *The curriculum: Problems, politics and possibilities* (pp. 177–198). Albany, NY: SUNY Press.

Ziman, J. (2000). *Real science: What it is, and what it means*. Cambridge, UK: Cambridge University Press.

Zoras, B., & Bencze, L. (2014). Utilizing social media to increase student-led activism on STSE issues. In J. L. Bencze & S. Alsop (Eds.), *Activist science & technology education* (pp. 435–449). Dordrecht, The Netherlands: Springer.

Part II
Documentaries

Overview

The chapters in this section provide accounts of STEWISE frameworks in practice. Authors include teachers, graduate student researchers and me. These accounts were produced in relation to analyses of data collected in action research intended to understand efficacy of STEPWISE in promoting research-informed and negotiated actions to address socioscientific issues/problems of interest to students, along with factors affecting such outcomes. In many cases, authors integrated various theoretical perspectives in their descriptions and analyses of teaching/learning situations in schools, in an after-school context and in the context of a science teacher education course

The Chapters

Chapter 3
School Science Ruling Relations and Resistance to Activism in Early Secondary School Science

Darren Hoeg, Tanya Williamson, and Larry Bencze

3.1 Introduction

A central challenge for teachers aiming to develop activist education is enacting socially just teaching practices that may be poorly supported by the existing school culture. Acquisition of these practices requires a supportive *environment*, or what Roderick Watts, Nat Williams, and Robert Jagers (2003) term an *ecology*. An ecology effective for activism is a local context that enables sustained, repeated engagement with relevant social issues that is affirmed through meaningful social change (Watts, Griffith, & Abdul-Adil, 1999). A supportive ecology enables individuals to acquire knowledge about social inequities, develop awareness and empathy, and enact new behaviours to change relationships reproducing these inequities (Watts, Williams, & Jagers, 2003). School ecology is communicated through complex institutional networks that often value particular, reproducible types of science teacher practices, such as formulaic laboratory practica, and common science inquiry activities, that can constitute an ecology that is resistant to activism (Hoeg, 2016). This chapter describes experiences of Amy, teaching in a grade 10 science classroom in Ontario, Canada, as she encounters both supportive and resistant school ecologies while developing and implementing activist science education experiences.

D. Hoeg (✉)
Department of Secondary Education, University of Alberta, Edmonton, AB, Canada
e-mail: dhoeg@ualberta.ca

T. Williamson
Halton District School Board, Burlington, ON, Canada
e-mail: williamsont@hdsb.ca

L. Bencze
Ontario Institute for Studies in Education (OISE), University of Toronto, Toronto, ON, Canada
e-mail: larry.bencze@utoronto.ca

L. Bencze (ed.), *Science and Technology Education Promoting Wellbeing for Individuals, Societies and Environments*, Cultural Studies of Science Education 14,
DOI 10.1007/978-3-319-55505-8_3

3.2 Science and Technology Education Promoting Wellbeing for Individuals, Societies and Environments (STEPWISE)

Socioscientific issues (SSIs) have been described as social issues related to practices and products of science and technology (Zeidler, Sadler, Simons, & Howes, 2005). Many of these issues, such as debates about climate change and loss of bio-diversity due to human settlement and deforestation, require citizen attention and action. According to Derek Hodson (2003), one of the aims for science education incorporating SSI's is to engage students in sociopolitical action, or activism, which he describes as "acquiring the capacity and commitment to take appropriate, responsible, and effective action on matters of social, economic, environmental and moral-ethical concern" (p. 658). This aim supports arguments for what we term *socioscientific activism* to be taught. Socioscientific activism is supported in school science through curricular developments such as Socioscientific Issues (SSIs) and Science, Technology, Society and Environment (STSE) education. However, these components are often taught, if at all, as simply add-on content (Hodson, 2003), limiting their potential to engage students in relevant socioscientific issues. The open-ended nature of knowledge constructed as students engage in addressing socioscientific issues may be difficult to implement in school education increasingly committed to preparing students for high stakes exams (Calabrese-Barton, 2012).

STEPWISE is a pedagogical framework that orients student learning and activity toward activism by encouraging and enabling them to use at least some of their science education to address SSI through actions intended for common good. The research and learning activities described in the framework include constructivist and student-directed activities, resulting in students addressing social and ecological problems related to decisions made by powerful people and groups pertaining to fields of science and technology. A key feature of STEPWISE is to engage students in self-directed primary (e.g., their own studies) and secondary (e.g., internet searches) research to help inform their decisions about and actions toward SSIs, often culminating with self-directed actions to address SSIs. STEPWISE is represented in a tetrahedral model in Fig. 3.1.

This model emphasizes an idyllic nature of learning science leading toward activism that is an organic process that cannot be predicted, and one that responds to different other actants as they change. In theory, a teacher can start at any point, because each element is co-dependent. However, this model of learning may be difficult to envision and enact by teachers socialized in science teacher cultures that privilege ruling relations based on progressively gaining expertise and confidence in science content-driven instruction and the authority of science knowledge, making the cyclical and multi-directional nature of the tetrahedral STEPWISE model of enactment an unfamiliar and potentially uncomfortable pedagogy.

In order for activist education such as that described by STEPWISE to be effective, it appears to require a supportive ecology that is responsive to both planned as well as organic and unpredictable learning events (Hoeg, Lemelin, & Bencze,

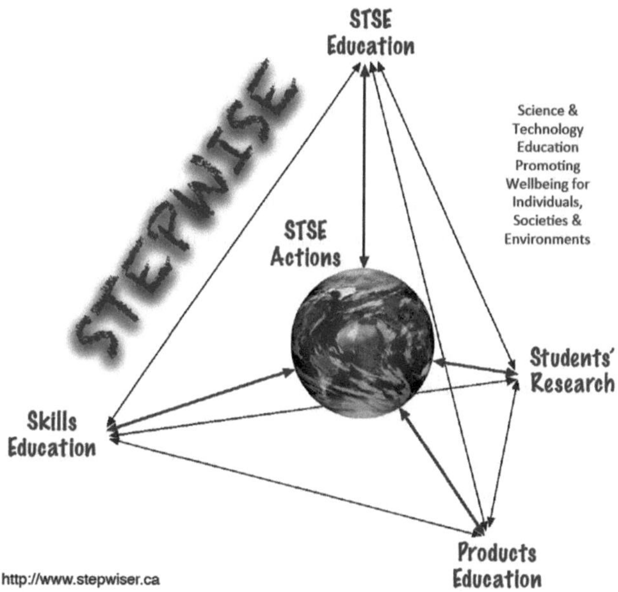

Fig. 3.1 The STEPWISE theoretical framework

2015). Schools, like other social institutions, are characterised by particular ecologies sustained through repeated and empowered social practices articulated through social relations. Social relations can be described as the person-to-person or, more commonly, text-to-person interactions that occur in daily life (Smith, 1999). In order for institutions to maintain their characteristic features in diverse localities, social relations must be coordinated trans-locally. This can be achieved through text and discourse (Smith, 1999), or what might be thought of more holistically as a form of institutional culture. Dorothy Smith (1987) describes institutional culture as being defined by *ruling relations* — enacted social relations that reproduce patterns of activity. Ruling relations are understood as activities conducted by individuals, from which connections to and patterns in the activities of other people, often in distant locations but in the same or similar institutions, can be identified.

Schools can be thought to be organised by ruling relations pertaining to, for example, age (grade levels), ability within age groups (streaming), authority (the teacher) and performance (report cards) (Ball, 2006). Yet, more subject-specific ruling relations also exert their influence on orthodoxies of professional practice of teachers. In school science, for example, ruling relations appear to reproduce authoritative science knowledge — prescribed science knowledge determined by the powers of authority (scientists) — that is taught as a product that must be acquired by students at the expense of their autonomy or democratic determination (Apple, 2004). Authoritative knowledge is intended to be learned as is, rather than interpreted or constructed by teachers and students, resembling the problematic 'banking' type of education described by Freire (1997). Activist science education, however, is largely predicated upon the notion that students are enabled to produce their own knowledge about social phenomenon related to processes and products of science and technology that may not be explicitly articulated in the curriculum. This necessitates forms of science teacher practice that are novel, and may not be supported by existing school science ecologies, requiring teachers to challenge the very ruling relations that are privileged in schools and school science itself. In light of these tensions, we aimed to evaluate the extent to which Amy was able to implement experiences based on STEPWISE, as well as how this was enabled and resisted by school science ruling relations.

3.3 Context and Methodology

Our study examines experiences of Amy as she implemented STEPWISE in her grade 10 Applied Science classroom. The Applied Science course is designed to "focus on the essential concepts of a subject, and develop students' knowledge and skills through practical applications and concrete examples" (Ministry of Education [MOE], 2008). The course provides opportunity for students to learn science used in their everyday lives; thus, making connections between science, technology, societies and environments (STSE), and extended projects that use science to solve problems are frequently enacted by students in this course (Interview, January 19, 2009). Because the course is not preparation for post-secondary science, there is apparently less pressure on teachers to impart the large volume of factual knowledge that is often associated with science and university science preparation courses (Interview, January 19, 2009). In addition, STSE is the first of three learning goals in the Ontario Science Curriculum, and "provides the context for developing the related skills and conceptual knowledge necessary for making connections between scientific, technological, social, and environmental issues" (MOE, 2008, p. 16). The practical nature of the applied course, a lack of emphasis placed on exams, and foci on social and environmental implications of science and technology inherent in STSE, make the Applied Science course, in theory, ideal for the enactment of STEPWISE based science education — although there can be issues of student engagement in science with such students (Aikenhead, 1996).

Our research approaches stem from concepts and practices of Institutional Ethnography (IE). IE is widely attributed to Dorothy Smith, and investigates interactions, or ruling relations, between daily human activity and broader social structures in society (Smith, 1999). In IE, investigations often focus on "what people are doing and experiencing in a given local site that are at the same time hooked into sequences of action implicating and coordinating multiple local sites where others are active" (Smith, p. 52). Researchers using IE frequently are interested in how extra-local social phenomena that act to similarly organize people's actions are also necessarily produced by those very actions. An IE perspective allows us to understand how reproductive ruling relations occurring in Amy's science classroom enabled and resisted implementation of STEPWISE-based science experiences.

As in other ethnographic approaches, we used observations, interviews and analysis of documents as our primary sources for claims. Interviews allowed us to explore how Amy came to understand and develop STEPWISE science experiences through her discussions with Larry, who acted as a process facilitator and researcher, and her observations about its implementation. Initial interviews were open-ended, involving discussion about the STEPWISE framework and its potential enactment, enabling Amy to make personal choices in developing units and lessons based on this framework. Short interviews/discussions focusing on Amy's views about the success of particular activities also occurred immediately after the completion of lessons. Several more formal, semi-structured, interviews were conducted after the semester, to evaluate Amy's views on the STEPWISE-based pedagogy she developed. Classroom observations were conducted each week during the second semester (January–June) when the majority of STEPWISE experiences occurred. Lesson plans, student activity sheet/lesson outlines, and students' work also were examined. In addition, three students were evaluated for their commitment to activism; this occurred through an interview and evaluation of the STEPWISE-based work they completed throughout the semester.

Data analysis in IE attempts to understand how common daily activities of people are aligned with similar practices of people elsewhere, in order to illuminate ruling relations characteristic of this common activity. Since we observed only a single teacher, we consulted the literature to provide broader context for the teaching practices with which Amy may be aligned. Through interpretations of text and raw data, concepts or themes were derived corresponding to ruling relations that influenced Amy's enactment of STEPWISE. Our analysis involved a repetitive and circular coding procedure — deriving, defining and modifying coding categories while reading, rereading and assigning responses to the categories (Wasser & Bresler, 1996), in a manner that corresponds to a general inductive approach (Thomas, 2006).

3.4 Findings and Discussion

Findings of this study illuminate a rocky and uneven pedagogical ecology characterised by ruling relations that appear to have pipelined Amy's teaching practice down familiar and well-worn paths, often resisting new routes and ideal forms of activist science education. Ruling relations appeared to be derived from various spheres of activity, including Amy's past experiences in science and school science, an existing culture of science teaching in which she was engaged, and her views of ideal forms of science education.

Teacher Background and Connections to Activism Whatever social relations exist in schools that tend to cause reproduction of particular practices, individual motivation and interpretation of social settings work within these relations, reifying and potentially changing them, resulting in enactment (Sawchuk & Stetsenko,

2008). Accounting for the individual, what has often been termed agency in the social sciences is, therefore, an important consideration in understanding Amy's development of STEPWISE experiences. Amy completed a B.Sc. in Applied Science with a focus on Biology, Geography, and Geology. She later went to teacher's college eventually becoming a high school science teacher — which she had been doing for 9 years by the time of this study. Amy recently completed a Masters of Arts (education) degree, during which she examined impacts of environmental place-based education on student perceptions of school and learning and how experience impacted their knowledge and attitudes about environments.

Amy's dispositions toward science, technology and society (STS) were obtained during targeted interviews. Amy disagreed with placing science as separate and self-contained from technology and society, stating "Society is better informed by science," but this is "a two directional process, where science affects society and society affects science. This is also reflected in the researchers who get funded based on what research is deemed socially important" (Interview, January 9, 2009). Amy also suggested that a layer of environment needed to be added to an STS diagram she was shown (Adapted from Figure 1 in Ziman, 1984, p. 4), with arrows indicating how the four realms directly inform each other. "Both science and technology have a positive and a negative effect on society" (Interview, January 19, 2009) — these beliefs suggest Amy is critical of some of the products of science and technology that are inherent to social issues. Such criticism may be necessary for a teacher's sustained commitment to activism (Hoeg, Lemelin, & Bencze, 2015).

Amy's commitment to activism was influenced by past personal experiences and her desire to provide more student-centred and socially-relevant pedagogy. Amy discussed tree planting as a student, which she now considers a form of activism. During her undergraduate degree, Amy went to Taiwan to participate in research on dolphin conservation. "I really credit a lot of my interest and commitment to the environment and conservation to my time working and living in Asia." Amy said she was affected by the bursting, densely populated cities of Asia, which she described as very polluted and toxic to the natural environment. These experiences are reflected in how she "wants to see kids view the effects of humans on the environment" (Interview, January 19, 2009) after doing lessons and activities in her class. Amy also indicated an STSE course she took during her graduate degree modelled discussion based, student centred and project oriented (a class debate) approaches, and these influenced her desire to explore these more frequently in her teaching.

Despite these goals, Amy expressed considerable insecurity about her ability to change as a science teacher to a degree necessary to implement non-traditional approaches, such as STEPWISE. This suggests an awareness of both her personal limitations as well as some of the ruling relations in schools and school science that might resist these approaches. For example, Amy lacked confidence in her suitability to be involved in STEPWISE, claiming, "I would not really say that I am an activist. And, if I don't have confidence in myself as an activist, how can the kids have confidence in themselves?" (Interview, January 19, 2009). Her lack of conviction about being an activist appears to be contradicted by her involvement in tree planting, suggesting uncertainty about what activism might entail. Despite these inconsistencies, Amy was determined to change her teaching practice and try new approaches. Amy was critical of her mentors during teachers' college who were quite traditional in ways in which they approached science education, motivating her to separate herself from those traditions. Through trying a critical thinking approach, she hoped to get away from the more didactic approach of "[Amy imagining another teacher talking] 'Here's the textbook, let me write the lesson on the board and now off you go.' I want to involve more critical thinking and hands-on activities in my lessons." (Interview, Feb 27, 2009).

Implementing STEPWISE Amy met with Larry six times prior to and during the 2008–2009 school year to discuss science education, activism, STSE, and STEPWISE. During these discussions, a major decision about STEPWISE occurred — largely due to Amy's struggles with bridging the tetrahedral version of the framework with dominant pedagogical perspectives and, as she said, "realities of the classroom" (Interview, Jan. 19, 2009). In particular, while Larry emphasized that "there should be a point in which the kids are able to make decisions about their research where they don't need the teacher anymore, and take action," Amy responded that she was unsure how that could be accomplished "without [teachers] having some control over what they do" (Interview, Jan. 19, 2009). Amy's discomfort with the tetrahedral version of STEPWISE (Fig. 3.1) led Larry to develop — first as a hand-written sketch sitting with Amy—a re-arranged version of it, like that in Fig. 3.2, that they felt might be more practical for teachers and for students.

The timeline represents learning as occurring in a more linear-progressive manner, with a definite start and finish, that models the "unit of learning" conceptualisation of education dominant in schools (Roth, 2013). Amy was able to implement several learning experiences based on this more practical, linear, version of STEPWISE during the 2008–2009 school year. The activities for which data were collected are presented in Table 3.1.

Although these activities do not each represent a start to finish completion of the timeline in Fig. 3.2, each represents at least one part of it. For example, Amy used the helicopter drop experiment as an apprenticeship to teach students about doing science inquiry — the intention is that they will later use these skills in more student-guided experiments related to activism. Amy was thoroughly impressed by the students' levels of engagement in the STEPWISE activities. She felt the students were excited by the chance to do hands-on experiments and she noticed a positive change in classroom behaviour; students were 'acting-out' less because they were concentrating on their work. For example, Amy felt the student posters developed around the question of "Why is learning chemistry important?" were effective because "students are more engaged in science if they know why/how the knowledge is relevant to their everyday lives" (Interview, Feb. 10, 2009). Students' posters listed reasons such as "to further enhance your knowledge," "to know which chemicals are good or bad," "for fun!," "to learn what stuff is made of," "to know what to do if a bad reaction happens," "more job opportunities, "new discoveries," and "to know where to dispose harmful chemicals." Such relevancy is especially important to students in an Applied science course, said Amy, because "they will not be doing

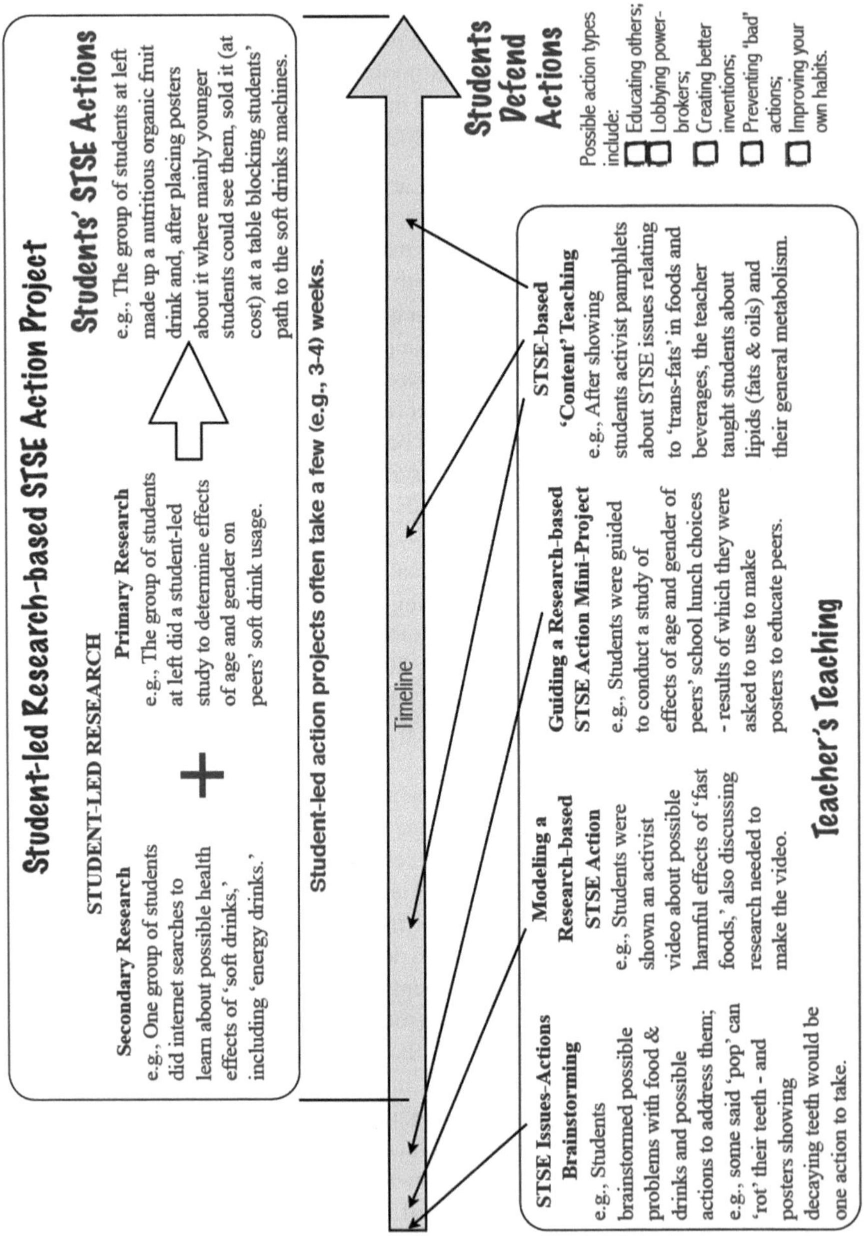

Fig. 3.2 The STEPWISE education timeline

Table 3.1 Stepwise units and activities enacted by amy

Unit	STEPWISE activity	Description of activity
Chemistry	Why study chemistry? (Brainstorming)	This activity provides a diagnostic assessment on student attimdes towards chemistry before and after the unit. Students were encouraged to answer this question in their journal before the unit began. Students answered the same question once the unit was complete and responses were compared to see how perceptions changed over the course of the unit.
Chemistry/E cology	The Story Of Staff (Modelling research based STSE activism)	Students were given an anticipation guide to answer before and after watching the movie. A discussion ensued regarding the effect of 'stuff' on the environment and how student choices (i.e. what they buy) impacts the environment and others in the world.
Motion	No texting while driving (Students STSE Actions)	Tins investigation created awareness on how distractions affect reaction time. Students attempted to catch a metre stick with and without texting or talking on the phone. Students found that talking and texting increased the amount of time it took to catch the ruler. The results were applied to the issue of distracted driving.
Motion	Parachute Drop activity (Apprenticeship)	Students investigated what variables affect the drop time or precision of a parachute drop. Results were used to discuss how parachutes are used in military and rescue operations.
Ecology	Soy Bean growth experiment (Apprenticeship)	Tins activity examined the impact of 'acid rain' on the growth of bean plants. Students used varying amounts of acid to water their bean plants.
Ecology	Determining the Effects of Pollutants on Freshwater Invertebrates (Daphnia) (Apprenticeship)	Tins activity examined the effects of various chemical and pollutant on the growth and survival of Daphnia. a fresh water invertebrate. Live Daphnia were provided to students, who then designed inquiry based activities to test the effects of various pollutants on the survival of the Daphnia,
All year	Science Journal (reflections)	Students were encouraged to write their thoughts and opinions related to what they learned in science class. Several entries were made throughout the semester.

science in university; and, my kids need science to be accessible because they may have had bad experiences with it in the past" (Interview, Feb 24, 2009). After being taught chemical concepts, students completed a journal entry, in which several commented that they would want to use their new knowledge to help people. Amy was pleased with successes of this activity, commenting, "I would like to do this every unit, and it was a good assessment piece. The idea that some of them tied in education, to use chemistry to teach other people, I mean it's a form of activism, teaching other people" (Interview, Feb. 27, 2009). After gaining some technical expertise implemented STEPWISE experiences, she remarked, "I have no idea how I was teaching before! I'm embarrassed to think about how I was teaching because I feel so strongly that this [STEPWISE] is the way I want to organize my classes" (Interview, July 21, 2009). These positive results indicate, to some extent, development and enactment of STEPWISE educational experiences, although certainly, these still appeared to be limited by personal and institutional factors.

3.5 Resistances to STEPWISE-Informed Activism

Although Amy implemented some STEPWISE-based education practices, she experienced significant challenges to their enactment. Dominant ruling relations that sustain conservative practices connected to teaching prescribed knowledge in school science appeared to resist enactment of STEPWISE, and re-oriented Amy's teaching practice back towards more conservative conceptions of teaching and learning. These ruling relations are discussed in the proceeding sections.

The "Unit" Model of Teaching and Learning A dominant pedagogical structure of school science is the discrete unit of study, defined by a clear beginning, a progressive increase in students' skill and knowledge, terminating with some product that will be assessed/evaluated (Ball, 2006). Teachers may see their jobs as developers and deliverers of such 'units' of education (Wallace, 2012), and seldom has this taken-for-granted metric of teaching and learning been problematized (Roth, 2013). The unit can be seen, however, as a reductive, and awkward, if not unnatural and oppositional, organization of human learning (Stetsenko, 2012).

The requirement of using the unit as a metric of education is communicated implicitly in the Ontario curriculum through its division of related content into the traditional disciplines of Biology, Chemistry, Earth and Space Science, and Physics — each of these might be seen as a discrete unit of study, propagating a fragmented view of science learning. The curriculum also advocates for a common method of unit planning, known as 'backward design,' an approach supported in *Understanding By Design* (UBD) (Wiggins & McTighe, 1998). In such unit planning, teachers start with the learning outcomes (content) they expect students to achieve, and plan backward towards assessment tools to measure, and then the specific lessons that will lead to student acquisition of skills and knowledge articulated in the curriculum. The STEPWISE pedagogical framework in Fig. 3.2 makes a compromise between a more ideal form of activist learning (e.g., Fig. 3.1) and realities and structures of contemporary school science that can, unfortunately, also be seen to continue traditions of representing learning as occurring through a unit of study.

A conceptualisation of learning based on the discrete unit was apparent throughout the data. For example, each activity presented in Table 3.1 was part of a larger unit of study. During initial unit planning, Amy stated, “I am struggling to come up with a WISE related skills activity for my grade 10 motion *unit* for an applied class” (Amy, email communication, 2009). This suggests that viewing activism as occurring in and through a unit was problematic for Amy, as unit content often did not provide obvious and clear direction for potential activism. This struggle is a common one for teachers committed to providing activist opportunity for students — how can activist education, requiring unknown lengths of time and unpredictable experiences, be represented as a cohesive body of knowledge, skills and activity (the unit) that is conveyed to students and evaluated? Several features of this incongruity are apparent in Amy’s struggles, and suggest other school science ruling relations connected to the discrete content-unit that resist activist educational frameworks such as STEPWISE.

Linear-Progressive Conceptions of Learning Learning is traditionally thought of as a linear process in which the learner progressively accumulates the skills and knowledge required to achieve some desired outcome (Treagust & Duit, 2008). This process initiates from the students’ existing mental frameworks, conceptions or discourses that, assisted by teachers through the activity they construct, lead to some new state of knowing that matches predetermined learning goals. New knowledge is theorized as the logical outcome of students’ progressive actions that are the consequences (effects) of the learning intentions (causes) of teachers (Roth, 2013). Yet, alternative *process philosophies*, such as that of Deleuze, discussed by Jesse Bazzul and Shakhnoza Kayumova in this book (Chap. 30), conceptualise learning as an unknown “event in-the-making” (Roth, 2013, p. 2), diffuse and sporadic (Watts et al. 2003), challenging linear-progressive views of learning that appear to dominate school and teacher cultures (Wallin, 2012). Process models of learning suggest learning as something unfinished and therefore as something that *cannot* be grasped because it does not yet exist. Learners and teachers do not know what is to be learned; therefore, they cannot intentionally orient to constructing future knowledge or discourse (Roth, 2012).

A model of learning based on a linear progression, however, remains a cherished way of thinking about curriculum and learning (Roth, 2013), and works toward restructuring or, to use Deleuze and Guattari’s term (1987), ‘reterritorialising’ new forms of science education. *Reterritorialization* is a process by which practices are re-appropriated, re-habitualized, or held together such that acting and thinking differently, such as is required by a teacher attempting to implement activist education, is more difficult. Evidence of a linear progressive conceptualization of learning that reterritorialized Amy’s teaching practice was apparent in the data we collected. For example, when asked what she would change about her planning of activities, Amy answered: “I would change the whole approach to it…..I found that I jumped all over the place, and there was really no *progression*” (Interview, Feb. 24, 2009). This demonstrates Amy’s dissatisfaction with the temporal logic (progression) of many activities; she felt they were “disorganized and very chaotic lessons” (Interview, Feb

24, 2009), suggesting the influence of dominant ruling relations that work to structure Amy's conceptions of learning and teaching as occurring along a linear progression. During the parachute drop activity, Amy felt students' skills of accurately measuring distance and correctly timing the drop were weak and that she had not adequately prepared students for the activity. As a result of their poor progress, she felt the results were unreliable, causing her to regress back to an earlier point along the linear progression of the activity, so students could complete a successful parachute drop. Finally, when asked about activism, Amy said "Action is the last step. Students would progress toward that" (Interview, Feb. 24, 2009). This suggests a conception of a linear progression toward a terminal point, a known event, characterised by a learned act of activism. These tensions were in large part the reason for the development of the pedagogical version of STEPWISE (Fig. 3.2), itself a linear-progressive model of enactment for activist science education upon which Amy's activities were based. Yet, the linear progressive model of education may be problematic for the development of socio-political orientations required for activism, as these appear to occur more diffusely through repeated engagements with social issues and encountering new events, during which students construct new personal knowledge about these issues (Watts et al. 2003). Thus, activist education may better be served through models of education based on process philosophies of learning.

Content-Driven Instruction A requirement of linear progressions of learning is that students and teachers have a learning goal toward which progress occurs — usually toward prescribed learning outcomes (Wallin, 2012), what is often considered *content*. Content is prescribed by the Ministry of Education in Ontario (e.g., MOE, 2008), and is a priority for school science, potentially compromising teachers' development of learning experiences that might lead to unknown events, such as student-centered experiences in which they construct knowledge (Bencze, Sperling, & Carter, 2012). Thus, *covering the content* through authority-centric practices, such as didactic teacher lecture and formulaic labs, seem to have become dominant 'ruling relations' in school science (Hodson, 2009).

The normalisation of *covering the content* was apparent when speaking to Amy. She demonstrated frustration about a culture of school that prioritised content teaching, stating "schools only want to focus on exam preparation, they won't do the [STEP]WISE focused experiments" (Interview, February 27, 2009). Amy also expressed concern that both classes had to write the same exam, stating, "It's (the exam) so traditional, multiple choice and content based, it doesn't reflect the hands-on approach I want to be taking with them (her students)" (Interview, February 27, 2009). A requirement to prepare her students for the exam made the STEPWISE activities an extra assignment for Amy's students, who questioned her about why they had to do more assignments than their peers in other classes. This content-driven instruction can be seen as a ruling relation that Amy resisted in order to implement STEPWISE lessons, yet it imposed an institutional expectation that restricted the amount of time Amy was availed to enact these lessons. Despite valuing a hands-on, more open-ended and activist approach to learning, Amy herself demonstrated an internalisation of a content-driven ruling relation, stating "although I love how engaged students are, I wonder if they are learning the science they really need to know" (Interview, Feb. 27, 2009). The fact that the culture of science teaching at her school valued content driven instruction suggests Amy, and her students, were attempting to enact activist education in a seemingly non-supportive ecology.

Content-driven pedagogy influences enactment of teaching and learning practices in which tendencies of conformity are developed, prevent students and teachers from drawing their own conclusions, and critiquing knowledge and those who control it (Wood, 1998), making it a poor choice for activist-based science education. A systemic school-wide preoccupation with providing students with knowledge to consume potentially reconstitutes activist education as simply an exercise in acquiring another type of authoritative content knowledge (Santos, 2009). Dominance of content-drive instruction suggests it is a ruling relation that reterritorialises novel ways of teaching and learning that are required for social justice education such as activism. This objectification of socio-political and socio-scientific practices that are the target of activist education could lead to ambivalence towards the very communities and environments oppressed through these practices (Beyer, 1998).

Counter-Cultural Practice Amy's implementation of novel STEPWISE based experiences seemed to be resisted by established practices, or ruling relations, that are part of science teacher "cultural memory" (Handa & Tippins, 2011, p. 2), constituted in a repository of tried-and-true approaches that have become institutionalised in school science (Hoeg, 2016). Rethinking these practices is difficult because it requires a "nomadic way of thinking curriculum" (Reynolds & Webber, 2004, p. 16), in which familiar concepts, cause–effect relations, and "bifurcated opposition[s]" (p. 16) are suspended. These challenges may be expected, as teachers have developed pedagogical content knowledge, knowledge about *how* to teach various science concepts, that has been structured by dominant, content-oriented teaching practices. As an example of Amy's difficulty in envisioning *counter-cultural practices* (Hoeg & Bencze, 2014), she said, "I just can't think of any sort of [STEP]WISE activity that can go along with the mousetrap cars. I haven't taught this course in a while, but the students' summative [final assessments] in classes past is to design a mousetrap car. But what sort of action can they take once they gather data?" (Interview, Feb. 10, 2009). Amy mentioned that finding connections to larger STSE issues was tricky when exploring a topic like chemical naming/formulas, which she viewed as fact-based content that is typically imparted to students. Even Larry, a former science teacher himself, was not immune to the challenges in overcoming cultural memory, when he commented about planning activist experiences with Amy, stating, "I admit this is difficult .... but, what a sense of relief when you come up with an idea!" (Larry, e-mail communication, 2009). Amy mentioned she was looking forward to a unit on weather or ecosystems, where STSE issues were more obvious, and established teaching practices were already better aligned with STEPWISE.

Activism education such as that described in STEPWISE likely requires a re-orientation, in which potential educational experiences start, for example, from an issue, or problem, rather than specific content knowledge. The change in perspective about how to create educational experiences for students creates a challenge for teachers that takes considerable creativity, courage, and counter-cultural perspectives to overcome.

Individualism Learning commonly is theorized in terms of individual and social construction of concepts or discourse (Scott, Mortimer, & Aguiar, 2006). Yet, teaching approaches based on the linear-progressive models of learning valued by many science teachers are designed to impart and evaluate knowledge acquired by the *individual* student, which may be oppositional to communitarian intentions of activist education (Bencze & Carter, 2011). Additionally, quick delivery of scientific content likely makes students (and teachers) become individually competitive (Eisenhart, Finkel, & Marion, 1996). Each educational experience designed by Amy was to be completed by and evaluated for each individual student. Practices privileging the individual in lessons and assessments likely have a subliminal persuasive effect (Hardt & Negri, 2009), imparting values of individuals onto students. Their sense of individualism as it related to learning chemistry can be seen in students' comments, such as "to further enhance your (my) knowledge", "to know which chemicals are good or bad for you (me)", "for (my) fun!", and, "more job opportunities (for me)". The taken-for-granted value of the individual in school was demonstrated by Amy, when she commented, "it was quite successful in my opinion because students love to talk and they enjoy connecting science to their *own* lives" (Interview, Feb. 29, 2009), rather than connecting science to society or the community.

Student's Commitment to Activism Several of Amy's students were evaluated for their commitment to activism, based on the following criteria developed during previous STEPWISE research: (1) Passionate indications (e.g., statements about issue and/or action) to address issues; (2) Clear indication of intention to implement actions; (3) Confidence in effectiveness of action ('outcome expectancy'); (4) Student self-efficacy (i.e., that s/he feels capable of effectively implementing the action); (5) Detailed analysis/planning of action; and (6) Number and variety of actions (Bencze, Sperling, & Carter, 2012). Evaluation was based on students' work throughout the course, students' written reflections, and student interviews. Students could score from 1 to 5 for each of the indications of activism identified in Table 3.2, below, with 1 being the lowest level of indication, and 5 being the highest.

Students generally were not very emotionally engaged with the issues they studied, nor were they enthusiastic about taking action. Students' intention to take action, however, was relatively well articulated. For example, Sandy, who's primary research was determining the oxygen level in water in relation to temperature, identified "organizing rallies, giving announcements, creating posters, telling people what you learnt" (Interview, June 9, 2009) as actions she would like to carry out to reverse global warming. These remained more of a theoretical proposition, however, as she demonstrated no clear intent to carry out these actions, stating, "It's

Table 3.2 Students' commitment to activism

Indications of activism/3	Sandy	Melita	Ryan
Passionate indications	1	1	1
Clear indication of intention to implement action	2	2	2
Confidence in effectiveness of action	1.5	2	1
Self-efficacy	1	2	1
Detailed analysis/planning of action	1	2.5	1
Number and variety of actions	0	1	0
Mean commitment to activism	1.1	1.7	1

kinda hard…because since global warming is such a big issue…you can't really go around the world and change the water temperature". In her final reflection, Sandy wrote, "Hopefully, the actions you take will get people to stop littering and polluting the environment" (Reflection), indicating uncertainty about how effective would be this action. Ryan, who's final project involved adding an acid to a small bean plant to determine its effects on soybean plant growth (finding that the acid stunted growth), identified "turning off lights and electronic things when they are not in use" and not "burning certain things" (Final reflection) as two actions he could take to save energy and reduce nitrate emissions. Ryan, however, also demonstrates little clear intent to carry out these actions, stating, "I *guess* I could use what I learned in this [acid project] about acid rain affecting plant growth to not leave on lights and electronic devices," suggesting a lower level of intent.

Student self-efficacy was also relatively low among the three students. In an anticipation guide before watching the Story of Stuff (storyofstuff.org), Ryan circled "Disagree" in his answer to the question: "I can make a difference in creating a more sustainable and equitable society". In her final reflection, Melita, who collected data on the number of plastic bottles consumed each week by students (finding 300 plastic bottles on average were consumed), stated, "I *could* help the environment by not using plastic bottles" (Interview), suggesting that she is not completely confident that her personal actions will be effective. What may be most noteworthy about students' commitment to activism was their inability to take actions to change the issues they investigated. Only Melita took any observable form of action, producing a petition to oppose the selling of bottled water, for which she obtained 60 signatures. However, once the petition was complete, no further action was taken.

The low level of action among students could be explained in part by the fact that Amy introduced this project in the last week of school. She stated in her final interview that the projects were rushed: "students did not have enough time to do more elaborate projects, to reflect on more creative forms of action, and to actually carry out those actions" (Interview, 2009). Although lack of time is certainly a feasible explanation for students' low commitment to activism, we suggest that "running out of time" in large part stems from a pedagogical model based on the linear progressive unit of study. With a definite beginning and end based on acquisition of required content, time becomes a subsuming factor. The discrete, linear-progressive model of learning appeared to limit how Amy (and likely other teachers) perceived possibilities and boundaries of science teaching practice. This model is likely validated and empowered by the linear model of STEPWISE (Fig. 3.2), potentially providing limitations to a more ideal enactment of activist education that leads to more significant commitment to activism among students (Hoeg et al., 2015).

3.6 Conclusion

This study highlights several relevancies of theory and practice pertaining to enactment of activist education by school science teachers. Curricular innovations, such as STSE and SSI, can be seen as enabling a more supportive ecology for the enactment of activist science education. These reforms may allow for changes in traditions of teaching cultures, and characteristic *ruling relations* that comprise them. Yet, extant and apparently dominant ruling relations connected to individualism, linear-progression models of learning, and discrete units of study based on content acquisition, appeared to limit students' sustained engagement with social issues that may lead to unplanned and *process* oriented learning more characteristic of sociopolitical development (Watts et al., 2003). We suggest these ruling relations may act to re-structure, or reterritorialise, novel social justice oriented pedagogical approaches, such as those required by STEPWISE, back towards conservative, dominant teacher practices. Although the linear STEPWISE timeline does not explicitly forbid a sustained and process oriented exposure to an issue or topic with unknown outcomes, we suggest its clear resemblance to a discrete, linear-progressive unit model of education that already may be reproduced through extant ruling relations make this a problematic representation for STEPWISE activism. Science teachers using the STEPWISE timeline may need considerable professional support, and to be provided with successful examples of its enactment in similar institutional ecologies, to develop activist science education that can lead to significant sociopolitical development of students.

References

Aikenhead, G. S. (1996). Science education: Border crossing into the subculture of science. *Studies in Science Education, 27*, 1–52.

Apple, M. W. (2004). *Ideology and curriculum* (3rd ed.). New York: Routledge Falmer.

Ball, S. J. (2006). *Education policy and social class*. Oxon, UK: Routledge.

Bencze, J. L., & Carter, L. (2011). Globalizing students acting for the common good. *Journal of Research in Science Teaching, 48*, 648–669.

Bencze, J. L., Sperling, E., & Carter, L. (2012). Students' research-informed socio-scientific activism: Re/visions for a sustainable future. *Research in Science Education, 42*, 129–148.

Beyer, L. E. (1998). Schooling for democracy: What kind? In L. E. Beyer & M. W. Apple (Eds.), *The curriculum: Problems, politics, and possibilities* (pp. 245–263). Albany, NY: SUNY Press.

Calabrese Barton, A.M. (2012). Citizen(s') science. *Democracy & Education, 20*(2). Article 12. Available online at http://democracyeducationjournal.org/home/vol20/iss1/12

Deleuze, G., & Guattari, F. (1987). *A thousand plateaus: Capitalism and schizophrenia*. London: Bloomsbury Publishing.

Eisenhart, M., Finkel, E., & Marion, S. F. (1996). Creating the conditions for scientific literacy: A re-examination. *American Educational Research Journal, 33*(2), 261–295.

Freire, P. (1997). *Pedagogy of the oppressed* (New Rev. 20th-Anniversary ed.). New York, NY: Continuum.

Handa, V. C., & Tippins, D. J. (2011). Cultural memory banking in preservice science teacher education. *Research in Science Education, 42*(6), 1–17.

Hardt, M., & Negri, A. (2009). *Commonwealth*. Cambridge, MA: Belknap Press of Harvard University Press.

Hodson, D. (2003). Time for action: Science education for an alternative future. *International Journal of Science Education, 25*(6), 645–670.

Hodson, D. (2009). *Teaching and learning about science: Language, theories, methods, history, traditions and values*. Rotterdam, The Netherlands: Sense Publishers.

Hoeg, D. (2016). *Constitutions of nature by teacher practice and discourse in Ontario grade 9 and 10 Academic science*. Unpublished doctoral dissertation, University of Toronto, Toronto, ON.

Hoeg, D., & Bencze, J. L. (2014). Counter cultural hegemony: Student teachers experiences implementing STSE-activism. In S. Alsop & J. L. Bencze (Eds.), *Activist science and technology education* (pp. 575–596). Dordrecht, The Netherlands: Springer.

Hoeg, D., Lemelin, N., & Bencze, J. L. (2015). Sociopolitical development of private school children mobilising for disadvantaged others. *Cultural Studies of Science Education, 10*(4), 1155–1174.

Ministry of Education, Ontario. (2008). *Grade 9 and 10 Science curriculum*. Retrieved July 11, 2014, from http://www.edu.gov.on.ca/curriculumcouncil/education.html

Reynolds, W. M., & Webber, J. A. (2004). Introduction: Curriculum dis/positions. In W. M. Reynolds & J. A. Webber (Eds.), *Expanding curriculum theory: Dis/positions and lines of flight* (pp. 1–18). Mahwah, NJ: Erlbaum.

Roth, W.-M. (2012). Mathematical learning, the unseen and the unforeseen. *For the Learning of Mathematics, 32*(3), 15–21.

Roth, W.-M. (2013). To event: Toward a post-constructivist of theorizing and researching the living curriculum as event*-in-the-making. *Curriculum Inquiry, 43*(3), 388–417.

Santos, W. L. P. (2009). Scientific literacy: A Freirean perspective as a radical view of humanistic science education. *Science Education, 93*(2), 361–382.

Sawchuk, P., & Stetsenko, A. (2008). Sociological understandings of conduct for a noncanonical activity theory: Exploring intersections and complementaries. *Mind, Culture, and Activity, 15*(4), 339–360.

Scott, P. H., Mortimer, E. F., & Aguiar, O. G. (2006). The tension between authoritative and dialogic discourse: A fundamental characteristic of meaning making interactions in high school science lessons. *Science Education, 90*, 605–631.

Smith, D. E. (1987). *The everyday world as problematic: a feminist sociology*. Toronto, ON: University of Toronto Press.

Smith, D. E. (1999). *Writing the social*. Toronto, ON: University of Toronto Press.

Stetsenko, A. (2012). Personhood: An activist project of historical becoming through collaborative pursuits of social transformation. *New Ideas in Psychology, 30*(1), 144–172.

Thomas, D. R. (2006). A general inductive approach for analyzing qualitative evaluation data. *American Journal of Evaluation, 27*, 237–247.

Treagust, D., & Duit, R. (2008). Conceptual change: A discussion of theoretical, methodological and practical challenges for science education. *Cultural Studies of Science Education, 3*, 297–328.

Wallace, C. (2012). Authoritarian science curriculum standards as barriers to teaching and learning: An interpretation of personal experience. *Science Education, 96*(2), 291–310.

Wallin, J. J. (2012). Bon mots for bad thoughts. *Discourse Studies in the Cultural Politics of Education, 33*, 147–162.
Wasser, J. D., & Bresler, L. (1996). Working in the interpretive zone: Conceptualizing collaboration in qualitative research teams. *Educational Researcher, 25*(5), 5–15.
Watts, R., Griffith, D., & Abdul-Adil, J. (1999). Sociopolitical development as an antidote for oppression: Theory and action. *American Journal of Community Psychology, 27*, 255–272.
Watts, R., Williams, N. C., & Jagers, R. (2003). Sociopolitical Development. *American Journal of Community Psychology, 31*(1), 185–194.
Wiggins, G., & McTighe, J. (1998). *Understanding by Design*. Alexandria, VA: Association for Supervision and Curriculum Development.
Wood, G. H. (1998). Democracy and the curriculum. In L. E. Beyer & M. W. Apple (Eds.), *The curriculum: Problems, politics and possibilities* (pp. 177–198). Albany, NY: SUNY Press.
Zeidler, D. L., Sadler, T. D., Simmons, M. L., & Howes, E. V. (2005). Beyond STS: A research-based framework for socioscientific issues education. *Science Education, 89*, 357–377.
Ziman, J. (1984). *An introduction to science studies: The philosophical and social aspects of science and technology*. Cambridge, UK: Cambridge University Press.

Chapter 4
Activating Students' Conceptions and Positions on STSE Issues in Preparation for Socio-political Activism

Varsha Patel

4.1 Introduction

Ontario secondary school science teachers are expected to meet 'STSE' (Science, Technology, Society, and the Environment) teaching/learning expectations set forth by the province's Ministry of Education, including assessing students on their ability to critically analyze various socioscientific issues (MoE, 2008). Through such critical analyses, their reflections and research may lead them to actions aimed at positive social and/or environmental changes. The ability to relate science to technology, society, and the environment is required in order to function in a scientifically-literate world where science manifests itself in all facets of life. Its placement as the first set of teaching/learning expectations for all content strands provides the context within which to develop scientific literacy as science knowledge and skills to be learnt in ways anchored meaningfully within a social, technological and environmental framework. The inter-dependence amongst the four elements of STSE makes science education much more interdisciplinary, with controversies potentially arising from these relationships.

Controversies can surface through the activation of students' preconceived notions on a given STSE topic, a tack recommended from fundamental constructivist learning theory (Osborne & Wittrock, 1985). Such preconceived notions are brought to conscious awareness by engaging students in meaningful dialogue with themselves, with each other and through interactions with media (e.g., movies, news articles, images, videos). In contrast to traditional approaches, often involving transmission of information between individuals through a teacher-talk and student response format of instruction, constructivist theory suggests that an individual's understanding of the world around him/her is constructed by integrating incoming

V. Patel (✉)
Toronto District School Board, Toronto, ON, Canada
e-mail: varsha.patel@tdsb.on.ca

L. Bencze (ed.), *Science and Technology Education Promoting Wellbeing for Individuals, Societies and Environments*, Cultural Studies of Science Education 14,
DOI 10.1007/978-3-319-55505-8_4

information into his/her existing schema, a mental model of how s/he views the world (Llewellyn, 2005). New knowledge is best developed by linking it to what the individual already believes to be true. Sometimes, students are not aware of their beliefs, but by activating their schema, such beliefs can move to the forefront of their consciousness and, in so doing, be constantly revised and refined in light of new information. When embedded within one's own schema, newly-acquired information can be evaluated according to that schema. At the same time, such reconstructions can be highly complicated by social interactions and discourse among peers and others (Vygotsky, 1978).

Given that the first set of curricular expectations of each unit in Ontario curricula are STSE-based, teachers can structure entire units in terms of socioscientific issues as students come to grasp new concepts from the sciences. However, given traditional preferences—by textbook publishers, teachers and others—to address knowledge expectations in the curriculum, attention to the STSE issues are often limited to engaging students in secondary research and summary reporting, wherein they are not encouraged to take research-informed and negotiated action(s) that could contribute towards creating a better world. Such a world would house citizens that act in ways that promote a healthy existence by positively influencing decisions made by stakeholders and not acting in ways that could bring harm to societies and environments. Accordingly, there are needs for curricular approaches that promote critical views of STSE relationships and personal and social actions to address controversies perceived by students. The 'STEPWISE' (Science & Technology Education Promoting Wellbeing for Individuals, Societies & Environments) framework is one such approach. Originating in Ontario (see Chap. 2 in this volume), it arranges teaching/learning goals in ways encouraging and enabling students to 'spend' at least some of their science literacy (e.g., knowledge and skills) on efforts to bring about a better world. More specifically, its prime goals are to help students to: (i) understand power relationships in STSE relationships; (ii) conduct student-led primary, as well as secondary, research; and (iii) conduct socio-political actions based on their research and education (Bencze, 2013a). Through STEPWISE approaches, students are not just encouraged to be aware of opposing positions in various STSE issues but to become life-long active citizens who care enough about the world in which they live, enough to cause them to be continuously critical and active in trying to bring about even a small-scale change. If more individuals became active citizens, their cumulative actions could invariably bring about large scale changes as their scope of influence widens. This is why it becomes fundamentally important to infuse principles like those inherent to STEPWISE into the educational system.

Although the original STEPWISE framework was in the form of a tetrahedron, with learning expectations for knowledge, skills, STSE relationships and student-led research outcomes around the periphery, all pointing to (and drawing from) students' socio-political actions, field-tests with teachers led to development of a more linear version of the schema that they found more practical (Bencze & Carter, 2011). This pedagogical framework suggests providing students with one or more sets of 'apprenticeship' lessons and activities that would help them to develop expertise, confidence and motivation for eventually self-directing research-informed and negotiated action (RiNA) projects to address controversies they perceive in STSE relationships. These apprenticeships, which are based on basic constructivist learning theory (refer above), begin by asking students to 'express' (e.g., state, write, draw) some of their pre-instructional conceptions about STSE relationships. To ensure all students have more equal access to useful conceptions, teachers are then encouraged to more directly share with students knowledge and perspectives about STSE relationships and RiNA projects others have conducted. Students are then invited to develop, perhaps with teacher support, small-scale RiNA projects to address STSE controversies of interest to them. Depending on various factors, including students' age and previous experiences with these kinds of perspectives and practices, teachers may have to provide additional apprenticeship lessons and activities before asking students to self-direct their own RiNA projects. Strategies introduced in this chapter can facilitate activation of students' pre-instructional schemas about such issues as stepping stones towards moving them further along an apprenticeship towards student-led research-informed and negotiated actions on STSE issues.

4.2 Introductory Activity Aimed at Eliciting Image-Generated Student Reactions

The first activity to discuss is one I used with students in my grade 11 biology course during the 2012–2013 academic year. They were given an opportunity to respond to images taken from *Evaluating Technologies* resource document from the STEPWISE website (www.stepwiser.ca). The activity involved several components. Students responded to images of various commodities, with one image displayed on each paper and all papers taped throughout the classroom. Figure 4.1 depicts an image similar to one used in my activity.

Fig. 4.1 Fast foods (Source: Commons. wikimedia.org/)

They each had to freely, but quietly, respond on the paper itself, with each student at separate stations. Students had to visit at least 5 stations (2 min per station) from a total of 27, and write down pros and cons about the technology that was depicted in the image. They also were asked these questions: Identify individuals and/or groups who would be supportive of your claims (who would agree with pros, with cons?). Which side are you more in favour of and why? How does the image portrayed on the card relate to biology? This image-response activity allowed for concentrated individual reflection. The students were quite engaged as no one was talking and all were writing. The second component involved uploading of individual student responses to his/her image (either selected from an envelope or self-chosen) onto Moodle (https://moodle.org), a learning management system. Thirdly, students responded to their peers' viewpoints using a discussion forum I created on Moodle. A discussion forum was a quick way for students to reply to their peers' posts.

If I were to do the activity again, students would be encouraged to visit as many stations as time allowed, as some wrote faster (ideas came quicker) than did others. I would place a time limit so that slower writers—not necessarily bad thinkers—would be able to provide meaningful reflections, albeit, fewer in number. Some students asked if they had to write their name next to their reflection. I think it would have increased student participation if they were held accountable to their contribution by way of writing their names down. However, I did not want them to feel that I was judging them in any way and wanted to encourage them to be freely expressive; so, they were told not to write their names, as the activity was intended to access their preconceived notions, to allow them to see other students' viewpoints, and to allow those viewpoints to trigger other ones. I did tell them that the goal of the exercise was to expose them to other students' viewpoints so as to facilitate their own thinking surrounding the issues. The idea was to generate as many viewpoints as possible as they relate to all images. Another modification I would make to this activity would be to expose them only to specific biotechnical images (www.nwabr.org/teacher-center/stem-cell-research#lessons) related to a specific unit of study in the course (see Fig. 4.2) rather than images that portray a wide range of technologies that do not necessarily tie in directly to the curriculum but may relate to biology in general.

For example, Fig. 4.2 portrays GMO clones that relate to the Genetic Processes unit. This could then be used as a unit opener activity, to be revisited towards the end of the unit in order to see if there had been any refinement in their views regarding genetic engineering. Such views, in turn, could be pivotal in motivating them to take research-informed and negotiated actions. Throughout the unit, students could be introduced to actions people have taken by showing videos and/or having them analyze case studies.

Students were graded on their responses to their selected card (uploaded to *STSE Image Response Assignment* site on Moodle) and their response to at least two classmates' written opinions of their image cards (posted in the discussion forum). They were marked on how well they communicated their thoughts and on the depth of connections they made between biology, technology, society and the environment

Fig. 4.2 Genetically-modified plant seedlings in test tubes (Source: Sepp Hasslberger, licensed under CC BY-NC-SA-1)

Table 4.1 Image response rubric

	Level 1 (0.5–1.5)	Level 2 (2–2.5)	Level 3 (3–3.5)	Level 4 (4–5)	Total
Communication Expression and organization of ideas and information (*e.g. clear expression, logical organization*)	Expresses and organizes ideas and information with limited effectiveness	Expresses and organizes ideas and information with some effectiveness	Expresses and organizes ideas and information with considerable effectiveness	Expresses and organizes ideas and information with a high degree of effectiveness	
Making connections between biology, technology, society and the environment (*e.g.* assessing the impact of biology on technology, people and other living things, and the environment)	Makes connections Between biology, technology, society, and the environment with limited effectiveness	Makes connections between biology, technology, society, and the environment with some effectiveness	Makes connections between biology, technology, society, and the environment with considerable effectiveness	Makes connections between biology, technology, society, and the environment with a high degree of effectiveness	

Source: www.edu.gov.on.ca/eng/curriculum/secondary/2009science11_12.pdf

(see Table 4.1). The rubric could have included a criterion relating to how effectively students were able to identify relevant stakeholders based on issues that a given image elicited. However, I was content in simply evaluating students using a generic rubric since the intent of the assignment was met; namely, getting students to freely express their ideas without judgment. All student ideas should be encouraged and not be judged; so, an evaluation rubric cannot evaluate students based on how accurate their ideas are. Student ideas are meant to be open-ended and, as such, should be evaluated based on how well they can express those ideas and the amount of effort put into expressing them in terms of richness of connections they make. Some connections may or may not be grounded in reality, but they are connections nonetheless, to be *valued* by the teacher. Some teacher direction can be used in cases where students may need to be primed in order to initiate a reaction to a given issue.

If students felt the need to conduct research in order to solidify their views on the topic(s), they were welcome to do so; but, it was not a requirement—since the assignment was intended to address preconceived notions students had regarding the relevant issues. I gave them this research option because students may not necessarily have ideas of their own, especially if they never had any experience with a given piece of technology, were unaware of issues related to an image or had a very vague idea of what the image conveyed. I found that students who did conduct secondary research had a stronger, more powerful, response than those who did not conduct research. The research seemed to have guided their thinking, although, of course, they made choices on where and what to search and what to synthesize. I could have extended this assignment further by expecting all students to conduct secondary research on their topics and develop a simple action, such as development of a poster or PowerPoint™ slide show that may be placed in a public place of their choice where the action would have the greatest influence (e.g., image of fast food items linked to posting an action-based product near the vending machine in the school cafeteria rather than on a classroom door).

4.3 Using Cooperative Learning Structures to Get Students to Express Ideas and Learn New Ideas About STSE Issues

A 'jigsaw' cooperative learning structure was used to expose students to techniques used to extract stem cells and controversial issues tied to each technique (www.nwabr.org/teacher-center/stem-cell-research#lessons). This structure allows expert groups to acquire new information on a given technique and then share it with another expert group. Members of each expert group formed the jigsaw group, within which sharing of different techniques and associated controversies took place. Each student in a group of three received the following materials:

1. One white letter-sized sheet: This had specific questions to answer individually after reading the handout silently. This allowed each student to be accountable for his own learning and have something to contribute once the group began its discussions.

2. One white legal-sized sheet: Each group had to collaboratively fill in the section that related only to the technique to which they were assigned. The following were section headings relating to each technique: How is it done? Origin of stem cell. Points of controversy. These were to be filled in the respective boxes of the table corresponding to their technique. On the other side of the sheet were four boxes and they had to diagram the technique based on what they read about it.
3. One coloured sheet: One side contained a description of the technique with a diagram; the other side was a news article related to a real-life application of it. Both sides contained controversial issues.

Each small group was identified by the colour of the paper onto which the reading material was photocopied, for a total of four different groups (four different techniques). Every student was given the reading material. So, there were *blue, green, pink* and *yellow* groups. Once they completed their same-coloured group-based work, they had to form new groups containing only one representative of each *colour*, such that all groups would then have one *blue*, one *green*, one *pink* and one *yellow* representative. I found this to be quite useful because I could quickly tell whether the groups were correctly constructed. Students had no difficulty as they knew where to go. In the newly constructed heterogeneous group, each student shared his/her own work (summarized on legal-sized sheet) based on the technique to which s/he was assigned. I really enjoyed this activity and I believe the students did, as well. They were fully engaged, were on task and eager to communicate with other students. All in all, the work got done and students were explaining the techniques/controversies to other students, so it was a real learning experience for them. They were then ready to conduct research to learn more about stem cell science and technology by searching for and reviewing an article. For details, please see page 75 of the document at this link[7]. I was contemplating on showing them a 15-min NOVA ScienceNOW video[1] prior to having students carry out the in-class activity described above, but I knew the activity would take a full class period. This video shows an animation of some stem cell techniques, their potential uses, and ethical and political issues over legalizing the cloning of embryonic stem cells. In retrospect, I would sequence my lessons differently by first posting the link to the video on Moodle and assigning students to watch the video as homework so that when they came to class the next day, they were better equipped to complete the cooperative learning activity in a timely manner. Even though I did specify how much time to spend on each part of the activity, I found that some groups had struggled to complete the in-class work and rushed through it towards the end of the period. Alternatively, this video could be scheduled for viewing in class the day before the cooperative learning activity was to be conducted. This would allow time for post-video discussion and possibly gauging where students stand on the notion of whether or not they value the use of cloning embryonic stem cells and for what purpose(s). This way, I could get students to position themselves on a line according to how strongly they felt about the

[1] http://www.pbs.org/wgbh/nova/body/stem-cells-research.html

stem cell research—from strongly supporting it to strongly opposing it. Once the line was formed, students of opposing views could face each other after the line was folded in half. For example, the first person who strongly supports cloning would face the last person who strongly opposes it. This could give opportunities for students with opposing views to discuss their reasoning and possibly try to sway their opponent or face-partner to see their perspective. When students are engaged in lively discussions, they are likely to have a better appreciation of why they feel the way they do on a given issue because they have to defend their arguments. Their preconceived ideas on any given issue manifest themselves through the natural course of talking. It would have also been interesting to see if, after the cooperative learning activity and secondary research assignment was completed, whether or not their positions on the *value* line would change and, if so, what caused that change to arise. In other words, it would be interesting to see how their post-instructional ideas compared to their pre-instructional ones after students interact with learning objects. After all, it was Piaget who stated that an individual adapts his/her own conceptions by either assimilating or accommodating new ones to which s/he is exposed (O'Loughlin, 1992). In the case of accommodation, a person experiences cognitive dissonance in that his/her own preconceptions deviate greatly from incoming conceptions from the outside world; so, in order to make sense of such discrepant events, s/he may need to greatly change his/her schema by creating new categories into which the new information can be placed. If external conceptions reinforce existing internal conceptions, then they can easily be assimilated into or placed into existing categories of their schema so that the schema is only slightly modified, ensuring that new and existing conceptions are congruent.

4.4 Introducing Correlational Studies to Students

A powerful way to induce a lasting change in students' schema that embodies the STEPWISE philosophy is to have them understand correlational studies so that they can eventually conduct one of their own design, the results of which can provide evidence and motivation for positive change in society; that is, be beneficial to WISE (wellbeing of individuals, society and the environment). When students conduct research of their own design, they can learn more about STSE issues. Due to the unethical nature of experiments' manipulation of certain variables in society (e.g., giving narcotics to subjects to see its effect on amount of sleep subjects experienced), studying the effect of a naturally changing variable on another variable might best be accomplished indirectly by conducting a correlational study (Bencze, 1996). While scientists often conduct quantitative and qualitative studies and experiments, the former may be a good choice for investigating STSE issues, since, unlike experiments, they do not involve purposeful changes to variables that could generate harmful effects. Rather, they involve collection of naturally-changing variables and then looking for correlations between pairs of variables.

I introduced correlational studies to my students by reading an excerpt from an article entitled "In praise of fat" (Graham, 2012), which describes how Ancel Keys used government funding to conduct a study to support his theory that *fat makes you fat*. Out of the 22 countries he visited, he only used data collected from 7 countries, as only those countries supported his viewpoint. He failed to notice the correlation evident in countries (like France) that had high fat diets also had low obesity rates. I chose to use this example as way of stressing the importance of using a large sample size (not ignoring any portion of it that refuted one's thinking) to make valid correlations. I then introduced a hypothetical correlational study that might look at the relationship between hours of time spent in daycare and level of aggression demonstrated in children who enter kindergarten. I stressed to students that any correlation found between these two variables does not automatically mean that a naturally changing variable (i.e., daycare hours) causes a definitive effect on another variable (i.e., level of aggression), since other variables must be factored in that could potentially influence the relationship you are investigating. A correlational study must aim at keeping other variables that might affect the relationship to be investigated constant (e.g., children from same ethnic background, living in same area, watching same amount of T.V., at least comparable amount of it, viewing similar programs on T.V. etc.). Scientists can manipulate variables in a controlled way (one variable at a time while other variables are kept constant) to see what effect changing one manipulated variable has on another variable. Selected pages (pp. 20, 27, 24, 25, 19, and 26) from the *Skills Apprenticeship* resource (Bencze, 2013b [goo.gl/tPILNi]) were completed by students working in small groups. Students enjoyed discussing connections they thought might exist between **possible** *cause and* **possible** (not definitive) *result* variables. Such connections may bring to light views related to STSE topics about which they had not been aware.

As a follow-up assignment, students were asked to write a reaction paper to a reading assignment that included two correlational studies.[2,3] This assignment stressed importance of paying attention to the reliability and validity of sources. Such articles can be misleading if they were shown in a newspaper or on a morning news show. However, there are studies of smoking, for example, that have made some very important findings and developments. Such findings hold merit as they would be vetted within a scientific community. As an extension activity, students were asked to propose a plan that would allow them to conduct a correlational study of interest to them. As a way of encouraging them to express their prior conceptions about STSE issues, they were asked to predict and explain the outcome of their proposed study. If time permitted, I would have conducted a whole-class discussion based on students' responses so as to allow students to see the merits of their work and lead the way towards a discussion of possible executable courses of action. Perhaps, of the correlational studies proposed, one could be selected for the purpose of instructing students on how best to go about creating a survey/poll as way of gauging public opinion.

---

[2] http://news.bbc.co.uk/go/pr/fr/-/2/hi/health/3622817.stm

[3] http://news.bbc.co.uk/go/pr/fr/-/2/hi/health/3086013.stm

Students can be assigned homework requiring them to come up with survey questions targeting a particular audience and these questions can be later discussed and refined. An agreed-upon Google™ survey can be completed by classmates alone, just for the purpose of generating data that can then be analyzed as a class. Brainstorming survey/poll questions as a class provides students opportunities to evaluate such questions in terms of their validity based on the purpose of the preliminary study of interest. When designing a poll or survey, it is important to think about what you want to determine, why you need this, who you need to ask, how to word the question(s), how to analyze and represent the data (e.g., in tabular and/or graphical form), what the data allows you to conclude and what course(s) of action would be suitable to address such conclusions. If students collectively are guided through this entire process, then they would be well on their way towards successfully completing their own RiNA project. An online discussion board could be set up for the purpose of teacher and peer feedback. A component of their assigned project mark might include how actively they used the discussion board as a way of encouraging its usage. This adds an intrapersonal component for those students who might shy away from a whole-class discussion but would welcome the opportunity to respond within the safety of a discussion forum, especially those who learn best through this form of multiple intelligence.

4.5 Integrating the Case Methods Approach with Cooperative Learning Structures in Order to Trigger Student Reactions Towards STSE Issues and the Eventual Creation of Products of Science and Technology

A smorgasbord of activities/assignments were prepared, all with the intention of getting students to express their preconceived ideas related to products of science and technology, some of which involved them producing products of their own. One such activity involved the case method. A case is any documentary of an STSE issue and a method is a set of student activities developed to further their ideas around it and to help them establish their own opinions/views. Based on this definition, the image-based activity related to uses of technology described earlier in this chapter would qualify as a good example of a case method as the images themselves would be the documentaries. However, this section looks at treating cooperative learning activities as case methods' *activities* that promote student expression of ideas to cases (i.e., video and articles) that expose them to STSE issues. In fact, case methods can teach new ideas too, and so, can border between expressing and learning ideas about STSE/RiNA, both of which were evident in the responses students gave to the cases to which they were exposed. This time, student-centered but teacher-directed cooperative learning structures were set up over a span of 2 days to allow students to analyze controversies surrounding water pollution, such as assessing the

Fig. 4.3 Placemat template

1	4
2	3

impact of human intervention on the health of fish and on the health of drinking water and STSE issues related to drinking water by comparing bottled water to tap water. By varying the activities, student engagement is increased as is the multitude of perspectives and responses students bring in. On the first day, students viewed a 15 min video[4] that highlighted concerns centering on water pollution, usage and treatment (Patel, 2013a). Following this, students were arranged into placemat groups of four, each group identified by a letter. The placemat itself is essentially a recording sheet wherein each individual in a group records his/her response in a given section of it. This is followed by the sharing of ideas amongst members of the same placemat group—usually in the middle of the sheet. The placemat activity[5] promotes thinking and discussion by first allowing students to think individually on a question, record their responses on a quadrant of the placemat (i.e., one section of a paper), then share their ideas with the small group prior to sharing ideas with the whole class. The purpose of this structure is to give students an opportunity to reflect individually on an issue, engage with others, and extend their thinking by building upon other students' ideas. Ideas common to all members of the group can be written by a nominated scribe in the centre of the placemat for the purpose of sharing to the whole class (Bennett & Rolheiser, 2001). If I had time for whole class discussion, then I would have added a central oval to the placemat shown in Fig. 4.3 where common group ideas would be recorded.

The class was divided in half with placemat groups situated on the left side receiving a set of video questions that differed from those groups situated on the right side. Each placemat group folded a very large blank sheet of paper (larger than legal size) to form four squares, with each square identified by a number that corresponded to a video question (see Fig. 4.3). For example, on the square numbered one, student #1 responds quietly to video question #1, student #2 to video question #2, and so on. A sample video question was: What actions can people take to minimize water consumption? Alternatively, students can be asked to pose a specific question on the placemat that the video answers with a different student in a placemat group answering it. This way, students are expected to critically analyze the video and identify the key issues that were presented. Students can be encouraged to come up with as many questions as possible based on the content of the video. By doing so, more enriched placemat and group discussions can be fostered.

All students simultaneously responded to their designated video question in the respective square of the placemat. After 3 min, students rotated the placemat

[4] http://dev.conservationontario.ca/source_protection/indexswpeducate.htm

[5] http://www.eworkshop.on.ca/edu/pdf/Mod36_coop_placemat.pdf

clockwise such that student #1 responded to video question #2, student # 2 to question #3 etc. This process continued until all students responded to all four video questions. Individuals within each placemat group were paired off and summoned to share their responses with each other. After completing the placemat activity, an individual from a placemat group was randomly selected to share the group's response to a given video question. For example, Placemat A/individual #1 might be called out by the teacher to respond. Encouraging students to participate in whole class discussions in this manner makes all learners equally accountable as anyone can be called upon to respond to a question. Having the class divided in half could set the stage for debates whereby students situated on the left side respond to one side of an issue and students on the right side would defend the opposing viewpoint. Switching the set of questions after students have responded to the first set can force students to look at both sides of an issue. The video can be looked at through a new set of lens to see if it can lend itself to debatable questions.

In preparation for the next day, students were assigned reading homework that entailed generating questions and answers based on the reading article. A sample question taken directly from the Ministry of Education document[6] (referenced above) that one of the reading articles addresses includes: What public health concerns are associated with the consumption of water bottled in plastic containers? The articles were posted on a class wiki, in numerical order. Each student in class was assigned a reading article number that matched the posted article by number (Abrahams, 2011; Fry, 2008; Hill, 2011; "Male bass," 2009). On the second day, students conducted a three-step interview within their homogenous (i.e., same) reading group, followed by a jigsaw activity within a heterogeneous reading group (Patel, 2013b). Each reading group received an instructions sheet (see Fig. 4.4).

Students were then given a project to do that related to issues they were exposed to during these 2 days (see Fig. 4.5).

The only thing that I would add to this assignment that would make it more action-based would be to have students publicize their products, at least within the school community so as to expose other students to their views and get their reactions by means of collecting and analyzing responses gathered from a survey or poll. For example, a comic strip could be added to the school newspaper and a link to a Google Forms™[6] be provided where student responses can be polled. A song could be played over the school's PA system, allowing the entire school body to hear and hence, react to it. Such displays of student work can be presented during Earth week, which my school holds. I feel that the greatest reaction can be generated during Earth week than during any other part of the school year. I have much to think about in the future in terms of finding ways to at least bring about some change within the school community. This would be the first step towards expanding this to include a wider community.

[6] https://support.google.com/docs/answer/87809?hl=en

1. Form same reading groups of 2 or 3. For example, assemble with students who have read the same article as you. This is a homogenous reading group.
2. For 3 minutes, individually come up with as many questions as possible. When prompted by the teacher, move onto the next step as outlined below.
3. Assign a role to each member of reading group:
   A. Interviewer – asks the questions
   B. Interviewee – answers questions posed by interviewer
   C. Scribe- records pertinent details (suggested format: question followed by answer) onto recording sheet given by teacher.
4. If you are a group of 2, then one person is assigned roles A and C as described above, and one person is assigned role B.
5. When teacher signals end of the first 3-step interview round, rotate roles clockwise so that interviewer (person A) is now the interviewee (person B) and the scribe (person C) is now the interviewer; i.e. A →B, B→C, C→A. The above process is repeated until teacher signals end of second round of 3-step interview.
6. When teacher signals end of second round of 3-step interview, roles again rotate clockwise so that each member performs a role he/she has not yet performed.
7. After three rounds, all scribed work is shared by all members of the same-reading group so that every member has a record of the three rounds of interviews.
8. Return to your home or "placemat' group that was formed yesterday. The home group comprises of only one member from each reading group and so, can be referred to as a heterogeneous reading group. Assign a number to each member according to the reading article each person was expected to read for homework. For example, if you were assigned to read article #1, then you are individual #1. Thus, the heterogeneous reading group would comprise of individuals #1 to #4.
9. Individual #1 summarizes his/her information based on article #1 while individuals #2 - #4 listen attentively and record the information in the appropriate section of handout entitled *Summary of All Reading Articles*. In other words, each person fills in summary sheet except for the person speaking. Continue the process (i.e. #2 now speaks and everyone listens/records etc.) until all members have shared and recorded their work.
10. At the end of the period, every student must submit the *Summary of All Reading Articles* handout to the teacher. Each student's scribed work from the three-step interview must be added to the relevant portion of the table corresponding to his/her assigned reading article.

Fig. 4.4 Three-steps interview and JIGSAW Cooperative Learning Activity

4.6 Summary and Conclusions

This chapter summarizes some ways in which students can be encouraged to express their preconceived ideas related to STSE issues and extend them into constructing new ideas through various constructivism-informed approaches embedded within the STEPWISE framework. In order to continuously revise one's own schema, it must first be activated before actions can be taken that can impact the wellbeing of individuals, society and the environment. An individual's schema can be activated and invariably be refined through engaging him/her in 'expression' activities. Many of the 'expression' activities highlighted in this chapter link back to case methods in one way or another. Students were exposed to different kinds of documents, such as images, videos, and articles. They had to then interact with such documents in a meaningful way such as partaking in small/large group online or class discussions and participating in a variety ofcooperative learning structured activities (e.g. placemat, jigsaw, three-step interview, and value line formation).

Water and Society

Overarching Idea: You will analyze economic, social, and environmental issues related to distribution, purification, or use of drinking water (e.g., the impact on the environment of the use of bottled water).

Something to think about, when conducting your research:

- In developing countries, thousands of people, many of them children, die every year from drinking contaminated water. Many of these countries cannot afford to build water treatment plants. In North America, where safe water is generally available, we spend millions of dollars on bottled water, draining sources of fresh water and challenging waste-disposal systems.

Some sample questions you might want to answer:

- What are the economic costs of building, maintaining, and monitoring water-purification plants?
- What are the social and environmental costs if these plants are not properly maintained and monitored?
- How effective are municipal wastewater treatment processes at removing pharmaceuticals such as hormones and antibiotics from our drinking water?
- What public health concerns are associated with the consumption of water bottled in plastic containers?

(Source: http://www.edu.gov.on.ca/eng/curriculum/secondary/2009science11_12.pdf)

You are to use the following chart to determine what your project will look like. If you have additional suggestion(s) for the ROLE, AUDIENCE and/or FORMAT, please confirm with the teacher before proceeding with your project.

ROLE	AUDIENCE	FORMAT	TOPIC
Government	**Consumers**		

Fig. 4.5 Description of Student Water Project Assignment

Throughout the chapter, I have given suggestions for future growth such as getting students involved in conducting surveys or polls within the classroom environment with the intention of preparing them to conduct their own correlational studies. Effectively, I would like to focus my efforts on getting students to be actively involved in conducting research-informed and negotiated actions related to STSE issues about which they are passionate in pursuing, rather than simply expressing their ideas. In other words, I want students to take their ideas further by acting on them in a way that will improve the wellbeing of individuals, societies and environments, however small such improvements may be. How students go about influencing public opinion is not relevant, so long it is done ethically and with a purpose in mind. One such way might involve having them create a blog that allows the public to follow their ideas as followers. This would encourage followers to offer their own opinion(s) and suggest ways in which the students can exercise greater influence by providing contacts for students to pursue. This is but one kind of technological tool I can explore further as there are others in which I am interested. Effective uses of technology will require me to figure out ways in which different tools can be used together to meet a common goal. One possibility might include having students watch a video on an STSE issue, such as deforestation, while, at the same time, communicating with one another in a chat room like TodaysMeet™. Students could pose questions related to a particular video, enter words of interest, and/or other associations they make with the content of the video in the chat room as they pertain to science and technology. From this online conversation, groups can be assigned to add one question or issue into a shared Google™ document set up using Google Drive™. From this shared Google™ document, groups can sign up for an issue they would like to explore further. This can even lead to a whole host of *global* collaboration projects as the possibilities are endless.

References

Abrahams, M. (2011, August 8). *Studies reveal that water tastes like water.* Retrieved from http://www.guardian.co.uk/education/2011/aug/08/bottled-water-taste-research

Bencze, J. L. (1996). Correlational studies in school science: Breaking the science-experiment-certainty connection. *School Science Review, 78*(282), 95–101.

Bencze, L. (2013a). *STEPWISE: Science and technology education promoting wellbeing for individuals, societies and environments.* Retrieved from http://stepwiser.ca

Bencze, L. (2013b). *STEPWISE: 'Expressing Ideas: e.g., Encouraging students to say what they know.* Retrieved from http://webspace.oise.utoronto.ca/~benczela/ExpressingIdeas.html

Bencze, L., & Carter, L. (2011). Globalizing students acting for the common good. *Journal of Research in Science Teaching, 48*(6), 648–669.

Bennett, B., & Rolheiser, C. (2001). *Beyond Monet: The artful science of instructional integration.* Toronto, Canada: Bookation.

Conservation Ontario. (2009). *The ripple effect: The role of Ontario's conservation authorities in protecting source water.* Retrieved from http://dev.conservationontario.ca/source_protection/indexswpeducate.htm

Create a Survey Using Google Forms. (2015). Retrieved from https://support.google.com/docs/answer/87809?hl=en

Eating Pizza 'Cuts Cancer Risk.' (2003, July 22). Retrieved from http://news.bbc.co.uk/go/pr/fr/-/2/hi/health/3086013.stm

Fry, C. (2008, March 21). *Tread lightly: Give up bottled water.* Retrieved from http://www.guardian.co.uk/environment/2008/mar/21/water.pledges

Graham, J. (2012, July/August). In praise of fat. *Optimyz.*

Hasslberger, S. (2011, October 1). *Polish minister says: Ban GM across Europe.* Retrieved from http://www.newmediaexplorer.org. Photo available under a Creative Commons Attribution-Noncommercial license.

Hill, R. (2011, April 21). *Real water – With added electrons!* Retrieved from http://www.guardian.co.uk/science/blog/2011/apr/21/real-water-added-electrons

Llewellyn, D. (2005). *Teaching high school science through inquiry: A case study approach* (pp. 38, 40–41, 51, 52). Thousand Oaks, CA: Corwin Press.
'Luckiest People Born in Summer.' (2004). Retrieved from http://news.bbc.co.uk/go/pr/fr/-/2/hi/health/3622817.stm
Male Bass with Female Features Found in U.S. (2009, September 15). Retrieved from http://www.cbc.ca/technology/story/2009/09/15/tech-biology-fish-sex-organs.html for CBC News
Ministry of Education [MoE]. (2008). *The Ontario curriculum, grades 9 and 10: Science*. Toronto, Canada: Queen's Printer for Ontario.
Moodle. (2014). Moodle Pty Ltd. Retrieved from https://moodle.org/
Northwest Association for Biomedical Research. (2014). *Lesson two: Techniques for obtaining stem cells.* Retrieved from https://www.nwabr.org/teacher-center/stem-cell-research#lessons
O'Loughlin, M. (1992). Rethinking science education: Beyond piagetian constructivism toward a sociocultural model of teaching and learning. *Journal of Research in Science Teaching, 29*(8), 791–820.
Ontario Ministry of Education. (2008). *The Ontario Curriculum Grades 11 and 12: Science.* Retrieved from http://www.edu.gov.on.ca/eng/curriculum/secondary/2009science11_12.pdf
Osborne, R., & Wittrock, M. (1985). The Generative Learning Model and its implications for science education. *Studies in Science Education, 12*, 59–87.
Patel, V. (2013a). Post-video reflection on water pollution, usage and treatment using placemat cooperative learning structure. *Crucible Online, 45*(2), 1–3.
Patel, V. (2013b). The origins and effects of water pollution: An investigation of the economic, social, and environmental issues related to drinking water. *Crucible Online, 4*(1), 1–7.
Place Mat. (2006). Retrieved from http://www.eworkshop.on.ca/edu/pdf/Mod36_coop_placemat.pdf
Stem Cells: Early Research. (2005). Retrieved from http://www.pbs.org/wgbh/nova/body/stem-cells-research.html
TodaysMeet. (2008). Retrieved from https://todaysmeet.com/.
Vygotsky, L. (1978). *Mind in society*. Cambridge, MA: Harvard University Press.

Chapter 5
Both Sides Now: Exploring the Art of Persuasion to Enhance Actions Within a STEPWISE Framework

Angeliki Grundy

5.1 Introduction

I joined the teaching profession as a second career, having worked in research in both academic and industrial settings. A question I grappled with when embarking on this new adventure was, "What is my role as a teacher?" An unconditional answer to this question is that a teacher is there to promote learning. But then the question evolves into more complicated ones, such as: "*What* should be learned?" and "For what purpose?" In Ontario, the focus of the curricula for the various science disciplines is to create citizens who are scientifically literate; that is, possessing "the scientific knowledge, skills, and habits of mind required to thrive in the science-based world of the twenty-first century" (MoE, 2008, p. 3). While not everyone will move on to become a scientist, everyone will be surrounded by information on science and technology from a variety of sources (for example, conventional media, politicians, a neighbour's blog) and need to be able to critically evaluate this information in their decision-making processes. Because science does not exist in a vacuum, concepts should be considered in terms of influences of technological advancements, environmental considerations and economic, social, and political perspectives. The Ontario curriculum (e.g., MoE, 2008) has adopted STSE (Science, Technology, Society and Environment) as a form of science education that allows students to engage in issues pertaining to the impact of science in everyday life. A key goal of STSE is to help students realize the significance of scientific developments in their daily lives and foster a voice of active citizenship (Pedretti & Forbes, 2000). However, I have wrestled with the stumbling block of my STSE lessons, simply becoming another thing that the students felt they had to learn. I did not want to simply be the font of all knowledge and decision-maker of what was 'right' or

A. Grundy (✉)
Toronto District School Board, Toronto, ON, Canada
e-mail: angeliki.grundy@tdsb.on.ca

L. Bencze (ed.), *Science and Technology Education Promoting Wellbeing for Individuals, Societies and Environments*, Cultural Studies of Science Education 14,
DOI 10.1007/978-3-319-55505-8_5

'wrong'. I felt that my place as the promoter of learning was to give my students tools they needed to arrive at their own conclusions.

In my fourth year of teaching, I was invited to join a STEPWISE focus group of educators led by Larry Bencze. 'STEPWISE' is the acronym for Science & Technology Education Promoting Wellbeing for Individuals, Societies & Environments. It is a curricular and pedagogical framework that organizes teaching/learning goals in ways that acknowledge reciprocal relationships among them, but prioritizes student-led actions to address problems students perceive about relationships among fields of science and technology and societies and environments (To learn more about STEPWISE, refer to its website: www.stepwiser.ca). In that context, I learned about the framework of STEPWISE as discussed in Chap. 2. I was intrigued with the notion of having students shift from being passive recipients of information, whether from a teacher or from their own research, to becoming active, engaged citizens of the world around them. I was also excited to find a new way to promote individual thought and to try to have students relate their scientific learning to their experiences.

5.2 Towards a Focus in STEPWISE Application

I was introduced to STEPWISE when I was teaching a Grade 12 'College' Chemistry course. This is a course for students who are heading for a community college as opposed to university. As such, the emphasis is more on 'applied' chemistry and inquiry techniques, as opposed to more theoretical chemistry. There was a wide range of skills and understanding within the class of just under twenty, as some students came to the course from the 'Academic stream' (pre-university) in Grades 9 and 10, while others arrived from the 'Applied' stream (college- or work-bound). There also were differences in their motivations for taking the course. I took an anonymous poll of my students and found that, while a few were there due to interest, some were there because the course was required for their post-secondary area of study, and others were there only because they had few other options available in order to graduate high school. I was trying to find a way to provide those who needed knowledge and concepts of this course for their future endeavours while still trying to maintain interests of those who would rather be anywhere but in a chemistry class. I decided to try to incorporate some of the STEPWISE pedagogy in order to have these students develop a personal interest in their learning.

Since a major goal of the STEPWISE framework is to have students perform research-informed and negotiated action (RiNA) projects to address STSE issues, I thought about what sort of skills and information they would need to be successful. For RiNA, the students would obviously need to practise their research skills. Another important component for this action would be to understand opposing viewpoints on their issue—to provide a focus and increase the likelihood of success in convincing others to make a change. There is no improvement by preaching to those who are already of like opinion. While an individual can make an impact by altering her/his own habits, a greater one can be made by persuading others to change theirs as well. Robert Cialdini (1993) notes in his book, *Influence: The Psychology of Persuasion,* that "[a] well-known principle of human behavior says that when we ask someone to do us a favor we will be more successful if we provide a reason. People simply like to have reasons for what they do" (p. 3). Thus, I felt that students would need to have skills in persuasion in order to be able to ask others to change their ways.

5.3 Setting the Stage for Persuasive Argumentation

Persuasion essentially "involves careful preparation, the proper framing of arguments, the presentation of vivid supporting evidence, and the effort to find the correct emotional match with your audience" (Conger, 1998, p. 3). An important part of persuasion, therefore, is some ability to take perspectives of someone else, even if it is not your own. There have been studies that have shown that perspective-taking ability increases with age (Clark & Delia, 1976); so, I wanted the first activity I did with the class to give me an idea of how easily my students could see multiple perspectives on a variety of issues, as well as gauge their pre-existing knowledge and understanding. The first thing I did was introduce the idea that all the parts of STSE influence each other. I started with a simple diagram (from STEPWISE focus group meetings):

$$\text{Science} + \text{Technology} \longleftrightarrow \text{Society} + \text{Environment}$$

After defining the terms, we talked about examples of this influence, such as influences of technology on society through fashion (pads on gloves to be able to use a touchpad while keeping the fingers warm) and the reverse of fashion influencing technology (the size of earphones over time). I emphasized that, in the end, these are all driven by human beings making decisions. We discussed how a decision is made, in the abstract. The class gave suggestions, such as looking at information and facts and getting advice from others. Delving deeper, we talked about rules, laws, the rights of the individual versus society and costs of the decision to be made. I offered up a scenario where the students were to think of an owner of a house who wants to do a renovation and asked, "Who should be involved in this decision?" Initial responses involved something like, "The owner of the house." Then, I asked them to imagine that they were a neighbour whose view would now be obstructed, or whose garden will now be perpetually in the shade due to the renovation—does the decision not also affect them? After some discussion, it was agreed that perspectives of the neighbours could be taken into account with the decision. This triggered additional discussion regarding effects on traffic and parking (imagine if the renovation resulted in multiple apartments), local green spaces (imagine if front and back yards covered in cement), increased taxes to city due to addition, and so on. Now, I felt they were starting to think beyond themselves as a single individual and understand that there were other stakeholders, or people who had even a remote vested interest and therefore holders of a viewpoint, in every situation.

Table 5.1 Controversial positions about fast food

Issue/technology—fast food	
Positive	**Negative**
Fast and easy to obtain	Not nutritious
Cheap to buy	Waste from fast food containers
Stakeholders in favour	**Stakeholders against**
Corporations like McDonalds™	Medical professionals
Busy families	People trying to lose weight

The class then performed an activity to help them exercise their thinking skills. I printed a number of cards with the names and pictures of various technologies or products (e.g., fast food, nuclear power, genetically-modified organisms, cell phones, diamonds—32 in total). Each student would choose one card from the pack that I had face down and fanned out. For their technology, they had to come up with two statements in favour of the technology along with two statements against, as well as two stakeholders who would be in favour and two who would be against [see Table 5.1 for a sample answer]. When they completed their first one, they could either trade cards with another student or come back to the fanned out cards, pick a new one and replace their old card.

They continued to do this until they had performed the task for four different technologies, writing down their answers. During this activity, I specifically asked them not to search out any of the topics on their personal electronic devices. I wanted to see what background information they already had or what sort of preconceived notions they were bringing with them to class. While there were some topics about which the students found it easy to find positive and negative statements (e.g., nuclear power), there were others that they found difficult. For example, one student picked up a card with cigarettes and felt frustrated because he could not think of anything positive to say. Further discussion brought out the fact that some students felt that they should not need to come up with positive statements about something that was 'known' to be negative. I asked him to think about the fact that people still smoked and cigarettes were still legal to sell—if there is nothing positive to be said about them, why would this be the case? I found it intriguing to see how students who took one side did not see the purpose to understanding another viewpoint.

An interesting side effect of this activity was seeing just how much of an effect the teacher's opinion would influence the students. In order to help encourage their thinking processes, I engaged them in discussion as a 'devil's advocate'—taking whatever the opposing viewpoint that they had for their particular card. The students listened, even when I was talking to their classmates, because at one point as I was helping one student by taking the 'negative' side of their topic, one of the others exclaimed, "But Miss! You gave me good reasons for [this topic], why are you giving him bad ones?" When I explained that I was just helping her to think of the other side, the students started wanting to know what my stance was. I just smiled and said nothing. I did not want to influence them, to have them fall on one side or another just because I was in a position of authority. I did, however, continue to ask them questions that helped them get their minds thinking (for example, branded clothing—does the company pay for advertising commercials, billboards, etcetera? Do they pay you to advertise for them? Do you pay more or less for something with a logo on it?).

I noticed that all the students had at least one of the topics that they had difficulty finding stakeholders that were in opposition to their stance, and more often than not, it would be because they did not have enough knowledge about the technology or issue. Cell phones, for example, were thought to be generally a good thing and a positive tool in their lives, but they were unaware of issues associated with them, such as environmental destruction causing loss of gorilla habitat (Lovgren 2006) or loss of privacy through monitoring of cell phone activity (Editorial Board 2013). I decided to continue the RiNA apprenticeship with the next step to incorporate research and create an informed action.

5.4 Encouraging and Enabling Youth to Engage in Persuasive Letter-Writing

I felt that this would be a good opportunity for me to design an activity which could both advance the STEPWISE goal of having the students consider an 'action' that could be taken to cause a change, and would allow me to assess what sort of research skills the students had developed through their respective paths through high school. There are a variety of approaches that have been used by those with varying levels of activism in order to try to initiate a change. I decided to have all the students use the same method in order to create a uniformity of product and simplify the process. My choice of action was that of the letter-writing campaign, specifically a letter that would be written by the student to someone (real or invented) that would be a stakeholder holding the opposite viewpoint of the student. I wanted each student to have her/his own topic or issue for which s/he would formulate a stance and develop a persuasive argument; so, I returned to the issue cards used in the previous activity. The students had already worked with these cards and had used them to begin to develop their ability to identify multiple points, so I hoped that this would help involve them in the assignment.

While I did not give direction in terms of research skills, as the government teaching/learning expectation for a Grade 12 course would be that the students already had exposure to gathering information from a variety of sources and being able to references these sources, I did give them a brief introduction to the concept of letter writing. Although they have written formal essays, these were students from a technological generation where personal communication would involve use of emails and texts, as opposed to actual letters. Although an email could replace a physical piece of mail, there is still an expectation of formality if an email is to be written to someone else as a persuasive document. Within their letter, I required them to make sure they had evidence supporting their positions and that they felt would be persuasive to their choice of audience. In order to help them think about the idea of persuasion, we discussed in class about how they would go about convincing their friends to go out with them on Saturday. I would describe a friend and their personality and ask how you could persuade this person to go out. For example, one invented friend was cheap and did not like spending a lot of money; the students offered up arguments in favour of going out by suggesting activities that would not be expensive. As the discussion progressed, the students started to hone their arguments better, trying to focus on those that would appeal the most to their targeted 'friend.' For their assignment, I provided them with an outline to help them focus their thoughts and to help keep them on task:

Issue: ____________________
Position: Pro or Con
Opposing Stakeholder: _____________________

I. Introductory statement ____
____

A. Claim #1 ____
____

a. Evidence ____
____

B. Claim #2 ____
____

b. Evidence ____
____

C. Claim #3 ____
____

c. Evidence ____
____

II. Concluding statement ____
____

I also went over the parts of a formal letter. I gave them the formatting example in Fig. 5.1, identifying the various parts, and within the body of the letter offered some advice in the writing:

The students were provided with two classes in the library to give them access to research materials for the guided action of this RiNA apprenticeship. Because of the diversity of academic behaviours exhibited in the class, I tried to provide an incentive to have the assignment completed by the deadline. While many of my students were quite happy to take part in discussions, there were some who tended to disregard certain aspects of academic expectations, such as completing assignments and studying for tests. At the time, marks were going to be submitted for the first term report card and there were a number of students with a less than stellar average for this reporting period. I informed the class that the assignment would be included among the marks to be reported, hoping that this would galvanize the more

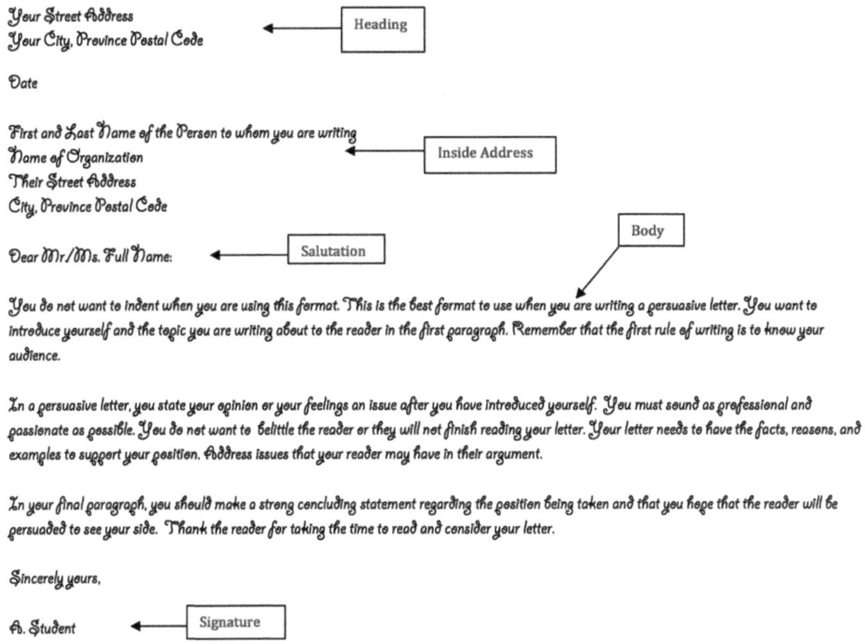
Your Street Address
Your City, Province Postal Code

Date

First and Last Name of the Person to whom you are writing
Name of Organization
Their Street Address
City, Province Postal Code

Dear Mr/Ms. Full Name:

You do not want to indent when you are using this format. This is the best format to use when you are writing a persuasive letter. You want to introduce yourself and the topic you are writing about to the reader in the first paragraph. Remember that the first rule of writing is to know your audience.

In a persuasive letter, you state your opinion or your feelings an issue after you have introduced yourself. You must sound as professional and passionate as possible. You do not want to belittle the reader or they will not finish reading your letter. Your letter needs to have the facts, reasons, and examples to support your position. Address issues that your reader may have in their argument.

In your final paragraph, you should make a strong concluding statement regarding the position being taken and that you hope that the reader will be persuaded to see your side. Thank the reader for taking the time to read and consider your letter.

Sincerely yours,

A. Student

Fig. 5.1 Persuasion letter form

indifferent students into action. I was heartened to receive the majority of the letters within a day of the due date, and almost all within a few days.

I found that the results were variable for this particular class. All the students were able to identify or create a stakeholder who could reasonably hold a position opposite that of the student. I was pleased about this, as I felt this would make it easier for them to target actions for other projects in the future by being able to identify their audience. Because the purpose of this RiNA apprenticeship was to allow the students independence of their own opinion based on their research, I had no plan to make judgements on the position taken, and evaluation would be based solely on appropriateness of the recipient of the letter and the arguments put forth by the student in favour of their position. Most students were also able to create a research-informed opinion and to provide evidence that would help solidify their position. While there were a couple who took a less academic approach by having an opinion and then providing little to back it up, I was generally pleased with the students' abilities to synthesize their researched information and be able to take a stand on their particular issue.

There also were a few students who were able to use what we had discussed in class about targeted persuasion to use arguments specific to their stakeholder. For example, one student had the issue of nuclear power. When he randomly selected his topic, he confided in me that he knew very little about the specifics of nuclear power generation and that he did not know how he felt about it. I told him that the first step of an educated person should be to not have an opinion about something until they have learned a bit about the topic. During the time we spent in the library, he remained quite focussed on his research task. He addressed his letter to the Minister of Energy in Ontario because there had been a recent push for green energy sources that often focused on wind power generation; the student argued in favour of nuclear power. For his arguments, he focused on the current percentage of energy used by Ontario that is created through nuclear power generation and the equivalent number of wind turbines that would be necessary to fill the gap. He focussed on costs, on land use and efficiency of energy transformation; all arguments that would be of interest to a government official who would need to make a decision on this issue, and thus would be more persuasive to the recipient.

5.5 Summary and Way Forward

Overall, I was pleased with the results of the basic RiNA apprenticeship described here. My students showed a heightened interest in the activities involved, compared to the daily tasks of learning the curriculum for the course. I was able to observe some changes in the way they thought about the world around them, and their place in it, as they tried to relate to opinions and experiences that were outside themselves. Although initially it was difficult for them to think past their own personal bubble, by asking them questions and guiding discussion, I was able to help them consider an issue from multiple points of interest. Within the framework of STEPWISE, I was also able to separate myself from the students' processes of arriving at their own opinions, acting as a guide to help them come to their own research-informed (and, often, negotiated) conclusion—as opposed to acting as the authority designating the stand to be taken. While secondary research was used here, as it was only a basic apprenticeship, I was able to give them feedback regarding choosing appropriate sources, because appropriate design of primary research can only grow out of an understanding of what has already been discovered. I was also able to introduce the concept of persuasion to my students, and provided them with an example of a common action taken to try to change the way an individual or an organisation interacts with the world around them in the form of letter-writing. The thought process of considering another point of view and tailoring persuasive arguments to suit the stakeholder can be used in almost any form of informative action that can be taken, from something as simple as personally creating a Facebook™ group or Twitter™ feed, to conducting nation- and world-wide campaign for an established or newly-created not-for-profit organisation.

> Vision without action is merely a dream. Action without vision just passes the time. Vision with action can change the world
>
> —(Joel A. Barker, 1993)

References

Barker, J. A. (Director & Producer). (1993). *The power of vision* [DVD]. Burnsville, MN: Charthouse International Learning Corporation.

Cialdini, R. B. (1993). *Influence: The psychology of persuasion* (revised). New York: HarperCollins eBooks.

Clark, R. A., & Delia, J. G. (1976). The development of functional persuasive skills in childhood and early adolescence. *Child Development, 47*(4), 1008–1014.

Conger, J. A. (1998, May). The necessary art of persuasion. *Harvard Business Review*, 84–95.

Editorial Board. (2013, June 10). Canadians need proper debate about monitoring of phone and Internet. *The Globe and Mail*. Retrieved from http://www.theglobeandmail.com/globe-debate/editorials/canadians-need-proper-debate-about-monitoring-of-phone-and-internet/article12445818/

Lovgren, S. (2006, January 20). Can cell phone recycling help African gorillas? *National Geographic News*. Retrieved from http://news.nationalgeographic.com/news/2006/01/0120_060120_cell-phones.html

Ministry of Education [MoE]. (2008). *The Ontario curriculum, grades 11 and 12: Science*. Toronto, Canada: Queen's Printer for Ontario.

Pedretti, E., & Forbes, J. (2000). STSE education: From curriculum rhetoric to classroom reality. *Orbit, 31*(3), 39–42.

Chapter 6
Learning About Youth Engagement in Research-Informed and Negotiated Actions on Socio-scientific Issues

Mirjan Krstovic

6.1 Reforming Science Pedagogy Through Research-Informed and Negotiated Activism

I started teaching ten years ago in one of the most populated, ethnically diverse and 'tech savvy' high schools in the Peel District School Board, the second largest school district in Canada. Like many beginner teachers, I learned to manage my classroom, establish daily routines, design mostly teacher-centred lessons, prepare 'recipe-style' lab activities, gain expertise and confidence with classroom technology, keep track of students' learning and perform a variety of other curricular and co-curricular duties. Different professional development opportunities at the school, district and provincial level triggered deeper critical thinking about my role as a twenty-first century science teacher and fuelled my passion for on-going professional growth.

My interest in processes of teaching and learning, education research and praxis inspired me to pursue a Masters of Education at the Ontario Institute for Studies in Education (OISE), University of Toronto. Many graduate courses, particularly the history, philosophy and sociology of science (HPSS) course taught by Larry Bencze, the editor of this book, encouraged profound reflection and discourse on dominant science pedagogy. I questioned if my students gained realistic conceptions about the nature of science and technology (NoST). Was I doing enough to encourage awareness of the complex interactions among science, technology, society and environment (STSE)? Were my students learning how to apply their scientific knowledge and skills in meaningful and purposeful ways?

I concluded that my early years of practice promoted mostly development of conceptual and theoretical knowledge, with one or two assignments that encouraged

M. Krstovic (✉)
Peel District School Board, Mississauga, ON, Canada
e-mail: mirjan.krstovic@peelsb.com

L. Bencze (ed.), *Science and Technology Education Promoting Wellbeing for Individuals, Societies and Environments*, Cultural Studies of Science Education 14,
DOI 10.1007/978-3-319-55505-8_6

students to make connections between science and other disciples. My past experiences resonated with readings in my graduate courses (e.g., Abd-El-Khalick & Lederman, 2000; Bencze & Carter, 2011; Crawford, 2007; Hodson, 2003), which suggest that school science systems compromise learning of STSE issues, stagnates development of science inquiry and misrepresents the nature of science and technology. I felt that many in-class activities, especially the labs, could be more *authentic* to allow for student-led, open-ended inquiries that are contextualized in real world applications. Actions to address STSE controversies never made it into my course plan until three years ago, when I began implementing research-informed and negotiated action (RiNA) projects in my tenth grade 'academic' science class.

After I completed a final research paper on school science reform in the HPSS course, I asked Larry if he could facilitate an action research project in my tenth grade 'academic' science class on students' progress of NoST knowledge. Larry agreed; however, he suggested that we should study students' expertise and motivation for self-directing RiNA on STSE issues, and that students' NoST views would relate to their actions. I was eager to begin our collaborative inquiry and learn about Larry's STEPWISE instructional framework.

STEPWISE offers an approach for STSE education that enables students to self-direct primary (e.g., correlational studies and experiments) and secondary (e.g., Internet searches) research as bases for developing and implementing plans of action to address a variety of socio-scientific issues. To help students to develop relevant expertise and motivation, we provide students (as reviewed in Chap. 2) with two successive 'apprenticeships.' This begins with 'basic' apprenticeship activities that first includes a teacher-guided project (e.g., I model student-led action projects), and moves towards a more 'advanced' apprenticeship that includes a second teacher-guided RiNA project (e.g., I guide students through secondary and primary research). After such guidance, students often are then ready to self-direct RiNA projects, in which they have opportunities to apply their learning. Derek Hodson (2014) supports the 'apprenticeship' model, stating that "students can gain experience of action, and thereby learn *through* action and learn *from* action, via the familiar 3-phase apprenticeship approach: modelling, guided practice and application" (p. 87).

I have had the privilege of observing various outcomes of an issues-based and action-oriented science education on students' academic and personal growth. This chapter offers some insights into the following three questions, which have guided my practice in ways that would allow my students to propose and implement research-informed and negotiated actions to address critical STSE issues in their school and community:

- What would our classroom look like, and feel like, if we let our students voice their opinions and positions on critical STSE issues?
- How do we set conditions in which students would be learning science and *doing* science in the context of real issues and gain a greater sense of purpose through education, other than merely earning grades?

- How do we equip our students with the capacity and commitment to take responsible and effective actions on matters of social and environmental and moral-ethical concern?

6.2 Building a More *Balanced* Science Curriculum

Science scholars and educational jurisdictions have been promoting STSE activities for over forty years (Pedretti & Nazir, 2011). With the latest elementary and secondary science curriculum revisions in Ontario, STSE education is given increased priority (MoE, 2008). With more attention to STSE education, students would develop a broader understanding of science; they would develop better critical thinking and decision making skills; and, they would be better prepared for active and responsible citizenship, now and in the future. Therefore, instructional frameworks that encourage exploration of socio-scientific issues are worthy of attention – given many serious social and ecological problems that humanity faces. This form of critical pedagogy is a more radical approach that politicizes science education, challenges dominant relations of power and positions students as agents of positive change in their schools and the wider community.

Experiences that encourage civic-mindedness, like socio-political activism in school science, are *authentic* learning phenomena that go beyond 'academic learning' (sometimes called 'book smarts'), which some students associate with boredom. I am not undermining the importance of 'academic learning' of scientific knowledge that underlies many important socio-scientific issues. Instead, I am suggesting that a more '*balanced*' approach to science education should triumph over the traditional 'concepts-mostly' education. Learning science (and technology) concepts should be balanced with *doing* science and technology, learning *about* science and technology, and *engaging* in socio-political action. Derek Hodson (2003) used these four broad learning domains to define science literacy, which is the overall aim of the secondary science programme in many jurisdictions.

Over the last 3 years, I learned that students can – and want – to make a difference in our world, and that a balanced approach is more equitable (Krstovic, 2014). Related to equity, we learned that RiNA promotes '*street smarts*,' the idea that practical intelligence and experiential education stand in contrast to traditional 'book smarts' (Chap. 13: Phillips, Krstovic & Bencze, this volume). This finding has implications to social justice, in that RiNA leads to increased participation of students who normally do not do well in traditional 'book smart' environments that seem to dominate many science classrooms. Having said this, many students are conditioned by traditional 'book smarts' environments that rely heavily on direct instruction, or what some call 'teach-test-teach' approaches. In this chapter, I discuss why some students experience difficulties with RiNA, and what factors contribute to student success.

6.3 Significant Theories and Philosophies That Have Inspired My Journey Through RiNA

While I was a graduate student in education, I learned about work of many transformative educators (e.g., Dewey, 1938; Foucault, 1991; Freire, 1970; Latour, 2005; etc.) whose theories have affected my practice and continue to shape my philosophy of education. Paulo Freire's (1970) *Pedagogy of the Oppressed* reaffirmed my belief that students should be co-creators of knowledge and not empty vessels waiting to be filled. According to Freire, if teachers wish to develop students' critical literacy skills, then they need to encourage students to question issues of power. I am drawn to Freire's ideas that students sometimes become teachers, that teachers and students construct knowledge together, and that together they interrogate significant life issues. I am the 'lead learner' in my classroom, and it is important for me that my students see me as a learner.

John Dewey's (1938) *Experience and Education* has made me think more critically about the learning experiences that I create for my students, or that students co-create with me, both inside and outside the classroom. Dewey (1938) asked several important questions about the 'character' of students' experiences in schools:

> How many [students] came to associate the learning process with ennui and boredom? How many found what they did learn so foreign to the situations of life outside the school as to give them no power of control over the latter? How many came to associate books with dull drudgery, so that they were 'conditions' to all but flashy reading matter? (p. 27).

Dewey's questions relate well to the present reality of twenty-first century school science. Teacher-centred, didactic and textbook driven methods may no longer meet the needs and reflect the experiences of our students. If we want to engage our students in learning processes, then the role of the teacher needs to change in response to the rapid changes in our society. Alsop and Bencze (2009) remind us that:

> [o]ur practices cannot afford to repeat the same experiments over and over again, mixing those same chemicals, when everything else has changed around us; we should not let our sphere of influence slip to a semi-historical re-enactment of our own educational experience – reducing our remit to efficiently covering dislocated facts and leaving all matters of concern to the politicians, the popular media and other moralizers (p. ii).

Science teachers should set conditions that allow growth in an 'experiential continuum,' which represents "the kind of present experiences that live fruitfully and creatively in subsequent experiences" (Dewey, 1938, p. 28). The stepwise nature of RiNA apprenticeships gradually builds experiences that allow students to develop knowledge, skills and attitudes required to understand and address STSE issues of their interest. This process is both continuous (i.e., extends throughout the course) and progressive (i.e., students' expertise and confidence grow with experience).

It is apparent that we live in a largely *neoliberal* capitalist society. Larry Bencze and Lyn Carter (2011) use Michel Foucault's (1991) notion of neoliberal *governmentality* to explain how people may believe that they are self-governed in this system, while

> [t]heir 'choices' may lack real agency and be fully congruent with aims of neoliberalism due to repeated exposure to messages from business-controlled news, sports, and entertainment media promoting such virtues as *individual responsibility, competition, excellence, efficiency, standardization, privatization,* and *commodification*" (p. 650).

School science (and technology) may be significantly influenced by the 'neoliberal pedagogy' that seems to promote practices that encourage individual competiveness (among other traits) and that advantage students who already possess considerable *cultural capital* (Bourdieu, 1986). Under this system, relatively few students with sufficient cultural and social capital reap benefits of such neoliberal education. By encouraging students to examine powerful networks, they learn to interrogate societal issues and challenge dominant power relations associated with various products and services of science and technology.

Related to students' understanding of dominant power relations, Larry Bencze introduced me to *Actor Network Theory* (ANT) (Latour, 2005). Students use ANT to explore living, non-living and semiotic 'actants' (or 'components,' as I refer to them in class) involved in everyday commodities. Students' understanding of various relationships among actants, through development of *actor-network maps*, allows students to consider new actants as they plan and implement their actions to address controversial issues (e.g., producing an activist video to address impossible standards of beauty set by popular media). Their actions become new 'actants' (or groups of actants) within the network that can challenge dominant semiotic messages by governing powers (e.g., students develop an activist video that challenges the idea that wearing brand name cosmetic products will make young women as beautiful as the models in popular magazines).

Increasing civic participation is good for a democratic society. Our young citizens should be involved in local, national and international decision-making. Hodson (2014) reminds us that:

> [b]y engaging in public issues at the local level, students see democratic process in action and learn how to engage in and negotiate them. By working alongside others, they learn about the demands and difficulties of taking action and learn to develop effective coping strategies. Research suggests that participation in these kind of activities in childhood and adolescence is associated with levels of civic participation, community service and political activism in adulthood up to four times higher than the norm (p. 86).

Students should not wait until they become adults to learn about civic engagement on issues that affect personal, social and ecological wellbeing. Where else, if not in our schools, can we train the most number of students to develop civic-mindedness? Research-informed and negotiated action projects represent one avenue for teachers and students to learn about the most pressing socio-scientific issues and ways in which we can collectively address them. Understanding of power, especially with the help of actor-network theory, will help increase students' capacity to take responsible and effective actions now, and in the future. The remainder of this chapter provides more specific examples of practices broadly described above.

6.4 'Basic' Apprenticeship Activities to Introduce Students to STSE Issues and RiNA Projects

6.4.1 *The Card Exchange Game*

Based on constructivist learning theories, students can benefit from exploring their pre-instructional ideas, attitudes and beliefs about STSE issues before learning about views of various stakeholders and developing (and hopefully implementing) actions to address problems that relate to the wellbeing of individuals, societies and environments (WISE). Previously, we have suggested a 'card exchange game' to help students explore a potpourri of STSE issues (Krstovic & Bencze, 2012). In this 'game, students evaluate various STSE issues by expressing their positions and sharing them with their peers. Teachers write one STSE issue per cue card (e.g., Governments should encourage development of nuclear energy). They distribute four, or more, different cue cards randomly to each student. The students play the 'exchange game,' trading statements with which they least agree with those about which they most agree. Students gain appreciation for various perspectives as they discuss current STSE issues with each other. Under teacher guidance, they start to think about possible actions or solutions in response to real world problems. We suggest this activity as a first step towards research-informed activism.

6.4.2 *Exploring Controversial Statements: The Four Corners Tactic*

The Four Corners tactic is one of my favourite techniques to get student to react to various controversies, prior to guiding them into the first research-informed action project. I label each corner of the room with 'strongly agree,' 'somewhat agree,' 'strongly disagree,' and 'somewhat disagree.' I project a controversial statement on the white screen. Students first think about the statement individually, then I instruct them to move to the corner that represents their opinion. For example, I share the following statement with my tenth grade 'academic' students: "*Be it resolved that governments should allow competing companies to decide how much greenhouse gases to emit each year.*"

When the students get to their corners, they pair up, or form a group of three. The students have a few minutes to discuss their opinions with likeminded peers. A whole class discussion begins with a volunteer from each corner sharing his/her opinion. The teacher ensures that students' voices are heard by reinforcing active listening and paraphrasing. Judgement and debate are suspended in lieu of hearing different positions. Teacher can facilitate discussion around possible actions that citizens can take to minimize any negative consequences on individual, social and ecological wellness.

This activity can be repeated with another controversial statement. The activity takes about 15–30 min, depending on the level of student engagement with the statement. After I finish with this activity, I hand out a list of possible issues for the first RiNA project. Students form groups of three to four. They select an issue that interests them as they are more likely to develop deeper attachment and commitment to the project. What follows next is another stage of expressing pre-conceived ideas, but this time students brainstorm what they know in their groups using a place mat, mind map or another tactic. (Placemat tactic involves groups of students working both alone and together around a single piece of paper to simultaneously involve all members. Students may write or draw their ideas on a larger piece of paper that is divided into three or four sections, depending on the size of the group.)

6.4.3 Brainstorming Ideas About STSE Issues in Small Groups

Each science unit is accompanied by STSE issues suggested in the official secondary science curriculum document. Teachers can add more issues, if they wish. For example, this is a list of water-related issues that I gave to my eleventh grade 'university level' Chemistry students for their RiNA project in the Solutions and Solubility unit:

- Social, economic and environmental implications of using plastic water bottles
- Specific toxins present in water (e.g., industrial, pharmaceutical)
- Sanitation issues in developing nations/or in developed nations
- Privatization of water resources
- Oil spills and oil dispersants
- Water conservation technologies (e.g., roof tanks, etc.)
- Other...students suggest an issue related to global water supplies

Students begin by exploring what they already know. A placemat or a mind map can be used to demonstrate their collective knowledge and understanding prior to beginning any secondary research to learn more about the issue (see Fig. 6.1 for a student-generated mind map of oil spills and oil dispersants).

Teachers should offer some guidance to students as they start to express their ideas. For example, teachers may ask students to express positive and negative consequences on individuals, societies and environments associated with their issues. At this stage, teachers should encourage students to think about the positions of various stakeholders (e.g., governments, corporations, parents, youth groups, social and environmental activists, etc) and possible power relations.

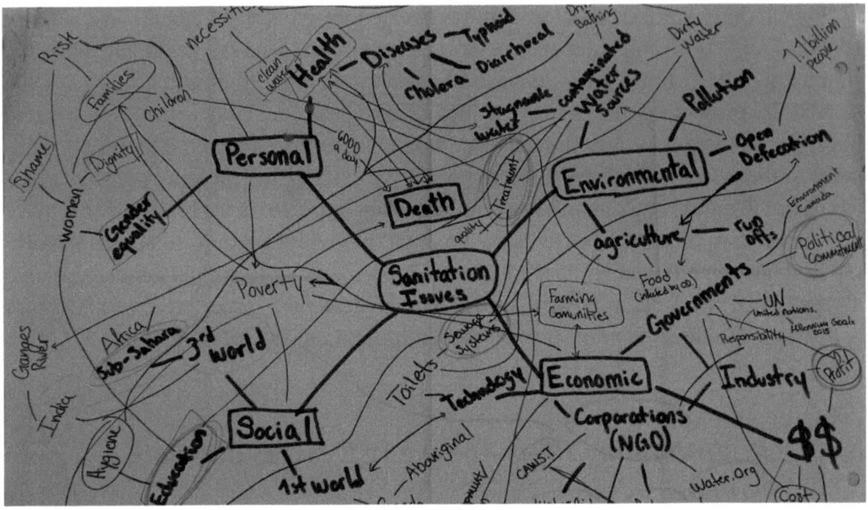

Fig. 6.1 Students' mind map of their preconceived ideas about the impact of water sanitation issues on the well-being of individuals, societies and environments

6.4.4 Guiding Students into 'Secondary' Research

Research is challenging in the age of digital information overload, especially considering numerous networks associated with socio-scientific issues and myriad effects on individuals, societies and environments. Research is a skill that students acquire over time. Teachers could work with the teacher librarian to develop a lesson on conducting proper secondary research before the students access the Internet or other sources. One of the most important aspects of secondary research is confidence in the validity of information. Students need to be able to discern reliability of sources, like personal wikis, blogs, popular magazine articles, etc. Related to this, students need to learn to reference information properly in their final RiNA report using an appropriate format, such as the APA style. A list of questions helps guide the students through the secondary research for the first RiNA project, and for other activist projects in the course. I recommend the 5Ws+How sample questions in Table 6.1.

Students may be given two to three 75-min periods to work on their secondary research with their group. Students should divide work evenly to ensure fairness. Teachers should monitor students' research progress by setting deadlines and scheduling student conferences to discuss achievement. Evidence of progress and achievement should be collected throughout the project, and these data should be considered in teachers' professional judgement for the final evaluation. Students are also expected to put in additional time for the secondary research outside of the classroom.

Table 6.1 List of 5W+how guiding questions for secondary research

Question	Sample guiding questions
What	What is the issue? State the controversy clearly. Remember an issue has two sides – pros and cons – and it can be debated. What are the key science concepts that you need to know to understand the issue?
Who	Who are the key stakeholders and powerful decision makers? Who benefits and who might be harmed?
When	When has the issue become a concern for the well-being of individuals, societies and environments? Review the historical timeline, and perhaps, significant historical events that may coincide with the controversy
Where	Is the STSE issue of a local, national or a global concern? Is it specific to a region (e.g., the school, municipality, etc.)?
Why	Why should citizen learn about this issue?
How	How do we address some of the negative consequences associated with the issue? Think of possible actions that you might take to address the issue

6.5 Challenges of Contextualizing Learning of Science Concepts

Since most RiNA projects are unit-specific, although they do not need to be, it follows that specific science concepts can be contextualized in various socio-scientific issues. For example, when students learn about functions of major organs and organ systems, teachers can help relate functions of the organs to specific issues. For example, consumption of high sugar foods can impact one's pancreatic function and lead to various health concerns, such as diabetes and obesity. Sometimes, the concepts that teachers are required to teach may not directly relate to the issue(s) that the students are studying. For example, the major focus on the tenth grade 'academic' curriculum in the Chemistry unit is on chemical reactions. However, the RiNA project that I facilitate is based on personal hygiene and beauty products. There is no *immediate* connection between the types of chemical reaction, balancing, and the law of conservation of mass, for example, and effects of everyday commodities on wellbeing of individuals, societies and environments. However, each commodity (e.g., shampoo, soap, deodorant, etc.) involves a series of complex chemical reactions that sometimes use controversial chemicals with possible undesirable effect on WISE. In this case, the big idea that teachers need to convey is that chemical reactions may have negative impacts on individuals, societies and environments. Students' actions should be targeted towards potentially negative effects of such commodities. These effects do not necessarily need to be related to the chemicals inside the products. The negative effects on WISE can be related to incredulous claims by advertisers, unethical testing of a product on animals, unfair wages paid to workers (often in poor countries), improper disposal of waste, and many other social and ecological justice issues.

In addition to acquiring knowledge and understanding of science concepts relevant to the issue, students need to learn about complex interactions among science and technology with society and environment. For example, throughout the RiNA learning cycle, students should consider individual health concerns, economic impacts, various ethical and moral considerations, political decision-making, corporate motives, power relations, media's influences, and various ecological concerns related to the issue of their interest.

6.6 'Advanced' Apprenticeship Activities to Facilitate Application of Actor Network Theory to RiNA

A year and half into my work with RiNA, I started to infuse actor-network theory to help students develop 'a big picture' view of the issues by considering various living, nonliving and semiotic actants related to their STSE issues (Pierce, 2013). Although I am still learning how to best introduce students to this complex theory without overwhelming or confusing them, a few strategies have worked relatively well. The secondary research that students compile should help them identify various actants, although teachers need to explicitly model this. The objective is to construct an actor-network map to show the co-dependence of many actants (watch videos in Fig. 6.2 for examples of students' actor-network maps). Infusion of actor-network theory would be considered a more 'advanced' apprenticeship activity, as it pushes higher order thinking skills such as critical analysis of networks and evaluation of actants' alignment to support dominant semiotic messages. Teachers should not name the theory. They should keep the language simple and easy for students to understand. We discuss the findings of our research into ANT in Chapter 9. Here, I outline some practical strategies that I use to help infuse ANT into RiNA.

Fig. 6.2 QR codes to two videos in which: (i) I discuss the actor-network map about 'smart phones,' and (ii) students discuss their network maps illustrating relationships among various entities connected to different consumer products

6.6.1 Trojan Horse Metaphor for Teaching Students About 'Hidden' Actants

The Trojan horse metaphor is a powerful analogy that can be used to help students understand that there is more to everyday commodities (e.g., cell phones, computers, cosmetics, pesticides, water bottles, toys, etc.) than what our eyes see. Colourful packaging, enticing ads, 'sexy' designs, feel-good messages, and other external features and semiotic messages can occlude potentially negative effects associated with certain products and services of science and technology.

In my tenth grade 'academic' science class, the students analyze and evaluate some of the potentially negative impacts of personal hygiene and beauty products on wellbeing of individuals, societies and environments. As part of the 'basic' apprenticeship prior to conducting secondary and primary research, I bring to the class several hygiene products and ask the students how these products might be like a Trojan horse. Students brainstorm some positive and negative effects of various products (e.g., shampoo, deodorant, lip balm, etc.) and a whole class discussion follows about some 'hidden' components. I extend students' thinking beyond just the chemicals inside the product and their effects on health, which is what students first consider. I ask questions like: 'Where do these raw materials come from?,' 'Who extracts the raw materials?,' 'Where are they extracted?,' 'What is the packaging made of?,' 'How is the product marketed?,' 'What features of this product make it attractive for the buyers?,' 'How is the product disposed?,' etc. The students start to consider the 'big picture' and various living, non-living and semiotic components. These questions serve to guide students' thinking and they set the stage for *The Story of Stuff* video by Annie Leonard.

The Story of Stuff video takes students through the life cycle of a product from extraction of raw materials to production, distribution, consumption and disposal. The video is useful to show when the STSE issues relate to everyday commodities. As students watch the video, they record significant points about each stage. The teacher discusses key ideas with the class after the video. The discussion should be framed in the context of actor-network theory by asking questions such as: 'What are some living components in each stage (e.g., miners, factory workers, truck drivers, animals, plants, etc.)?,' 'What are some non-living components (e.g., raw materials, technologies, etc.)?,' and 'What might be some semiotic messages associated with each stage (e.g., miners' work feels dangerous, the product makes you feel 'trendy,' 'sexy,' etc.)?' Teachers should help students develop some understanding of negative impacts on 'WISE' resulting from each stage in the life cycle of a product.

Together with the Trojan horse metaphor and *The Story of Stuff* video, students gain an appreciation of some 'hidden' components/actants associated with everyday commodities and controversial STSE issues. This prepares them well for using their secondary research to construct a study for their 'primary' research and to inform their action plans.

6.6.2 Modelling Construction of a Network Map for a Cell Phone

Teachers should model creation of an actor-network map. After the students watch *The Story of Stuff* video, I pick an everyday product, such as a cell phone. I start constructing a network map showing different actants associated with this product (Scan the QR codes in Fig. 6.2 to view the videos of an actor-network map and students working on their network maps for various consumer products). I engage the students in creation of the network map by asking them to name the components of the cell phone that they see. Our discussion moves to components that we do not see, but that are part of the cell phone (e.g., Coltan miners, phone engineers, corporations that employ the engineers, advertisers and marketers, drivers who distribute the product, technology used to assemble the phone, consumers who buy the product, cost of the product, etc.). The students see how I connect various components and how each component (or actant) is co-dependent on other actant(s). For example, miners cannot mine without technology, and technology allows extraction of raw materials without which a phone (or another product) cannot exist. Additionally, I highlight components that may align to support the dominant semiotic messages. For example, I circle human organizations that encourage positive feelings associated with cell phones. I circle with a different colour living and non-living components that 'suffer' for us to have cell phones (e.g., miners, exploitation of Earth's natural resources, threatened species, etc.). The final network map serves as an exemplar for students to construct their own network maps.

6.7 Apprenticeship Activities for 'Primary' Research

6.7.1 'Basic' Apprenticeship Activities: Helping Students Understand Correlational Studies

An aspect of science inquiry that often does not receive enough attention in high school science is use of correlational studies – as opposed to experiments – for attempting to understand phenomena in nature. Correlational studies are inquiries in which investigators try to find relationships between variables that change *naturally*. Experiments, on the other hand, require that the investigator forces an independent variable to change and then measures changes in the resulting dependent variable. There is an apparent bias in science education towards experimentation and away from correlational studies (Bencze, 1996). Curriculum guidelines, as well as science textbooks, emphasize the experimental nature of science through the 'scientific method' (Gott & Duggan, 1995). However, experiments are not always ideal when students are asked to explore STSE issues. With correlational studies, unlike experiments, the choice to induce potentially negative outcomes (e.g., cancer) would not be in the hands of the investigator but, rather, be left to others (e.g.,

Table 6.2 Exercise for developing students' understanding of studies vs. experiments

Cause variables	Result variables
Vegetarianism	Teenagers' hearing
Rock music	Teenagers' learning
Exercise	Plant height
Hormones	Yeast fermentation
Drugs	Aggressive behaviour
T.V. watching	Physical fitness
Temperature	Muscle strength

smokers). We learned that students' results and conclusions from correlational studies can be used to inform activism to address STSE issues (Krstovic & Bencze, 2012). But students need a proper introduction and guidance to these types of science inquiries as they embark on the first RiNA project.

Teachers need to introduce students to correlational studies through various apprenticeship activities before the students conduct their own studies. Teachers should start by sharing examples of correlational studies while contrasting them with experiments. For example, when studying effects of smoking on lung cancer, it would not be ethical to force any group of people to smoke and determine if they develop lung cancer in comparison to non-smokers. Teachers would explain to students why a correlational study is more appropriate in this case. As an extension to this example, teachers can give students a list of 'cause' (independent) and 'result' (dependent) variables (see Table 6.2) and ask the students to match one cause variable to one result variable. Students should come up with five examples of correlational studies and two examples of experiments. For example, students can match T.V.-watching with teenagers' learning as an example of a correlational study, and temperature with yeast fermentation as an example of an experiment. Teachers should probe for deep understanding by asking students to justify why a correlational study or an experiment is most appropriate for each matched pair of variables. In both correlational studies and experiments, it is important to discuss control variables.

As a follow-up exercise, teachers can ask students to decide if they would conduct a correlational study or an experiment in several inquiry cases, such as these:

- Different types of light on the growth of a plant
- Time spent in front of the computer and quality of sleep
- The effect of pH on the amount of corrosion
- One's gender and their reaction rate

Students can justify their method of choice in small groups, and the teacher can take up each case with the whole class before guiding the students into a mini correlational study as a form apprenticeship activity prior to conducting a study for the first RiNA project. In terms of timing, one class period is sufficient for the above exercises, followed by another 75 min period to guide the students through a mini in-class correlational study.

6.7.2 Guiding Students Through a Mini Correlational Study

A purpose of guiding students through a mini correlational study is to engage students in small-scale data collection and analyses. For example, students can determine if there is a correlation between gender (independent variable) and one of the following dependent variables: preference between meat or veggie diet, reaction time, memory, tongue rolling ability, resting heart rate, extracurricular involvement in school, favourite school subject, etc. Gender is an easy choice of independent variable for a class with about equal boys to girls ratio; however, it is not a continuous variable, and therefore, students are limited to producing bar graphs only. Also, if the class is too small, teachers could give students ready-made data to analyze and evaluate. An important part of the discussion with students should focus on 'bias,' 'validity,' and 'reliability' of the results. Students need to understand that their mini study is biased, and that in order to increase the validity and reliability of their results, students would need a larger sample size and, if possible, repeat their study several times.

6.7.3 Preparing a Correlational Study for the First RiNA Project

With the abovementioned apprenticeship activities, students should be prepared to design a small correlational study for their first RiNA project. Students will require the teacher's guidance and facilitation to select questions for the survey and to analyze the data. As mentioned earlier, gender, age, and grade level are easy independent variables to select. Age and grade level are continuous variables and will allow for a line graph instead of a bar graph. But teachers should encourage students to explore a range of variables suitable to their STSE issue.

Selecting 'good' questions to put on surveys is one of the first challenges students face. Teachers should show a sample of study questions, like the one below (See Fig. 6.3). Students should conduct some secondary research prior to designing their studies as this helps with the selection of questions for the survey.

Students should have about four or five questions to allow enough time to analyze the results. Teachers should encourage students to divide the work evenly in their group. Students collect data in their classes or during lunch time. Usually, two days are sufficient for students to collect all the data. There may be one or two groups that might need extra time. Teachers should use their discretion when deciding how much time to give students for data collection and analysis. I ask students

Grade 10 Climate Change Correlational Study

1. Are you male or female? □MALE □FEMALE

2. What mode of transportation do you use most often to get to school?
   a) Bus b) Car c) Bike d) Walk

3. What kind of food do you prefer to eat most often?
   a) meat b) vegetables c) meat and vegetables

4. How often do you eat out at fast food restaurants?
   a) once a month b) once a week c) twice a week
   d) more than two times per week

5. What source of water do you drink most often?
   a) tap water (filtered or not filtered) b) bottled water

6. How long do you shower for on average?
   a) 0 to 5 minutes b) 5 to 10 minutes c) 10 to 15 minutes
   d) 15 minutes or more

Fig. 6.3 Sample survey for a tenth grade climate change correlational study

to collect and tally the data on their own time, and I give them one class period to graph the results. I check their progress on agreed-upon dates and I conference with students regularly, especially with groups that might be experiencing challenges.

6.7.4 Modeling Different Forms of Activism

Modeling different ways in which citizens can engage in activism is of paramount importance. Students can be motivated and inspired when they hear about activist work by teenagers and others in their community. Table 6.3 lists some examples that I share with my students.

Activism is not only about organizing rallies, protests and chaining oneself to a metal post. It is a type of public action that takes many different forms, such as

- collecting petitions to ban the sale of energy drinks to Minors
- writing letters to the editors of local or national newspapers and magazines
- developing and posting educational YouTube™ videos
- lobbying the school administration and teachers to save energy
- organizing games in school to encourage more recycling
- promoting a 'Walk/Bike to School Day'
- planning and implementing new technology designs that consider social and ecological justice issues, etc.

Table 6.3 Examples of actions that I show to my students to model different forms of activism

Title	Web address/QR code	Description
Teens Against the Privatization of Water	http://tapthatwater.tumblr.com	Teens raise awareness of the impacts of bottled water and water privatization
TEDxTeen – Natalie Warne – Anonymous Extraordinaries	http://www.youtube.com/watch?v=FszSc7Fb8ss	As a volunteer for Invisible Children, Natalie Warne talks about her journey to expose Africa's longest running war involving child soldiers
TEDxSIT – Sam Stevens – Moving Youth Towards Action and Activism	http://www.youtube.com/watch?v=ALqzWs9gjGI	Sam Stevens has participated in various youth activism project both locally and globally. This video will motivate youth to use their education for make this world a better place for all
Piano stairs – TheFunTheory.com	http://www.youtube.com/watch?v=2lXh2n0aPyw	How do we get more people to take the stairs over the escalator? The Fun Theory suggests that 'the easiest way to change people's behaviour for the better is by making it fun to do.'
The Life Cycle of Foundation	http://www.youtube.com/watch?v=WhN6PS1GT9c	This student developed YouTube™ video exposes and educates the viewers about some hidden actants in cosmetic products

Proposing and implementing actions is probably the most exciting and memorable part of RiNA projects. Students are also motivated by what they hear and learn from their peers. Two years ago, one of my Grade 10 students commented in an interview that:

> [T]hese projects are more fun because you get to do more, when something is more fun you will remember it…Social awareness, I think that's what makes people remember it [the STSE issue] the most, hearing it from a friend you're more likely to listen than hearing it from your parents or reading it in a book (Student interview, April 2013).

I am pleased to hear that students associate the learning process with fun and not boredom, and that they learn from each other as they become young activists.

6.8 Celebrating Successful RiNA Projects and the Publication of Youth Issue of JASTE

Over the last 3 years, I have seen a number of successful RiNA projects that have benefitted students, societies and environments. A collection of ten successful projects was published in the first youth issue of the *Journal for Activist Scienceand Technology Education* (JASTE). The on-line journal is available at bit.ly/1t3B4XI or simply scan the QR code below (Fig. 6.4). This publication is a celebration of students' commitment, responsibility and solidarity to make the world a better place. JASTE is also an important actant that may bring about fundamental changes in science and technology education. Hopefully, more teachers will re-position science education as a vehicle for social and environmental transformations.

Although I have seen many great projects, two are worthy of mentioning here: 'No Car Day,' and 'Concerns over X-rays.' Both of these projects resulted in a positive contribution to the community and increased student confidence.

6.8.1 'No Car Day:' Reducing Our Carbon Footprint

A group of three students learned that transportation contributes the most carbon dioxide to our atmosphere. They learned about the impacts of climate change on wellbeing of individuals, societies and environments. The students identified powerful actors, such as oil companies and governments, which support uses of fossil fuels. They also learned about groups that oppose the use of fossil fuels and researched alternative energy sources.

After conducting some secondary research, the students conducted a study to determine if there is a correlation between gender and modes of transpiration. They found that 57% of students come to school by a car with no significant difference between genders. They also counted number of cars that dropped off students in the morning prior and post their proposed action.

Fig. 6.4 QR code to the first youth issue of the *Journal for Activist Scienceand Technology Education*

In response to their research, the students created posters to promote a 'No Car Day' at the school. They stood in the parking lot in the morning and after school with the signs to encourage parents and students to use public transportation, carpool, bike or walk to school. They counted 156 cars before their action, and 115 on the 'No Car Day' event. The students felt that they made a small difference in helping to reduce the carbon footprint of the school.

6.8.2 Concerns over X-rays

A group of four girls in grade 10 'academic' science learned about X-rays as one type of medical imaging technology. They researched advantages and disadvantages of this technology, focusing on how often doctors prescribe them and amounts of radiation released for certain areas of the body. One student wrote the following in her final research report: "[W]e discovered while researching that teenagers tend to receive the highest amount of X-rays in a year, which is why we have decided that teenagers should be our target audience" (Student Report, 2014). The same student reflected on her personal experience:

> I had six X-rays in total for the one injury all within six weeks. Doctors still insisted that I go for three more X-rays even though I knew nothing would be found. Finally after nine X-rays of the same hand, I was sent for a different procedure which found the solution… Having too many X-rays can make you sick especially at a young age (Student Report, 2014).

For their primary research, the girls designed a simple study asking their peers three questions:

(i) Have you had any X-rays in the past three years? If yes, what part of the body?
(ii) Are you aware of the amount of radiation released by X-rays?
(iii) What part of the body received the greatest dose of radiation? ~Arm ~Foot ~Dental ~Pelvis ~Skull ~Abdomen (liver level) (circle one)

The girls learned that about equal number of boys and girls (19/30, and 17/30, respectively) had X-rays within the last three years. Dental, arm and leg were the highest X-rayed body parts for both genders. Almost all surveyed boys and girls were unaware of the level of radiation exposure, with most believing that skull would receive the highest amount of radiation.

For their action, the group wrote a letter to the Head of Diagnostic Imaging at a local hospital informing the doctor about their research. They received a two-page reply from the doctor, who thanked the girls for their detailed inquiry, and who wrote in his letter: "I applaud your efforts to educate yourselves, your students and your community and I look forward to statistically lower dose rates in the future as patients become more educated and involved in their healthcare and I thank you for your role in contributing to this."

Fig. 6.5 Sample photos from the first STSE fair with the Grade 11 University level Chemistry students

6.8.3 Organizing an STSE Fair at the School

We are all too familiar with science fairs, at which students showcase their projects and judges evaluate their work based on pre-determined criteria. But rarely, if ever, has anyone organized a fair to showcase students' research-informed and negotiated action projects. The setup of an STSE fair resembles a 'traditional' science fair, but on a closer look, students' displays do not necessarily follow the rigid 'scientific method.' Instead, students explain controversies surrounding their issues, display actor-network maps, present results of their correlational studies, show their ready-to-use action materials and discuss the impact and effects of their implemented actions.

I organized the first STSE fair in the school library and invited several classes to visit, one at a time (See photos in Fig. 6.5). Teachers and students visited students' displays and asked questions about the projects. Handing out a list of questions to the visiting classes may be a good idea to prepare students and increase accountability for participation. If more teachers decide to run STSE fairs, the main foyer or the school cafeteria could be excellent settings for these events. I encourage teachers to organize similar events at their schools.

6.9 Factors That Contribute to Successful RiNA Projects and the Related Impacts on Students' Academic and Personal Growth

Student success with the RiNA projects depends on myriad factors. I identified the following factors as the most important in helping the students achieve success:

- students' (and teachers') sustained interest and motivation to learn about contemporary socio-scientific issues
- students' positive and productive group dynamics
- teacher's knowledge and skills in guiding students through apprenticeship activities and collaborative work
- opportunities for students to self-direct their research and actions
- students' explicit cycles of reflection on the nature of research and actions

We discuss the last two factors in more detail in the eighth Chapter. These six factors are mutually dependent. Teachers play a central role in helping students understand and develop expertise and confidence in conducting research-informed and negotiated activist projects.

I learned from my unsuccessful groups that students' lack of success resulted from their lack of understanding of the purpose of RiNA. It would be unrealistic to say that all students enjoy these projects and that they are all equally successful. Some students prefer direct instruction, recipe-style labs and individual projects, which all serve a purpose. But the students' roles in the tradition-bound science classroom is different from students' roles in a more student-led, open-ended and collaborative environment. Teachers' guidance towards greater student autonomy for conducting RiNA may require explicit discussion with students about purposes of RiNA projects, their place in the science curriculum and students' expectations and factors that contribute to success. Parents should also be informed about these learning activities.

Outcomes of student-led research-informed and negotiated actions on socio-scientific issues are vast. I have observed students who continue to be engaged in activist projects and leadership activities beyond the classroom. After his 'No Car Day' event at the school, one student ('Robert') led a group of 40 students for the World Vision 30 Hour Famine event. The students raised over $400 for World Vision. Robert started the Interact club at the school for his peers who want to tackle the issues in their community and internationally through various service projects. He explains that RiNA projects in Grade 10 inspired him to continue to make a positive difference in his school and the community. I am proud of students who use their activist science education to develop into young leaders that strive to make this world a better place for all.

6.10 Moving the System Forward: Promoting Issues-Based and Action-Oriented Science Curriculum

I have been a persistent learner in every setting in which I have worked. In many cases, I have taught 'against the grain' and stayed optimistic despite resistance. As one might imagine, school and/or system wide change comes slowly. One of the keys to change is building a collaborative climate in which everyone pursues change together. And this is precisely what I have been doing over the last four years. I started to promote issues-based and action-oriented science education by first working with a small group of teachers, one or two, in my science department to build their instructional capacity and professional efficacy. After I had left the first school in which I worked, teachers continued to implement issues-based and action-oriented curriculum. One of the teachers I worked with went on to become the department head of science at another school where she promotes RiNA projects with her science staff. In addition to working with small groups of teachers, I have promoted RiNA projects at various district and provincial conferences (e.g., Science Teacher's Association of Ontario conference, my school district's Environmental Education Conference, and my school district's Social Justice Innovation Day, etc.). However, sometimes it is difficult to measure the exact impact of these presentations on teachers' practice. With the recent publication of the JASTE issue, each high school in the my district received a hard copy of the journal. The journal serves as an actant to promote activist projects for social and ecological justice in science. To recognize the importance of social and environmental activism, I have initiated an award at my school called "Action for Social and Ecological Justice Award" to recognize students who make positive contributions to the community through their RiNA projects. Most recently, I have invited Larry Bencze as the Keynote speaker at my school district's first Science and Technology Inquiry Symposium, at which he spoke about application-based inquiry. We are in the process of initiating a 'STSE Controversy and Political Actions' group in with elementary and secondary teachers who are interested in embedding social and environment justice issues in their classes. We are creating a positive movement with the support from several system leaders in the district. I am excited and motivated by the impact that we can have as we engage and empower our students to take actions on contemporary socio-scientific issues. I feel confident in our ability to galvanize change and I am humbled by the possibilities of a more active and politicized science education.

References

Abd-El-Khalick, F., & Lederman, N. G. (2000). The influence of history of science courses on students' views of nature of science. *Journal of Research in Science Teaching, 37*, 1057–1095.

Alsop, S., & Bencze, L. (2009). Editorial: Beyond the confines of matters of fact. *Journal of Activist Science and Technology Education, 1*(1), i–v.

Bencze, J. L. (1996). Correlational studies in school science: Breaking the science-experiment-certainty connection. *School Science Review, 78*(282), 95–101.
Bencze, J. L., & Krstovic, M. (2013). *Students' social studies influences on their socioscientific actions*. A presentation at the annual conference of the American Educational Research Association, San Francisco, CA, April 27–May 1, 2013.
Bencze, L., & Carter, L. (2011). Globalizing students acting for the common good. *Journal of Research in Science Teaching, 48*(6), *648–669*.
Bourdieu, P. (1986). The forms of capital. In J. G. Richardson (Ed.), *The handbook of theory: Research for the sociology of education* (pp. 241–258). New York: Greenwood Press.
Crawford, B. A. (2007). Learning to teach science as inquiry in the rough and tumble of practice. *Journal of Research in Science Teaching, 44*(4), 613–642.
Dewey, J. (1938). *Experience and education*. New York: Collier/Macmillan Press.
Foucault, M. (1991). Governmentality. In G. Burchell, C. Gordon, & P. Miller (Eds.), *The Foucault effect: Studies in governmentality* (pp. 87–104). Hemel Hempstead, UK: Harvester Wheatsheaf.
Freire, P. (1970). Pedagogy *of the oppressed.* New York: Continuum.
Gott, R., & Duggan, S. (1995). *Investigative work in science curriculum*. Buckingham, UK: Open University Press.
Hodson, D. (2003). Time for action: Science education for an alternative future. *International Journal of Science Education, 25*(6), 645–670.
Hodson, D. (2014). Becoming part of the solution: Learning about activism, learning through activism, learning from activism. In J. L. Bencze & S. Alsop (Eds.), *Activist science and technology education* (pp. 67–98). Dordrecht, The Netherlands: Springer.
Krstovic, M. (2014). Preparing students for self-directed research-informed actions on socio-scientific issues. In L. Bencze & S. Alsop (Eds.), *Activist science and technology education* (pp. 399–417). Dordrecht, The Netherlands: Springer.
Krstovic, M., & Bencze, J. L. (2012). Using a 'card exchange' game to help students explore STSE issues. *Crucible Online, 43*(3), 1–8.
Latour, B. (2005). *Reassembling the social: An introduction to actor-network-theory*. Oxford, UK: Oxford University Press.
Ministry of Education [MoE]. (2008). *The Ontario curriculum, grades 9 and 10: Science*. Toronto, Canada: Queen's Printer for Ontario.
Pedretti, E., & Nazir, J. (2011). Currents in STSE education: Mapping a complex field, 40 years on. *Science Education, 95*(4), 601–626.
Pierce, C. (2013). *Education in the age of biocapitalism: Optimizing educational life for a flat world*. New York: Palgrave Macmillan.

Chapter 7
Students' Social Studies Influences on Their Socioscientific Actions

Larry Bencze and Mirjan Krstovic

7.1 Introduction

School systems in many jurisdictions have, especially over the last few decades, been encouraging students to learn about so-called 'socioscientific issues' (SSIs); which often appear to involve controversies around potential problems facing individuals, societies and/or environments associated with fields of science and technology. This aspect of science education has not, however, been able to make significant inroads into dominant goals and practices. School science systems have tended to prioritize instruction and evaluation in 'products'—such as laws and theories—of fields of science and technology. Elaborate attention to potential problems associated with these fields may detract from dominant programmatic agendas. There does, however, seem to be hope for further implementation of SSI education by engaging students in conduct of research that may yield results that motivate them to learn more about the issues and, perhaps, take actions to address them. In this report, we share results and conclusions of efforts to enhance research-informed and negotiated action experiences of students by encouraging them to base actions on *correlational studies*, forms of investigation that often appear not to have been promoted to a great extent in school science.

L. Bencze (✉)
Ontario Institute for Studies in Education (OISE), University of Toronto,
Toronto, ON, Canada
e-mail: larry.bencze@utoronto.ca

M. Krstovic
Peel District School Board, Mississauga, ON, Canada
e-mail: mirjan.krstovic@peelsb.com

L. Bencze (ed.), *Science and Technology Education Promoting Wellbeing for Individuals, Societies and Environments*, Cultural Studies of Science Education 14,
DOI 10.1007/978-3-319-55505-8_7

7.2 Socioscientific Issues Education in Schools

It seems clear that there are many potential harms to wellbeing of individuals, societies and environments associated with decisions made by powerful people and groups affecting fields of science and technology. People are concerned, for instance, about possible harms from various commercial products and services, such as: household cleaning and hygiene products (Leonard, 2010); pesticides (Hileman, 1998); tobacco (Barnes, Hammond, & Glantz, 2006); and, pharmaceuticals (Angell, 2004). Many people also expect serious personal, social and environmental degradation associated with dramatic increases in average global temperatures often linked to excessive fossil fuel uses (Klein, 2014). For various reasons, however, not everyone agrees about causes and/or seriousness of such real or perceived problems. Indeed, it seems that some people and groups have actively campaigned to cast doubt on such concerns. There are strong suggestions, for example, that business-sponsored groups have acted—sometimes by employing reputable scientists—to discredit harmful effects of such commodities and, indeed, trustworthiness of scientists who have published negative findings about them and, moreover, trustworthiness of whole fields of science (Oreskes & Conway, 2010).

Since about 1970, educators and school systems have given some priorities to engaging students in learning about controversies such as those noted above (Pedretti & Nazir, 2011), falling under different names in various jurisdictions, including: socioscientific issues (SSIs) in the USA and elsewhere (Sadler, 2011) and STSE (relationships among fields of science & technology and societies & environments) issues in the UK, Canada and elsewhere (Pedretti & Nazir, 2011). Despite its relatively long and broadly-distributed history, implementation of STSE (the term used here, due to its use in local curricula) education has been, while extensive and deep in some contexts, generally *modest* (Hodson, 2011). Among many barriers to its implementation, a prominent and persistent one has been traditional emphases in school science systems—including governments, educational researchers, textbook publishers, school and school system administrators, science teachers, etc.—on instruction in 'products,' such as laws and theories, of fields of science and technology. Such teaching is relatively easy, in the sense that it involves presenting to students, in various ways, convenient 'packets' of information (e.g., at regular intervals, the human heart pumps oxygenated and de-oxygenated blood out to the body and lungs, respectively); and, students' knowledge of such products can then be relatively-easily assessed and evaluated. Moreover, attention to STSE issues can question idealized images of the nature of science—e.g., that it is highly-efficient, unbiased and unproblematic for societies and environments (Allchin, 2003)—that often are associated with didactic foci on products education (Hodson, 2008). Indeed, it is apparent that efforts to enlighten students about potentially problematic aspects of the nature of science often are resisted (Hodson, 2008).

There are, undoubtedly, many factors limiting STSE education. The nature of its implementation in many contexts around the world, though, implicates one general inhibiting factor. Often, the aim of STSE education emphasizes student learning of evidence and arguments for conflicting positions, which they are then encouraged to use—frequently in social situations—to develop, in highly logical ways, *personal* positions on issues (Levinson, 2013). On the one hand, there appear to be clear benefits of such personalized and argumentative approaches, including: development of *socioscientific reasoning skills* (Sadler, Barab, & Scott, 2007) and learning of products of science (e.g., laws & theories) (Venville & Dawson, 2010). On the other hand, emphases on personal decision-making suggests an orientation towards societal *individualism*—a trend away from collectivism that has, apparently, been increasing in the last few decades (Putnam, 2000). This is a mark of a broader ideological trend that seems to have increasingly dominated societies globally since the mid-1970s. According to McQuaig and Brooks (2010), after the second world war and subsequent recovery that involved major increases in government spending and intervention in economies, it seems that the share of wealth of the richest 1% of the population dropped from pre-war levels around 24% to post-recovery levels near 10%. This, apparently, led them to re-invent the pre-war policy of economic liberalism—a view advocating individuals' and groups' (e.g., corporations) rights to engage in economic markets free (liberated) from government intervention. This renewed policy, known as *neo*[new]-liberalism, intensified and re-invented old economic liberalism (McMurtry, 2013). For example, new 'extra-national' organizations, like the World Bank, World Trade Organization and International Monetary Fund, were formed to represent economic and social elite—free from obligations to any country or its people. Apparently in cooperation with governments and extra-national organizations, like banks and 'think tanks,' policies and agreements—such as tax laws and international trade agreements—are arranged in ways to favour for-profit activities by individuals and groups (Robertson & Verger, 2012). Reminiscent of the *survival-of-the-fittest* phenomenon attributed to Charles Darwin, neoliberalism has, apparently, contributed to significant increases in possessive individualism within cultures—which, in turn, seems to have contributed to dramatic worldwide increases in gaps between rich and poor (McQuaig & Brooks, 2010). Associated with this wealth concentration has, apparently, been significant damage to wellbeing of many individuals, societies and environments (WISE); hence, STSE issues like the ones noted above.

Possessive individualism is, perhaps, epistemologically and ethically unsound. Based on social epistemology (e.g., Fuller, 2002) and actor-network theory (e.g., Latour, 2005), all people, other living things, artefacts they produce, abstract images and messages and inanimate objects, etc. all, to varying degrees, co-influence each other. One implication of such a worldview is that many of our 'positive' (and 'negative,' etc.) attributes (e.g., attitudes, skills, knowledge [ASK]) are given to us by other actants in our networks. From a *communitarian* perspective, we can recognize such connectedness and promote positive actions on networks—to reciprocate benefits received and/or to realize further benefits from a more 'positive' network (Peters, 2011). People acting in positive ways within networks, however that may be defined by them, could, in principle, contribute to general improvements to WISE.

There appears to be a particular need for promotion in and through school science of positive student actions on actor-networks. Fields of science and technology, particularly in the neoliberal age, seem to have been heavily-influenced by profit motives and this, in turn, seems to have led to many compromises to their topic choices, methods and dissemination and application decisions (Ziman, 2000). Krimsky (2003), for instance, has documented many such compromises in biological and biomedical fields that are associated with such problems as birth defects, cancer, diabetes, cardio-vascular diseases and other preventable illnesses. Arguably to protect positive images of science (and technology), however, school science systems rarely refer to problems associated with business-science partnerships (Carter, 2005). Indeed, promotion of any kinds of actions—such as petitions to power-brokers—to address STSE issues, regardless of perceived causes (e.g., business-science partnerships), are rare in school science. As Hodson (2003, p. 657) said, "[i]t is almost always much easier to *proclaim* that one cares about an issue than to do something about it!" (emphases added). Nevertheless, at least because we need more activist societies to address the many serious potential and realized problems for WISE associated with decisions made by powerful people and groups about fields of science and technology, activism should be promoted in and through school science and technology (Hodson, 2011; Levinson, 2010; Dos Santos, 2009).

7.3 Research Context & Methods

Since 2006, the first author of this chapter (Larry) has been supporting teachers in efforts—based on the 'STEPWISE' (Science & Technology Education Promoting Wellbeing for Individuals, Societies & Environments) pedagogical framework—to encourage and enable students to take socio-political actions, at least partly based on their research, to address STSE issues. STEPWISE is elaborated in Chap. 2 in this volume and at: www.stepwiser.ca. In the spring of 2011, the second author (Mirjan) of this paper was a member of a graduate course in history, philosophy and sociology of science for which Larry was the instructor. After the course ended, Mirjan asked Larry if he could facilitate his study of progress in students' views of the nature of science (NoS). As reported below, however, subsequent discussions about this request led us to conclude that, instead of focusing only on NoS, Larry would serve as researcher-facilitator of Mirjan's efforts to encourage and enable students to develop and implement research-informed and negotiated actions (RiNA) to address issues. We agreed that students' NoS views may relate to their actions.

In this chapter, we report Mirjan's efforts to encourage and enable students in his first-semester (Sept. 2011-Jan. 2012) tenth-grade 'academic' (for university entrance) science class to conduct socio-political actions on issues. We also aimed to understand factors that may influence students' actions.

Given it was Mirjan's first time promoting actions on SSIs, he chose to design activities that aligned well with the pedagogical sequence in Fig. 7.1, which many teachers working with Larry had found appropriate for their contexts. Although it may not be the most philosophically-sound approach, given that it assumes students

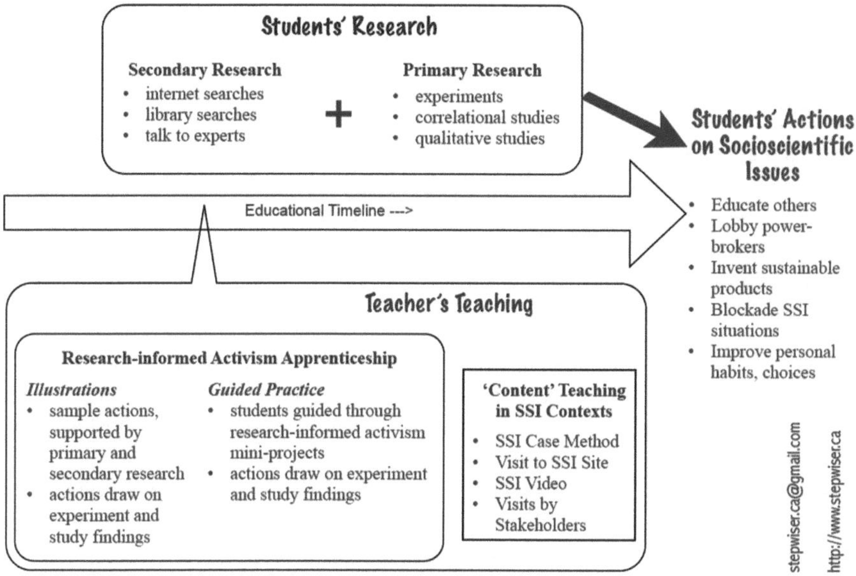

Fig. 7.1 A pedagogical framework for promoting sociopolitical actions on socioscientific issues

lack significant expertise and motivation for actions on SSIs, many teachers believe that such a 'deficit' approach is necessary because of students' and teachers' relative inexperience with such educational approaches (Bencze & Carter, 2011).

Primarily between Oct. and Dec., 2011, Mirjan developed and implemented a program with his tenth-grade students that, in synchrony with instruction about 'products' (laws & theories for chemistry, physics, biology and earth & space science), featured the following kinds of activities:

- Students expressed (e.g., discuss, write about) their pre-instructional views about SSIs;
- Mirjan modelled/demonstrated and discussed various cases of research-informed actions on SSIs taken by citizens (e.g., Occupy Toronto [via video taken by Mirjan]; Problems With Take-out Coffee on YouTube™; *The Story of Stuff* video and resources [www.storyofstuff.org/]);
- Student groups conducted secondary research (e.g., Internet searches) to learn more about one of three chemistry issues (i.e., *household cleaners*, *acid rain* or *oil spills*) suggested to them;
- Mirjan conducted lessons and activities to help students to develop expertise and confidence with controlled experiments and correlational studies;
- Mirjan asked students to consider various types of socio-political actions, such as *educating others*, *lobbying power-brokers* and *developing more sustainable inventions*, and then required them to develop actions to address their selected chemistry issue;

- Mirjan guided students in implementing a whole-class correlational study about climate change, using students' questions about relationships like gender vs. students' shower lengths; and then asked students to develop actions addressing results of a survey.

To achieve our research agenda, we conducted data-collection and analyses having *rationalistic* and *naturalistic* characteristics (Guba & Lincoln, 2011). Rationalistically, we focused, for example, on students' ongoing motivation to act on issues. Naturalistically, we collected data that enabled emergence of unexpected situational outcomes. Data collected from 29 students (ages 14–18) and from Mirjan included:

- Project Work Artefacts: Samples of products generated by students were collected, including: issue descriptions, research plans, data collected, written reports, project reflections, action plans and forms of action (e.g., posters, petitions, videos);
- Project Instructional Materials: Copies of Mirjan's pedagogical plans and instructional materials (e.g., paper handouts, videos, PowerPoint™ presentations, and internet site web addresses) were made;
- Digital Recordings of Students' Project Work: Photographs (23) and videos (16 for each of the 7 groups) were produced depicting youth presenting and defending their forms of action in public fora (e.g., to fellow students within and outside of class).
- Semi-structured Interviews: Five volunteering students were interviewed three times, near the beginning, at the end and 4 months after the course. Questions focused on their views about issues, research and actions. Mirjan was interviewed 11 times, for about 60 min each, about project progress. All interviews were audio-recorded and later transcribed.

Regarding analyses, each of us coded data for categories and then developed encompassing themes—using constant comparative methods based on constructivist-grounded theory (Charmaz, 2014). Categories and themes were then negotiated between to achieve consensus (Wasser & Bresler, 1996). Member checks with participants were conducted to help ensure *trustworthiness* of claims, each of which was based on at least three supporting data sources.

7.4 Results & Discussion

7.4.1 Preamble

The RiNA projects developed by students in Mirjan's tenth-grade science class resulting from his first efforts to promote such thinking and activism seemed broadly comparable to those conducted by students in other teachers' cases (e.g., Bencze & Sperling, 2012; Bencze, Sperling, & Carter, 2012). Guided by the

STEPWISE pedagogical framework in Fig. 7.1, all groups of students in Mirjan's class generated and implemented multiple actions—based on secondary and primary research about which they had significant control—to address STSE issues of their concern. In association with their project work, they also developed expertise in the different STEPWISE domains and, moreover, significant commitments to critical research on SSIs and actions to address them. Development of these outcomes mainly spanned two curriculum units; that is, *chemical change* and *climate change* (MoE, 2008)—each lasting about six weeks. As shown in Table 7.1a, student groups conducted secondary research into various chemistry issues and developed corresponding actions to address them; while in the climate change unit, as shown in Table 7.1b, they conducted secondary and primary research as sources of claims for actions on issues of their choice.

In the following sub-section, some specific examples from the class studied are provided in support of the above claims. Such results appear to be influenced by multiple—and, often, interacting—contextual variables. This complex of factors is discussed in the second sub-section below. Particular focus in this discussion is placed on roles of students' uses of *local data* as bases for corresponding actions in *local contexts*.

7.4.2 Young Activists in the Making

Given the relatively student-led nature of research-informed and negotiated actions, it is difficult to impose precise criteria on student 'achievement' regarding such projects. By their very nature, they can—and, perhaps, should—be highly idiosyncratic and situated. Moreover, given that they are a relatively new phenomenon in science education, formal evaluation criteria are lacking. Hodson's (2011) categories for 'success' in SSI education, however, seem useful—particularly in terms of their *breadth*. He suggests that students should develop expertise in each of the following domains (not, necessarily, in the following order): *understanding* of SSI issues; awareness of *power relations* in SSIs; establishment of *personal SSI positions*; and, *sociopolitical actions* on SSIs. Given that students in our research had very little prior experience with SSI education, primary research and development and implementation of social actions to address SSIs, their 'expertise' by the end of the course seemed well-progressed. Their initial views of 'STSE Issues' could be described as *technicist*; that is, seeing 'issues' not as controversies but as 'problems.' For example, as shown in Fig. 7.2, students' views of cigarette smoking emphasized their harmful effects (e.g., cancer)—although there was some allusion to controversy (e.g., peer pressure to smoke, depicted by the eyes and 'Do it!' advice). Eventually, however, their discourse about issues increasingly included reference to controversies regarding decisions by people in powerful positions. Generally, although there was no mention of sometimes problematic influences of capitalists on scientists and/or engineers, as some have described (Krimsky, 2003; Ziman, 2000), they did tend to name 'people with money' as major actants in STSE networks. When asked, for instance, who were the opponents in climate change issues, 'Zoë' said:

Table 7.1a Summary of students' first RiNA projects

Team	Socioscientific issue	Secondary research	Primary research	Action(S)
3 boys	**Oil Spills** (Topic posed by teacher, chosen by students)	Student-led, mostly using the internet	N/A	Created Facebook™ group, with regular oil spill cases added, supported by messages on Tumblr™ and Twitter™
3 girls	**Cigarette Smoking** (Topic posed by teacher, chosen by students)	Student-led, mostly using the internet	N/A	Brochures distributed around local community centre; Posters posted around school; made and wore t-shirts with anti-smoking messages
3 boys	**Cigarette Smoking** (Topic posed by teacher, chosen by students)	Student-led, mostly using the internet	N/A	Produced two large anti-smoking posters they placed where people smoke; and produced and distributed an anti-smoking brochure to teachers and students
4 girls	**Cigarette Smoking** (Topic posed by teacher, chosen by students)	Student-led, mostly using the internet	N/A	Produced an anti-smoking lecture, which included four anti-smoking posters, they conducted with a ninth-grade class in the school
1 boy	**Acid Rain** (Topic posed by teacher, chosen by students)	Student-ied, mostly using the internet	N/A	Activist video placed on YouTube™
3 boys	**Household Cleansers** (Topic posed by teacher, chosen by students)	Student-led, mostly using the internet	N/A	Produced a website about household cleaners, included other's activist video
1 girl	**Household Cleansers** (Topic posed by teacher, chosen by students)	Student-led, mostly using the internet	N/A	Produced activist video, featuring she and her cousin in a role-playing scenario
3 girls	**Household Cleansers** (Topic posed by teacher, chosen by students)	Student-led, mostly using the internet	N/A	Produced an activist video posted to YouTube™, and posted flyers the cleaners' isle of a local store
3 boys	**Personal Hygiene Products** (Topic posed by teacher, chosen by students)	Student-led, mostly using the internet	N/A	Produced PowerPoint™ presentation, which they planned to present to a junior class

> People who know it [waste] is a problem, but they choose to think it is not as important because of the cost of it. … People who are rich, like politicians, they don't want to help [address] climate change because they feel it will do them more harm than good. … [Opposed to them are] the protesters [laughter]!" (Dec. 16, 2011).

Most students seemed to develop relatively strong critical views of large companies. 'Brent,' for example, who investigated students' computer uses, said: "Computer manufacturers would lose a lot of money if they made computers that were more efficient and don't break down" (Dec. 16, 2011). Similarly, 'Sean,' who had

Table 7.1b Summary of students' second RiNA projects

Team	Socioscientific issue	Secondary research	Primary research	Action(s)
2 boys	**Consumption of food in different categories by boys and girls**	Student-led, mostly using the internet	Part of whole-class study designed by teacher & students	Lobbied male gym. teachers to encourage boys to eat less meat and more vegetables and fruit, using a brochure they produced
2 boys & 2 girls	**Computer usage by boys and girls**	Student-led, mostly using the internet	Part of whole-class study designed by teacher & students	Produced activist video posted to YouTube™
2 girls & 2 boys	**Water consumption (esp. shower lengths) by bays and girls**	Student-led, mostly using the internet	Part of whole-class study designed by teacher & students	Produced activist PowerPower™ slideshow, which contained information about shower uses and photos of them talking to fellow students as they distributed their brochures about shower uses.
2 boys & 2 girls	**Water consumption (esp. shower lengths) by boys and girls**	Student-led, mostly using the internet	Part of whole-class study designed by teacher & students	Produced activist PowerPower™ slideshow, shown to their class, and they produced a website about water uses. They also made a poster, distributed around the school
1 girl & 2 boys	**Uses of bottled vs. tap water by students of different ages in the school**	Student-led, mostly using the internet	Part of whole class study designed by teacher & students	Produced activist PowerPower™ slideshow, shown to their class. They also produced an activist poster they used to speak to other students in the school about bottled water
2 girls & 2 boys	**Playing of video games among girls and boys**	Student-led, mostly using the internet	Part of whole-class study designed by teacher & students	Produced activist PowerPower™ slideshow, shown to their class. They also produced 2 large activist posters that they planned to place outside the boys' washrooms
2 boys & 2 girls	**Recycling practices by students of different ages**	Student-led, mostly using the internet	Part of whole-class study designed by teacher & students	Produced activist PowerPower™ slideshow, shown to their class. They also produced *A* posters, to be placed in cafeteria neartrash bins. They also handed out brochures they produced

(continued)

Table 7.1b (continued)

Team	Socioscientific issue	Secondary research	Primary research	Action(s)
2 girls & 2 boys	**Recycling practices by boys and girls**	Student-led, mostly using the internet	Part of whole-class study designed by teacher & students	Produced activist PowerPower™ slideshow, shown to their class. They also lobbied the school principal using a letter they wrote and read an announcement over the school PA system to encourage recycling in different ways

Fig. 7.2 Some students' initial conceptions of STSE issues

investigated students' plastic water bottle uses, said that, despite knowing about hazards (e.g., Bisphenol A), companies persist in marketing them because of "money. They really don't care about anything other than their money" (Apr. 23, 2012). At the same time, some shared their views with peers that 'average' citizens/consumers may not be entirely blameless. Brent, for example, seemed to feel humans, generally, were abusing their/our intellectual gifts:

> There is this magical thing that happened to the evolution of Man [sic], which was that we were allowed to think. … This has allowed us to make different things and be different than other organisms. [But,] we are just using our knowledge to destroy each other because, basically, we just want more money (Dec. 16, 2011)!

In researching information about STSE issues, students seemed to develop some relevant expertise. Mirjan said students had developed a 'healthy' skepticism, for example, about information retrieved from websites—generally distrusting websites with corporate sponsors, preferring government-sponsored sites, and regularly triangulating claims; that is, only drawing conclusions when at least two sites they believed to be reputable corroborated each other's claims. Such skepticism, however, seemed to exist in students' minds prior to the course we studied. What was new to these students, on the other hand, were conceptions about *correlational studies*; that is, empirical investigations in which researchers attempt to find correlations between naturally-changing variables. Correlational studies are used in many fields of science, but apparently seldom used or emphasized in school science (Bencze, 1996). In interviews, students confirmed that they had never heard about correlational studies prior to the course about which we are reporting. Sean exclaimed, for instance, "I told my friends in other classes that we are doing correlational studies in science. They said, 'What's that?' " (Nov. 16, 2011). By about the mid-way point of the course, however, most students seemed to understand them reasonably well. Two typical statements given by students about them were: "You can't really experiment with people with smoking. My teacher was saying, 'You can't really make half of the class smoke and the other half of the class not smoke' " (Zoë, Nov. 16, 2011); and, "In an experiment, you are trying to alter one of the variables; whereas, in a correlational study, you can't alter any of them [variables]. You have to observe the people to see what patterns occur" (Brent, Nov. 16, 2011). While this seemed to be a positive outcome of the course, we are not sure about students' *expertise* in conducting correlational studies—mainly because they were only part of one such study in this course, and that was largely teacher-guided (refer below for elaboration). As a class, with teacher help, students generated findings for possible correlations for several variable pairs—which, as shown in Table 7.1b, included links between age and gender and variables like: shower length; food group eating; plastic water bottle use; and video game-playing.

Once students felt they had gathered sufficient information from their research, given their available time, they negotiated, developed and often implemented actions to address their chosen STSE issues (see Tables 7.1a and 7.1b). All actions drew, to varying extents, on their secondary and primary research (refer below for details); and, all took particular stances on personal and/or group decisions (e.g., 'Girls might reduce shower times'). As indicated in Tables 7.1a and 7.1b, there was a range of breadth and depth of student groups' actions. Difficulties some students experienced may, as discussed below (*Factors Affecting Actions*), be somewhat attributable to this was Mirjan's first use of the STEPWISE framework and, moreover, he had only started using it in the second month of a 5-month course. While not necessarily typical, actions generated by one student group may give readers a sense of the potential of this kind of student project work. Their main actions, which are depicted in Figs. 7.3a–c, illustrate that they combined findings from their primary and secondary research in producing an informational website (Fig. 7.3a), which they promoted through their Facebook™ and Twitter™ contacts and an

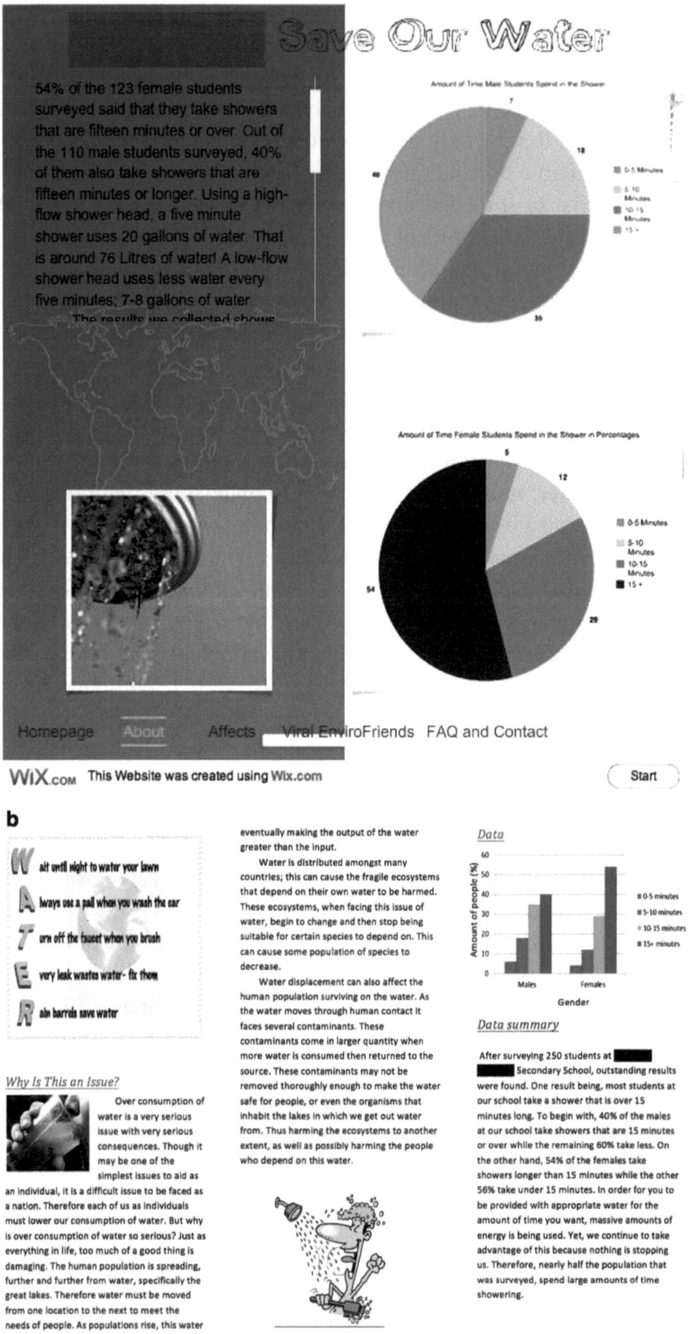

Fig. 7.3 (**a**) Excerpt from students' activist website, (**b**) excerpt from students' activist pamphlet, (**c**) a student teaching a peer

c

Fig. 7.3 (continued)

informational pamphlet (Fig. 7.3b), which they used as a basis for conversations in which they engaged students in hallways within their school (Fig. 7.3c).

In reviewing students' various actions (refer to Tables 7.1a, 7.1b and 7.2), it became quite apparent that they tended to be restricted to local/familiar contexts—such as peers and teacher(s) in class or in the school—or remote, but somewhat impersonal, contexts like contacts on YouTube™, Facebook™ or Twitter™. Overall, apart from one case in which a student group actively lobbied the school principal and janitors to move recycling bins to prominent locations in the school cafeteria, most actions could, arguably, be considered 'safe.' Most students felt comfortable interacting with familiar people not in an authoritative role. Reflecting most students' views, 'Brent,' an otherwise confident student, said:

> I am OK with talking to people in my age group. I just don't like it when I explain my ideas to older groups of people. I feel like they don't take me seriously. I feel like they say, '[You] didn't go to university. [You] don't have all these credentials. … I feel that when I am talking to my own age group, I am more safe and more comfortable talking to them – because I can connect with them on a personal level (Dec. 16, 2011).

While it was apparent that students were most comfortable with *safe* actions, there seemed to be, nevertheless, some depth to their commitments to actions on STSE issues. Several students offered that their focus on STSE issues and actions led them—and, sometimes, family members—to change their personal behaviour regarding several issues. 'Hal' said, for instance, that he doesn't litter any more—adding: "I try to stop my friends [from littering]" (Apr. 23, 2012). 'Paul' concurred, saying that he and his parents have tried to reduce their car use now that he has become more aware of climate change (Apr. 23, 2012). Moreover, there appeared to be some durability to students' commitments to research-informed and negotiated actions. Although few said they thought they would conduct primary research

Table 7.2 An analysis of students' qualitative considerations for actions

Quotation(s) and summaries	Knowledge (Capital)	Possible changes
Recycling project Adrienne said they planned to get the recycling bins located in the cafeteria in places where they knew students would see and perhaps use them (Dec. 7, 2011). Karla said, "Usually at lunch time, people are too lazy to go to the back of the cafeteria to use the recycling bins. So, it's better if the recycling bins are next to the trash cans; so that they [students] will sort it [recyclables] out" (Dec. 7, 2011). Paul added that "Usually, they [students] just leave it [containers, etc.] on the tables or throw it in the garbage [bin].... There is, like [approximately], one trash can for every two tables. Even with two trash cans per table, people are just lazy and leave their stuff on the table" (Dec. 7, 2011). Zoe added that "A lot of students put cans in the trash can, which is wrong" (Dec. 7, 2011)	Students' typical movements in a cafeteria	Recommendations for placement of recycling bins
	Students' tendency to believe that recycling is Not 'cool' (Note: obtained through personal communication with a teacher)	Hope that students use recycling bins
Adrienne: "We ... decided to talk to the janitors about the lack of recycling bins ... to see if they could help by possibly putting out more or moving them to a more accessible location" (Dec. 12, 2011). Adrienne's group chose to place their poster about placement of recycling bins in front of the school office "so that the principal or teachers would see it and get inspired [to recycle]" (Dec. 7, 2011). Zoe's group seemed keen to get their recommendation of having recycling bins moved; they chose to take their letter to the principal, rather than just sending it to her. They felt they could explain it better in person (Dec. 16, 2011)	Students' tendency to not clean up after themselves	Hope that students do not place recyclables in trash cans
	Location of trash cans	
	Number of trash cans pertable	
	Students uses of trash cans	
	Familiarity with the school's janitors	
	Awareness of principal's and teachers' positions of power	
	Understanding that personal contact can be better than through a medium	

(continued)

Table 7.2 (continued)

Quotation(s) and summaries	Knowledge (Capital)	Possible changes
Shower time project When Mirjan asked students how they could convince other students to reduce water and energy use regarding showers, Karla said: "When we were talking, we said it would be hard to tell people, 'Just reduce your [shower] time!' They would say to your face, 'Oh, whatever...' and walk away. So, that's why we created these 'Did You Knows' [e.g., 'low flow shower heads use about 46L less per 5 min. of showering' on their poster]. It would get people mentally thinking; and it's [sic] facts they wouldn't know. So, maybe it would take to them [sink in] and they would actually start thinking about reducing [water use]" (Class Presentation, Dec. 7, 2011)	Understanding of students' listening practices/priorities	Hope that students reduce shower times
	Comfort in speaking to peers	Hope that students would use reasoning for their choices
Hal: "We made a brochure [about water usage] and Mr. K. made a bunch for us. We went around the halls [of the school] handing them to and talking to people. We went [asked] 'How long do you shower?' They gave us an answer. [They congratulated students who took short showers.] We asked them why [they take such long showers]. We encouraged them to read the brochure. We told them how it harms the environment and how it affects global warming. A lot of people did listen, but some people [ignored us].... We [suggested] they take shorter showers" (Dec. 16, 2011)		Speaking to peers about shower uses

beyond their science course, most indicated that their curiosity for secondary research had been piqued and their orientation towards actions on STSE issues had been deepened. Hal reflected this view when he said:

> I will continue with actions. I like researching things. If I don't know something, I want to know it! Just the other day, I was researching about photosynthesis. … I was thinking maybe people won't have to change. Maybe, as we are adapting to [increasing' CO_2 emissions, maybe we can have this piece of machinery that I am trying to invent. [It would] capture CO_2 and put oxygen [into the air]. … I'm going to 'high-five' myself if I do it! … I am [also] going to inform people about issues that *I* feel strongly about; just try to spread the word (Dec. 16, 2011).

In support of this claim for depth of commitment, most students agreed with this recommendation: "I think a lot of schools should do this. Not only did we learn about the aspects of science that are part of the grade 10 curriculum, but we learned it and we made a change to it. So, you actually feel like you did something while learning" ('Sidney,' Apr. 23, 2012).

7.5 Factors Influencing Actions

Myriad Factors Promoting RiNA projects to address STSE issues should be relatively easy in the province of Ontario, in which Mirjan conducted his work and which makes provision for most aspects of such projects in its latest curricula (e.g., MoE, 2008). Mirjan acknowledged that such curricular support was, indeed, a prime driver behind his promotion of RiNA projects (Written reflection, July 10, 2012). On the other hand, based on actor-network theory (Latour, 2005), we must assume myriad, often co-affecting, factors may influence outcomes. In some ways, we might imagine concentric circles around teaching/learning in classrooms, at least including influences from fellow students, teachers, parents, administrators, school district officials and the larger society (e.g., with advertisements). Nevertheless, despite such complexity, evidence from Larry's studies of Mirjan's work in comparison to that of other teachers with whom he has worked suggests that student outcomes like those reported above would not have been possible without a teacher at least somewhat like Mirjan. Prior to our collaboration, it is apparent he already was active as a critical reflective practitioner—having, for instance, participated in educational research and dissemination about a decade earlier (Percy & Krstovic, 2001). At the time of this research, he was a graduate student studying for a Masters of Education degree focusing on science education. He participated in and led several workshops in science education, is a member of a 'professional learning community' within his school district (regularly collaborating with other teachers) and maintains an online educational blog highlighting his ongoing thoughts about education. Arguably related to this background, his philosophical positions seemed congruent with principles and practices for teaching about problematic STSE relationships and promoting student-led RiNA projects. On the *Scientific Theory Profile* (Loving, 1991), for instance, he indicated that he adheres to 'Naturalist-Antirealist' views about science—which, in contrast to 'Rationalist-Realist' views, suggests that knowledge generation in the sciences may be affected by factors beyond logic, including: societal cultural values, socio-economic factors, a scientist's gender, religion, theoretical perspectives, personal biases, etc. Such views about science seem conducive to encouraging students to challenge authority of scientists and others in STSE decisions. On the other hand, teachers may sometimes struggle implementing such views if immediate colleagues are not, at least, supportive of a teacher exercising some critical reflective practice in his/her teaching—support that Mirjan said, indeed, contributed to his students' successes with RiNA projects.

Social Studies Findings An overwhelming focus of our collaboration became the role of correlational studies as sources of data/findings that may motivate and guide student actions. Larry had been advocating for their use in science education for many years (Bencze, 1996). Mirjan was not, like many other science teachers, initially familiar with them. School science textbooks, which often guide teachers' instructional perspectives and practices, tend to valorize experimentation as the way to truths and, moreover, as <u>the</u> defining feature of science separating it from other ways of knowing (Knain, 2001). In our discussions, Mirjan agreed with Larry that it seemed reasonable to encourage students to conduct correlational studies in investigating STSE issues—largely on *ethical* grounds; because, unlike experimentation, studies were unlikely to generate problematic outcomes for living things. Accordingly, after considering the apprenticeship framework in Fig. 7.1 (Bencze & Carter, 2011), along with instructional resources provided by Larry (Bencze, 2000), Mirjan developed teaching/learning materials for helping students to develop expertise, confidence and motivation for conducting correlational studies. An excerpt from these resources, in which students are guided through a short in-class correlational study, is shown in Fig. 7.4a. After such introductory activities, Mirjan then worked with students to co-generate a large-scale, whole-class, correlational study relating to climate change—in the second unit of the course, but the first time students were introduced to primary research as a contributor to decisions about actions on STSE issues. Outside of class, students first conducted some secondary research—via the Internet—about climate change and, then, in class, brainstormed many possible contributors to it. This work then allowed them, in groups, to suggest possible questions for correlational studies. At Mirjan's suggestion, many such questions used age and gender as possible independent variables. One group's questions, for example, were: " 'How does gender affect the amount of times you wash your clothes per week?' and 'Does your age affect your transportation to and from school?' " (Nov. 22, 2011). After some whole-class negotiation, the survey questions in Fig. 7.4b were generated. Students then collected survey responses from 250 students in their school and, with Mirjan's help, organized them into a large data table. Each of the seven student groups were then asked to pick results from one variable combination (e.g., students' shower times varying by age). Groups' choices are given in Table 7.1b. They were then asked to use the results from the class survey and their secondary research on their chosen topic (and on climate change, generally) as bases for actions they were to develop and implement that would address their issue.

A special feature of students' correlational studies that appeared to contribute to the depth of commitment to actions on STSE issues was their *local* nature. In short, despite the remoteness of much information about climate change, there was a close connection between the context of their primary research (correlational studies of school peers) and actions (e.g., pamphlets given to school peers). For example, 'Karla' reported that results of the study of shower lengths found that: "54 percent of the 123 females surveyed take showers 15 minutes or over" (Dec. 12, 2011). She said, in addition, that it was their understanding that results from this local study logically implied local action: "The website [and posters] is directed to the students at ['Central High'] because the study was done at the school" (Dec. 12, 2011). The contexts, in other words, for their research and actions were much the same. This can be understood, perhaps, in terms of the schema in Fig. 7.5a—which is an adaptation of a theoretical relationship between 'science' and 'technology' described by Roth (2001). Accordingly, students conducted 'Science'/Research by collecting data about fellow students' shower lengths and representing that information as graphs (e.g., as seen in Fig. 7.3b). Once phenomena of the World are translated into representations (Signs), they are part of the cognitive and social milieu and, therefore, may be *manipulated*. Typically, as described above (e.g., Tables 7.1a and 7.1b), a

Part III) Perform a small correlational study in class!

You will perform a small scale correlational study in class today. You will examine if there is a correlation between GENDER (male/female) and one other variable that you choose from the list below:

- preference between meat or vegetables
- attitudes towards reusing, reducing and recycling
- extracurricular involvement in school
- walking to school, biking or getting dropped off in a car
- staying home to watch a movie or going out to a theater

Question:

Is there are correlation between gender and ____________________?

Hypothesis:

I think that ____________________ because ____________________

Method:

Describe how you will obtain your data.

Results:

Construct a table in which you will record your results:

Ex: The following is an example of data collected by one student who wanted to determine if there was a correlation between gender and preference for math, science or art.

	# of Male students	# of Female students
I prefer art	23	7
I prefer math	15	15
I prefer science	13	23

Fig. 7.4 (**a**) Excerpt from correlational study activity, (**b**) Class-developed correlational study

b

Grade 10 Science: Climate Change Survey

Age: ________ Gender: M or F Initials: ______

Circle the option that best represents your action(s).

1, What mode of transportation do you use to get to school most often?
a) bus b) walk c) bike d) car

2. What kind of food do you prefer to eat most often?
a) meat b) vegetables c) meat and vegetables

3. How often do you eat out at fast food restaurants?
a) once a month c) twice a week
b) once a week d) more than two times per week

4. What source of water do you drink most often?
a) Tap water (filtered or not filtered) b) Bottled water

5. How often do you (or your parents/guardians) wash your clothes?
a) once in two weeks b) once a week c) 2 times a week d) 3+ times a week

6. How long do you shower for on average?
a) 0 to 5 minutes b) 5 to 10 minutes c) 10 to 15 minutes d) 15 minutes or more

7. Which of these electronic devices do you use for more than 1 hour a day?
a) computer b)video game console c) television d) radio/mp3 player e) cell phone

8. After eating lunch in the school, how often do you recycle materials such as glass, aluminum cans or plastics ?
a) always b) only sometimes c) rarely ever

Fig. 7.4 (continued)

students then use their pre-conceived notions, primary and secondary research findings, etc. to develop *new* 'representations' (Signs) of the world—which they then expand into their social milieu (e.g., among peers), as 'Technology'/Actions, to try to bring about changes to the World (e.g., 'Changes in Teens' Shower Lengths'). Presumably—and hopefully—students then engage in series of cycles of research-informed and negotiated actions in school and beyond.

A key to the apparent contribution of 'the local' to students' engagement in and attachments to RiNA projects appears to be 'the personal.' In an analysis preceding, but parallel to that in Roth's (2001) paper, Wenger (1998) claimed that the more learners are personally-engaged in decisions about reciprocal relationships between phenomena and representations (the latter he said must be *reified*) the deeper and more committed would be their learning—at least because personal choice seems associated with personal identity and positive self-esteem development. There are

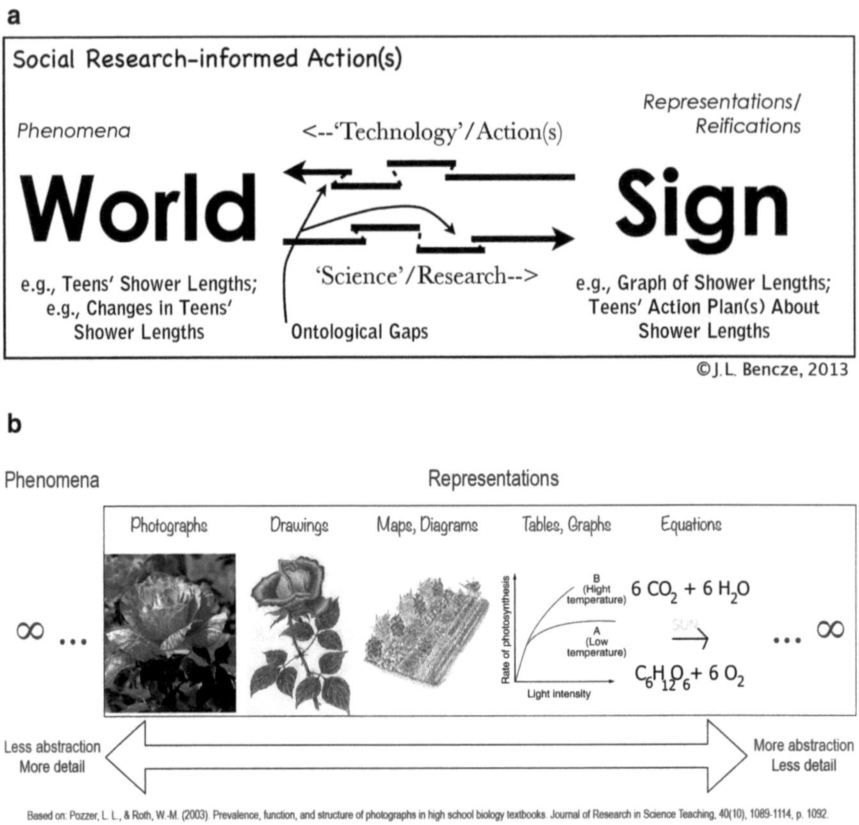

Fig. 7.5 (**a**) Schema for research-informed action projects, (**b**) relationships between phenomena and their representations

several examples in this study in which this applied to the research and to action sides of the relationship:

Research While the RiNA projects conducted about climate change were the first such projects by these students, which meant some teacher modeling and guidance seemed appropriate (Fig. 7.1), Mirjan gave students considerable control of decisions throughout the process. In terms of 'research,' as described above, the students worked with Mirjan to generate potential study questions and, although guided through the actual study, were allowed to choose which 'cause-result' relationship (e.g., shower lengths by gender) on which to conduct secondary research. They then had freedom about how to display their selected data (from the class study and their secondary research) and conclusions to draw from them; e.g., as given in Fig. 7.3b. Although we suggest that the ultimate goal for such projects is full (within safety limits) student control of decisions (perhaps after an apprenticeship like that provided here), it seemed clear that students found this research much more engaging than they had previously experienced as school science students:

> Last year, my science teacher would give us [STSE] assignments, but he wouldn't let you choose the topic. He'd just give it to you and you'd be expected to kind-of follow through.... and, it's like, it was all right. I did good and stuff, but I wasn't interested in it. I didn't have that much of a motivation to really go further. He [Mr. K] has given us a choice and said it should be something you have an interest in. I think that is better, because I am motivated to do more (Nov. 16, 2011).

With such freedom of choice, it is apparent students chose—to a great extent—findings from the class study that had considerable personal meaning for them as stimuli for action. 'Adrienne,' a member of one of the two groups investigating recycling in the school, exclaimed on two occasions how disappointed they were in 'discovering' that many of their peers did not recycle packaging in the context of their lunch in the cafeteria. On Dec. 6, 2011, she said that the fact that most students don't recycle is "startling to find out. So, we decided to take action against it. We believe that we should put up posters and educate people, but also have more recycling bins right there in the cafeteria, where students will be." Later, she stated this another way: "While we discovered that there is not really a strong correlation between age and the amount of recycling that is occurring, we discovered that a shocking 72% of both seniors and juniors only sometimes or rarely ever recycle after eating lunch at school" (Dec. 12, 2011).

Actions Although students had only perhaps 'minor' control over the whole-class correlational study, mainly because it was the first one most, if not all, had conducted, regulation of decision-making was ceded to them in using research (and other influences) for actions to address their issues. As indicated in Table 7.1b and Figs. 7.3a–c, this seemed to lead to considerable variety and creativity in actions. This personalization may have contributed to the level of commitment to RiNA discussed above because of its influence on student identity and self-esteem. This, indeed, seemed evident in students' comments about freedom they experienced, particularly in acting on research findings. Zoë, for instance, made this highly appreciative comment about the action component of their science education: "The difference with Mr. K. is that he wants us to take action. He doesn't just want us to just make a report" (Nov. 16, 2011). 'Sidney' concurred: "Mr. Krstovic gives us a lot of freedom about how we want to take action. That's another thing we like about him" (Nov. 16, 2011).

'Holistic' Correlational Studies Although contributions of students' social correlational studies to motivation and direction for actions in the same local contexts seems significant, it may be that its influence was even greater than what we imagined. On the one hand, students' actions were not solely determined by correlational study results. Some students indicated that their decisions drew on both secondary and primary research findings. About his group's actions about shower lengths, for example, Hal said: "The longer the shower you take, you are, basically, wasting energy to make the water warm and you are taking water from the lakes and contaminating and, sometimes, they can't get the contaminants out, right? So, that was our main problem" (Dec. 16, 2011). Similarly, regarding her group's actions on recycling, Zoë said: "I used to think there was just one pop [metaphorically] can in the ocean, but it is so serious, you know?" (Dec. 16, 2011). Nevertheless, it seems clear that Mirjan placed considerable priority on use of correlational study findings as a source of direction for actions. This was evident in Mirjan's instruction sheet for the RiNA project assignment: "Whatever action you take, for example if it's an educational brochure, it should include information from your secondary research (i.e., the Internet, textbook and other resources) as well as your results from your correlational study with proper bar graphs or other graphs that visually display your data" (Nov. 22, 2011). Moreover, nearly fifty percent of the grade on the final RiNA assignment report was associated with students' use and reference to correlational data. Indeed, it was apparent that students worked under the assumption that actions based on correlational data was a priority. Sidney said, for example:

> [Our findings will help us] decide how to take actions. Suppose a lot of people don't really know about global warming. Our *responsibility* would be to put up signs around the school, to educate people on global warming. .. Whatever our findings, we take actions on those results (Nov. 16, 2011).

On the other hand, it is apparent that students' motivation and direction for action on STSE issues was broader than from correlational data. One might ask, 'To what extent do people fully experience the world through correlational studies?' The answer appears to be that they may be a 'thin slice' of reality. Correlational studies, like experiments, are *reductionist* in nature. Out of the complex world of variable interactions, pairs of them are isolated for consideration. Such abstraction, however, seems to be prioritized in the sciences. However, in the study reported here, it is apparent that freedom of choice that Mirjan had afforded students for their RiNA projects led them to explore and act on the world in more expansive ways than just through correlational studies. In short, their RiNA projects were more *holistic* than might be expected by projects supposedly driven by correlational data. In this regard, to supplement the quantitative character of correlational studies, there was a significant *qualitative* character to their projects. This was apparent in both the research and action phases:

Research Despite being encouraged to conduct and use quantitative correlational studies, but perhaps because of the freedom they were given, students chose to add *qualitative* observations to their research. As Zoë explained, many of them felt that the quantitative studies needed to be supplemented with personal interactions with peers:

> We went around [during a full class period] to find out why they didn't recycle. … We got the percentages [of student recyclers from their survey], but we wanted to know [why]. That's how we figured out it was because a lot of them [students] didn't know where they were [located] or that the cafeteria even had recycling [bins]. … We couldn't really find a solution unless we knew what the problem was (Dec. 16, 2011).

Hal corroborated this claim in the context of his group's study of peers' shower times. While conducting his survey, he asked spontaneous supplementary questions to determine students' reasons for longer showers. He said that the "girls had better reasons. The guys would say [things like], 'Oh, I just stand in the shower [for whatever time]'; whereas the girls would say, 'Oh, I shave my legs, and I have to wash my hair nicely and I have to condition it [their hair] …' They gave us all these reasons" (Dec. 16, 2011). Additionally, he said: "We found that a lot of people [about

70% of students], regardless of gender and age, take showers longer than 15 minutes. ... Some people I talked to take showers of 2 hours!" (Dec. 16, 2011).

Actions By nature, actions on the world may be considered 'technological'; that is, purposeful actions that may bring about changes desired by individuals and/or groups. Creating technologies/inventions is a complex activity, often involving simultaneous consideration of many often interacting variables. It does draw on information from the world (World → Sign), but then needs to use that information in ways that achieve the desired outcomes (Sign → World). Because technologies tend to be developed for specific contexts of use (e.g., wheel chairs for people of certain sizes, etc. and meeting certain legal standards), those carrying out technological change/action need to know the world into which the technology/invention/action will be placed. Because these students chose local contexts for their research, they did—indeed—know a great deal about the context of use; that is, *their* school. As illustrated in Table 7.2, they appeared to draw, to a great extent, on both *cultural* and *social* capital (Bourdieu, 1986)—e.g., *what* you know and *who* you know, respectively—while making decisions about possibly-effective actions.

7.6 Conclusions and Scholarly Significance

It may be expected that a science teacher would allow more student-led technological activities. There appears to be a tradition in school science of allowing *applications* of principles of science in terms of technology design challenges (e.g., Fortus, Dershimer, Krajcik, Marx, & Mamlok-Naaman, 2004)—despite arguments that technological developments often have preceded well-developed science knowledge (Layton, 1993). Arguably, science education systems allow such student-directedness in technological design as long as canons of the sciences (e.g., widely-accepted laws & theories) are protected as part of an apparent *Platonic Legacy* (Lewis, 1995)—perpetuation of dependence on potentially self-interested abstract rulings from so-called *Philosopher Kings*. Nevertheless, contributions of such student-led, holistic, technology design/action development to the depth of students' commitments to research-informed and negotiated actions must be appreciated. Several students in this project acknowledged this contribution. About depth of learning, for example, Hal said that self-directed learning and being able to use his personal knowledge is about "being able to pioneer your own ideas. ... Right now, I can grab my science textbook and learn everything about grade 10 curriculum, but I am not going to know it a year from now. In this class, we learned to take the knowledge and act, make a difference, talk to people, help" (Dec. 16, 2011). Sidney added that "[i]t sticks with you" (Dec. 16, 2011) and, later, that: "[e]xperience always stays with you, whereas book [learning] always goes away after you are done the subject [course]. It's better to [use book and personal knowledge], which is what we did with these action projects" (Apr. 23, 2012). Indeed, as Sidney suggested, associated with depth of learning was *personalization*—which, again, makes sense in terms of our arguments regarding the importance of learner engagement in decisions about reciprocal relationships between phenomena and their representations (Wenger, 1998). Several students made comments in this regard. Hal said, for instance, that RiNA projects allowed them to "take in knowledge and then … do something with it,[which he felt was less likely than when just learning from books]" (Apr. 23, 2012). Concurring, Zoë added: "In school, we most learn subjects that are dedicated to facts and a system. There are always certain rules you must follow, whether it is with math or with English [courses]. Whereas, with this project, we got more freedom—and we could use our own personal knowledge" (Apr. 23, 2012).

What was not expected in this case was the contextualization of primary research that occurred in Mirjan's class, despite his emphasis on quantitative correlational studies, and its apparent contribution to the depth of students' commitment to research-informed and negotiated actions we perceived. Evidence cited above suggests that students' more holistic take on primary research—such as by drawing on their personal discussions with students about their shower times—was a significant contributor to their motivations. There are, perhaps, many ways this effect may be explained. However, a schema again drawing on ideas about relationships between phenomena and their representations seems to help. As depicted in Fig. 7.5b, it may be that the deepest and most committed learning occurs in the 'research' phase (World ➔ Sign in Fig. 7.5a) when representations are the *least abstracted* and, therefore, have the most detail. Although it is not shown in the schema, by this logic, results of correlational studies (as 'Tables, Graphs') may engender less commitment and depth of learning than various, student-led, qualitative descriptions and anecdotes—such as that girls may take longer showers to allow for shaving of legs, etc.

Overall, findings of this study contribute to our growing understanding of factors influencing students' research-informed and negotiated actions on STSE issues. Students' *social* correlational studies in a school science context appeared to significantly influence their motivation to act and their action foci. But, perhaps more importantly, the relative freedom of choice students were afforded in their projects seemed to lead them to expand the scope of their research beyond the relatively narrow foci of quantitative correlational studies. Students' motivation to act seemed to be enhanced by their decision to spontaneously use prior notions and collect qualitative information about their contexts of actions; that is, their school. This conclusion appears to challenge the traditional emphases in school science on quantitative, reductionist, science research; and emphasis that may be part of the ancient adherence to conceptions of society depicted in Plato's *Republic*.

References

Allchin, D. (2003). Scientific myth-conceptions. *Science Education, 87*(3), 329–351.

Angell, M. (2004). *The truth about the drug companies: How they deceive us and what to do about it*. New York: Random House.

Barnes, R. L., Hammond, S. K., & Glantz, S. A. (2006). The tobacco industry's role in the 16 Cities Study of Secondhand Tobacco Smoke: Do the data support the stated conclusions? *Environmental Health Perspectives, 114*(12), 1890–1897.

Bencze, J. L. (1996). Correlational studies in school science: Breaking the science-experiment-certainty connection. *School Science Review, 78*(282), 95–101.

Bencze, J. L. (2000). Procedural apprenticeship in school science: Constructivist enabling of connoisseurship. *Science Education, 84*(6), 727–739.

Bencze, J. L., & Sperling, E. R. (2012). Student-teachers as advocates for student-led research-informed socioscientific activism. *Canadian Journal of Science, Mathematics & Technology Education, 12*(1), 62–85.

Bencze, L., & Carter, L. (2011). Globalizing students acting for the common good. *Journal of Research in Science Teaching, 48*(6), 648–669.

Bencze, L., Sperling, E., & Carter, L. (2012). Students' research-informed socioscientific activism: Re/visions for a sustainable future. *Research in Science Education, 42(1), 129–148.*

Bourdieu, P. (1986). The forms of capital. In J. G. Richardson (Ed.), *The handbook of theory: Research for the sociology of education* (pp. 241–258). New York: Greenwood Press.

Carter, L. (2005). Globalisation and science education: Rethinking science education reforms. *Journal of Research in Science Teaching, 42*(5), 561–580.

Charmaz, K. (2014). *Constructing grounded theory* (2nd ed.). New York: Sage.

Dos Santos, W. L. P. (2009). Scientific literacy: A Freirean perspective as a radical view of humanistic science education. *Science Education, 93*(2), 361–382.

Fortus, A., Dershimer, R. C., Krajcik, J., Marx, R. W., & Mamlok-Naaman, R. (2004). Design-based science and student learning. *Journal of Research in Science Teaching, 41*(10), 1081–1110.

Fuller, S. (2002). *Social epistemology* (2nd ed.). Bloomington, IN: Indiana University Press.

Guba, E. G., & Lincoln, Y. S. (2011). Competing paradigms in qualitative research. In N. K. Denzin & Y. S. Lincoln (Eds.), *The SAGE handbook of qualitative research* (pp. 105–117). Thousand Oaks, CA: Sage.

Hileman, B. (1998). Industry's privacy rights: Is science shortchanged? *Chemical & Engineering News, 76*(17), 36.

Hodson, D. (2003). Time for action: Science education for an alternative future. *International Journal of Science Education, 25*(6), 645–670.

Hodson, D. (2008). *Towards scientific literacy: A teachers' guide to the history, philosophy and sociology of science*. Rotterdam, The Netherlands: Sense.

Hodson, D. (2011). *Looking to the future: Building a curriculum for social activism*. Rotterdam, The Netherlands: Sense.

Klein, N. (2014). *This changes everything: Capitalism and the climate*. Toronto, Canada: Simon & Schuster.

Knain, E. (2001). Ideologies in science textbooks. *International Journal of Science Education, 23*(3), 319–329.

Krimsky, S. (2003). *Science in the private interest: Has the lure of profits corrupted biomedical research?* Lanham, MD: Rowman & Littlefield.

Latour, B. (2005). *Reassembling the social: An introduction to actor-network-theory*. Oxford, UK: Oxford University Press.

Layton, D. (1993). *Technology's challenge to science education*. Milton Keynes, UK: Open University Press.

Leonard, A. (2010). *The story of stuff: How our obsession with stuff is trashing the planet, our communities, and our health – And a vision for change*. New York: Free Press.

Levinson, R. (2010). Science education and democratic participation: An uneasy congruence? *Studies in Science Education, 46*(1), 69–119.

Levinson, R. (2013). Practice and theory of socio-scientific issues: An authentic model? *Studies in Science Education, 49*(10), 99–116.

Lewis, T. (1995). From manual training to technology education: The continuing struggle to establish a school subject in the USA. *Journal of Curriculum Studies, 27*(6), 621–645.

Loving, C. C. (1991). The Scientific Theory Profile: A philosophy of science model for science teachers. *Journal of Research in Science Teaching, 28*(9), 823–838.
McMurtry, J. (2013). *The cancer stage of capitalism: From crisis to cure* (2nd ed.). London: Pluto.
McQuaig, L., & Brooks, N. (2010). *The trouble with billionaires*. Toronto, Canada: Viking.
Ministry of Education [MoE]. (2008). *The Ontario curriculum, grades 9 and 10: Science*. Toronto, Canada: Queen's Printer for Ontario.
Oreskes, N., & Conway, E. (2010). *Merchants of doubt*. London: Bloomsbury Press.
Pedretti, E., & Nazir, J. (2011). Currents in STSE education: Mapping a complex field, 40 years on. *Science Education, 95*(4), 601–626.
Percy, J. R., & Krstovic, M. (2001). Later life learners: A significant and receptive audience for introductory astronomy. *Journal of the Royal Astronomical Society of Canada, 95*, 205–207.
Peters, M. A. (2011). *Neoliberalism and after?: Education, social policy and the crisis of Western capitalism*. New York: Peter Lang.
Putnam, R. (2000). *Bowling alone: The collapse and revival of American community*. New York: Simon & Schuster.
Robertson, S.L., & Verger, A. (2012). Governing education through Public Private Partnerships, published by the Centre for Globalisation, Education and Societies, University of Bristol, Bristol BS8 1JA, UK at: http://susanleerobertson.com/publications/
Roth, W.-M. (2001). Learning science through technological design. *Journal of Research in Science Teaching, 38*(7), 768–790.
Sadler, T. (Ed.). (2011). *Socio-scientific issues in the classroom: Teaching, learning and trends*. Dordrecht, The Netherlands: Springer.
Sadler, T. D., Barab, S. A., & Scott, B. (2007). What do students gain by engaging in socioscientific inquiry? *Research in Science Education, 37*(4), 371–391.
Venville, G. J., & Dawson, V. M. (2010). The impact of an argumentation intervention on Grade 10 students' conceptual understanding of genetics. *Journal of Research in Science Teaching, 48*(8), 952–977.
Wasser, J. D., & Bresler, L. (1996). Working in the interpretive zone: Conceptualizing collaboration in qualitative research teams. *Educational Researcher, 25*(5), 5–15.
Wenger, E. (1998). *Communities of practice*. Cambridge, UK: Cambridge University Press.
Ziman, J. (2000). *Real science: What it is, and what it means*. Cambridge, UK: Cambridge University Press.

Chapter 8
Epistemic Contributions to Students' Autonomous Socioscientific Actions

Larry Bencze and Mirjan Krstovic

8.1 Introduction

A major trend in science education over the last several decades has been to engage students in consideration of *socioscientific issues* (SSI); which may be considered controversies around potential problems facing individuals, societies and/or environments associated with fields of science and technology. This movement has not, however, been able to make significant inroads into dominant goals and practices in science education. School science systems have tended to prioritize instruction and assessment/evaluation in 'products'—such as laws and theories—of fields of science and technology. Elaborate attention to potential problems associated with these fields may detract from dominant programmatic agendas. There does seem to be hope for further implementation of SSI education through, for example, engaging students in conduct of research that may yield results that motivate them to learn more about issues and, perhaps, take actions to address them. In this chapter, we share results and conclusions of efforts—based on the 'STEPWISE' pedagogical framework—to enhance research-informed and negotiated action projects of students by engaging them in reciprocal relations between action projects and meta-analyses of them.

L. Bencze (✉)
Ontario Institute for Studies in Education (OISE), University of Toronto,
Toronto, ON, Canada
e-mail: larry.bencze@utoronto.ca

M. Krstovic
Peel District School Board, Mississauga, ON, Canada
e-mail: mirjan.krstovic@peelsb.com

L. Bencze (ed.), *Science and Technology Education Promoting Wellbeing for Individuals, Societies and Environments*, Cultural Studies of Science Education 14,
DOI 10.1007/978-3-319-55505-8_8

8.2 Socioscientific Issues Education and Actions

Although it continues to be taught alone, particularly in secondary schools, there have been efforts over several decades to encourage teachers to integrate other subjects with science. One such movement, 'STSE' education, involves lessons and activities to relate fields of science and technology and societies and environments (Pedretti & Nazir, 2011). Students may, for instance, be taught that norms of practice in the sciences—including skepticism and objectivity, for instance—have sometimes been influenced by non-scientist individuals/groups (Ziman, 2000). Perhaps the most popular movement in relating science to other fields, however, is to engage students in considering so-called *socioscientific issues* (Sadler, 2011)—such as controversies regarding government (de-)regulation of food manufacturing companies. Very often, while negotiating decisions with peers, students are expected to weigh conflicting data and claims and develop personal reasoned defences of their position(s) on issues (Levinson, 2013). There are numerous socio-scientific controversies, and evidence suggests that engaging students in making personal reasoned decisions about them has led to some significant learning gains—including, for instance, development of *socioscientific reasoning skills* (Sadler, Barab, & Scott, 2007); learning of products of science (e.g., laws & theories) (Venville & Dawson, 2010) and, learning about the nature of science (Khishfe & Lederman, 2006).

Asking students to make personal choices about SSIs may benefit them in various ways, but it seems limiting on a social level. In *representative* democracies, citizens educated about socio-scientific controversies may be in good positions to, for instance, vote for candidates who would represent their informed positions (Wood, 1998). Such a system seems appropriate and, clearly, has numerous examples worldwide. However, there are suggestions—from scholars and others—that societies (and environments) could benefit from more *participatory* forms of democracy. Rationale for greater citizen engagement in public affairs are complex and varied; but, in terms of science and technology, major concerns often emphasize issues surrounding private sector financial agreements with scientists, engineers and other related professionals (along with institutions employing them). At least in part, scientists and engineers with financial ties to private sector interests have been known to compromise topic choices, research methods and dissemination approaches (Krimsky, 2003; Mirowski, 2011; Ziman, 2000). Associated with such compromises, in turn, appear to be numerous potential (and realized) harms linked to various commercial products and services—such as: genetically-modified foods, etc. (Kleinman, 2003; household cleaning and hygiene products (Leonard, 2010); pesticides (Hileman, 1998); tobacco (Barnes, Hammond, & Glantz, 2006); and, pharmaceuticals (Angell, 2004). Many people also expect serious potential personal, social and environmental problems associated with dramatic increases in average global temperatures that often are linked to excessive fossil fuel uses (Klein, 2014). While such potential and realized harms linked to government-supported business-science partnerships seem to exist, it is sometimes difficult for citizens in representative democracies to make informed choices regarding them. There is evidence to suggest, for instance, that private sector interests (e.g., financiers & companies) have paid scientists and engineers and others to discredit findings of other scientists and engineers that would problematize various commercial products and services, including those relating to petroleum, nuclear energy, pesticides/herbicides, tobacco and weaponry (Oreskes & Conway, 2010).

In light of such concerns as government-sponsored private sector ties with fields of science and technology, many scholars and others are urging school science systems to encourage and enable youth to develop expertise, confidence and motivation to critically evaluate fields of science and technology and their relationships with societies and environments and to develop and implement informed actions to address potential/realized harms they perceive (Bencze & Alsop, 2014; Hodson, 2011; Levinson, 2010; Santos, 2009). Promoting such critical and activist science education is, however, likely to be very difficult. School science systems—including governments, educational researchers, textbook publishers, school and school system administrators, science teachers, etc.—tend to prioritize instruction in 'products,' such as laws and theories, of fields of science and technology. Such teaching is relatively easy, in the sense that it involves presenting to students, in various ways, convenient 'packets' of information (e.g., the human heart pumps oxygenated and de-oxygenated blood out to the body and lungs, respectively, at regular intervals)—students' knowledge of which can then be relatively-easily assessed and evaluated. Moreover, attention to STSE issues can call into question idealized images of the nature of science; e.g., that it is highly-efficient, unbiased and unproblematic for societies and environments (Allchin, 2003), which often are associated with didactic focus on products education (Hodson, 2008, 2011). Indeed, it is apparent that efforts to enlighten students about potentially problematic aspects of the nature of science often are met with resistance (Crawford, 2007; Hodson, 2008). Arguably to protect the positive image of science (and technology), for example, school science systems rarely make reference to problems associated with business-science partnerships (Carter, 2005). Consequently, promotion of many kinds of actions—such as petitions to power-brokers—to address STSE issues, regardless of the perceived causes (e.g., business-science partnerships), are rare in school science. As Hodson (2003) said, "[i]t is almost always much easier to *proclaim* that one cares about an issue than to do something about it!" (p. 657; emphases added).

8.3 Research Context & Methods

8.3.1 Research Context

Since 2006, research groups led by Larry (first author here) have studied educators' efforts to encourage and enable students to develop and implement actions to address STSE issues based, in part, on their self-directed research. This work has drawn, to a great extent, on the 'STEPWISE' (Science & Technology Education

Promoting Wellbeing for Individuals, Societies & Environments) theoretical framework. Refer to Chap. 2 here (and www.stepwiser.ca) for more detail about this framework. Briefly, while acknowledging co-dependence of various learning domains (i.e., 'Products Education' ←→ 'Skills Education'), this framework prioritizes 'STSE Actions'—which, like all of the other domains in the framework, depends on (and affects) all other domains in it. Among other things, this implies that, in essence, students would—when taking social actions intended to improve wellbeing of individuals, societies and environments be—'spending' some of their cultural and social capital (e.g., science literacy) on actions to bring about a better world. This is intended to orient STEPWISE towards altruism.

Two major features of STEPWISE should be stressed; that is, its emphasis on: i) *critical STSE education*, particularly in terms of encouraging students to consider potential problems for WISE due, at least in part, to decisions made by powerful people and/or groups about fields of science and technology; and ii) student-led primary, as well as secondary, research to possibly inform their decisions about social actions to address STSE issues. Justification for the first feature is given above and in other publications (e.g., Bencze & Carter, 2011), which also provide rationale for encouraging student-led research as a basis for actions. However, some additional rationale given here for this latter feature may help readers. Because there is not strong traditions of promotion of student-led socio-political actions in school science (refer above), students are likely to need added incentive to carry out such activities. In our work, we have found that students can become motivated to act on STSE issues if they have been encouraged and enabled to self-direct research ('primary' [e.g., correlational studies] and 'secondary [e.g., Internet searches]) to learn more about the issues (Bencze, Sperling, & Carter, 2012). This tack is based on *knowledge duality theory* (Wenger, 1998); that is, that students are likely to develop deep commitments to issues and actions if they strongly influence reciprocal relations between *phenomena* (e.g., citizens' views of consumer products) and *representations* of them (e.g., survey data; and, later, educational posters).

Among teachers working with Larry to explore implementation possibilities of STEPWISE, Mirjan Krstovic (second author here) stands out in terms of his longevity with the project and close working relationship with Larry. After taking an online course instructed by Larry in history, philosophy and sociology of science, Mirjan asked to work with him to study students' conceptions of the nature of science. After meeting each other face-to-face for the first time, we agreed that one approach to this goal may be to encourage and enable students to, eventually, self-direct research-informed and negotiated action (RiNA) projects to address STSE issues of their concern. Our reasoning was, in part, that the nature of science clearly includes research and negotiation; but, as well, could include socio-political actions informed by research findings. We began our collaboration in the September of 2011, with Mirjan developing lessons and student activities surrounding the STEPWISE pedagogical schema given in Fig. 8.1. This schema was derived from work with other teachers, who found its linear approach more feasible in terms of teaching/learning conditions in schools than the arguably more theoretically sound tetrahedral version

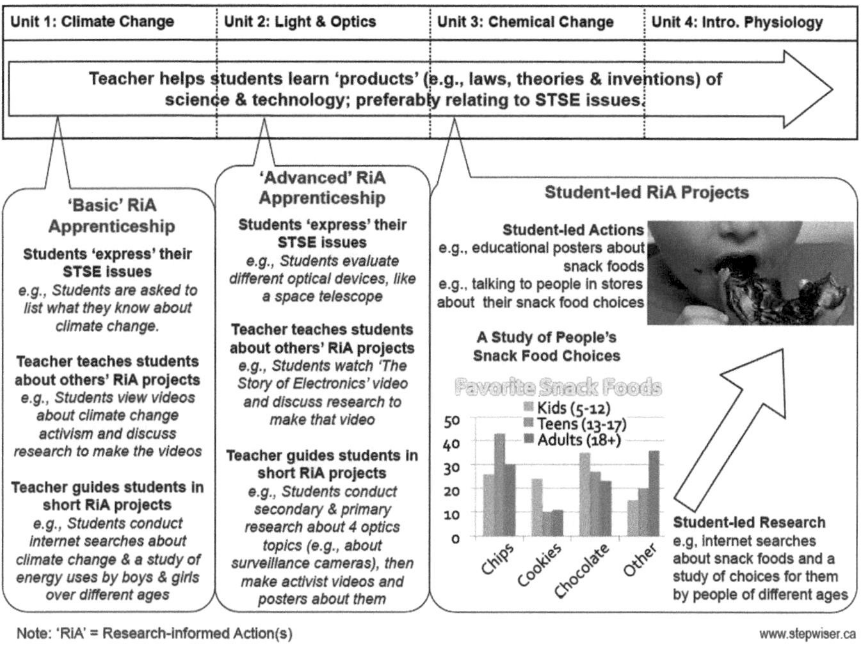

Fig. 8.1 STEPWISE pedagogicalschema

(Chap. 2, in this volume) of STEPWISE (Bencze & Carter, 2011). Briefly, the schema involves providing students with 'apprenticeship' lessons and activities that are aimed at helping them develop expertise, confidence and motivation for self-directing RiNA projects. Further details of this approach are provided in Chap. 2 of this volume. As Mirjan used this approach to encourage student-led RiNA projects, Larry served as 'researcher/facilitator,' asking Mirjan about his instructional plans and corresponding student achievements—and, at Mirjan's request, providing him with teaching/learning suggestions.

The study reported here draws from our experiences with implementing the pedagogical schema in Fig. 8.1 during the second semester (Feb. – June, 2012), which was Mirjan's second attempt at using this approach. In his first attempt, he had decided that his students were not ready to self-direct RiNA projects—partly because he had started the process late in Sept. of 2011 and partly because it was his first effort at such teaching/learning practices. Consequently, for his second attempt, he committed himself to developing pedagogical strategies that may help students to self-direct RiNA projects. Approaches that Mirjan developed, in consultation with Larry, are discussed below—under 'Factors Contributing to Outcomes.' Our research goals were to document and evaluate the extent of students' self-directed RiNA projects and explain such outcomes through reference to relevant theories.

8.3.2 Research Methods

To achieve our research goals, Larry conducted data-collection and analyses having *rationalistic* and *naturalistic* characteristics (Guba & Lincoln, 2011). Rationalistically, for example, students' ongoing motivation to act on issues was tracked. Naturalistically, data were collected that enabled emergence of unexpected situational outcomes. Data collected from 14 students in a tenth-grade 'academic' (for university qualification) stream science class and Mirjan included:

- Project Work Artefacts: Samples of products generated by students were collected, including: issue descriptions, research plans, data collected, written reports, project reflections, action plans and forms of action (e.g., posters, petitions, videos);
- Project Instructional Materials: Copies of Mirjan's teaching materials (e.g., handouts, videos, PowerPoint™ presentations, and internet site web addresses) were made;
- Digital Recordings of Students' Project Work: Photographs (23) and videos (16 for each of the 7 groups) were produced depicting youth presenting and defending their forms of action in public fora (e.g., to fellow students within and outside of class).
- Semi-structured Interviews: Five volunteering students were interviewed twice, near the beginning and end of the course. Questions focused on their views about issues, research & actions. Mirjan was interviewed 11 times for about 60 minutes each about project progress. All interviews were audio-recorded and later transcribed.

Regarding analyses, each of us coded data for categories and then developed encompassing themes—using constant comparative methods based on constructivist grounded theory (Charmaz, 2014). Categories and themes were then negotiated between to achieve consensus (Wasser & Bresler, 1996). Member checks with participants were conducted to help ensure *trustworthiness* of claims, each of which was based on at least three supporting data sources.

8.4 Results & Discussions

8.4.1 Preamble

Mostly adhering to the framework in Fig. 8.1, Mirjan was able to get students in his tenth-grade academic-stream science class to complete what he considered very successful self-directed social actions projects—based, to a great extent, on finding of their primary and secondary research—to address STSE issues of their concern. Again, this appears to be an exceptional achievement (Bencze & Carter, 2011). In collaboration with Larry, Mirjan found ways to cede considerable amounts of control of research-informed and negotiated activism to students. In the following sub-section, examples of students' RiNA projects as they progressed through the course are provided—to illustrate their achievements and as a basis for analyses of factors apparently contributing to them, the latter of which is provided in the second sub-section below.

8.4.2 *Progress Towards Student-Led Research-Informed Actions*

As illustrated in Fig. 8.2, Mirjan began his efforts to promote student-led research-informed and negotiated actions on STSE issues with apprenticeship activities intended to help students to develop (or enhance) expertise, confidence and motivation for such autonomous projects. Although he felt that the science department in which he worked had a history of practices aligned with some STEPWISE principles, he also believed his students had relatively little experience in several aspects of it—particularly in terms of self-directed primary research, a situation commonly-found in many secondary school contexts. Like many secondary school science teachers, those conducting other sections of the same tenth-grade science

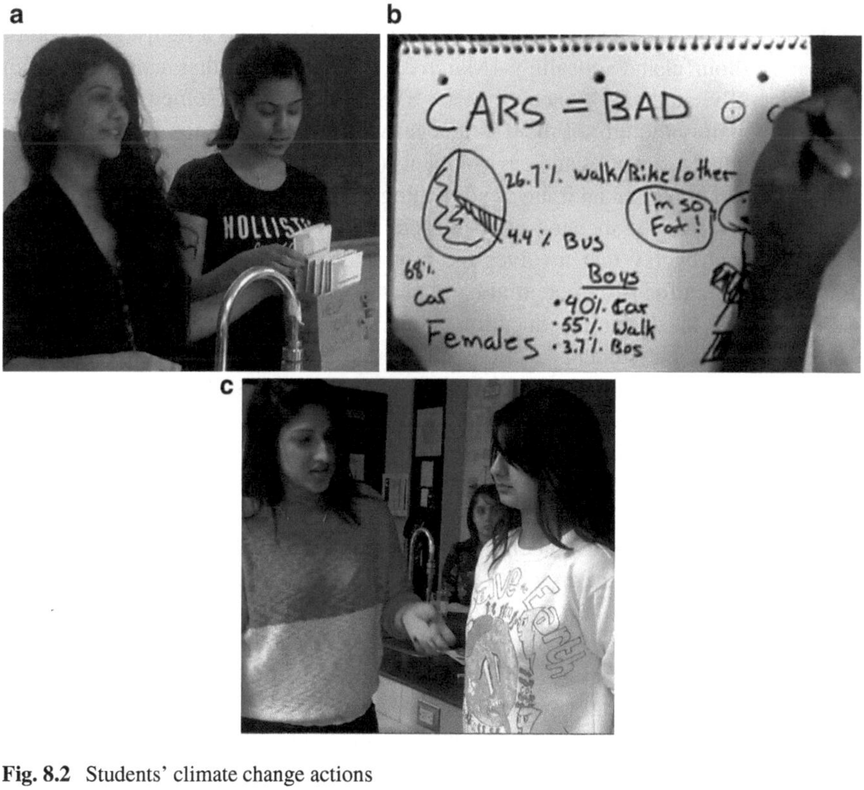

Fig. 8.2 Students' climate change actions

course that semester in his school were not promoting RiNA projects on STSE issues. By the end of the first unit—i.e., *Climate Change* (MoE, 2008)—of the course, however, seven student groups (each with 3-4 members) had completed a RiNA project to address particular potential contributors to climate change. Topics addressed included: garbage accumulation/incineration and limits to recycling (2 groups); non-local food sources; students' internet usage; students' school-home transportation practices (2 groups); and, students' water consumption practices. Both groups of students who studied peers' transportation choices found, for example, that girls are more likely to be driven to and from school than boys. Using this finding, one group developed, for example, an informational pamphlet provided in a colourfully-decorated box (Fig. 8.2a) that they situated in the school's students' services office, while the other group produced an informational video (Fig. 8.2b) modelled after those generated by RSA-animate (www.thersa.org/events/rsaanimate). Meanwhile, included among another student group's actions was a colourfully-designed 't-shirt' (Fig. 8.2c) with messages urging fellow students to consider their amounts of Internet use.

In reviewing/evaluating students' projects (from their classroom presentations and written reports), Mirjan indicated that he was very satisfied with students' uses of primary and secondary research (both with teacher support) in informing actions relating to climate change; the latter as being very student-led, he felt were very creative. About the t-shirt with the messages about Internet uses, for example, he said this to the class: "Did everyone see that shirt? They did a really great job of making it colourful and appealing" (March 23, 2012). On the other hand, Mirjan felt that, generally, students' conceptions of STSE issues were somewhat limited—since they mainly thought of them in a negative way, as 'problems' for the wellbeing of individuals, societies and environments (WISE). For instance, he said: "Many students understood that an issue as something that is a 'problem' and that there are negative side-effects associated with an issue. … [N]o one stated explicitly that an issue is a controversy that involves the pros and cons" (July 10, 2012). Perhaps associated with a focus on potential harms for WISE, most of the students' actions were relatively local; that is, education aimed at fellow students in the school. One exception was some students' actions aimed at 'friends' connected to social network websites, like those on Facebook™, or general users of YouTube™ (in the case of students' video actions). Overall, their actions were local, mainly aimed at peers and friends—and, associated with this, not much aimed at people or groups apparently strongly-influencing decisions linked to potential or realized climate change (Klein, 2014).

Somewhat satisfied with students' initial (guided) research-informed action projects, Mirjan—in consultation with Larry—made a few adjustments for the second set of RiNA apprenticeship activities. Addressing the curriculum unit, *Light & Geometric Optics* (MoE, 2008), students were given brief outlines of four issues from which to choose; that is, (i) Privacy invasion (e.g., as it related to surveillance cameras), (ii) Commercial advertizing (e.g., involving use of semiotic messages), (iii) Laser eye surgery (e.g., with possible negative side-effects) and iv) Consumer electronics (e.g., involving qualities of flat-panel TVs). Students were asked to conduct secondary (first) and primary research and consider using findings to develop actions to address issues they determined and they were given considerable freedom in their choices. Students were, then, asked to report results of their work in two contexts; that is, (i) in-class, with opportunities for classmate and peer feedback, and (ii) in a school hallway, with summaries of their work given on display boards (as used in science fairs) and opportunities for feedback from teachers and students (beyond their class). In reviewing the project reports, Mirjan had three broad conclusions. First, with regards to STSE issues, many of the students did seem to emphasize controversies resulting from decisions by powerful individuals or groups. For example, one of the students whose group had explored issues associated with advertising said, "[What advertizers say] "is not true and they know it is not true. [It is meant to get consumers to] buy the product" (Class presentation, March 26, 2012). Meanwhile, a member of the group who studied common uses of surveillance cameras in public places, made reference to the concept of the *Panopticon* discussed by Foucault (1977) as perhaps problematic control of citizens by authorities. Mirjan subsequently asked his class if their presence indicates less societal trust in people. A student responded: "It is a controversial issue. Some people would agree with you that it is a problem [that we have such surveillance. They might ask,] 'Why do we need cameras watching our every movement? It is my right not to be seen by the government.' But, other people have safety concerns[, thus approving of surveillance]." One girl offered that the increased level of violence in movies has caused society to "be paranoid[, thus accepting surveillance]" (March 26, 2012).

While Mirjan felt that students in his class had developed more sophisticated conceptions of STSE issues, he concluded, however, that their expertise in primary research and action choices still needed development. Regarding primary research, Mirjan said: "Although most felt that they had confidence in their secondary research, not many students understood what is meant by a valid study. The concepts of independent, dependent, control variables and sample size needed to be revisited before the students engaged in the last STSE action project. On the positive side, most students understood the difference between a study and an experiment" (July 10, 2012). For example, after hearing about students' research into peers' interpretations of advertizing that have had product-identifying information removed (Fig. 8.3), Mirjan said: "It is a very biased study—right?—with only 10 males and 10 females. How would we make this more valid and reliable?" (March 25, 2012). Meanwhile, despite students' relatively sophisticated views about STSE issues, acknowledging controversies involving decisions made by powerful individuals/groups, their actions—although often creative, such as in production of attractive and informative posters and videos (including another in 'RSA Animate' style)—continued to be largely aimed locally and not at powerful others. Referring, for example, to a very large poster made by a group who had studied semiotic messages in advertizing, Mirjan said in his written feedback to them: "Actions were adequate (posters). I wonder if they made a difference in any way" (March 27,

Fig. 8.3 Students' study object for learning about advertizing

2012). By his latter comment, he was alluding—as he stated in a later interview (May 10, 2012)—to students' lack of attention to influential individuals/groups. While this may be a legitimate concern, Gramsci (2007) suggests that 'counter-hegemonic' actions—such as alerting the public to ways in which they may be unconsciously conditioned to accept perspectives and preferred practices of dominant classes—often are necessary in democratic societies.

To address his concerns regarding students' progress in self-directing effective research-informed and negotiated actions to tackle STSE issues of their concern, Mirjan—in consultation with Larry—implemented a set of 'nature of RiNA' reflection activities. These occurred after students had completed two apprenticeships (on climate change and optics), at the beginning of the third (basic chemical change) of four course units. In connection with these activities, Mirjan introduced the final RiNA project—one that was to be self-directed by students and dealing with the fourth unit of the course (*Tissues, Organs, and Systems of Living Things*). Along with other events and phenomena, these reflective activities appeared to contribute to students' eventual successes with self-led RiNA projects discussed below (under *Factors Influencing Outcomes*). Briefly, a major factor contributing to students' successes with RiNA projects appeared to be implicit and explicit epistemic attention to such projects. Prior to this discussion, however, we provide a summary of the nature and extent of students' final course RiNA projects and learning outcomes associated with them.

By the end of the course, students in Mirjan's tenth-grade 'academic' science class had all—in seven teams—completed and defended an apparently much more independent research-informed and negotiated action project to address an STSE issue of their choice. The nature and extent of these projects, along with learning outcomes associated with them, are discussed below under the three broad RiNA categories emphasized by Mirjan with students. Some overall findings also are presented and discussed.

8.4.2.1 Conceptions of STSE

Instead of thinking of STSE issues as 'problems' that, to a great extent, need *technical* fixes (e.g., moving recycling bins to a more prominent location), as they had initially, most of the students came to view issues as controversies surrounding decisions by powerful people/groups. In several cases, students focused on power residing in companies and governments, as this quote from a student's final report indicates:

> Why aren't people who have the power to actually do something about [side-effects of multivitamins] not doing anything at all to help others from harm? The answer is simple[; that is,] profit. People in power are the people who produce multivitamins and the government. … Companies will not improve their product by finding a better way to produce these products since they are making money. …If [governments] wanted to, they could stop the production of bad products. Although they have the power to do so, they don't make the effort to since they're benefiting from these companies as well as since these companies share their profit with the government. … The reason why I think some vitamins harm people is because companies usually want to find the cheapest and fastest way to produce their product. When they find a way to do so, they don't concentrate on the negative effects bit[;] instead think about the profit they will make. … For example, instead of making all-natural vitamins, companies decide to play around with chemicals and find a solution that's much cheaper than extracted vitamins from real foods ('Randi,' June 19, 2012).

8.4.2.2 Research Expertise

In addition to the usual secondary research, about which they had considerable expertise (e.g., knowing to triangulate claims from the Internet), students conducted correlational studies with reasonably-large sample sizes. As 'Nancy' said in her final RiNA report of her team's study of peers' understanding of issues surrounding ultrasound diagnostic tests, "[o]ur survey had a good sample size of about 100 people, 50 males and 50 females" (June 19, 2012). One of the two graphs produced by her team from this study is provided in Fig. 8.4. Their conclusion, which Mirjan evaluated to be reasonable, was: "There is no real co-relation between gender and perception of ultrasound … [but] … a significant number of females (38% were aware of the long term effects of ultrasound than males (16%)" (Final Report, Nancy, June 19, 2012).

8.4.2.3 Action Expertise

Compared to projects these students completed earlier and those completed by students in Mirjan's first semester, actions students developed for their last project were more *diverse* (e.g., pamphlets & YouTube™ videos), *less local* (e.g., beyond the school) and more often *aimed at powerful actors* (e.g., drug companies). A brief summary of each team's actions is provided in Table 8.1. For example, rather than just displaying posters within their school, as had been their practice in the previous

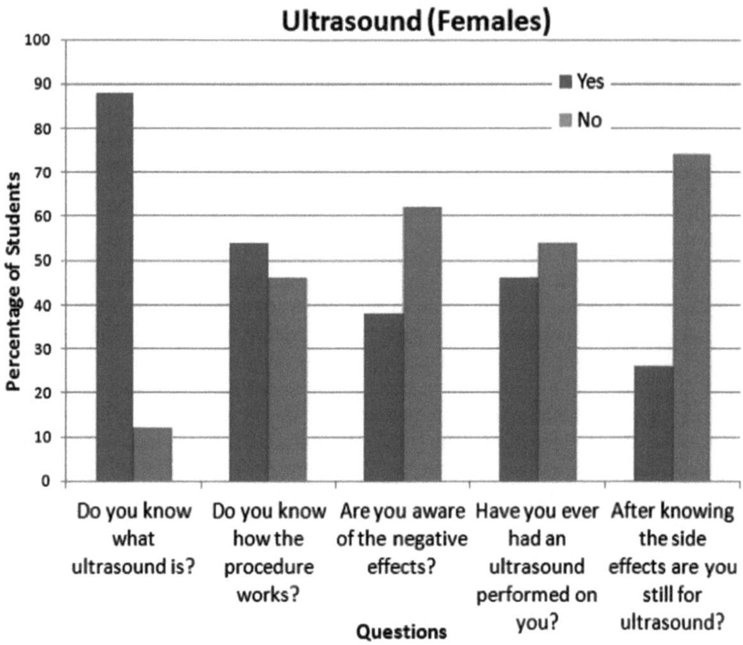

Fig. 8.4 Students' research results about ultrasound

Table 8.1 Students' third RiNA project topics

Ultrasound	Smoking I	Smoking II	Snack food	Energy drinks	Multi vitamins	GM foods
Poster, which they placed in the school	Poster	Developed 2 posters and distributed it to where people buy cigarettes	Made 3 posters and placed them in different local food outlets; and talked to people about snack foods	Petition with 500 signatures sent to Energy Drink companies	2-part video on YouTube (role-playing)	Letter to Greenpeace
Different letters sent to a website used by expectant mothers	Petition sent to Mayor	Developed an anti-smoking Facebook page	Talked to peers and families about snack foods	Created a lesson plan for PhysEd teachers	Brochure that they used to talk to peers and people at drug stores	YouTube video (like RSA Animate)
Brochure, which they used in speaking to women at maternity stores		Talked directly to students and store 0patrons		Sent letter to Health Canada	Wrote suggestions on Centrum's Facebook page	

two projects, the group that studied fast foods, displayed them in stores where people purchase the items and, moreover, discussed consumers' purchases with them. One of the students related a story, for instance, of a reason given by a child (who was with his parent) for his snack food choice; that is, that the item's container displayed a Spider Man™ image. When discussing issues with people in such commercial contexts, some students seemed to note that an issue regarding power was that 'front-line' actants—such as store clerks—may not be willing participants in generation of potential problems, perhaps because of their alienation from major decisions about products 'they' sell. A student who had asked a store clerk to speak on video camera about energy drinks, said:

> They said 'no' because, they don't know what they are selling, they don't know anything about energy drinks, they don't know how it effects the body. So, basically, they have no idea about what they are selling (June 18, 2012).

Meanwhile, students who had taken their actions more directly to powerful decision-makers, seemed to experience alienation of a different kind; that is, lack of response to their critical comments/questions. Little occurred, for example, in response to a group's note on the Centrum™ Facebook™ page promoting exclusive use of natural vitamins in multivitamins and in response to five different letters sent—which the group thought would be more effective than a single letter—by another group to a website (BabyCenter, Canada [www.babycenter.ca], a private company) used by expectant mothers. Similarly, the Ministry of Health in Canada did not respond to students' petition with 500 signatures on it recommending the following:

> Stop advertising to youths and cut down the production of energy drinks. Many adolescents are being impacted, and some have died because of overdose of energy drinks[,] mostly from Monster[™] and Rockstar[™]. Also, many people suffer from horrible with-drawl [sic] symptoms from the drink, as well as negative effects both mentally and physically (Report, June 19, 2012).

The main responses students received from their actions were, besides that from people in the school or in commercial outlets they talked directly to about issues, from people who viewed activist videos several of these student uploaded to YouTube™, such as the role-playing scenario (www.youtube.com/watch?v=EaiNh6ys8XQ) a group of students produced about multivitamin uses (with nearly 80 'views') (refer to Fig. 8.5).

In association with the expertise demonstrated by many of the students in Mirjan's class was the apparent general view that research-informed and negotiated actions on STSE issues was a worthwhile component of a science course. A student working in a group investigating and acting on cigarette smoking, for example, summarized his group's sentiment about their project by saying: "More people [students] should do this [RiNA] because … What's the point of having knowledge [as in book learning] if you can't do nothing [sic] with it?" (Interview, June 20, 2012). Moreover, given such positive views about RiNA, it seemed that students had—over the course of three such projects—developed significant self-efficacy (Bandura, 1997) towards conduct of these kinds of projects. Such feelings of autonomous control did seem to be manifested in students' apparent new-found identities—as

Fig. 8.5 Students' role-playing action video

expressed by 'Nancy', whose group had investigated ultrasound, in the beginning of the letter she wrote to a website aimed at expectant mothers: "I am a student and a young activist" (June 19, 2012).

8.4.3 Factors Influencing Outcomes

8.4.3.1 Myriad Factors

In theory, promotion of research-informed and negotiated actions on STSE issues should be relatively-easy in schools within the Canadian province of Ontario—the jurisdiction governing Mirjan's teaching. In its latest science curriculum document (MoE, 2008), STSE education is listed first (instead of 3rd, as in the previous curriculum) among its three overall goals—which also makes provision for student-focused science investigations and social actions on issues. Mirjan acknowledged that this official mandate for RiNA was essential for his promotion of it (Written reflection, July 10, 2012). Having said that, it seems clear teachers' instructional decisions depend on myriad situational variables—often co-affecting each other (Latour, 2005). For instance, we must consider the nature of the larger society, the curriculum, the teacher, textbook publishers, school district administration, the school and its infrastructure and colleagues and staff, parents, students, etc.

Despite the complexity and associated uncertainty in precisely-determining factors affecting student learning outcomes, it seems clear that a major factor enabling them to develop and implement RiNA projects to address critical STSE issues is the teacher in this case. At the time of this research, Mirjan was a graduate student studying for a Masters of Education degree focusing on science education. He had participated in and led several workshops in science education, was a member of a 'professional learning community' within his school district (regularly collaborating with other teachers) and (still) maintains an online educational blog highlighting his ongoing thoughts about education. His views about society and science and technology, moreover, seemed highly congruent with perspectives and preferred practices inherent to STEPWISE. On the *Scientific Theory Profile* (Loving, 1991), for instance, it is apparent that he holds positions that adhere to 'Naturalist-Antirealist' (NA) views about science—which, in contrast to 'Rationalist-Realist' views, suggests that knowledge generation in the sciences may be affected by factors beyond logic, including: societal cultural values, socio-economic factors, a scientist's gender, religion, theoretical perspectives, personal biases, etc. and that scientists' claims can only be approximations of reality. With such views, teachers may be more comfortable encouraging students to challenge the integrity of fields of science (and technology) than if they adhered to Rationalist-Realist views—since NA views acknowledge the fallibility of science (and technology), as seems apparent in business-science partnerships (Krimsky, 2003; Ziman, 2000). While this may be the case, it also is apparent that teachers sometimes struggle enacting their perspectives about science when working with colleagues holding opposite views (Hodson, 2008). Mirjan assured Larry, however, that his teaching colleagues were very supportive of his views and his efforts to improve his science teaching. This gave him considerable confidence in enacting a major new framework with students.

8.4.3.2 Epistemic Factors

As discussed above, Mirjan implemented two sets of apprenticeship activities, following the schema in Fig. 8.1, before asking students to conduct a self-directed RiNA project to address an STSE issue of their choice. This approach has been used—with some successes—in educational contexts with teachers and students (Bencze et al., 2012; Sperling & Bencze, 2010) and with student-teachers (Bencze & Sperling, 2012). It seemed to, again, work for students in Mirjan's class—all of whom were able to complete *generally* (refer below for qualifications to this claim) self-directed RiNA projects. Indeed, students indicated that the apprenticeship activities did help them to become more comfortable with RiNA project work. When asked if he and classmates were nervous without much teacher guidance for the final project, for example, 'Ken' said, "We did it [RiNA projects] enough times before [the final project] that we knew how to do,[for example,] a survey, knowing what kinds of questions to ask" (Interview, June 20, 2012).

The apprenticeship that Mirjan conducted during the semester studied here had at least one important modification from previous uses of it—a modification that may have influenced students' expertise and motivation for research-informed and negotiated actions on STSE issues. Broadly, new emphasis was placed on students' *epistemic agency* (Damsa, Kirschner, Andriessen, Erkens, & Sins, 2010); that is, students' self-directedness of decisions and actions associated with understanding of the nature of their thoughts and actions. This is a process similar to *metacognition*; that is, thinking about and planning one's thinking/learning (Niemi, 2002). In that sense, it involves cycles of these reciprocal relations: RiNA Projects ←→ Nature of RiNA Projects. With this broad conception in mind, we can consider it in more detail as depicted in Fig. 8.6. In this theoretical framework, 'RiNA Projects' is

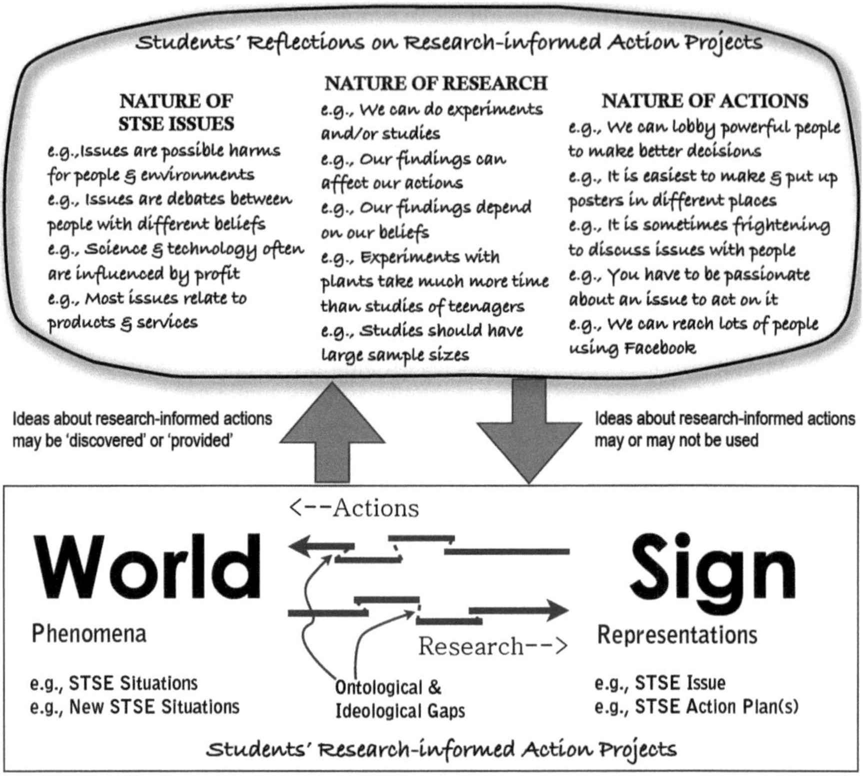

Fig. 8.6 Detailed model for RiNA projects ←→ nature of RiNA projects

expanded to suggest that they involve cycles of reciprocal relationships between phenomena of the 'World' and representations ('Signs') of them (Roth, 2001). More specifically, research into STSE issues would involve translations from phenomena (e.g., an STSE situation, like a business-science contract) into representations of them (e.g., drawings, written statements, and abstract concepts, like 'efficient,' etc.). Actions would, in turn, first involve thinking of new representations (e.g., a poster and video about business-science contracts that may circulate among people and groups to bring about revisions to such contracts; that is, 'New STSE Situations'). New STSE Situations could, in turn, be represented and the above cycle repeated. This series of cycles may not, however, change significantly without an additional second loop; that is: RiNA Projects ←→ Nature of RiNA Projects. Like 'double-loop learning,' metacognition in which people examine the nature of their thinking and acting and possibly revise assumptions about them—possibly leading to improved learning (Argyris, 2002). For example, as illustrated in Fig. 8.6, students may find (or be told) that one can do experiments or studies when researching STSE issues (RiNA → Nature of RiNA Projects). Possibly, such awareness may lead them to conduct investigation types (e.g., studies) that had not previously-considered

(Nature of RiNA Projects → RiNA Projects). As also depicted in Fig. 8.6, such metacognitive acts can occur within different learning domains, including, but not limited to: *Nature of STSE Issues*; *Nature of Research*; and, *Nature of Actions*. We found examples of these in this research, some of which are given in Fig. 8.6, while others are given below in the context of discussions surrounding learner control.

A key element of epistemic agency is, by definition, *learner control*. Greater epistemic agency implies more learner control, particularly in terms of the learner's conceptions of the nature of that learning (Damsa et al., 2010). This is supported by *knowledge duality theory*, as described by Wenger (1998)—who suggested that people may develop deep understanding and commitment to their learning if they are personally-engaged in decisions about reciprocal relationships between phenomena of the world and representations of them. Therefore, it may be that the extent to which students had control over decisions about reciprocal relations in Fig. 8.6 could be key to their epistemic agency. In reviewing data from this study, although students' third RiNA project (World ←→ Sign) was meant to be largely, if not wholly, student-led (Fig. 8.1), it is apparent that students' control of decisions indicated in Fig. 8.6 was shared—to varying extents—with the teacher (Mirjan). This appeared to be the case for both RiNA projects (with World ←→ Sign translations) and for RiNA Projects ←→ Nature of RiNA Project translations. Given it is impossible to document or share them all, examples of decisions mainly made by the teacher and students for each kind of translation are given in the sub-sections below:

RiNA Projects (i.e., World <----> Sign)

In reflecting on progress of students in his class, Mirjan concluded that some students had not learned as much about RiNA as he hoped and/or had not conducted their projects soon enough or in sufficient depth. In this regard, he wrote:

> What became apparent from the first two research-informed STSE action projects was that there was some variation in terms of the depth (and breadth) of learning amongst students in the same group and between different groups. This makes sense since the students had control over what information to learn; however, it was also easy for some students to do less work than intended, and as a result, gain less from these projects (July 10, 2012).

Consequently, Mirjan provided more 'scaffolding' than he perhaps would have liked—given his earlier-stated intention—for his second attempt at implementing STEPWISE—of helping his students to achieve self-directed RiNA projects. For example, to ensure topics had a critical edge to them, he gave students a list of (7) topics, including: Energy Drinks; Snack food labels; Smoking; Ultrasound; Multivitamins; Vaccines; Sun and skin cancer. Each topic title came with a brief description, such as this one for 'Snack Food Labels':

> As consumers are increasingly mindful of their health, many companies are attempting to make their products seem more nutritious by including catch phrases like "whole grain", "low fat", "made with real fruit" and "organic" on their packages. A more careful look at ingredients and nutrition information reveals that many of these products are not as healthy as they might appear (Biology RiNA Assignment Supplement, June 4, 2012).

These and the other six descriptions were written to help the students to focus on controversies, often between companies and those who may oppose some of their decisions. To help ensure such a focus, he then arranged a structured online forum for students to post results of their research that contained five fields, each with a leading research question or instruction:

- Clearly describe the STSE issue and the possible causes.
- Who are the powerful actors (people in power) that are influencing the issue? Explain.
- Clearly describe what you and your group already know about the issue.
- Briefly describe the science and technology involved in …Topic...
- List and describe 2–3 negative effects on the [separate sections for individuals, societies & environments]. Include references that you used to learn this information.

This guidance seemed to ensure students chose to focus on controversies regarding decisions about products and services often linked to companies and science and technology, examples of which are given above.

After they had completed their secondary research, students were asked to conduct primary research designed by them. These, too, involved some scaffolding. Attached to the general description of the RiNA assignment was a rubric for evaluation of the completed assignment. Although students had considerable choice on details, this rubric was designed to help ensure their studies (or experiments, which none conducted) had features commonly-associated with professional inquiry, including this criterion: “Your investigation is valid and reliable (sample size is large enough, wide range of independent variables used where possible, etc.)” (June 7, 2012). Such guidance, in addition to that described below regarding the RiNA reflection class, seems to have contributed to Mirjan’s conclusion that all student correlational studies had reasonably-high sample sizes—a major contributor to reliability and validity in studies.

In terms of actions, the evaluation rubric also provided guidance. Sample criteria included: “Two different actions are proposed to deal with the STSE issue; … They target various groups including people in power” June 7, 2012). These were, as above, guidance in terms of setting parameters for RiNA while leaving decisions about details to students. Support for this claim beyond Mirjan’s statement of this in interviews was seen during students’ presentations about their final RiNA projects. Mirjan seemed genuinely surprised—not in any detectable evaluative way—at students’ particular choices regarding STSE issues, their research about them and actions to address them.

RiNA Projects ←→ Nature of RiNA Projects

As discussed above, the apprenticeship (Fig. 8.1) that Mirjan provided during the semester about which we are reporting here featured a special focus on encouraging students to think about and possibly apply aspects of the nature of RiNA projects—a possible contributor to their epistemic agency and ability and motivation to conduct RiNAs. Apart from the evaluation criteria provided in the assignment sheets, along with Mirjan's evaluative feedback on completed projects, a particular innovation that may have contributed to students' expertise and motivation for RiNA may have been the 90-minute class devoted to student collaborative reflection on the nature of their previous two projects and ways in which their third project might be improved. This class began with students individually completing a survey, containing questions (with 5-point rating scale) for each of the three phases of a RiNA project, such as: "How confident are you that you understood the controversy involved in the issue you studied? [STSE Issues]; … How confident are you in the results of your study? [Research]; … To what extent do you feel that your actions were targeted towards people in power? [Actions] (May 21, 2012). Afterwards, students were asked to share their responses in small groups. This, in turn, was followed by asking students to share their ideas about RiNA in a 'community circle' (a tack used by Mirjan on other occasions, for different purposes). Finally, students were given time to consider topics for the final RiNA project—during which they also were invited to consider possibly applying ideas about the nature of RiNA gleaned from the earlier part of the class period.

Overall, students' access to and application of Nature of RiNA Projects (refer to Fig. 8.6) were, as with their RiNA projects (World ←→ Sign), partly teacher-scaffolded, but student-led. These reflection-action cycles seemed to pertain to each of the three phases of RiNA defined by Mirjan (in consultation with Larry); that is, STSE Issues; Research; Actions. Students also explored ideas about the overall nature of RiNA.

One of the key aspects of the nature of RiNA that Mirjan wanted students to consider was possible problems due to decisions made by people/groups with power. During the reflection class, for example, he asked about this using an example: "Who would be in favour of laser eye surgery?" Janice replied, "The doctors and surgeons who make the money [favour laser eye surgery]" (May 24, 2012). Later in this discussion, while students were applying their reflections to plans for their third project, it seemed clear that students were quite conscious of such issues of power. The following debate occurred among the students: Henry said, "Let's say you've been studying all night and you forgot to sleep! You had a test the next day. You can use that energy drink to stay awake." Carl retorted, "Say you are in university, you have projects, tests and stuff [for which to prepare] every night for, like, two months – and you're drinking those things [energy drinks] every night. Then, you have to look at the side-effects of that... It wouldn't be good for you" (May 24, 2012). Similarly, with regards to discussions about research, Mirjan was interested in ensuring students had what he considered useful conceptions of what he considered key aspects of it. The class emphasizing reflections on the nature of RiNA led, for example, to this response when Larry asked when students might choose studies over experiments:

> I don't know how to explain it. But, say you wanted to find out how smoking affects lung cancer, for a study, we would find people who smoked and we would ask them how many cigarettes they smoke a day. Whereas, if we did an experiment, we would give people the

> smokes and say, 'Smoke this [cigarette], and we will see if you get cancer later on.' That would be the difference. ...[We wouldn't do an experiment in that case] because it is unethical to give them cigarettes and see if they get cancer. ... We don't always know the effects for certain experiments. So, that is why we do studies (Carl, May 24, 2012).

There were many other comments from students about the nature of research stemming from the survey and Mirjan's questions. But, a comment from a student that allowed Mirjan to transition into a discussion about the nature of activism seemed very enlightened and enlightening. In response to another student's comment on the importance of the research, Carl said:

> It's not a research project. It's an action project. The research is just the base of it; but, the actual project is to take the action. ... We did [research into] consumption of water. Once we researched it, we realized that, if humans, as a society, keep going [consuming] at this rate, we are not going to have any water left. It's our responsibility to take action to survive" (May 24, 2012).

Regarding reflections on and possible applications of ideas about activism, meanwhile, there were many important discussions. As part of the transition in discussion about research into that about actions, a student made yet another useful point. When Mirjan asked if students' actions were mainly based on results of their studies that suggested that more girls were being dropped off at school than boys, Jeff said: "It all depends. You have to look at the larger world. If more cars are being used, then more CO_2 [will be emitted]" (May 24, 2012). While this seemed very helpful, Mirjan's particular focus was to encourage students to consider actions aimed at decision-makers. He did so particularly in feedback on their second RiNA project, in the May 24, 2012, reflection class, and in the context of the evaluation scheme. This is evident, for example, in Carrie's comments: "When we talked about the projects we did before, we realized that none of us really did [actions towards] people in power. That actually drove us to do something for people power this time" (Interview, June 20, 2012). She added that it wasn't just the May 24 class, but "all of it" (Interview, June 20, 2012). Indeed, students seemed to have little trouble engaging this concept. At first, some of their discussions in this regard were limited to local decision-makers. When asked by Mirjan about addressing power, Nancy replied, for example, that they could target "people in authority ... [such as] the mayor of Brampton [their city]. [We could] write a letter to her." Henry added, "Rather than sending just one letter, let's say this entire class writes letters. So, there would be 24 letters!" (May 24, 2012). However, other members of class did highlight for peers possibilities for actions on more distant decision-makers. Towards the end of their third project, for example, Charlie said:

> [W]e realized that these things are not just local. People do these things, like smoking, everywhere. ... We discussed in our group that, if we target an older group, maybe a change will happen. We wanted to write a letter to the bigger companies, but we discussed that they wouldn't really care. They would just throw the letter away. They just want to make a profit. They are likely, like, billionaires already (Interview, June 20, 2012).

A sense of action 'targets' went beyond those—such as companies—they considered possibly responsible for issues. Mirjan also asked them to consider enlisting celebrities to possibly rally significant public support for their causes. In response, Alan said, for example, "[It's] amazing, eh?! If Lady Gaga tweets something, then people are all over it." Mary then said that this logic was much like how the Kony 2012 video was so massively–viewed [As of Aug. 11, 2012, it had 92,380,955 views]. "It was all over Facebook™, all over Twitter™, all over everything!" She added, "Teenagers can do a lot of things, as long as there is a large group behind them" (May 24, 2012).

In addition to ideas about 'separate' (which may be co-dependent) aspects of the nature of RiNA projects (e.g., issues vs. actions), Mirjan (and Larry) encouraged student discussion about and possible application of overall aspects of the process. A suggestion from Larry that introverts and extroverts may choose different kinds of actions generated an interesting discussion that may or may not have ultimately affected students' projects. Alan, for example, said, "I am not really a social person. But, after this project, [being involved in] talking to people, I developed more leadership skills" Jeff added "For some actions that people took, you have to be more of an extrovert. You have to go up to people and give them brochures. It can be kinda nerve-wracking (May 24, 2012). Similarly, students seemed intrigued (although a bit puzzled, at first) with Larry's suggestion that a person's emotional state—and not just knowledge and logic—might affect RiNA decisions. Janice, for example, said emotions "didn't really affect me, at first; but when you research it, you kinda get knowledge about it. ... [I found that] about 50% of girls in America, and they are like 10 years-old, are worried about becoming bulimic" (May 24, 2012). Jeff added: "To try to pursue an issue and to try to help people, you have to have some compassion; have some humanity to help people for the greater good." Meanwhile, Mary said: "I think it [level of emotion] depends on the topic. For my group, we did [research on] privacy and that [decision] was based on how people felt about being watched or, like, video-taped. It wasn't so much logical as emotional" (May 24, 2012).

Finally, students indicated that different pedagogical acts initiated by Mirjan to stimulate reflection on the nature of RiNA projects led them to some overall conclusions about 'the' (not likely singular or certain) process. Ken said, for example, that "it opened your eyes to some things. [He would tell himself,] 'I should get a bigger sample size. I should try something new[, like] a letter to something [e.g., the mayor].' A lot of new ideas came out... It also gave us confidence because [we knew] everybody was doing it, because we sat in a circle [to share ideas]" (Interview, June 20, 2012).

Overall, it seems that teacher scaffolding/guidance in both RiNA projects and reflections on and applications about them (RiNA Projects ←→ Nature of RiNA Projects) can be a necessary and helpful aspect of promotion of RiNA in secondary schools. Based on fundamental (psychological) constructivist learning theory (Osborne & Wittrock, 1985), for example, it is difficult for learners to discover abstract ideas without at least having a notion of them in their heads. In terms of social constructivism, moreover, it is apparent that learning can be enhanced through collaboration in familiar and friendly contexts. For instance, Carrie said: "Since you are sitting there with your friends, it makes it a lot more comfortable to talk about what you did. … You [also] listen more because you are more relaxed" (Interview, June 20, 2012). Regarding self-efficacy theory (e.g., Bandura, 1997), meanwhile, learners can be assisted—in part—in developing attachment to and confidence in tasks (of various sorts) when having them modeled for them as, in a sense, a vicarious experience. In this case, such modeling was provided by the teacher when discussing and illustrating, with examples, the nature of RiNA. If they also receive modeling from peers, additional encouragement may arise from those with whom they share identities and experiences.

As argued above, perhaps the deepest, most meaningful, learning—in many, if not all, domains—occurs when students have full control over decisions. There was, indeed, evidence of this occurring with regards to students' reflections on and/or in RiNA. We saw this primarily through students' relatively (e.g., without specific questions in interviews) spontaneous statements about how the projects contributed to their views on them and, possibly, uses of such claims. Through a trial-and-error process, for instance, students decided that experiments seem to take longer than studies. Charlie said his group thought of an experiment in which they watered plants with water that had had cigarettes soaking in it but decided this would take too long "because the plants grow so slowly" (Interview, June 20, 2012). Meanwhile, Penny gave this definition of validity of research after her experiences with it: "A valid study would help you to come up with information for it [your topic]; whereas, another study may not contribute to what you are studying" (May 24, 2013).

Finally, in addition to epistemological conceptions, students' reflections on the nature of RiNA projects seemed to affect their overall attitudes/comfort-level/self-efficacy with regards to such projects. When asked, for example, why most of the projects seemed to move outside the school, Carrie said, "Because we had experience. It's not the first time doing this type of project. We sort-of know what to do, and what we want to do" (Interview, June 20, 2012). This comment seemed to emphasize their increased understanding of processes, but embedded within that may have been some affective gains. Indeed, it seemed that, generally, students felt a sense of independence grow as they moved through the three RiNA projects, feedback on them and embedded reflections. Alan said, for example, that the third project

> is a different way to learn. He [Mirjan] is showing that, in the future, there is not going to be someone showing us what to do. We have to know what to do ourselves, right? We went through two [partly guided RiNA projects] and that was, like, the experience. [After that,] we knew what to do, what was expected" (Interview, June 20, 2012).

In support, Carrie said, when asked how she got her confidence to confront people in drug stores, that "after doing multiple [research-informed action] projects, I got more confident" (Interview, June 20, 2012). Related to that, she said, seemed to be the degree of openness of the assignment: "By leaving it open, it gives [us] a lot of freedom, [allowing us to] think up our own ways [of doing things]" (Interview, June 20, 2012).

8.5 Conclusions and Implications

The apprenticeship framework used in the study reported here seemed to generate considerable expertise and motivation for the conduct of research-informed and negotiated actions (RiNA) to address STSE issues. A major contributor to such actions seemed to be a special innovation introduced here; that is, reciprocal relationships between conduct of RiNA projects and reflections about the nature of RiNA projects. In using this scheme, we reasoned that a key aspect of this influence should be significant degrees of student control of decision-making. Here, we found that, despite his intention to enable such student control, he felt a need to continue to provide some guidance for students' imagined self-directed RiNA projects. There are many possible reasons to explain this result. Perhaps the teacher's development of expertise in promoting self-directed RiNA needs continued development. On the other hand, perhaps this teacher's experiences with this aim is an indication of a fundamental—and long-standing—problem in science education; that is, those controlling school science have created curriculum and assessment schemes that overly-emphasize teaching and learning of products of fields of science and technology at the expense of students' opportunities to develop expertise and confidence for self-determining their own conceptions of the world and ways in which to change it.

References

Allchin, D. (2003). Scientific myth-conceptions. *Science Education, 87*(3), 329–351.

Angell, M. (2004). *The truth about the drug companies: How they deceive us and what to do about it*. New York: Random House.

Argyris, C. (2002). Double-loop learning, teaching, and research. *Academy of Management Learning and Education, 1*(2), 206–218.

Bandura, A. (1997). *Self-efficacy: The exercise of control*. New York: W. H. Freeman.

Barnes, R. L., Hammond, S. K., & Glantz, S. A. (2006). The tobacco industry's role in the 16 Cities Study of Secondhand Tobacco Smoke: Do the data support the stated conclusions? *Environmental Health Perspectives, 114*(12), 1890–1897.

Bencze, J. L., & Alsop, S. (Eds.). (2014). *Activist science & technology education*. Dordrecht, The Netherlands: Springer.

Bencze, J. L., & Sperling, E. R. (2012). Student-teachers as advocates for student-led research-informed socioscientific activism. *Canadian Journal of Science, Mathematics & Technology Education, 12*(1), 62–85.

Bencze, L., & Carter, L. (2011). Globalizing students acting for the common good. *Journal of Research in Science Teaching, 48*(6), 648–669.

Bencze, L., Sperling, E., & Carter, L. (2012). Students' research-informed socioscientific activism: Re/visions for a sustainable future. *Research in Science Education, 42(1), 129–148.*

Carter, L. (2005). Globalisation and science education: Rethinking science education reforms. *Journal of Research in Science Teaching, 42*(5), 561–580.

Charmaz, K. (2014). *Constructing grounded theory* (2nd ed.). New York: Sage.

Crawford, B. A. (2007). Learning to teach science as inquiry in the rough and tumble of practice. *Journal of Research in Science Teaching, 44*(4), 613–642.

Damsa, C. I., Kirschner, P. A., Andriessen, J. E. B., Erkens, G., & Sins, P. H. M. (2010). Shared epistemic agency: An empirical study of an emergent construct. *Journal of the Learning Sciences, 19*(2), 143–186.

dos Santos, W. L. P. (2009). Scientific literacy: A Freirean perspective as a radical view of humanistic science education. *Science Education, 93*(2), 361–382.

Foucault, M. (1977). *Discipline and punish: The birth of the prison*. New York: Pantheon Books.

Gramsci, A. (2007). *Prison notebooks, Volumes 1–3*. (J.A. Buttigieg, Trans.). New York: Columbia University Press, 1992–2007.

Guba, E. G., & Lincoln, Y. S. (2011). Competing paradigms in qualitative research. In N. K. Denzin & Y. S. Lincoln (Eds.), *The SAGE handbook of qualitative research* (pp. 105–117). Thousand Oaks, CA: Sage.

Hileman, B. (1998). Industry's privacy rights: Is science shortchanged? *Chemical & Engineering News, 76*(17), 36.

Hodson, D. (2003). Time for action: Science education for an alternative future. *International Journal of Science Education, 25*(6), 645–670.

Hodson, D. (2008). *Towards scientific literacy: A teachers' guide to the history, philosophy and sociology of science*. Rotterdam, The Netherlands: Sense.

Hodson, D. (2011). *Looking to the future: Building a curriculum for social activism*. Rotterdam, The Netherlands: Sense.

Khishfe, R., & Lederman, N. G. (2006). Teaching nature of science within a controversial topic: Integrated versus nonintegrated. *Journal of Research in Science Teaching, 43*(4), 395–418.

Klein, N. (2014). *This changes everything: Capitalism and the climate*. Toronto, ON: Simon & Schuster.

Kleinman, D. L. (2003). *Impure cultures: University biology and the world of commerce*. Madison, WI: University of Wisconsin Press.

Krimsky, S. (2003). *Science in the private interest: Has the lure of profits corrupted biomedical research?* Lanham, MD: Rowman & Littlefield.

Latour, B. (2005). *Reassembling the social: An introduction to actor-network-theory*. Oxford, UK: Oxford University Press.

Leonard, A. (2010). *The story of stuff: How our obsession with stuff is trashing the planet, our communities, and our health – And a vision for change*. New York: Free Press.

Levinson, R. (2010). Science education and democratic participation: An uneasy congruence? *Studies in Science Education, 46*(1), 69–119.

Levinson, R. (2013). Practice and theory of socio-scientific issues: An authentic model? *Studies in Science Education, 49*(10), 99–116.

Loving, C. C. (1991). The Scientific Theory Profile: A philosophy of science model for science teachers. *Journal of Research in Science Teaching, 28*(9), 823–838.

Ministry of Education [MoE]. (2008). *The Ontario curriculum, grades 9 and 10: Science*. Toronto, ON: Queen's Printer for Ontario.

Mirowski, P. (2011). *Science-mart: Privatizing American science*. Cambridge, MA: Harvard University Press.

Niemi, H. (2002). Active learning—A cultural change needed in teacher education and schools. *Teaching and Teacher Education, 18*(7), 763–780.

Oreskes, N., & Conway, E. (2010). *Merchants of doubt*. London: Bloomsbury Press.

Osborne, R., & Wittrock, M. (1985). The Generative Learning Model and its implications for science education. *Studies in Science Education, 12*, 59–87.

Pedretti, E., & Nazir, J. (2011). Currents in STSE education: Mapping a complex field, 40 years on. *Science Education, 95*(4), 601–626.

Roth, W.-M. (2001). Learning science through technological design. *Journal of Research in Science Teaching, 38*(7), 768–790.

Sadler, T. (Ed.). (2011). *Socio-scientific issues in the classroom: Teaching, learning and trends*. Dordrecht, The Netherlands: Springer.

Sadler, T. D., Barab, S. A., & Scott, B. (2007). What do students gain by engaging in socioscientific inquiry? *Research in Science Education, 37*(4), 371–391.
Sperling, E., & Bencze, J. L. (2010). 'More than particle theory': Citizenship through school science. *Canadian Journal of Science, Mathematics and Technology Education, 10*(3), 255–266.
Venville, G. J., & Dawson, V. M. (2010). The impact of an argumentation intervention on Grade 10 students' conceptual understanding of genetics. *Journal of Research in Science Teaching, 48*(8), 952–977.
Wasser, J. D., & Bresler, L. (1996). Working in the interpretive zone: Conceptualizing collaboration in qualitative research teams. *Educational Researcher, 25*(5), 5–15.
Wenger, E. (1998). *Communities of practice*. Cambridge, UK: Cambridge University Press.
Wood, G. H. (1998). Democracy and the curriculum. In L. E. Beyer & M. W. Apple (Eds.), *The curriculum: Problems, politics and possibilities* (pp. 177–198). Albany, NY: SUNY Press.
Ziman, J. (2000). *Real science: What it is, and what it means*. Cambridge, UK: Cambridge University Press.

Chapter 9
Students' Uses of Actor-Network Theory to Contextualize Socioscientific Actions

Larry Bencze and Mirjan Krstovic

9.1 Introduction

A significant component of science education research and development in the last few decades has been attention to so-called *socioscientific issues*. While this work has had some successes in terms of implementation of pedagogical practices in schools in many parts of the world, there appears to be a certain 'timidity' in this regard. Arguably, issues may not be treated as critically as some recommend and, related to this, students often are not encouraged and enabled to take strong social actions to address them in ways that may bring about a better world. In our previous work, we seem to have had some successes in this regard through promotion of student-led primary (e.g., studies and experiments) and secondary (e.g., Internet searches) research as some bases for such activism. With increasing awareness of actor-network theory as a way of conceiving of phenomena, however, including in terms of helping to 'democratize' students' conceptions of socioscientific issues (Pierce, 2013), we decided to pursue infusion of aspects of actor-network theory into our ongoing programme of promotion of research-informed and negotiated actions on issues informed by the 'STEPWISE' curriculum framework developed in 2006. After a review of literature relating to implementation of socioscientific issues in science education, we describe and analyze one teacher's efforts to promote activism through uses of actor-network theory in science education.

L. Bencze (✉)
Ontario Institute for Studies in Education (OISE), University of Toronto, Toronto, ON, Canada
e-mail: larry.bencze@utoronto.ca

M. Krstovic
Peel District School Board, Mississauga, ON, Canada
e-mail: mirjan.krstovic@peelsb.com

L. Bencze (ed.), *Science and Technology Education Promoting Wellbeing for Individuals, Societies and Environments*, Cultural Studies of Science Education 14,
DOI 10.1007/978-3-319-55505-8_9

9.2 Neoliberal Critique of Science Education

For at least four decades, governments, researchers and others have promoted perspectives and practices in science education that acknowledge that fields of professional science are not isolated from—but, rather, have significant relationships with—fields of technology and societies (including special interest groups) and environments (Pedretti & Nazir, 2011). Over that time period, although there have been various foci, a major one has been to emphasize *controversies* regarding relationships among fields of science and technology and societies and environments (STSE)—often called, *socioscientific issues* (SSIs). Different 'stakeholders'—such as politicians, community activists, business representatives, general citizens, and others—debate merits, for example, of various products of science and technology, including: nuclear reactors, cosmetic surgery, plastic containers, coal energy sources, 'fast' and manufactured foods, genetically-modified organisms, computer 'data mining' and sun tanning beds. In engaging students in such SSIs, educators have used a range of goals and practices. It appears, however, that many of them largely involve uses of SSIs as contexts for promotion of *individual student choice and achievement* (Levinson, 2013). Such emphases have, apparently, led to some positive learning outcomes for students, including achievements in: *products of science* (e.g., laws & theories) (Venville & Dawson, 2010), *socioscientific reasoning skills* (Sadler, Barab, & Scott, 2007); and, the *nature of science* (Khishfe & Lederman, 2006). As a multidisciplinary focus, it seems logical that SSI education can promote learning in a wide range of domains.

Encouraging and enabling youth to debate details of issues and arrive at personal decisions seems reasonable. In *representative* democracies, in which citizens regularly vote to determine leadership of their societies by politicians and experts (e.g., scientists, lawyers, etc.) (Wood, 1998), outcomes like those noted above should be useful (Sadler, 2011). On the one hand, decisions may not be easy. There are so many issues that individuals can struggle to learn enough to make informed decisions, some issues pertain to new fields (e.g., nanotechnologies) about which less information is available and, of course, decisions often involve emotional debates frequently driven by competing ideological stances. Nevertheless, citizen engagement in such controversies must be upheld in democracies (Pedretti & Nazir, 2011). It seems, however, that preparing citizens for debate and personal decision-making—which appears to be very common in science education—may be inadequate. Apart from difficulties cited above, it is apparent that citizens need to not only understand and make personal decisions about issues but also be prepared to take socio-political actions to address potential harms associated with them in what may be considered more *participatory* forms of democracy (Bencze & Alsop, 2014; Dos Santos 2009; Hodson, 2003, 2011). Reflecting this stance, Levinson (2010) recently has analyzed SSI education, concluding that—with reference to Table 9.1—current SSI practices tend to frame students in terms of either 'Deficit' (dependency on experts and leaders) or 'Deliberative' (discussions with, but still deference to, experts/leaders) models of citizenship. Given various societal and environmental

Table 9.1 Variations in citizenship in SSI education

Framework	Socio-epistemic relations	Epistemology	Controversy and participation	Pedagogy	Implications for democratic participation
Deficit	Knowledge flow is from scientist-teacher-student	Science is the corpus of knowledge	Ability to engage is constrained by access to technical knowledge	Knowledge for addressing an issue can be brought to the attention of the student.	There is a socio-epistemic inequality between the scientist/teacher and students which limits ability to bring about political change from below but does not preclude influential specialists making a political impact.
Deliberative	Knowledge flow is predominantly from scientist to the teacher and students, the latter two might be working in concert.	Science is understood to be uncertain and fallible.	Dialogue is open. Lay participants are informed but often lack the political means to bring about change. In schools, students might have opportunities for deliberation through group work and school councils but action might be constrained depending on the democratic nature of the school.	Emphasis on critical thinking and understanding of scientific methods and procedures.	Participation is real but often ineffectual in generating democratic change because participants do not have the 'clout' to make crucial decisions.

(continued)

Table 9.1 (continued)

Framework	Socio-epistemic relations	Epistemology	Controversy and participation	Pedagogy	Implications for democratic participation
Science education as praxis	Knowledge is distributed and emergent.	Knowledge is situated. Students become inducted into communal ways of knowing through legitimate peripheral participation in particular but changing contexts.	All participants work with a shared sense of social purpose.	Knowledge is provided on a need to know basis. The teacher is not epistemologically privileged.	Active and egalitarian participation to enhance change which might assume political literacy.
Dissent and conflict	This can be variable but is likely to have similar characteristics to science education as praxis.	What is known is contextualised by socio-political concerns.	Political action.	Knowledge provided on a need to know basis with an emphasis on political literacy.	Political understanding and action for change are foregrounded

Levinson, R. (2010). Science education and democratic participation: An uneasy congruence? *Studies in Science Education, 46*(1), 69–119; Table 9.1, pp. 83–84 harms faced by humanity (refer below), he—like authors quoted just above—suggests SSI education needs to include more 'Praxis' (critical, reflective, practice) and 'Dissent and Conflict' (critique and actions).

Reasons for promoting socio-political activism through school science are complex, but it seems a prime consideration must be current socio-economic conditions. Although capitalism has existed for centuries, many scholars and others suggest that societies are currently under immense—and, in many ways, extremely harmful—influences from *neoliberal capitalism*. Although the term is controversial, neoliberalism seems to be a socio-economic system that uses government *intervention* in markets in ways that prioritize private sector interests with, perhaps, the view that benefits will 'trickle down' to the masses in societies (McMurtry, 2013; Springer, Birch, & MacLeavy, 2016). Neoliberal capitalism appears to be extremely powerful, particularly in terms of its globalized (and global*izing*) network of actants aligned to its causes (e.g., individual competitiveness). Apparently assimilated into the 'Global Capitalist Network' (GCN) are, for example, individual financiers, banks, transnational corporations, advertizing agencies, telecommunications and shipping networks, supportive governments and universities, 'think tanks' (e.g., Atlas Liberty Network) and, crucially, supra-national economic organizations (e.g., World Trade Organization and the International Monetary Fund) that negotiate transnational trade agreements to isolate business activities from local/national government controls (Ball, 2012; McMurtry, 2013).

Apparently, key agents contributing to enormous power associated with the GCN are fields of science and technology (S&T) and their educational counter-parts (e.g., science education). Although their foci cannot be reduced to one emphasis, prime among them appears to be *consumerism*; that is, encouraging individuals to engage in cycles of consumption and disposal of goods, services, ideas, perspectives, feelings, etc. Promotion of such consumerism can be understood in terms of the schematic in Fig. 9.1, which is adapted from Wolff-Michael Roth's (2001) analysis of relationships between 'science' and 'technology.' In this light, it seems that fields of 'technoscience' (i.e., viewing S&T as co-influencing each other's nature) often assist neoliberal capitalists through creation of virtual *Trojan horses*; that is,

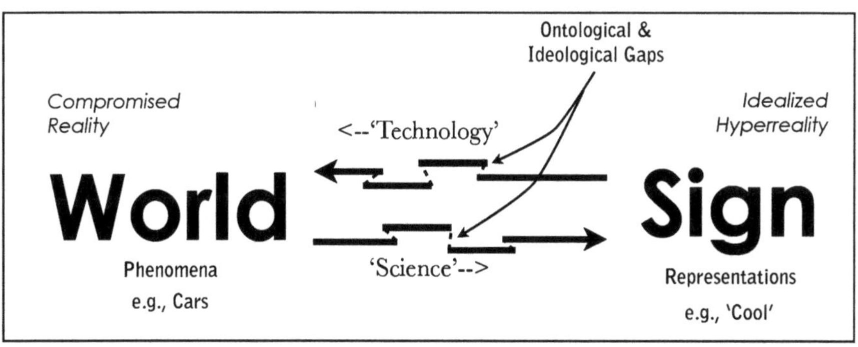

Fig. 9.1 Neoliberalism-inflenced technoscience

commodities that appear idealistically positive (and desirable) on the outside (e.g., 'classy') while hiding or ignoring many potential problems on the inside (e.g., government support for petroleum industries). In terms of Fig. 9.1, this may be accomplished, for example, when there are large *ideological* gaps (purposeful mistranslations)[1] in World → Sign relationships that create idealized hyperrealities (illusions of truth, such as 'classy') that may incite desires for the commodity (e.g., a car); while, at the same time, because of companies' legal rights to *externalize* their costs, compromising the quality of their commodities (e.g., reducing costs of labour, materials, etc.). Such externalities may be interpreted as ideological gaps in Sign → World translations that, in turn, lead to numerous potential problems for wellbeing of individuals, societies and environments (WISE) (Bencze & Carter, 2015). Among such commodities linked to WISE harms are: genetically-modified foods, etc. (Kleinman, 2003; household cleaning and hygiene products (Leonard, 2010); pesticides (Hileman, 1998); tobacco (Barnes, Hammond, & Glantz, 2006); pharmaceuticals (Angell, 2004); and, petroleum products (Klein, 2014).

As suggested above, supporting neoliberalism-influenced technoscience activities to promote consumerism appears to be—among networks of entities—science education. Firstly, it is apparent that a major component of the GCN is a sub-network, sometimes called the *Global Education Reform Movement* (GERM), that seems to be orchestrating educational actants to support neoliberalism—including through: *curriculum standardization*, *international competitiveness*, *testing and reporting* and *emphases on 'core' literacies* (e.g., language(s), mathematics, science and information technology) (Sellar & Lingard, 2013). Within this sub-network, given key roles for fields of technoscience, as discussed above, a major element seems to be school science. Briefly, as elaborated elsewhere (e.g., Bencze & Carter, 2011), it is apparent that school science often functions as a mechanism for achieving two capitalist goals; that is, for: (i) selecting and educating relatively few students who may become technoscientists and other professionals, such as business managers, who may establish and manage the scenario in Fig. 9.1; and, (ii) generation of large masses of citizens who may serve as compliant followers of labour instructions from professionals and as enthusiastic and unquestioning consumers of for-profit commodities. The first goal appears to be often achieved, for instance, through rapid-fire instruction in abstractions (e.g., a point mass occupies no space)—a tack that seems to favour students with considerable cultural and social capital (Bourdieu, 1986), which appears to be important for functioning in elite careers (Reich, 2007). Creating consumers, meanwhile, often involves use in science education, as above, of the concept of the Trojan horse. Fields of science and technology are, for instance, portrayed to be idealistic; e.g., made to appear efficient, unbiased and unproblematic in terms of adverse effects on individuals, societies and environments, such as by avoiding discussions (refer above) of neoliberal influences on fields of technoscience (Carter, 2005). Without being educated about such problematic situations involving fields of technoscience, students may struggle making

[1] Ontological gaps are mistranslations in World ←→ Sign relationships due to differences in the nature of ontological entities involved (e.g., tree vs. picture of tree).

appropriate decisions surrounding socio-scientific issues. This problem seems to be contiguous, moreoever, with contexts in their larger communities. There is evidence to suggest, for example, that companies have frequently hired reputable scientists to discredit science, in general and, more specifically, science claims about such commodities as tobacco, pesticides, weapons, nuclear power and possible causes and effects of climate change (Oreskes & Conway, 2010).

In light of influences of the GCN and GERM on societies, generally, and science education, more particularly, promoting critical and activist science seems daunting. Indeed, recent, but quickly-emerging, 'STEM' (Science, Technology, Engineering & Mathematics) education movements—which emphasize instruction in 'products' (e.g., laws, theories & innovations) and skills/methods (e.g., engineering design) of STEM fields, apparently in order to select and educate more STEM professionals so that jurisdictions (e.g., states/provinces, countries) can be successful in global economic competitions—seem to avoid education that might problematize STEM fields (Gough, 2015; Zeidler, 2016). Accordingly, approaches to science education are needed that may enlighten students about problematic STSE relationships and help them to develop expertise, confidence and motivation for taking socio-political actions to address socio-scientific issues concerning them.

9.3 Research on Critical & Activist Science Education

9.3.1 Research Context

Since 2006, based on the 'STEPWISE' (*Science & Technology Education Promoting Wellbeingfor Individuals, Societies & Environments* [www.stepwiser.ca]) curricular and pedagogical framework, the first author of this paper has been working with teachers and student-teachers to promote personal and social actions to address STSE issues of their concern. This framework organizes broad learning domains, such as 'Products Education' (e.g., learning laws & theories) and 'Skills Education' (e.g., skills for science inquiry) into a tetrahedral shape, with 2-way arrows between all domains, including with 'STSE Actions' in the centre. A general prioritiy is that students' education (e.g., via Products Education) and research findings would be used by them to promote wellbeing for (other) individuals, societies and environments (WISE)—generally, an *altruistic* orientation. A main feature of this approach is to encourage and enable students to self-direct *primary* (e.g., correlational studies), as well as secondary (e.g., Internet searches), research as contributions to their decisions about actions to address STSE issues of their choice. Given the apparent opposition to activist science education (refer above), it seems that students—and, perhaps, teachers and others—need compelling reasons to engage in such challenging activities as socio-political actions. A prominent claim in this regard is that students need deep, personal, attachments to issues to be motivated to act on them (Hodson, 2011). In this light, we have found that affective engagement with SSIs and actions may develop if students have leading roles in decisions regarding primary and secondary research into issues (e.g., Bencze, Sperling, & Carter, 2012). Findings can motivate and direct actions they might take to address SSIs. This tack is premised on assumptions, based on *knowledge duality* theory (Wenger, 1998), that students should develop deep commitments to actions if they are personally-engaged in reciprocal relations between *phenomena* (e.g., citizens' views of consumer products) and *representations* (e.g., survey data) of them. Such student-led research-informed and negotiated action (RiNA) projects align well, we think, with 'Science education as praxis' and 'Dissent and conflict' (refer to Table 9.1), as recommended by Levinson (2010) and others. Moreover, we have developed considerable evidence to suggest that secondary school students are able to self-direct successful and personally-meaningful projects, depending on such influences as: primary vs. secondary research (Bencze et al. 2012); the social nature of their correlational studies (Bencze & Krstovic, Chap. 7 here); reflections on RiNA projects (Bencze & Krstovic, Chap. 8 here); and, their 'street smarts' (vs. 'book smarts') (Phillips-MacNeil, Krstovic, & Bencze, Chap. 13 here).

An aspect of STEPWISE that we have felt still needs signficant work pertains to the critical nature of students' conceptions of STSE issues—and how that may influence their socio-political actons. In previous work, we had mainly asked students to consider positions of different human 'stakeholders,' such as members of government, scientists (and their data), people in business and activist groups. Recently, however, we were particularly influenced by the work of Clayton Pierce (2013) and Stephen Ball (2012)—who, among others (e.g., Latour, 2005), sensitized us to complexities of the Global Capitalist Network and Global Education Reform Movement. Pierce (2013) provides, as depicted in Fig. 9.2, an example of an actor-network theory (ANT) (Latour, 2005) analysis of STSE issues associated with genetically-modified salmon. In pointing out complexities of such networks, he stresses benefits to students in terms of providing them with more *realistic*—and, therefore, more *democratic*—conceptions of STSE relationships than is commonly-emphasized in school science. In integrating actor-network theory, along with Ball's (2012) analysis of global capitalist networks, into our efforts to encourage and enable students to self-direct research-informed and negotiated action projects, we imagined students developing deep conceptions about a wider range of actants regarding issues, and perhaps corresponding commitments to actions considering many of them. We understand RiNA projects as depicted in Fig. 9.3. By encouraging students to develop and progressively-revise (as more ideas, perspectives, etc. arise) actor-networks to depict STSE issues of their interest and actions to address them, we expected that their 'Actions' (Sign → World) translations may be based on more complex and realistic Signs they derived from 'Research' (World → Signs).

Mirjan Krstovic, a teacher involved in this project since Sept. 2011 and the second author of this chapter, agreed to encourage students in his tenth-grade science class to use actor-network theory to focus on issues associated with consumerism during the Feb. – June 2013 semester. We reasoned that, given its emphases on a range of kinds of actants—including living and nonliving things and semiotic messages—within networks, such a focus may provide students with access to richer

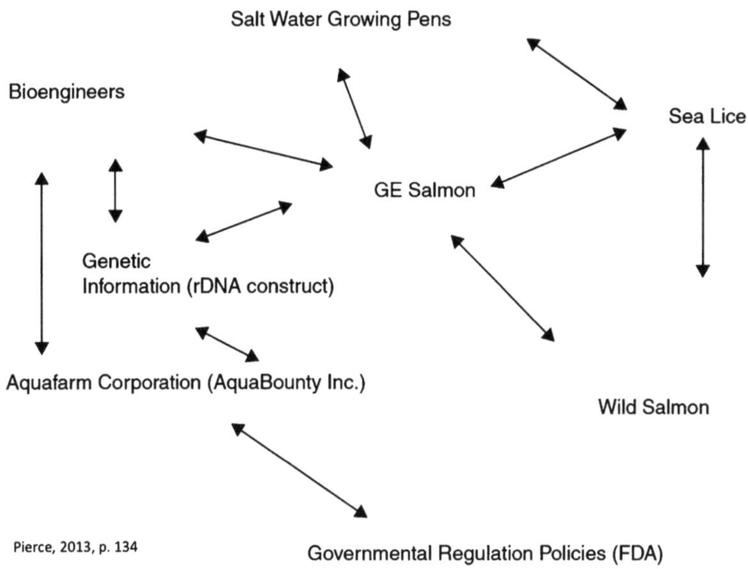

Fig. 9.2 GM Salmon network (Pierce, 2013, p. 134)

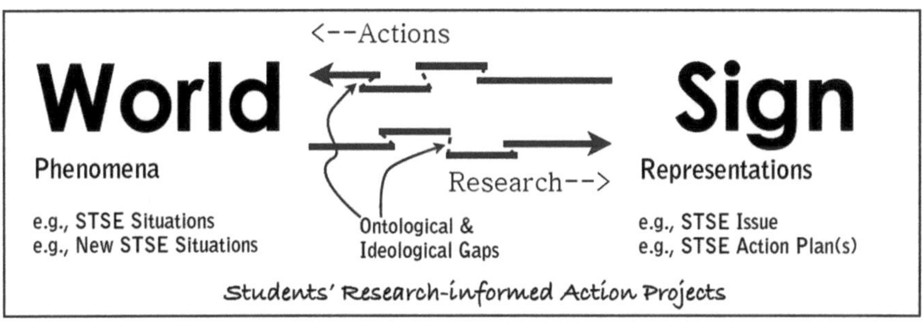

Fig. 9.3 A model of research-informed actions on STSE issues

contexts relating to their issues of choice (Pierce, 2013). Especially when students are given considerable control of decisions, ANT-infused RiNA projects may allow students to develop more personally-meaningful and contextually-rich conceptions of socioscientific issues that, in turn, may motivate them to act on such issues (Wenger, 1998). Such an education also would help students advance to Levinson's (2010) 'Science education as praxis' and 'Dissent and conflict' levels of citizen engagement in SSIs.

As he had done in three previous semesters of implementating the STEPWISE instructional framework, Mirjan used the schema shown in Fig. 9.4, which emphasizes (partly teacher-guided) apprenticeships aimed at helping students to develop expertise, confidence and motivation for self-directing RiNA projects. He used the first apprenticeship activities (in the Climate Change unit) to introduce students to

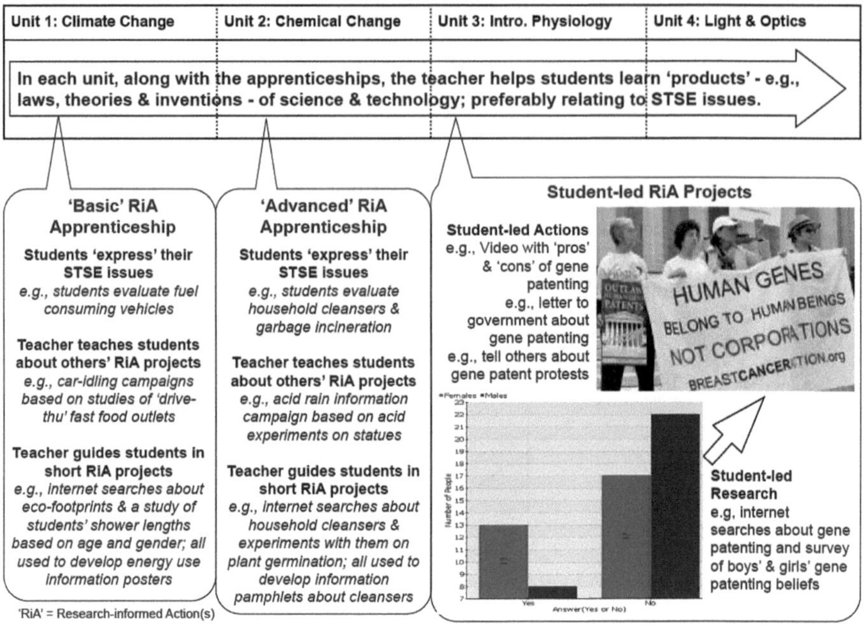

Fig. 9.4 STEPWISE instructional framework

conduct of RiNA projects before introducing ANT to them in the Chemical Change unit. Principles of ANT shared with students included the following points:

- Individual actants are *heterogeneous*, composed of influences from other actants;
- Types of actants include: *materials* (e.g. living & non-living things, inventions, inscriptions) and *semiotic messages*;
- Actants may co-affect each other, with effects that constantly change;
- Actants can align, particularly under influences from powerful actants, so that a common semiotic message is supported by all; and,
- Activism may involve introducing new actants and re-orienting existing ones so that dominant semiotic messages change.

To teach students about such ANT principles, Mirjan used lectures, whole-class and small-group discussions that emphasized the Trojan horse metaphor for consumer products. A copy of his full lesson sequence for this is given in Appendix A. During this sequence, he combined a Socratic lesson based on an actor-network (Fig. 9.5) that he had developed concerning cell phones. During this lesson, he emphasized the ANT principles above—with special focus on the Trojan horse metaphor, in the sense that dominant actants can be aligned to support of an idealized message about the commodity that can hide actants which, if made more prominent, would support a quite different, likely more problematic, message. To supplement

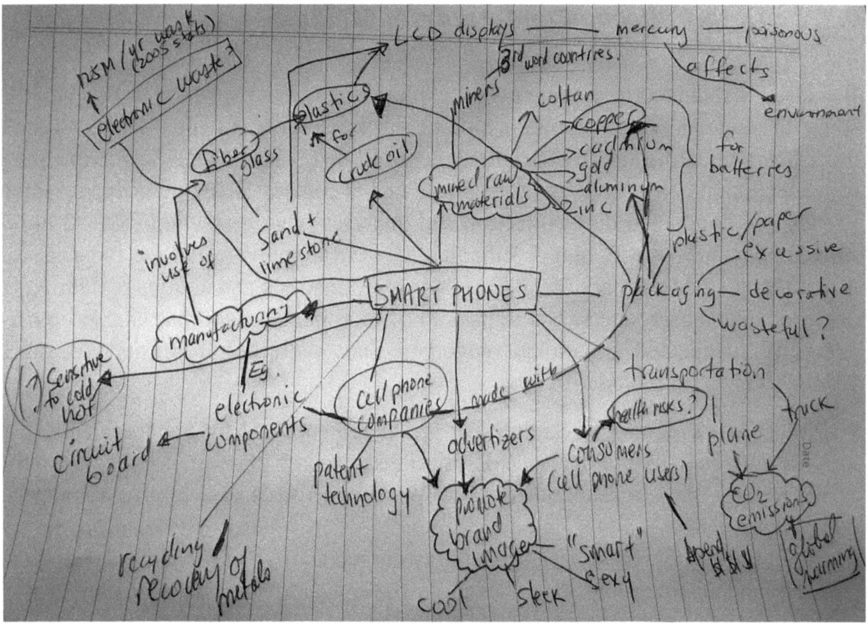

Fig. 9.5 Mirjan's model actor-network about cell phones

this lesson, he showed students the activist video, *The Story of Stuff* (SoS [www.thestoryofstuff.org]), which is structured around the 'materials economy,' tracking commodities in this sequence: Extraction → Production → Distribution → Consumption → Disposal. We felt that this video would effectively expose students to a wide range of actant types, including many that are often hidden or, as Latour (2005) suggests, 'black-boxed.' This is a feature emphasized by Pierce (2013) regarding his ANT study of genetically-modified salmon:

> [S]cientific literacy needs to be radically rethought in an age where genes of an Ocean Pout (an eel fish) are spliced with those of a Chinook (king) salmon, implanted in Atlantic salmon eggs, and a corporation patents this process *and* the new species of the fish itself … (p. 113; emphasis in the original).

Associated with exposing students to a breadth of actants, the SoS emphasizes two major aspects of consumerism noted above; that is, perceived (e.g., through marketing) and planned (e.g., through engineering) obsolescence.

In light of the context description above and arguments about influences of the global capitalist network earlier, the study reported here intended to explore the extent to which secondary school students could develop more democratic modes of citizenship, as defined by Levinson's (2010) criteria (Table 9.1).

9.3.2 Data-Collection and Analyses Methods

Because we intended to document and explain one teacher's efforts to broaden the scope of students' conceptions of a particular broad concept (i.e., consumerism), this research qualifies as an *instrumental* case study (Stake, 2000)—a documentary of a specific situation to be compared to a larger context. To achieve our research agenda, we conducted data-collection and analyses having *rationalistic* and *naturalistic* characteristics (Guba & Lincoln, 2011). Rationalistically, we focused, for example, on students' uses of ANT in their investigations and actions pertaining to personal hygeine products. Naturalistically, we collected data that enabled emergence of unexpected situational outcomes. Data collected from students (ages 14-16) and Mirjan included:

- Project Work Artefacts: Samples of products generated by most students (57) were collected, including: issue descriptions, research plans, data collected, written reports, project reflections, action plans and forms of action (e.g., posters, petitions, videos);
- Project Instructional Materials: Copies were made of all of Mirjan's pedagogical plans and instructional materials (e.g., paper handouts, videos, PowerPoint™ presentations, and internet website addresses);
- Digital Recordings of Students' Project Work: Photographs and videos were produced depicting youth presenting and defending their forms of action in public fora (e.g., to fellow students within and outside of class).
- Semi-structured Interviews: Eight volunteering students were interviewed twice, near the beginning and at the end of the course. Questions focused on their views about issues, research & actions. Mirjan was interviewed 11 times, for about 60 minutes each, about project progress. All interviews were audio-recorded and later transcribed.

Regarding analyses, each of us coded data for categories and then developed encompassing themes—using constant comparative methods based on constructivist grounded theory (Charmaz, 2014). Categories and themes were then negotiated between us (Wasser & Bresler, 1996). Member checks with participants were conducted to help ensure *trustworthiness* of claims, each of which was based on at least three supporting data sources.

9.4 Towards Critical & Activist Science Education

9.4.1 Preamble

Despite a broader context that appears dominated by the global capitalist network (GCN) and which appears to prioritize formation of subjects conducive to capitalist enrichment, it is apparent that pedagogical practices implemented by Mirjan enabled and encouraged students to achieve more democratic levels of citizenship. In terms of the framework developed by Levinson (2010), it appeared to us that most students in Mirjan's tenth-grade science class demonstrated considerable expertise and comfort in engaging in 'Science education as praxis' and 'Dissent and conflict.' Having made this claim, it is important to recognize that its 'external validity' (e.g., applicability to outside contexts) can only be made by each reader relating the description of the instrumental case described below to his/her context (Guba & Lincoln, 2011).

9.4.2 Students and Networks

Mirjan's first effort at infusing actor-network theory for promotion of students' research-informed and negotiated actions seemed relatively successful. All student teams (1-4 members/team) completed actor-network maps to illustrate their preconceived and ongoing (via secondary and primary research and collaboration) notions of issues surrounding a consumer product relating to chemistry and then developed socio-political actions to address the issue. Particularly for the chemistry unit, when ANT was introduced and emphasized, many student projects accommodated most aspects of ANT listed above. Students only seemed to have difficulties developing deep, explicity-stated, understanding of the abstract concept that each actant is heterogeneous. For their final projects, although few took time to draw actor-networks, most projects infused several ANT principles.

To support our claims that Mirjan's programme corresponds to the two more engaged levels of citizenship in Levinson's (2010) framework (Table 9.1), we now present a brief overview of one student's research-informed action project—supplemented by a few other examples—completed during the Chemical Change unit (Fig. 9.4). This case gives great priority to exposing citizens to actants and semiotic messages frequently hidden from consumers, often because of the prominence of idealized semiotic messages associated with commodities. As she learned more about liquid foundation makeup from her secondary research, 'Connie' developed the actor-network map shown in Fig. 9.6 to represent it. Although some actant types, such as think tanks and transnational advocacy groups (Ball, 2012) were absent, her network included a wide range of actant forms, including: living things (e.g., [human] teens, cheap labour, rabbits), human organizations (e.g., companies [e.g., Maybelline™], factories), technologies (e.g., editing [software]), non-living things (e.g., aluminum, inscriptions (e.g., fake pictures), and semiotic messages (e.g., feel prettier, feel grown up). In illustrating relationships among them, moreover, she makes relatively-explicit reference to the Trojan horse metaphor—indicating prominent pro-capitalist actants (e.g., happy companies, advertising, planned obsolescence) perhaps distracting consumers from such negative effects as: [human] depression, fake results [of animal testing], non-renewable energy use. To supplement her investigations, she surveyed fifty teenaged girls in her school regarding

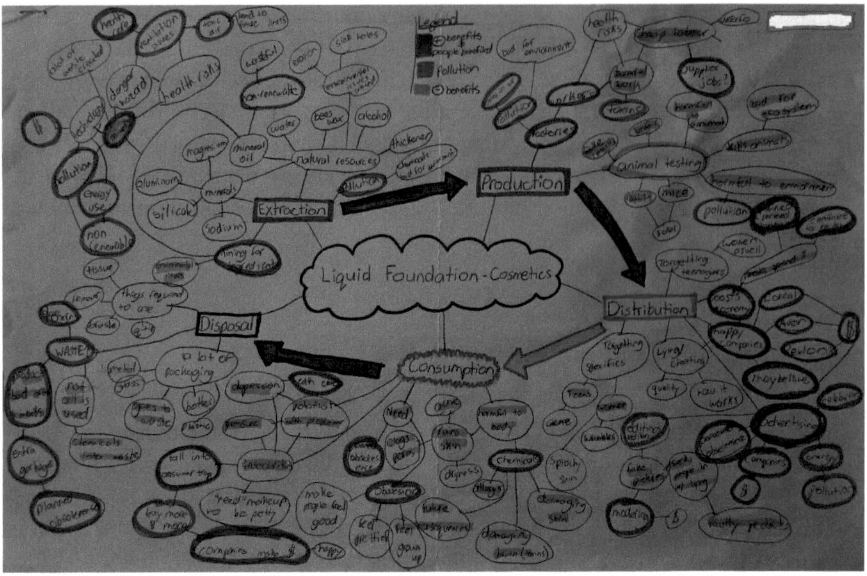

Fig. 9.6 A student's actor-network map about foundation makeup

their uses of makeup. This study also made reference to different actants, as indicated by her summary of study results:

> I can conclude that the media does have a generally large impact on teenage girls makeup usage. Out of the 86% of girls who started wearing makeup at a pre-teen age 81% of them said that the media was the influence that got them interested in it. Also 93% of the girls who started wearing makeup at a young age say they wear makeup for more than half of the week. Only 30% of girls are not influenced from the media about their makeup choices. So the media has a very large impact on teenage girls makeup consumption, by impacting 70% of them (June 3, 2013).

In particular, in addition to reference to makeup and girls' ages, she focused on influences from (although not specifically) advertizers. Other students' research-informed actor-networks also indicated signficant consciousness regarding a range of actants associated with the consumer product they studied. The comment below is somewhat representative of classs members' conceptions and perspectives in this regard:

> Most of these corporations [regarding electronic games] have American [sic] or Japanese ownerships (www.zdnet.com). The most powerful groups in this field are console developers such as Sony, Panasonic, and Microsoft (ca.ign.com). These are the massive corporations that have other branching corporations which manufacture the video games. The biggest few video game manufacturers are: Square Soft, Sega, Capcom, SONIC Team, and Infinity Ward (ca.ign.com). These massive corporations completely control the media related to the gaming community. Massive advertisement companies have billion dollar contracts with the above mentioned companies. Therefore, one can assume that only positive things about video games are being advertised - not any of the negative stuff is being

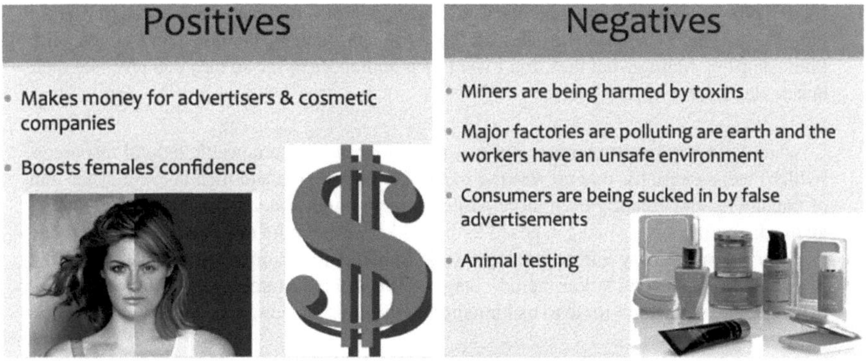

Fig. 9.7 Excerpt from a student's activist video

> mentioned. Massive corporations that make millions of games and gaming consoles are mostly based in China. This is due to the fact that gaming consoles can be manufactured at a very cheap labour cost; this is because of China's massive population and high unemployment rate. In China, safe working conditions are not guaranteed either, shifts of 12 hours a day are in effect. But these weak stakeholders can not do much because of the fact that thousands of people are waiting in line for the same job. An average worker in these companies is paid about $1.36 (www.huffingtonpost.com). These are important stakeholders to this network of video game manufacturing, but this group is certainly less privileged and has basically no power compared to the massive corporations that employ them (Miguel, Project Report, June 13, 2013).

Actants discussed here at least include: corporations; countries; media/advertizers; consoles; games; money; contracts; labour; working conditions; power; priviledge. Uses of actor-networks have been shown, as noted above, to provide students with these kinds of more realistic insights into socioscientific issues than is apparently being encouraged by the GCN (Pierce, 2013).

Based on students' reflections on their pre-conceived notions and on their secondary and primary research, they seemed to develop deep, meaningful, concerns for wellbeing of individuals, societies and/or environments and, accordingly, develop and implement plans of action to address their concerns. For example, based on her research, Connie developed and posted to YouTube™ an excellent educational video (www.youtube.com/watch?v=WhN6PS1GT9c) advising viewers (now with over 500 'views') of 'positive' and 'negative' aspects of liquid foundation consumption—such as those depicted in Fig. 9.7 (screen shots from her in-class PowerPoint™ presentation). The video, which is structured around the stages of the materials economy from *The Story of Stuff*™ (refer above), again (like her actor-network) emphasizes the Trojan horse metaphor—featuring a range of actant types, some aligned to support positive messages while others that may support negative messages are, largely, hidden. Evidence for these claims can be seen in the following set of excerpts from the script of her video:

> [Foundation is about] hiding what we don't want others to see because we are scared to get judged. … This is what advertizers do [showing a woman's picture being edited with Photoshop™], hiding what they don't want their audiences to see so they can promote their businesses as best as possible;
>
> … when hard-working miners are put on the job to mine non-renewable natural resources, [which] we, essentially, end up wasting in a bottle on our skin and then take off at the end of the day … and creating even more waste. Being a miner is a very hazardous job, due to all the dangerous toxins being exposed to the workers. Also, a lot of pollution is created due to all of the machinery and technology used. Some of the main resources needing to be extracted to make foundation include magnesium, silicon, aluminum, alcohol and many petro-chemicals that are toxic to us humans and the environment. …
>
> Some big problems with factories [where raw materials are used to make cosmetics] are harmful toxins in the air, the low working wages, and the long hours. … [regarding product-testing]. Many of the big well-known make-up companies unfortunately test their products on animals, like rabbits and sometimes cats. This is an issue due to the harming of animals but, unfortunately, many big companies do not care. They only care about profit.
>
> [This is when] the advertising starts to try to make money. They suck the consumers in with many common tactics, like telling teens that it will clear or heal their acne and making wrinkle-free, anti-ageing, products for [women]. Cosmetic products get falsely-advertized through the excessive amount of editing to make the models look flawless. The product ends up not working as well as people expect and, then, blame it on themselves for not being pretty enough; but, really, it is the product's fault. So, [it is] no wonder why over the past decade self-esteem has decreased in teenaged girls; and, about 70% of teenaged girls won't leave the house without being all 'dolled-up' and, still, feel self-conscious and insecure about themselves. And, this is mostly due to fake advertising and too much pressure.
>
> During this stage, many harmful toxins that are created during the production stage are spread all over the consumer's face. People are tricked into thinking that foundation is to help feel beautiful and make your skin clear, but it is not [the reality]. Although the product claims to keep your skin clear, hydrated, prevent acne, and not clog your pores, all of these claims mean nothing - because the advertizers find ways around the laws. Consumers will, eventually, start to become addicted to the product and continuously buy more, which is exactly what the companies want to make profit.
>
> Once the foundation is done, it goes to the last stage of [its] life, disposal. Since the foundation should only be used for 3 months and then thrown away, many people are not able to finish the bottle and the chemicals are getting thrown out.
>
> So, this is foundation's life. Just like a human, it goes through many experiences that people would never know about. Just like a book, never judge it by its cover (April 25, 2013).

Again, like Connie, many other students seemed to hold conceptions (although not explicitly stated by them) that the consumer product they studied may be considered a complex entity, 'something' that has agency through its relations with other actants. These realizations, it seems, led many of them to write and/or speak with considerable passion about injustices associated with deception apparently inherent in advertizing for many consumer products. 'Brenda,' for example, whose team studied mascara, noted that advertisements used names of major cities for their products—such as: Maybelline, New York™; Rimmel, London™; Lancôme, Paris™. About this, while presenting her actions to her classmates, she said: "I couldn't figure out why a lot of mascara products have a city or something after the names, but I think that it is something that is trying to make it [the product] more interesting and intriguing … because cities, like Paris, London, New York .. Everyone wants to go there.." (April 26, 2013). In thinking about this, though, she seemed to have a fairly strong conception of roles for semiotics in consumer product networks:

> [Regarding a Maybelline advertisement,] the ideas and emotions that it creates towards the product are that mascara is exciting, because they use all of these bright colours and all of these bold letters, [use of] capitals .. and making it so expressive … And, they use phrases like, … 'Super-sonic JumboBrush' to make it sound so intense … to make people want to buy this product. [Meanwhile,] what is not said in this ad is the fact that mascara is tested on rabbits for eye irritation, as you can see here [showing a picture of a rabbit with irritated eyes and mascara]. It's eye is all blistery and red and swollen because they take drops of mascara and drop it into the eyes [of the rabbit]. They kill the rabbits after they do the tests (Class presentation, April 26, 2013).

Overall, in light of the above examples, it seems clear that students engaged in Levinson's (2010) 'Science education as praxis' and, consequently, developed 'Dissent' (concerns about issues) and, to some extent, 'conflict' (social actions).

Finally, it should be noted that, contrary to many criticisms of a focus on SSIs, inquiry and socio-political actions, it seems that such educational experiences need not compromise students' 'achievement' in traditional 'content' knowledge. A remarkable finding from work with Mirjan, something he confirmed several times throughout the two years of our collaboration, was that promotion of student-led research-informed and negotiated actions on STSE issues seemed not, apparently, to have detrimental effects on students' overall grades. Rather, he said,

> I have had a lot of success with the STEPWISE framework. Students are more engaged on average. My weaker students (ie. those who generally do poorly on knowledge-based tests) have done a lot better in class with the STEPWISE. Students like discussing socio-scientific issues and being empowered to act and make a difference in our society. Students' inquiry projects (experiments and correlational studies) have more meaning since they are contextualized (Blog entry, Nov. 23, 2011).

This result has ramifications for GCN goals for science education, a priority of which appears to be reflected in this statement (from a quotation above): "…4 percent of the nation's workforce is composed of scientists and engineers; this group disproportionately creates jobs for the other 96 percent" (NRC, 2011, p. 2). Under the current system, as argued above, most of the 4% referenced above will be comprised of advantaged students. If Mirjan's finding has merit, and two years of trials seem to support this claim, then more disadvantaged students may become knowledge builders. This would represent a significant social justice improvement over systems governed by the GCN.

9.4.3 Teachers and Networks

In analyzing data collected, it was apparent that—as with other instructional situations—myriad factors likely influenced results like those reported above. Moreover, in light of actor-network theory (Latour, 2005), it is likely none of these factors act in isolation; but, rather, act as networks of dynamically-changing co-affecting actants. Accordingly, while it often seems convenient to consider individual factors, we must not lose sight of their collective agency.

While we may claim several factors, such as Mirjan's promotion of actor-network theory (the focus of this paper), as contributors to students' orientations towards 'science education as praxis' and 'dissent and conflict,' it seems that a signficant limiting/facilitating actant in this context pertains to Ontario's curriculum prioritization (at least in principle) of STSE education (including in terms of promoting student development of plans of action to address issues) and research. Mirjan said it was only after introduction of the revised Ontario curriculum (MoE, 2008), which had placed STSE education as the first of three overall curriculum goals, that he began to seriously implement this educational component. Expressing his rationale for this kind of education, he said: "[W]ith regards to my focus on academic, social and moral development, it [STSE] fits in nicely" (Interview, Dec. 17, 2012).

As elaborated elsewhere (Bencze & Krstovic, Chap. 11, this volume), much of the success of the ANT-based activism activities noted above would not have been possible without Mirjan—arguably a quite exceptional teacher. He has a number of basic positive characteristics that would serve any teacher well, but he has grown significantly in his perspectives and practices in the last few years in ways very much conducive to promoting student-led research-informed and negotiated actions on socioscientific issues. He is extremely energetic, which most of us know is so essential for effective work as a teacher. Perhaps most importantly, he is a *reflective*—and, essentially, an 'entrepreneural'—*practitioner*. This characteristic, which is likely essential for innovative teaching, much like what would be required to infuse ANT into high school science teaching, appeared to arise, more or less, by chance. While an undergraduate student, he worked as a research assistant for a science professor with a focus on education. This work led to a publication—fortuitously, about lifelong learning—in a refereed journal (Percy & Krstovic, 2001), an event that perhaps led Mirjan to then conduct an action research project dealing with gender differences regarding his students' interests in astronomy which, in turn, led to another refereed journal article[2]. About this research, he said: "I saw this kind of inquiry work … as being a natural part of teaching and learning—something that my previous experiences reinforced" (Written reflection, Aug. 6, 2013). Because of his commitment to continual reflection and change, he then enroled in a graduate programme leading to a Masters of Education degree—a move that appeared to lead to at least two major changes that seem congruent with teaching that would generate

[2] The citation for this publication is omitted here because it would reveal the identity of Mirjan's school and students involved in our study.

results reported above. Firstly, many of the courses he took gave him new perspectives on education—some of which he attempted to integrate into his teaching. For example, Mirjan was chosen by his school's vice-principal to lead a government-funded project to study implementation effects of pedagogical techniques—such as 'placemat,' 'think-pair-share' & 'community circle'—relating to 'instructional intelligence' about which he learned (Bennett & Rolheiser, 2001). As a result of this work, Mirjan concluded: "When various instructional tactics are integrated in class[,] students' communication skills as well as their knowledge and understanding of the key concepts improve" (Project Reflective Journal, Krstovic, 2009, p. 29). Buoyed by his successes, he then sought a collaborative relationship with the instructor (Larry) of an online graduate course dealing with the history, philosophy and sociology of science. Throughout that course, Mirjan had shifted his views about science towards the 'Naturalist-Antirealist' quadrant of Loving's (1991) *Scientific Theory Profile*—positions that challenge the certainty of methods and conclusions in the sciences. Naturalists, in contrast to Rationalists, assume that conduct of science is highly situational and idiosyncratic, depending on various factors, including psychological, social, cultural and political influences. Antirealists, opposing Realists, claim that human knowledge cannot precisely match the nature of phenomena. After discussing possibilities, including a focus on the nature of science, we agreed Larry would work as 'researcher-facilitator' for Mirjan's efforts to promote research-informed and negotiated actions (RiNA) to address socioscientific issues. So, it seems clear that this collaboration contributed to the nature and extent of Mirjan's teaching.

> Since I have started implementing the STEPWISE framework in my Grade 10 Academic classes, I have had a lot of success. One of the ways in which I define that success is by the extent to which my students experience the joy of learning that goes beyond the traditional learning of knowledge and understanding of certain fundamental science concepts. I have also experienced the joy of teaching (and learning) critical STSE issues and ways of implementing STSE activism.

Despite Mirjan's apparent successes, in several ways, conclusions about this should be tempered with the realization that STEPWISE implementation seemed to significantly depend on his teaching and learning contexts. Much of his overall success appears attributable to perhaps ideal conditions in the first school in which he had worked (Sept. 2006 – June 2012). Mirjan stressed, at different times, that his orientation towards reflective practice arose because, to a large extent, the department head—supported by the school's administrative team (e.g., principal)—in his first school set a tone of exploration for department colleagues. Mirjan said, for example: "The leader … was, I would say, a more forward-thinking teacher. Leadership is important. He wanted to innovate in the science department by introducing … cross-curricular assignments" (Interview, July 30, 2013). Although he noted that most of his colleagues, while innovative, did not take their instruction beyond levels #1 and 2 in Table 9.1, he felt habits of reflective practice were reinforced from the very beginning of his teaching career. Particularly in the context of larger schools, Mirjan suggested that an essential condition enabling him to develop and explore new approaches was his freedom to 'step aside from collaboration' with colleagues teaching other sections of the same course. This was a routine practice for any teacher who wanted to innovate in the first school. It also was, though, a circumstance he experienced in the first year of his new, more conservative, school in the second year of the project. He was hired there with recognition that he was performing differently than most teachers and, accordingly, was allowed to teach courses for which there were no other sections—which meant he could freely explore different approaches in his teaching.

Over the three semesters of our collaboration (since Sept. 2011), Mirjan had developed and implemented numerous lessons and activities based on the pedagogical framework shown in Fig. 9.4. By the point of this study, he was able to implement it relatively fully—with two apprenticeships for RiNA prior to encouraging and enabling students to, to a great extent, self-direct RiNA projects. As discussed above, a major factor enabling him to implement this framework was the decision of the science department head to allow Mirjan to independently teach courses in his first year at the school. From previous studies of use of this framework, we concluded that this seemed to be an effective approach to enabling and promoting student-led research-informed and negotiated actions to address SSIs of student concern/interest (e.g., Bencze et al. 2012). In the study reported here, as outlined above (and in Appendix A), Mirjan chose to infuse ANT into this framework in the second unit (chemistry), rather than the first one, so that students would first become generally aware of and confident with RiNA prior to being introduced to a new and complex sets of concepts surrounding actor-network theory. The approaches Mirjan chose to use during the second unit appeared to be particularly effective in helping students to understand ANT, to develop critical conceptions of their chosen issues and to engage, apparently, in more critical actions to address them—thus, achieving Levinson's (2010) 'dissent and conflict' level of citizen engagement. There is considerable evidence to support these claims. In examining the actor-network developed by 'Connie' (Fig. 9.6) and her subsequent video (refer to Fig. 9.7 and transcript notes, above), for example, there is very clear evidence of Mirjan's use of *The Story of Stuff* and *The Story of Cosmetics*, along with the Trojan horse metaphor. In other words, while the actor-network drawing and video are both organized in terms of the materials economy (i.e., Extraction → Production → Distribution → Consumption → Disposal), they also are broadly organized into groups of actants supporting positive semiotic messages (e.g., Foundation makeup is 'good') that appear to occlude actants that, if made more prominent, may support much less flattering semiotic messages (e.g., 'problematic'). Most student teams in Mirjan's class included these aspects in their networks and actions.

Finally, given Mirjan's ANT-infusion project occurred in the context of a science department where teaching has been very traditional (e.g., focused mostly on instruction in products of science and technology using relatively didactic approaches), it should be no surprise to learn students entering his course were relatively conditioned to common pedagogical practices. Consequently, it was important for Mirjan to gradually re-orient students' conceptions of and comfort with new and more diverse pedagogical practices. In the following excerpt from an email message he sent to Larry to describe his experiences with students, it is apparent that he seemed to be handling this complexity well:

> [S]tudents are coping with a two-[or 3]-tier instructional program in my class. Let me explain: as you know, traditionally students listen to a lesson, take notes, then they do questions from the textbook, which the teacher may or may not take up the next day; may be they do a closed-ended, scripted procedure style lab that proves a concepts (ie. law of conservation of mass), they may write a small lab report, then they write a unit test. Somewhere in there, an STSE reading is assigned (or a research project) which the students hand in as a written report (and it only gets read by the teacher and no one else). I call this a one-tier instructional program (for the lack of better words). However, my students are learning to cope with a multi-tiered system. So not only are we handling content in also a very 'instructionally intelligent' way (I use a lot of instructional tactics/strategies from *Beyond Monet*, by Barrie Bennett and Carol Rolheiser of OISE), but we are also adding a layer of apprenticeship activities (ie. how to conduct correlational studies, learning about activism, photojournalism, etc.) which are part of the research-informed activism instructional tier. As much as I try to merge the two tiers (i.e. use RiNA as the CONTEXT to drive the learning of the CONTENT, that does not always happen) some of the kids MAY feel (and this is my hypothesis) that their learning environment is messy. To add to these two tiers, there is the third tier of learning additional skills necessary to execute 'creative' actions, such as how to use Windows Movie Maker™ software, how to publish a video on YouTube™, how to use Prezi™, etc. (Oct. 9, 2012; italics and Trademark logos added).

9.5 Uses of Actor-Network Theory in Critical & Activist Science Education

There appear to be numerous serious concerns for wellbeing of individuals, societies and environments associated with decisions made by powerful people and groups regarding directions and practices in fields of technoscience. Potential and realized harms seem to include social and environmental devastation and disruption due to climate change, various illnesses linked to manufactured foods and beverages and poverty for billions of people working under poor labour conditions to mine and manufacture goods and services for relatively few advantaged individuals. In science education in many contexts, such challenges often are treated, however, as controversial—often urging students to determine logical and evidence-based personal decisions about them. On the one hand, such approaches appear to have merits, including in terms of helping students to develop logical reasoning competencies. However, according to several scholars (e.g., Hodson, 2011; Levinson, 2010; Dos Santos, 2009), students also need to engage in knowledge generation concerning issues/problems and critical analyses and social actions to address them—forms of citizenship Levinson (2010) calls, respectively, 'science education as praxis' and 'dissent and conflict.' While our previous research appears to indicate successes in this regard as a result of various strategies encouraging and enabling student-led research-informed actions to address socioscientific issues (SSIs) of their interest/concern, we felt that further approaches were needed to enhance such ends.

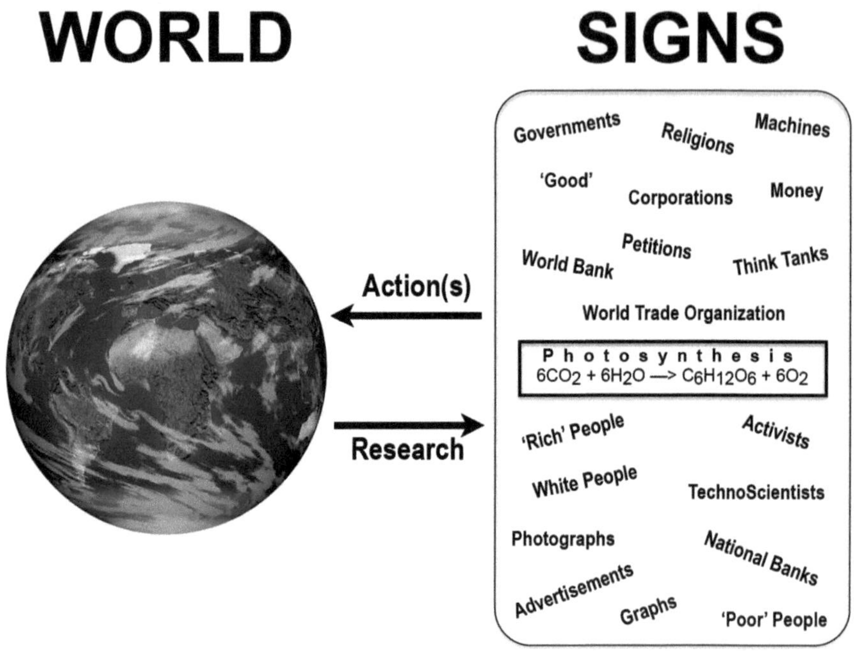

Fig. 9.8 ANT-based RiNA projects on STSE issues

Drawing largely from the work of Stephen J. Ball (2012) and Clayton Pierce (2013), along with principles of actor-network theory (ANT), particularly from the Bruno Latour (2005), the study reported here was premised on the idea that commitments towards research-informed and negotiated activism could be enhanced by encouraging students to consider a broad spectrum of actants and relationships among them associated with socioscientific issues. As indicated in Fig. 9.8, instead of developing relatively reductionist conceptions of the 'World,' such as the equation for photosynthesis, students may develop much more holistic and realistic ones, including critical conceptions of powerful actants such as corporations and The World Bank and, in turn, then develop strongly-motivated actions to address their concerns relating to such potentially-problematic actants. Data from this study appear to provide support for this claim. Students seemed able to understand and use at least five major principles of ANT in developing critical conceptions of power in relations associated with their SSIs that many of them then seemed to use as bases for their actions to address the issues.

In light of actor-network theory, a great variety of actants must align to enable uncommon practices like ANT-infused student-led research-informed and negotiated actions to address highly-contentious SSIs of students' choices. It seems clear a teacher must contend with situations in which many actants are aligned to support highly didactic instructional practices that, in terms of attention to SSIs, tends to favour more teacher-led logical decision-making by students with regards to arguably less-critical—e.g., in terms of avoidance of attention to matters of political economy—socioscientific issues. In the study reported here, while existence of official curricular sanctioning of attention to SSIs, research and actions appeared to help, it also seems clear that student learning outcomes reported here (e.g., regarding 'praxis' and 'dissent & conflict') may not have been very possible without the energy, talents and perspectives of a very special teacher, Mirjan, whose persistent desire for instructional reflecton and improvement appeared to contribute greatly to his implementation of an ANT-infused activist programme. At the same time, it also seems clear his association with a researcher-facilitator (Larry) seemed to contribute greatly to apparently successful pedagogical decisions, such as: a teacher-led Socratic discussion about actor-network theory, aided by use of a personally-drawn actor-network depiction of an SSI (i.e., regarding cell phones), the Trojan Horse analogy/metaphor and videos from *The Story of Stuff*™ activist series.

Given the very narrow context of the study reported here, including in terms of its status as an instrumental case study of one teacher's instructional experiences with a class of tenth-grade students, the external validity/transferability of our claims must be considered limited—largely dependent on interpretations of individual readers (Guba & Lincoln, 2011). Moreover, the likelihood of many other science teachers implementing activities depicted here seems limited—given the extent to which a range of actants would have to be re-aligned, such as in terms of a teacher's professional development, particular curricular change, availability of alternative theoretical frameworks (e.g., ANT), etc. Nevertheless, we suggest that ways need to be found to re-align actants to enable such practices exhibited here. As argued earlier, for at least the reason that many socioscientific issues are associated with potentially serious personal, social and/or environmental problems, we and others believe societies need to become more activist in nature. Our study suggests, following Pierce (2013), especially, that infusion of principles of actor-network theory into apprenticeship activities promoting research-informed and negotiated actions to address critical socioscientific issues can be effective, along with alignment of several other actants, in allowing students to achieve Levinson's (2010) 'science education as praxis' and 'dissent and conflict' levels of citizen engagement in socio-political issues relating to science and technology.

The approaches noted here, while prioritizing students' self-determination, were not entirely neutral. No formal educational experience can, of course, be unbiased. Indeed, through uses of actor-network theory, we have introduced students to potentially-problematic actants associated with commodities (e.g., Fig. 9.5). This tack is recommended for bringing more 'realism' and, therefore, democracy, to students' engagement in socioscientific issues (Pierce, 2013). This also aligns with Levinson's (2010) call for significant 'dissent and conflict' (Table 9.1) in science education as a way of democratizing citizen engagement. It could be argued, however, that such overt efforts to, in effect, disrupt the GCN is, in itself, anti-democratic. We agree that this is, indeed, a tension of the approach offered here. On the other hand, it also could be argued that, given the extent to which the GCN appears to be overtly and covertly attempting to instill capitalism-friendly perspectives and practices into the minds of increasingly younger children, through, for example, advertizing (Acosta-Alzuru & Lester Roushanzamir, 2003; Bakan, 2011), perhaps offering youth alternatives to this programme may be justified. At the same time, it should be clear that, while dissent and conflict are urged here, student-led decision-making also was prioritized. We encouraged, for instance, students to make judgements they considered ethical in World ←→ Sign translations—including regarding ideological gaps in them (Fig. 9.3).

Appendix A: Grade 10 Academic Science: Chemistry Research-Informed STSE Action Project

Each lesson presented here is intended for a 75 minute period

Lesson 1: Expressing your ideas

Introduction

Most of our everyday products that we, in developed and industrialized countries, use and/or consume involve the use of chemicals and chemical reactions. From the moment that you get up from your bed in the morning you start to interact with everyday substances such as soaps, toothpaste, shampoos, body sprays, etc. You may be taking multivitamin supplements, wearing wrinkle resistant shirts or drinking carbonated (or non-carbonated) beverages. All of these products contain chemicals.

Group Activity

1. In a group of three to four students, do a two to three minute brainstorm to outline a list of everyday products containing chemicals that you use at home.
2. Create a chart to show at least one **positive** and one **negative** consequences of each product on the well-being of societies and/or environments.

3. In another column of the same chart list any information you would need to research in order to state the positive and negative consequences of the product.
4. Consider the image of a Trojan horse. The Trojan horse was a great example where warriors were hidden inside a giant horse statue and delivered into the heart of the city. The town's people saw how great a gift the horse was but didn't realize what was lurking inside. Discuss with your group how the Trojan horse metaphor is relates to the products you listed in your table.
5. Consider the following statement: *Be it resolved that companies should reduce the use of harmful chemicals in their products even if it compromises the product effectiveness.* Decide if you strongly agree, somewhat agree, strongly disagree or somewhat disagree with this statement. Move to the corner of the classroom that best represents your view. Discuss your position with a partner or in a group of three when you get to your appropriate corner of the room.

Lesson 2: Lifecycle of a Product

The whole **life cycle of a product** (from its creation through to its disposal) needs to be taken into account when considering its impact on the well-being of individuals, societies and environments. This is where we begin with a deeper exploration of various chemical products, which on the surface may appear to be a simple product, but in reality there are many components that link together in the life cycle of a product.

Activity

1. Watch "The Story of Stuff" with Annie Leonard.
2. Fill in the table below stating at least one negative consequence on the well-being of individuals, societies and environments during each stage in the lifecycle of a product.

Stage of the lifecycle of a product	Negative consequence on individuals	Negative consequence on societies	Negative consequence on environments
Extraction			
Production			
Distribution			
Consumption			
Disposal			

3. Discuss the consequences with your group, then with the whole class (*teacher guided discussion*). Ensure that your table is complete as you will need this information for the next stage of your research-informed action project.

Lesson 3: Creating a Mind Map

Many products come with negative features which are often masked to make the product appear a certain way. This is analogous to the Trojan horse we explored earlier. The negative aspects of many personal hygiene products are usually kept away from the general public. We only see what the companies want us to see so they attract the consumers and generate as much profit.

Goals of the Lesson and Your Task

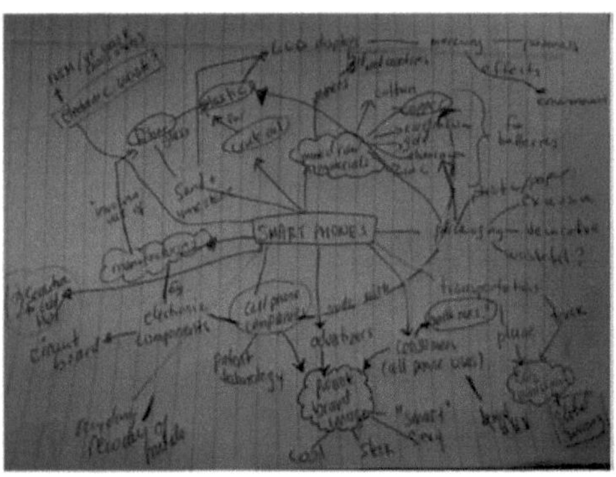

1. Your teacher will create a mind map for a 'smart phone' showing some of the most important components (living and nonliving) during the lifecycle of this product. We will begin to uncover some of the hidden social and environmental costs that are often hidden from the general public.
2. You will pick **one personal hygiene product** (e.g., shampoo) and start developing a similar mind map. You can start by expressing what you know already

based on what you learned from "The Story of Stuff" and the table that you completed in class. You are required to do some additional secondary research for your first mind map.
3. It's important that you show the many links to various components (living/non-living and hidden messages) for your product as well some of the social and environmental costs associated with your product.
4. Another great video to watch is 'The Story of Cosmetics' by Annie Leonard.

The Story of Cosmetics (2010)

Lesson 4: Revising the Mind Map

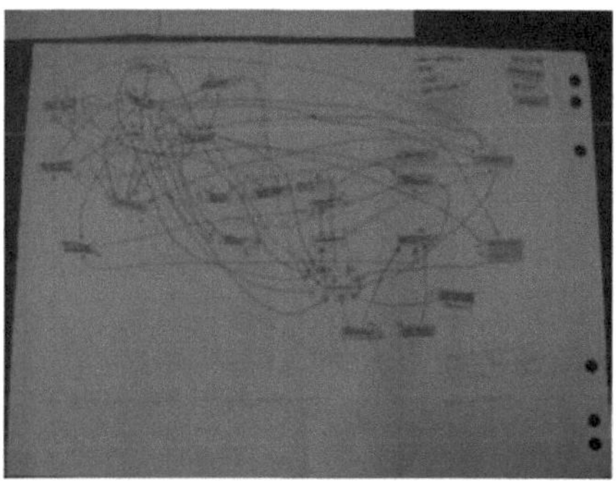

Since the goal of this project is to address an STSE issue (or a few related STSE issues) in connection to the personal hygiene product you selected, you will first need to identify an issue, or issues, you feel is/are most relevant and most important to address through research-informed activism. You will need to revise your original mind map according to the instructions below.

Your Goal Is to

1. Revise your first mind map so that you group different components according to overall goals they have. For example, one might represent: advertisers, cell phone companies, people who often are seen using cell phones, stores that sell

cell phones, etc. All of these may be aligned to support such common *hidden messages* like, cell phones are 'cool,' 'sleek,' 'powerful,' etc. You may put circles around groups of components that operate as a unit, or color code these components.

2. With another circle, or another colour, show living and non-living components that might tell a different story, **which is often less prominent in society**. For example, if we consider cell phones again, this might include: miners in developing countries, living things adversely affected by mining, people with cancer from cell phone energy waves, landfills with heaps of cell phones in them, etc. The components in this circle/or with this colour code—if it were more prominent - might send messages like, 'cell phones are harmful/toxic,' 'they separate people from each other,' etc.
3. **Identify less prominent messages that may be important in society about your selected chemical product.** For example, many personal hygiene products, like shampoos, contain some possible carcinogenic substances or neurotoxins which can affect our health. One or more of these less prominent messages will become the STSE issue(s) that you will address with your group through your choice of actions.

Lesson 5: Conducting primary research into your chemical product

Purpose To plan and conduct an original investigation (e.g., a correlational study) about a personal hygiene product of your choice. It would be best if your investigation relates to one or more of the less prominent messages/STSE issues you identified.

Getting Started

List of possible investigations that your group can do:

- Design an investigation to determine what smells/fragrances of soaps/shampoos are preferred by teenage boys and girls and how fragrance seems to be used as a marketing tool.

- Evaluate various labeling claims used on several brand name products as *acceptable* or *unacceptable* (according to Heath Canada Guidelines for Labeling Claims: http://www.hc-sc.gc.ca/cps-spc/pubs/indust/cosmet/index-eng.php#s2).
- "*If you can hook teens when they're young, you have a customer for a lifetime*," said Matt Britton, chief of brand development at Mr. Youth, a marketing firm. Design an investigation to study various methods that markers/advertisers use to entice teenage boys and girls to buy a particular product. For example, you may look at particular ads and determine if there is a difference in the way that boys vs. girls perceive these ads.
- Design a controlled experiment to test stability of personal hygiene products (see http://www.intertek.com/beauty-products/testing/cosmetic-stability/). For example, for your product you may look at pH, viscosity, appearance/colour and odour at various conditions such as different temperatures, different amounts of light and/or free-thaw conditions.
- Design and carry out a correlational study to determine how hypgeine products for girls and boys differ in hidden messages; e.g., how they make the user feel (refer to: http://www.beautypackaging.com/articles/2003/11/semiotics-research-deciphering-packaging-codes-rev)
- Your idea…discuss it with the teacher and get it approved before beginning the investigation.

Developing a Method and Preparing for Your Investigation

1. After you decide which investigation you'd like to perform with your group, you need to first come up with method to conduct your investigation. Plan it so that its results may be 'trustworthy'; e.g., valid and reliable.
2. Make sure that you show your method to your teacher before beginning the investigation.
3. As part of your preparation to conduct an investigation, you need to have all your materials ready. For example, if you conduct a study, you need to have your survey questions developed. If you are studying particular ads for your product, you need to have these ready as well.

Collecting and Analyzing Data

1. Remember that your data can be both qualitative (descriptive, no numbers used) and quantitative (usually numerical data). The type of investigation you chose will determine whether your data is qualitative and/or quantitative.
2. All data needs to be summarized in a properly labeled Table.
3. Quantitative data should be shown graphically—as a bar, line or pie graph.
4. Graphs should have titles with properly labeled axis.

Interpreting Data and Drawing Conclusions

1. You should make sense of what your results show by interpreting your results. Can your original focus question be answered using the data you collected? What does your data suggest? What conclusions can you make from your primary research?
2. What are some possible weaknesses in your investigation and how can they be addressed?

Lesson 6: Preparing for and taking action: Consumer Activism!

The last stage of your research-informed action project is to prepare for and take action to address an STSE issue related to the personal hygiene product you selected. You will be given one class period in which you will work with your group to propose and prepare for your actions. Your actions should be informed by both your secondary and primary research.

Your Goal Is to

1. Propose an action that your group can take to address the issue you identified.
2. Develop ready-to-use action materials. You will need to spend additional time outside of the designated class time to work on your ready-to-use action materials.
3. Present and defend your actions during a 10-muinute class presentation.

Ideas for Actions

- A public service announcement about the health and environmental effects of chemicals in shampoos, or lead additives in cosmetics
- An exposition/narrative about the lives of miners/factory workers affected during the extraction or production stages of development of your selected product
- A letter to powerful groups (e.g., Health Canada) asking for better regulations of chemicals in everyday products

- Students may propose safer/healthier and greener alternatives to some personal care products and develop a campaign to promote their use over other products
- Other actions of your choice—get them approved by your teacher

What's Assessed and Evaluated at the End of This Project:

At the end of this research-informed action project, you will be assessed and evaluated on the following components:

(i) Initial and revised mind maps
(ii) Results and conclusions of your primary investigation
(iii) Ready-to-use action materials/actions you propose and take
(iv) 10 minute presentation defending your actions

Mirjan Krstovic, 2013

References

Acosta-Alzuru, C., & Lester Roushanzamir, E. P. (2003). Everything we do is a celebration of you!: Pleasant company constructs American girlhood. *The Communication Review, 6*(1), 45–69.

Angell, M. (2004). *The truth about the drug companies: How they deceive us and what to do about it*. New York: Random House.

Bakan, J. (2011). *Childhood under siege: How big business targets children*. Toronto, ON: Allen Lane.

Ball, S. J. (2012). *Global Education Inc.: New policy networks and the neo-liberal imaginary*. Abingdon, UK: Routledge.

Barnes, R. L., Hammond, S. K., & Glantz, S. A. (2006). The tobacco industry's role in the 16 Cities Study of Secondhand Tobacco Smoke: Do the data support the stated conclusions? *Environmental Health Perspectives, 114*(12), 1890–1897.

Bencze, J. L., & Alsop, S. (Eds.). (2014). *Activist science & technology education*. Dordrecht, The Netherlands: Springer.

Bencze, L., & Carter, L. (2011). Globalizing students acting for the common good. *Journal of Research in Science Teaching, 48*(6), 648–669.

Bencze, J. L., & Carter, L. (2015). Capitalists' profitable virtual worlds: Roles for science & technology education. In P. P. Trifonas (Ed.), *International handbook of semiotics, vol. 1 & 2* (pp. 1197–1212). Dordrecht, The Netherlands: Springer.

Bencze, L., Sperling, E., & Carter, L. (2012). Students' research-informed socioscientific activism: Re/Visions for a sustainable future. *Research in Science Education, 42*(1), 129–148.

Bennett, B., & Rolheiser, C. (2001). *Beyond Monet: The artful science of instructional integration*. Toronto, ON: Bookation Inc.

Bourdieu, P. (1986). The forms of capital. In J. G. Richardson (Ed.), *The handbook of theory: Research for the sociology of education* (pp. 241–258). New York: Greenwood Press.

Carter, L. (2005). Globalisation and science education: Rethinking science education reforms. *Journal of Research in Science Teaching, 42*(5), 561–580.

Charmaz, K. (2014). *Constructing grounded theory* (2nd ed.). New York: Sage.

Dos Santos, W. L. P. (2009). Scientific literacy: A Freirean perspective as a radical view of humanistic science education. *Science Education, 93*(2), 361–382.
Gough, A. (2015). STEM policy and science education: Scientistic curriculum and sociopolitical silences. *Cultural Studies of Science Education, 10*(2), 445–458.
Guba, E. G., & Lincoln, Y. S. (2011). Competing paradigms in qualitative research. In N. K. Denzin & Y. S. Lincoln (Eds.), *The SAGE handbook of qualitative research* (pp. 105–117). Thousand Oaks, CA: Sage.
Hileman, B. (1998). Industry's privacy rights: Is science shortchanged? *Chemical & Engineering News, 76*(17), 36.
Hodson, D. (2003). Time for action: Science education for an alternative future. *International Journal of Science Education, 25*(6), 645–670.
Hodson, D. (2011). *Looking to the future: Building a curriculum for social activism*. Rotterdam, The Netherlands: Sense.
Khishfe, R., & Lederman, N. G. (2006). Teaching nature of science within a controversial topic: Integrated versus nonintegrated. *Journal of Research in Science Teaching, 43*(4), 395–418.
Klein, N. (2014). *This changes everything: Capitalism and the climate*. Toronto, Canada: Simon & Schuster.
Kleinman, D. L. (2003). *Impure cultures: University biology and the world of commerce*. Madison, WI: University of Wisconsin Press.
Latour, B. (2005). *Reassembling the social: An introduction to actor-network-theory*. Oxford, UK: Oxford University Press.
Leonard, A. (2010). *The story of stuff: How our obsession with stuff is trashing the planet, our communities, and our health – And a vision for change*. New York: Free Press.
Levinson, R. (2010). Science education and democratic participation: An uneasy congruence? *Studies in Science Education, 46*(1), 69–119.
Levinson, R. (2013). Practice and theory of socio-scientific issues: An authentic model? *Studies in Science Education, 49*(10), 99–116.
Loving, C. C. (1991). The scientific theory profile: A philosophy of science model for science teachers. *Journal of Research in Science Teaching, 28*(9), 823–838.
McMurtry, J. (2013). *The cancer stage of capitalism: From crisis to cure*. London: Pluto.
Ministry of Education [MoE]. (2008). *The Ontario curriculum, grades 9 and 10: Science*. Toronto, ON: Queen's Printer for Ontario.
National Research Council [NRC]. (2011). *Successful STEM education: Identifying effective approaches in science, technology, engineering, and mathematics*. Washington, DC: National Academies Press.
Oreskes, N., & Conway, E. (2010). *Merchants of doubt*. London: Bloomsbury Press.
Pedretti, E., & Nazir, J. (2011). Currents in STSE education: Mapping a complex field, 40 years on. *Science Education, 95*(4), 601–626.
Percy, J. R., & Krstovic, M. (2001). Later life learners: A significant and receptive audience for introductory astronomy. *Journal of the Royal Astronomical Society of Canada, 95*, 205–207.
Pierce, C. (2013). *Education in the age of biocapitalism: Optimizing educational life for a flat world*. New York: Palgrave Macmillan.
Reich, R. B. (2007). *Supercapitalism: The transformation of business, democracy, and everyday life*. New York: Knopf.
Roth, W.-M. (2001). Learning science through technological design. *Journal of Research in Science Teaching, 38*(7), 768–790.
Sadler, T. (Ed.). (2011). *Socio-scientific issues in the classroom: Teaching, learning and trends*. Dordrecht, The Netherlands: Springer.
Sadler, T. D., Barab, S. A., & Scott, B. (2007). What do students gain by engaging in socioscientific inquiry? *Research in Science Education, 37*(4), 371–391.
Sellar, S., & Lingard, B. (2013). The OECD and global governance in education. *Journal of Education Policy, 28*(5), 710–725.

Springer, S., Birch, K., & MacLeavy, J. (Eds.). (2016). *The handbook of neoliberalism*. New York: Routledge.
Stake, R. E. (2000). Case studies. In N. K. Denzin & Y. S. Lincoln (Eds.), *Handbook of qualitative research* (2nd ed., pp. 435–454). Thousand Oaks, CA: Sage.
Venville, G. J., & Dawson, V. M. (2010). The impact of an argumentation intervention on Grade 10 students' conceptual understanding of genetics. *Journal of Research in Science Teaching, 48*(8), 952–977.
Wasser, J. D., & Bresler, L. (1996). Working in the interpretive zone: Conceptualizing collaboration in qualitative research teams. *Educational Researcher, 25*(5), 5–15.
Wenger, E. (1998). *Communities of practice*. Cambridge, UK: Cambridge University Press.
Wood, G. H. (1998). Democracy and the curriculum. In L. E. Beyer & M. W. Apple (Eds.), *The curriculum: Problems, politics and possibilities* (pp. 177–198). Albany, NY: SUNY Press.
Zeidler, D. L. (2016). STEM education: A deficit framework for the twenty first century? A sociocultural socioscientific response. *Cultural Studies of Science Education, 11*(1), 11–26.

Chapter 10
Science Students' Ethical Technology Designs as Solutions to Socio-scientific Problems

Larry Bencze and Mirjan Krstovic

10.1 Introduction

Much of the world seems to be in dire straits—although many of us living in certain advantaged parts of the world and/or in certain advantaged segments of societies seem relatively oblivious to difficulties around us. Despite increases in the Middle Class in China and India, for instance, overall global differences between rich and poor have recently increased—and, apparently, due to capitalism, are destined to continue to increase (Piketty, 2014). Associated with massive poverty increases, apparently, are government-supported efforts to make work more *precarious*; that is, more part-time (if not without any guaranteed work hours) and with minimal labour benefits (e.g., health care and environmental protections) (Hardt & Negri, 2009; McMurtry, 2013; Pierce, 2013). Apparently strongly-linked to wealth accumulation by small fractions of societies, meanwhile, are considerable harms to health of many people, other living things and environments. Arguably at the top of the list in this regard is devastating climate change (Klein, 2014), but harms from a range of consumer products—such as cigarettes, pesticides, household cleansers, radiation and food additives—also plague us (McMurtry, 2013).

Given that fields of science and technology are implicated, to some extent, in many of the harms noted above, and given that school science has, for many years, included mandates (at least) to educate citizens about possible harms associated with relationships amongst fields of science and technology and members of societies and environments, there seems to be some hope for a better world. On the other

L. Bencze (✉)
Ontario Institute for Studies in Education (OISE), University of Toronto, Toronto, ON, Canada
e-mail: larry.bencze@utoronto.ca

M. Krstovic
Peel District School Board, Mississauga, ON, Canada
e-mail: mirjan.krstovic@peelsb.com

L. Bencze (ed.), *Science and Technology Education Promoting Wellbeing for Individuals, Societies and Environments*, Cultural Studies of Science Education 14,
DOI 10.1007/978-3-319-55505-8_10

hand, recent dramatic emphases on identification and education of workers in STEM (Science, Technology, Engineering & Mathematics) fields, apparently largely for economic reasons, seem to threaten efforts to educate citizens about 'science-linked' personal, social and/or environmental problems and provide them with expertise, confidence and motivation to address such problems in direct ways. In this chapter, however, we report some successes in encouraging and enabling youth to develop technology designs that, in addition to being relatively functional, take into consideration potential for wellbeing of individuals, societies and environments. In light of the seemingly hegemonic nature of STEM education emphases, the approach reported here may offer citizens a form of STEM education that uses fields of science and technology (and related disciplines) for wellbeing of larger fractions of societies and environments.

10.2 Theoretical Background

Although fields of science and technology appear to have contributed many benefits to societies, not the least of which perhaps include increases in human health and longevity related to developments in medical and agricultural sciences and technology, there also are apparent causes for concern. Additives to manufactured and 'fast' foods, for instance, often are linked to human health problems like cardio-vascular disease, cancer and diabetes (Weber, 2009). Perhaps most worrisome, however, are many devastating potential (and realized) harms to individuals, societies and environments due to climate change often linked to excessive petroleum uses (Klein, 2014). Reasons for such harms are, undoubtedly, complex. In light of actor-network theory (Latour, 2005), for instance, all living and nonliving things (and symbolic entities, like 'prestige') are linked—to varying degrees—to each other. Accordingly, sharing 'culpability' with fields of science and technology may be, for example, governments and laws they enact and many citizens indicating desires and needs for products and services of such fields.

Having acknowledged distributed 'blame,' however, underlying many harms appears to be intense and pervasive *consumerism* promoted by capitalists and others. In his book, *Consumed*, Benjamin Barber (2007) says consumerist enticements are: *ubiquitous* (everywhere); *omnipresent* (always there); *addictive* (creates reinforcements); *self-replicating* (spreads 'virally'); and, *omnilegitimate* (self-promotional). Although those of us doing most of the consuming may feel better through our purchases, enjoyment often is fleeting; and, through various means, we often discard commodities, only to replace them with 'new' ones—often at the expense of various harms to wellbeing of individuals, societies and environments (Leonard, 2010). Again, while not alone, by any means, major contributions to consumerism and associated harms appear to be fields of science and technology. A way to envision their roles seems to be through Roth's (2001) depiction of relationships between science and technology, as illustrated in Fig. 10.1. Before discussing their roles in consumerism, however, it may be helpful to point out some fundamental

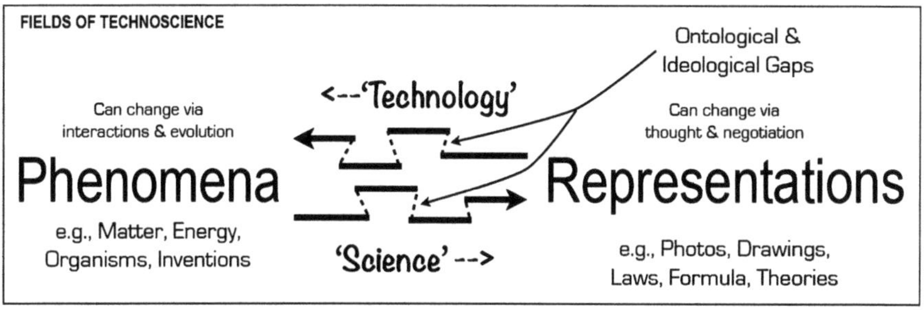

Fig. 10.1 Depiction of relationships between science & technology

features of relationships between the two processes. Firstly, although science and technology may be considered opposite translations (i.e., 'Science' = [World → Sign]; 'Technology' = [Sign → World]), perhaps because they co-affect each other (are reciprocal) they should be thought of as one combined entity, sometimes called, *technoscience* (Sismondo, 2008). Secondly, while there may be translation inconsistencies ('ontological gaps' due to differences in translated entities) between World and Sign (in both directions), negotiations among technoscientists and others over many years may help reduce these (Ziman, 2000).

In terms of consumerism, the schematic in Fig. 10.1 includes consideration of possible ideological gaps; that is, *purposeful* inconsistencies in translations between World and Sign. Such mistranslations often seem to occur when technoscientists are funded by the private sector—which seems to apply to those working directly for companies and, as well, university-based 'academic' scientists. While academic scientists have used private money for at least the last century, private funding of university-based scientists became legalized in about 1980 with passage of the Bayh-Dole Act in the USA—a practice that then spread to other countries (Mirowski, 2011). At about the same time, recognizing that companies and fields of technoscience had developed dramatic increases in production capabilities (more commodities ['Phenomena' in Fig. 10.1]), capitalists began emphasizing generation of consumer *desires* for commodities—prioritizing repeating consumption-disposal cycles among consumers with few needs (mostly in 'advantaged' places). This is largely accomplished through product design and advertising, which (referring to Fig. 10.1) emphasize invention of abstract idealizations ('Representations') that are associated with commodities. Such representations of commodities as 'cool,' 'slick,' 'powerful' and 'sexy' may be highly detached (large [ideological] gaps) from the Phenomena (commodities) they are to represent, yet consumers apparently often associate them with the real commodities. These representations are said to be *hyperreal*, meaning that people confuse them with real phenomena (Baudrillard, 1998). With consumers distracted by abstract idealizations in their purchases, companies can then compromise the quality of commodities people are consuming. In such cases, there would be large gaps between the idealized representations and the compromised phenomena/commodities. Governments have given companies the right to compromise the quality of products (and labour used to make them) for the sake of profit—by *externalizing* their costs (transferring them to others) (McMurtry, 2013). Examples of commodities that are sold on the basis of idealized representations while having compromised qualities include: cigarettes (Barnes, Hammond, & Glantz, 2006), pesticides (Hileman, 1998), fast foods and other manufactured foods (e.g., Weber, 2009), household cleansers (e.g., Vasil, 2007), petroleum products and petroleum-powered vehicles (Klein, 2014) and biotechnology (Kleinman, 2003). Pharmaceutical companies may compromise drug quality by, for instance, testing of 'new' drugs that have only had minor modifications against placebos and by testing them with young subjects, who are less likely to experience negative side-effects (Angell, 2004). In this light, many commodities seem to function as Trojan horses—highly attractive on the outside, but housing various hidden dangers inside. In terms of actor-network theory (ANT), commodities are—in other words—*punctualized* (Callon, 1991); that is, made to appear as isolated entities (e.g., 'smart' phone), while actually being connected to a range of (possibly problematic) ones, such as transnational corporations, poor labourers and banks (McMurtry, 2013).

Given various personal, social and/or environmental harms linked to fields of technoscience (associated with many other entities), particularly in terms of their roles in consumerism, it seems clear citizens must be educated about such possible harms, networks associated with them and possible ways to address relevant concerns. There is, indeed, potential for such consumer education. Jurisdictions around the world have—for at least the last four decades—included in curricula opportunities for students to understand (possibly problematic) relationships among fields of science and technology and societies (including financiers) and environments (STSE) (Pedretti & Nazir, 2011). It is apparent, however, that potential problems associated with STSE relationships often are treated as *controversial*. Often known as education in socioscientific *issues* (SSIs) (Sadler, 2011), students frequently are asked to consider disputed data and differing opinions of various 'stakeholders' (e.g., scientists, citizens, government officials, etc.) relating to products and services of science and technology and develop argumentative personal decisions about them. Zeidler, Sadler, Applebaum, and Callahan (2009), for example, who have had significant influences on the nature of SSI education, summarize the approach this way: "Central to this approach is the concerted effort to provide opportunities for students to reflect on issues in order to evaluate claims, analyze evidence, and assess multiple viewpoints regarding ethical issues on scientific topics through social interaction and discourse" (p. 75). On the one hand, common practices reported in research papers suggest that focus on SSIs can lead to several important learning outcomes, including those relating to: *products of science* (e.g., laws & theories) (Venville & Dawson, 2010), *socioscientific reasoning skills* (Sadler, Barab, & Scott, B, 2007) and the *natureof science* (e.g., non-linear nature of inquiry) (Khishfe & Lederman, 2006). Despite such apparent gains, relatively individualized approaches to SSI education seem limiting. According to Levinson (2010), such approaches provide either 'Deficit' or 'Deliberative' models of citizenship, both of which seem to reserve ultimate decisions about controversies to experts (e.g., scientists) and/or those with power (e.g., governments and corporations). Education of this type seems appropriate for citizenship in *representative* democracies, within which politicians promise to represent citizens' wishes through periodic campaigning for their support (Wood, 1998). However, there are concerns that people with power cannot always be trusted. Indeed, there are suggestions that members of the private sector have paid some scientists and engineers to cast doubt on (create 'issues' about) science that would reveal problems with various commercial products and services, such as cigarettes, petroleum, pesticides, and weaponry (Oreskes & Conway, 2010). Some pro-industry governments, meanwhile, seem to have taken steps to either limit such science or prevent relevant scientists from reporting results that would discredit commodities (Turner, 2013).

In light of concerns about capitalists' influences on governments and fields of technoscience, it seems that we need greater citizen vigilance and, where appropriate, actions to address their concerns (Bencze & Alsop, 2014; Hodson, 2011; dos Santos, 2009). Ralph Levinson (Levinson, 2010), in his review of socioscientific education approaches, agrees; suggesting schools need to provide students with models of citizenship involving 'Praxis' (e.g., reflective engagement) and 'Dissent and Conflict' (e.g., critique and protest). Very recently, however, there are concerns that such approaches to science education may be severely limited by fast-emerging and broadly-accepted 'STEM' (Science, Technology, Engineering & Mathematics) education initiatives. On the one hand, there appears to be much to celebrate about STEM education movements. By their very nature, in the sense of promotion of interrelationships among and/or integration of the four STEM fields, they are much more realistic than highly reductionist single-subject teaching/learning (Rennie, Venville, & Wallace, 2012). The initiatives' frequent promotion of greater inclusivity—e.g., encouraging more participation by females and people of colour—in STEM fields also seems laudable. On the other hand, several scholars suggest that STEM education is extremely reductionist—in terms of its apparent pre-occupation with identifying and training potential workers in STEM fields with the hope they will help jurisdictions (e.g., provinces/states and countries) compete in international economic markets (Bencze, Reiss, Sharma, & Weinstein, in press; Pierce, 2013). Apparently largely because of foci on jobs and economic productivity, STEM education tends to de-emphasize studies of socioscientific issues, including by limiting students' consideration of perspectives from the humanities (Zeidler, 2016) and, perhaps more importantly (given the discussion above about problems associated with politician-supported capitalism), by its tendency to present students with highly apolitical conceptions of goals, methods and outcomes of STEM fields (Gough, 2015; Hoeg, & Bencze, 2017).

Given the immense power of highly reductionist, idealized, STEM education initiatives, stopping them seems highly unlikely—and reforming them seems … *difficult*. 'Attacking' (i.e., critiquing) them also can be problematic. As Paulo Freire (1997) suggested, liberators may take on characteristics of their oppressors through dialectical relationships between them. Consequently, Michael Hardt and Antonio Negri (2009) advised that it may be better to develop alternatives. Such a tack, moreover, seems to have merit in light of Kuhn's (1970) conclusion that dominant paradigms can only be replaced (or evolved) if alternatives exist. One such alternative is discussed through the research and development project summarized below.

10.3 Research Context and Methods

Since 2006, Larry (first author here) has been supporting teachers in efforts to encourage and enable students to self-direct research-informed and negotiated action (RiNA) projects to address their concerns about socioscientific issues—such as the extent to which governments should regulate various businesses that may be contributing to problems (e.g., climate change, species losses, warfare and human diseases) for wellbeing of individuals, societies and/or environments (WISE). The basis for this work has been the 'STEPWISE' (Science & Technology Education Promoting Wellbeing for Individuals, Societies & Environments [see: www.stepwiser.ca]) theoretical framework, which organizes four teaching/learning domains (i.e., *STSE Education*, *SkillsEducation*, *Products Education* & *Students' Research*) into a tetrahedron, with reciprocal relationships among all four domains and between them and *STSE Actions*. A basic premise of this framework is that students are encouraged to 'spend' some of their cultural, social and other forms of capital (Bourdieu, 1986) on actions for common benefits—which seems antithetical to education based on neoliberal capitalism, which tends to prioritize individual possessiveness (McMurtry, 2013). The framework also is meant to be highly interdisciplinary, particularly with its emphasis on *critical* STSE education—encouraging students to consider a wide range of actants that may relate to such fields as sociology, politics, economics, history, philosophy, as well as fields of technoscience. Accordingly, STEPWISE may be considered an alternative form of STEM education.

Among teachers engaged in work with STEPWISE, Mirjan (second author here) stands out as having had considerable relevant successes (Chaps. 6 & 11, this volume; Krstovic, 2014). He has implemented—with Larry serving as 'researcher-facilitator'—STEPWISE-informed activities in at least one science course each semester (Sept.-Jan; Feb-June) since Sept. 2011. At that time, he was beginning his seventh year of teaching and was enrolled in a graduate programme leading to a Masters of Education degree (which he has now completed). During the time that he worked on the STEPWISE project, he developed considerable confidence and expertise (as well as numerous relevant instructional resources) for STEPWISE implementation.

Up until the time of the study reported here, Mirjan's previous instructional activities based on the STEPWISE framework had mainly emphasized strategies for helping students to gain expertise and confidence regarding the nature of STSE issues, primary (e.g. correlational studies) and secondary (e.g., Internet searches) research and uses of such information for developing (and, sometimes, enacting) personal and social actions to address issues. In terms of actions to take, Mirjan had

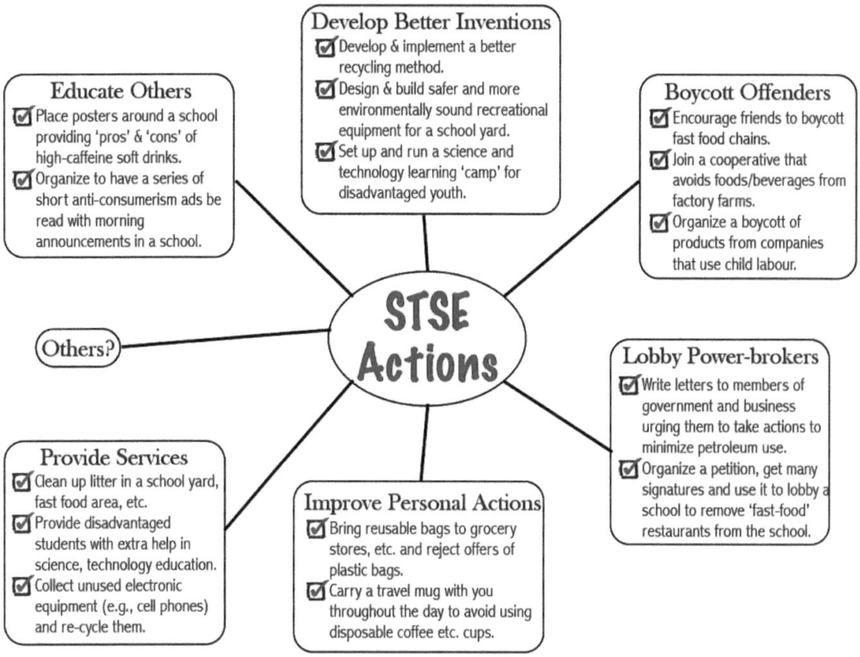

Fig. 10.2 STSE action types

shared possibilities (with relevant student and citizen sample actions) like those depicted in Fig. 10.2. For the most part, when given a choice (which students invariable were), most students chose to develop and implement the following action types: 'Improve Personal Actions' (e.g., less water use); 'Educate Others' (e.g., posters, pamphlets, YouTube™ videos, Twitter™ feeds, etc.) and 'Lobby Power-brokers' (e.g., letters to government and businesses). We have not studied reasons for their choices. Just prior to the project reported here, however, we decided that students could benefit from learning to 'Develop Better Inventions' (Fig. 10.2). Mirjan had taken a graduate course conducted by Larry that dealt with history, philosophy and sociology of science (and, to some extent, technology) and, consequently, he was familiar with relationships between fields of science and technology—as discussed above (also, refer to Fig. 10.1). In discussing future revisions to his teaching based on STEPWISE, Larry suggested that Mirjan teach students in at least one of his classes about possibilities for technology design as forms of actions on STSE issues—including teaching them some things about the *nature* of technology design. Although he supported the idea of teaching about technology design, his background was in science (Hon. B.Sc., Molecular Biology) and he had only taught school science. So, Larry shared with him an article he had written several years ago (Bencze, 2001) that described teachers' efforts to encourage and enable students to conduct technology design projects, along with several relevant 'skills education' resources (goo.gl/tPILNi). In line with our previous work, Mirjan

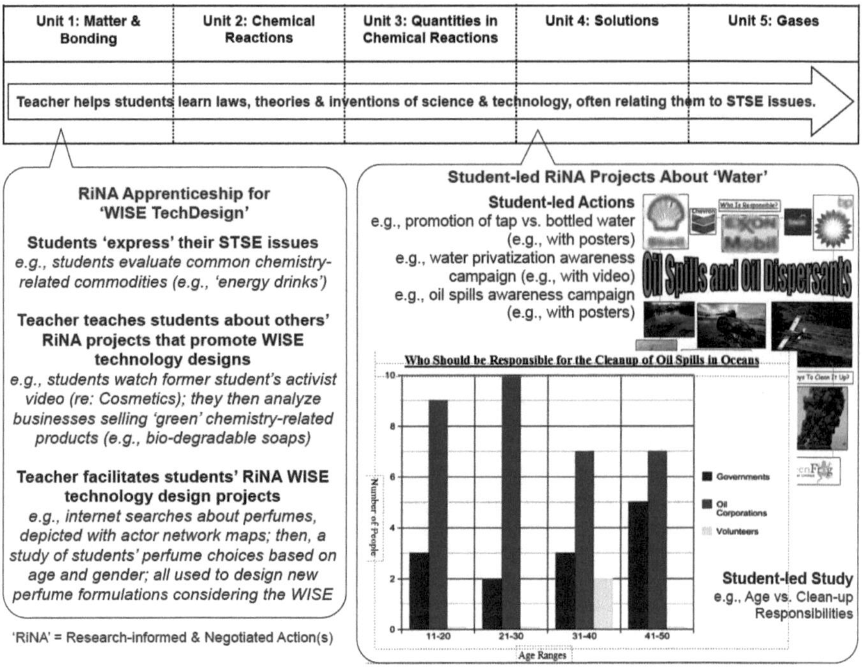

Fig. 10.3 Pedagogy for promoting student-led RiNA projects

aimed to develop instructional strategies that would help students to eventually self-direct research-informed and negotiated actions (RiNA) that may help to improve the wellbeing of individuals, societies and environments (WISE). In light of his decision to encourage and enable students to develop and implement technology designs that may promote WISE, we called the project about which we are reporting here 'WISE TechDesign.'

To implement the WISE TechDesign project, Mirjan incorporated relevant instructional strategies into the pedagogical framework he had been using for STEPWISE implementation since Sept. 2011. This framework, illustrated in Fig. 10.3, is a linear version of the tetrahedral STEPWISE framework. While it may not be as theoretically-sound as the tetrahedral version (which is more holistic), most (if not all) teachers have elected to use this more linear version (Bencze & Carter, 2011). The basic premise of this approach is that students are provided with 'apprenticeship' lessons and activities that, eventually, will give them sufficient expertise and confidence enabling them to self-direct RiNA projects to address STSE issues of their concern/interest (STSE education is the first of three goals of science curricula in Ontario [e.g., MoE, 2008], the site of this research). The apprenticeship lessons and activities are informed by constructivist learning theory, encouraging learners to first 'express' their pre-instructional attitudes, skills and knowledge (regarding STSE issues and RiNA), then be provided (by the teacher)

Fig. 10.4 A model for RiNA projects

with examples of STSE issues and RiNA projects conducted by others before conducting (with, perhaps, some teacher guidance) small-scale RiNA projects on STSE issues of their concern/interest.

Because it was the first time he had ever taught students about technology design, Mirjan chose to implement the WISE TechDesign project with a class (during the Feb-June 2014 semester) of grade 11 'academic' (university-qualifying) chemistry—some of whom had experienced STSE-RiNA apprenticeships and conducted RiNA projects when they were enrolled in a grade 10 science class taught by Mirjan. Appendix A provides outlines of his lessons and activities. His plans allowed him to introduce students to a RiNA project assignment that was to feature WISE technology design actions early in the course, with the hope that students would not be troubled by basic RiNA project concepts and, accordingly, could concentrate on learning about WISE technology designs.

Much of the focus of the apprenticeship (Fig. 10.3) can be thought of in terms of the theoretical schema for RiNA projects in Fig. 10.4. Mirjan began by first asking students, indirectly, to express their pre-conceived notions about STSE issues (World → Sign)—by asking them to evaluate some commodities linked to chemistry (e.g., 'sports' drinks, snack foods, and e-cigarettes). Students also were, however, asked to brainstorm pre-conceived notions about technology design (about revising Signs; and, about Sign → World). Mirjan then taught students about STSE issues (Signs) while also teaching them about actor-network theory (ANT) (Latour, 2005), partly through having students answer RiNA-oriented questions (World ←→ Sign) about an activist video produced by a student in his grade 10 science class the previous year. In Chap. 9, we describe earlier work with students in Mirjan's classes to use ANT with regards to RiNA projects. With their ANT-informed conceptions of STSE issues and technology design, students were then asked to analyze and evaluate some 'eco-friendly' businesses (e.g., SunOpta™ [www.sunopta.com], providing organic foods). Associated with this, he used Twitter™ to share some WISE technologies developed by others, such as cardboard furniture (www.youtube.com/

watch?v=cOa1kHEiIpg). An ensuing class discussion helped to solidify students' conceptions of STSE issues and actions in terms of actor-network theory. This enabled students to then develop actor-network maps (Signs), as recommended by Pierce (2013), to depict a chemistry-related commodity (technology) of their interest. They were then expected to brainstorm possible positive and negative effects of the commodity on WISE (World ←→ Sign) and, based on that, propose an alternative technology design (revised Signs). At this point, Mirjan provided students with a summary of some major points about STSE issues and RiNA (World ←→ Sign). Students were then given the assignment of developing a technology/commodity that took into consideration various aspects of WISE (revised Signs → World). As part of this process, they also were asked to conduct some secondary (learning Signs) and primary (World → Sign) research about their commodities. Based on their findings, they were asked to revise their technology designs/commodities (Signs) and then defend them in presentations to panels of 'experts' (other teachers, senior students, and Larry) (revised Signs → World).

In light of the above arguments and strategies, our research goals were to document and explain the nature and extent of students' orientations towards WISE technology designs as forms of action to address socioscientific issues of their interest. To achieve that research agenda, we conducted data-collection and analyses having *rationalistic* and *naturalistic* characteristics (Guba & Lincoln, 2011). Rationalistically, we focused, for example, on students' orientations towards accommodating WISE considerations into their technology design projects. Naturalistically, we collected data that enabled emergence of unexpected situational outcomes. Data collected from students (ages 16–17) and Mirjan included: Project Work Artefacts: Samples of products generated by most students (28) were collected, including: issue descriptions, research plans, data collected, written reports, project reflections, action plans and forms of action (e.g., design plans, posters, videos); Project Instructional Materials: Copies of all of Mirjan's pedagogical plans and instructional materials (e.g., handouts, videos, PowerPoint™ presentations, and internet site web addresses) were made; Digital Recordings of Students' Project Work: Photographs and videos were produced depicting youth presenting/defending their actions in public fora (e.g., to fellow students within and outside of class); Semi-structured Interviews: Five students were interviewed twice, near the beginning and end of the course. Questions focused on their views about issues, research and actions. Mirjan was interviewed 11 times (~ 60 min.) about project progress. All interviews were audio-recorded and later transcribed. Regarding analyses, each of us coded data for categories and themes—using constant comparative methods based on constructivist grounded theory (Charmaz, 2014). Categories and themes were then negotiated between us. Member checks with participants were conducted to help ensure *trustworthiness* of claims, each of which was based on at least three supporting data sources.

10.4 Results and Discussion

10.4.1 Outcomes

Mirjan's first effort at encouraging school science students to design technologies/commodities that accommodated WISE considerations seemed relatively successful. While only one of the nine student groups (3–4 students/group) *implemented* their designs (Sign → World, Fig. 10.4), the rest successfully *designed* technologies (revised Signs, Fig. 10.4) that considered various aspects of WISE; including: a cologne for males; deodorants for males and females (2 groups); sustainable road deicer; and, a board game for WISE choices about batteries.

Perhaps the most significant characteristic of students' technology designs was that they were *networked*. While they understood 'technical' aspects of technology design, such as that changing one ('independent') variable may lead to both positive and negative effects on desirable 'result' (dependent) variables, their designs also transcended physical products/services, encompassing a range of actants. Consider, for instance, design features used by the group that created a new men's cologne, based—to a great extent—on their actor-network map for cologne, given in Fig. 10.5:

> The cologne itself is made from primarily pure and natural ingredients these include: Bergamot essential oil, cedar wood essential oil, lemon, cinnamon sticks, green tea leaves, and absolute vodka. Many of our ingredients, if not grown in Canada do not contain harsh chemicals so the people working in other countries to harvest cinnamon for example, would not be exposed to dangerous fumes. … [T]he production process of cork is less harmful to the environment then making a glass or plastic lid, the cork is stripped off the trunk of the tree every 9 or 10 years, this does not kill the tree. … Our product would appear more attractive to a consumer because they are able to read and identify all seven ingredients and maybe even have majority of them in their household (March 21, 2014).

A particularly salient feature of all student groups' networks appeared to be awareness that many commodities were produced by poor people working under inadequate labour conditions in far-away places:

> [I]f you have a product and you are producing it in a Third World country, where you are not giving people the proper amount of pay and they are living in a low ['destitute'] place, then that society is not doing well – and that is partly due to your product … [and] … [W]e strongly believe in fair wages, unlike other manufacturing plants around the world, and will only employ legal and adult workers. … What makes our product more attractive to a consumer than the original antiperspirants is that it hides nothing behind closed doors. … [I]t is a 'what you see is what you get' type of product … (Antiperspirants Group, March 21, 2014).

Prior to instructional activities in this unit, when they were asked about common commodities, they made statements like: "It's okay to eat processed foods (snacks, fast food) because companies are always finding ways to make them more nutritious (e.g. adding vitamins)."

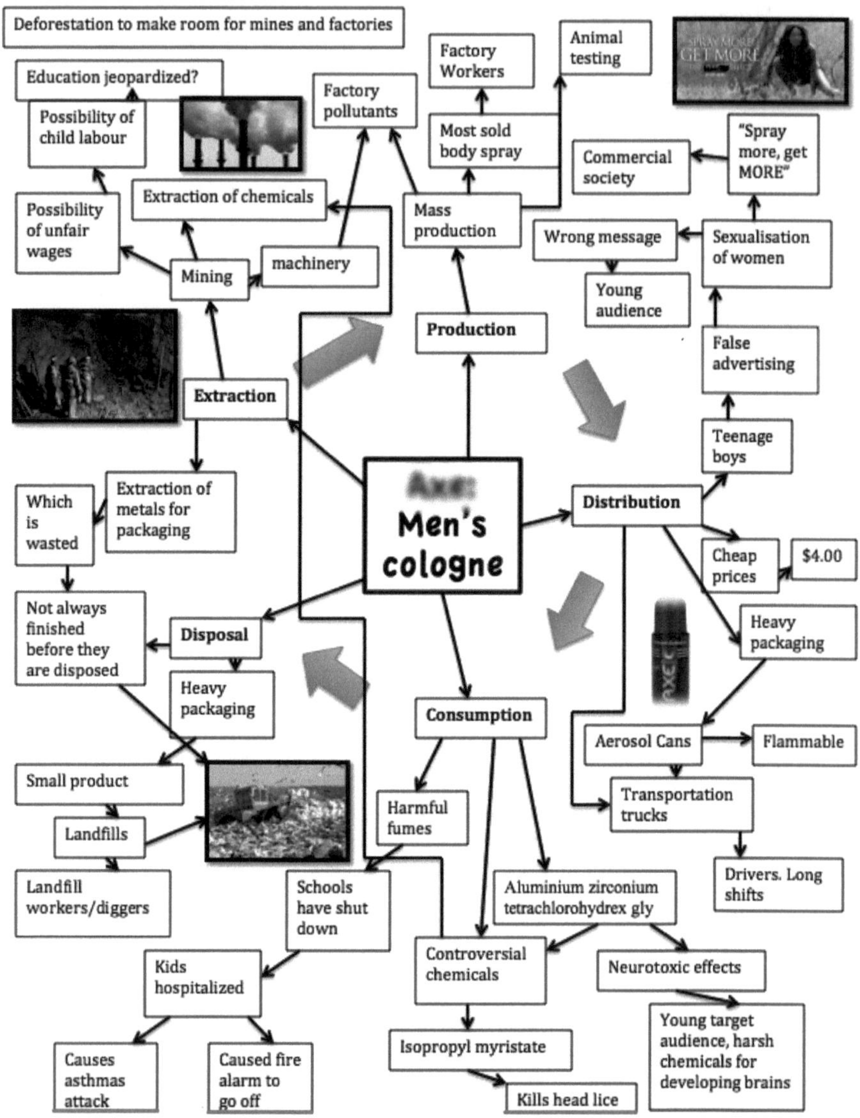

Fig. 10.5 Students' actor-network map for 'Cologne'

The significance of students' designs can largely be understood in terms of the schematic in Fig. 10.4. In expressing their pre-conceived notions of commodities, especially those that became the subject of the technology designs (e.g., cologne) and conducting research (secondary and primary), students were able to negotiate representations (e.g., Signs, like the network map in Fig. 10.5) of commodities, which were—to a great extent—also influenced by their *anticipation* of effects these commodities may have on WISE (e.g., use of aerosol cans vs. glass bottles for environmental wellbeing). Although most students may not have brought about better Worlds that they perhaps hoped with their designs, they *imagined* them (Sign → World is imagined). Perhaps the most significant feature of their designs, however, was that they had de-*punctualized* the commodities being re-invented. Punctualization is a process by which a complex actor-network is 'black-boxed; that is, made to appear as a single actant, while still being dynamically connected to many other actants (Callon, 1991). Often, for-profit commodities are punctualized and, with marketing, appear—like a Trojan horse—positive on the outside (e.g., sexy') while harbouring (when black-boxed) potentially-harmful actants on the inside (e.g., unfair labour practices). De-punctualization exposes complex networks of which apparently single actants are composed and connected—such as those students suggested were connected to a cologne they studied (Fig. 10.5). In terms of the schema in Fig. 10.4, the much greater complexity and 'critical authenticity' (Pierce, 2013) of students' Signs and *imagined* Worlds suggests that the ideological gaps in (mostly imagined) translations from Sign to World are much more ethical than often is the case when capitalists influence such translations (as discussed above, under Theoretical Background).

In developing research-informed and negotiated 'WISE' technology designs as socio-political actions, it is apparent that most of the students in this study achieved Levinson's (2010) 'Science education as praxis' and 'Dissent and Conflict' levels of citizen engagement, which appears to surpass many current programmes promoting interactions with socioscientific issues (Gough, 2015; Levinson, 2010; Pedretti & Nazir, 2011; Zeidler, 2016).

10.5 Contributing Factors

In analyzing data, it was apparent that—like many teaching/learning contexts—myriad factors likely influenced results like those reported above. In terms of actor-network theory (Latour, 2005), no one actant can be identified as an individual factor contributing to outcomes reported above. To do so would be to punctualize networks of which the above (and other) outcomes are a part. Using an ANT lens, one way of explaining the above results is in terms of *dispositifs*; that is, Michel Foucault's (2008) concept of "coupling of a set of practices and a regime of truth [that] form an apparatus" (p. 19). A dispositif is, in other words, a set of actants that are aligned to achieve particular purposes—such as, in this case, student orientations towards envisaging technology designs/commodities that prioritize, in addition to their functionality, wellbeing of individuals, societies and environments. Ironically, a dispositif could, like any network, become punctualized/black-boxed, if not analyzed/de-constructed. Punctualization can be problematic if, as conditions change, people are unaware of internal workings of the apparatus. Accordingly, there may be some merit in analyzing/de-punctualizing the dispositif that appeared to contribute to students' orientations towards WISE commodity designs.

It is difficult to know where to start when discussing a dispositif, which is considered a dynamic and networked 'phenomenon.' However, as has been reported elsewhere, a starting point for this work appears to be Ontario's curriculum for secondary school science. Although the STEPWISE framework is not acknowledged by Ontario curricula (e.g., MoE, 2008), all of the major curricular elements are addressed by the framework. Perhaps most notably, STEPWISE prioritizes STSE education and actions. This may account for considerable success of the dispositif noted here. Some teachers, including Mirjan (Krstovic, 2014), seemed to gain a sense of 'freedom' to enact STSE-based education after the STSE learning domain was listed as the first overall curriculum goal in 2008 (Bencze, Sperling, & Carter, 2012), a movement that was initiated in Canada about two decades earlier (CMEC, 1997). On the other hand, it is clear that not many science teachers in Ontario are still attending to STSE education to the degree allowed by/promoted in curricula (Hodson, 2011; Pedretti & Nazir, 2011). Therefore, some other factors may be involved in the dispositif explored here.

It seems quite clear that student achievements regarding WISE technology/commodity designs like those reported above were possible, to a great extent, because of characteristics of their teacher (Mirjan). He appears to hold views about the nature of science (and technology), for instance, that seem to align with more critical and comprehensive conceptions of STSE relationships that became apart of students' designs (e.g., Fig. 10.5). Through our various interactions (Bencze & Krstovic, Chap. 11, this volume), it is clear that Mirjan relatively strongly supports Naturalist-Antirealist positions on Loving's (1991) *Scientific Theory Profile*—positions that contrast with Rationalist-Realist positions and, for instance, acknowledge that scientists' decisions are not only based on logic and evidence, but may also include influences from various internal and external factors, such as the profit motive. Sources of such views, although complex, seem to include his course work in graduate studies leading to a Masters of Education degree (which he now holds)—one of which was a course (taught by Larry) exploring ideas from history, philosophy and sociology of science. It was, indeed, at the end of this course (taught online) when Mirjan requested participation in action research led by Larry (whom he had not met in person at that time). Although evidence suggests Mirjan already was an innovative reflective practitioner (Bencze & Krstovic, Chap. 11, this volume), he seemed particularly motivated to pursue research in STSE education after this course. With our collaboration, Mirjan has developed considerable expertise and confidence for enacting teaching and learning approaches based on STEPWISE (Krstovic, 2014). At the time of implementing the WISE TechDesign project reported here, he had developed and enacting STEPWISE-informed teaching/learning materials in five previous semesters (with at least five different classes of students). Not only did Mirjan have such expertise and confidence, however. In successive interviews (March 21 & June 10, 2014), students in the class under study here indicated they already had some expertise/confidence—given that they had taken a course the previous year taught by Mirjan that involved research-informed and negotiated action projects. Moreover, due to his persistent and relatively obvious involvement in promotion of RiNA projects, including in collaborating with teaching colleagues in his school, teachers and members of administration and support staff in Mirjan's school had become relatively accepting of his efforts in this regard (Bencze & Krstovic, Chap. 11, this volume).

With the above actants aligned, it appears that Mirjan was then able to implement pedagogical practices that relatively effectively enabled students in his eleventh-grade chemistry class to develop research-informed and negotiated technology designs as actions that may contribute to improvements in wellbeing of individuals, societies and/or environments. As with our previous explorations regarding STEPWISE (e.g., Bencze et al., 2012; Krstovic, 2014), the constructivism-informed apprenticeship (Fig. 10.3)—which provides exemplars and guided practice—appeared to be an important contributing factor to students' general understanding of STSE relationships and expertise, confidence and motivation for RiNA projects.

The focus of this chapter, however, deals more specifically with students' implementation of technology designs that took into consideration not only their functionality but, moreover, aspects of wellbeing of individuals, societies and environments. An important, fundamental, outcome of the apprenticeship seemed to be students' fairly-effective understanding of general processes of technology design. In interviews with students during the time that they were learning about technology design and WISE considerations, many of them demonstrated considerable expertise in this regard. A key concept they all seemed to understand was that changes in one variable may be insufficient due to effects of other variables. When discussing with them design of a better pen, for instance, one student said: "If you only have enough funds to make the product environmentally friendly, maybe you are going to have to sacrifice some of the quality. Maybe the smudging [of the ink] will be a problem" (March 21, 2014). At the same time, such understanding of complexities (e.g., compromises in) technology design seemed, perhaps more importantly, related to students' developing applied understandings of different kinds of actants associated with commodities (refer above). When asked about design features of lined paper that happened to be nearby during an interview, students noted numerous actants relating to paper: "Going back to WISE, it came from trees; how they managed to get the ink; a group of people that had to produce it; the machines that cut the paper; the idea of how much space people need to record their ideas; the paper could be recycled; it's complementary to the binder [in which it may be stored]." Moreover, these conceptions seemed to be clearly linked to the metaphor of the Trojan horse often associated with commodities. In explaining the cologne they were analyzing and revising, for instance, a group member said:

> This product could be compared to the Trojan horse because from the outside it looks like a sexy, effective product that will attract many pretty women but we don't know what harmful chemicals are on the inside. For example, I doubt that everyone who sprays the cologne considers the allergic reactions it could cause or the factory workers, miners and landfill workers that are all involved in the product life cycle of axe. The "good" smell masks the negative side effects of the product and also creates a very unrealistic, even toxic, expectation of cologne for young boys (*Is it Toxic?*) (March 31, 2014).

When asked how they came to understand commodities in such networked ways, students named two key aspects of the apprenticeship. The first of these were videos that Mirjan used to teach about commodities through ANT. During the WISE TechDesign apprenticeship, students were soon exposed to the Story of Stuff (storyofstuff.org) video, which uses the Trojan horse metaphor (without naming it) in depicting a number of actant types. Associated with this, students also said they found the activist video (www.youtube.com/watch?v=WhN6PS1GT9c)—using the Trojan horse metaphor in depicting 'foundation' cosmetics—prepared by one of Mirjan's former students helpful. However, hearing about ANT analyses of commodities from others was not enough. Students also mentioned the importance of analyses of 'green companies' Mirjan asked them to conduct. One student, for instance, said: "I found the [analyses] of the eco-friendly businesses [to] be very helpful, as it gave me ideas about my product and how to endorse it. It also showed me how many negatives can disappear [be black-boxed] through simple manageable ways" (April 3, 2014). Their analyses of 'green' companies, indeed, indicated significant understanding of ANT and the Trojan horse metaphor, one example of which was:

> Green Beaver [www.greenbeaver.com] does not have many pollution related actants because majority of their ingredients are organically based so as a result no heavy machinery is used in the initial stage of extraction. … The advertisement of this company or one of the strong semiotic messages is "natural = healthy". And the use of 'mother nature' rather then harmful chemicals" (Student, speaking for her group of 4, March 3, 2014).

Finally, although students did not mention it, it seems logical to assume that Mirjan's decision to visit local eco-friendly businesses and send Twitter™ feeds of examples of sustainable products to students would help—demonstrating that he, too, is committed to WISE technology/commodity design. Examples of such 'green' commodities—in this case, from *Planet Organic* (planetorganic.ca)—are given in Fig. 10.6.

Not only did the activities in the WISE TechDesign apprenticeship seem to help students to understand commodities as part of networks, it also seemed to help them develop emotional commitments to acting to address relevant concerns exposed through such de-punctualization. 'Cary,' for example, said:

> I think people purposely forget [about hidden actants]. If they are using a product they know is made in a Third World country where they are not being paid [well] and stuff, and if they really like the product and they want to keep using it, they kind-of pretend they didn't hear [about the poor workers] [They might say,] "It's not happening in our backyard" (March 21, 2014).

To this, she added: "Some companies are not going to say, 'People in poor countries make our shoes!" Her group member, 'Matthew,' concurred: "They only promote the good stuff around it." In such de-punctualization work, students seemed to become particularly concerned for workers in poor countries labouring under poor working conditions: "A lot of companies are being more environmentally friendly, but they are still using those factories that have unsafe conditions and unfair wages. So, is it worth it to buy it [products]? You are still causing harm to people" ('Mandy,' March 21, 2014). With such realizations, students tended to be very critical of companies—and capitalism, more generally. Apparently speaking for his group, 'Kerry' said:

Fig. 10.6 Sample 'Green' commodities 'Tweeted' by Mirjan

> A lot of technologies these days are just focusing on the individuals, not the society and environments; because, if you [manufacturer] affect the society and environment, it doesn't really affect your pay [profit]. You [a company] really just want to worry about the consumer, if they are going to buy it [products] or not (March 21, 2014)

To this, he added: "Businesses don't really care about the environment and stuff [social justice?]; they just care about profit" (March 21, 2014). His group member, 'Margaret,' later concurred: "Everything is about money" (June 10, 2014). Indeed, students' motivation to act seemed to largely arise from a sense of *subterfuge* they felt about companies:

> Technology is great in many ways; it has saved lives, time, granted "convenience", and has started to become the fuel powering our world. This mainly applies to first world countries.

> We have a tendency to forget the third world countries that are engulfed by the negative effects of technology without getting the chance or opportunity to use it as other societies do. For example: child labour, pollution, waste; we do not see it much in our society yet it surrounds others in their day to day lives. … To some, it's their life, their future; and the key to innovation. To others, it's the thing responsible for countless storms, floods, it could be the things that claimed their child's life while mining. … To be brutally honest, I am surrounded by technology mainly by choice due to the way it is portrayed in media. I fall victim to the idea that after a phone is a year old, it is useless and "behind". I always want the newest technology and tend to waste a great amount of it. Hardly do I ever consider the work that went into making that piece of technology. In a way, it is almost like the phrase, "ignorance is bliss". I felt rather disturbed when I learned more about the industry that I along with almost all teens in North America support. … In the future, technology should take into consideration the working conditions of the people producing it in the different stages and the environmental consequences such as pollution" ('Erin,' March 31, 2014).

Such realizations seemed to move students to want to educate others about companies' trickery. 'Cary' reflected this well when he said: "You have to do this [promote WISE commodities] in small steps. You can't change everything at once. I think you need to start small and work your way up. … If everyone is aware of this kind of stuff [WISE issues], then corporations will aim to do that [appeal to people's WISE interests], because that is what will raise their profits" (March 21, 2014). Related to this, as they said to Larry in the final interview (June 10, 2014), they recognized that their actions had to be networked. Indeed, as indicated above, their technology/commodity designs did attempt to accommodate effects on multiple actants—including many problematic ones, such as poor labourers in distant countries, that are normally hidden.

10.6 Summary and Conclusions

In light of the apparent power of what appear to be capitalist-friendly STEM education movements, it seems that opportunities for citizens to gain access to critical perspectives about relationships among fields of science and technology and societies and environments (informed by history, sociology, politics, economics, etc.), expertise for critically analysing/evaluating them and expertise and motivation for addressing them seem to be significantly compromised. In this chapter, we suggest that promotion of research-informed and negotiated technology designs (along with other actions) to address apparent socio-scientific problems linked to commodities familiar to youth may be one viable alternative to apparent capitalist-friendly STEM education movements. Indeed, the focus here on encouraging and enabling students to develop designs for commodities that consider possible wellbeing of individuals, societies and environments (WISE) seems very necessary—given capitalists' current emphases on encouraging relatively rich people in the world to repeatedly consume and discard for-profit products and services, often at the expense of wellbeing of individuals, societies and environments (often in the poorest parts of the world).

In the study reported here, it was apparent that eleventh-grade chemistry students developed critical understandings of and commitments towards STSE relationships involving familiar commodities and, based on their secondary and primary research, developed technology designs for commodities that they felt would be functional but also may improve wellbeing of individuals, societies and environments. Such outcomes appeared to be made possible because of development (including elements prior to the study reported here) of a *dispositif*—i.e., a network of actants aligned to support a common purpose. Fundamental to this dispositif was, apparently, Ontario's curriculum prioritization of STSE education, research/inquiry and, to some extent, activism. The teacher promoting WISE technology designs, however, also was key to the programme's success, having various congruent epistemological and political commitments, experience with RiNA project work, and having influenced his immediate milieu to support work of this type. Such findings are, however, not particularly new. We have previously seen these factors influence student research-informed and negotiated actions. In this study, however, there was evidence to suggest that a combination of use of actor-network theory (aided by various media, such as *The Story of Stuff* videos) and analyses of examples of sustainable technologies/commodities greatly assisted students in developing outcomes like those described above. Although the dispositif apparently responsible for these outcomes cannot ever be repeated, this study suggests that alternatives to STEM education that prioritize WISE are possible. From a Kuhnian (1970) perspective, however, it may be that use of such alternatives to this hegemonic system may require one or more 'crises' with STEM education—perhaps in the form of efforts to expose its apparently problematic aspects—along with offering educators viable alternatives.

Appendix A

Grade 11 University Chemistry

Unit 1: WISE-Technology Design Project

Introduction

The main purpose of this project is for students to propose plans (and possibly implement and develop prototypes) for technology that considers the well-being of individuals, societies and environments (WISE). You will learn about the relationships between science and technology as you explore effects on WISE of some common, but potential hazardous, household chemicals.

Task You will propose WISE-Technology designs that will lessen the harmfulness of various chemical products that we find at home, school or workplace. Your creative ideas can be put to a test and your WISE inventions may contribute to a better world.

Part I – Expressing Your Pre-conceived Ideas

Four Corners Activity

Consider the following statements and decide if you SOMEWHAT AGREE, STRONGLY AGREE, SOMEWHAT DISAGREE or STRONGLY DISAGREE.

Statement 1 *"Sports drinks are unnecessary and counterproductive for most people."*

Statement 2 *"It's okay to eat processed foods (snacks, fast food) because companies are always finding ways to make them more nutritious (e.g. adding vitamins)."*

Academic Controversy

In groups of 4–5 students, you will be assigned to either the PRO or CON side of an argument. The teacher will guide you through the academic controversy surrounding the following statement:

E-cigarettes, a new form of technology, should be banned

Reflection Journaling

In your reflective journal, write about your perspective regarding the nature of technology design. Discuss what technology is and what kind of things could be considered technology? What are some important considerations when making new technologies?

Part II – WISE Tech Design

In Part I of WISE Tech Design project you expressed your views on several controversial socio-scientific issues such as sports drinks, processed foods and electronic cigarettes, all which could be considered forms of technology. You reflected in your journal on a broader view technology and important considerations that need to be made when designing new technology.

For Part II of WISE Tech Design, we will discuss various types of *actants* involved in creating new technologies and considerations we need to make when designing alternative 'WISE' technology.

Video Group Discussion

The Life Story of Foundation

http://www.youtube.com/watch?v=WhN6PS1GT9cv by a grade 10 student from last year

Watch the video above and in your group discuss the following:

(i) What are the stages in the life cycle of a product?
(ii) What are three types of *actants* involved in each stage of the life cycle of foundation? *(Actants* can be *material* (living and non-livings things) and semiotic messages (e.g., cool, pretty, sexy, etc)
(iii) Considering cosmetics as an example of technology, which *actants* can align to support common *semiotic messages* about cosmetics.
(iv) What new actants can be introduced to change dominant semiotic messages about cosmetics/foundation?

Eco-friendly Businesses

Your tutorial group (TG) will be assigned ONE of the top ten eco-friendly businesses. You will visit their website and report of the following. Individual member of your TGs has to answer all of these questions. As a group you will discuss your answers.

(i) What does the company specialize in and what do they value? Check out the 'About Us' section of their website (100–150 words)
(ii) What specific Products and/or Services do they provide? (50 words)
(iii) How are their products and/services examples of technology that considers WISE? Discuss in terms of various *actants* that this company considers to make the new product. (150 words)

WISE Technology Design – Part III

Actants and Eco-friendly Businesses Group/Class Discussion

Discuss with your group the following points and be prepared to share with the rest of the class:

1. Brainstorm a list of all the actants that your group members identified (living, non-living and semiotic) that were explicitly (or implicitly) mentioned in Cathy's video. Use the chart paper provided.
2. Circle actants that align to support a dominant semiotic message about cosmetics/liquid foundation.
3. Suggest new actants that can be introduced to change the dominant semiotic message.
4. Share the eco-friendly business that you researched with your group members. Discuss how are their products and/services examples of technology that considers WISE. Discuss in terms of various *actants* that this company considers to make the new product.
5. Be prepared to share 1–4 above with the rest of the class during whole group discussion.

Actor-Network Map

In the past, and perhaps still in the present, chemists/technologists often develop products without considering their full effects on the well-being of individuals, societies and environments. The practice of 'WISE Technology' involves the inventions, design and use of products and processes that have minimal personal, social and environmental impacts.

You and your group will pick **one everyday product** (cleaning, personal hygiene, beauty, medical, etc) and you will work individually to create an actor-network map showing as many actants involved in each stage of the life cycle of this product. Begin by brainstorming what you already know with your members, before you begin your independent research.

By next week Monday (March 3, 2014) you need to have your Actor-Network Maps completed with references showing where you got your information.

Critical Analysis of the Technology

Analyze and critique the technology for which you created the actor-network map. Consider the following questions for your critical analysis:

1. To what extent does your selected product ensure sustainability and safety? In other words, does it use renewable materials, are toxic chemical used and produced and how is the health of individuals and environments impacted? (200 words)
2. How might the product be like Trojan Horse? Consider the actants from your Actor-Network Map in your analysis. (100 words)
3. Propose an alternative technology design that not only looks at functionality but also considers other actants (not only the chemicals inside the product). What are the advantages and some disadvantages of the new, alternative technology that considers WISE? (250 words)

Your critical analysis should be double space and 12 point font. It should include at least two-three references in proper APA format. It is due Monday, March 3, 2014.

WISE Technology Design – Part IV

Summary of What We Accomplished

We started the project by expressing our views about some controversial technologies such as sports drinks, processed foods, e-cigarettes, etc. You watched several green technology videos that I posted to Twitter™ (@MrKrstovic) to get an idea of what may constitute sustainable technology development. However, just because technology is eco-friendly does not mean that it considers issues of social justice (e.g., fair labour practices). In your technology designs, you need to be mindful of personal, social and environmental issues.

We watched Cathy's video about Liquid Foundation/Cosmetics and identified various actants involved in each stage of in the life cycle of Liquid Foundation. We discussed how dominant semiotic messages about cosmetics and beauty can be changed by inserting new actants. You learned that actants can be living, non-living or semiotic. You explored one eco-friendly business and discussed how it considers the well-being of individuals, societies and environments.

Finally, in Part III of WISE Technology Design, you selected one common, but potentially hazardous, household product and created an actor-network map. You were expected to research various actants related to your selected product. You wrote a critical analysis about this product and proposed an alternative with overview of advantages and disadvantages.

What's Next?

Focus Group Discussion:

You and your group will compare your actor-network maps and focus your discussion on your answers to the critical analysis questions.

Developing Alternative Technology Designs That Consider WISE

You will plan, develop, design and market a product that considers WISE. You will be purposeful in your design. Your product will illustrate your group's commitment to the well-being of individuals, societies and environments. You will put the product to the ultimate test: convincing investors to support its production.

Secondary Research

In your group, you may need to conduct more research to learn more about

- The principles of green chemistry
- The principles of social justice as it relates to technological production
- What makes technology products that consider WISE appealing to investors and consumers

Primary Research

You will need to conduct some original market research (e.g. market surveys) on your product prior to defending your WISE tech design.

Describe you design briefly to a selection of potential shoppers (e.g. students, parents/teachers) and judge responses in terms of gender, age, approximate family income, etc. Your survey should include no more than five questions. Choose one dependent variable (either gender, age, or other). Gather, analyze and interpret the data prior to your product defence presentation.

Identify Solutions and Make Decisions

Now you will design an alternative to the conventional product that you selected. You may challenge yourself and be creative with your WISE Technology designs. For example, you may design a board game, or a computer game, which when played allows consumers to learn about alternatives to your selected product while also becoming conscious about many of the actants related to the conventional product.

The 'WISE' Product

Develop explanatory responses to the following questions:

- How does your product and the production process consider WISE? (250 words)
- What makes your product more attractive to a consumer? (100 words)

Dragons' Den!

You may have seen the CBC television program *Dragons'Den,* in which investors, entrepreneurs, and product designers try to sell their ideas to a panel of potential investors. You will take turn in the dragons' den! In your group, plan and present your 'pitch' to a panel of investors (selected classmates, teachers, admin, etc). Your pitch must include

- Attention grabbing introduction
- Information about your 'WISE' product (contrasted with the conventional options)
- A clear and concise evaluation of your products considerations for WISE
- A wrap-up to convince the investors to back your product/design

After listening to your pitch, the investors will confer and give their answer.

References

Angell, M. (2004). *The truth about the drug companies: How they deceive us and what to do about it*. New York: Random House.

Barber, B. R. (2007). *Consumed: How markets corrupt children, infantilize adults, and swallow citizens whole*. New York: Norton.

Barnes, R. L., Hammond, S. K., & Glantz, S. A. (2006). The tobacco industry's role in the 16 Cities Study of Secondhand Tobacco Smoke: Do the data support the stated conclusions? *Environmental Health Perspectives, 114*(12), 1890–1897.

Baudrillard, J. (1998). *The consumer society*. London: Sage.

Bencze, J. L. (2001). 'Technoscience' education: Empowering citizens against the tyranny of school science. *International Journal of Technology and Design Education, 11*(3), 273–298.

Bencze, J. L., & Alsop, S. (Eds.). (2014). *Activist science & technology education*. Dordrecht, The Netherlands: Springer.

Bencze, L., & Carter, L. (2011). Globalizing students acting for the common good. *Journal of Research in Science Teaching, 48*(6), 648–669.
Bencze, L., Reiss, M., Sharma, A., & Weinstein, M. (in press). STEM education as 'Trojan horse': Deconstructed and reinvented for all. In L. Bryan & K. Tobin (Eds.), *Thirteen questions in science education* (pp. xx–xx). New York: Peter Lang.
Bencze, L., Sperling, E., & Carter, L. (2012). Students' research-informed socioscientific activism: Re/Visions for a sustainable future. *Research in Science Education, 42*(1), 129–148.
Bourdieu, P. (1986). The forms of capital. In J. G. Richardson (Ed.), *The handbook of theory: Research for the sociology of education* (pp. 241–258). New York: Greenwood Press.
Callon, M. (1991). Techno-economic networks and irreversibility. In J. Law (Ed.), *A sociology of monsters: Essays on power, technology and domination* (pp. 132–161). London: Routledge.
Charmaz, K. (2014). *Constructing grounded theory* (2nd ed.). New York: Sage.
Council of Ministers of Education, Canada [CMEC]. (1997). *Common framework of science learning outcomes K-12*. Ottawa, Canada: CMEC.
dos Santos, W. L. P. (2009). Scientific literacy: A Freirean perspective as a radical view of humanistic science education. *Science Education, 93*(2), 361–382.
Foucault, M. (2008). *The birth of biopolitics: Lectures at the Collége de France, 1978–1979* (M. Senellart, Ed.). New York: Palgrave Macmillan.
Freire, P. (1997). *Pedagogy of the oppressed* (New Revised 20th-Anniversay ed.). New York: Continuum.
Gough, A. (2015). STEM policy and science education: Scientistic curriculum and sociopolitical silences. *Cultural Studies of Science Education, 10*(2), 445–458.
Guba, E. G., & Lincoln, Y. S. (2011). Competing paradigms in qualitative research. In N. K. Denzin & Y. S. Lincoln (Eds.), *The SAGE handbook of qualitative research* (pp. 105–117). Thousand Oaks, CA: Sage.
Hardt, M., & Negri, A. (2009). *Commonwealth*. Cambridge: Belknap Press of Harvard University Press.
Hileman, B. (1998, August 17). Industry's privacy rights: Is science shortchanged? *Chemical & Engineering News, 76*, 36.
Hodson, D. (2011). *Looking to the future: Building a curriculum for social activism*. Rotterdam: Sense.
Hoeg, D., & Bencze, L. (2017). Values underpinning STEM education in the USA: An analysis of the Next Generation Science Standards. *Science Education, 101*(2), 278–301.
Khishfe, R., & Lederman, N. G. (2006). Teaching nature of science within a controversial topic: Integrated versus nonintegrated. *Journal of Research in Science Teaching, 43*(4), 395–418.
Klein, N. (2014). *This changes everything: Capitalism vs. the climate*. Toronto, ON: Simon & Schuster.
Kleinman, D. L. (2003). *Impure cultures: University biology and the world of commerce*. Madison, WI: University of Wisconsin Press.
Krstovic, M. (2014). Preparing students for self-directed research-informed actions on socio-scientific issues. In L. Bencze & S. Alsop (Eds.), *Activist science and technology education* (pp. 399–417). Dordrecht, The Netherlands: Springer.
Kuhn, T. S. (1970). *The structure of scientific revolutions* (2nd ed.). Chicago: University of Chicago Press.
Latour, B. (2005). *Reassembling the social: An introduction to actor-network-theory*. Oxford, UK: Oxford University Press.
Leonard, A. (2010). *The Story of Stuff: How our obsession with stuff is trashing the planet, our communities, and our health – and a vision for change*. New York: Free Press.
Levinson, R. (2010). Science education and democratic participation: An uneasy congruence? *Studies in Science Education, 46*(1), 69–119.
Loving, C. C. (1991). The Scientific Theory Profile: A philosophy of science model for science teachers. *Journal of Research in Science Teaching, 28*(9), 823–838.
McMurtry, J. (2013). *The cancer stage of capitalism: From crisis to cure*. London: Pluto.

Ministry of Education [MoE]. (2008). *The Ontario curriculum, grades 9 and 10: Science*. Toronto, ON: Queen's Printer for Ontario.

Mirowski, P. (2011). *Science-mart: Privatizing American science*. Cambridge, MA: Harvard University Press.

Oreskes, N., & Conway, E. (2010). *Merchants of doubt*. London: Bloomsbury Press.

Pedretti, E., & Nazir, J. (2011). Currents in STSE education: Mapping a complex field, 40 years on. *Science Education, 95*(4), 601–626.

Pierce, C. (2013). *Education in the age of biocapitalism: Optimizing educational life for a flat world*. New York: Palgrave Macmillan.

Piketty, T. (2014). *Capital in the twenty-first century*. Cambridge, MA: Cambridge University Press.

Rennie, L., Venville, G., & Wallace, J. (Eds.). (2012). *Integrating science, technology, engineering and mathematics: Issues, reflections and ways forward*. New York: Routledge.

Roth, W.-M. (2001). Learning science through technological design. *Journal of Research in Science Teaching, 38*(7), 768–790.

Sadler, T. (Ed.). (2011). *Socio-scientific issues in the classroom: Teaching, learning and trends*. Dordrecht, The Netherlands: Springer.

Sadler, T. D., Barab, S. A., & Scott, B. (2007). What do students gain by engaging in socioscientific inquiry? *Research in Science Education, 37*(4), 371–391.

Sismondo, S. (2008). Science and technology studies and an engaged program. In E. J. Hackett, O. Amsterdamska, M. Lynch, & J. Wajcman (Eds.), *The handbook of science and technology studies* (3rd ed., pp. 13–31). Cambridge, MA: The MIT Press.

Turner, C. (2013). *The war on science: Muzzling scientists and wilful blindness in Stephen Harper's Canada*. Vancouver, Canada: Greystone.

Vasil, A. (2007). *Ecoholic: Your guide to the most environmentally friendly information, products and services in Canada*. Toronto, ON: Vintage.

Venville, G. J., & Dawson, V. M. (2010). The impact of an argumentation intervention on Grade 10 students' conceptual understanding of genetics. *Journal of Research in Science Teaching, 48*(8), 952–977.

Weber, K. (Ed.). (2009). *Food Inc.: How industrial food is making us sicker, fatter, and poorer – and what you can do about it, A participant media guide*. New York: Public Affairs.

Wood, G. H. (1998). Democracy and the curriculum. In L. E. Beyer & M. W. Apple (Eds.), *The curriculum: Problems, politics and possibilities* (pp. 177–198). Albany, NY: SUNY Press.

Zeidler, D. L. (2016). STEM education: A deficit framework for the twenty first century?: A sociocultural socioscientific response. *Cultural Studies of Science Education, 11*(1), 11–26.

Zeidler, D. L., Sadler, T. D., Applebaum, S., & Callahan, B. E. (2009). Advancing reflective judgement through socioscientific issues. *Journal of Research in Science Teaching, 46*(1), 74–101.

Ziman, J. (2000). *Real science: What it is, and what it means*. Cambridge, UK: Cambridge University Press.

Chapter 11
Resisting the Borg: Science Teaching for Common Wellbeing

Larry Bencze and Mirjan Krstovic

11.1 Introduction

> The great aim of education is not knowledge, but action! (Herbert Spencer).

> The aim (of education) must be the training of independently acting and thinking individuals who, however, can see in the service to the community their highest life achievement (Albert Einstein).

Despite increases in our understanding of science education and perspectives and practices that may transform it in ways benefitting individuals, societies and environments, problematic traditions in school science seem highly resistant to reform. Arguably among the more pressing aspects of science education needing attention are *socioscientific issues*. Many such issues are associated with possible personal, social and/or environmental harms linked to various commodities, such as electronic communication technologies, petroleum for fuel, plastics and other uses, and genetically-modified foods. Consequently, many scholars, politicians and others are recommending that school science, along with other subjects, contribute to development of citizens more prepared to be proactive about addressing such issues. To understand how we might have more success in promoting a more activist citizenry through science education, it may be that we can learn a great deal from teachers who manage, despite significant opposition, to 'teach against the grain' (Cochran-Smith, 1991). In the instrumental case study described below, we illustrate ways in which Mirjan Krstovic, the second author of this chapter, achieved

L. Bencze (✉)
Ontario Institute for Studies in Education (OISE), University of Toronto, Toronto, ON, Canada
e-mail: larry.bencze@utoronto.ca

M. Krstovic
Peel District School Board, Mississauga, ON, Canada
e-mail: mirjan.krstovic@peelsb.com

L. Bencze (ed.), *Science and Technology Education Promoting Wellbeing for Individuals, Societies and Environments*, Cultural Studies of Science Education 14,
DOI 10.1007/978-3-319-55505-8_11

considerable successes in helping students to design and conduct research-informed and negotiated actions to address socioscientific issues of their concern—school science practices that are, generally, highly discouraged.

11.2 Theoretical Background

11.2.1 Inertial Science Teaching

Humanity appears to be facing many serious threats to wellbeing of individuals, societies and environments associated with fields of technology and science (or, 'technoscience'). People are concerned, for instance, about possible harms from various commercial products and services, such as: genetically-modified foods, etc. (Kleinman, 2003); household cleaning and hygiene products (Leonard, 2010); pesticides (Hileman, 1998); tobacco (Barnes, Hammond, & Glantz, 2006); and, pharmaceuticals (Angell, 2004). Many people also expect (and dread) serious problems associated with dramatic increases in average global temperatures that often are linked to excessive fossil fuel uses (Klein, 2014). Not everyone is convinced, however, of the seriousness (in terms of outcomes) or likely causes of such potential problems. There are many reasons for such disagreements, one of which appears to be that some business-sponsored groups have acted—sometimes by employing reputable scientists—to discredit findings about causes and apparently-harmful effects of commodities and, indeed, trustworthiness of scientists who have published such findings and, moreover, trustworthiness of whole fields of science and science as a discipline (Oreskes & Conway, 2010). Because of controversies surrounding potential problems like those noted above, and because of their association with fields of technoscience, they are known as *socioscientific issues* (SSI).

For various and complex reasons, many scholars and policymakers have—since at least 1970—attempted to infuse such issues into science education (Pedretti & Nazir, 2011). Although there have been some notable successes, it is apparent, however, that actual implementation of SSI education in schools remains relatively modest. In many school situations, it is not addressed at all; while, where there is attention to it, scholars continue to call for reforms/improvements (Hodson, 2011). Similar frustrations have been expressed by scholars regarding related reform movements. Crawford (2007), for example, lamented that teachers are having difficulty "creating classroom environments that are inquiry-based, and that support their students in developing informed views of scientific inquiry and the nature of science" (p. 613). Given difficulties with fuller implementation of SSI education, it is as if there were an 'invisible hand' inhibiting its infusion into school science. There likely are many factors contributing to this inertial effect. However, because many of the potential problems involve for-profit commodities, such as automobiles that appear to contribute to climate change and other environmental problems, it may be that resistance to SSI education may be significantly-related to *capitalism*. It seems clear that capitalism has come to dominate the *zeitgeist* of many societies—despite periodic exposure of its weaknesses, such as the 2008 global financial crisis (McMurtry, 2013). Following the rebuilding period after World War II, capitalists' share of the wealth apparently had dropped significantly—largely because of costs of infrastructure (e.g., buildings, roads, power generators, etc.) and social programmes (e.g., health care, unemployment insurance, etc.). To return to their earlier share of wealth, it seems they resurrected and re-invented traditional *economic liberalism* (McQuaig & Brooks, 2010)—an ideology favouring freedom (liberation) from government intervention in individuals' efforts to maximize their private wealth. The new form of economic liberalism, called *neo*-liberalism, appears to be much more strategic (or *opportunistic*) and highly globalized (i.e., further integrated throughout the world). These changes have, apparently, made neoliberalism extremely powerful and resistant to significant revision. Much of this strength appears to stem from the vastness and complexity of cooperative relations established among a wide variety of living and nonliving entities (or 'actants') that form a veritable 3-dimensional web surrounding and infiltrating into nation states (Ball, 2012; McMurtry, 2013; Springer, Birch, & MacLeavy, 2016). Types of actants involved in this global capitalist network (GCN) include various human individuals and groups, such as scientists and engineers, universities, transnational corporations (e.g., Coca Cola™, Nike™, Walmart™), 'think tanks' (e.g., Cato Institute), 'transnational advocacy networks' (e.g., Atlas Liberty Network) 'philanthrocapitalists' (e.g., Bill and Melinda Gates Foundation), national and extra-national banks (e.g., World Bank), extra-national trade organizations (e.g., World Trade Organization), national (e.g., Achieve™) and international (e.g., Programme for International Student Assessment) testing 'services', etc., along with numerous non-human entities, like the sea scallops in Callon's (1986) analyses of fishermen in St. Brieuc Bay and various products of genetic engineering, like the AquaAdvantage™ genetically-modified salmon, and salt water growing pens in which they are contained (Pierce, 2013).

With such great complexity and breadth, the GCN seems to exercise immense power over school systems—although acknowledging that every teaching/learning situation depends on myriad contextual variables and is difficult to accurately predict (Barnett & Hodson, 2001). Nevertheless, McLaren (2000), for one, has suggested that influences from the GCN have been so powerful that one might think that "the major purpose of education is to make the world safe for global capitalism" (p. 196). Some people have called this powerful sub-set of the GCN the Global Education Reform Movement ('GERM') (Hargreaves et al., 2001). GERM appears to be influencing governments to increasingly structure their educational systems in ways supporting neoliberalism. Indeed, in many jurisdictions around the world, it is common to see priorities like the following emphasized: curriculum standardization, international and local competitiveness, testing and reporting and emphases on 'core' literacies (e.g., language(s), mathematics, science and information technology) supporting economic activities (Sellar & Lingard, 2013). In some places, making such changes has taken on a militarist fervour. In a veritable economic 'war' with countries like China and India, a conflict sometimes known as the 'neo-Sputnik' race (Pierce, 2013), US education, for example, has being engineered—according to the foundational document, *Rising Above the Gathering Storm: Energizing and Employing America for a Brighter Future* (NAS, 2007)—to help further develop the USA's economic and political security. In light of the power of the GCN-GERM, it seems unsurprising that many educational reform agendas struggle for success.

Given prominent roles played by various fields of science and technology in businesses, it seems logical to suppose that science education systems must also be involved in this regard. Indeed, the National Research Council in the USA (NRC, 2011), for example, which helped facilitate development of that country's revised national science education standards (*Next Generation Science Standards*, Achieve, 2013), stated:

> The primary driver of the future economy and concomitant creation of jobs will be innovation, largely derived from advances in science and engineering. . . . 4 percent of the nation's workforce is composed of scientists and engineers; this group disproportionately creates jobs for the other 96 percent (p. 2).

From such statements, it seems very clear that school science systems may (and, likely, have been) under immense pressure to structure themselves in ways conducive to national successes in international economic competitions. The statement also makes clear that school science systems should concentrate on identifying students with aptitudes for becoming knowledge producers—such as those who would work in fields of science, technology, engineering and mathematics (STEM) (Pierce, 2013). Along with other professionals, like lawyers, accountants and business managers, STEM workers can provide *immaterial labour* (e.g., abilities to manipulate symbols, including words, concepts, numbers and graphics), useful for inventing commodities and ways to process and market them (Lazzarato, 1996). Such labour seems very powerful, setting tasks for many other workers (as the above quote notes). Although various mechanisms may be used for this filtering process, a priority on *abstractions* seems congruent with the goal. Indeed, many scholars have noted that school science systems focus "almost exclusively on the well-established products [e.g., laws & theories] of science and cookbook approaches to laboratory exercises, using authoritarian teaching modes" (Bell, 2006, p. 430). Many of these 'products' are abstract and 'delivered' quite quickly to students, apparently serving as a mechanism for selecting students who can quickly comprehend such abstractions (e.g., that a point mass occupies no space), often in the absence of self-directed applications (Bencze, 2010). Such an emphasis may be convenient for teachers, allowing them to teach relatively discrete 'packets' of information (e.g., the human heart pumps oxygenated and de-oxygenated blood out to the body and lungs, respectively), which can then be fairly-easily assessed and evaluated. This intense instructional approach may, however, be unfair for many students—favouring those who are advantaged in terms of cultural (and social) capital (Bourdieu, 1986), because of their abilities to rapidly comprehend abstractions. Nevertheless, these abstract thinkers may be useful to capitalists, developing and managing production and consumption of for-profit innovative commodities on their behalf.

The GCN-GERM not only appears to benefit from school systems' generation of knowledge producers, but it seems they also get citizens conditioned to function—to a great extent—as *knowledge consumers*. In the neoliberal age, increased value is placed on consumption, as opposed to production, of innovative for-profit commodities. After Jean Baudrillard's (1998) seminal work on *the consumer society*, several authors (e.g., Bakan, 2011; Usher, 2010) have advised that capitalists increasingly emphasize creation of positive semiotic messages—like 'cool,' 'powerful' & 'sophisticated'—associated with commodities rather than on high quality in products and services. Simply, it seems that 'image sells.' This is a powerful tack for capitalists, as it is much easier to re-design, for repeating cycles of consumption, images about commodities than to revise/invent them. De-emphasizing production can be profitable, allowing capitalists to minimize costs of labour, materials, transportation, etc., reductions that can generate negative side-effects (e.g., disease, pollution) of commodities (McMurtry, 2013).

In several ways, school science under GCN-GERM appears to provide capitalists with citizens oriented towards consumerism—including, for example, citizens conditioned to follow labour instructions and unquestioningly purchase for-profit commodities (Giroux & Giroux, 2006). This has been discussed at length elsewhere (Bencze & Carter, 2015) but, broadly, at least three—likely related—mechanisms appear to be used for this conditioning:

- Continuous Re-identification: Guy Claxton (1991) once recounted a description by his daughter about her experiences in school science. She said it was like riding in a windowless train, not knowing where you are going, but stopping periodically to look around—all the while not being informed by the conductor how each stop related to the others. This seems to be similar to what students experience when, for example, their science course consists of a series of perhaps significantly-different units; such as changing from *Sustainable Ecosystems* to *Atoms, Elements, and Compounds* (MoE, 2008). By being asked, in essence, to briefly focus on a planned sequence of often unrelated science topics, it seems that students must frequently adjust their identities—not unlike cycles of commodity purchases, as each 'new and improved' product/service is introduced (Barber, 2007; Usher, 2010).
- Technoscientism: Although it is difficult for school science to fully represent the nature of fields of technoscience, there is considerable evidence that they are significantly misrepresented in various ways. Broadly, we can think of such representation in terms of Ziman's (1984) 'internal' and 'external' sociology of science, with the former concerning practices of technoscientists *within* their professional communities and the latter dealing with their interactions with people and groups *outside* of them. From an internalist perspective, there are numerous reports that technoscience practitioners often are portrayed as highly-efficient, objective and collegial—usually strictly adhering to Merton's (1973) stereotypical norms of practice, such as originality and skepticism (Hodson, 2008). Idealized images of technoscientists can lead citizens to overly trust these professionals, perhaps disregarding possible negative effects of their relationships with other societal members. It seems clear, for example, that school science systems tend to avoid one of the most contentious externalist relationships; that is, between technoscientists and capitalists (Bencze, 2010; Carter, 2008). Evidence suggests that business-technoscience relationships in many fields often are

problematic for wellbeing of individuals, societies and environments—since, for instance, for-profit motives often compromise topic choice, research methods and free dissemination of research findings (Mirowski, 2011). Although there are numerous reports in scholarly literature suggesting that teachers—often in research contexts—have engaged students in learning about relationships among fields of technoscience and members of societies, it is apparent that such SSI education often lacks a critical, political, edge (e.g., with references to problematic business-technoscience relationships [also refer below]) (Levinson, 2010; Pedretti & Nazir, 2011).

There are, undoubtedly, various general ways of analyzing how school science may depict fields of technoscience. Based on the above examples, however, it seems clear school science tends to portray fields of professional technoscience as relatively Rationalist-Realist on Loving's (1991) *Scientific Theory Profile* (STP)—depictions suggesting that these fields are highly efficient, unbiased and relatively (or nearly) certain about conclusions (Hodson, 2008). Science and technology studies, meanwhile, suggest that they are more Naturalist-Antirealist in terms of the STP—positions often avoided in school science. Without a breadth of perspectives about fields of technoscience, students may be less able to make informed judgements about such fields and their relationships with societies and environments. Such a limited view of possibilities seems undemocratic (Pierce, 2013).

- <u>Alienation</u>: Great injustices occur when people are, in various ways, separated—or 'alienated'—from their world and/or their being (Marx, 1990 [1867]). Science education systems—particular when greatly influenced by the GCN-GERM—appear to accomplish this, in at least two ways. The first is related to the degree to which students' learning is controlled. It may seem paradoxical, for example, but it is apparent that students rarely are allowed to 'do' science in school science. In other words, individuals (e.g., teachers) and/or groups (e.g., those associated with textbooks) control decisions regarding knowledge building and/or uses (Hodson, 2008). Even with popular inquiry-based learning approaches, which are said to promote (more) independent knowledge construction, students' decisions often are 'scaffolded' to ensure—apparently above all else—that they arrive at conclusions supported by mainstream fields of technoscience (Bencze & Alsop, 2009). Such excessive guidance also seems to apply, to some extent, to SSI education. Although there are, likely, many exceptions, SSI education appears to emphasize student learning of teacher-provided evidence and arguments for conflicting positions, which they are then encouraged to use—frequently in social situations—to develop, in highly logical ways, personal positions on issues (Levinson, 2010).

There are several ways to evaluate such SSI approaches. A number of these appear to relate to Levinson's (2010) analysis of citizen participation regarding SSIs, depicted in Table 11.1. In his comprehensive review, he concludes that most forms of SSI education are limited to his first two categories; that is, 'deficit' and 'deliberative' approaches. Consequently, it seems clear that students are being asked to consider science information given to them and negotiate personal positions regarding any controversies among data interpretations. On the one hand, there appear to be clear benefits of such argumentation-based approaches, including:

Table 11.1 Types of citizenship engagement in school science

Summary of four frameworks for democratic participation in a school context

Framework	Socio-epistemic relations	Epistemology	Controversy and participation	Pedagogy	Implications for democratic participation
Deficit	Knowledge flow is from scientist-teacher-student	Science is the corpus of knowledge	Ability to engage is constrained by access to technical knowledge	Knowledge for addressing an issue can be brought to the attention of the student.	There is a socio-epistemic inequality between the scientist/teacher and students which limits ability to bring about political change from below but does not preclude influential specialists making a political impact.
Deliberative	Knowledge flow is predominantly from scientist to the teacher and students, the latter two might be working in concert.	Science is understood to be uncertain and fallible.	Dialogue is open. Lay participants are informed but often lack the political means to bring about change. In schools, students might have opportunities for deliberation through group work and school councils but action might be constrained depending on the democratic nature of the school.	Emphasis on critical thinking and understanding of scientific methods and procedures.	Participation is real but often ineffectual in generating democratic change because participants do not have the 'clout' to make crucial decisions.
Science education as praxis	Knowledge is distributed and emergent.	Knowledge is situated. Students become inducted into communal ways of knowing through legitimate peripheral participation in particular but changing contexts.	All participants work with a shared sense of social purpose.	Knowledge is provided on a need to know basis. The teacher is not epistemologically privileged.	Active and egalitarian participation to enhance change which might assume political literacy.
Dissent and conflict	This can be variable but is likely to have similar characteristics to science education as praxis.	What is known is contextualised by socio-political concerns.	Political action.	Knowledge provided on a need to know basis with an emphasis on political literacy.	Political understanding and action for change are foregrounded

Levinson (2010), Table 1, pp. 83–84

development of *socioscientific reasoning skills* (Sadler, Barab, & Scott, 2007) and learning about the nature of science (Khishfe & Lederman, 2006). On the other hand, limiting students to such personal decision-making on the basis of received data and claims seems alienating in terms of students' control of learning. Despite options for thought and negotiation, it is apparent that authority tends to rest with scientists and other professionals in these educational contexts (Levinson, 2010).

Without significant, if not full, control over knowledge development and dissemination as at least part of their education, the depth of students' learning—in various learning domains—seems limited. For deep learning, Etienne Wenger (1998) suggests that learners need significant control over decisions in Phenomena (e.g., tree) ←→ Representation(s) (e.g., drawing of tree) translations. Otherwise, they may become alienated from them and, therefore, be amenable to manipulation by those with more control over such relations. They may, for instance, be more amenable to consuming commodities (with semiotic messages) provided by elite capitalists.

A second way in which alienation may occur through school science pertains to the degree to which students have access to complete contexts in Phenomena ←→ Representation translations. There appear to be various cases in school science limitations in the regard, but a particularly problematic domain for this relates, again, to their exposure to complete SSI contexts. According to Pierce (2013), for example, SSI education tends to ignore many actants that may problematize many for-profit commodities recommended for SSI education in school curricula. As an example, he summarizes corporate ownership of living entities—an aspect of genetic engineering often omitted in schools:

> [S]cientific literacy needs to be radically rethought in an age where genes of an Ocean Pout (an eel fish) are spliced with those of a Chinook (king) salmon, implanted in Atlantic salmon eggs, and a corporation patents this process *and* the new species of the fish itself ... (Pierce, 2013, p. 113; emphasis in the original).

Overall, the above arguments and examples suggest that school science systems are, perhaps, greatly influenced by or congruent with the GCN-GERM and, as noted in the above quotation from the US National Research Council (NRC, 2011), structured/engineered to generate a relatively small group of knowledge producers and large groups of compliant citizens that, together, may enrich the very small fraction of the world's population controlling capitalism.

11.3 Towards a Just Science Teaching

To increase social justice and environmental sustainability through school science, it is apparent from the above arguments that we need *counter-hegemonic* (Cohn, 2005) actions to provide citizens with alternative perspectives and practices from which to choose. Transforming school science systems in ways conducive to significant improvements to wellbeing of individuals, societies and environments is not likely to be easy, however, given the aforementioned powerful nature of the global capitalist network (GCN) and related global education reform movement (GERM). A major part of the difficulty in challenging the GCN-GERM is its vast and diffuse nature. It is, as described above, comprised of a great range of actants, operating between and within nations (and states & provinces, etc.). There is no one person or group to challenge. The 'target' is more like an idea—or *ideology*—than a physical thing, person, etc. Although it is very powerful, transforming it is not as straightforward as defeating a dictator and his/her military—although that also is, clearly, not easy. A major aspect of the difficulty seems to derive from the apparent fact that neoliberal ideological perspectives have infiltrated the sub-conscious of great fractions of the human population. Often without knowing about it, people think and act in ways supportive of neoliberalism—in the form, for example, of individual competitiveness and struggles with empathy/concern for those less 'successful' and for slowly-degrading environments. Gramsci (2003 [c1929–1935]) discussed this generally as *cultural hegemony*, infiltration into a population of a set of (albeit dynamic) perspectives that may be more aligned with powerful others than individuals in the population. Foucault (1991) discussed much the same concept regarding neoliberalism, calling it *neoliberal governmentality*. His work emphasized that, while believing they are self-governing, individuals may be enacting powerful others' perspectives. Ball (2003) discussed this in terms of '*performativity*'; that is, the idea that subjects perform (e.g., through their speech and actions) perspectives that are not, necessarily, based on their deep personal identities, and that such performances serve to spread such perspectives across populations/cultures. A key aspect of cultural hegemony, governmentality and performativity is the idea that subjects are frequently monitored, assessed and evaluated and, consequently, may begin to *self-regulate* their thoughts and actions without particular pressure from powerful others. Such self-regulation is very much related to the concept of the *Panopticon*—a prison structure used as a metaphor for governmentality (Foucault, 1977). Ball (2000) expressed this well:

> Performativity is a technology, a culture and a mode of regulation, or a system of 'terror' in Lyotard's words, that employs judgements, comparisons and displays as means of control, attrition and change. The performances (of individual subjects or organisations) serve as measures of productivity or output, or displays of 'quality', or 'moments' of promotion (there is a felicititous ambiguity around this word) or inspection. They stand for, encapsulate or represent the worth, quality or value of an individual or organisation within a field of judgement. 'An equation between wealth, efficiency, and truth is thus established' (Lyotard 1984, p. 46) (p. 1).

In practical terms, we might find teachers performing educational acts (e.g., choice of topics and pedagogical approaches) that reinforce ideological perspectives inherent to curriculum policy documents and textbooks produced by powerful entities. In the USA, Pearson Educational Inc., probably the world's largest educational business, controls—in concert with an educational assessment company (Achieve, 2013)—much of the curriculum and textbook market (Ball, 2012) and may, ultimately, be strongly influencing teachers' performances.

Clearly, with deep and complex global capitalist networks holding hegemonic influence over many teachers, it is likely to be very difficult for them to 'escape' the mainstream science teaching 'paradigm' and establish new perspectives and practices. Cochran-Smith (1991) said that, for teachers daring to be different, it would be like 'teaching against the grain'; that is, there is likely to be considerable opposition, not the least of which may arise from colleagues who have been assimilated into 'the' (acknowledging its dynamism) GCN-GERM paradigm. Comparable to Kuhn's (1970) advice regarding paradigm shifts, she suggested that teachers need to arrive at veritable 'crises' (deep questioning of the mainstream paradigm) in order to develop motivation for change; but, additionally, they need access to alternative perspectives and practices that may 'relieve' the crises they feel. In her words, she suggested teachers need 'critical dissonance' (like a crisis) and have access to alternatives through what she called 'collaborative resonance'—working with a critical friend (Cochran-Smith, 1991). These tacks seem similar to that recommended by McLaren (2000) who, in turn, was drawing upon work of Che Guevara and Paulo Freire; that is, revolutionary change may occur if participants experience *conscientization* (e.g., critical consciousness of problems and alternatives), but—to avoid further oppression—arrive at solutions through a process of *praxis*, which implies working through cycles of critical, reflective, practice. These processes are, in turn, not dissimilar to emancipatory and participatory action research; that is, collaborative efforts to bring about informed changes (actions) that may provide (research) insights into the context of study. Accordingly, in this paper, we describe results of an action research project, in which a teacher (Mirjan Krstovic, the second author here) experienced some successes 'teaching against the grain' (collaborating with Larry Bencze, the first author here) in ways that may enlighten us about alternatives to the GCN-GERM as it pertains to science education.

11.4 Research Context and Methods

11.4.1 Context

Mirjan was a secondary school science teacher with six years teaching experience and enrolled in a graduate programme in education, working towards a Masters of Education degree. In early Sept. 2011, soon after having taken an online graduate course in history, philosophy and sociology of science (HPSS) for which Larry was the instructor, Mirjan asked Larry—whom he had not met face-to-face—if he would help facilitate his efforts to infuse perspectives about the nature of science (NoS) into his secondary school science teaching. After some discussion, Mirjan expressed interest in incorporating his focus on NoS into efforts to encourage and enable students in his science classes to self-direct research-informed and negotiated actions (RiNA) to address socioscientific issues—a central feature of the 'STEPWISE' (Science & Technology Education Promoting Wellbeing for Individuals, Societies

& Environments [see: www.stepwiser.ca]) curriculum and instructional framework, to which Mirjan had been introduced during the HPSS course. We also decided Larry would serve as researcher/facilitator (R/F) for Mirjan's efforts in this regard. At the time of this writing, Larry and Mirjan have collaborated on this project for two successive years, involving four separate semesters and four classes of students. As noted above, this facilitation was to employ the two main principles discussed by Cochran-Smith (1991)—i.e., *critical dissonance* and *collaborative resonance*—in helping teachers teach against the grain (TAG). Larry had worked in a similar way with several other teachers (in elementary & secondary schools) and, as indicated below, few were as successful as Mirjan in encouraging and enabling students to self-direct RiNA projects. Moreover, as demonstrated below, his facilitated teaching seemed very much 'against the grain' with regards to mainstream/traditional science teaching (locally and beyond).

11.4.2 Research Methodologies, Data Collection & Analyses Methods

To illustrate a teacher's efforts to teach against the grain, we constructed an *instrumental* case study (Yin, 2003)—which involves collecting data to provide details about a particular situation, emphasizing its significance relative to a broader issue. Clearly, the issue of note here is the extent to which teachers can teach against the grain—particularly in light of the broader milieu of the GCN-GERM—and what factors may influence such teaching. In choosing instrumental cases, it is important to ensure that particular patterns have been established and, consequently, trust that a relatively-specific phenomenon is being illustrated. Here, data and analyses of Mirjan's case were conducted over two years, including regular discussions throughout the school years and during the holiday seasons. This process was, in essence, a form of *participatory action research* (Kemmis & McTaggart, 2000) that allowed us to learn from cycles of collaborative reflection on actions. Action research has much potential to conduct socially-meaningful research and it can "valid knowledge, theoretical development, and social improvements that the conventional sciences have promised" (Greenwood & Levin, 2000, p. 87). Participatory action research is a democratic form of inquiry, one in which professional researchers *collaborate* with local stakeholders to seek and enact solutions to problems that are important to stakeholders. Although Larry continually acted as a critical friend, posing alternatives to theory and practice, Mirjan retained ultimate control of decisions—although it must be acknowledged that every relationship is *dialectic* (Hegel, 1977 [1806]) and, consequently, we influenced each other.

To achieve our research agenda, we conducted data-collection and analyses having *rationalistic* and *naturalistic* characteristics (Guba & Lincoln, 2011). Rationalistically, we focused, for example, on Mirjan's efforts to teach against the grain—in comparison to immediate colleagues (e.g., in the science departments in which he worked) and teachers more generally, as reflected in scholarly educational literature. Naturalistically, data were collected enabling emergence of unexpected situational outcomes. Data collected included the following items from Sept. 2011–Aug. 2013:

- Project Instructional Materials: Copies of Mirjan's teaching materials (e.g., handouts, videos, PowerPoint™ presentations, and internet site web addresses) relating to STEPWISE implementation were collected;
- Project Work Artefacts: Samples of products generated by students in four tenth-grade classes were collected, including: issue descriptions, research plans, data collected, written reports, project reflections, action plans and forms of action (e.g., posters, petitions, videos, PowerPoint™ slides);
- Digital Recordings of Students' Project Work: Photographs and videos were produced depicting youth presenting and defending their forms of action in public fora (e.g., to fellow students within and outside of class).
- Semi-structured Interviews: Each semester, about four volunteering students were interviewed twice, near the beginning and end of the course. Questions focused on their views about issues, research & actions. Mirjan was interviewed at least once per month (sometimes in person and sometimes via Skype™), for about 90 minutes each, about project progress. Four of these interviews focused exclusively on his perceptions around being considered to be 'teaching against the grain.' All interviews were audio-recorded and later transcribed.
- Email Exchanges: Copies of all email exchanges between us (at least once per week) were collected.
- Mirjan's Written Reflections: During the first semester in which he implemented STEPWISE, Mirjan kept an online journal, which amounted to 19 pages of text and graphics (e.g., pictures of students' work). He also maintains an online educational blog and 16 entries pertaining to his STEPWISE experiences were copied. He also wrote a 7-page reflection on his status as a teacher 'teaching against the grain' (received Aug. 6, 2013). Finally, a copy of a report he wrote for a graduate course that summarized his interview with a senior colleague's positions about SSI education was retained.
- Inscriptions of Mirjan's Positions: Mirjan completed several 'inscriptions' to depict his positions on a range of topics. These included: (i) *Political Compass*: After answering a series of multiple-choice questions (www.politicalcompass.org), a grid depicting a person's views about economic and social matters is generated. Figure 11.9 depicts the grid generated for Mirjan; (ii) *Scientific Theory Profile*: This is described above. In addition to indicating his position on it, Mirjan wrote a 2-page description of his rationale for his choices; (iii) *Repertory Grid*: Mirjan placed a series of teaching strategies relating to STEPWISE on a series of continua, after which RepGrid 5.0 software was used to generate a 'FOCUS' display of his positions (Gaines & Shaw, 1993) (refer to Fig. 11.7); (iv) *Word Cloud*: Text from the Mirjan's first-semester journal was entered into the Wordle website (www.wordle.net), which generated a word cloud depiction of his reflections.

- Artefacts of Mirjan's Work: Copies of the following items representative of his work in this project were collected, including: (i) the school newsletter from Jan. 18, 2013, including a note about STSE-Action Fair he organized; (ii) an 18-page summary of his experiences at his first academic conference (American Educational Research Association, 2013), Mirjan produced for colleagues; (iii) two publications and an unpublished report of research he conducted prior to our collaboration.
- Letter from a Student: A copy of an unsolicited 1-page letter written by a twelfth-grade student to Mirjan at the end of the 2012–2013 school year (her graduating year) was retained.

Regarding analyses, each of us coded data for categories and then developed encompassing themes—using constant comparative methods based on constructivist grounded theory (Charmaz, 2014). Categories and themes were then negotiated between to achieve consensus (Wasser & Bresler, 1996). Member checks with participants were conducted to help ensure *trustworthiness* of claims, each of which was based on at least three data sources.

11.5 Results and Discussion

11.5.1 Mirjan's Teaching Against the Grain

As described in the next sub-section, it was apparent to us that Mirjan was 'teaching against the grain' (TAG) before our collaboration. However, after two years of our work together, he indicated that his TAG was considerably advanced—and likely to continue to do so. Depending on how one views 'teaching against the grain,' either as ideal or realized practice, it also varied across contexts. In other words, his actual implementation of TAG varied across time and contexts, although there seemed to be steady growth in his *priorities* for and expertise and comfort with it. Effects of such situational variables are discussed below, under Factors Influencing Mirjan's Teaching Against the Grain. The representation (case) below of Mirjan's work is a construction of TAG practices in what appears to be relatively ideal conditions (relative to those in 'mainstream' contexts), with an elaborated discussion of inhibiting and enabling factors to follow.

Although there are various ways of defining or evaluating teaching against the grain, it seems that many of his practices represented significant alternatives to those supportive of the GCN-GERM. Given our collaboration, he worked over the two years in which we have, so far, been engaged to more-deeply develop and implement the STEPWISE framework (refer above). By the fourth semester of his work with this framework, he had implemented a pedagogical sequence given in Fig. 11.1—which is, generally, derived from the tetrahedral, more theoretical, framework (Bencze & Carter, 2011). Among the many aspects of STEPWISE he implemented, the following elements, with examples, seem sufficient to illustrate

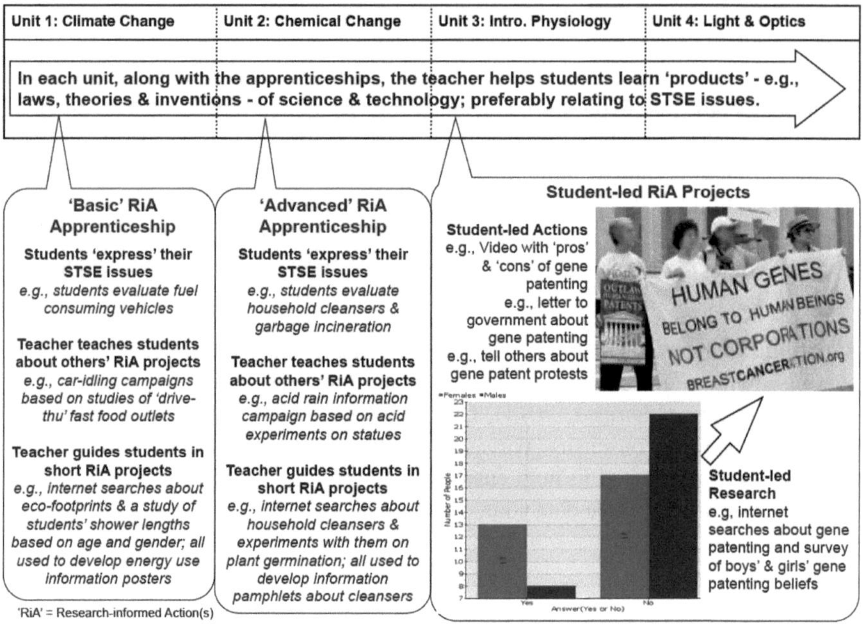

Fig. 11.1 STEPWISE Pedagogical Framework

his 'teaching against the grain.' Most of these activities were largely student-led and open-ended (Lock, 1990), thus more personalized, but they also provided students with broader, more critical, conceptions of the world (Pierce, 2013).

11.5.1.1 Expressing SSIs

Recommended based on basic constructivist learning theory (Osborne & Wittrock, 1985), students were asked to 'express'—in various forms—their pre-instructional views about a range of socioscientific issues. Doing so, can not only inform the teacher about gaps in students' learning, but also allow students to become more conscious of their own positions—which they might later use as bases for self-directed explorations. An example of this activity includes the 'card exchange game,' in which students evaluate and exchange cards containing social action statements regarding SSIs (Krstovic & Bencze, 2012a). An example of a student group's positions on four such statements is given in Fig. 11.2. Students had considerable freedom of expression here.

Fig. 11.2 Students' views about SSI action statements

11.5.1.2 Critical SSI Education

Based on arguments provided above regarding benefits of infusion of actor-network theory into science education (e.g., Pierce, 2013), students were asked to develop and revise (based on secondary research and peer negotiations) networks relating to a particular SSI of their concern/interest. An example of a student group's actor-network, here relating to hairspray, is reproduced in Fig. 11.3. Students demonstrated their growing awareness of roles of powerful actants, like Target™ and Walmart™ (Fig. 11.3), relating to SSIs.

11.5.1.3 Student-LedPrimary Research

In light of Wenger's (1998) claim that deep, personally-meaningful, learning is promoted the more students have control over reciprocal relationships between phenomena and their representations (Phenomena ←→ Representation(s)), students were encouraged to design and conduct primary research to generate data and claims that may inform their sociopolitical actions (refer below). This tack also appears to align well with Dewey's (1916) advice that "inquiry has to serve common social needs" (Levinson, 2010, p. 74). Most students conducted correlational studies, which tend to be de-emphasized in school science (Bencze, 1996), with results from one student group's study given in Fig. 11.4.

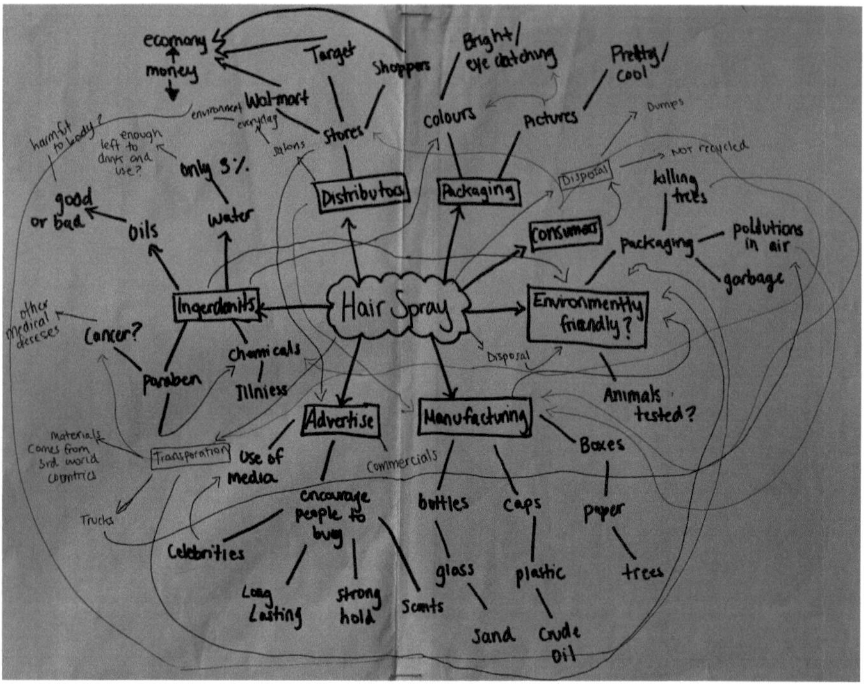

Fig. 11.3 A student group's actor-network about hairspray

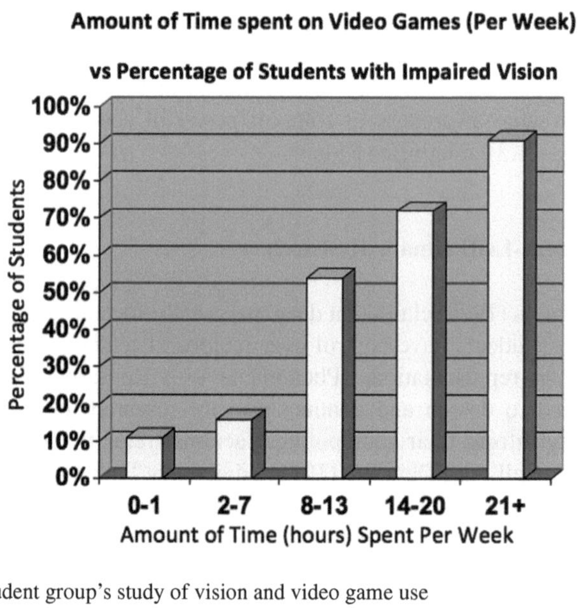

Fig. 11.4 A student group's study of vision and video game use

11.5.1.4 Socio-political Activism

Because of potential seriousness of problems for wellbeing of individuals, societies and environments (e.g., Hodson, 2011) and because of many more immediate learning benefits, students were encouraged to develop and implement forms of personal and social actions that they believed would address SSIs concerning/interesting them (Bencze & Carter, 2011). Students' actions took many forms, ranging from personal changes to their perspectives and practices through local social actions aimed at anyone noticing their posters, pamphlets and videos, etc. to actions directed specifically at those they felt may affect changes. A typical poster developed by a student group is given in Fig. 11.5, while a typical letter to an advocacy group is given in Fig. 11.6.

Perhaps not unsurprisingly, teaching such a complex programme requires simultaneous teaching in several learning domains. Students were exposed to instruction/learning in each of the three curriculum domains, including: STSE (SSI), Investigative Skills and Products (MoE, 2008). Particularly with STSE education, science education can be—and was—broadened from traditional practices to include such domains as: technology education, psychology, sociology, politics and economics. Additionally, his instruction branched into the visual arts—with our collective aim to understand the role of various forms of visual representation in students' orientations towards RiNA projects. Among perspectives students in Mirjan's class experienced was exposure to the activist artistry of Chris Jordan (www.chrisjordan.com), who uses aspects of our apparently self-destructive world—such as

Fig. 11.5 A student group's poster about alcohol abuse

Dear BCA,

We are four students from Ontario, Canada. Recently, we finished our final science project in biology which was on the gene patent controversy with Myriad Genetics. While researching on this issue we came across your great website and were able to inform ourselves more about this topic.

After researching both sides of the argument, we believe that this patent is invalid and are against it. It's shocking to believe that thousands of women who have mutations in the BRCA 1 and BRCA 2 genes have almost triple the chances of developing breast and ovarian cancer. We think it's wrong that Myriad Genetics claims it should have the patent on the BRCA 1 and 2 genes because they 'isolated' them. However, we do recognize that Myriad spent $500 million in 17 years to discover the genes and to discover a method of diagnosing patients with the BRCA mutations but that is not does make something patentable. We believe only new inventions or procedures should be patentable.

As part of our Biology project we created an informational video that informs others on gene patenting. This video will be uploaded on social media networks such as Facebook, Twitter and YouTube. We hope to get more people involved in making a difference to stop companies from patenting human genes. This letter is also part of our final action. We hope to reach out to Breast Cancer Action and find out more about your organization.

Lastly, we'd like to thank you for reading our letter and we hope to take positive action against patenting laws.

Sincerely,

Note: Identifying Names Deleted

Fig. 11.6 A student group's letter to a breast cancer advocacy group

nuclear power, plastic packaging and disposable products—to sensitize viewers to many potential problems for WISE. Such a multi-disciplinary curriculum seems democratizing, as Levinson (2010) suggests: "If science education is both to serve democracy and to be a source of democratic values then a picture of an interdisciplinary problem-solving curriculum reflecting wider social and global matters begins to emerge" (p. 75). For students unaccustomed to such diverse teaching and learning, however, it can be disorienting. Early into his second year with the project, Mirjan described such struggles:

> [S]tudents are coping with a two-[or 3]-tier instructional program in my class. Let me explain: as you know, traditionally students listen to a lesson, take notes, then they do questions from the textbook, which the teacher may or may not take up the next day; may be they do a closed-ended, scripted procedure style lab that proves a concepts (ie. law of conservation of mass), they may write a small lab report, then they write a unit test. Somewhere in there, an STSE reading is assigned (or a research project) which the students hand in as a written report (and it only gets read by the teacher and no one else). I call this a one-tier instructional program (for the lack of better words). However, my students are learning to cope with a multi-tiered system. So not only are we handling content in also a very 'instructionally intelligent' way (I use a lot of instructional tactics/strategies from *Beyond Monet*, by Barrie Bennett and Carol Rolheiser of OISE), but we are also adding a layer of apprenticeship activities (ie. how to conduct correlational studies, learning about activism, photojournalism, etc.) which are part of the research-informed activism instructional tier. As much as I try to merge the two tiers (i.e. use RiA as the CONTEXT to drive the learning of the CONTENT, that does not always happen) some of the kids MAY feel (and this is my

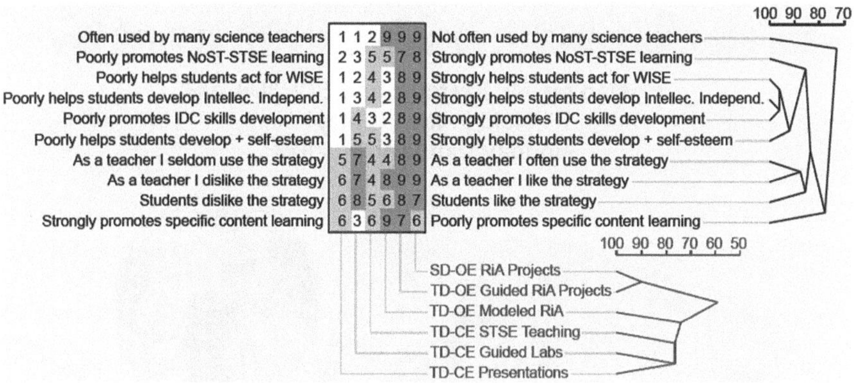

Fig. 11.7 Repertory grid completed by Mirjan

> hypothesis) that their learning environment is messy. To add to these two tiers, there is the third tier of learning additional skills necessary to execute 'creative' actions, such as how to use Windows Movie Maker™ software, how to publish a video on YouTube™, how to use Prezi™, etc. (Email message to Larry, Oct. 9, 2012; italics and Trademark logos added).

Overall, it seems clear Mirjan was 'teaching against the grain.' It is apparent his teaching aligns well with levels 3 and 4 of the framework for citizen engagement depicted in Table 11.1, which Levinson (2010) suggests is rare in science teaching and, moreover, contrary to approaches aligned with neoliberalism. The above quotation in a message from Mirjan indicates his awareness of his distinctive practices. Indeed, there were many indications that he was very much conscious of his TAG status. The repertory grid (Fig. 11.7) that he completed in January of 2013 indicates that he perceived his teaching to be significantly different from that of his immediate departmental colleagues. For example, he "likes" and "often uses" several strategies—such as SD-OE RiA Projects and TD-OE Guided RiA Projects—that he indicated most other teachers avoided. Indeed, Mirjan said that his perspectives and practices were so different from department colleagues that one of them said he "was *terrified* to teach with me" (Interview, Jan. 14, 2013; italics added to indicate inflection).

Mirjan's teaching against the grain was not covert. He did not simply close the door to his classroom and teach differently in a private way. Rather, even in the context of the more conservative school in which he worked in the second year of this study, he was proactive in making public his alternative practices. It was part of his routine, for instance, to display examples of students' work in public spaces. The poster shown in Fig. 11.5, for example, was one of several displayed on the wall just outside his classroom soon after the activity. This may have sent subtle messages to his colleagues that his teaching was distinct from the norm. More directly, Mirjan said that, particularly for the first course unit (Fig. 11.1), in which he had to use considerable amounts of time helping students to develop attitudes, skills and knowledge (ASK) for conducting RiNA projects, he would not finish units at the same time as colleagues teaching other sections of the same course. As discussed

Fig. 11.8 Students' RiNA project displays during the 'STSE-Action Fair'

below, this was not so much a problem for teachers in the first school in which he worked, but it became a concern for teachers in the second school. Despite the more traditional nature of the second school in which he worked (refer below), however, he was quite overt in ensuring colleagues were aware of his unique teaching approaches. An arguably quite overt message to colleagues was his use of the quote from Gandhi—that is, 'First they ignore you, then they laugh at you, then they fight you, then you win'—that he used as his computer's screensaver. Perhaps the most elaborate overt messages to colleagues, however, was the 'STSE-Action Fair,' photographs from which are shown in Fig. 11.8. During a regular class period in his second school, teachers and students were invited to the school's library to discuss with students in Mirjan's class RiNA projects they had completed.

Soon after Mirjan's first effort with STEPWISE, he became an advocate for the project, generally, and promotion of research-informed and negotiated actions to address SSIs, more specifically, in various interactions with teachers. His enthusiasm for encouraging and enabling students to conduct correlational studies to inform socio-political actions was, indeed, infectious—in at least the first school in which he worked. About this, he said several of his colleagues adopted his approach during the second semester during his first year of this project:

> They liked it so much that, in second semester, everybody teaching grade 10 did [worked with students to] the climate change action project! ... Teachers of grade 11 chemistry picked up on this project, as well. I think, overall, there were nine classes [promoting a RiA project]. And, we did it in grade 12 chemistry in that same semester. ... [So, t]he work [we did in first semester] went 'viral' in the second semester! (July 30, 2013).

Partly from this success, Mirjan developed an article about this strategy for teachers, published in the professional journal, *Crucible* (Krstovic & Bencze, 2012b).

There were several more instances in which Mirjan shared aspects of his work on STEPWISE with teachers in his school and beyond, but two more of these seem sufficient to highlight his pro-active defence of approaches he had been developing. After he had conducted a 2-hour workshop with student-teachers enrolled in Larry's pre-service science teacher education methods course (an elective), several student-teachers expressed concern that students' primary research results may not be sufficiently valid to warrant socio-political actions. Mirjan acknowledge this concern, saying he would adjust his approaches, but also defended it in a blog entry he asked Larry to share with student-teachers: "Although some have used their primary results to motivate their actions, many have erred on the side of secondary research findings. This suggests that fine-tuning apprenticeship activities is important so that students feel more confident in carrying out valid primary research" (Dec. 3, 2012). This was a challenge because, although these student-teachers had chosen this elective course, it was the only one in science education available and, moreover, most seemed conditioned to prioritize instruction in 'products' of science and technology. The second 'incident' related to Mirjan's participation in an international, refereed, education conference; the 2013 AERA conference. As a result of his experiences, Mirjan produced an elaborate summary of what he considered highlights of his visit to San Francisco, the site of the conference. In short, most of the content of this report represented a challenge to the dominant paradigm in education, with special attention to issues of social justice. He quoted speeches by prominent educators, including one by Kent McGuire (Southern Education Foundation): "We have this highly stratified education system in this country, with schools historically and legitimately taking pride in serving who? Women, working class people and minorities?—none of whom have enjoyed much status in this country." In support of this contention, he spoke with and photographed several homeless people in the city—which he also highlighted in his report. Because his report contains many indicators of his stance as a critical and creative science educator, a copy is provided in Appendix A of this paper.

11.5.2 Factors Influencing Mirjan's Teaching Against the Grain

There are, likely, numerous factors possibly accounting for Mirjan's instructional approaches and, moreover, their counter-culture nature. As noted above, for every teaching and learning situation, there are likely to be myriad—possibly interacting—contextual variables affecting it, such as the nature of the curriculum, teacher, students, parents, administrators, textbooks, etc. (Barnett & Hodson, 2001). Some of the more salient factors from this study are discussed below, in light of relevant data.

11.5.2.1 The Official Curriculum

According to the above analyses, while Mirjan appeared to be teaching against the grain relevant to colleagues and, moreover, the GCN-GERM, his teaching was aligned—to a great extent—with the local jurisdiction's curriculum, which officially sanctioned such practices as: STSE education, primary research, and socio-political actions to address SSIs (MoE, 2008). Mirjan indicated that this sanctioning had major motivating influences on his decision to pursue STSE education—including with regards to research-informed and negotiated action projects led by students. He said it was only after introduction of the revised Ontario curriculum (MoE, 2008), which had placed STSE education as the first of three overall curriculum goals, that he began to seriously implement this educational component. Expressing his rationale for this kind of education, he said: "[W]ith regards to my focus on academic, social and moral development, it [STSE] fits in nicely" (Interview, Dec. 17, 2012). Nevertheless, it seems that many teachers are not fully implementing this curriculum.

11.5.2.2 About Mirjan

Given difficulties many teachers, including several of Mirjan's colleagues, face in implementing educational practices like those described above, it follows that he likely possesses personal characteristics aligned with them. In light of actor-network theory, 'Mirjan' may be considered a *complex* entity, composed of—and, to a certain extent, representing influences from—many other actants (each of which also is complex) (Latour, 2005). In our numerous discussions about what makes Mirjan unique from other teachers, a major influence appears to be his orientation towards continuous and highly-ambitious educational exploration, innovation and learning.

There are many ways such orientations can be described, but it was apparent to us that he can be considered to be at least a 'double-loop' learner—that is, teachers who question fundamental assumptions surrounding their practices and attempt to adjust them accordingly (Argyris, 2002). When asked if he felt he fit into this category, Mirjan said: "I think that one sustains inquisitiveness throughout adulthood by developing meta-cognitive skills; it is in this ability to reflect as an adult that one finds a deeper purpose for one's work and life in general" (Email message to Larry, Aug. 2, 2013). By contrast, he said: "Some teachers don't [ironically] welcome learning. It's bizarre. ... I think there is research that says that teacher learning is closely-linked to student learning" (July 30, 2013). He felt that many teachers with whom he had worked (in two secondary schools and through associations with teacher groups outside of school) were, at the most, 'single-loop' learners; that is, those who may revise their practices based on findings about student learning outcomes, but who are unlikely to question fundamental premises of their strategies. Such critical and creative reflective practice (CCRP) may lead to dramatic changes in teacher practices and, arguably, corresponding changes in student learning.

We speculated that Mirjan's proclivity for CCRP may have begun as a result of his status as an *immigrant*, having moved to Canada from war-torn Yugoslavia when he was 12 years old (1993) (not long after his father was killed during the war). Indeed, he said that "[t]his marked a very difficult period in my life, but ... I found comfort in the academic life. I was determined to succeed in whatever I chose to do in life" (Written reflection, Aug. 6, 2013). The extent to which such experiences contributed to his tendencies towards CCRP is unclear, but it seems that an event congruent with CCRP during his undergraduate years may have been pivotal. Needing funding for his education, he pursued a part-time job with a professor, assisting him in research to determine abilities of older people (after schooling years) to learn about astronomy. Fortuitously, the study placed great emphasis on benefits of people who consider themselves 'lifelong learners' (Percy & Krstovic, 2001). This work seemed prescient, given our view that a key reason for his successes with practices aligned with levels #3 and 4 in Table 11.1 was his status as a lifelong learner. Indeed, Mirjan confirmed this regarding his comments about research he had conducted in collaboration with high school students during his third year of teaching—research that led to another publication-. About this research, he said: "I saw this kind of inquiry work in teaching as being a natural part of teaching and learning—something that my previous experiences reinforced" (Written reflection, Aug. 6, 2013). Apparently in light of his orientation towards CCRP, Mirjan was then chosen by his school's vice-principal to lead a government-funded project to study implementation effects of pedagogical techniques—such as 'placemat,' 'think-pair-share' & 'community circle'—relating to 'instructional intelligence' (Bennett & Rolheiser, 2001). Mirjan felt that this action research project was very successful, leading him to conclude: "When various instructional tactics are integrated in class[,] students' communication skills as well as their knowledge and understanding of the key concepts improve" (Krstovic, 2009, p. 29).

Mirjan's action research project seems to have been significant with regards to the project reported here. It became apparent to us that, in addition to the fact he had been teaching for six years when we began our collaboration and, therefore, had developed pedagogical expertise for teaching in ways acceptable in his school context, his action research—which involved efforts to implement a series of teaching strategies (refer above) that prioritize student engagement—appeared to help him to more-easily implement STEPWISE instructional practices, which prioritize student engagement in such components as: expressing pre-instructional views, conducting partially-guided and student-led science inquiries and designing and implementing plans of action to address socio-scientific issues. In interviews and blog entries, Mirjan mentioned that part of his success with STEPWISE seemed to be related to his relatively well-developed expertise with corresponding "pedagogical content knowledge" (Interview, July 30, 2013). He also noted, in contrast, that a senior colleague he had interviewed for a graduate course assignment had told him that, although he *valued* what Mirjan was doing regarding promotion of research-informed and negotiated actions on SSIs, he felt he lacked necessary *expertise* for enacting congruent pedagogical practices (Feb. 26, 2013).

Although we feel that an association between more student-centred pedagogical expertise and enactment of STEPWISE practices is a very significant finding, there also appeared to be several other ramifications of Mirjan's status as a critical and creative reflective practitioner. Although he had positive feelings about his work with instructional intelligence, he also felt a strong need to continue his learning; and, accordingly, in Sept. 2010, he enrolled in a graduate studies programme leading to a Masters of Education degree (now completed). In addition to learning about STEPWISE through a graduate course he took from Larry, he felt that his other courses made significant contributions to his teaching approaches—particularly in the way they enabled him to 'teach against the grain.' In our discussions about this, true to a supporter of constructivist learning theory, it was unclear to him the extent to which his studies added to his ASK, etc., as compared to helping them to emerge from inside of him. Larry suggested this reminded him of a biography of Michelangelo, who was to have said that his task as a sculptor was to chip away at a rock in such a way that the inner figure is revealed. He said the MEd programme played a huge role in motivating him to focus on Critical STSE, primary research and social actions. Several writers, including Hodson and others, "help[ed] bring out something buried in me somewhere" (July 30, 2013). Regardless of their origins, it is apparent to us that many of Mirjan's perspectives have considerable congruence with counter-hegemonic practices in science education. Among these are the following, each with some explanation and evidence:

- <u>Views About Science</u>: Regarding Loving's (1991) *Scientific Theory Profile* (STP), Mirjan appeared to hold views somewhere in the middle of the 'Naturalist-Antirealist' quadrant. He said, for instance, this about his position:

  > I believe that construction of knowledge depends more on socio-cultural contexts rather than strictly adhering to Merton's institutional imperatives and pure logic. I support the idea that one's psychological make up would interact with one's logical reasoning, thus affecting one's judgment. In addition, I hold the position that reaching the 'truth' about knowledge is a matter of consensus amongst professionals and that strict scientific methods may not necessarily lead us to the truth about laws and theories governing physical and natural phenomena (Written Statement About Science, Dec. 12, 2012).

  Holders of such positions seem amenable to exposing students to possible roles for powerful actants within networks involving products of fields of technoscience, which is something Mirjan frequently did during the two years of this study. This may be particularly important for science educators who, through idealized

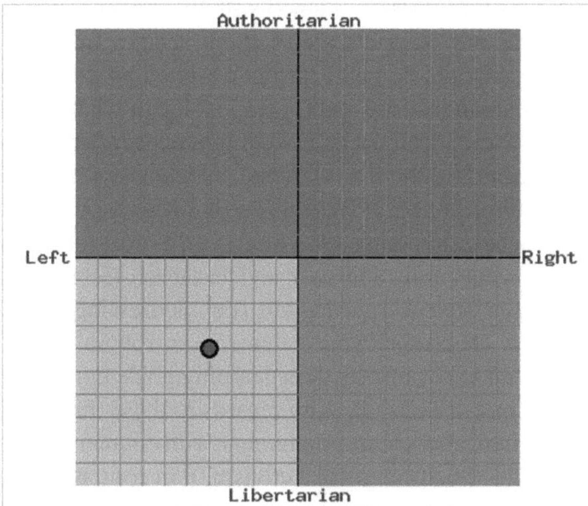

[Economic Left/Right: -4.00
Social Libertarian/Authoritarian: -4.05]

Fig. 11.9 Mirjan's position on the 'Political Compass'

portrayals of the nature of science (refer above), may be unaware that school science can, in effect, behave like a Trojan horse (Bencze & Carter, 2015); that is, hide less positive realities about science—including those involving business-technoscience partnerships (e.g., Mirowski, 2011).

- Views About Society: Mirjan's views of 'society' seemed strongly aligned with his views about science, particularly in terms of his conceptions of roles for powerful actants. Although its accuracy is subject to dispute, he indicated he held relatively strong '(Economic) Left—(Social) Libertarian' views on the *Political Compass* online tool (refer to Fig. 11.9)—which suggests he believes in some government regulation to control powerful economic actants and believes in contextual decision-making in many social matters. About capitalist systems, he said, for instance:

> You can probably argue that the level of control of the capitalist system does not exist, or [at least] not very much. … We live in a capitalist system controlled by a very few. We are subjects of that system. … You call it a democracy, but I cannot protest [as a teacher, under a current law] any more (Interview, Jan. 14, 2013).

Such views seemed congruent with his decision to attend the Occupy Toronto (www.occupyto.org) encampment and share his photographs and stories about it with his students. He explained this move with this response: "I am on the side of the socialists when it comes to the growing gap between the rich and the poor" (Interview, Jan. 14, 2013). Associated with such views, he seemed to have a strong orientation towards social justice—as reflected in this blog entry: "We all can be that disruptive force that works together to bring about change in parts of our world that are plagued with injustice, poverty, hunger, lack of education, violence and inequality" (May 8, 2012).

- Views About Education: Perhaps aligned with his tendency for critical and creative reflective practice, Mirjan's views about science and society seemed to translate, to a considerable extent, into corresponding educational views and practices for students in his care and for teachers with whom he interacts. Many of his educational views contrasted, as we might expect, with those of several of his colleagues. Regarding the interview he conducted with a senior department member, he said, for instance: "I was known [in his previous school] for some of my radical approaches and perhaps some anti-conformist ideas (e.g., challenging the traditional 'concepts' only approach and test-teach-test method) in aspiration of a more *'balanced'* science curriculum [i.e., addressing all 3 learning goals [MoE, 2008]" (Feb. 26, 2013). Other statements corroborated this claim, such as when he said he could not rank the three learning goals in importance, believing them to be equal (Dec. 17, 2012). With regards to these three learning domains, his educational beliefs also seemed congruent with counter-hegemonic practices outlined above (e.g., "Socio-political Activism"). In contrast to traditional orientations towards use of SSIs as instruments for achieving other learning goals, as argued above, he seemed to have strong commitments towards also addressing them as potential problems for wellbeing of individuals, societies and environments. For instance, he said that "the traditional way it [SSI education] is done is …many creative ways…but without the added bonus, if you will, of activism" (Interview, Jan. 14, 2013). He was not, however, one to impose his views on others. Rather, he prefers a more democratic approach in education—such as through sharing of alternative perspectives while allowing learners to self-determine their personal decisions. This applied to his views about teaching of students and about giving other teachers opportunities to implement STEPWISE.

  Regarding his school science teaching, it seems very clear that he valued student participation and a range of educational outcomes above many other considerations. This is quite clear in the word cloud reproduced in Fig. 11.10, which was created from the journal he kept about STEPWISE implementation in the first semester (Sept. 2011–Jan. 2012) of this project. There were several statements and other indictors to corroborate this claim, including a statement in his blog: "Nowadays teenagers want most things around them to be 'customizable.' Classroom experiences are still traditional, for the most part. They are not very customizable by the students. Students should be able to 'customize' their learning experiences by tailoring them to their interests" (Jan. 12, 2012). In a similar vein, he often spoke about promoting his perspectives and practices among colleagues (local and distant), but in a gentle way that honoured their professional integrity. A typical comment in this regard was: "I expect these changes to be slow, but with the right amount of pressure and support, mobilization and advocacy we can initiate a systemic change" (Email message to Larry, Apr. 4, 2013). Indeed, he agreed with the senior colleague he interviewed, who suggested that systemic change will be like trying to change the direction of a supertanker; that is, slowly, with considerable effort (Feb. 26, 2013).

  An interesting tension arose in the context of the many discussions we had to try to understand reasons for Mirjan's teaching against the grain. On the one hand, as indicated above, he seemed very sincere about wanting to share his

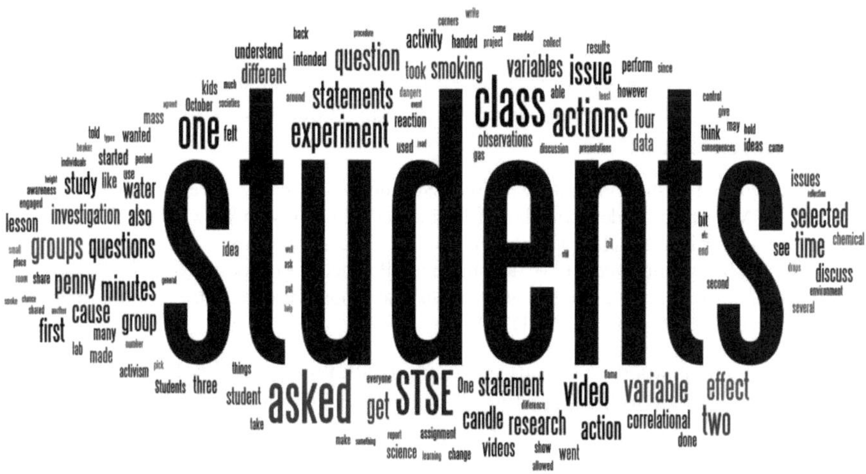

Fig. 11.10 Word cloud created from Mirjan's project journal

ideas and approaches with colleagues. Indeed, he did so on many occasions. On the other hand, there appeared to be some reluctance in this regard. This centred around his clear and frequent claim that he was driven, to a great extent, by his apparent need to be seen as "inimitable" (Dec. 17, 2012) and, perhaps, "part of history" (Jan. 14, 2013). Related to this, he wished that the teaching profession, like fields of medicine, had formal specialties (Blog entry, June 7, 2012). In other words, he wanted to be an educational 'engineer' or 'entrepreneur.' Moreover, he felt such need to be acknowledged for one's notoriety in teaching was normal. He asked, for instance: "Don't we all want to leave a mark. … It is what distinguishes us that drives us to do what we do" (Interview, July 30, 2013). About this, he added:

> Call me crazy, but we have to have that inimitability factor, to define who we are as teachers. … So, do I want everyone in my school [department] doing STEPWISE? [The answer is,] No! [Let me put it this way,] If all of your colleagues did everything the same way [as you], it would really take away from students being in your course (Dec. 17, 2012).

This finding begs the question, 'To what extent do educators need to *believe* in the changes they promote?'

11.5.2.3 Success Breeding Success

Teaching against the grain would, likely, be much more difficult for teachers if its results were not successful. Mirjan, as might be expected, experienced considerable successes using approaches aligned with STEPWISE. A particularly important finding that he noted on several occasions was that, "I have found that my Grade 10 students scored at, or above, average on unit tests and final common exams compared to other Grade 10 classes. This important finding may suggest that even content learning (concepts education) may be improved when the course is contextualized in real, meaningful and relevant STSE issues" (Blog entry, Dec. 3, 2012). Moreover, in addition to noting successes in various learning domains, he also emphasized that the approaches seemed to promote successes among students who often struggled with traditional science teaching:

> I have had a lot of success with the STEPWISE framework. Students are more engaged on average. My weaker students (ie. those who generally do poorly on knowledge-based tests) have done a lot better in class with the STEPWISE. Students like discussing socio-scientific issues and being empowered to act and make a difference in our society. Students' inquiry projects (experiments and correlational studies) have more meaning since they are contextualized (Blog entry, Nov. 23, 2011).

Additionally, there appeared to be considerable 'sticking power' with his approaches. After listening to an audio-recording of student interviews conducted by Larry four months after their course had ended, he remarked how pleased he was that students had retained significant portions of the instruction regarding research-informed and negotiated activism: "This is exactly what I want these kids to be able to do: to talk intelligently about their STSE issues and defend their action months after they have completed them!" (Email message to Larry, Nov. 16, 2012). Finally, although many more examples could be provided, a letter from a twelfth-grade student indicated the degree to which some students were appreciative of Mirjan's work:

> I have always envisioned my life turning out different. I had a few ideas, here and there, of what I would like to dedicate my life to but I was still doubtful. You have given me a place to start with activism and fuelling my passion and motivation into dedicating myself to such causes. It has made me feel beyond ecstatic and energize[d] to have meaning and passion re-enter my life. After reflecting on the past year, it is funny how you came at just the right moment (June 26, 2013).

Such deep, personalized, attachments to education may be explained in various ways. However, again, Wenger's (1998) knowledge duality theory seems helpful here. In other words, with reference to Fig. 11.11, students' engagement in reciprocal relations between phenomena of the 'World' and representation(s) ('Signs') of them may have deepened their commitments to issues, research findings and actions and, more broadly, the education experience they perceived (e.g., regarding the letter from a graduating student to Mirjan, as reflected in the above quote).

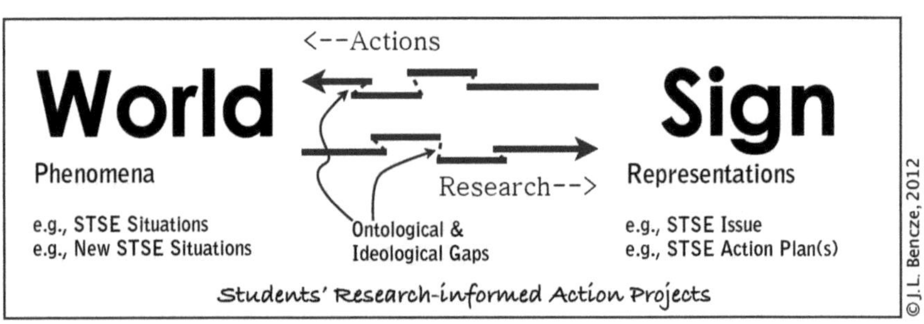

Fig. 11.11 Educational personalization through STEPWISE

11.5.2.4 Educational Contexts

Despite Mirjan's apparent successes, in several ways, conclusions about the degree to which he was able to 'teach against the grain' should be tempered with the realization that STEPWISE implementation significantly depended on his teaching and learning contexts. Much of his overall success must be attributed to the perhaps ideal conditions in the first school in which he had worked (Sept. 2006–June 2012). Mirjan stressed, at different times, that his orientation towards reflective practice arose because, to a large extent, the department head—supported by the school's administrative team (e.g., principal)—in his first school set a tone of exploration for the department colleagues. Mirjan said, for example: "The leader … was, I would say, a more forward-thinking teacher. Leadership is important. He wanted to innovate in the science department by introducing … cross-curricular assignments" (Interview, July 30, 2013). Although he noted that most of his colleagues, while innovative, did not take their instruction beyond levels #1 and 2 in Table 11.1, he felt habits of reflective practice were reinforced from the very beginning of his teaching career.

Particularly in the context of larger schools, Mirjan suggested that an essential condition enabling him to develop and explore new approaches was his freedom to 'step aside from collaboration' with colleagues teaching other sections of the same course. This was a routine practice for any teacher who wanted to innovate in the first school. It also was, though, a circumstance he experienced in the first year of his new, more conservative, school in the second year of the project. He was hired with the recognition that he was performing differently than most teachers and, accordingly, was allowed to teach courses for which there were no other classes of the same course.

11.5.2.5 The Research/Facilitator-Teacher Relationship

Just as collegial support—even in the form of avoidance of collaboration, as had been the case in Mirjan's new school—appeared to be a significant factor in enabling Mirjan to teach against the grain, he believed that our collaboration was a significant factor. While, in his written reflection about what contributed to his abilities to teach against the grain (Aug. 6, 2013), he named several people who had served as "mentors" throughout his teaching career, data from this study clearly indicate effects of Larry's perspectives and practices on those of Mirjan. Mirjan—and, in many cases, students in his classes—routinely used phrases from the STEPWISE project developed by Larry, including: 'wellbeing for individuals, societies & environments,' 'research-informed activism' and 'primary and secondary research.' Moreover, with each successive scholarly paper developed from studies of his teaching, he incorporated ideas from them into his teaching in the next semesters. On the other hand, there also was considerable evidence that Larry's perspectives and practices were affected by his interactions with Mirjan. A clear example of this pertained to the pedagogical framework in Fig. 11.1, which was continuously-revised from an earlier version as new findings emerged from our successive inquiries into Mirjan's practices.

11.6 Summary and Conclusions

Up against a veritable 'Borg' (from the Star Trek™ entertainment series) in the form of the global capitalist network and accompanying Global Education Reform Movement that has, apparently, structured education in ways conducive to maximizing profits for relative few global individuals, some teachers seem to manage to avoid 'being assimilated'—'teaching against the grain,' in the words of Marilyn Cochran-Smith (1991). Mirjan Krstovic appears to be such a teacher. Many of his personal qualities seemed to embolden him to teach in counter-culture ways, including by prioritizing student choice and context, enlightening them about powerful actors apparently arranging their world to suite their interests, and encouraging and enabling them to take actions to bring about a better world. Among these qualities, it seems that his perpetual drive to investigate his teaching, to interrogate fundamental premises influencing his decisions and his energetic and creative spirit, all aimed at helping every student develop to her/his potential, stood out as essential. Indeed, it seemed his penchant for lifelong learning and, moreover, an apparent need to be acknowledged as unique and successful may over-shadow, to a degree, his desires to improve the world through science education.

Although Mirjan embodies many characteristics highly suitable for teaching against the grain towards a better world, we must not consider him as having acting alone in this regard. Based on actor-network theory (Latour, 2005), each 'separate' actant (e.g., human) can be considered to be a 'complex'; that is, an entity composed of influences from many other entities (actants) in dynamic (fluctuating) ways. In other words, we can assume Mirjan represented just one entity within a network that, collectively, enabled and 'defined' (for the context) 'teaching against the grain.' Having acknowledged this complexity, however, the research here suggests some actants within this network were particularly influential. With little doubt, the Ontario government's decision (itself influenced by many actors) to prioritize education about socioscientific issues, students' research and socio-political actions was a major contributing factor; and one that perhaps needs to be celebrated and emulated. Additionally, however, that Mirjan began this project both having previously conducted action research and having been nurtured (e.g., by administrators, senior teachers and mentors) in the early years of his teaching career into habits of critical and creative reflective practice seemed pivotal; and, yet another reminder of benefits of provisions (e.g., time and infrastructure) in school systems for ongoing professional development. Associated with this was an interesting finding that more personally-meaningful learning—like that encouraged by STEPWISE—can be facilitated, in part, by teachers who have developed expertise in a range of student-focused pedagogical practices, such as those inherent to *instructional intelligence* promoted by Barrie Bennett and Carol Rolheiser (2001). At the same time, it seemed clear that a merging of Mirjan's personal characteristics and attitudes, skills and knowledge generated through educational research—here in the form of his graduate studies in education and research collaboration with Larry—was an essential ingredient for the apparently highly-successful teaching against the grain that may contribute to improvements to the wellbeing of individuals, societies and environments.

Appendix A

Sharing the Learning

Mirjan Krstovic, OCT

This year I had the privilege of attending the AERA conference in San Francisco where I presented two research papers with several graduate students and faculty members from the Ontario Institute for Studies in Education (OISE). It was an incredible professional development experience and I'd like to share with you some of the highlights from this conference from my point of view. Over 20,000 people attended the event with many world experts sharing their expertise and wisdom in education. A good friend of mine once said: "What good is your knowledge if you don't share it with others?" So, I hope that you find this document somewhat useful in your attempt to build a more issues-based and action-oriented curriculum that builds our students' character.

E-mail: mirjan.krstovic@peelsb.com
Twitter: www.twitter.com/MKrstovic
ClassWiki: http://mrkrstovic.wetpaint.com
EduBlog: www.mkrstovic.edublogs.org

Favorite Quotes from the Featured Presidential Sessions

	"We are moving form a national, analog, industrial economy to a global, digital, information economy. Every one of our social institutions was created for the former. Schools need to be re-fitted for new society." *Arthur Levine, Woodrow Wilson Fellowship Foundation*
	"Do something with what you know! It is knowing the structure of the discipline, how to use it and how to make it useful that is important to convey to our students—not just the facts and concepts of the discipline." *Sharon P. Robinson, American Association of Colleges for Teachers of Education*
	"We have this highly stratified education system in this country. With education schools historically and legitimately taking pride in serving who? Women, working class people and minorities—none of whom have enjoyed much status in this country." *Kent McGuire, Southern Education Foundation, Inc.*

Please e-mail me (mirjan.krstovic@peelsb.com) **if you would like FULL videos featuring these four prominent speakers from the conference.**

Food for Thought !

Although the featured speakers that I listened to are from the USA, what they are saying (as captured by the quotes above) can be easily applied to our country and much of the 'developing' world, too. As one listens to these dynamic speakers, it's hard not to think about the multitude of ways in which we can better prepare students for the future, especially marginalized and 'at-risk' students. The following questions came to mind after I reflected carefully on the content of each speech:

- What changes need to occur in our practice so that our students are better prepared for this 'new society'?
- How might our practice evolve in response to global and local concerns?
- Our subject disciplines do not occur in some kind of a 'sociocultural vaccum.' How can develop better cross-curricular assignments that allow our students to experience the interdisciplinary nature of education (for citizenship)?

Arne Duncan, US Secretary of Education Controversial Quotes?

Photo by: Mirjan Krstovic

"Today federal state and local governments spend billions of dollars each year on professional development for teachers, yet we know surprisingly little about the effectiveness or return on investment of professional development."

"I've seen devastating effects of poverty…But give me the poorest kid, from the toughest community and the toughest family and put them in a high quality setting, have them go to elementary, middle school and high school, have them go to college—I'm actually optimistic about that child."

"I never said poverty doesn't matter. I said that poverty should never be destiny!"

"We have to close the opportunity gap before we can close the achievement gap. All children need opportunity to be successful."

"Talent, time and resources—that's what will close the opportunity gap." [He is referring to putting talented teachers in tough communities with underprivileged children]

Food for Thought !

The special invited address entitled *"Choosing the right battles: remarks and conversation"* by Arne Duncan, the ninth U.S. secretary of education drew the largest audience at the 2013 AERA conference. Secretary Duncan is a very good orator who also handles controversial questions rather well. Although he made several good points in his speech (e.g., poverty should not be destiny), his policies on 'school reform' supported by President Obama and other Democrats have many *for-profit* educational industries benefiting, particularly technology firms, publishing and testing corporations, test prep and monitoring centres, educational management companies, investment bankers, venture philanthropy and think tanks. Also, servants of power and flexians (people who move between foundations), private companies and universities, as well as human capital economists are all benefitting financially from "Race to the top" reform. So are Secretary Duncan and his allies really serious about reducing inequality gap when his governments' policies are clearly privileging those who are already at the top? Should our country's governments (both federal and provincial) be taking advice from our neighbors? What position do we take if, or when, our governments (and large corporations) start to oppress the less privileged and/or the middle class in our society?

What the Tweets Say About Arne's Speech #AERASec

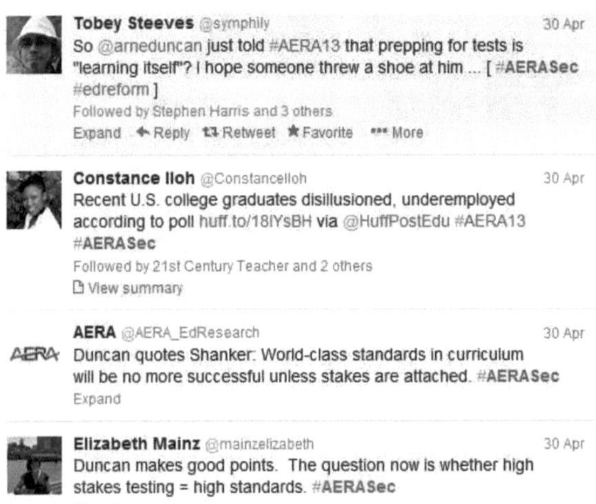

For more Tweeter updates, please visit #AERASec

Photos from Occupy AERA—Protesting Secretary Duncan's Visit to AERA

Photos by: Mirjan Krstovic Videos are also available upon request. Please e-mail me (mirjan.krstovic@peelsb.com) if you are interested for the video footage.

Teacher Leadership for Educational Quality and Equity

What makes education in Finland so unique and highly effective?

This is what I learned from the Dean of University of Helsinki, Finland, Dr. Hannele M. Niemi, at a roundtable session chaired by Dr. Ann Lieberman of Stanford University.

In Finland:

- There is a **high level of respect** for teacher.
- Every teacher is a **leader**.
- It is **very competitive** to get into teacher's college.
- Teachers have high level **pedagogical thinking** but also **pedagogical content knowledge**.
- Teachers understand how students are learning at each stage.
- Teachers are not only responsible for next level of education but also for how students can go forward in their lives.
- There are **no national tests**; that leaves lots of freedom but also lots of responsibilities for teachers.
- Teachers decide how they organize their work in the school in cooperation with other teachers and principals.
- Every teachers must have a **5 year Master's level degree**.
- Teachers must have **high level analytical thinking** and also credibility to make decisions based on evidence.
- Teachers go in many different places in Finland - they are highly sought after professionals in Finnish society.
- Teachers are proud of their profession in Finland.

- Principals are also teaching.
- Beginner teachers have high level of support and mentoring.
- Collaboration with outside partners is an area of need for professional development for teachers.

Audio recording of Dr. Nieme's talk is available upon request. Please e-mail: mirjan.krstovic@peelsb.com

Learning with Dr. Ann Lieberman

Ann and I after a roundtable discussion about teacher leadership. Ann Lieberman was previously a senior scholar at The Carnegie Foundation for the Advancement of Teaching and is Professor Emeritus of Education at Columbia University. She is also a senior scholar at Stanford University. I met Ann four years ago when I was part of the Teacher Learning and Leadership Program (TLLP) for experienced teachers. Ann was the key note speaker at the TLLP conference. I kept in touch with Ann. Her work has greatly contributed to a very successful cross-curricular *Leadership for Learning* team that was formed five years ago at Fletcher's Meadow Secondary School. Ann joined us via web conference during one of our professional learning team meetings. It was a pleasure seeing her again in San Francisco where we shared our learning about teacher leadership.

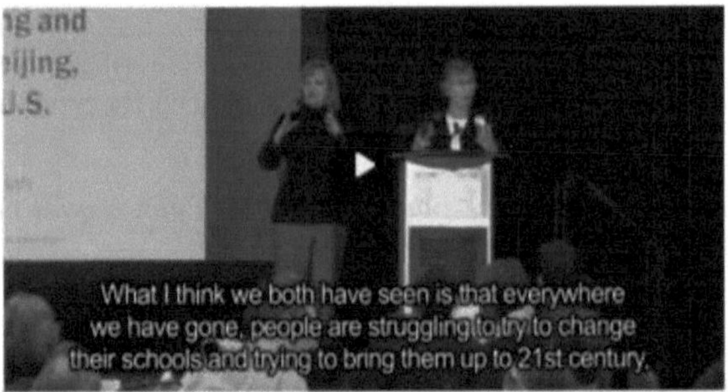

Learning from Dr. Pedro Noguera

Dr. Pedro Noguera, is the Professor of Education at New York University. He is a remarkable speaker who certainly knows how to motivate and inspire the crowd!

Key Learning Points from Dr. Noguera's Speech:

- We give the most to children that have the most, and **we don't give enough to those who need it the most.**
- We need to **empower kids as learners**! We need to spend more time on empowering kids and motivating them to learn!
- Failure is not an option! **Fear is not a motivator**! What kids need is hope!
- Racial inequality is still an issue—how do we reduce these disparities?
- So much of what is wrong has to do with the politics of education and the policies of education. The fact that we continue to view education as a vacuum, that we have not created a more integrated strategy connected to health, housing, community development… So much of what's wrong is that **we continue to blame teachers for problems that they do not create**.
- We need to move away from blaming to thinking more deeply about the nature of the problems.

Key Ingredients for School Improvement

1. Coherent instructional guidance system
2. Development of the professional capacity of its faculty
3. Strong parent and community school ties
4. A student-centered learning climate
5. Leadership that drives learning

Questions for Thought

1. How do we build and sustain a climate for learning in our schools?
2. What additional resources do we need to help sustain this climate for learning?
3. How do we foster community partnerships to support our climate for learning?
4. Which of the above ingredients for school improvement needs more attention at our school?

Teacher Activists and Character Education

Three characteristics of teacher activists:

(i) Vision of a more socially just world
(ii) Work to enact this vision
(iii) Stand up to oppression

Vincent Harding's Wisdom

"If we teach youth to run away from the darkness, rather than to open up the light in it, then we are doing great harm to them."

"We human beings are meant to be sources of light for each other."

"How do we talk together in ways that will open up our best capacities and our best gifts?"

Resources

http://www.youtube.com/watch?v=ouudXr-csZg

*"We have to re-think the education.**Education has to be for social and political action!**"* - Dr. Chris Emdin

Great TED Talks on Ecological Justice

http://www.youtube.com/watch?v=EzZzZ_qpZ4w

These are truly inspirational videos!

http://www.youtube.com/watch?v=gQ-cZRmHfs4

Learning from the Homeless on the Streets of San Francisco

As I walked out the Hilton Hotel I came across a queue of homeless people who were waiting rather patiently to get lunch from a local church organization called "Glide". I asked for permission to interview a few of them to learn more about their experiences since the theme of the AERA conference was Education and Poverty.

People flock here [San Francisco]…there is so much homelessness here. Here you get a tiny room with shared bathroom for $1000 per month. There are not enough spaces in shelters to accommodate everyone….A lot of homelessness as well is mental illness, drugs, alcoholism, all this stuff. I've been on the streets for more than three years.

I've been on a waiting list for five different apartments for the past six months. The average waiting list is between two to four years for single room occupancy.

It's also safe to say that, by now, many Bay Area residents have realized that every trip to the grocery store and the BART station will necessitate walking past between two and twenty people with outstretched hands, shadowed by at least twice that number in severe, specific, and immediate need.

This isn't the full picture. This is just *their* picture, the picture that colors their neighborhood, their day, their sense of community and fairness, and whether or not the world is a good place to be.

It's safe to say most of them have hit the breaking point and can no longer imagine that handing out dollars and dimes represents any kind of solution to poverty. It's safe to say that most of them recognize that a radical change in housing policy is not just a civic, but also a moral obligation.

Excerpt from the Street Spirit newspaper given to me by one of the homeless gentlemen.

Check out this newspaper. A number of homeless people were handing it out on the street in exchange for some money. Articles from this newspaper can be used to stimulate discussion in class around social justice. Visit www.thestreetspirit.org

"The Housing Circle of Life." Artwork by Carol Denney

Social Justice in MATH Class!

Dr. Indigo Esmonde, Assistant Professor from OISE/UT, specializes in learning to teach mathematics for social justice! I was impressed with her presentation and her enthusiasm for a 'different' kind of math education. She introduces us to the opportunities for learning math in the context of community activism. Who would have thought these two fields can ever connect?

I took a photograph of the Abstract of her poster:

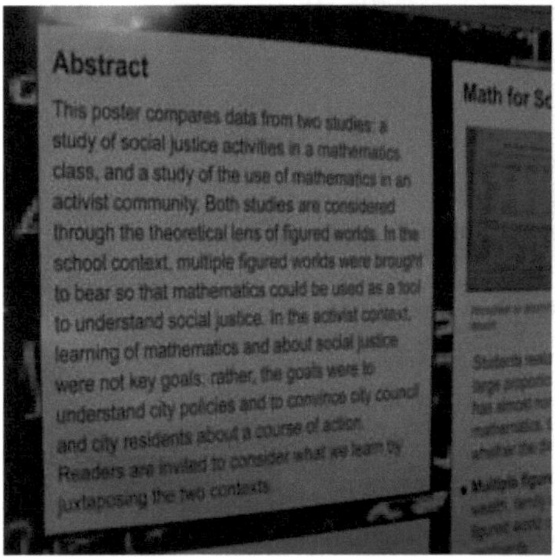

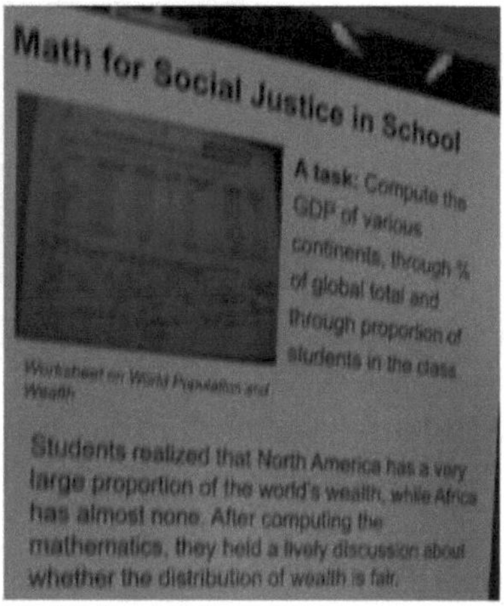

For the math teachers interested in social justice, contact Dr. Esmonde at iesmonde@oise.utoronto.ca. I spoke with Indigo on the airplane on the way back from San Fran and she is very cool, and looks forward to connecting with teachers who want to adopt social justice perspective in their math classes. Imagine the possibilities!

My Contributions to AERA

Students' Social Studies Influences on Their Socioscientific Actions

A Presentation at the annual conference of the *American Educational Research Association* April 27–May 1, 2013, San Francisco, CA, USA

Theme; "Education and Poverty; Theory, Research, Policy, and Praxis"

J. Lawrence Bencze & Mirjan Krstovic

For interested staff, full article that Dr. Bencze and I presented can be found here:http://webspace.oise.utoronto.ca/~benczela/AERA2013_Bencze-Krstovic.pdf

Christina Phillips and I during our poster presentation (See Abstract of our paper below)

Students' Socioscientific Actions: U Sing & Enhancing Their ' 'Street Smarts'

Christina Phillips, Mirjau Krstosvic and J. Lawrence Beneze Ontario Institute for Studies in Education. University of Toronto

Abstract
In the context of promoting actions to address socioscientific issues, we found that student-led research projects were effective 'border-crossing' instruments that enabled students to increase their scientific street-smart savvy. Arguably, increased street smarts provided by research-informed student activism allow a greater degree of scientific knowledge to be internalized as students have been permitted to explore the discipline as personalized 'experts' These approaches may be instrumental in increasing student success and transcending scientific power dynamics as they enable students to view-science as a relevant participatory activity. This may aid in the diminishment of intellectual, economic and moral poverty where future citizens arc more capable of making scientifically-literate decisions.

THANK YOU!

I would like to give special thanks to the following people:

- Principal **Sue Turner** for her support, interest in the conference and approval of my PD leave
- VP **Chris Lane** for arranging my supply coverage

- **Kathryn Dertinger** for her support in taking care of the STPDL forms
- **OSSTF** for the supply coverage money
- **Erindale Science** staff for checking up on my classes while
- I was away ☺
- Colleagues from OISE/UT, especially **Dr. Bencze** for his scholarship and mentorship
- My ***Leadership for Learning Team***—too many to name ☺

References

Achieve, Inc. (2013). *Next generation science standards*. Washington, DC: Achieve Inc..

Angell, M. (2004). *The truth about the drug companies: How they deceive us and what to do about it*. New York: Random House.

Argyris, C. (2002). Double-loop learning, teaching, and research. *The Academy of Management Learning and Education, 1*(2), 206–218.

Bakan, J. (2011). *Childhood under siege: How big business targets children*. Toronto, ON: Allen Lane.

Ball, S. J. (2000). Performativities and fabrications in the education economy: Towards the performative society? *Australian Educational Researcher, 27*(2), 1–23.

Ball, S. J. (2003). The teacher's soul and the terrors of performativity. *Journal of Education Policy, 18*(2), 215–228.

Ball, S. J. (2012). *Global Education Inc.: New policy networks and the neo-liberal imaginary*. Abingdon, UK: Routledge.

Barber, B. R. (2007). *Consumed: How markets corrupt children, infantilize adults, and swallow citizens whole*. New York: Norton.

Barnett, J., & Hodson, D. (2001). Pedagogical context knowledge: Toward a fuller understanding of what good science teachers know. *Science Education, 85*(4), 426–453.

Barnes, R. L., Hammond, S. K., & Glantz, S. A. (2006). The tobacco industry's role in the 16 Cities Study of Secondhand Tobacco Smoke: Do the data support the stated conclusions? *Environmental Health Perspectives, 114*(12), 1890–1897.

Baudrillard, J. (1998). *The consumer society*. London: Sage.

Bell, R. L. (2006). Perusing Pandora's box: exploring the what, when, and how of nature of science instruction. In L. B. Flick & N. G. Lederman (Eds.), *Scientific inquiry and nature of science: Implications for teaching, learning, and teacher education* (pp. 427–446). Dordrecht, The Netherlands: Springer.

Bencze, J. L. (1996). Correlational studies in school science: Breaking the science-experiment-certainty connection. *School Science Review, 78*(282), 95–101.

Bencze, J. L. (2010). Exposing and deposing hyper-economized school science. *Cultural Studies of Science Education, 5*(2), 293–303.

Bencze, J. L., & Alsop, S. (2009). A critical and creative inquiry into school science inquiry. In W.-M. Roth & K. Tobin (Eds.), *The world of science education: North America* (pp. 27–47). Rotterdam, The Netherlands: Sense.

Bencze, J. L., & Carter, L. (2015). Capitalists' profitable virtual worlds: Roles for science & technology education. In P. P. Trifonas (Ed.), *International handbook of semiotics, vol. 1 & 2* (pp. 1197–1212). Dordrecht, The Netherlands: Springer.

Bencze, L., & Carter, L. (2011). Globalizing students acting for the common good. *Journal of Research in Science Teaching, 48*(6), 648–669.

Bennett, B., & Rolheiser, C. (2001). *Beyond Monet: The artful science of instructional integration*. Toronto, ON: Bookation Inc..

Bourdieu, P. (1986). The forms of capital. In J. G. Richardson (Ed.), *The handbook of theory: Research for the sociology of education* (pp. 241–258). New York: Greenwood Press.

Callon, M. (1986). Some elements for a sociology of translation: Domestication of the scallops and the fishermen of St-Brieuc Bay. In J. Law (Ed.), *Power, action and belief: A new sociology of knowledge?* London: Routledge and Kegan Paul.

Carter, L. (2008). Globalisation and science education: The implications for science in the new economy. *Journal of Research in Science Teaching, 45*(5), 617–633.

Charmaz, K. (2014). *Constructing grounded theory* (2nd ed.). New York: Sage.

Claxton, G. (1991). *Educating the inquiring mind: The challenge for school science*. London: Harvester Wheatsheaf.

Cochran-Smith, M. (1991). Learning to teach against the grain. *Harvard Educational Review, 61*(3), 279–310.

Cohn, T. H. (2005). *Global political economy: Theory and practice*. New York: Pearson Education.

Crawford, B. A. (2007). Learning to teach science as inquiry in the rough and tumble of practice. *Journal of Research in Science Teaching, 44*(4), 613–642.

Dewey, J. (1916). *Democracy and education: An introduction to the philosophy of education*. New York: MacMillan.

Foucault, M. (1977). *Discipline and punish: The birth of the prison*. New York: Pantheon Books.

Foucault, M. (1991). Governmentality. In G. Burchell, C. Gordon, & P. Miller (Eds.), *The Foucault Effect: Studies in Governmentality* (pp. 87–104). Hemel Hempstead, UK: Harvester Wheatsheaf.

Gaines, B. R., & Shaw, M. L. G. (1993). Knowledge acquisition tools based on personal construct psychology. *Knowledge Engineering Review, 8*(1), 49–85.

Giroux, H. A., & Giroux, S. S. (2006). Challenging neoliberalism's new world order: The promise of critical pedagogy. *Cultural Studies ↔ Critical Methodologies, 6*(1), 21–32.

Gramsci, A. (2003 [c1929–1935]). *Selections from the Prison Notebooks Of Antonio Gramsci* (Q. Hoare, & G. Nowell Smith, Eds. & Trans.). New York: International Publishers.

Greenwood, D. J., & Levin, M. (2000). Reconstructing the relationships between universities and societies through action research. In N. K. Denzin & Y. S. Lincoln (Eds.), *The handbook of qualitative research* (2nd ed., pp. 85–106). Thousand Oaks, CA: Sage.

Guba, E. G., & Lincoln, Y. S. (2011). Competing paradigms in qualitative research. In N. K. Denzin & Y. S. Lincoln (Eds.), *The SAGE handbook of qualitative research* (pp. 105–117). Thousand Oaks, CA: Sage.

Hargreaves, A., Earl, L., Shawn, M., & Manning, S. (2001). *Learning to change: Teaching beyond subjects and standards*. San Francisco, CA: Jossey-Bass.

Hegel, G. W. F. (1977 [1806]). *Phenomenology of spirit* (A.V. Miller, Trans.). Oxford: Oxford University Press.

Hileman, B. (1998, August 17). Industry's privacy rights: Is science shortchanged? *Chemical & Engineering News*, *76*, 36.

Hodson, D. (2008). *Towards scientific literacy: A teachers' guide to the history, philosophy and sociology of science*. Rotterdam: Sense.

Hodson, D. (2011). *Looking to the future: Building a curriculum for social activism*. Rotterdam, The Netherlands: Sense.

Kemmis, S., & McTaggart, R. (2000). Participatory action research. In N. K. Denzin & Y. S. Lincoln (Eds.), *Handbook of qualitative research* (2nd ed., pp. 567–606). Thousand Oaks, CA: Sage.

Khishfe, R., & Lederman, N. G. (2006). Teaching nature of science within a controversial topic: Integrated versus nonintegrated. *Journal of Research in Science Teaching, 43*(4), 395–418.

Klein, N. (2014). *This changes everything: Capitalism and the climate*. Toronto, ON: Simon & Schuster.

Kleinman, D. L. (2003). *Impure cultures: University biology and the world of commerce*. Madison, WI: University of Wisconsin Press.

Krstovic, M. (2009). *Improving subject area literacy learning for grade 9 applied science students. An unpublished report for the Teacher Learning and Leadership Program*. Mississauga, ON: Peel District School Board.

Krstovic, M., & Bencze, J. L. (2012a). Using a 'card exchange game' to help students explore STSE issues. *Crucible*, *43*(3), Online at: http://www.stao.ca

Krstovic, M., & Bencze, J.L. (2012b). The use of correlational studies to encourage activism in science. *Crucible*, *43*(4), Online at: http://www.stao.ca

Kuhn, T. S. (1970). *The structure of scientific revolutions* (2nd ed.). Chicago: University of Chicago Press.

Latour, B. (2005). *Reassembling the social: An introduction to actor-network-theory*. Oxford: Oxford University Press.

Lazzarato, M. (1996). Immaterial labor. In P. Virno & M. Hardt (Eds.), *Radical thought in Italy: A potential politics*. Minneapolis, MN: University of Minnesota Press.

Leonard, A. (2010). *The Story of Stuff: How our obsession with stuff is trashing the planet, our communities, and our health - and a vision for change*. New York: Free Press.

Levinson, R. (2010). Science education and democratic participation: An uneasy congruence? *Studies in Science Education, 46*(1), 69–119.

Lock, R. (1990). Open-ended, problem-solving investigations: What do we mean and how can we use them? *School Science Review, 71*(256), 63–72.

Loving, C. C. (1991). The Scientific Theory Profile: A philosophy of science model for science teachers. *Journal of Research in Science Teaching, 28*(9), 823–838.

Lyotard, J.-F. (1984). *The postmodern condition: A report on knowledge*. (G. Bennington, & B. Massumi, Trans.). Minneapolis, MN/Manchester, UK: University of Minnesota Press/ University of Manchester Press.

Marx, K. (1990 [1867]). *Capital: A critique of political economy* (Vol. I). London: Penguin Books.

McLaren, P. (2000). *Che Guevara, Paulo Freire, and the pedagogy of the revolution*. Lanham, MD: Rowman & Littlefield.

McMurtry, J. (2013). *The cancer stage of capitalism: From crisis to cure*. London: Pluto.

McQuaig, L., & Brooks, N. (2010). *The trouble with billionaires*. Toronto, ON: Viking.

Merton, R. K. (1973). The normative structure of science. In R. K. Merton (Ed.), *The sociology of science: Theoretical and empirical investigations* (pp. 256–278). Chicago: University of Chicago Press.

Ministry of Education [MoE]. (2008). *The Ontario curriculum, grades 9 and 10: Science*. Toronto, ON: Queen's Printer for Ontario.

Mirowski, P. (2011). *Science-mart: Privatizing American science*. Cambridge, MA: Harvard University Press.
National Academies of Sciences (NAS). (2007). *Rising above the gathering storm: Energizing and employing America for a brighter economic future*. Washington, DC: National Academies Press.
National Research Council [NRC]. (2011). *Successful STEM education: Identifying effective approaches in science, technology, engineering, and mathematics*. Washington, DC: National Academies Press.
Oreskes, N., & Conway, E. (2010). *Merchants of doubt*. London: Bloomsbury Press.
Osborne, R., & Wittrock, M. (1985). The Generative Learning Model and its implications for science education. *Studies in Science Education, 12*, 59–87.
Pedretti, E., & Nazir, J. (2011). Currents in STSE education: Mapping a complex field, 40 years on. *Science Education, 95*(4), 601–626.
Percy, J. R., & Krstovic, M. (2001). Later life learners: A significant and receptive audience for introductory astronomy. *Journal of the Royal Astronomical Society of Canada, 95*, 205–207.
Pierce, C. (2013). *Education in the age of biocapitalism: Optimizing educational life for a flat world*. New York: Palgrave Macmillan.
Sadler, T. D., Barab, S. A., & Scott, B. (2007). What do students gain by engaging in socioscientific inquiry? *Research in Science Education, 37*(4), 371–391.
Sellar, S., & Lingard, B. (2013). The OECD and global governance in education. *Journal of Education Policy, 28*, 710–725.
Springer, S., Birch, K., & MacLeavy, J. (Eds.). (2016). *The handbook of neoliberalism*. New York: Routledge.
Usher, R. (2010). Consuming learning. In J. A. Sandlin & P. McLaren (Eds.), *Critical pedagogies of consumption: Living and learning in the shadow of the "Shopocalypse"* (pp. 36–46). New York: Routledge.
Wasser, J. D., & Bresler, L. (1996). Working in the interpretive zone: Conceptualizing collaboration in qualitative research teams. *Educational Researcher, 25*(5), 5–15.
Wenger, E. (1998). *Communities of practice*. Cambridge, UK: Cambridge University Press.
Yin, R. K. (2003). *Case study research: Design and methods* (3rd ed.). Thousand Oaks, CA: Sage.
Ziman, J. (1984). *An introduction to science studies: The philosophical and social aspects of science and technology*. Cambridge, UK: Cambridge University Press.

Chapter 12
Giving Meaning to STSE Issues Through Student-Led Action Research: An Actor-Network Theory Account of STEPWISE in Action

Neil T. Ramjewan, Brandon Zoras, and Larry Bencze

12.1 Introduction

In this chapter, we use Actor-Network Theory (ANT), a disparate assemblage of material-semiotic tools (Law, 2009), to describe our implementation of STEPWISE in a Grade 9 pre-International Baccalaureate class. STEPWISE is a pedagogical framework for science education that promotes wellbeing for individuals, societies and environments in hopes of developing reflective, critical, and sustainable relationships between these domains by resisting neoliberal logics that instrumentalize science and technology education, research, and practice as a profit centers within a free market paradigm (Bencze & Carter, 2011). STEPWISE tries to reposition and reimagine science and technology education such that its participants can envision themselves as ethical society makers, changers and activists (Bencze, Sperling, & Carter, 2012).

This study highlights merits of determining a personally-relevant STSE issue and ambivalence in said process in terms of framing and moving between localities and the global. Furthermore, we consider the role of a digital virtual social media space, 'the Wiki,' and its role in constituting subject positions and shaping capacities for action or, dare we say, agency. Finally, we consider effects of engaging in activist science projects and their roles in displacing students from determinism inherent to traditional science pedagogies into ambivalent and indeterminate crossroads of reflexive society making.

N.T. Ramjewan (✉) • L. Bencze
Ontario Institute for Studies in Education (OISE), University of Toronto, Toronto, ON, Canada
e-mail: neil.ramjewan@mail.utoronto.ca; larry.bencze@utoronto.ca

B. Zoras
Toronto District School Board, Toronto, ON, Canada
e-mail: brandonzoras@gmail.com

L. Bencze (ed.), *Science and Technology Education Promoting Wellbeing for Individuals, Societies and Environments*, Cultural Studies of Science Education 14,
DOI 10.1007/978-3-319-55505-8_12

12.2 Theoretical Background

12.2.1 Actor-Network Theory in Abstraction

ANT is employed methodologically in this study as a means of doing what some suggest ANT does best — tell interesting and *descriptive* stories (Law, 2009) using the vocabulary of ANT in search of new descriptors to narrate how our network(s) wax(es) and wane(s). But, first, the question of whether ANT can be abstracted and, thus, whether it warrants nomenclature of being a theory at all has been a source of ongoing controversy that must be considered momentarily (Calás & Smircich, 1999). Though the focus of this section is not to engage in that debate, perhaps attempting to state what *is* ANT is somewhat a participation in that debate; but, like John Law (2009), who makes a definitive claim that ANT can be abstracted and reified as a coherent assemblage of critical ideas and methods, we attempt to briefly do the same here. By implication, this text in its attempt to state what is ANT, as Law (2006) reminds us, is a translation of ANT through our lens and situation and, thus, is immediately a betrayal of the nature of ANT.

With origins in the sociology of science and technology, ANT, also known as the "sociology of translation" (Law, 1992, p. 380), is concerned with power and relationships. ANT is interested in *the* social but as a descendant of postmodernism and poststructural theory, ANT is invested in particularities of social phenomena, or multiplicities of social *orders.* Law (1992) notes, "…there is no such thing as … "*the* social order" with a single center, or a single set of stable relations. Rather, there are orders, in the plural" (p. 386, italic in original). As such, ANT is invested in describing how multiple social localities come into being, interact, and fall out of being.

In a seminal study, Michel Callon (1986a) asserts that ANT is founded on three principles: agnosticism, generalized symmetry and free-association. These principles suggest that ANT, in its telling of small stories, is impartial to its constitution of actors, which can be human or non-human, or materially heterogeneous, and can in turn be ontologically social, natural and technological (Wong & Tatnall, 2010). This posture has the effect of flattening the social by way of not privileging human actors and acknowledging dependence of social orders on all forms of materiality (Latour, 2005). To further elaborate the concept of an actor, Arthur Tatnall and Stephen Burgess (2002) draw on Law (1992), noting: "An actor is seen not just as a 'point object' but rather as an association of heterogeneous elements, themselves constituting a network. Each actor is thus itself also a simplified network" (p. 183). Hence, we get the Actor-Network hyphenation, ambiguity, and fluidity that the name "ANT" suggest in the constitution of its object of study.

By implication, networks are black-boxed, or hidden, as they are semiotically represented resulting in an oversimplification of said networks in the form of punctualized actors, perceived as singular, stand-alone entities. Callon (1986b) considers that which becomes black-boxed followed by a commonly used strategy of un-black-boxing, or thick descriptions of actors and relations that bring them into being, in turn exposing other critical and relational actors. Law (2009) further elaborates the importance of such relations indicating that ANT is "a disparate family of material-semiotic tools, sensibilities and methods of analysis that treats everything in the social and natural worlds as a continuously generated effect of the webs of relations within which they are located" (p. 141) outside of which there is no discernible reality.

Marta Calás and Linda Smircich (1999) note that early ANT scholarship conceived networks as manifest out of research related analytical structures created by analysts. As such, early ANT scholars were interested in ways in which actors became centered as part of a network. Works by scholars such as Callon (1986a) with this goal in mind call attention to four moments in sociological translation that provide a vocabulary for analysis that we utilize in this study.

The first moment of translation, *problematization*, involves key actors re-defining problems in terms of constructed solutions made possible only by said actors, thus rendering the actor indispensable. This usually involves a double movement (Callon, 1986a) in which said actor(s) determine other actors in the network and their identities in relation to the solution proposed, an effect that translates said actor into an obligatory point of passage (OPP) in the network, or a node which other actors must encounter to exist in the network (Callon, 1986a). The second move, *interessement*, involves an actor (A) becoming 'interested' or coming between an entity (B), which is brought into relational existence by (A), and other entities (C, D, E,...n) trying to coopt (B) into problematic relations. The function of this move further stabilizes the network and its actors via the imposition of roles onto surrounding actors (Wong & Tatnall, 2010), effectively locking allies in their place (Callon, 1986a). The third moment is *enrolment*, or the coordination of imposed roles. However, of critical distinction is that roles are not only imposed through the process of problematization and interessement, but now they are 'yielded-to' in enrolment, a structural willingness again strengthening network relations. Finally, *mobilization*, refers to network spread and acceptance accomplished by 'spokespersons' in which some actors speak on behalf of other actors thereby punctualizing, or black-boxing, certain actors, which are actually networks in themselves (Callon, 1986b).

In more recent scholarship, those using ANT have not only been interested in this centering but also means by which actors become decentered, how networks fail and "oscillations" that seem to occur, a concept referred to as "ontological choreography" (Calás & Smircich, 1999, p. 663). These studies have included concepts such as ambivalence, indeterminacy and multiplicity (Singleton & Michael, 1993) that help to describe fluidity and ever- changing characteristics of networks that contribute to their ebb and flow and account for maintenance of networks, while highlighting the precarious nature of networks (Law, 1992). These notions play in the realm of uncertainty, in which all relations are mediated and are extensively considered through analyses in this study.

Lastly, but certainly not least, is the notion of reflexivity. Calás and Smircich (1999) note that ANT is a reflexive methodological paradigm, or a system of knowledge production that "both constitutes and describes its objects of interest" (p. 663). ANT traditions have centered this reflexivity thereby offering a criticism of positivism and relatedly empirical studies in general, which operate under the same terms of reflexivity, but manage to decenter this self-referentiality by privileging a distinct ontological separation between observer and object. For ANT scholars and analyses, this point of reflexivity is not one that is elided but, rather, embraced and centered with two ends in mind.

First, acknowledging this self-fulfilling form of knowledge production, ANT analyses irrevocably bind the observer and its objects, yielding an infinity of intersectionality and complexity in construction of descriptions of the world and phenomena. Consequently, this makes ANT particularly capable of telling detailed, fluid, and dynamic stories of reality from multiple perspectives, but in a way that "defamiliarizes what we may otherwise take for granted" (Calás & Smircich, 1999, p. 663). Second, this centering of reflexivity implicates ANT *as an actor* and thus, an Actor-Network in the networks it seeks to describe (Singleton & Michael, 1993). This point is crucial to the framing of this study since it was rooted in elaborating interconnectedness of actors in our context. Consequences of this involve possible dissolution of ANT as a cohesive Actor-Network that will inevitably come with the passage of time as 'it,' as an assemblage of practices, is translated into different intellectual tools and practices seeking to make sense of our realities (Law, 2009).

12.3 Why ANT?

Pedretti and Nazir (2011), in their comprehensive report on the last 40 years of STSE education, point to a critical moment in science education's recent history that involves making connections among fields of science, technology, societies and environments (STSE). This study and STEPWISE are part of that tradition in many ways. In continuing efforts to make connections, enabling students to see STSE relationships of which they are a part, inevitably uncovers relationships and exposes networks of interrelation. In fact, one of the earliest activities that high school students engaged in involved identifying entities connected to specific technologies keeping the acronym STSE in mind (see Fig. 12.1). Building on these connections and using the STEPWISE framework, we were interested in helping students situate themselves in these networks with the intention of realizing that their actions can ultimately disrupt said relations, leading to possible change and resolution of STSE issues. We believe that these efforts to build and 'see' networks surrounding our technoscientific contemporary lives make obvious connections to ANT and its concern with network translation, choreography, and oscillation, making ANT a suitable theoretical lens for understanding STEPWISE in action.

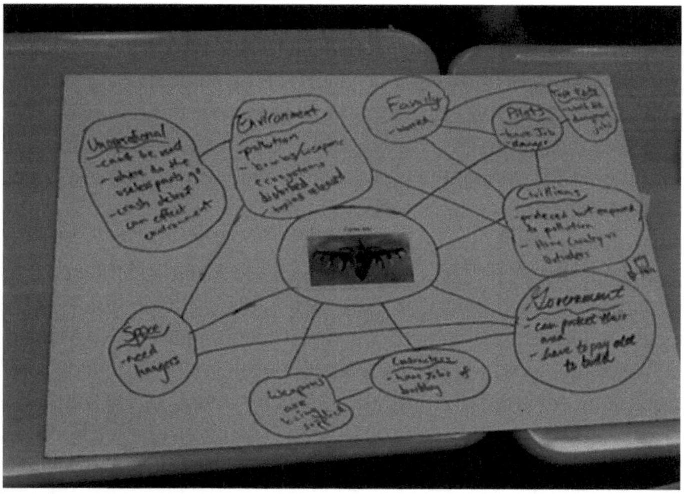

Fig. 12.1 Student work sample of a technology (fighter jet) located in a network of relations

12.4 Research Context and Methods

The study reported here takes place primarily at a large urban public high school, specifically in a course titled 'Grade 9 Academic Science' that is part of the International Baccalaureate (IB) program. We implemented STEPWISE over the course of a semester from September 2013 to January 2014. The IB Diploma Program (DP) officially begins in Grade 11, but Grade 9 and 10 are transition years, or "pre-IB" with a less dense curriculum that involves close ties with the IB system, curriculum, standardized testing, and external marking protocols (ibo.org/en/programmes/diploma-programme) and so on. The pre-IB student participants in this study were screened by the school to determine eligibility for the IB program based on both academic and non-academic criteria. The high school teacher, Mr. Zoras (author two here), had 6 years of teaching experience, although this was his first time teaching in the IB program. Mr. Zoras had previously participated and implemented STEPWISE with another group of high school students (see Zoras & Bencze, 2014).

As a pedagogical framework that tries to disrupt and decenter neoliberal logics, STEPWISE values student knowledge and interest. Thus, the framework promotes student-directed, as opposed to teacher-directed, learning pathways and open-ended versus closed-ended rigid curricular outcomes and expectations (Bencze, Alsop, & Bowen, 2009). High school students were taught, via an accelerated apprenticeship model, how to engage in research-informed and negotiated action (RiNA) projects that address STSE issues. In general, students were taught to conduct secondary research, primary research and, finally, take action to engage STSE issues.

Secondary research involves using sources about others' research, such as existing reports and articles on STSE issues accessible via the library or mainly the Internet. This skill was directly tied to identification of an STSE issue, which was student-directed in the sense that it was determined by students, but somewhat closed in the sense that it needed to relate to the IB unit of study (Ecology). Primary research, or research in which students collected data, typically via creation of digital surveys using Google Forms™, was analyzed by students to inform them about how people in their daily lives related to the STSE issue they chose to address. This information was then utilized to design and conduct actions to address, disrupt, or make better their STSE issue, the outcomes of which were unpredictable and thus, open-ended by design.

The final phase of the study involved high school students, with the gained expertise of doing RiNA projects, mentoring Grade 5 and 6 elementary school students to conduct RiNA projects of their own. This collaboration involved digital communication and interaction across a social media platform — Wikispace™ — hereafter "the Wiki" to which it was referred throughout. All interactions with the elementary school occurred digitally either synchronously or asynchronously using either Google Talk™, the Wiki, or YouTube™. The elementary school teacher, Mr. Romano, did not teach students how to conduct STEPWISE projects, rather supported their use of technology, thus enabling high school students to be teachers of activist science. The elementary school is part the same large urban school board as the high school, but is geographically distinct and culturally and linguistically diverse.

Unfortunately, the course of collaboration between classes was fraught with institutionalized schedules leading to temporal conflicts and ultimately, *failure* of this aspect of the research to fully materialize. But why mention it here as opposed to just leaving it out of this text? The primary reason for this is because students were informed of this being a possible aspect early in the course of the project and, consequently, imposed social relations thereby producing social positions, perceptions, and actions. As well, high school students created teaching resources for the elementary school students yielding instructional artefacts for analysis in this study. In fact, the elementary school students did manage to use and implement these resources in their own RiNA projects; however, due to network constraints, such as end-of-school year and frequency of project engagement by elementary school students, sufficient data were not collected to fully comment on potential outcomes of these relations in terms of mentorship. However, because of digital mediums and resulting virtual collaborative spaces (i.e., Wiki/social media) within which said artifacts are produced (teaching resources such as videos), we do consider this relationship in ANT terms since these artefacts are representative of said relations and are, thus, actors *in* network relations.

12.5 Data Collection and Analyses Methods

Data were created, collected, and mined in a variety of ways using in a number of sources, including:

- *Semi-structured interviews:* As a team, researchers devised questions to learn about processes of doing RiNA projects and teaching said process to younger students and how this relates to one's social position in society. Seven high school students were interviewed on three occasions for approximately 20 minutes each session. Three elementary school students were interviewed once for approximately 20 minutes each. Interviews were recorded and transcribed. The transcriptions were coded and themes were allowed to emerge from the transcripts.
- *Artefacts of students' work:* As part of the STEPWISE project, students were required to produce certain products for teacher evaluation. These products were created on a variety of web platforms (Facebook™, Twitter™, YouTube™, word processor) but were stored or presented in the class/project Wiki site organized by groups and further into STSE issue secondary research, primary research (mobilized by Google's™ suite of analytic tools), and actions. As well, each student was responsible for three reflections using an Attitudes, Skills, Knowledge (ASK) framework to help guide their reflections when thinking about STEPWISE, engaging as activists, and thinking about social position.
- *Anecdotal records/Observational notes:* Significant portions of in-class time was allotted for the projects, thus, field notes were used to supplement the picture of our story.

Using constant comparative methods premised on constructivist grounded theory, we discussed the emerging data regularly to dialectically construct the direction of the project and to make meaning of what was coming into focus. Data were codified, categorized and themes were allowed to emerge. Themes were negotiated between researchers/teachers to improve the trustworthiness of claims (Charmaz, 2000). This process continued until saturation or until no new themes emerged or were apparent (Corbin & Strauss, 2008). For triangulation purposes, multiple sources were examined to validate themes, including transcripts, student reflections (ASK), and other student artefacts, including social media screenshots, videos, images and researcher field notes (Lincoln & Guba, 2000). As a result, categorical analysis of mentorship, motivation, and social location revealed key patterns of concern by high school students, including: the process of determining an STSE issue and ambivalence towards activist efforts and outcomes. Meanwhile, an ANT theoretical lens brought the conditions of social position into focus, which were supported by interview accounts of one's changing social position while doing STEPWISE along with an articulation of one's sense of agency.

12.6 Commentaries

Law (2006) notes that ANT is not only rooted in empirical case study, but also is better performed as a means of explaining its practices, rather than trying to represent it under apprehension that "faithful translation" (p. 48) is aloof. Thus, this

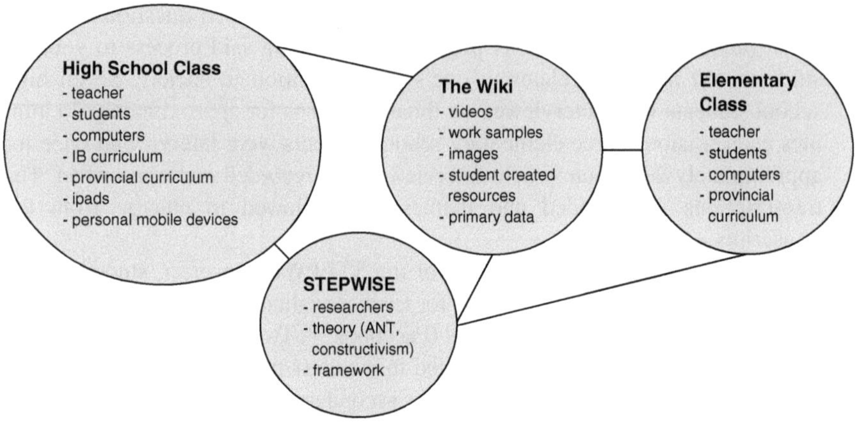

Fig. 12.2 Actor-world's and actor-network

section is ANT- in-action or, more specifically, a description of the actor-worlds of high school students and elementary students as they engage in STEPWISE, assuming a multiplicity of roles, including but not limited to that of students, mentors, mentees, teachers and researchers.

In trying to make sense of an infinite array of relations, the initial analysis of the data yielded a graphical representation (see Fig. 12.2) as described by Tony Bryant and Kathy Charmaz (2007), which expounds usage of grounded theory in situational analysis, and which is a commonly-used strategy to tell ANT stories (see Callon, 1986b). We have decided to use this strategy and the resulting graphic as a starting point in the telling of this particular story of STEPWISE. As one can imagine, this representation has undergone countless permutations through this process of analysis and writing, is essentially imprecise and, as a translation, is unfaithful to the ontologies that it attempts to describe, but we feel serves as a good starting point in this story. We have reflexively tried to justify and amend this depiction throughout analysis using ANT concepts, including problematization, interessement, enrolment, mobilization and ambivalences, to consider how certain actors get centered and to understand how the network waxes and wanes.

Tatnall and Bugress (2002) suggests identifying actors in the network as a starting point. Of course, this list can never be comprehensive, in that all actors are in themselves networks of actors. Ones of interest include high school students, elementary school students, researchers, teachers, the IB curriculum, the STEPWISE framework, STSE issues, secondary research data, primary research data, and various forms of technological actors, such as: iPads™, computers, personal mobile devices, websites, social media (Facebook™, Twitter™, YouTube™, the Wiki), and student artefacts, such as videos, digital documents, and ASK reflections. From here, we follow these actors and see how they relate and where they lead.

12.7 Problematization: Processes of Determining an STSE Issue

An STSE issue is problematized in terms of a possible solution derived by way of research-informed and negotiated actions, as defined by the STEPWISE framework, which students learn and carryout. In this approach, students are provided with 'apprenticeship' lessons and practice activities helping them to eventually self-direct RiNA projects. Consequently, roles of actors surrounding a problematic STSE node are quickly translated in relation to the solution(s) proposed. Students are defined as indispensable actors that select STSE issues of interest, conduct research, and perform actions to address the problem. The Internet, websites, information, and computers are actors that make secondary research possible, which contribute to secondhand accounts of the STSE issue. These same actors, as well as others such as Google™ analytic programs, Facebook™ and Twitter™ are now instrumental in terms of gathering and analyzing data to describe how high school student peers, friends, family members, and community members are connected to said STSE issues. These networks of human actors are defined in relation to the STSE issue of concern. For example, one group of high school students decided to address problems with electronic waste, which is of environmental concern. They petitioned their classmates to understand what kind of e-waste they had accumulating in their homes and, from this, determined that mobile phone e-waste amongst their peer group was a significant problem. In doing this, they defined their peers as e-waste producers. Ultimately, their action was to collect said waste for recycling with an incentive offered by a recycling company for the most recyclable content collected. However, this account of problematization is not complete and, thus, calls for student accounts of the process.

After informing high school students about what constitutes an STSE issue through transmission style teaching, including YouTube™ video's from the 94 Elements series (www.94elements.com/elements/29/), the classroom teacher, Mr. Zoras directed students to the Internet to search for an STSE issue of interest. *This was no easy task.* Phase 1 interviews reveal the selection, and thus production, of STSE issues as a critical actor that set the tone of possible success or failure of individual projects.

During Phase 1, after taking action to confront STSE issues, students were asked about what one would tell younger students in terms of identifying an STSE issue; one student notes: "I would ask them, 'What do you care about?,' just trying to get them interested from the start by trying to see what their passions are." Meanwhile, another student notes:

> First of all, it's really important to choose a topic that you really care about because if you just have something that you don't care about then you are not going to help other people care about it or lead anything that would make a change about it. So I would tell them to first just find a topic that they really feel interested in, [and then] they are bound to make a difference about it, it's not just that — okay I have to find a topic by tomorrow — I have to like it too and I have to feel strongly about it.

Here, we see multiple participants placing importance on identifying an STSE issue about which one is passionate in a way that intertwines the issue with one's sense of self, a move that makes the student indispensable to the solution(s) that follow. This move is tantamount in terms of the project being meaningful beyond institutional demands such as timelines and deadlines. This allows students to develop a sense of agency that warrants convincing other human actors to contribute to efforts of change, or a process of interessement that will be described shortly. Law (1992) outlines ANT's position on agency that is important to visit here to situate the idea of the agent in ANT terms. He states, "social agents are never located in bodies and bodies alone, but rather that an actor is a patterned network of heterogeneous relations, or an effect produced by such a network…an actor is also, always, a network" (p. 384). This framing of an agent as a network interprets students development of agency as an effect tethered to their increasing entanglement in the network of relations imposed by the STEPWISE framework.

In Callon's (1986a) seminal ANT study that suggests an unlikely, but empirically-supported, relationship between "scallop anchorage," a problem determined by scientist, and fishermen in St. Brieuc Bay, the notion of an obligatory point of passage (OPP) is introduced. He notes that in the process of problematization, certain "movements and detours…must be accepted as well as…alliances that must be forged keeping in mind that actors in the network are "fettered,"" (p. 70) or interdependently constrained into being; thus, actors "cannot achieve what they want by themselves. Their road is blocked by a series of obstacles problems" (p. 70); thus, are dependent on other actors to be brought into contextual being. Similarly, students articulate the importance of not just *selecting* an STSE issue but, rather, *determining* a personally-relevant STSE issue about which one cares and to which one is connected. This translation marks the construction of an OPP, or an obstacle that is determined by one set of actors (high school students) and imposed onto another set of actors (elementary school students) as one that must be overcome, or traversed, in order to solve the problem; in this case, the very STSE issue itself. This act of determining a personally-relevant STSE issue about which one deeply cares can be interpreted as the initial step in translation of global STSE issues into local ones that can be acted on in meaningful ways, which we address in the next section.

12.8 The Ambivalence of Problematization: Localizing the Global

The goal of mentoring younger students was not to establish the roles of mentor-mentee as made possible by the Wiki, the interesting entity that makes possible relations between high school and elementary school students. Rather, the general aim, as envisaged by teachers and researchers, was for high school science students, with gained expertise doing RiNA projects in Phase 1, to teach younger students to do similar projects by sharing their expertise in hope of effecting change for the betterment of society and environments.

Numerous high school students, after doing their first project in Phase 1, lamented on their inability to *really* change the world (have a large and lasting impact), despite being very excited about taking action to resolve an STSE issue. In many cases, high school students transitioned from having high hopes, to disappointment and, finally, through reflection, to being ambivalent about the actions they took which, in turn, affected their articulations of what younger students needed to know about taking research-informed and negotiated actions. Interestingly, students tied the efficacy of actions to the scope of the STSE issue as it intersects with one's own life. For example, one student who was addressing the issue of deforestation notes in response to a question about potential actions:

> ... many animals...lose their homes and they are going to die and then we are going to lose what we need to survive as well [oxygen], and the reason for all this is because of the things that we need to pay for...We are pretty much taking advantage of those forest and those organisms but we are not realizing that it is going to come back to us. I learned more about myself and how we all take advantage of things sometimes, so it really goes in depth more than the topic and so it makes you learn more about yourself.

In the same line of questioning, the same student, when asked about what younger students needed to know about doing a STEPWISE project, responded:

> ...the first thing I would tell them is not to think of anything too big that you know you are not going to be able to handle because one or two people can't change the world, but they can take little steps...it's not going to happen in a day because that's what I thought too, oh I'm gonna do something so big, like if I do something big I'm going to get the most marks... don't dream too big. When you have that feeling that it's not going to get accomplished, but still you dream big, but work for it and take little steps and think little before you start thinking big.

Despite advising others to not think "*too big,*" this participant contradicts her/his advice in saying "*but you still dream big,*" thereby characterizing the conflict between *local* and *global*.

Our participant behaves as does the dissatisfied sociologist that Bruno Latour (1999) describes as she/he obsessively moves back and forth between the micro and the macro, or the local 'face-to-face' and the 'far away' intangibility that frames the situations of our lives (culture, society, norms, context, structure and so on). At a glance, this seems like a reframing of the global in local terms; however, this reinforces the dichotomy of the local versus global perspectives that is problematic in assuming that a global perspective is knowable outside of localities and vice versa. Latour (1999) notes:

> Actor and network...designates two faces of the same phenomenon, like waves and particles, the slow realization that the social is a certain type of circulation that can travel endlessly without ever encountering either the micro-level—there is never an interaction that is not framed—or the macro-level—there are only local summing up which produce either local totalities ('oligoptica') or total localities (agencies). (p. 18–19)

This ambivalence appears to be at work in student thinking in a way that enables one to conceive 'small' local actions that are interwoven into 'large' STSE issues, as neither large and/or small, local and/or global, micro and/or macro but, rather, entertains an ontology of circularity (Latour, 1999). This state of indeterminacy allows movement between narrow localities (oligoptica), which together cast impressions of the global (Latour, 2005), or what David Bigham (2013) calls a "panorama of localities" (p. 18). Despite feelings of frustration and defeat related to project outcomes in discord with initial expectations, high school students felt it was important to try to act in ways to address global STSE issues, but reflexively redefined their expectations, actions and strategies to make potential outcomes more locally oriented; thereby establishing pragmatic social bonds to maintain activist networks *in between* notions of the local and the global.

12.9 Mobilization, Problematization, Interessement, and Enrollment: The Wiki and Conditions for Subject Positions

Above, we discussed the problematization of STSE issues because interview data, namely student reflections, reflect these concerns. However, in revisiting the data, what became evident was that the primary repository of information and means of inter-class collaboration and communication-the Wiki-is punctualized and, thus, black-boxed, in part due to its intangible nature as a digital virtual space; thereby concealing its role as an actor. This section attempts to open the black-box by problematizing the Wiki as a solution to geotemporal discord between elementary and high school classes to consider what actors are semiotically translated for the sake of social cohesion, or inter-class collaboration. In this effort, we also consider the other moments of translation, including interessement, enrollment, and mobilization to account for our use of technology to mediate collaboration and activism.

In the past, Mr. Zoras and Mr. Romano had used a similar Wiki forum like the one used in this study as a means of inter-class collaboration on another project that set the stage for this particular union. Mr. Zoras and Mr. Romano discussed the prospect of collaboration, problematizing geographical distance, as well as scheduling conflicts between classes; thus, necessitating a communicative solution and necessarily traversed node, or an OPP (problematization). The Wiki was quickly integrated to accommodate these misalignments, bringing classes together, as well as being a storehouse for student work (interessement and enrollment).

The final moment of translation detailed by Callon (1986) is mobilization, or the spread of the network by 'spokespersons,' who speak on behalf of actors, effectively turning them into an opaque entity whose function is "reduced to a few well-defined parameters" (Callon, 1986b, p. 29), or a black-box. However, unlike Callon's (1986a, b) seminal works, which appears to position mobilization as final moment in the process of translation, mobilization of the Wiki happened prior to its role interesting and enrolling elementary school students, partly because it was already in use by high school students in Phase 1. More specifically, when the interclass collaboration was proposed to elementary students, the classroom teacher spoke on behalf of the Wiki (black-boxing) presenting it to students as the medium of communication that they *must* traverse (an OPP), which he would help them navigate as their technology teacher.

The Wiki can be considered as a form of social media, or as part of "a group of Internet-based applications that build on the ideological and technological foundations of Web 2.0, and that allow the creation and exchange of User Generated Content" (Kaplan & Haenlein, 2010, p. 61). All high school student generated content — project word documents, Facebook™ and Twitter™ screenshots, ASK reflections, teaching resources (videos and documents) — was made accessible for exchange with other users, namely elementary school students, teachers, and researchers, via the Wiki.

With the black-boxed Wiki in the relational mix, it quickly locked certain relationships in their place. Recall, interessement, the second moment of sociological translation, is generally characterized by an actor (A), becoming 'interested,' or coming between an entity, (B), which is brought into relational existence by (A), and other entities (C, D, E,...n) trying to coopt (B) into 'problematic' relations (C----A===B) (Tatnall & Burgess, 2002); thus, locking allies into place (Callon, 1986a). (We have represented these relationships here linearly but we encourage the reader to think of them in nonlinear, uncertain, spatiotemporal flux. As well, we have depicted the various relational bonds using "---" and "===" to simply differentiate the kind of bonds, not to depict strength, but again the fragility and obduracy of these bonds are never static and always in dynamic contention with other fluctuating actors.) The Wiki as the interesting actor "impose[s] the identities and roles defined in the problematisation on the other actors" (Tatnall & Burgess, 2002, p. 185). It accomplishes this in terms of bringing into allied relation the actor worlds of high school and elementary school students. Phase 2 of the project involved high school students engaging in a second RiNA project, as well as creating resources for elementary school students to learn to do such a project. In our scenario, the Wiki mediates relations between high school and elementary school students, but in a way that forges new possible identities; namely, that of the *mentor-mentee* relationship between high school and elementary school students respectively. Thus, we can think of the Wiki as the actor that interests, or attracts, the elementary school actor-world by coming between it and the high school actor-world and in the process semiotically translating the roles of actors within these worlds.

Again, the third moment of sociological translation is enrolment, or the coordination of imposed roles that are 'yielded-to,' which was made possible through structures of relationality established during problematization and interessement, ultimately strengthening network relations (Tatnall & Burgess, 2002). Interviews reveal that elementary students, having no prior experience doing activist science projects, eagerly learned from high school students via the use of technology and, in general, deferred to the expertise and teachings of high school students. Thus, we can say that elementary school students are enrolled into the network through the OPP of the Wiki as such (Elementary----the Wiki===High school) establishing social relations, positions, and capacities for action.

12.10 Agency, Multiplicity and Ambivalence

Final interviews exploring high school students' perceived social roles in their schools and communities continue to be ambivalent and contradictory. Even though students did not articulate a shift in one's social position from that of a student to an activist, numerous participants expressed a sense of agency, or ability to act in meaningful and purposeful ways to affect their world and the society in which they live. Our urge is to discuss agency as an outcome developed and as a measure of empowered students; however, to do this would be a breach of our theoretical commitments. ANT is not particularly interested in the structure/agency debate and tends to bypass this binary (Latour, 1999). Rather, we consider agency in loose terms; not as an esoteric element of the social in binaristic tension to structure, but as an effect of social entanglements and commitments related to one's intersubjective positioning, existing within multiple, narrow, experiential potentialities or, as Latour (1999) puts it, "total localities (agencies)" (p. 19). The intimacy of these myopias that one is connected to and reflexively constitutes, yields a space in which the effects of one's actions, be it intentional or not, affect one's immediate relations, leaving the mark of freedom and independence of one's actions.

As mentioned, the Wiki enabled communication between high school and elementary school students, as an interesting entity, an essential position that defined the roles of other actors. Specifically, human actors on either side of the chasm were brought into being either as mentors (high school students) or as mentees (elementary students). When asked about why they continued to mentor younger students past the end of the course, high school students commented on their commitment to younger students as, "*the next generation*," and feeling responsible to them, since the wellbeing of others was caught up in the web of relations of STSE issues, in which they themselves were consciously entangled. In this particular social ordering, one high school student, connected to others via STSE issues and through the Wiki, began to see the world in multiplicitous ways, and used a mountain metaphor to describe this:

> When you look at a mountain you're just looking at the mountain from that one side and you just see a bunch of rock, we don't think about the different sides we could see it from. We could see it from a bird's eye view, we could go under the crust and see how the mountain is formed and that's the thing, and since I have been looking at things from different perspectives it's definitely changed the way I grasp things in my head and the ways I make decisions now.

Interestingly, the same student when asked about how she/he sees oneself in society after engaging in activist science projects notes:

> In society, I am still playing the same role that I did before…I have learned more about the issues going on, but compared to society I feel that I am still standing in the same place that I did before, although in my heart I know that I have tried to make a change…

This student's mountain metaphor can be understood through the contradictions of Latour's (2005) notion of *panoramas*. Latour uses the metaphor of a panorama to address the micro/macro separation, noting that the panorama is unlike the oligoptica, in that it can "...see everything. But they also see nothing since they simply show an image painted (or projected) on the tiny wall of a room fully closed to the outside" (Latour, 2005, p. 187). The resulting coherent assemblage of projections gives a sense of the whole, but is not the whole; rather, is a mosaic effect that constructs an impression of the whole. Similarly, our participant notes being able to see the whole, but in local, multivariate, ways that stand on their own; but, at the same time, constitute the whole, which has an impact on the way that he/she makes decisions.

Despite being locked into a position of obligation, or responsibility, this student make clear the uncertainty of one's role in the world, even within the many landscapes of the social; thereby, exemplifying the circularity of the social. This circularity, or state of constant motion within particular structures, relativizes one's position; thus, changing our network commitments while concomitantly establishing them. This double move of localizing and totalizing, enables the capacity to act, which either maintain or undo the relations of which we are part, in turn affecting said capacities.

12.11 Conclusions: A Final Betrayal

The fragility and obduracy of networks is related to the establishment of problems to which solutions can be conceived and to which actors can be made indispensable in relation. Within these snares, STSE issues ought to be chosen with care and passion, such that one's sense of self is entangled with solutions proposed. There is, of course, more to choosing than mere choice. STSE issues are pervasive and encompassing. These problems are overwhelmingly too out there, too large, and disconnected from the practitioner, begging the question, 'Is change even possible?' In working towards change, perceived failure is not a source of individual weakness, but a congealment of a disparate array of ambivalences into a focused pattern of resistances (Singleton & Michael, 1993, p. 259). Thus, effects of agency being felt and internalized become diminished, weakening bonds of necessity that define roles and subjectivities of actors and, ultimately, destabilize networks of resistance against neoliberal ideologies.

And, so, the question, 'Can the global be localized?' comes into focus. The scope of STSE issues ought to be considered and reconsidered, reflecting along the way, asking 'Where does the global meet the local?' in hopes of connecting lives, activities, and every day experiences of practitioners to pressing issues of our time. But is this localization? Any entity is a network in a network, relationally situated from without and is overwhelmed by uncertainty, not necessarily because of the inadequacy of signifiers to signify but, rather, because of vicissitudes of circumstances. Hence, the actor and its location is never fixed and always in the ebb and flow of the circular tides that little orders of the social stir up, and which the congealment of like uncertainties concertedly overflow. Maybe now one can act but, again, agencies are not without their certainties, leaving us fettered, in tension, and endlessly bound.

Our contemporary entanglements are mediated by the technical, without which said networks simply fail, or never come into being. Thus, careful consideration is recommended in terms of how said mediations are implemented. Technical actors, like the Wiki, can be walls upon which localities are scattered to yield panoramas of an ungraspable totality, or the cavern into which scenes of the everyday can simply slip away. But even the oligopticon is imprecise, or as Bruno Latour (2005) reminds, is easily blinded by the "tiniest bug" (p. 181). As such, students can easily perceive reifications in digital mediums as faithful representations of reality, as truth, when, in fact, an ontological loss has occurred in the translation of representation leading to a moment of ethical questioning through the ontologies of, 'What is lost?,' 'Who/what might be brought into being through these mediations?,' 'What is my role in and through these mediations?' This moment positions STEPWISE as an actor that structures capacities for ethical engagement through praxis. Thus, in the process of getting tied up in STSE issues, as well as locked into pedagogical relation, we have moved into capacities of action, into total localities of relation, and then in the following moment, into another, and another, and so on, infinitely, yielding a panorama of myopias, which, if looked at with squinted eyes, might just blur into focus shadows of the futures that we hope to create, futures of resistance in the form of reflective social action, or praxis.

Acknowledgements We would like to thank the wonderful class for all your hard work and commitment and in particular those students and parents that gave their consent to be represented by us in this text. This chapter could not have been written without the passion and dedication that you all have. Additionally, we are grateful to the Social Sciences and Humanities Research Council, Canada, for their generous funding of research about this case.

References

Bencze, J. L., Alsop, S., & Bowen, G. M. (2009). Student-teachers' inquiry-based actions to address socioscientifc issues. *Journal for Activist Science and Technology Education, 1*(2).

Bencze, J. L., & Carter, L. (2011). Globalizing students acting for the common good. *Journal of Research in Science Teaching, 48*(6), 648–669.

Bencze, J. L., Sperling, E., & Carter, L. (2012). Students' research-informed socio-scientific activism: Re/visions for a sustainable future. *Research in Science Education, 42*(1), 129–148.

Bigham, D. J. II. (2013). *Hearing connections: An actor-network study of online guitar activities* (Order No. 3557205, University of California, San Diego). ProQuest Dissertations and Theses (pp. 1–213).

Bryant, A., & Charmaz, K. (2007). Grounded theorizing using situational analysis. In A. Bryant & K. Charmaz (Eds.), *The SAGE handbook of grounded theory* (pp. 362–398). London: Sage.

Calás, M. B., & Smircich, L. (1999). Past postmodernism?: Reflections and tentative directions. *The Academy of Management Review, 24*(4), 649–671.

Callon, M. (1986a). Some elements in a sociology of translation: Domestication of the scallops and the fishermen of St. Brieuc Bay. In J. Law (Ed.), *Power, action and belief* (pp. 196–223). London: Routledge.

Callon, M. (1986b). The sociology of an actor-network: The case of the electric vehicle. In M. Callon, J. Law, & A. Rip (Eds.), *Mapping the dynamics of science and technology: Sociology of science in the real world* (pp. 19–34). London: Macmillan.

Charmaz, K. (2000). Grounded theory: Objectivist and constructivist methods. In N. K. Denzin & Y. S. Lincoln (Eds.), *Handbook of qualitative research* (pp. 509–535). Thousand Oaks: Sage.

Corbin, J., & Strauss, A. (2008). *Basics of qualitative research (3rd ed.): Techniques and procedures for developing grounded theory*. Thousand Oaks, CA: Sage.

Kaplan, A. M., & Haenlein, M. (2010). Users of the world, unite!: The challenges and opportunities of social media. *Business Horizons, 53*(1), 59–68.

Latour, B. (1999). On recalling ANT. *The Sociological Review, 47*(S1), 15–25.

Latour, B. (2005). *Reassembling the social: An introduction to actor-network-theory*. Oxford, UK: Oxford University Press.

Law, J. (1992). Notes on the theory of the actor-network: Ordering, strategy, and heterogeneity. *Systems Practice, 5*(4), 379–393.

Law, J. (2006). Traduction/Trahison: Notes on ANT. *Convergencia. Revista de Ciencias Sociales, 42*(13), 47–72.

Law, J. (2009). Actor network theory and material semiotics. In B. Turner (Ed.), *The new Blackwell companion to social theory* (pp. 141–158). West Sussex, UK: Blackwell.

Lincoln, Y. S., & Guba, E. G. (2000). Paradigmatic controversies, contradictions, and emerging confluences. In N. K. Denzin & Y. S. Lincoln (Eds.), *Handbook of qualitative research* (pp. 163–188). Thousand Oaks, CA: Sage.

Pedretti, E., & Nazir, J. (2011). Currents in STSE education: Mapping a complex field, 40 years on. *Science Education, 95*(4), 601–626.

Singleton, V., & Michael, M. (1993). Actor-networks and ambivalence: General practitioners in the UK cervical screening programme. *Social Studies of Science, 23*(2), 227–264.

Tatnall, A., & Burgess, S. (2002). Using actor-network theory to research the implementation of a BB portal for regional SMEs in Melbourne, Australia. In *15th Bled Electronic Commerce Conference-'eReality: Constructing the eEconomy'*. Bled, Slovenia: University of Maribor.

Wong, L., & Tatnall, A. (2010). Factors determining the balance between online and face-to-face teaching: An analysis using actor-network theory. *Interdisciplinary Journal of Information, Knowledge, and Management, 5*, 167–176.

Zoras, B., & Bencze, L. (2014). Utilizing social media to increase student-led activism on STSE issues. In L. Bencze & S. Alsop (Eds.), *Activist science & technology education* (pp. 435–449). Dordrecht, The Netherlands: Springer.

Chapter 13
Students' Socioscientific Actions: Using and Gaining 'Street Smarts'

Christina Phillips-MacNeil, Mirjan Krstovic, and Larry Bencze

13.1 Introduction

Our world is facing many serious problems, such as socio-economic disparity, conflict and violence and environmental destruction. Climate change is one of the most pressing issues of our time. Our societies continue to 'borrow' from futures of our children and grandchildren as we deplete precious environmental resources—including clean air, water and carbon sinks, such as forests and oceans—with little payback. While fields of science and technology have offered societies many benefits, 'progress' appears in some cases to be devoid of forethought and, consequently, seem to be at least partly implicated in many of the above-mentioned problems. As Edward O. Wilson (1998) stated, "We are drowning in information, while starving for wisdom" (as cited in Peter & Wals, 2013, p. 300).

To deal with problems like those noted above, some school science systems have, for some 40 years, been promoting 'STS[E]' education—which involves helping students to understand and perhaps attend to problematic relationships among fields of science and technology and societies and environments (Pedretti & Nazir, 2011). A related movement, known as *socioscientific issues* (SSIs) education, encourages students to negotiate a range of controversial data sources and claims about them relating to science and technology (Sadler & Zeidler, 2005). Such curricular components promote exploration of issues and topics from a number of subject areas not traditionally associated with science, such as "social, technological, cultural, ethical, and political contexts" (Pedretti & Nazir, 2011, p. 602). Moreover, as Catherine

C. Phillips-MacNeil (✉) • L. Bencze
Ontario Institute for Studies in Education (OISE), University of Toronto, Toronto, ON, Canada
e-mail: chrisphillips1202@gmail.com; larry.bencze@utoronto.ca

M. Krstovic
Peel District School Board, Mississauga, ON, Canada
e-mail: mirjan.krstovic@peelsb.com

L. Bencze (ed.), *Science and Technology Education Promoting Wellbeing for Individuals, Societies and Environments*, Cultural Studies of Science Education 14,
DOI 10.1007/978-3-319-55505-8_13

Barrue and Virginie Albe (2013) say, they may allow young people to "see themselves as full citizens and not as citizens-in-the-making" (p. 1094).

For at least the last decade, perhaps in light of the seriousness of the many problems mentioned above, there have been numerous calls encouraging school science systems to help students/citizens to not only become aware of issues and problems but, as well, to take informed and responsible actions to try to rectify them (Hodson, 2011). Among approaches to implement such a mandate have been those based on the 'STEPWISE' curriculum and pedagogical framework (refer below). Since its inception in 2006, several teachers, student-teachers and youth educators have been using variations of this framework to encourage young people to critically evaluate products of science and technology and, where problems are perceived, to take research-informed and negotiated actions to address them. This framework has had some successes in this regard, partly, it seems, due to recent listing of STSE education as the first of three (rather than the third in that list in curricula from a decade earlier) curriculum goals in the province of Ontario (MoE, 2008), the context for development and implementation of STEPWISE.

In this chapter, we explore general factors that may affect the nature and extent of students' research-informed actions on STSE/socioscientific issues. Our findings suggest—among several outcomes—that such projects may encourage students to draw on and perhaps enhance their 'street smarts,' along with more traditional 'book smarts.'

13.2 STSE and SSI Education

Socioscientific issues education, which may be considered a sub-set of arguably more neutral STSE education (Zeidler, Sadler, Simmons, & Howes, 2005), appear to have dominated global research, curricula and practice in terms of relating science to other disciplines (Levinson, 2013). Frequently in such approaches, students are presented with conflicting data and claims pertaining, for instance, to merits of various products (e.g., evolutionary theory or nuclear vs. wind power) of science and technology and, often in small groups, and encouraged to develop reasoned arguments in defence of their personal positions of such issues (Sadler & Zeidler, 2005). There are, undoubtedly, various ways to analyze and evaluate such argumentation-based personal-choice approaches. An apparently helpful framework in this regard, however, has been developed by Ralph Levinson (2010), who—as indicated in Table 13.1—suggests that, while citizenship education can involve at least four levels of sophistication, argumentation-based approaches tend to be limited to deficit and deliberative experiences, both of which he suggests often place significant, if not all, influence in the hands of authorities (e.g., politicians and scientists). Such emphases on logical decision-making surrounding socioscientific issues seem appropriate in *representative* democracies (Wood, 1998), in which citizens mainly are asked to elect leaders every few years and, otherwise, be able to make personal decisions on matters pertaining to science and technology. Indeed,

Table 13.1 Summary of four frameworks for democratic participation in science

Framework	Socio-epistemic relations	Epistemology	Controversy and participation	Pedagogy	Implications for democratic participation
Deficit	Knowledge flow is from scientist-teacher-student.	Science is the corpus of knowledge.	Ability to engage is constrained by access to technical knowledge.	Knowledge for addressing an issue can be brought to the attention of the student.	There is a socio-epistemic inequality between the scientist/teacher and students which limits ability to bring about political change from below but does not preclude influential specialists making a political impact.
Deliberative	Knowledge flow is predominantly from scientist to the teacher and students, the latter two might be working in concert.	Science is understood to be uncertain and fallible.	Dialogue is open. Lay participants are informed but often lack the political means to bring about change. In schools, students might have opportunities for deliberation through group work and school councils but action might be constrained depending on the democratic nature of the school.	Emphasis on critical thinking and understanding of scientific methods and procedures.	Participation is real but often ineffectual in generating democratic change because participants do not have the 'clout' to make crucial decisions.

(continued)

Table 13.1 (continued)

Framework	Socio-epistemic relations	Epistemology	Controversy and participation	Pedagogy	Implications for democratic participation
Science education as praxis	Knowledge is distributed and emergent.	Knowledge is situated. Students become inducted into communal ways of knowing through legitimate peripheral participation in particular but changing contexts.	All participants work with a shared sense of social purpose.	Knowledge is provided on a need to know basis. The teacher is not epistemologically privileged.	Active and egalitarian participation to enhance change which might assume political literacy.
Dissent and conflict	This can be variable but is likely to have similar characteristics to science education as praxis.	What is known is contextualised by socio-political concerns.	Political action.	Knowledge provided on a need to know basis with an emphasis on political literacy.	Political understanding and action for change are foregrounded.

From Levinson (2010, Table 1, pp. 83–84)

studies suggest that students gain a range of personal benefits from them, including abilities to use logic in practical situations (Sadler & Zeidler, 2005).

Argumentation-based decision-making surrounding socioscientific issues may, despite apparent personal benefits, be somewhat problematic, in the sense that students might lack skills to think critically about information presented by 'experts'—which seems essential in light, for example, of adverse effects of private sector funding of science and technology (Mirowski, 2011) and, associated with that, private sector creation of doubt and controversy surrounding negative science findings about commercial products and services (Oreskes & Conway, 2010). Consequently, it seems clear that, in addition to developing understandings of complex STSE relationships and being able to make well-reasoned personal decisions about controversies in them, students/citizens need to be prepared to critique socioscientific data and claims and, where they perceive problems, be ready and willing to take personal and social actions to address them. In Levinson's (2010) schema for citizenship in this regard (refer to Table 13.1), in other words, students should be prepared to engage in praxis (reflective practice) and dissent and conflict (critique and actions). These kinds of approaches may be aligned with more participatory forms of democracy (Wood, 1998). In support of this kind of recommendation, Derek Hodson (2003) suggests that science education need to include all four levels of his schema for socioscientific issues education, as given below:

1. Appreciating the societal impact of scientific and technological change and recognizing that science and technology are, to some extent, culturally determined.
2. Recognizing that decisions about scientific and technological development are taken in pursuit of particular interests and that benefits accruing to some may be at the expense of others. Recognizing that scientific and technological development is inextricably linked with the distribution of wealth and power.
3. Developing one's own views and establishing one's own underlying valuepositions.
4. Preparing for and taking action. (p. 655)

There appear to be several reasons for including more 'activist' goals (such as in #4, above) in science education. As with the first three levels of involvement above, students' actions on socioscientific issues can lead to improvements in their knowledge, skills and attitudes about them; but, as well, it seems clear that the seriousness of many potential and realized problems (e.g., climate change) warrant more immediate actions. Nevertheless, although such approaches may be beneficial, it appears very unlikely that students will engage in actions associated with their selected SSI or STSE topic (Lester, Ma, Lee, & Lambert 2006). Larry Bencze and Erin Sperling (2012) remark that much of science education tends to focus on Hodson's first three outcomes above, as opposed to student engagement in actions. Consequently, it is apparent that new perspectives and practices are needed to help educators prepare more citizens for more critical and activist democratic participation.

13.3 STEPWISE

In 2006, to help citizens to take on more participatory roles on matters pertaining to science and technology, the 'STEPWISE' (Science & Technology Education Promoting Wellbeing for Individuals, Societies & Environments) curriculum and pedagogical framework was developed. It is a framework that organizes teaching and learning in science and/or technology education in ways encouraging and enabling students to use at least some of their literacy (especially through science education) to try to bring about a better world. Broadly, this framework provides students with constructivism-informed 'apprenticeship' lessons and activities aimed at helping them to develop expertise, confidence and motivation for eventually self-directing research-informed and negotiated action (RiNA) projects to address problems they perceive in relationships among fields of science and technology and societies and environments (STSE). Refer to Chap. 2 in this volume for a fuller description of the STEPWISE framework and educational approaches. Mirjan Krstovic's (2014) chapter in an earlier book about activism in science education also provides some relevant practical and theoretical suggestions for implementation of STEPWISE.

Since its inception about a decade ago, several teachers, student-teachers and youth educators (in informal learning contexts) have developed and implemented various teaching/learning strategies drawing from the STEPWISE framework and relevant principles (e.g., apprenticeships, self-directedness, authenticity and altruism). Although many teachers avoid use of this framework, perhaps because of tendencies of school science systems to prioritize teaching/learning of products (e.g., laws, theories and inventions) of science and technology, the framework has been attractive to some teachers and students (Hodson, 2011). There are, perhaps, numerous and varying reasons for teachers to work with STEPWISE perspectives and practices. On a more philosophical level, people seem to crave participation in initiatives where they feel that that one's actions matter. RiNA projects may help to fulfill some of these objectives for both teachers and students as they are making a difference regarding their selected SSI. RiNA projects may, in that vein, be connected to Jürgen Habermas' (1996) discussion of the 'life world.' In that regard, Hyslop-Margison and Thayer (2009) state:

> The life world for Habermas consists of those fundamental human experiences and interactions that generate a sense of inner peace or individual wellbeing, and provide the necessary community space, such as liberal learning institutions, for democratic discussion. (p. 7)

The enactment of a science curriculum that fuses STSE/SSI objectives and activism (i.e., RiNA projects), such as the STEPWISE approach, may help to build Habermas' 'life world' in our classrooms where wellbeing for everyone is enhanced as we participate in more democratic discussion and provide necessary autonomy for our students to engage in their projects and subsequent actions.

Something Was Missing… Both feeling that, despite having been successful teachers of science for several years, something was missing from their teaching practice, 'Amanda' and Mirjan (second author here) agreed to work with Larry (third author here)—serving as a 'researcher/facilitator'—to develop and implement teaching/learning strategies based on the STEPWISE framework. Although their educational backgrounds and work situations were somewhat different, they both felt that this framework might fill in some gaps in their respective professional and pedagogical practices. Amanda had just returned to teaching from a 2-year leave of absence to complete a master's degree and understood, from both an academic and practical perspective, social and political complexities that are associated with science and wanted to more authentically approach research and STSE in her teaching practice. Some of her previous practices had emphasized knowledge-based curricular outcomes of science through more teacher-directed approaches that often are not connected with the outside world. She was dissatisfied with her approaches, especially given the serious nature of the environmental issues that she had spent the last 2 years investigating and acting to protect. She wanted her students to understand complexities of issues pertaining to science and technology and to realize that they can take action on issues that concern them. It seemed to her that the STEPWISE approach fused these objectives in a practical and meaningful way, and she was enthusiastic about incorporating this framework in her science classroom.

Mirjan's introduction to STEPWISE was quite different than that of Amanda. He learned about it toward the end of a graduate course in history, philosophy and sociology of science for which Larry was the instructor. He had been thinking critically and profoundly about his role as a science educator and purposes of science education. Mirjan reflected on his pedagogical practices and felt that he had rarely addressed SSIs or asked students to propose any practical courses of action, let alone take socio-political action and felt that something profound was missing from his teaching. He did not feel that there were any contextualized or lasting connections made to the outside world even after engaging in many creative teaching strategies with his classes.

STEPWISE Implementation After discussions with Larry about characteristics of the STEPWISE framework and common relevant teaching/learning strategies, Amanda and Mirjan proceeded to develop lessons and activities that they could use in their respective science teaching contexts. Both were teachers in public secondary schools in the Greater Toronto Area, with student populations representing diverse ethno-cultural and gender compositions typical of the region. Amanda chose to implement STEPWISE in a 'college-level' (mostly preparing students for entrance into community non-degree granting colleges or directly into the job world) chemistry class while Mirjan did so with a class of tenth-grade students in a course (Feb. – May 2012) meant for future university students. Broadly, both teachers implemented lessons and activities based on the 3-phase STEPWISE apprenticeship schema, shown in Fig. 13.1, involving 1–2 cycles of these phases: (i) students reflect on and express (e.g., write, draw, state) existing attitudes, skills and knowledge ('ASK') about STSE relationships, research and actions; (ii) the teacher teaches students examples of RiNA projects that others have conducted to address problematic STSE relationships (e.g., an anti-smoking video); and, (iii) students are

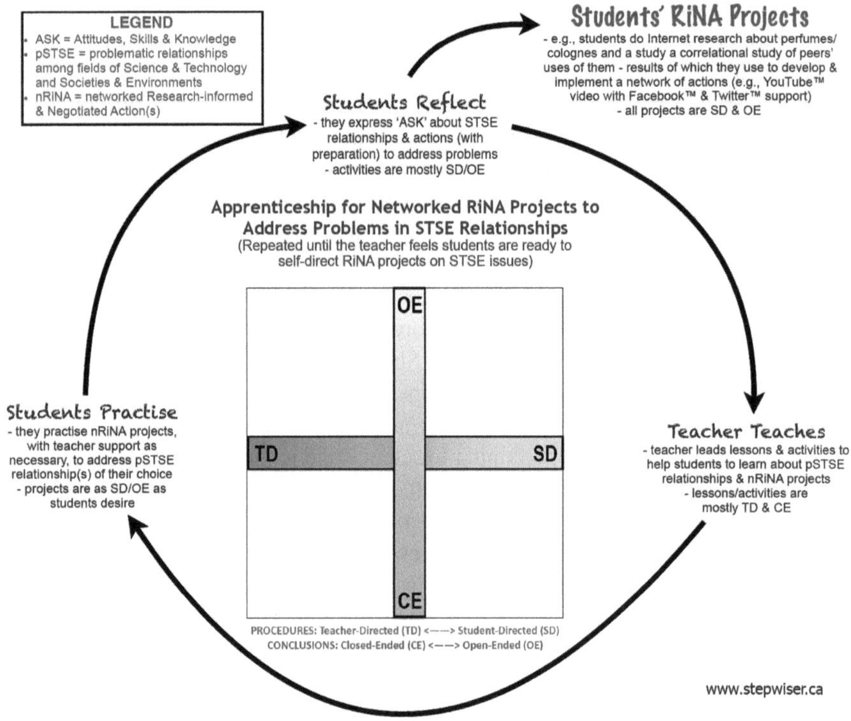

Fig. 13.1 STEPWISE pedagogical framework

asked to co-design and conduct RiNA projects to address problems they perceive in STSE relationships (e.g., posters to advise schoolmates of climate change problems linked to excessive hot water shower use), with teacher support upon student request. As each course proceeded, the teachers ceded to students more and more control of decisions for most aspects of projects. Details regarding each teacher's approaches are elaborated below in the context of descriptions of student results.

Data-Collection and Analyses Methods As teachers developed and implemented activities, sometimes in consultation with Larry, qualitative ethnographic data-collection approaches were used, attempting to balance naturalistic and rationalist research perspectives (Guba & Lincoln, 2011). Amanda and Mirjan were interviewed (for about 90 min each) before, during and after their courses had ended. Semi-structured interview questions focused on their views about the nature of science and teaching/learning approaches and outcomes based on STEPWISE. Students in Mirjan's classes (only) were interviewed (in small groups, for about 30 min each) during and after their RiNA projects. Larry asked Mirjan's students questions about purposes, methods, findings and significance of their RiNA projects, with special focus on results of their primary and secondary research, along with the nature of and rationale for their action choices.

Many types of documents were collected to construct a case for each teacher, including: copies of teachers' lesson plans, instructional slideshows, student handouts and samples of students' work (e.g., secondary research, study graphs, and action posters, etc.) relating to their RiNA projects. For data analyses, each of us independently and repeatedly examined and reflected on the data for relevant categories and then developed various themes using constant comparative methods based on constructivist grounded theory (Charmaz, 2014). Categories and themes were then negotiated between us until we were in agreement with our findings.

13.4 Students' Expertise, Confidence and Motivation for RiNA Projects

Preamble Although there appeared to be clear variations across the two very different school science contexts (e.g., tenth-grade university-qualifying science vs. college-qualifying chemistry in two different schools) involved in this study, both teachers concluded that many students—including many who do not normally 'succeed' in school science—were highly engaged in their RiNA projects to address problems they perceived in STSE relationships. In support of these claims, we provide, below, brief descriptions of the three major aspects (power-related STSE relationships, student-led primary research and social actions) of RiNA projects developed and implemented in both classes.

Some RiNA Projects from Amanda's Class After Amanda had worked with students to have them consider possible problems linked to chemicals in a range of common consumer products, such as cosmetics, plastic containers (e.g., for bottled water) and personal hygiene products, taught them about characteristics and practices involved in primary research (e.g., correlational studies vs. experiments) and actions (educational pamphlets) people take to address socioscientific issues, she escorted her class of students outside near a local body of water (pond) to conduct various studies of water content after first obtaining parental/guardian and administrative permissions. She supplied students with a water testing kit, which enabled them to monitor acidity and content of a range of ions (e.g., phosphates) in the water, along with other measuring devices, such as meter sticks and water volume containers. While such resources amounted to provision of some guidance, students then were advised by Amanda to self-determine independent variables to sample—such as distance from the roadway (which typically contains salt used to melt ice) and distance from groupings of geese (which tend to leave large amounts of fecal material in and around the water). Such levels of choice were new to many of her students, who typically are asked to carefully follow teacher inquiry procedures. About such sudden freedom, Amanda noted the following in her journal:

> [S]ome students appeared to be a bit nervous about this: they looked quite surprised-other students looked happy—to me it is the ultimate in treating them as adults, as many of them

> are adults in my class, why not morph into this role where they seek the information and I am present to guide them?

While students seemed initially apprehensive about such relative freedom of choice, students in this course seemed to thrive in this environment—enthusiastically and carefully completing their studies, something Amanda had not seen to the same extent with these students when conducting teacher guided inquiries. At the same time, perhaps because it was the first time they had self-directed some aspects of inquiries, Amanda mentioned that several of these students indicated that they were somewhat skeptical of their findings and, consequently, felt that any actions they might take needed to be informed by better research.

Some RiNA Projects from Mirjan's Class In what was his second semester of implementing the STEPWISE pedagogical framework (Fig. 13.1), Mirjan indicated that students in his class seemed to develop considerable expertise, confidence and motivation regarding such projects. As recommended by the framework, students developed and implemented two RiNA projects with support, as they requested, from Mirjan, prior to conducting a self-directed project. It is difficult to summarize the major projects completed by students, but there were a few notable outcomes indicating 'considerable' student achievements. In the first unit of the course, one dealing with climate change, socioscientific issues investigated and acted upon seemed quite varied and novel. One group, for instance, investigated variations in fellow students' diets, particularly comparing the extent of vegetarianism (vs. meat-eating) as affected by gender. Finding that boys tended to eat more meat than girls and that much of this may be shipped to them over long distances, they developed posters and a pamphlet to educate fellow students about relationships among diet, transportation and climate change. Another group, meanwhile, investigated the extent and nature of peers' uses of electronic devices, such as cell phones, tablets and computers—attempting to get a sense of youth electricity uses as possible contributions to climate change. In terms of actions, they developed a page on Facebook™ to inform 'friends' of such contributions—including a suggestion that girls may use more electricity this way—and they developed artistic designs on 'T-shirts,' urging peers to reduce their electricity use. Yet another group, investigating relationships between petroleum-based transportation and climate change, developed a creative video (goo.gl/o5FC38)—featuring techniques used by 'RSA Animate' (goo.gl/dcluRR)—they posted to YouTube™, advising viewers to consider more sustainable transportation forms.

In the second unit, students' projects—as one might expect—deepened. Many of their topics in studies relating to light and optics seemed novel. One group, for instance, investigated the extent and nature of uses of surveillance technologies (e.g., hidden cameras) to contribute—in light of the *panopticon* concept (a prison design, requiring few guards who, by their periodic and random appearance, persuade inmates to self-regulate their behaviour)—produced a large poster they displayed in a main school hallway to educate people about relative merits of surveillance. A group member's comment about this controversy appears to indicate students had more networked conceptions of science and technology, seeing the seen and unseen about them:

> It is a controversial issue. Some people would agree with you that it is a problem [that we have such surveillance. They might ask,] 'Why do we need cameras watching our every movement? It is my right not to be seen by the government.' But, other people have safety concerns[, thus approving of surveillance]" One girl offered that the increased level of violence in movies has caused society to "be paranoid[, thus accepting surveillance].

Indeed, much focus was placed by students in this class during the optics unit on a broader conception of 'optics'; that is, one based on semiotic/symbolic conceptions. One student group, for instance, explored advertizers' adjustments to photographs using image-editing software that may lead viewers (often youth) to conjure up idealized and/or stereotypical conceptions of 'beauty,' 'success,' 'normal,' etc.

Another group had a similar idea, but their approach was to explore peers' roles of semiotics in attracting consumers to the products/services while perhaps distracting them from problematic elements. Their research tack was to show students in the hallways of their school pictures of advertisements (minus the product [a shower gel] name) and then ask the students what product is being advertised and what the advert. is conveying. The advertisement that this student group opted to show other students depicted a young woman wearing a flesh-toned bikini and covered in dirt. On her lower abdomen the phrase, "Wash Me" appeared to be hand-written in the dirt covering her body. A student in the group suggested that their conclusion, particularly about this image, is that: "[s]o, you are buying this [sexuality] over this [the product]" (June 19, 2012). These research discussions, which were video-recorded by students in the group, became their actions—as they eventually revealed to fellow students the targeted product in each advertisement and engaged peers in discussions about merits of such tactics inclusive of the overt sexism toward women that was present in these depictions.

Although Mirjan was, generally, satisfied with the quality of student projects during the 'apprenticeship' phases reflected above, he indicated even more satisfaction with those students 'self-directed' (admittedly partially limited by assessment/evaluation criteria) in the last unit of the course. A particularly salient feature of students' final projects in this course was their more community-based nature—both studying and acting on those beyond peers and the school. A group investigating the nature and uses of 'energy drinks' (e.g., Red Bull™, Monster™, etc.) visited several local commercial outlets (e.g., WalMart™) and asked clerks if they would talk on video camera about energy drinks. The clerks refused. One of the students said, "They said 'no' because, they don't know what they are selling, they don't know anything about energy drinks, they don't know how it effects the body." After also surveying fellow students' uses and knowledge about energy drinks, these students then chose to create a petition, including many peers' signatures, and letter to advocate for healthier drinks. A group member explained rationale behind their actions:

> We chose to send the letters to big companies and Health Care Canada because we believe that it is important for adolescents like us to support a good cause and to help our peers

understand the seriousness of this issue. We believed that if we sent our letters and petitions to Monster Inc. and Rockstar Inc then they would understand our perspective and respect the fact that their dedicated consumers are harming themselves, and should cut down their production [of energy drinks].

A group of four girls, meanwhile, chose to explore the merits of ultrasound technologies—including for such diverse uses as fetal monitoring and submarine scanning. After secondary research to learn more about it, finding that there appear to be several often-unreported negative side-effects, they conducted a survey of 50 boys and 50 girls in their school, results of which are given in Fig. 13.2—a pamphlet in which they summarized their research findings. To extend influence of their findings, however, they also chose to send the pamphlet—along with four letters, one written by each student—to a maternal health website that they felt was not adequately informing women of possible negative side-effects of ultrasound scans. Moreover, they chose to visit shops for maternity clothing and baby accessories, pass out their brochures and talk to women. Such direct actions seemed to reinforce these students' research and actions. In their final report, they stated: "This one lady wasn't totally sure if she wanted to do [have] an ultrasound [examination] but, when we gave the brochure to her, she said she would think about not doing it" (June 19, 2012).

13.5 Factors Affecting Outcomes

Preamble

The way kids learn to make good decisions is by making decisions, not by following directions
(Alfie Kohn, 2005, p. 169).

The cross-case analysis reported here suggests that progressively ceding learning control to students in the context of research-informed and negotiated actions to address socioscientific issues of their concern/interest deepens their learning and commitments to it. This can be explained, apparently, in terms of students' access to control of various decisions. However, as our work progressed in this area, it became apparent that a related factor contributing to student understanding and commitments was their uses (and, perhaps, increases in) 'street smarts' (vs. 'book smarts'). The two, not necessarily mutually-exclusive, broad factors are discussed below. We base our findings on, in part, our *personal practical knowledge* (Clandinin & Connelly, 1992). This is acknowledgment that teachers have unique knowledge of students because of their close proximity to teaching and learning situations and because their personal evaluations of student achievement may have validity that others cannot replicate.

Epistemic Agency It seems clear from others' scholarship that the more control students have over decision-making in their learning, not forgetting that significant teacher control also can help alleviate differences in students' cultural capital

Long Term Negative Effects

Ultrasound is a very non invasive, inexpensive common procedure however there are many unknown possible negative long term effects that it has, most of which include:

- It can increase the blood flow in the womb and the temperature
- Can create gas bubbles that put pressure on the tissue.
- Mechanical effects like movement of the fluid that surrounds your cells, which can also put pressure on local tissues
- Studies show that children who have been exposed to ultrasound while in their mothers' bellies have a greater chance of suffering from dyslexia and other speech and learning problems than children who have not been exposed to ultrasound.
- Ultrasound waves in laboratory experiments have been known to damage chromosomes, produce internal cellular heat which damages cells, retard the normal development of cells, and many other phenomenon.
- Ultrasound also has some very serious emotional and psychosocial side effects.

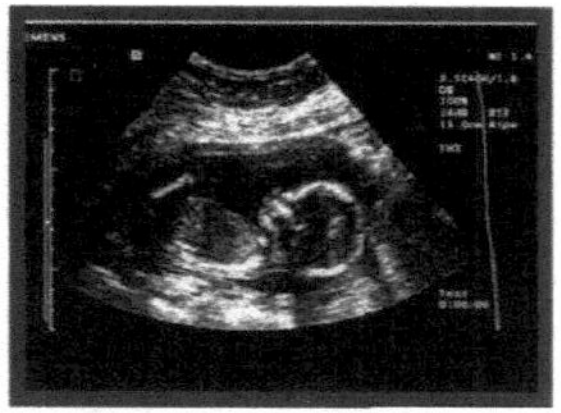

Studies

Our studies show that there is not a co-relation between gender and the perception of ultrasound. However we found that a significant number of females were aware of the long term negative effects of ultrasound whereas males were less informed.

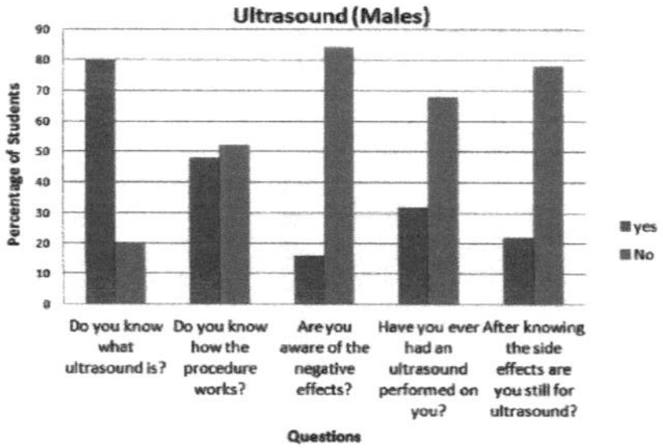

Figure 0.1:This graph displays the percentage of males

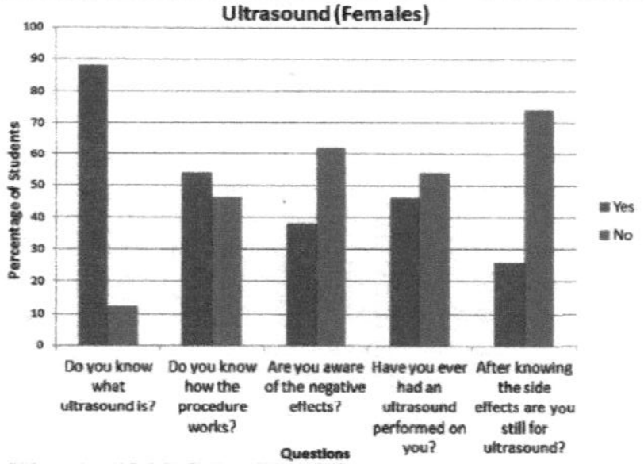

Figure 0.2: This graph displays the percentage of females

Interesting Facts

- Ultrasound is also referred to as Sonography
- Ultrasound does not involve the use of X-ray, so there is no radiation.
- The speed of ultrasound does not depend on its frequency; it depends on what material or tissue t is traveling in.
- Ultrasound travels faster in dense materials and slower in compressible materials.
- Travels freely through fluid and soft tissues but is reflected back as 'echoes' when it hits more of solid/dense surface
- Often time's individuals are instructed not to eat at least 12 hours before an ultrasound procedure. However a full bladder is almost always a necessity in order to obtain a better image.

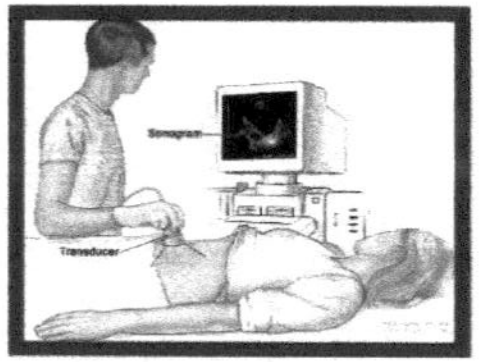

- 3D and 4D ultrasounds have improved the images obtained by ultrasound machines. The images viewed by these ultrasounds can offer better views of the body.
- Typically every pregnant woman will have an ultrasound preformed between the 18th and 25th month of pregnancy. The ultrasound test can detect abnormalities in the fetus's development, multiple birth pregnancies, sex of the fetus, and placenta location

Fig. 13.2 Students' pamphlet addressing research and actions about Ultrasound

(Bencze & Alsop, 2009), the deeper their learning and the greater they may become attached (e.g., emotionally) to it. Damsa, Kirschner, Andriessen, Erkens and Sins (2010) refer to such learning control and attachments as *epistemic agency*, suggesting, in essence, that it has two dimensions; that is, a *knowledge-related dimension* (i.e., creating awareness, alleviating lack of knowledge, creating shared understanding, generative collaborative actions); and a *process-related dimension* (i.e., *projective* [e.g., creating joint plans]; *regulative* [e.g., reflecting on actions]; and, relational [e.g., negotiating social conflict]) (p. 175). This seems comparable to *knowledge duality theory*—which suggests that depth and commitments to learning increase as learners have more control over reciprocal translations in Phenomena ← → Representation(s) relationships (Wenger, 1998). Such conceptions seem, in turn, to align with translations in RiNA projects; that is, while acknowledging their reciprocal nature, 'research' involving Phenomena (e.g., cell phone uses) → Representation(s) (e.g., graph of student cell phone uses) translations, while 'negotiated actions' involving Representation(s) (e.g., wise energy use T-shirts) → Phenomena (e.g., less cell phone energy uses) translations. From Larry's interviews with students in Mirjan's class, there is ample evidence to support our claims about student engagement in both kinds of translations. In a small-group interview towards the end of their semester, students made statements like the following about both kinds of translations: "…we do something to change the world, whether it is something big or small," "What we learn, we actually use it ourselves," "Most people got affected by what they did and wanted to help other people understand" and "Before I learned about it, I didn't really care about it…" (May 3, 2012).

As discussed above, student learning control of research and action translations appeared to help them deepen their understanding of and commitments to their learning in science education. Students often, for instance, used emotionally-charged terms in discussing their work—such as: "For my group, we did [research on] privacy and that [decision] was based on how people felt about being watched or, like, video-taped. It wasn't so much logical as emotional" (May 24, 2012) and "…we're **having fun** while we were doing studying" (June 20, 2012). Such emotional investment in their SSIs projects may be an important part of engagement with activism initiatives. Stephen J. Gould seems to concur when he eloquently states that, "[w]e cannot win this battle to save species and environments without forging an emotional bond between ourselves and nature as well—for we will not fight to save what we do not love" (cited in Orr, 2004, p. 43). Associated with such attachments are, as argued above, various learning outcomes in the subject of inquiry and action. Robert Sternberg and Elena Grigorenko (2004), "[w]hen students think to learn, they also learn to think" (p. 275). At various times throughout Mirjan's work with students, they made comments congruent with this claim, such as: "We have more skills. We are learning to be more interactive [with ideas, etc.] with our projects." (Nov. 16, 2011) and "…it helps us to understand what we are learning, 'cause then you are doing something instead of just sitting down, instead of just doing something for the sake of passing the course" (May 3, 2012).

Given our emphases on authority concessions to students regarding research and negotiated actions relating to socioscientific issues, it follows that some of this deep learning would align with Levinson's (2010) conceptions of more active citizenship; that is, as summarized in Table 13.1, in terms of promotion of 'praxis' (reflective practice, such as in terms of student-led research) and 'dissent and conflict' (e.g., critical views of STSE relationships and socio-political actions, respectively). Indeed, in an email message to us, Mirjan wrote that a student who had been in his tenth-grade science class and conducted RiNA projects had been recognized as a 'youth community leader' by their school's guidance department—a sign, he felt, of enduring effects of student control over research and negotiated actions on socioscientific issues of interest to them.

Ontological Relevance Associated with control over research and negotiated action, as we claim above, appeared to be increased uses of local and personal contexts for both of these. We suggest this may be referred to as *ontological relevance*; that is, uses of conceptions of being that are more relevant to students' personal experiences. Awareness of this phenomenon arise naturalistically in this research. In the course of an interview with Amanda while she was working with students in her chemistry class, she mentioned that the projects seemed to appeal to these students' 'street smarts.' Larry appreciated this comment, but it wasn't until an interview with students in Mirjan's tenth grade science class that its significance became more apparent. While discussing motivation for their groups' projects, a student said:

> This is a hip way of saying it, but the way other teachers teach us, just through a textbook, that is called being 'book smart,' because you are just learning everything from the textbook. Basically, what happens to me is that when I learn stuff from a textbook I won't always remember it; but, the way Mr. K. teaches us, it is called 'Street Smart' (Nov. 16, 2011).

Immediately, Larry recalled that Amanda had mentioned roughly the same thing about the relative role of 'street smarts,' as opposed to 'book smarts,' in her students' engagement in RiNA projects. Broadly, from our subsequent review of relevant literature, although there appear to be some disagreements on their meanings, it seems that 'street smarts' often refers to more local and personal contextualized knowledge, while 'book smarts' is more about abstract, de-contextualized, knowledge claims developed by historical communities of practitioners in various specialized fields (e.g., Hatt, 2007). These constructs should not, however, be considered 'binary' in nature, assuming that people either use one or the other. To make this point, Gerald Graff (2001), for instance, discusses the case of Marilyn Monroe's divorce from famous baseball player, Joe DiMaggio, someone perhaps considered strong in 'street smarts,' and marriage in 1956 to the playwright, Arthur Miller, someone thought to be strong in 'book smarts,' there was some general public disenchantment with her choice—apparently largely due to prevalence of anti-intellectualism of the day. Miller had actually come from a poor family that, for instance, lost their small business during the Great Depression and, so, may have had some 'street smarts' ("Arthur Miller Biography," 2014) and Monroe, with very little formal education, had an extensive and varied library and perhaps had considerable 'book smarts' (Keogh, 2010, p. 143). Nevertheless, there is the view that 'street smarts' is a set of practical capabilities associated with informal education, often obtained in 'every-day' (non-formal) situations, often linked to more spontaneous problem-solving—as one might experience while conducting 'mental math' in personal financial transactions in local markets (Saxe, 1988). Students in our study did seem to have such a view of 'street smarts,' as evidenced by other comments made by some of them in interviews:

> For street smarts, you're actually experiencing something…experience always stays with you…book smarts might go away after you're done the subject" (April 24, 2012).

> [T]his whole idea of taking action at an early age is a good idea and lots of schools should participate in this because not only did we learn about the aspects of science that we are suppose and that are part of the curriculum in Grade 10, we made a change, so you actually feel like you did something while learning…like, the term comes again '*street smarts*,' [emphasis added] not only book smarts (April 24, 2012).

> [I]t's like a real-life lesson to you, right? And you're actually going to remember it. I remember everything we did right (May 3, 2012).

> [We participated in] taking the science to another level and really, like applying it and making it real" (May 3, 2012).

> [W]ell, when you go out and you talk to people, it's sort of like how you approach them, that's street smarts. If you use all books you might not know how to act so it's like a combination of both [book smarts and street smarts] (June, 2012).

Students' suggestions about relationships between RiNA projects and 'street smarts,' as opposed to 'book smarts,' appears to relate to a long-standing tension in science education—that is, between proponents of so-called 'Vision I' and 'Vision II' versions of science literacy (Roberts, 2011). The former prioritizes preparation of future scientists and engineers (etc.) and emphasizes relatively reductionist foci on knowledge and skills prominent in such fields; whereas, the latter places significant emphases on students' roles as citizens and, accordingly, their education is said to require more holistic experiences—something, for instance, prioritizing understanding and actions regarding relationships among fields of science and technology/engineering and societies and environments (STSE). It seems clear that school science systems tend to place greater emphasis on the former than the latter, part of what Theodore Lewis (1995) called a 'Platonic Legacy'; that is, high status associated with abstract, de-contextualized, knowledge that could, in theory, be applied across many contexts by 'Philosopher Kings.' There appears to be, moreover, renewed (if not continuing) emphases on such reductionism with advent of STEM (science, technology, engineering & mathematics) education initiatives (Gough, 2015). A problem with such narrow foci, however, seems to be that most students' backgrounds, interests and abilities do not align with Vision I perspectives and practices. Victoria Costa (1995, p. 316) claimed, for instance, that many science classes are composed of a range of student 'types,' varying from *Potential Scientists* through to *Outsiders*—most of whom either preferred not or whose backgrounds limited their opportunities to pursue careers in the sciences (and engineering, etc.). For most students, Costa (1995) claimed, school science was 'another world' than their everyday worlds. In other words, in terms of the Phenomena ← → Representation(s) dialectic described above, it may be that many students (including in both classes studied here) may be investigating (Phenomena → Representation(s)) and acting on (Representation(s) → Phenomena) more holistic ontological worlds than often is prioritized in professional science and requires considerable 'book smarts.' Sternberg and Grigorenko (2004), for instance, suggest that, when educators teach for more contextualized learning or (what they call) "successful intelligence, …it enables children to capitalize on their strengths and to correct or to compensate for their weaknesses, and it allows children to encode material in a variety of interesting ways" (p. 302).

Among benefits for students when teachers provide educational experiences drawing on both 'street smarts' and 'book smarts' appears to be much greater *inclusivity*—a claim that both teachers supported, in that each noted that students who frequently struggled in the past with science education more aligned with 'book smarts' characteristics appeared to thrive when given opportunities to engage in RiNA projects, which draw significantly on 'street smarts.' For these students, at least, one major repercussion may be increases in so-called 'fate control' (Rowe, 1978); that is, a feeling that, rather than living under assumptions that others controlled their lives, making them dependent on 'fate,' they could take more control over decisions—here, in terms of research and actions.

Although, as argued above, ceding learning control to students in Phenomena ← → Representation(s) translations (i.e., via RiNA projects) seems to enable students to draw on and use 'street smarts,' thus increasing overall class engagement, it may be that other approaches used by these teachers may also have contributed to such outcomes. Amanda, for instance, drawing on her background in research (i.e., through her Master's degree), prioritized outdoor experiences for students in her class. About such real-life experiences, Emilia Fägerstam (2014), for instance, claimed the following.

> Students communicate and participate in the classroom too, but when students engage in practical outdoor activities in collaboration with others they learn by doing and participating in a concrete 'real-life' context. This differs from the more abstract classroom situation (p. 58).

At the same time, data here suggest that both teachers also attempted to engage students with the outside world by bringing citizens' representations of them to students—such as in terms of use of activist videos from *The Story of Stuff* (storyofstuff.org) project, which highlights citizens' roles in various stages of production, consumption and disposal of consumer products and services.

13.6 Some Concluding Thoughts

Perhaps beyond—or along with—improvements to the wellbeing of individuals, societies and environments by working to cede control of decision making (e.g., through self-led research) about relationships among fields of science and technology and societies and environments (STSE) and socio-scientific issues/controversies within them and, moreover, to encourage and enable students to self-direct research-informed and negotiated actions to address relevant perceived problems, data from this cross-case analysis suggests that such perhaps more 'self-less' acts by teachers can significantly increase student/citizen engagement in learning (and acting). Traditional science education seems to prioritize selection and education of students who are most like practising scientists (Costa, 1995). Acknowledging tensions regarding emphases on Roberts' (2011) Vision I versus Vision II versions of science literacy, while acknowledging students' rights to prioritize their own ontological priorities (e.g., 'street smarts'), perhaps our research here requires us to re-emphasize Glen Aikenhead's (2000) advice about honouring students' life-worlds while giving them access to dominant perspectives and practices associated with 'book smarts.' Graff's (2001) statement below about this appears to be eloquent reinforcement of this recommendation:

> Bridging this gulf is not a matter of turning "them" into mini-versions of "us," or of asking students to give up their language in favor of our academic discourse. It is a matter of finding points of convergence and translation, moments when student discourse can be translated into academic discourse and vice versa, producing a kind of "bilingualism" on both sides of the student-teacher divide (p. 23).

Given dominance of perspectives and practices in science education aligned with Vision I conceptions of science literacy, it seems more educational research may be needed to explore and promote students' 'street smarts'—drawing, for example, from work surrounding the *funds of knowledge* (González, Moll, & Amanti, 2005); that is, knowledge students gain from their family and cultural backgrounds to make their classrooms more inclusive.

References

Akenhead, G. S. (2000). Students' ease in crossing cultural borders into school science. *Culture and Comparative Studies, 85*(2), 180–189.

Barrue, C., & Albe, V. (2013). Citizenship education and socioscientific issues: Implicit concept of citizenship in the curriculum, views of French middle school teachers. *Science & Education, 22*, 1089–1114.

Bencze, J. L., & Alsop, S. (2009). A critical and creative inquiry into school science inquiry. In W.-M. Roth & K. Tobin (Eds.), *The world of science education: North America* (pp. 27–47). Rotterdam, The Netherlands: Sense.

Bencze, J. L., & Sperling, E. R. (2012). Student teachers as advocates for student-led research-informed socioscientific activism. *Canadian Journal of Science, Mathematics and Technology Education, 12*(1), 62–85.

Biography.com Editors. (2014). *Arthur Miller biography*. Retrieved 09:12, October 1, 2014, from http://www.biography.com/people/arthur-miller-9408335
Charmaz, K. (2014). *Constructing grounded theory* (2nd ed.). New York: Sage.
Clandinin, D. J., & Connelly, F. M. (1992). Teacher as curriculum maker. In P. W. Jackson (Ed.), *Handbook of research on curriculum* (pp. 363–401). New York: MacMillan Publishing.
Costa, V. B. (1995). When science is 'another world': Relationships between worlds of family, friends, school, and science. *Science Education, 79*(3), 313–333.
Damsa, C. I., Kirschner, P. A., Andriessen, J. E. B., Erkens, G., & Sins, P. H. M. (2010). Shared epistemic agency: An empirical study of an emergent construct. *Journal of the Learning Sciences, 19*(2), 143–186.
Fägerstam, E. (2014). High school teachers' experience of the educational potential of outdoor teaching and learning. *Journal of Adventure Education & Outdoor Learning, 14*(1), 56–81.
González, N., Moll, L. C., & Amanti, C. (Eds.). (2005). *Funds of knowledge: Theorizing practices in households, communities, and classrooms*. New York: Routledge.
Gough, A. (2015). STEM policy and science education: Scientistic curriculum and sociopolitical silences. *Cultural Studies of Science Education, 10*(2), 445–458.
Graff, G. (2001). Hidden intellectualism. *Pedagogy, 1*(1), 21–36.
Guba, E. G., & Lincoln, Y. S. (2011). Competing paradigms in qualitative research. In N. K. Denzin & Y. S. Lincoln (Eds.), *The SAGE handbook of qualitative research* (pp. 105–117). Thousand Oaks, CA: Sage.
Habermas, J. (1996). *Contributions to a discourse theory of law and democracy*. Cambridge, UK: Polity Press.
Hatt, B. (2007). Street smarts vs. book smarts: The figured world of smartness in the lives of marginalized, urban youth. *The Urban Review, 39*(2), 145–166.
Hodson, D. (2003). Time for action: Science education for an alternative future. *International Journal of Science Education, 25*(6), 645–670.
Hodson, D. (2011). *Looking to the future: Building a curriculum for social activism*. Rotterdam, The Netherlands: Sense.
Hyslop-Margison, E. J., & Thayer, J. (2009). *Teaching democracy: Citizenship education as critical pedagogy* (Vol. 42). Rotterdam, The Netherlands: Sense.
Keogh, P. (2010). *Are you a Jackie or a Marilyn?: Timeless lessons on love, power and style*. New York: Gotham Books.
Kohn, A. (2005). *Unconditional parenting: Moving from rewards and punishments to love and reason*. New York: Atria/Simon & Schuster.
Krstovic, M. (2014). Preparing students for self-directed research-informed actions on socio-scientific issues. In L. Bencze & S. Alsop (Eds.), *Activist science and technology education* (pp. 399–417). Dordrecht, The Netherlands: Springer.
Lester, B. T., Ma, L., Lee, O., & Lambert, J. (2006). Social activism in elementary science education: A science, technology, and society approach to teach global warming. *International Journal of Science Education, 28*(4), 315–339.
Levinson, R. (2013). Practice and theory of socio-scientific issues: An authentic model? *Studies in Science Education, 49*(10), 99–116.
Levinson, R. (2010). Science education and democratic participation: An uneasy congruence? *Studies in Science Education, 46*(1), 69–119.
Lewis, T. (1995). From manual training to technology education: The continuing struggle to establish a school subject in the USA. *Journal of Curriculum Studies, 27*(6), 621–645.
Ministry of Education (MoE). (2008). *The Ontario curriculum, Grades 9–10*. Toronto, ON: Queen's Printer for Ontario.
Mirowski, P. (2011). *Science-mart: Privatizing American science*. Cambridge, MA: Harvard University Press.
Oreskes, N., & Conway, E. (2010). *Merchants of doubt*. London: Bloomsbury Press.
Orr, D. W. (2004). *Earth in mind: On education, environment, and the human prospect*. Washington, DC: Island Press.

Pedretti, E., & Nazir, J. (2011). Currents in STSE education: Mapping a complex field, 40 years on. *Science Education, 95*, 601–626.

Peters, S., & Wals, A. E. J. (2013). Learning and knowing in pursuit of sustainability: Concepts and tools for trans-disciplinary environmental research. In M. Krasny & J. Dillon (Eds.), *Trading zones in environmental education* (pp. 79–104). New York: Peter Lang.

Roberts, D. A. (2011). Competing visions of scientific literacy: The influence of a science curriculum policy image. In C. Linder et al. (Eds.), *Exploring the landscape of scientific literacy* (pp. 11–27). New York: Routledge.

Rowe, M. B. (1978). Science and soul: Some musings on fate control. In M. B. Rowe (Ed.), *Teaching science as continuous inquiry: A basic* (2nd ed., pp. 319–343). Toronto, ON: McGraw-Hill.

Sadler, T. D., & Zeidler, D. L. (2005). Patterns of informal reasoning in the context of socioscientific decision making. *Journal of Research in Science Teaching, 42*(1), 112–138.

Saxe, G. B. (1988). Candy selling and math learning. *Educational Researcher, 17*(6), 14–21.

Sternberg, R. J., & Grigorenko, E. L. (2004). Successful intelligence in the classroom. *Theory into Practice, 43*(4), 274–280.

Wenger, E. (1998). *Communities of practice*. Cambridge, UK: Cambridge University Press.

Wilson, E. O. (1998). *Consilience: The unity of knowledge*. New York: Vintage Books.

Wood, G. H. (1998). Democracy and the curriculum. In L. E. Beyer & M. W. Apple (Eds.), *The curriculum: Problems, politics and possibilities* (pp. 177–198). Albany, NY: SUNY Press.

Zeidler, D. L., Sadler, T. D., Simmons, M. L., & Howes, E. V. (2005). Beyond STS: A research-based framework for socioscientific issues education. *Science Education, 89*(3), 357–377.

Chapter 14
Tensions on Promoting Social Justice Through STEPWISE Pedagogies in an International Baccalaureate Preparatory Class

Neil T. Ramjewan, Brandon Zoras, and Larry Bencze

14.1 Introduction

School science education is a contested site of political and ideological struggle. It produces knowledge/power that maintains hierarchies entrenched through a recent history of colonialism and capitalism. As such, actors possessing appropriate cultural capital, according to Pierre Bourdieu (1986), are capable of investing that capital for the sake of producing more capital and, thus, accessing greater and greater conduits of power in society. 'STEPWISE' is, to a great extent, a response to such complex relationships. 'STEPWISE' is the acronym for Science and Technology Education Promoting Wellbeing for Individuals, Societies and Environments. As such, it is particularly interested in envisaging how these domains interact and reflexively construct each other in ways that reproduce power and maintain hierarchies of social relation. More importantly, it is committed to doing or acting on these relations to undo or at least resist neoconservative and neoliberal agendas that tend to prioritize capital before the wellbeing of people, communities, and environments. In other words, STEPWISE is an activist sciencepedagogical framework. It can be characterized by three main aspects but these should not be thought of linearly but, rather, reflexively in the course of taking sociopolitical action. First, STEPWISE is interested in making sense of the relations of power surrounding Science, Technology, Society, and Environment (STSE) issues, such as those regarding global warming, deforestation, and water contamination. However, being focused but not limited to school science, it is pragmatically oriented to enabling

N.T. Ramjewan (✉) • L. Bencze
Ontario Institute for Studies in Education (OISE), University of Toronto, Toronto, ON, Canada
e-mail: neil.ramjewan@mail.utoronto.ca; larry.bencze@utoronto.ca

B. Zoras
Toronto District School Board, Toronto, ON, Canada
e-mail: brandonzoras@gmail.com

L. Bencze (ed.), *Science and Technology Education Promoting Wellbeing for Individuals, Societies and Environments*, Cultural Studies of Science Education 14,
DOI 10.1007/978-3-319-55505-8_14

students to engage locally and immediately in their communities and everyday lives. Second, STEPWISE is committed to challenging roles of the learner as that of a consumptive subject to one of a productive capacity. As such, STEPWISE is rooted in a constructivist learning model that perceives the student as a holder of knowledge that is changed and produced as one interacts in social life. Finally, STEPWISE is interested in more than *just* analysis, but rather is committed to actually doing something about structural and social relationships in the world to affect the business as usual approaches to resolve contemporary STSE issues. This sort of reflection and action relationship has been described under the banner of praxis, which continues to be central to this pedagogical engagement.

The text that follows is organized with a few purposes in mind. First, we consider the background literature to frame STEPWISE in larger research contexts of science education. Secondly, we consider some theoretical conceptions that were influential from the onset of the teacher-student-researcher collaboration which, in turn, affected the shape of our teaching, our analytical tools, and the overall narrative that we construct here. Thirdly, we offer some practitioner oriented reflections on the implementation of STEPWISE, which we hope offer some insight into our processes and for future implementations. Finally, we critique STEPWISE and consider the discourses in which it is situated and the forms of capital that it may reproduce.

14.2 Background Literature in Science Education

In reviewing research literature, three interconnected themes seem to be of relevance in research on science education that frames STEPWISE and to which STEPWISE contributes both theoretically and pragmatically. Thus, in this section, we briefly consider scientific literacy, the nature of science (NOS) and humanizing science and *the* scientist through pedagogy.

A perennial goal of science education has been to prepare scientifically literate students (Cavagnetto, 2010). In technoscientific knowledge-based economies of today's increasingly globalized world, science literacy is arguably essential for economic access (Roberts, 2009) and democratic citizenship (Kolstø, 2008). However, the extent of democratic participation is increasingly limited to participation in neoliberal modes of relating to each other (i.e., in consumer terms) (Bencze & Carter, 2011; Giroux, 2008). Within science education scholarship there is a wide range of interpretations defining what it means to be scientifically literate but, in general, arguments include, but are not limited to, understanding of science concepts, processes of science inquiry (DeBoer, 2000), the nature of science (NOS) (Abd-El-Khalick & Lederman, 2000) and socioscientific interactions (Schwartz, Lederman, & Crawford, 2004) that together help to support scientific literacy (Cavagnetto, 2010).

These aims continue to be relevant in schools, are outlined in the introduction to the Ontario science curriculum (the context of this study), and are central to the STEPWISE framework. However, we find these aims are insufficient to understand science as more than a mere mode of knowing the world, in terms of concepts and interaction, or to imagine science as a dialectical relationship between the human subject, or the scientist, and its object of study, something in the world. Leopold Klopfer (1969) says that, "…understanding…scientific concepts and inquiry are without substance if students are unaware of the impact of science and related technologies on contemporary society" (p. 88). He draws our attention to the dialectical relationships between the epistemological enterprise that is science, and the societies in which it is situated and practiced. This tension is the space in which STEPWISE seeks to be located as a critical pedagogy which, through research informed and negotiated action, tries to disrupt, unravel, and possibly sever unjust relations of power that deepen varying forms of inequality.

Similar to science literacy, definitions for the NOS are varied and inconsistent. Nonetheless, Michael Clough (2006) indicates that the NOS refers to questions including, "What science is, how it works, the epistemological and ontological foundations of science, how scientists function as a social group and how society itself both influences and reacts to scientific endeavor" (p. 463). Despite this expansive definition of what science *is*, Hsingchi Wang and David Marsh (2002) indicate that school science education is commonly focused on established scientific facts and existing knowledge resulting in naïve understandings of the NOS; thus, deviating from expansive and integrated themes central to scientific literacy. Again, STEPWISE challenges these status quo relations between the learner and forms of knowledge by de-emphasizing scientific knowledge as static (facts) and orienting students towards the notion of knowledge as consequential of dynamic processes, relationships, and subject to ongoing debate.

Misrepresentations of scientific processes as primarily abductive forms of reasoning (see Bencze & Alsop, 2009) embedded in school science curricula and textbooks have mythicized science as a linear roadmap to knowledge and understanding (Allchin, 2003). This abductive reasoning is further exemplified by a reliance on experimental studies in schools, which Larry Bencze (1996) suggests is premised on practitioner belief that experimentation yields certain (versus uncertain) results representative of scientifically determined truths. Here, STEPWISE challenges these trends by working towards more authentic science pedagogies, which embrace uncertainty in scientific study. For example, Bencze (1995) notes that professional scientists often employ correlational studies that tends to utilize inductive and deductive reasoning such that explanations are open-ended and ultimately uncertain in terms of deciphering relationships between variables (covariance). This approach is in contrast to school science 'experimental studies' that generally aim to reproduce certain procedures and products to teach core concepts in a pedagogically engaging way. In general, school science 'experiments' begin with expected results that are reproduced methodically to yield expected observations. Students are then expected to speculate scientific explanations that account for the observed phenomena. These speculations are typically brought to a halt as the teacher intervenes to tell students the accepted scientific knowledge 'behind' their observations.

In processes of mythicizing scientific knowledge as predictable and graspable, the human actors that help bring science into being, scientists, have been valorized as infallible, unbiased freethinkers, pursuing truth and knowledge in the name of (hu)mankind, while the constraints of society and culture are dismissed (Hodson, 1998). This is, arguably, a form of dehumanization (Wang & Marsh, 2002), or reduction of human scientist to, as Barthes and Lavers (1972) put it the chapter titled, "The Brain of Einstein," a mythification of the scientist as a purely intellectual being, allegedly devoid of irrational impulses and cultural influences. Aside from implicating this vivisected form of being to declining student confidence and motivation (Matthews, 1994), the trope of the genius scientist functions to exonerate scientists, and science in general, from moral and ethical responsibilities by placing the scientist within the politically sterile trope of the laboratory. In research literature, efforts to resist these myths have been framed as the "humanization" of science, or the portrayal of science "as an organized activity of society" (Irwin, 2000, p. 8), within society, animated by fallible, irrational, emotional, political human actors, or scientists. STEPWISE explicitly takes science out of the classroom laboratory and into the world beyond the classroom as a means of exposing the political, social, and cultural bonds constitutive of modern Western science, which are ultimately acted upon via action to transform science, society and it various subjects.

14.3 Theoretical Influences: The Forms of Capital and Actor-Network Theory

First, given that we were working within a pre-IB/IB setting, we were conscious of a certain degree of privilege that students brought to the learning context. As such, we were interested in how students' cultural capital enabled them to reproduce certain forms of knowledge through STEPWISE, thus endowing them with more cultural capital as well as membership to a group through which social capital could be utilized for advancement in society. Second, given STEPWISE's concern for STSE issues, which tend to be large global problems, students tended to localize global issues in ways that they could act upon them in their daily lives. As such, we saw a connection to Actor-Network Theory (ANT), which insufferably wavers between the nodal actor and its network of constituent actors, all of which are networks of actors that are also networks—or a dialectic of unceasing local and global relations. However, these influences are not employed in a rigidly analytical manner since our aim here is to produce a text for practitioners to use in their daily work in science classrooms and to consider how the forms of capital impinge on social justice science education.

Bourdieu (1986) differentiates three general forms of capital: economic, cultural, and social. He says that economic, or the material form of capital can, through a process of *transubstantiation,* exist in immaterial forms—cultural and social capital, through various conversion processes that conserves value. These processes involve constitution of "purposeless finality of cultural and artistic practices and their products…the world of bourgeois man…the pure, perfect universe of the artist and the intellectual" (p. 16). The value of things represented and constituted within this realm, Bourdieu reminds us, is no less real than things themselves such that "priceless things have their price" (p. 16).

Cultural capital exists in three states: embodied, objectified, institutionalized, which resist the transferability of the economic form. In its embodied state, cultural capital is slowly acquired by self-investment towards some vision of a better self that enables the self to be more useful in society. As well, its acquisition occurs in subtle ways through contexts of upbringing, traditions, and is dependent on social class but not necessarily through inculcation. The objectified state refers to cultural artefacts through which value is reified in the form of art, books, or machinery. Finally, the institutionalized state is objectified cultural capital in the form of "academic qualification" (p. 20), which Bourdieu (1986) says neutralizes embodied properties of cultural capital. This objectification enables scrutiny of capital of the person, which is legitimized via 'the' academy and law, as distinct from the person, yet embodied. He says that "the power of instituting…impose[s] recognition" (p. 21) through the institutional network on the embodied *and* objectified cultural capital possessed by constituents of that network, or the human agent, such that "the academic qualification" (p. 21) can be compared and exchanged.

Social capital extends outward from the individual to the collective. It does so in such a way that *who* one knows, and the *quality* of one's social network, quality being distinguished in hierarchical class terms, determines one's social capital and mobility of the individual. Bourdieu (1986) says that social capital is the, "aggregate of the actual or potential resources, which are linked to possession of a durable network of more or less institutionalized relationships of mutual acquaintance and recognition, or in other words, to membership in a group" (p. 21). The aggregate of capital includes both cultural and economic forms, thus belonging to White bourgeois culture, generally affords the greatest social capital and its associated mobility in capitalist society. Membership implies obligation by the collective to its constituents, such that the individual is backed by collectively-held capital in such a way that credentials afforded by belonging, also affords credit, or a collective backing of economic and cultural wealth to members of the group. And so, as Bourdieu determines:

> The volume of the social capital possessed by a given agent thus depends on the size of the network of connections he can effectively mobilize and on the volume of the capital (economic, cultural or symbolic) possessed in his own right by each of those to whom he is connected (p. 21).

As such, bearing the appropriate social signifiers (intersections of family name, gender, race etc.) can provide access to economic capital held by the collective, without actually having economic capital as an individual. In addition, collectively held social capital has a "multiplier effect on the capital he [the agent] possess in his [sic] own right" (p. 21), or the gain of profit by the individual granted by mere belonging, thus resisting the reduction of economism applied to social capital, as we embrace the complexity of networks as a necessary form for social capital. As

Bourdieu (1986) acknowledges, these networks are not simply a "natural given" (p. 22) but rather a:

> …product of endless efforts at institution, of which institution rites – often wrongly described as rites of passage – mark the essential moments and which is necessary in order to produce and reproduce lasting, useful relationships that secure material or symbolic profits (p. 22).

Bourdieu (1986) refers to *reproducing* the group, which he continues to describe as a series of strategies by the agent or the collective, deliberately or inadvertently, which serves to *transform* relationships to yield more "durable obligations" (p. 22), or a strengthening of the network and, thus, profits of belonging. Here, the linkages to ANT are quite evident, referring to networks, mobilization, transformation (or translation in ANT), and durability. Though these concepts are similar, a key distinction is the effect of non-human elements on the obduracy and fragility of networks, which drastically changes the researcher gaze through said analytic looking glasses, and to which we now focus our gaze.

From the onset of the project, we were influenced by ANT, which draws on Michel Foucault's antihumanist philosophy. Sociologist and prominent ANT scholar John Law (1992) describes ANT as the "sociology of translation" (p. 380), which is particularly interested in the power relations between heterogeneous elements, or actors. It considers how certain actors exist in relationship with other actors which, in turn, shape those actors and bring into being networks of relations that are dynamic and fluid. Historically, ANT as a theoretical assemblage has been interested in how certain actors get punctualized, or centered, into a single node that obscures the infinite network of relations and actors that they constitute and are constituted by. More recent ANT scholars have been interested in how these relationship are constituted, change over time and space (context), and fall out of existence.

With these general ANT concepts in mind, one of the earliest activities in which students were asked to engage involved graphically representing 'actor-networks' to better understand the STSE issues that they were investigating and seeking to address via sociopolitical action. This mapping strategy was informed by ANT as a means of identifying various actors involved in and surrounding STSE issues (see Fig. 14.1). We found ANT to be very useful in depicting relationships, in part because of how it defines 'actors,' or the human, non-human, material, and immaterial elements that constitute networks. Arthur Tatnall and Stephen Burgess (2002) offer the following explanation of an actor, which we feel is useful: "An actor is seen not just as a 'point object' but rather as an association of heterogeneous elements, themselves constituting a network. Each actor is thus itself also a simplified network" (p. 183). In other words, an actor is simultaneously an actor *and* a network, implying that any depiction can theoretically expand *ad infinitum*. ANT has also been involved in the exploration of tension between structure and agency, which extends to its almost dizzying oscillation between the local and the global. Both these concepts were of importance in terms of students localizing large global issues, as well locating themselves within networks, or structures, positioning them

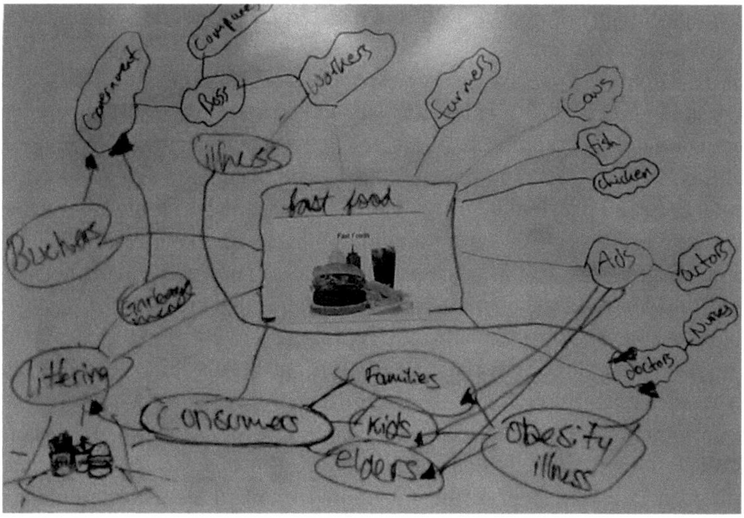

Fig. 14.1 Student explore ANT (Actor-Network Theory) and power relations through network mapping

as various actors that have an effect on the system (for an extended discussion on ANT specific to this context see Ramjewan, Zoras and Bencze, Chap. 12, this volume).

14.4 Research Context

The high school, comprising about 1000 students, in which our study was located is situated in a large urban centre in Canada. The International Baccalaureate (IB) program was relatively new to the school, being in its fifth year of implementation. Students were allowed to apply from across the school district. Consequently, many of the students were outside of the regular catchment area affecting the expected socioeconomic profile of the cohort with some student from working classes and others from intellectual classes. Though we did not explicitly gather empirical data to discern the economic background of participants, interview questions about student home life revealed some of these differences. The group of pre-IB Grade 9 students with whom we worked during this action research project entered into the program by taking a test, having an interview, and delivering a presentation to the school IB coordinator. This context, as well as the socioeconomic range, is not surprising given criticism of the IB program in the last few decades, followed by more recent efforts on behalf of the International Baccalaureate Organization (IBO) to diversify beyond elite institutions to inner city settings (Tarc & Beatty, 2012). However, this move does not exempt the program from an ableist ideology of "intellectual elitism" (Peterson, 1972, cited in Tarc, 2009, p. 255), which we address in the discussion section.

The IB Diploma Program doesn't officially start until Grade 11 and, thus, the Grade 9 student participants with whom we worked are considered "pre-IB." Pre-IB/IB students are grouped into cohorts and travel through the school's system together. As well, both pre-IB, IB, and regular students are tested and evaluated against the same standardized provincial curriculum. However, the IB program issues its own standardized tests as means of determining student ability to continue in the IB Diploma Program (as well as teacher accountability; however, this is not the focus of this discussion). By the end of Grade 10, students that are eligible must decide if they want to continue into the 'official' IB program or switch to the regular classes.

A major distinction of the IB program from the regular high school program is access to enrichment opportunities in and through their academic classes. It is for this reason that the classroom teacher, Brandon Zoras, was inclined to incorporate STEPWISE into his teaching and programing (however, STEPWISE has been implemented in a variety of settings including standard state curriculum settings; e.g., Bencze, Sperling, & Carter, 2012). In fact, Mr. Zoras saw a direct connection with STEPWISE's open-ended approach to addressing STSE issues and commitment to sociopolitical action with the pre-IB requirement for students to earn Creativity, Activity, Service (CAS) (International Baccalaureate [IB], n.d.a) credits towards the IB diploma and other elements of the IB learner profile (IB, 2013).

Mr. Zoras had been teaching for 6 years within the same school board, but at 4 different schools over those 6 years. He had completed his Bachelor of Education in Inner City Education and as well as a Masters of Education while teaching in urban settings. It is through social justice, equity, and information and communications technology that he approaches his science classes. His previous work with Larry Bencze examined student activism through social media, while doing school science (Zoras & Bencze, 2014). This was his first year teaching in the IB program, in which he taught Grade 12 Diploma Program Chemistry and Grade 9 pre-IB Academic Science, the latter of which participated in STEPWISE. He received a 3-day training program to be able to teach in the IB program and attended workshops and other professional development sessions relating to the IB program throughout the year.

14.5 Reflections on Implementing STEPWISE

This section is organized into three main parts that focus on how STEPWISE was implemented with the Grade 9 pre-IB students. The case is written mainly through the eyes and experiences of the classroom of Mr. Zoras; however, we use a plural pronoun to represent a degree of communication that we tried to achieve throughout the project. First, we reflect on how we first introduced STEPWISE and ANT to consider relations of power. Secondly, we consider texts that students generated with an emphasis on uses of the Wiki, or our virtual learning environment, as they conducted secondary and primary research and reflected in their journals throughout the process. Finally, we look back at how students used various texts to act in society, including the mentoring of elementary school students across the Wiki.

14.5.1 *Introducing STEPWISE and Science, Technology, Society, Environment Issues to Students*

Mr. Zoras introduced STEPWISE early in the semester to frame the course as being explicitly concerned with social justice. The project was presented as an opportunity to work with a teacher, graduate student, and an established scholar to examine conditions of inequality and respond to STSE issues. After considering cultural capital that pre-IB students brought to the learning context, we decided to do an accelerated apprenticeship to engage in action sooner and with greater frequency (multiple projects over multiple science units). Mr. Zoras provided students with 'apprenticeship' lessons and activities to prepare them for self-directed research-informed and negotiated action (RiNA) projects, to address problems they perceived about STSE relationships. The first of these apprenticeships was more teacher-led, while the second one gave students more control over most decisions. Consequently, students could complete two full sets of STEPWISE-informed activities, the latter of which involved the mentoring of a group of elementary school students through STEPWISE-related processes via the Wiki, on our online space for collaboration and knowledge production.

Students were exposed to many examples of social justice and STSE issues at the start of the course, mainly through YouTube™ videos and discussions. Over the years, Mr. Zoras has collected a range of media that examine STSE issues, power, and social action, which he integrates into the Ontario science curriculum. Keeping the end goal of sociopolitical action in mind, we wanted to not just expose students to STSE issues but enable them to critically examine and question why the issues exist at all. Mr. Zoras initially introduced students to large scale global issues, such as 'blood minerals,' using the documentary *Blood in the Mobile* (Poulsen, 2010). Though a process of thinking in actor-network terms, we were able to see how we are implicated in these large, and sometimes overwhelming environmental problems.

Consequently, for our chemistry unit, we looked at how minerals and mining affect the world but we also considered how we affect demand and uses of minerals. After watching the preview for the film, we examined power relations involved in mines in the Congo, which companies are utilizing these minerals, and the people who are creating demand for these products. We challenged students to think about how this problem was not just that of the Congolese people but as a problem created from outside of the Congo, in particular through Western appetites for electronics. Students were encouraged to show a web of actors (human and non-human) involved in mining. Students were also encouraged to look at who was taking action against this issue, what mode they took, and how that in turn enters the network as disruptive elements. We presented the idea that documentary makers can be thought of as activists who act through creation of media, which in turn acts upon us (the viewer) to affect our perspective and, thus, our relationship to the issue. This had the effect of getting students to reflect on their roles as consumers and, thus, connect their wants to mining in the Congo. As a consequence of having their consumer wants implicated in the process of mining, the class signed up for a recycling box through a local organization and ran a campaign within the school to get involved and spread the message.

In 2008, the Ontario Science Curriculum was revised and restructured, part of which involved moving teaching/learning expectations for STSE issues to the forefront of the curriculum document, while shifting knowledge and understanding of 'products' of science and technology (e.g., laws, theories & inventions) to the end. Similarly, STEPWISE is premised on the primacy of the politics of science discourse. As a consequence, we kept both the curriculum and the NOS in mind as we introduced various issues to students. In turn, it became evident that students started to realize that many of the chemistry issues stretched across biology, social sciences, law, and politics. Students even commented on other teachers in different disciplines who were surprised that they were addressing social issues in science class all while learning state-mandated scientific concepts, principles, and methods. Eventually students began to question why the schooling system itself is so compartmentalized when the issues under scrutiny involve massive networks of people and departments that are interconnected.

Many other examples were presented when introducing science knowledge and content throughout the semester. The 94 Elements online video series has a particularly moving piece that looks at the 'farming' of circuit boards from e-waste and the health and environmental consequences of this mining (Paterson, 2008). As well, the online *Story of Stuff* series has expanded to examine many issues related to STSE education, from bottled water, to cosmetics, to their newer *Story of Change*. Students were also inspired by many of the TED Talks™ and TED Ed lessons on environmental issues. They would often share with us new videos they found online and come into class with a blast of energy asking others if they have seen various videos. Other sources include the website *Upworthy* where we looked at effects of the molecule benzene on human health and how consumerism intertwines with health and science.

It was through these examples in conjunction with theoretical influences, such as ANT, discussed earlier, that students began to see that these far-off examples aren't so far away and that local issues are needing champions to create change. As mentioned, an early activity that students completed involved choosing an STSE issue around which they create an actor-network web to graphically represent various elements (actors) that constitute the issue. For their first attempt at mapping an actor-network, students were given a choice from a variety of STSE issues and within a group planned their map on a whiteboard. The group discussed below (see Fig. 14.1) chose fast food and looked at implicated elements (actors) and those most affected (typically human actors). As they were planning, we asked them to also think about who has power, who is driving the issue, and how we can interrupt the web.

As is evident from Fig. 14.1, students were able to, in a very short timeframe, expose some very sophisticated relationships to fast food industry, including relationships among government, industry and consumers (themselves), the impact on the environment, health and wellbeing, and the marketing machine proliferating images for visual consumption and shaping ideologies. Furthermore, actor-networks, such as the one in Fig. 14.1, represent an artefact of student knowledge created with their peers through dialogue drawing from their personal and collective experiences. These representations became a first crucial tool to creating further texts upon which to reflect and then act (i.e., praxis). Overall, we found that the notion of an actor-network was well understood by the students and central to how they examined issues and how they began to situate themselves in the issues being acted upon through the remainder of the STEPWISE project.

14.5.2 Making Socioscientific Activist Texts

The next phase of the project came naturally as the students felt it was not enough to just consume information about STSE issues, such that they were eager and enthusiastic to step in an affect change. This shift is a central component of the STEPWISE framework. To move students from content consumers to creators of content really empowered and engaged students as they began to gain a sense of agency, through situating themselves in the social structures in which they exist.

14.5.2.1 The Virtual Learning Space – The 'Wiki'

With advancements of the Internet and related technologies, people around the world are able to access information, communicate and collaborate on projects like never before. The Substitution, Augmentation, Modification, Redefinition (SAMR) model (Candace, 2013) shifts technology use from a simple substitution to redefining new tasks that were not possible without the technology. With this shift, we wanted to create a platform through blended learning where students could collaborate and communicate their projects to each other (as a group), to other groups, and to the teacher. The first STEPWISE action project was done through Wikispaces™. Wikis allow for students to collaborate in an online space that accommodate a mix of media, such as text and audio-video content, but also abilities to be fluid and progress over the semester. Students added to their projects as they went along and shared them to get feedback from peers, teachers and researchers. Students could view STSE issues and questions about distributions of power posted by us and work collaboratively to share about their chosen issues. This proved to be a useful tool for student collaboration and feedback as well as keeping us, as teachers and researchers, informed and organized (in terms of student work and projects) along the way.

For our second STEPWISE action project, we shifted from Wikispaces™ to Google Apps for Education™ through which we were able to leverage Google Sites™ as a learning platform—and as a platform for our students to mentor younger students at an elementary school. Students were able to embed various document types, interactive audio video content, and use this as a fluid site that was constantly evolving as they were learning and which younger mentees (discussed below) would use as a repository of information to learn about STEPWISE and act in their locality to address local/global issues.

Both platforms allowed for a blended learning space where students could log in from a variety of devices and locations. Through the Wiki/Google Sites™ platforms, high school students were positioned as content creators and curators who could shape their learning and the messages they wanted to convey to younger students learning about science and social activism.

14.5.2.2 Student Research

Secondary Research

Again, under the presumption that pre-IB students possessed a degree of privilege afforded by cultural and social capital, we agreed to, through transmission-style teaching, deliver an accelerated apprenticeship (also see Chap. 2, this volume) at the beginning of the unit. The students quickly came to their own conclusions that, before taking actions to address STSE issues, they would need to review and research the STSE issue in which they were interested via secondary research, or research using secondary sources. This typically involved understanding scientific concepts and technology involved, impacts on societies and environments, as well as contemporary efforts to address the issue. Students used various secondary texts (mainly online content in the form of documents and video) to find out as much as they could about the issue they chose. They were able to post this information to the Wiki as part of their ongoing work. Students also utilized email and various social media, such as Twitter™ and Facebook™, to contact companies typically involved in the network of the STSE issue, to ask about business practices, as well as about their knowledge and efforts to address the STSE issue to which they were contributing. In addition, students engaged government officials, activists, professional scientists, and not-for-profit organizations to get more information and better understand the STSE issue at hand.

One group of students constructed a page of information and resources for peers as they continued to learn about their STSE issue and before deciding on an action (see Fig. 14.2). The idea of a fluid space, such as the Wiki platform, that was able to change and grow as they learned was a key feature and reason why a digital platform was preferred over traditional paper-based media. They were able to embed videos, images and live links to relevant information, which eventually included videos of themselves either as an action, or as a mentoring resource and, in some cases, both. For example, the group represented in Fig. 14.2 collected a variety of

Fig. 14.2 Students use Wikispaces™ to create pages to organize and share their secondary research

e-waste resources and then links to help educate others; along with yielding forms of both secondary research and a social action to inform others of the problem at hand. As well, they linked to other organizations, like Greenpeace™—which provides an interactive map of where countries send their e-waste (Fig. 14.3).

Primary Research

Part of the apprenticeship offered to students involved teaching distinctions between experiments and correlational studies. Bencze (1996) highlights the preference of experimental investigations over correlational studies in school and college science, which persist in the current climate of high stakes testing, dense curriculum standards, and the rhetoric of being competitive within a free market economy. Despite extensive use of correlational studies by professional scientists, Bencze (1996) suggests that school science tends to opt for experimentation due to beliefs by teachers in potential of experiments to yield more certain results. This belief in the ability to discern firm, knowable, causal relationships between variables aligns with societal demands to advance via competition in a knowledge-based economy. In contrast to this belief, STEPWISE tries to challenge these ideologies by encouraging the use of correlational studies, which do not emphasize strict causal relationships, and dwells in the uncertainty of the (un)knowable and the open-endedness of acts of learning. Correlational studies also often are a more ethical choice than experimentation when studying possibly-harmful effects on living things—which is frequently the case when investigating STSE issues.

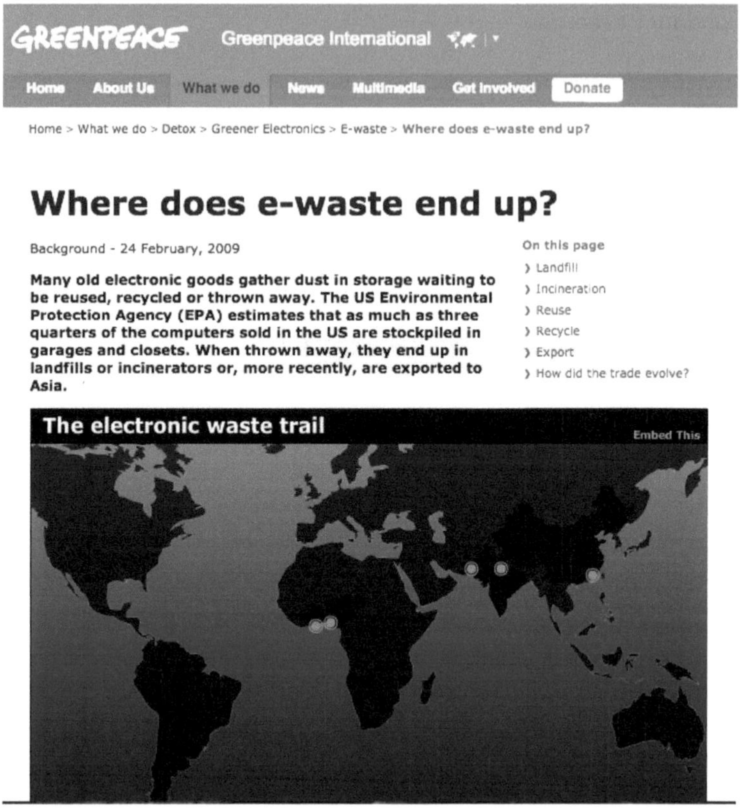

Fig. 14.3 Links to content are easily embedded in the Wiki. (http://www.greenpeace.org/international/en/campaigns/detox/electronics/the-e-waste-problem/where-does-e-waste-end-up/)

After having curated secondary information on their STSE issue, students were required to conduct primary research to learn more about it. Thus, students had to start asking critical questions about actors entangled in the network of their STSE issue that they were trying to resolve. Here, we encouraged students to think of who they felt were in positions of power and who they felt was most affected by the issue in order to devise actions that could yield the greatest potential impact towards some resolution. We encouraged students to think of ways to better understand their issue through voices and experiences of others. As such, we suggested gathering information via in-person interviews, phone, email, Twitter™, Facebook™, and through online surveys, such as through Google Forms™ and Survey Monkey™. In large part due to convenience, surveys were the tool of choice by the majority of groups, which was used to poll their social network (both in face-to-face in their classes as well as digitally to their social media networks), which resulted in data that were collected and analyzed by the students. Student data were primarily quantitative and enabled them to estimate frequencies of determined variables. They were able to utilize Google Site™ tools to generate graphical representations to compare their

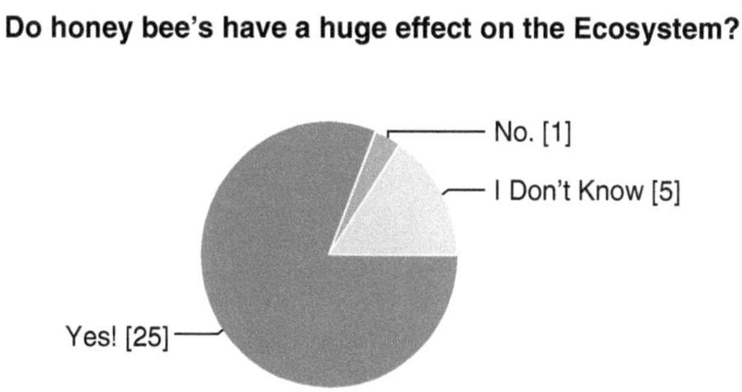

Fig. 14.4 Students publish data from their primary research using Google Forms™

variables of concern, giving them the ability to determine positive or negative relationships upon which to reflect and develop sociopolitical actions (Fig. 14.4).

14.5.2.3 Student Journals

We were very conscious to provide opportunities for students to reflect on relationships that they were representing through secondary and primary research, for the sake of making connections *between* these sets of information (and others), to enable sustained engagement with STSE issues, via research-informed and negotiated actions. By having students reflect multiple times over the semester, they could track how they are interpreting what they were learning as well as get formative feedback on their progress. Students were asked to write journal entries four times throughout the semester and use a chart to guide them. They considered their attitudes, skills, and knowledge (ASK) as they thought about their issue, the research they conducted, and the actions they took. A sample of a student's third journal can be found in Appendix 14.1. What is evident to us is that the student has become really committed to this work, developed a sense of agency, and is able to see key players and actions needed to influence change in the world. Since this was a digital journal, the other students and researchers also were able to see how everyone in the group was proceeding with their project.

14.5.3 Using Student Generated Texts to Act in Society

14.5.3.1 Actions

Having the students go through two rounds engaging STSE issues through RiNA projects, and reflecting through journals along the way, helped to shape the action process. We provided exemplars of previous student actions as part of the apprenticeship. We tried to emphasize the relationship between how STEPWISE activists were able to take their own secondary and primary research to form meaningful and,

Fig. 14.5 QR code placed on newsstands directing readers online for a financially and paper free copy

hopefully, impactful actions that disrupt networks of relations constitutive of the STSE issue. The first set of actions was very powerful from a group that was initially concerned with global issues of deforestation. First, through secondary research, they were able to localize the issue to paper waste. After doing secondary research, students were encouraged to think about actions to address their issue. Typically, this meant particularizing the issue to a specific context in which they could engage with the issue. The group considered here was annoyed by the fact that newspapers were littering their commuter route to school as newspaper companies distributed free copies of two local daily papers outside the subway. Again, they used secondary sources to determine the carbon footprint associated with printing and delivering papers meant for one time use.

Many groups chose to use digital mediums to educate their social networks on particular issues; however, this group decided to make a change in the subway to address issues linked to newspaper waste (and deforestation). They went to the subway to conduct primary research by polling commuters whether they preferred newsprint media or digital media. Using this information in conjunction with their secondary research, they decided to provide a Quick Response (QR) code which they placed on newsstands (see Fig. 14.5) and spent some time informing commuters of the impact of newspapers on deforestation and redirected them to the digital source. The QR code enabled readers to use their digital device to link to the online version of the paper, in turn potentially reducing paper waste and, thus, impacting large scale problems of deforestation. Of course, mining for precious metals that are necessary for production of mobile phones are implicated in global issues of deforestation and, so, there are many critiques that can be launched against the actions that we report here. However, we critique STEPWISE below in sociological terms to engage with these contradictions.

14.5.3.2 Mentoring

Mr. Zoras had partnered with an elementary school colleague in the past to test interclass communication through the Web. Mr. Zoras proposed extending the second RiNA project that his class was conducting to involve a similar interclass relationship with elementary school students. We thought that this would be a novel actor and network of relations in the form of mentorship of elementary school students by high school students, in itself representing an action on the world. We informed the Grade 9 pre-IB students of the plan to mentor younger students at the end of the first action project. Consequently, the prospect of being role models and mentors affected how Grade 9 students perceived their location in the network mediated by the Wiki, the means of communication with elementary students.

Grade 9 students were responsible for creating resources to post to the Wiki for elementary school students to learn about the STEPWISE process, including identifying STSE issues, conducting secondary and primary research and, finally, taking sociopolitical actions. They created entire modules through Google Sites™, including handouts, videos, and their own personal projects as examples of how they addressed STSE issues. As the Grade 9 students conducted their second STEPWISE project, they provided feedback and support for elementary students via the Wiki as elementary school students posted their work. Using exclusively high school student-made resources, elementary classes were able to identify issues in their own communities, conduct both secondary and primary research, brainstorm solutions, and start to take meaningful and authentic actions toward a better world. Although it took longer than expected for the elementary students to go through the process, it had lasting impressions on the way both high school and elementary students perceived the practice of science and their responsibility to others and the worlds they live in (for an extended discussion on the mentorship process see Ramjewan, Zoras and Bencze, Chap. 12, in this volume).

14.6 Discussion

Using both Bourdieu's (1986) notion of forms of capital (economic, cultural, and social) and ANT as theoretical lenses, we consider the tensions of doing social justice work in an IB program that aims to provide students with a competitive international advantage. First, using ANT, we consider how engaging in STEPWISE enables students to be positioned as more than mere consumers, resisting the illusion of market based agency, or choice, for a subjectivity of the social activist and, perhaps, a more authentic sense of agency, at least through resistance against capitalist modes of subjectification. Secondly, using Bourdieu's notions of different forms of capital, we consider how the IB program refracts and amplifies privileged students cultural capital to produce and reproduce network relations that yield personal and group profits in the form of social capital, thus maintaining hierarchies of privilege through a social justice oriented pedagogical framework such as STEPWISE.

In general, we found that as students expanded their networks (via pictorial representation) they eventually found themselves implicated in their STSE issue. Typically, students' entanglement in their STSE networks was in the form of consumers, or a subject of capitalist discourses, which was rarely blind to environment and material resources. It's familiar to interpret this as a cognitive *realization* by students that they were actors in the network of capitalist relations whose presence has an effect on the constitution and maintenance of that network. However, an interesting and powerful aspect of ANT is its tendency to implicate itself as an actor (and network) meaning what ANT *is*, is continually under revision, and possibly dissolution (Law, 1992). Through this almost obsessive self-reflexivity, and in this particular iteration, the actor STEPWISE (secondary research, primary research, reflection, sociopolitical action, teachers, researchers, students, virtual learning spaces, ANT, etc.) brought ANT (which constructs STEPWISE as an actor) into the pedagogical fold. In other words, STEPWISE and ANT were not mutually exclusive but, rather, coextensive elements that constituted the other in very particular ways in our context. Subsequently, students were *positioned* beyond the subjectivity of capitalist consumers, via discourses of science education, social activism, and the sociology of science which are not often presented in schools as interdependent. Consequently, students began to envision themselves as educators and social activists that could do more than merely 'choose' less harmful products as exclusively consumers in capitalist society because they were *brought into being* more than mere consumers. As educators and activists, they could act in strategic and deliberate ways to influence others such as business entities, individual workers, friends, family members, and other students to consider their role in certain STSE issues. However, despite becoming more than consumers, students were also in the process of becoming IB learners.

Recall that the Mr. Zoras saw a direct connection with the IB learner profile, the IB Creativity, Action, Service (CAS) requirement, and STEPWISE. Thus, the IB program was a structural actor that shaped the ontology of our social experiences. This text is not meant to be a critique of the IB program; however, we cannot ignore the literature on international curricula, since connections between the IB program and elite schooling is well established. Paul Tarc (2009) considers the issue of access in the IB program. He says the "IBO's de facto beginnings as an education for mobile elite were in tension with its dream of making a better world in an era of democratization" (p. 252) and, thus, found itself obliged to broaden access beyond an elite class. Tarc cites the IBO Director General's Report written by Alec Peterson (1972) in which Peterson tries to defend the charges of social elitism by distinguishing the IBO's concern for 'intellectual elitism' in a meritocratic society, which he believed to be worth developing towards a better and more democratic world. Here, Tarc cites Remillard (1978), who criticizes Peterson for being blind to the biological, social, and political inequalities overlooked by the ideology of meritocracy. Finally, Tarc notes that theories of socio-cultural reproduction of the 1970s, which detail how schools reproduce class relations, seems to have been in large part ignored by the IBO. As mentioned previously, the modern IBO is making efforts to broaden its contexts to inner-city settings (Tarc & Beatty, 2012), such as ours.

The IB is an internationally-recognized institutional program which, according to their mission statement, "aims to develop inquiring, knowledgeable and caring young people…[via]…challenging programmes of international education and rigorous assessment" (IB, n.d.b) with undeniable ideological roots in an elite social class (Tarc, 2009). More specifically, the IB community produces IB students that eventually attain an IB diploma, which represents cultural capital in its institutionalized state. According to Bourdieu (1986), then, the diploma, or graduating from said program, confers "academic qualification" (p. 20) objectified for the sake of separating it from the person such that it can be compared and exchanged for the sake of profits (in various forms). This form of cultural capital can be converted into social capital, which Bourdieu defines as the "aggregate of the actual or potential resources, which are linked to possession of a durable network of more or less institutionalized relationships of mutual acquaintance and recognition, or in other words, to membership in a group" (p. 21). As such, institutional programs such as the IB represent a group in which recognition of the agent as intellectual is imposed by its very distinction from the mainstream, non-internationally recognized program and learner. The need for exchangeability of the cultural commodity also imposes recognition on the embodied and objectified cultural capital possessed by the agent through the social network of instituting bodies or, in this case, the IB program and all institutions who recognize the IB program. As a consequence, despite efforts of equitable access through extension into urban contexts, the accumulation of social capital is maintained through the reproduction of IB students who are internationally and institutionally recognized as intellectually superior granting them access to forms of symbolic and material wealth.

As STEPWISE enters the market of exchange and circulation *through* the institution of the IBO, we must consider how it contributes to the reproduction of the social structures described above. As mentioned, Mr. Zoras saw connections between STEPWISE, the school, and the IB program. First, Mr. Zoras saw a connection between the social justice orientation of STEPWISE and the school, which promotes social justice through various tenets that they articulate on their website. Second, Mr. Zoras saw connections between the IB learner profile, which articulates fairness and justice, and the IB programs creativity, activity, service (CAS) requirements, to the student-led activist science framework of STEPWISE. However, despite these points of confluence, using Bourdieu's notion of cultural reproduction, we argue that STEPWISE, within an IB setting, is utilized to attain institutionally sanctioned cultural capital, which is reproduced to maintain capitalist modes of relation.

As mentioned, we were initially informed by three documents, *the IBO mission statement, the IB learner profile, and the CAS requirement*. The *IB learner profile* connects to STEPWISE concern for the wellbeing of individuals, societies, and environments, in that it aims to develop "internationally minded people who, recognizing their common humanity and shared guardianship of the planet, help to create a better and more peaceful world"(IB, 2013). As well, it specifically mentions justice as follows: "We act with integrity and honesty, with a strong sense of fairness and justice, and with respect for the dignity and rights of people everywhere" (IB,

2013). Here, the "IB learner profile" refers to the expectations of the individual learner while demanding the individual be concerned with the world beyond the self. In differentiating the individual "IB learner," the agent is discursively constructed. This agent is capable of possessing cultural capital, or skills that the IB program works to instill, such as inquiry, thinking, communication, open-mindedness, risk taking, and reflection in addition to striving to be caring, knowledgeable, and principled. These skills are recognized internationally by universities and colleges; thus, granting membership to an intellectual class or group. Furthermore, the social network, and thus the aggregated capital of its members, or social capital, is signified by the "We" that acts with integrity and honesty towards justice. In turn, the priority of access by the individual agent to an intellectual class is achieved through justice oriented activities. As a consequence, the reproduction of the intellectual is achieved through a collectively skilled effort by intellectually privileged (read able bodied/minded) students recognized by international institutions, to care for the wellbeing of the world. In the IB setting that we consider here, STEPWISE can be considered as a social justice instrument utilized in an institutionalized program that distinguishes itself from local institutions by reproducing intellectually superior agents constitutive of an international program recognized by other international institutions; thus, establishing an exchange value representative of said hierarchical distinctions that maintain capitalist networks of social inequality.

14.7 Conclusions

We have attempted to provide a reflective and practical text for practitioners to use for the sake of implementing STEPWISE in their classrooms and various learning environments, with a focus on the use of Web 2.0 technologies, or dynamic user created content. However, our reflections on our efforts are not without room for further reflection, a process of self-reflexivity that we have tried to foster and encourage though praxis with the students with whom we engaged. The purpose of this is to continually unsettle the self that settles in between moments of reflection for the sake of changing and adapting to social, economic, and political life (not that these dimensions are discrete). As such, we have also offered some more theoretical considerations of STEPWISE, as both a structure and as an instrument depicting the tensions of trying to teach social justice in a program that enables individualized upward mobility. Together, we hope these seemingly discordant texts work to encourage meaningful reflection to face the contradictions of institutional life.

Acknowledgements We would like to thank the wonderful class for all your hard work and commitment and in particular those students and parents that gave their consent to be represented by us in this text. This chapter could not have been written without the passion and dedication that you all have. Additionally, we are grateful to the Social Sciences and Humanities Research Council, Canada, for their generous funding of research about this case.

Appendix 14.1: Student Journal Entry

	Attitude	Skills	Knowledge
Issue	I feel like a big issue right now is the fact that consumerism and its harmful effects on the environment are being ignored.	Once again, I have to be good at deciphering the information I receive online between right and wrong, A small analysis of information can make a huge difference. I am helping to educate people about the effects of consumerism and how to take action against it. I definitely do not want to mislead them because I have the wrongs ideas about the issue. In order to inform others about this issue, I first need to understand it clearly and properly myself.	I need to know and understand this issue like the back of my hand. If I am going to be explaining this issue to others they may have questions and concerns. I need to have enough knowledge about the issue of consumerism to be able to answer most, if not all, of those questions accurately. This will be extremely important because I will be the one people are relying on and I need to give them the correct information they want and require to take action against consumerism.
	In fact, they are not even being ignored.		
	People just aren't aware of these effects.		
	I feel proud and happy that I am making a change by helping to make people aware about these effects and how they can do simple, small everyday things that can have an amazingly large impact		
Research	There has been extensive research done on the global issue of consumerism and electronic waste. I am grateful to all the people who have previously conducted research on this issue as it is helped my partner and I a great deal throughout our project. Furthermore, I am grateful to all the people who have stepped forward and taken action against consumerism and e-waste, thus making more people more aware of its harmful effects. I believe, that no one wants to cause harm to their home. Our earth is our home and therefore we should make sure that we are not harming it. I know that if people are more aware regarding the issue of consumerism they will take action against it and 1 day it will come to an end,	Like I mentioned above, I need to be able to tell the difference between right and wrong information because it can make such a big difference. It is crucial for me to make sure that the source I am receiving my information and research from is reliable.	Research done in the past on this issue is just stunning. Like I mentioned earlier so many people have looked into this issue and have researched on it. The research was not bias and it was accurate.
		Just as I am relying on a source to get more information on my issue, there are people relying on me to further explain this issue to them. I do not want to mislead them or even accidently give them the wrong information.	This made it a lot easier for me to conduct my research because I felt good about the sources I was getting the information from. There were visuals and contact information provided for further inquiry. This made me feel at ease because I was able to look at a chart and see how everything worked. If I felt like I had a question I was provided with contact information to get my answer.
		It is not right if I am trying to reduce the effects of an issue because again, one needs to be able to properly understand something to be able to explain it to others.	

(continued)

Appendix 14.1 continued

	Attitude	Skills	Knowledge
Action	I feel so proud and happy at this time. In fact, both my partner and I are extremely pleased with out action. We wanted to get the message of consumerism out there and we wanted people to start doing small things that will lead to big changes. Through the Recycle My Cell school challenge my partner and I held at our school, we got the students and staff of our school to understand the harmful effects of e-waste. In fact, we even got them to bring in their old chargers, cell phones and other e-waste into room 304 to put them in a box. We will send this box to Recycle My Cell this Monday. Although, we may not win the school challenge, I am happy that now so many people are aware of the effects of consumerism and are willing to take action against it by recycling their e-waste.	There were a lot of skills involved in the action portion of our project. We needed to be clear and confident with the information we were delivering to others. We had to take initiative to conduct our action portion in our school. A huge thank you to Mr. Zoras who helped and supported my partner and I a great deal throughout the course of our action. Without him our action would not have been possible. This is just one skill that we required, We also needed to be patient and calm with people who did not quite understand our project and the point of it. There are several other skills that I have mentioned in my earlier journal entries that were used and crucially needed to take action. One requires many skills to conduct a piece of action as large as this one.	In order to successfully conduct myaction I needed to find a new and creative way to engage my audience and get them to participate. While I was looking through Recycle My Cell's website, I found a school challenge and registered for it right away after confirming with Mr. Zoras if this would be alright. We set up a temporary box in our home form classroom while we waited for the cool designated box to arrive in the mail. We kept on announcing the challenge over the morning announcements to really get our message across as much as we possibly could. The 500$ prize for our school seemed like the motivation that everyone needed. Even though there was a prize involved, most people were just happy to have a place where they were able to give away the burden of e-waste and not feel guilty or unsure about the way it would be "recycled". Once again, I am extremely pleased with the results our action hashad. It was definitely a successful method seeing that many people participated in it and we have collected over 30 pieces of e-waste. Even though our chances of winning the 500$ are slim to none, I feel satisfied due to the fact that this e-waste will be recycled properly.
	I have had so much fun doing this project and would love to do something similar to this one in the future,	Thankfully, both my partner and I had these skills and Mr. Zoras was always there for us.	

References

Abd-El-Khalick, F., & Lederman, N. G. (2000). The influence of history of science courses on students' views of nature of science. *Journal of Research in Science Teaching, 37*(10), 1057–1095.

Allchin, D. (2003). Scientific myth-conceptions. *Science Education, 87*, 329–351.

Barthes, R., & Lavers, A. (1972). *Mythologies*. New York: Hill and Wang.

Bencze, J. L. (1995). *Towards a more authentic and feasible science curriculum for secondary schools* (Order No. NN02640). Available from ProQuest Dissertations & Theses Global. (304257613). Retrieved from http://search.proquest.com/docview/304257613?accountid=14771

Bencze, J. L. (1996). Correlational studies in school science: Breaking the science-experiment-certainty connection. *School Science Review, 78*(282), 95–101.

Bencze, L., & Alsop, S. (2009, June). *Towards communitarian science education: Epistemological & ethical arguments*. In Paper presented at the 10th biennial conference of the international history & philosophy for science teaching group, South Bend, Indiana.

Bencze, L., & Carter, L. (2011). Globalizing students acting for the common good. *Journal of Research in Science Teaching, 48*(6), 648–669.

Bencze, L., Sperling, E., & Carter, L. (2012). Students' research-informed socioscientific activism: Re/visions for a sustainable future. *Research in Science Education, 42*(1), 129–148.

Bourdieu, P. (1986). The forms of capital. In J. Richardson (Ed.), *Handbook of theory and research for the sociology of education* (pp. 241–258). New York: Greenwood.

Candace, M. (2013, May 30). *SAMR in 120 seconds*. [Video file]. Retrieved from https://www.youtube.com/watch?v=us0w823KY0g

Cavagnetto, A. R. (2010). Argument to foster scientific literacy: A review of argument interventions in K-12 science contexts. *Review of Educational Research, 80*(3), 336–371.

Clough, M. P. (2006). Learners' responses to the demands of conceptual change: Considerations for effective nature of science instruction. *Science Education, 18*, 463–494.

DeBoer, G. E. (2000). Scientific literacy: Another look at its historical and contemporary meanings and its relationship to science education reform. *Journal of Research in Science Teaching, 37*, 582–601.

Giroux, H. A. (2008). *Against the terror of neoliberalism: Politics beyond the age of greed*. Boulder, CO: Paradigm Publishers.

Hodson, D. (1998). Science fiction: The continuing misrepresentation of science in the school curriculum. *Curriculum Studies, 6*(2), 191–216.

International Baccalaureate. (n.d.a). *Creativity, activity, service*. Retrieved from http://www.ibo.org/programmes/diploma-programme/curriculum/creativity-activity-and-service/

International Baccalaureate. (n.d.b). *Mission*. Retrieved from http://www.ibo.org/en/about-the-ib/mission/

International Baccalaureate. (2013). *IB learner profile*. Retrieved from http://www.ibo.org/contentassets/fd82f70643ef4086b7d3f292cc214962/learner-profile-en.pdf

Irwin, A. R. (2000). Historical case studies: Teaching the nature of science in context. *Science Education, 84*, 5–26.

Klopfer, L. E. (1969). The teaching of science and the history of science. *Journal of Research in Science Teaching, 18*, 87–95.

Kolstø, S. D. (2008). Science education for democratic citizenship through the use of the history of science. *Science & Education, 17*(8–9), 977–997.

Law, J. (1992). Notes on the theory of the actor-network: Ordering, strategy, and heterogeneity. *Systems Practice, 5*(4), 379–393.

Matthews, M. R. (1994). *Science teaching: The role of history and philosophy of science*. New York: Routledge.

Paterson, M. (Producer/Director). (2008). *94/Elements* [Motion Picture]. London: PFILM

Peterson, A. D. C. (1972). *The International Baccalaureate: An experiment in international education*. London: G.G. Harrap.

Poulsen, F. P. (Producer/Director). (2010). *Blood in the mobile* [Motion Picture]. Copenhagen, Denmark: Koncern TV- og Filmproduktion.

Remillard, J. R. (1978) Knowledge and social control in a multinational context: An analysis of the development, content and potential of the International Baccalaureate. Unpublished doctoral dissertation, State University of New York, Buffalo

Roberts, S. A. (2009). Supporting English language learners' development of mathematical literacy. *Democracy and Education, 18*(3), 29–36.

Schwartz, R., Lederman, N. G., & Crawford, B. A. (2004). Developing views of nature of science in an authentic context: An explicit approach to bridging the gap between nature of science and scientific inquiry. *Science Education, 88*(4), 610–645.

Tarc, P. (2009). What is the 'International' in the International Baccalaureate? Three structuring tensions in the early years (1962–1973). *Journal of Research in International Education, 8*(3), 235–261.

Tarc, P., & Beatty, L. (2012). The emergence of the International Baccalaureate Diploma in Ontario: Diffusion, pilot study and prospective research. *Canadian Journal of Education, 35*(4), 241–275.

Tatnall, A., & Burgess, S. (2002). Using actor-network theory to research the implementation of a BB portal for regional SMEs in Melbourne, Australia. In *15th Bled electronic commerce conference-'eReality: Constructing the eEconomy',* Bled, Slovenia, University of Maribor.

Wang, H. A., & Marsh, D. D. (2002). Science instruction with a humanistic twist: Teachers' perception and practice in using the history of science in their classrooms. *Science & Education, 11,* 169–189. (305561906). Retrieved from http://search.proquest.com/docview/305561906?accountid=14771

Zoras, B., & Bencze, L. (2014). Utilizing social media to increase student-led activism on STSE issues. In L. Bencze & S. Alsop (Eds.), *Activist science & technology education* (pp. 435–449). Dordrecht, The Netherlands: Springer.

Chapter 15
'In the Eye of the Hurricane': Using STEPWISE to Address Urgent Socio-political Issues in Venezuela

Majd Zouda, Tomo Nishizawa, and Larry Bencze

15.1 Introduction

Our research began when the second author of this chapter—Tomo—deciding to involve her secondary school students in activist science education to address socio-scientific issues (SSIs). Tomo was a teacher from Japan who had her higher education in a Canadian university. At the time of this study, she was working as a biology teacher in an affluent, international high-school in Venezuela. Tomo was surfing the Web for different approaches to teach activist science when she encountered the 'STEPWISE' website (www.stepwiser.ca) and decided to use its pedagogical framework in her teaching. 'STEPWISE' is a curricular and pedagogical framework that encourages students/citizens to take informed actions to address perceived socioscientific problems/issues. The decision to use STEPWISE stemmed from its possible application as an alternative approach to conventional school science—which is argued to limit students' science literacy and role as critical and activist citizens (Bencze & Carter, 2011). A prime goal of this project was to encourage students in Tomo's classes to spend some of their cultural and social capital (Bourdieu, 1986) for attempting to improve the wellbeing of individuals, societies and environments. However, the types of student activism generated in the particular social and political context in Venezuela drew our attention to some subtle but powerful factors that seemed to affect students' motivation and perspectives. These factors may also shape students' activism in other contexts. In this chapter, we

M. Zouda (✉) • L. Bencze
Ontario Institute for Studies in Education (OISE), University of Toronto, Toronto, ON, Canada
e-mail: majd.zouda@mail.utoronto.ca; larry.bencze@utoronto.ca

T. Nishizawa
Faculty of Education, University of Victoria, Victoria, BC, Canada
e-mail: tomo.nishizawa@gmail.com

L. Bencze (ed.), *Science and Technology Education Promoting Wellbeing for Individuals, Societies and Environments*, Cultural Studies of Science Education 14,
DOI 10.1007/978-3-319-55505-8_15

briefly review the rationale for Tomo's promotion of student activism and for choosing the STEPWISE pedagogical framework. We also map the social and political context in Venezuela in which our research took place. Then we explore actions taken by affluent students to educate members of the public about local SSIs of high priority for Venezuelan society (e.g., food shortage and insecurity) and factors that may have affected students' perspectives and decisions regarding these actions. Finally, we conclude by suggesting rationale for incorporating real life scenarios in activist science education, while taking into consideration students' cultural backgrounds that may shape their perspectives on change. Challenges and recommendation for science educators also are included.

15.2 Rationale and Possibilities for Student Activism

On a daily basis, people negotiate their personal decisions. While many factors affect their choices, it could be argued that general 'attitudes to life' are largely shaped by our ontological perspectives. How we perceive elements of life would probably determine what we construct with them and our attitudes toward them. Science and education are examples of these elements. As science researchers/educators in this study, we hold an ontological perspective about science as a relatively naturalist field (Loving, 1991); that is, that science is a cultural construct that is affected by social, economic, political and other factors. Science, as described by Latour's (2005) Actor-Network Theory (ANT), is part of complex networks of human, nonhuman and semiotic actants. These networks appear to be dynamic and shaped by power relations. Education is also a crucial actant in these networks, and an important player in directing science. By placing science within complex social and environmental networks, we can better understand the reciprocal relations between them. We can also understand possible roles of science education in tackling personal, social and environmental problems associated with fields of science and technology.

Although there is much to celebrate about involvement of fields of science and technology in societies, people often need to take care in making personal choices about using their products. From quitting smoking to voting on public funding for space exploration, people refer to science as a main—although not the only (e.g., Sadler & Zeidler, 2004)—resort for making decisions. Accordingly, school science systems have—over at least the last 45 years (Pedretti & Nazir, 2011)—aimed to educate students about relationships among fields of science and technology and societies and environments (STSE). Among a range of approaches educators have used for STSE education, a prominent tack has been to provide students with controversial data and claims about them (regarding various products of science and technology) and encourage them to make personal decisions about them (Levinson, 2010). There are numerous such socioscientific issues (SSIs) for students to consider—including, for instance, alternative positions and data about uses of fossil fuels vs. various renewal energy forms. In making decisions about such controversies, students need to learn to undertake critical—well-reasoned—analyses of various data sets and claims about them (Levinson, 2010). Such abilities and orientations can prepare them well for making personal decisions in representative democracies, which—aside from periodic voting for governing bodies—generally involves personal decision-making about, for example, purchases of products and services (Wood, 1998). It is apparent, however, that being 'critical' must go beyond explaining existing norms to examining power relations and changing them (Fairclough, 2010). Hence, criticality is more than deciding whether genetically modified food, for example, is 'good' or 'bad.' It seems that citizens must examine how powerful agents, discourses, and relations collaborate and/or compete over spaces for social transformation (for the better or the worse); perhaps leading to needs to act to *transform* possible injustices (to societies and/or to environments).

While social and environmental injustices associated with SSIs may have been occurring for ages, facing them now seems more urgent than ever—at least because of the seriousness of problems and their globalized nature. Arguably of most concern is climate change, which has potential to increasingly disrupt lives a millions of people and other living things and alter—if not destroy—habitats (e.g., barrier reefs) over the next few decades unless immediate actions are taken (Klein, 2014). There also are, however, immediate health and social injustices linked to manufactured foods, for instance, that need intense attention in the near future (Weber, 2009). Although causes of such socioscientific problems are likely complex and uncertain, there are concerns that for-profit influences on fields of science and technology often are involved (Mirowski, 2011). At the same time, in light of actor-network theory (ANT), which conceives all living, nonliving and symbolic things ('actants') as being interconnected, it is difficult to associate culpability with any one actant. Nevertheless, under assumptions that fields of science and technology are associated, to some extent, with many personal, social and environmental problems like those outlined above, many scholars and others have recommended that school science systems place more emphases on encouraging and enabling students (as citizens) to take actions to address problems of their concern/interest (e.g., Hodson, 2011). Science education should help citizens perceive themselves as *responsible* and *capable* of participating in knowledge production, decision making and socio-political actions in what would be more participatory forms of democracy (Levinson, 2010).

When we advocate for citizens' participating in relevant knowledge production, we do not claim that the lay public can replace experts (e.g., scientists and technologists) in decision making around SSIs. Continuously emerging technologies, for instance, tend to give experts advantages in making decisions—perhaps also giving them more insights and possibilities for research and development. However, citizens can become co-investigators and co-producers of knowledge: They can critically question available information, further investigate them (theoretically and practically), produce legitimate knowledge, and participate in related political decisions (e.g., Pouliot, 2015). Public roles in questioning, investigating and producing relevant knowledge is essential for common wellbeing; especially with frequent incidences of manipulating science research for benefits of private industries (Krimsky, 2003). Therefore, unless citizens become partners in knowledge production, they will continue to hand their voices—such as about SSIs—to the relatively small number of experts to make decisions that do not, necessarily, keep in mind common wellbeing (Rudolph, 2005).

Along with hoped-for social and environmental benefits, there may be personal gains for students, too, through more proactive forms of citizenship. Students' activism may help them to develop better understanding of characteristics of the SSIs they are studying. While critically examining and investigating certain SSIs, students should construct new understandings based on their preconceptions, available information and past (and present) experiences, knowledge and skills (among other factors). However, these new conceptions are rarely tested. They are seldom challenged against students' social realities. Students do not practically experience complexities of SSIs, power dynamics that shape them, or possible 'points of entry' (Fairclough, 2010) for changes. Acting upon SSIs would provide students with possibilities to 'falsify' their new conceptions and adjust them in relation to complexities of their social realities. It brings these conceptions from abstract levels of representation into practicalities of the concrete world (Roth, 2001). As Scott Cook and John Seely Brown (1999) argue,

> [w]e act within a social and physical world, and since knowing is an aspect of action, it is about interaction with that world. When we act, we either give shape to the physical world or we affect the social world or both. Thus, 'knowing' does not focus on what we possess in our heads; it focuses on our interactions with the things of the social and physical world (p. 388).

By combining practical with theoretical elements, students get involved in cycles of deductive-inductive reasoning, and conceptions are continuously tested and adjusted into relatively newer ones. Here, we do not advocate for superiority of 'scientific' methods or for possibilities to reach particular realities. Rather, we argue that practical application—such as through activism—should provide essential cognitive experiences to learn SSIs from different angles. It also seems worth noting here that some forms of student activism, such as service learning, are argued to increase students' understanding of and engagement in subject-matter (Sherrod, 2006). Effective activism requires good understanding of topics involved and related subject-matter in order to justify and support certain actions and increase their effectiveness. For example, learning certain biological concepts, such as genetic engineering, DNA replication, or mutation, and their applications and consequences is expected for students leading a campaign regarding genetically modified food. Hence, students' socio-political activism is a necessity for individual and collective benefits: It allows student involvement in subject-matters, SSIs and democratic life. In the complex, interrelated networks of relations, collective societal and environmental benefits are inseparable from individual wellbeing.

15.3 A Novel Context for STEPWISE

For about a decade, the third author here has been using the 'STEPWISE' curriculum and pedagogical framework that he developed in mid-2006 to encourage and enable children/youth to critically evaluate fields of science and technology with regards to possible harms to wellbeing of individuals, societies and/or environments. Where they perceive problems, students would usually develop and implement plans of action to bring about a better world. Most of these efforts have occurred in and around Toronto, Canada, where Larry has worked as a science education professor since September 1998. When Tomo encountered the STEPWISE website, she wrote to Larry, asking if he could help facilitate her development and research into effectiveness of teaching/learning strategies. Part of her interest in the project stemmed from the fact that she was familiar with curricula in Ontario, the site of development of the framework. Larry then asked Majd, first author here, to work very closely with Tomo as a democratic facilitator (ultimately, deferring to Tomo) of curriculum development and as researcher.

The school in which Tomo was working could be considered an 'affluent' private school, catering to children of relatively prosperous Venezuelan and international families. Many of these families were owners of big businesses. A major attractive feature of the school was its provision of International Baccalaureate (IB) programmes, diplomas from which are recognized—with considerable prestige attached to them—around the world. Tomo chose to implement STEPWISE with students in her eleventh-grade IB biology course, considering that IB courses typically involve a student-led project.

Tomo was attracted to a major premise of STEPWISE, positing that students are likely to become attached to, and motivated to engage in, socioscientific issues, research and actions if they have increasing control over decisions about them. This principle is based on *knowledge duality* theory of Etienne Wenger (1998), who suggested that deep learning and affinities to learning develop when students have significant control over reciprocal decisions about translations between 'phenomena' (e.g., objects and energy) and representations (e.g., drawings, graphs, etc.) of them. Through STEPWISE, students are, ultimately, expected to control translations in 'research' (phenomena ➔ representations) and negotiated actions (representations ➔ phenomena).

Usually, students do not have significant expertise, confidence and motivation for self-directing research informed and negotiated action (RiNA) projects to address socioscientific issues of interest to them. Therefore, the STEPWISE pedagogical schema (Fig. 15.1) encourages teachers to engage students in one or more sets of 'apprenticeship' lessons and student activities until the teacher feels students are ready to self-direct projects. Typically, such apprenticeships involve three basic phases: (i) with some stimuli by the teacher, students reflect on and express (e.g., through writing, talking, etc.) their existing conceptions about socioscientific issues, including possible problems and actions to address them; (ii) the teacher actively helps students to learn about some different SSIs and research-informed and

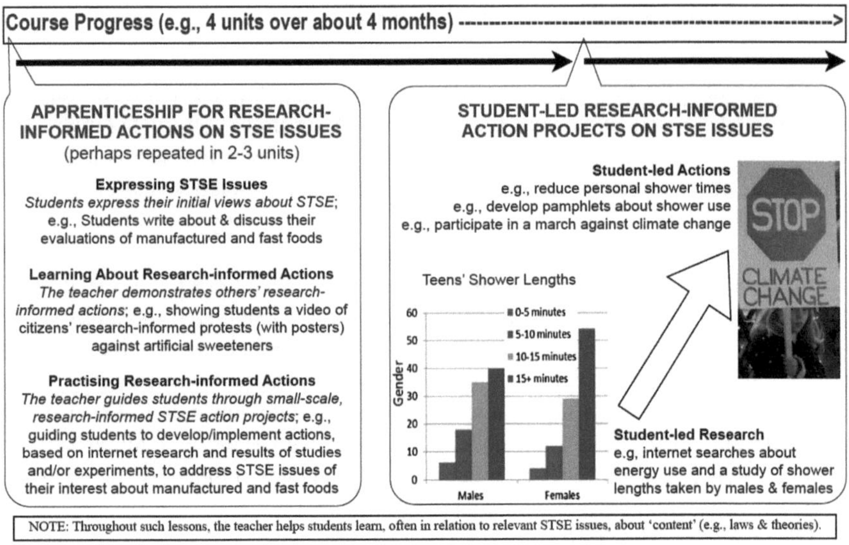

Fig. 15.1 STEPWISE pedagogical framework (Bencze & Carter, 2011)

negotiated actions people have undertaken to address perceived problems relating to them; and, (iii) the teacher facilitates students' small-scale research and action projects relating to a SSI of interest to them.

In helping Tomo to develop and implement apprenticeship lessons and students activities, it became apparent that at least two major aspects of our suggestions about STEPWISE implementation were new to her, as follows: (i) Because students (and other citizens) are often unaware of various, often problematic, actants, such as government deregulation of businesses, it has been recommended that students be encouraged to analyze and evaluate SSIs through construction of actor-network maps, which may—particularly through student research and teacher instruction—lead to awareness of such previously unknown actants (Pierce, 2013); and, (ii) Because experiments (forcing changes in independent variables) may lead to problematic outcomes when studying socioscientific issues, Larry Bencze and Mirjan Krstovic (Chapter 7, this volume) recommend students use correlational studies—in which investigators study relationships between naturally-occurring variables (e.g., cigarette smoking vs. cancer rates). With these in mind, Tomo developed and implemented two sets of apprenticeship lessons and activities—as outlined just below and elaborated later:

Apprenticeship 1: Focusing on Genetically Modified Organisms

- Before introducing the RiNA project assignment, students learned about commodities' life cycles, and some related socio-environmental issues, using *The Story of Stuff* (www.storyofstuff.org) video.
- Tomo then introduced actor-network theory, through teaching students to develop mind maps. Students, with the help of Tomo, developed a network map regarding

genetically modified (GM) products and expressed their preconceptions about them. The network focused mainly on possible positive and negative effects of GM products, and included living and nonliving actants.

- Students then learnt about controversies around GM products using different tools, such as the BBC documentary, *Playing God*, and the TED talk, *Waiters, there is a gene in my soup!* (these and others available on YouTube™). Students were also involved in a class debate regarding GM Golden rice (with increased Vitamin A production), and developed a more sophisticated actor-network map that encompassed different stakeholders, possible relations between them, and related semiotic messages.
- Finally, to conduct their first RiNA project, students (in groups) were required to pick a GM product of their choice and analyze its life cycle using secondary research (e.g., Internet searches) and actor-network theory. Then, with the help of Tomo, students conducted primary research—in the form of surveys—and developed their socio-political actions on bases of their findings.

Apprenticeship 2: Venezuelan Crisis

- Before the end of the first project, troubles started in different cities in Venezuela and students became emotionally and deeply involved in socio-political conflicts surrounding them. Maybe one of the most well-known things about Venezuela is that it is one of the main oil producing countries in the world and that it depends on oil revenue as a main source of national income. It may, however, be surprising to know that a leading oil producing country, such as Venezuela, is struggling from serious economic and social problems, such as scarcity of basic goods, electricity shortage, insecurity, and inflation (Cardenas, 2014). These urgent concerns are what Venezuelans were facing when protests burst out in major Venezuelan cities, such as Caracas and Maracaibo, in February, 2014. Accusations and blame emerged between the Leftist government led by the PSUV (United Socialist Party of Venezuela, led by President Maduro) and the opposition—whose main supporters have been members of the 'old ruling elites,' private sectors and, more recently, an increasing percentage of the middle-class (Wilpert, 2005). On one hand, the opposition argued that the main reasons for these problems are the government's socialist reforms (e.g., business nationalization and inhibiting private interests), the over-reliance on oil revenue and neglecting economic issues in favour of social ones (Cardenas, 2014; Durte et al., 2006). On the other hand, the government—which represented the legacy of late president Hugo Chavez and gained its legitimacy from support of the working-classes, the marginalized and, at different times, the middle-class (Wilpert, 2005)—blames the opposition for the current crisis. The government accuses the opposition of prioritizing their capitalist interests and intentionally creating problems (e.g., basic good scarcity) by hoarding goods, exporting them and manipulating price controls (among others) (Ellner, 2013). What the opposition considers as 'extreme' measures (e.g., government expropriation of private property deemed to be in the public interest) was viewed by the government as a response to the opposition's and private sector's maneuvers to avoid governmental regulations

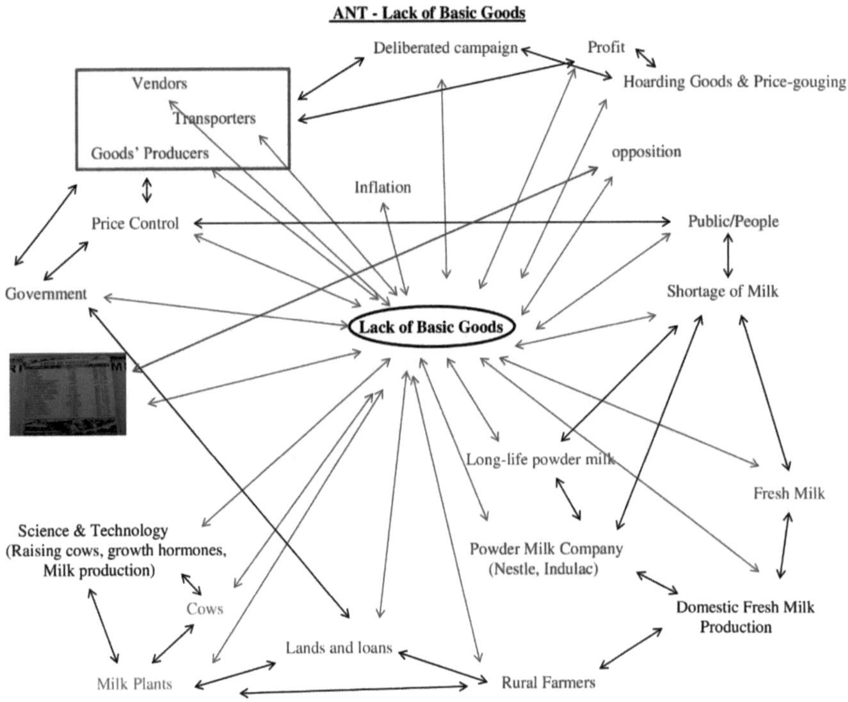

Fig. 15.2 Actor-network used by Tomo about lack of basic goods in Venezuela

(Ellner, 2013). Outbreaks of tensions relating to such differences led us to adjust the focus of students' second project. Students were asked to imagine that they are politicians with a strong background in fields of science. They were asked to pick and investigate a resource that they thought has potential to improve currently-divided political/economic/social situations in Venezuela and improve wellbeing of individuals, societies and environments. Students would then take actions to promote the resource and prepare a presentation to convince the rest of the politicians in the government (i.e., their peers) to support their initiative.

- In class, and before getting involved in the project, students brainstormed different reasons for the protests in Venezuela. Tomo then used an actor-network (Fig. 15.2) to teach students about complexities of one of these issues (i.e., lack of basic goods), and emphasized 'technoscience' (science and technology co-influencing each other; Sismondo, 2008) as an important actant.
- Students then proceeded in their projects; i.e., secondary research, primary research, and action.
- For this project, we encouraged students to place themselves and their families, in addition to technosciences, as essential actants in network maps.

15.4 Data Collection and Analyses

To study the nature of students' activism in their specific socio-political context, we used qualitative methods. These allow studies of "phenomena in terms of the meanings people bring to them" (Denzin & Lincoln, 2000, p. 3). They provide broad spaces for participants to express parts of their entities, and explain how they live and perceive these phenomena. The richness of the data allows researchers to better understand the cases under study.

For data collection, we used multiple methods to increase the trustworthiness of our research. This tack is a form of *triangulation*, which gives different insights into a phenomena (Denzin & Lincoln, 2000). While designing our data collection approaches, we considered developing them *naturalistically* and *rationalistically* (Guba & Lincoln, 2011). While the former allows the emerging of themes and theories, the latter is more predetermined and aims at examining pre-existing ones. In our research, we rationalistically focused on the nature of students' actions in specific contexts, and their understandings of complex power-relations in SSIs. Naturalistically, we collected data that allowed emergence of unexpected situational outcomes, such us the oppositional motifs in students' actions. Data collected includes:

- Projects Work Artefacts: Samples of students' work (12 students in groups), from the two RiNA projects, were collected and examined. These included: SSIs descriptions, data collected by students (through primary and secondary research), samples of actor-networks, written reports, projects reflections and forms of actions (e.g. posters, flyers, videos and Facebook™ and blog pages);
- Projects Instructional Materials: Copies of all Tomo's instructional and pedagogical materials (e.g., handouts, videos, and ANT maps) and Tomo's reflections on her teaching and student learning;
- Digital Recording of Students' Work: Audio recording of students' final presentations;
- Students' Questionnaire and Surveys: Seven students volunteered to complete a questionnaire at the beginning of the first RiNA project, and 12 students completed a Slavery (slaveryfootprint.org/) and an Ecological footprints (www.earthday.org/) surveys between the two RiNA projects; and
- Semi-structured Interviews: Seven volunteering students were interviewed near the beginning of the first RiNA project. Students were interviewed in groups of three and four. Each group interview lasted about 60 min (15–20 min/student). Three students (out of seven) were also interviewed at the end of the second RiNA project. The interview lasted about 15–20 min per student. All interviews were audio-recorded and later transcribed.

For data analysis, we used constant comparative methods based on constructivist grounded theory (Charmaz, 2014) as our methodological framework. Theories are expected to naturalistically emerge from data analysis through *interpretation* of available data, and knowledge is constructed through negotiation of meanings between researchers and participants. By using this method, different key-points in the data were coded. Initial codes were then constantly compared for refining and development of broader and more abstract categories, and then theoretical themes. The emerging themes were then examined by returning back to field work to collect further supporting/challenging data. During all these stages, categories and themes were continuously negotiated between the three researchers (Wasser & Bresler, 1996). Rationalistically, we used six indicators of students' commitment to activism. These indicators were developed by Larry Bencze, Erin Sperling and Lyn Carter (2012), and are: passionate concerns about an issue; clear intention to implement action; confidence in effectiveness of action; student self-efficacy; detailed analysis/planning of action; and number and variety of actions.

15.5 Results and Discussion

Our analysis of data collected revealed three main themes regarding the nature of students' socio-political actions on SSIs. In general, students' actions took educative forms to inform other citizens about potential problems and possible solutions related to certain SSIs. Perhaps more significantly, students' motivation to implement these actions was elevated when real life scenarios were integrated in their RiNA projects. At the same time, this increased agency for activism seemed to be shaped by students' affluent social class and the general political stance its members usually hold; which coloured students' actions with oppositional motifs. In the following, we discuss in detail these three themes, factors that may contribute to their development and their educational significance.

Educative Types of Actions

Throughout the two RiNA projects, students' forms of actions (e.g., posters, Facebook™ pages, YouTube™ videos, etc.) reflected their tendency to educate the public about different SSIs—rather than more 'aggressive' forms of actions, such as lobbying powerbrokers. Their main goals seemed to be to create awareness among the public, and to less extent, to encourage participation of the public in developing solutions for the issues or taking actions toward them. For example, when reflecting on her GMO project, 'Viki' wrote:

> The issue being faced in the daily markets when buying the product of GMO Rice is not knowing exactly what you're buying. The way my group decided to help this issue was taking the action of creating awareness poster. The poster consisted specifically of Golden Rice. In the poster, my group listed some of the many negative effects of consuming Golden Rice with a title in big words of "Are You Aware?"

Working in another group, 'Mariana' was concerned about the lack of public awareness regarding clothes containing GM cotton (a conclusion that she reached from surveying different students at her school). Her research-informed and negotiated action aimed at creating more public awareness and encouraging members of the public to make a change. In her reflection she wrote:

Fig. 15.3 Poster created by a student to educate and evoke public action

> After our survey, I noticed that I wasn't the only person that didn't know much about where my clothes come from…. This made me want to do something to inform people about the situations the workers labor in [sweatshops], and this is why I made the poster below with some facts about our clothes to hopefully inform people about everything….because I know if I had known that when I bought the only item I have from them [Abercrombie] I wouldn't have bought it, and so would many people I believe.

Below her reflection appeared a photo of her informative poster with some suggestions to the public to take actions against manufacturers and stores that carry GM clothes, Fig. 15.3.

Students' tendency toward educating the public and/or making personal improvements can be read in different, yet interrelated, ways. It is argued that schools should provide students with the tools to succeed in a 'culture of power,' teaching them its implicit and explicit rules (Delpit, 1988). However, as discussed earlier, dominant classroom SSI education practices limit students to discussion and personal decision making (Levinson, 2010). Taking transformative political actions is a tool to be learnt; it is a 'capital' that not too many students seem to be able to access. Hence, common practices in schools (i.e., debating and taking personal decisions) become students' guidelines and tools to act, and students may feel comfortable and confident in 'spreading the word' and convincing the public with what they know/need to know. This is not to ignore the easiness of creating a poster or a page on the social media, which may also direct students' choices about the types of their actions.

Another factor that may direct students' choices about what should be done seems to be their confidence in the effectiveness of their actions and its effect on their commitment to these actions (Bencze et al., 2012). Students may feel that educating the public is more effective and can evoke more tangible results than addressing people in power or decision-makers. Students may not perceive themselves as powerful agents in the decision-making process and, so, tend to gain more public support rather than acting individually. In other words, by educating the public, they implicitly attempt to create a 'collective identity' (Melucci, Keane & Mier, 1989): Shared interests and goals that can be used for further actions. The 'Likes' on their Facebook™ pages, or the 'followers' on their Twitter™ accounts may give students the impression that their actions are effective and that other people share their concerns and goals. Placing posters around the school/city to reach a lot of people may appear more effective than addressing an official in the government; particularly when considering the social class of our participants and their distrust of the government.

Having the tools for actions, as well as the confidence in the ability to act and the effectiveness of the actions, seem to be major factors in students' tendencies to educate others.

Increasing Agency for Activism in Real-Life Situations During the first RiNA project, students were involved in creative educative actions. However, their agency and motivation to implement 'effective' actions seemed to be more significant in the second project. When issues directly threatening people's lives (e.g., their consumption of basic goods or insecurity, etc.) were addressed in the second RiNA project, many students challenged their anxiety and fear of insecurity and headed toward public places to distribute handouts, surveys and flyers (e.g., Fig. 15.4) to educate the public and encourage their participation in finding solutions. This feverish public interaction occurred, despite the fact that some students had earlier declared that insecurity had restricted their social activities and recreations to spend their time in their friends' houses.

We argued earlier that students should have a leading role in research and decision making. In our projects, this usually begins by choosing the issues they want to investigate (which should be related to the unit they are studying). The freedom of choice usually ensures students' interests in their research and their willingness to act. However, when real-life scenarios were incorporated in students' research, their concerns were temporally and spatially located in their immediate milieu. The blackout, lack of basic goods, or insecurities were directly experienced by these students. The close proximity of these issues seemed to increase their urgency and, consequently, students' passionate concerns about them. For example, in response to an interview question about the second RiNA project, 'Fiona' said: "I really liked it because I'm experiencing these problems today. So, I think this was a very interesting project…. [I]t's my country and its problems that we face every day." Having passionate concerns about SSIs is argued to increase students' commitment to activism (Bencze et al., 2012); which, in our research, has translated into increased agency to act.

From a pedagogical perspective, two strategies (in the second RiNA project) may have contributed to passionate involvement of students in these real-life situations. The first strategy is asking students to place themselves and their families as actants in their SSI actor-network maps. For example, before developing and publicly distributing the handout in Fig. 15.4, students produced an actor-network map showing how their family members were affected by food shortages (Figs. 15.5a and 15.5b).

Fig. 15.4 Handout distributed by a group of students to the public. Translation: *Tired of all the queues?.... Problem of [goods] scarcity affect all Venezuelans. For that we shall all work together to solve it. The Solution for scarcity is by investing in the field of science and agriculture, especially agricultural technology to increase national production. If you want to share your solution, join us on Facebook...*

By asking students to place themselves as actants in their SSI-network maps, it appeared that we drew their attention to their personal stakes in these issues. Their personal interests became part of the larger societal wellbeing. Realization that improvements to wellbeing of others may lead to improvements in their own lives (Lehrer, 2001) seems to have motivated students to take 'effective' actions to address these issues.

The second strategy we used was emphasizing technoscience as a main actant in these socio-political issues, and designing a *solution-oriented* RiNA project. It is argued that when approaching a pressing socio-political problem, the implicit/explicit role of science should always be examined (Simonneaux & Legardez, 2010). This is a necessity regarding leading roles of fields of technoscience in today's world and complex relations they have with diverse actants in any societal

a

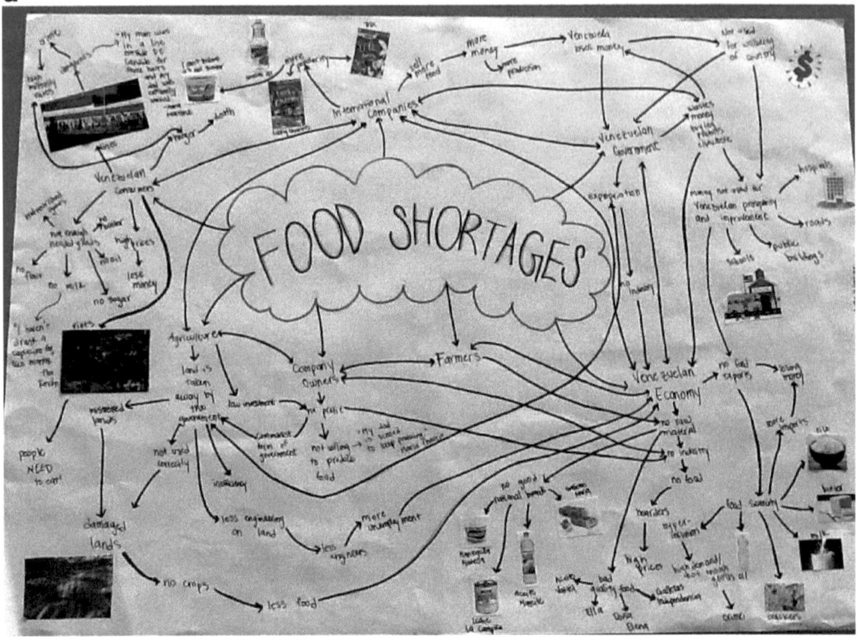

Fig. 15.5a Actor-network map including students & their families as actants

b1 b2

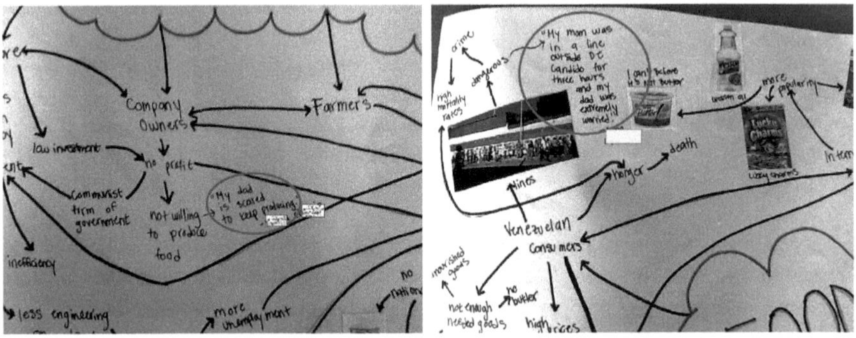

Fig. 15.5b Close captions on students' involvement as actants

network. We emphasized this point in the second RiNA project because these relations are less explicit in some socio-political problems and can easily pass unnoticed. For example, when examining scientific issues, such as GM products, technoscience seems a default key player; the same presupposition cannot be assumed in issues like insecurity or food shortage. Therefore, we explicitly asked students to include science and/or engineering in their actor-network maps. As a result, some students developed new perspectives about possible roles of technoscience in 'solving' socio-political issues. For example, during her second interview, and when asked how the second RiNA project may have changed her views about the Venezuelan conflict and the world, 'Lisa' said: "I thought that it was only related to politics and things like that and issues that happen in Venezuela, but through the project I was able to see also that science is one way to solve the problem [Food shortage] that we have right now." This new conception of the role of technoscience in their socio-political problems, and the tangible possibilities of finding solutions, may have increased students' confidence in the effectiveness of their actions and empowered them *to act* rather than feeling helpless. Believing that there is something that can be achieved, may have encouraged some of these students to enthusiastically apply their actions.

It is important to note that, at the beginning of the second RiNA project, most students struggled in connecting technoscience to the socio-political problems in Venezuela, and some of them continued to have vague understanding about it. Arguably, through science education, students are usually accustomed to examine technoscience products/services and how they affect the larger society; however, they rarely address an existing socio-political problem and investigate the underpinning roles of technoscience. Perceiving hidden roles of technoscience seems to be challenging for many students. Another important point to discuss here is that those who successfully developed a clearer perspective about possible relations between technoscience and socio-political issues viewed this relation in a positive way and mainly perceived technoscience as a part of the solution. This could be due to the solution-oriented way we structured the second RiNA project. While developing solutions seems to have encouraged students to act, it may have de-problematized technoscience; a point that should be challenged and taken into consideration when conducting similar types of projects.

Overall, increased agency for activism during the second RiNA project seems to be a result of many factors: Perceiving the personal stakes in urgent, immediate problems, and feeling capable of making a change.

Political Types of Actions Although most students in our study seemed to develop better and broader understanding about complexities of problems they face in Venezuela—especially by using actor-network maps—their emotional involvement and political resentment to the government did not seem to be affected and, indeed, tended to strongly-influence their actions. For example, when answering a question about how the second RiNA project changed her understanding, emotions, and views about the world, 'Fiona' said:

> The project changed my understanding but not my emotion.... Because at first I believed that it was only the government's fault of the food shortages, but after doing the project we learned that there are the hoarders and they are a huge cause of the problem and also we learnt that ... there [are] other major stakeholders. But my emotional understanding didn't change because I still resent the government. I think it's all the government fault. Even when the hoarders have a lot to do with it I think if the government had done something about it then hoarders wouldn't be doing what they are doing now...

Fig. 15.6a Oppositional comic posted on students' Facebook Page. Translation: *Maduro [The President]: "The newspapers exaggerate the little problems of the country"*

And when asked what features of the project brought these changes, she said:

> I think the secondary and the primary research because in the secondary research first of all, you got to know about all the stakeholders and everyone who is positively or negatively affected. First, I thought the government is positively affected from food shortage but they are not. They are negatively affected because they lose more money. Because there is no food, so they have to buy the food from other places. And in the primary research we got to survey people from different ages and genders and we saw the overall picture that all of us have these problems and all of us face them every day.

A basic premise of STEPWISE is that students/citizens may be willing to 'spend' at least some of their cultural capital (Bourdieu, 1986) (e.g., science literacy) to benefit other living and nonliving things. As members of affluent classes (and business owners) that have been negatively-affected by the government's 'socialist' procedures (Ellner, 2013; Wilpert, 2005); however, with at least one parent telling his child (in this class) that he is afraid to continue production for the fear of having his business nationalized, these students seem to hold strong oppositional stances and tended to emphasize them in their actions (in spite of the neutral political position held and stressed by their school). For example, Fig. 15.6a shows an oppositional comic posted on students' Facebook™ page as part of their action, and Fig. 15.6b shows an actor-network map (developed by a different group) with oppositional motifs.

Broadly, therefore, it seems that through the second RiNA project students started to see societal wellbeing as directly related to their own, personal wellbeing (a perception that tended to increase their agency for activism). Nevertheless, their views about this societal wellbeing seemed also to be shaped by their social class and the political stance they hold. The sincerity of students' concerns and their genuine desire to make a change is not questionable. However, students tended to perceive the prosperity of their society in ways that may guarantee their *status quo*.

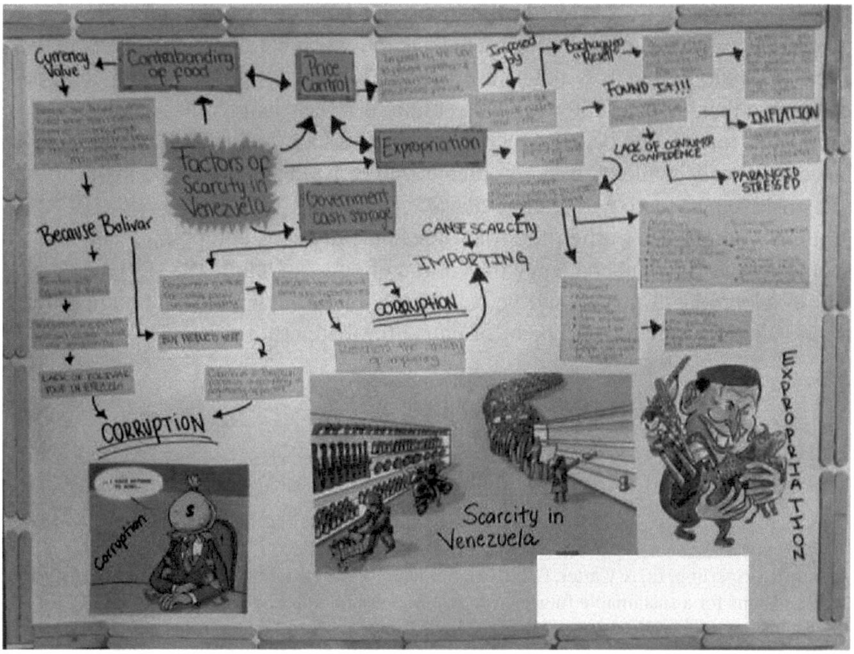

Fig. 15.6b Actor-network map with oppositional motifs

15.6 Conclusion and Recommendations

Overall, our findings suggest that, when involving students in research-informed and negotiated actions to address SSIs, teachers can integrate tangible real-life scenarios. Such contexts allow students to be personally and emotionally involved in SSIs, especially when students perceive themselves as significant stakeholders. This may motivate them to take actions beyond the expected. However, the motivation of students to act for societal wellbeing may also be greatly shaped by their cultural backgrounds (e.g., social class, political stance, religious beliefs, etc.), which tend to determine their conceptions of societal wellbeing. It is well argued that people rely on different factors, such as their emotions, intuition, personal experience, and family biases (among others) when making decisions about SSIs (Sadler & Zeidler, 2004). Our participants negotiated some of these factors when developing their actions. They aimed at a societal wellbeing that maintains their class interests.

As educators involved in activist science education, we are faced with some challenges: First, how do we present different SSIs as urgent real-life ones, and bring them to immediate concerns of students. Drawing attention to personal stakes using actor-network maps does appear to be an effective method. However, other pedagogies should be developed to motivate students' activism. Second, as 'scientific' reasoning is not the only factor that drives students' actions, new pedagogies should also be developed to incorporate, measure and challenge the other factors in play.

It is also important to consider how to manage a balance regarding the role of technoscience in socio-political issues. Students may struggle in connecting technoscience to socio-political issues; nevertheless, comprehending this relation may encourage them to search for possible solutions rather than feeling helpless. At the same time, the relation between technoscience and socio-political issues should also be continuously challenged and problematized to avoid being trapped in the normative positive perspective of technoscience.

Finally, as argued earlier, the role of schools includes equipping students with tools to succeed in and challenge the culture of power (Delpit, 1988). Hence, activism in its various forms should be part of school curricula, and examining power-relation in SSIs should be a 'point-of-entry' for that purpose.

References

Bencze, J. L., & Carter, L. (2011). Globalizing students acting for the common good. *Journal of Research in Science Teaching, 48*, 648–669.

Bencze, L., Sperling, E., & Carter, L. (2012). Students' research-informed socioscientific activism: Re/visions for a sustainable future. *Research in Science Education, 42*, 129–148.

Bourdieu, P. (1986). The forms of capital. In J. G. Richardson (Ed.), *The handbook of theory: Research for the sociology of education* (pp. 241–258). New York: Greenwood Press.

Cardenas, J. R. (2014). A nation divided: Venezuela's uncertain future. *World Affairs, 176*, 47.

Charmaz, K. (2014). *Constructing grounded theory* (2nd ed.). New York: Sage.

Cook, S. D. N., & Seely Brown, J. (1999). Bridging epistemologies: The generative dance between organizational knowledge and organizational knowing. *Organization Science, 10*, 381–400.

Delpit, L. (1988). The silenced dialogue: Power and pedagogy in educating other people's children. *Harvard Educational Review, 58*, 280–298.

Denzin, N. K., & Lincoln, Y. S. (2000). The discipline and practice of qualitative research. In N. Denzin & Y. Lincoln (Eds.), *Handbook of qualitative research* (2nd ed., pp. 1–28). Thousand Oaks, CA: Sage.

Durte, C., Ettkin, L. P., Helms, M. M., & Anderson, M. S. (2006). The challenge of Venezuela: A SWOT analysis. *Competitiveness Review, 16*, 233–247.

Ellner, S. (2013). Social and political diversity and democratic road for change in Venezuela. *Latin American Perspectives, 40*, 63–82.

Fairclough, N. (2010). *Critical discourse analysis: The critical study of language* (2nd ed.). Harlow, UK: Pearson.

Guba, E. G., & Lincoln, Y. S. (2011). Competing paradigms in qualitative research. In N. K. Denzin & Y. S. Lincoln (Eds.), *The SAGE handbook of qualitative research* (pp. 105–117). Thousand Oaks, CA: Sage.

Hodson, D. (2011). *Looking to the future: Building a curriculum for social activism*. Rotterdam, Netherlands: Sense.

Klein, N. (2014). *This changes everything: Capitalism vs. the climate*. Toronto, ON: Simon & Schuster.

Krimsky, S. (2003). *Science in the private interest: Has the lure of profits corrupted biomedical research?* Lanham, MD: Rowman & Littlefield.

Latour, B. (2005). *Reassembling the social: An introduction to actor-network-theory*. Oxford: Oxford University Press.

Lehrer, K. (2001). Indivadulism, communitarianism and consensus. *The Journal of Ethics, 5*, 105–120.

Levinson, R. (2010). Science education and democratic participation: An uneasy congruence? *Studies in Science Education, 46*, 69–119.
Loving, C. C. (1991). The scientific theory profile: A philosophy of science model for science teachers. *Journal of Reasearch in Science Teaching, 28*, 823–838.
Melucci, A., Keane, J., & Mier, P. (1989). *Nomads of the present: Social movements and individual needs in contemporary society*. Philadelphia: Temple University Press.
Mirowski, P. (2011). *Science-mart: Privatizing American science*. Cambridge, MA: Harvard University Press.
Pedretti, E., & Nazir, J. (2011). Currents in STSE education: Mapping a complex field, 40 years on. *Science Education, 95*, 601–626.
Pierce, C. (2013). *Education in the age of biocapitalism: Optimizing educational life for a flat world*. New York: Palgrave Macmillan.
Pouliot, C. (2015). *Quand les citoyen.ne.s soulèvent la poussière [When citizens raise dust]*. Montréal, QC: Carte Blanche.
Roth, W. M. (2001). Learning science through technological design. *Journal of Research in Science Teaching, 38*, 768–790.
Rudolph, J. (2005). Inquiry, instrumentalism, and the public understanding of science. *Science Education, 89*, 803–821.
Sadler, T. D., & Zeidler, D. L. (2004). The morality of socioscientific issues: Construal and resolution of genetic engineering dilemmas. *Science Education, 88*, 4–27.
Sherrod, L. R. (2006). *Youth activism: An international encyclopedia*. Westport, CT: Greenwood Press.
Simonneaux, J., & Legardez, A. (2010). The epistemological and didactical challenges involved in teaching socially acute questions: The example of globalization. *Journal of Social Science Education, 9*, 24–35.
Sismondo, S. (2008). Science and technology studies and an engaged program. In E. J. Hackett, O. Amsterdamska, M. Lynch, & J. Wajcman (Eds.), *The handbook of science and technology studies* (3rd ed., pp. 13–31). Cambridge, MA: The MIT Press.
Wasser, J. D., & Bresler, L. (1996). Working in the interpretive zone: Conceptualizing collaboration in qualitative research teams. *Educational Researcher, 25*, 5–15.
Weber, K. (Ed.). (2009). *Food Inc.: How industrial food is making us sicker, fatter, and poorer – and what you can do about it, a participant media guide*. New York: Public Affairs.
Wenger, E. (1998). *Communities of practice*. Cambridge, UK: Cambridge University Press.
Wilpert, G. (2005). Venezuela: Participatory democracy or government as usual. *Socialism and Democracy, 19*, 7–32.
Wood, G. H. (1998). Democracy and the curriculum. In L. E. Beyer & M. W. Apple (Eds.), *The curriculum: Problems, politics and possibilities* (pp. 177–198). Albany, NY: SUNY Press.

Chapter 16
Science for Citizenship: Using Prezi™ for Education About Critical Socio-scientific Issues

Darren Hoeg, Alexander DiGiacomo, Sarah El Halwany, Mirjan Kirstovic, Christina Phillips-MacNeil, Minja Milanovic, Tomo Nishizawa, Majd Zouda, and Larry Bencze

On the local news last night (August 31, 2015) citizens of Toronto were informed that much of the downtown core was closed to the public because the spire of the new Trump Hotel was loose, and may fall to the street. The incident stopped commerce, tourism, and citizen use of one of the busiest and most important parts of Toronto for a full day, as the hotel engineers examined the problem. Many late night variety shows in the USA made light of the issue, suggesting Donald Trump, a Republican candidate for the US presidential election, had the power to shut down Toronto. For Toronto citizens, however, this was no laughing matter. This same building shed dozens of panels of glass to the street on more than one occasion, closing the same busy section of Toronto. Donald Trump Jr., executive vice-president of the Trump Organization, told Reuters that one piece of broken glass doesn't "speak to the rest of the building" (www.theglobeandmail.com/news/toronto/glass-falls-from-trump-tower/article5306888/). Yet the growing laundry list of problems with the tower suggests otherwise. These incidents adversely affect the daily life of

D. Hoeg (✉)
Department of Secondary Education, University of Alberta, Edmonton, AB, Canada
e-mail: dhoeg@ualberta.ca

A. DiGiacomo • S. El Halwany • C. Phillips-MacNeil • M. Milanovic
M. Zouda • L. Bencze
Ontario Institute for Studies in Education (OISE), University of Toronto, Toronto, ON, Canada
e-mail: ajd508@gmail.com; sarah.elhalwany@mail.utoronto.ca;
chrisphillips1202@gmail.com; minja.milanovic@utoronto.ca; majd.zouda@mail.utoronto.ca;
larry.bencze@utoronto.ca

M. Kirstovic
Peel District School Board, Mississauga, ON, Canada
e-mail: mirjan.krstovic@peelsb.com

T. Nishizawa
Faculty of Education, University of Victoria, Victoria, BC, Canada
e-mail: tomo.nishizawa@gmail.com

L. Bencze (ed.), *Science and Technology Education Promoting Wellbeing for Individuals, Societies and Environments*, Cultural Studies of Science Education 14,
DOI 10.1007/978-3-319-55505-8_16

citizens of Toronto, and raise questions about policies and regulations that allow apparently substandard construction to take place. The effects of buildings such as the Trump Hotel to our common environment emphasises needs for citizen involvement in decisions about how our environment should be used, and who should be held accountable for detrimental environmental and social impacts of its misuse.

The Trump Hotel issue can be seen as an example of a Socioscientific issue (SSI) (Zeidler, Sadler, Simmons & Howe, 2005). SSIs can be described as social and environmental issues connected to the products and practices of science and technology. Controversies surrounding climate change, effects of food and drugs on human health, and industrial pollution are examples of SSIs. The ongoing problems with the Trump Tower can also be described as a citizen issue connected to products and practices of science, technology, engineering and mathematics (STEM). STEM is a science educational reform movement that is generally rationalised by seeing merit in integrating science with related disciplines of technology, engineering, and mathematics. The value of these reforms, claims some proponents (e g., Rennie, Venville & Wallace, 2012), is that this integration will make science more engaging, accessible to students, and more contemporaneously relevant to the increasingly science and technology oriented society in which we live, resulting in more pertinent science literacy for citizens. A second, but perhaps more influential rationalisation, is that STEM education is needed to prepare students for science and technology related jobs that are seen as the basis of the future economy (Pierce, 2013).

When viewed as an issue of citizenship, the Trump Hotel incident is an example of what many scholars (e.g., Gough, 2015) have reported as lacking in STEM education—an emphasis on the ethical and moral dimension of STEM knowledge and practice, and how these might be utilized by citizens in order to understand, and take action on, social issues that affect the communities and environments in which they live. In this chapter, we present a preliminary study in which we evaluated science students' experiences using the multi-media platform Prezi™, which presented information on various consumer-based commodities. We use this as context for discussion of the research team's thoughts on how Prezi™ might be used in science education designed to engage students in social action on SSIs.

16.1 Citizenship and Science Education

Citizenship has been a contested notion. Thomas Marshall (1998) frames citizenship as a nationally bounded set of universal legal and social rights and duties evolving out of the emerging historical and socioeconomic developments of post-war Keynesian states. Yet, as Bryan Turner (2001) suggests, a citizen within Marshall's progressive narrative is imagined largely as a *passive* recipient of rights rather than as an *active* political subject. Zygmunt Bauman (2001), Wendy Brown (2005) and Henry Giroux (2008) have analyzed the 'hollowing out' of civic life and subsequent colonization of citizenship by economic, market-based rationalities and practices. This conception of citizenship prioritizes individuals who strategize for themselves among various social, political and economic options, rather than striving with others to alter or organize these options to benefit the whole. Concurrent with these changes, there appear to be decreases in more traditional forms of democratic participation and citizenship.

Science education has struggled with issues of citizenship for decades (Calabrese Barton, 2012). Sociopolitical absences in past science curricula have limited opportunity to develop democratic orientations conducive to a citizenry that sees science as something they might utilize in the solution to community-based problems (Bencze, 2008). There exists a strong emphasis in science education reform that advocates for uses of science in ways that are relevant to students' lives and transformative to their communities (Mueller, Tippins & Bryan, 2012). A common view of what democratic citizenship in science education might look like is one in which students, teachers and communities work together to use science to understand and address community problems, as opposed to inaction or passivity that often characterizes what it means to teach or learn science (Calabrese Barton, 2012).

Contemporary science education faces ethical and moral questions related to citizenship—such as in terms of the extent to which dominant STEM initiatives may prepare students for specific types of employment and provide necessary student engagement in more socially participatory and community grounded science that may be required for social justice and citizenship education (Calabrese Barton, 2012). The ability to understand and use science is increasingly an everyday part of life for the average citizen. Joe Kincheloe, Shirley Steinberg and Deborah Tippins (1992) emphasize that teachers, students and parents come to view the "community as a mini-laboratory for democratic participation" (p. 223). These are compelling reasons to provide for citizenship and social justice considerations in science education. Unfortunately, participative forms of science that prepare students for democratic citizenship appear to be relatively non-existent in schools, and student participation (i.e., caring for a community) is tightly mediated by those already with authority—those who set up the questions, the tools and the resources for participation (Calabrese Barton, 2012). Angela Calabrese Barton (2012), among others, notes that place ought to serve as context for, subject of, and driving relationship framing the doing of science. Yet, teachers and students are rarely asked to identify with place (local environment/community) as a part of teaching and learning science.

Science education connected to citizenship, in the form of SSIs education, has existed for about 45 years (Pedretti & Nazir, 2011). SSI education *can be* a framework for development of skills and knowledge related to uses of science for democratic citizenship, yet this ideal is infrequently achieved (Levinson, 2010). Instead, SSIs education is often an afterthought, if there is enough time, after knowledge and theories, products of traditional science, are taught (Hodson, 2009). Pressure to prepare students for exams and assessment, appears to limit potential for leveraging political aspirations of students in actual decision making or advocating for social change related to SSIs (Activism) (Bencze & Alsop, 2014). An alternative for citizenship education incorporating SSIs is STEPWISE (www.stepwiser.ca). STEPWISE is a pedagogical framework that, among other things, encourages and enables students to self-direct primary (e.g., experiments) and secondary (e.g.,

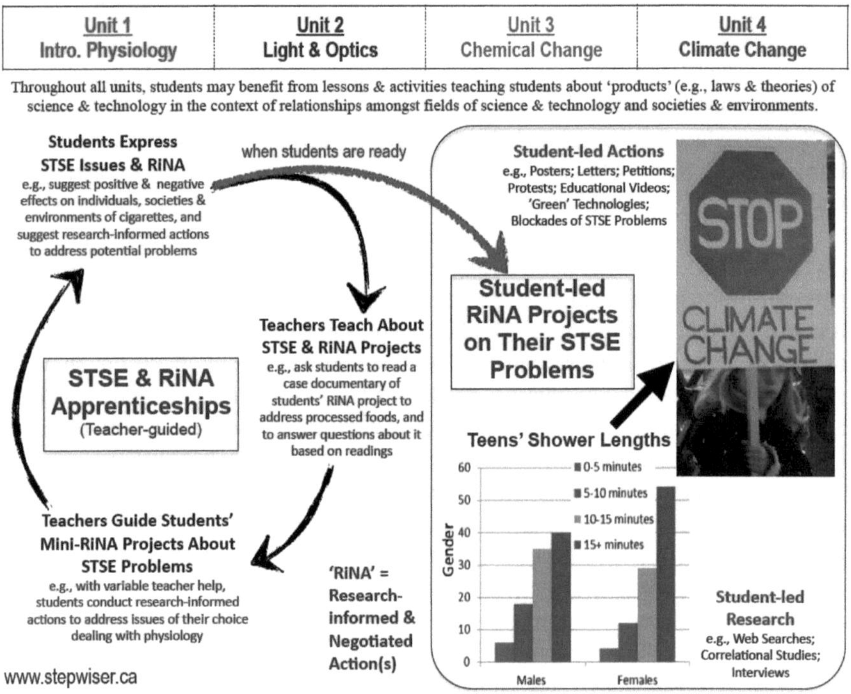

Fig. 16.1 STEPWISE educational framework

internet searches) research to help inform their decisions about SSIs and take action to address problems they perceive.

As illustrated in Fig. 16.1, the framework recommends that teachers provide 'apprenticeship' lessons and activities to help students to develop expertise and confidence to eventually self-direct research-informed and negotiated action (RiNA) projects to address SSIs of interest to them. These experiences may result in students learning about various stakeholders and entities ('*actants*') connected to SSIs under investigation, many of which (e.g., corporations, transnational trade agreements) often are not included in science education. This can be *democratizing,* because of the rich, perhaps more realistic, representations of issues they might provide (Pierce, 2013).

After nearly a decade of research, we have concluded that affective engagement with SSIs, resulting in taking actions, may develop if students have leading roles in determining personal relevance and roles in relationships associated with these issues, which can provide motivation to conduct primary and secondary research (e.g., Hoeg & Bencze, 2015). One of the challenges of engaging students in learning about SSIs is the elusive and hidden nature of the relationships between the stakeholders involved. The idea that relationships between stakeholders (actants) occurs through interconnected networks is a central tenet of actor-network theory (ANT) (Latour, 2005). ANT can be described as 'material-semiotic,' meaning it considers both material (people, groups, non-living material) and symbolic (e.g., the social meaning of luxury items) *actants* in a network, and how these come together to act as a whole. Through ANT, students may come to understand the world as one of a network, in which they and their consumer based-choices are linked to numerous other actants that influence these choices. Students' recognition of networks associated with the SSIs, and their role as actants within these networks, appears to be important to their development of socio-political (activist) orientations and actions (Watts, Williams & Jaggers, 2003). Yet, these networks are often difficult for students to see. Powerful actants, such as the producer of a cellular phone, act to align other actants toward certain ends, such as making the phone desirable to citizens. Actants are not always living; they can be non-living and semiotic (signs, symbols, language, memes). This positive-toned chorus enchants consumers, both implicitly and explicitly, about the necessity of having the latest cell phone, creating symbolic value beyond the phone's utilitarian purposes. Indeed, many of the actors involved in the production-consumption of the phone are immaterial, invisible or, purposely hidden from view, so that the phone itself becomes the actant that represents a much more complex network about which the average consumer is not aware. This is a phenomenon of actor-networks that Latour (2005) terms 'punctualisation'.

Educating students about the networks associated with SSI's, particularly so that they may be 'de-punctualised' should be, we suggest, one of the most important aims of SSIs education. We have found that explicit teaching of ANT to help students conceptualise relationships associated with SSIs is particularly effective in their development of socio-political orientations (Bencze & Krstovic, 2013a, 2013b).

16.2 Teaching Socioscientific Issues

SSIs are often taught through various forms of visual media, such as videos, PowerPoint™ slides, and case studies. These media are effective because they are cognitively appealing to students (Fleming, 1995). According to a cognitive approach to learning, there are a number of learning 'styles,' but three types are dominant in classrooms: auditory, visual and kinaesthetic (Ahola, 2004). Videos and electronic slideshows are visual and auditory media; case studies are visual and possibly kinesthetic. Yet none of these media engage all three learning styles simultaneously. These media also often portray a single perspective or point of view, and present topics in a linear fashion (Zukerman, 1999). These features can diminish the complexity and diffuse nature of the actants associated with SSIs. Criticisms of PowerPoint™, in particular, have increased in the past decade. Edward Tufte (2003), for example, argues that PowerPoint™ slides lead to over-reliance on a hierarchy of ideas, over-simplification, and linear thinking on part of the presenter and audience. In surveying classes with and without PowerPoint™ lectures, Dale Cyphert (2004) and Karl Kunkel (2004) found that there was no significant difference in student performance or understanding of material. These studies claim that PowerPoint™ usage stifled pedagogical creativity and led to poorer audience engagement.

Another issue is that videos, PowerPoint™, and case studies represent forms of media that may not be well suited to mental processing abilities of current school-aged students, a group often termed *digital natives*. The term 'digital natives' was first proposed by Marc Prensky (2001) to represent a generation of people born after 1980 who are immersed in digital technologies and therefore learn differently from previous generations of people. According to Prensky, digital natives reproduce a culture of connectivity and online creating and sharing. They have 'e-lives' that revolve around the Internet, where they access information and interact with others, for example blogging, playing online games, downloading music, purchasing and selling online and socialising via social media networks. Digital natives are active experiential learners who like receiving information quickly, are multi-taskers and parallel processers and prefer graphics over texts.

Alternatives to PowerPoint™ and other more linear representations of information have been sought that better appeal to contemporary students (Conboy, Fletcher, Russell &Wilson, 2012). One of these is Prezi™. Prezi™ is a free, cloud-based online visual presentation tool launched in 2009 that allows the audience to interact with its content by moving around and zooming in and out on a large canvas contained in separate *cells*, any number of which can be created (Fransson & Holmberg, 2012). Each cell can contain various forms of media, such as videos, text, and pictures, which can effectively represent one actor of an SSI network, not connected to others. Prezi™ is thought to be more engaging to digital natives due to its 'networked' capability and non-linear representation of information (Ng, 2012). Studies of Prezi™ are limited, although it has been reported, for instance, that its effectiveness is influenced by instructor style (Conboy et al., 2012). Others suggest a need for teachers skilled in using the software, and adequate preparation for classroom use, to optimize Prezi™ effectiveness (Virtanen, Myllärniemi & Wallander (2012).

Prezi™ has some apparent advantages over other forms of media used to explore SSIs; its non-linear presentation of information, and the discursive feature of separate and independent cells, may make it more suitable to encourage the viewer to self-determine SSI relationships, controversies, personal contributions, and decisions, so that exploration of the Prezi™ may be an open-ended, *inductive* experience (Lawson, 2005). At the same time, the Prezi™ identifies actants in SSI networks, and provides information about them, which may not be obvious to students, and may in fact be hidden (e.g., child labor workforce making cell phones); so, the learning experience is also instructive, or *deductive* (Lawson, 2005). The personal relevance established by such experiences can motivate secondary and eventually primary research that fuels RiNA (Hoeg & Bencze, 2014).

16.3 Research Methods and Context

The research reported here is based on an exploration of the use of Prezi™ to teach students about actor-networks associated with SSIs. This falls within the larger STEPWISE rationale of the need to develop democratic citizenship orientations and actions in students—ideally, forms of activism toward critical SSIs that are relevant to students and their communities. The initial phase of this project involved development of several 'multi-actant documentaries' (MADs) on Prezi™ platforms by members of the research team. Given the personalised and variable ways a Prezi™ can be created, specific criteria were provided for this construction, which included:

- The focus is on SSIs/STEM issues that are pertinent to students;
- Inclusion of actants that may not be obvious or well known by students;
- Content of each cell was to be a representation, but certainly not a full representation of individual actants;
- Each cell was a 'teaser', but aimed to give some semblance of the individual actants;
- Each cell contained mixed media; at a minimum, words and pictures. However, they could also include videos, links to other sites, etc.;
- There were to be no connections shown between actants (yet, at the same time, not to mislead students in terms of the networked nature of every cell—actants were not to appear as a single, unconnected point).
- The overall focus was on the issue—although the issues themselves are controversial, conclusions related to the controversy were not provided, and;
- Some positive support for each commodity (e.g. from companies…) was provided— although the whole presentation was intended to be generally critical of the commodity.

With these criteria in mind, thirteen Prezi™ based MADs were created (goo.gl/tRxwz0), based on the following topics: Cosmetics; Drugs; Fast and Manufactured Foods; Smoking; Tattoos; Automobiles; Cell Phones; Coffee; Popular Media; Pets; Fashion and Clothing; Fashion Accessories; and, Non-recyclable items. The MADs were developed as a way to provide a precise of information about a range of actant types (i.e., living, non-living, and semiotic) associated with actor-networks of commodities of science and technology/STEM. The MADs were meant to provide 'de-punctualisation,' an expansion and explication of actor-networks associated with the commodity about which students may have previously been unaware. Our view was that the MADs could be used to 'teach' students about these networks, and so they were seen as well-suited to the apprenticeship phases of STEPWISE, which typically involves more teacher-directed learning experiences.

We conducted research in two science classrooms. Since 2006, Larry has been supporting teachers in efforts to encourage and enable students to conduct RiNA projects to address SSIs. Among teachers engaged in this work, Mirjan stands out as having considerable successes along these lines (e.g., Krstovic, 2014). Mirjan Krstovic (author #4, here), a secondary school teacher of science with 8 years teaching experience, had promoted research-informed actions for three previous semesters when he decided to explore the use of MADs to teach his tenth-grade 'academic' (for university entrance) science class about SSIs. Tomo Nishizawa (author #7, here), the second teacher involved in this study, has been involved in STEPWISE since 2013, and has had noticeable success with her students in both years. Tomo's students, in grade 10–11 Biology at an affluent international school in Venezuela, have shown more agency to take actions and developed 'better understanding' of the complexity of STSE issues and the different stakeholders involved after education based on STEPWISE. Both teachers continued to use the STEPWISE framework depicted in Fig. 16.1 but, for this study, used MADs on fast food, cosmetics, cell phones, drugs, and automobiles, during the apprenticeship phase of a larger RiNA project. Students were instructed that the purpose of the activity was to learn about the actor-networks associated with some common commodities.

We collected data from a total of 28 students. Students were asked to complete a survey (Appendix B) on their perceptions of the MADs. The purpose of the survey was to get some initial feedback on being 'user-friendly' and interesting—along with the extent to which they enhanced perception of controversies and actor-networks associated with SSIs. The survey was based on the following constructs: Visual appeal and comprehensibility; clarity of controversy; relevance; clarity of relationships (network) between stakeholders (actants), and credibility/reliability of information. The survey consisted of a series of statements to which students indicated a level of agreement on a 10 point Likert-scale. In addition, survey questions were asked to provide more depth and detail to students' responses for each construct. Semi-structured interviews were also conducted with Mirjan and Tomo about their views on the use of MADS and Prezi™ in the STEPWISE apprenticeship. Data describing the evolution of the Prezi™ project is also present in the form of statements made by members of the research team during meetings. Data-collection and analyses had *rationalistic* (survey/interview) and *naturalistic* (research team member statements) characteristics (Guba & Lincoln, 1988). The data were coded for categories and then developed encompassing themes—using constant comparative methods based on constructivist grounded theory (Charmaz, 2014). Categories and themes were then negotiated between us (Wasser & Bresler, 1996).

16.4 Findings

The Prezi™ format appeared to be a welcome addition by Mirjan, Tomo and their students to the more typical forms of media that they have used during the apprenticeship stage of STEPWISE. Mirjan generally found Prezi™ s to be easy to use, engaging, and informative. One thing that stood out to him was that it "really provided depth and breadth to the issue, it really showed students the complex web of people, materials and ideas that make up these problems" (Interview, July, 2015), and Tomo stated "The Prezi™, more than other media, really caused a lot more

Table 16.1 Students responses to Prezi™

Prezi features	Average score /10	SD	Variance
Controversial nature	7.48	2.08	4.33
Relevancy of information	8.37	1.76	3.10
Clarity of relationship between stakeholders	6.94	2.72	7.39
Credibility/reliability of sources	8.48	1.57	2.49
Comprehensibility of presentation	8.71	1.67	2.8
Attractiveness of presentation	8.72	1.83	3.35

questions to be asked, and these drove students to ask questions about their contribution to SSIs" (Interview, June, 2015).

Table 16.1 shows a summary of data collected from all student surveys. Averages, standard deviation, and variance of student scores for each construct were calculated for each Prezi™. The same data for an individual Prezi™ can be found in Appendix A.

Overall, students rated *relevancy of information*, *credibility/reliability of sources*, *comprehensibility of the presentation*, and *attractiveness of the presentation* very highly. Students scored *controversial nature* and *clarity of relationship between stakeholders* relatively lower, although scores were still quite high for these constructs. Lower scores for these constructs might be expected, since each Prezi™ was designed so as not to define the controversy or relationships between stakeholders. These results, in addition to students' responses to questions related to each construct, are examined further in the following sections.

16.4.1 Appeal and Induction

Overall, students thought Prezi™ was highly comprehensible and visually appealing. When asked about how easy the media was to use and understand, there was strong agreement that each Prezi™ "was very well laid-out and pleasing to the eye. It presented a lot of information in an easy to follow Prezi™" (Sarah, survey). One advantage of the Prezi™ over some other forms of media is its ability to be viewed offline, and developed collaboratively, and it is therefore perhaps better suited to so called 'digital natives' (Prensky, 2001). Melissa commented "I liked that I could bring the Prezi™ home, it gave me more time to think about the issue and where I wanted to go with it" (Survey). Their networking ability is one of the reasons Prezi™, and other contemporary digital presentation media, such as Google Present™, appear to be more engaging than older forms of media such as PowerPoint™ (Ng, 2012).

On the other hand, each Prezi™ contains a significant amount of textual data, something about which many students were critical. For example, John said "bring information that is more visual" (Survey), Susan thought a Prezi™ could be improved by "not having slides that are so jumbled with content" (Survey), and

Cirah said "having less words crammed into each slide would make it better" (Survey). The amount of information on each Prezi™ was a key point of discussion among the research team. The provision of too much knowledge, can 'lead' the student, resulting in deductive processes such as knowledge acquisition and application rather than the perhaps more inductive process of knowledge construction (Roth, 2013). Knowledge acquisition as opposed to knowledge construction is central in debates about student learning, constructivism and inquiry in general (Bencze & Alsop, 2008). The amount of scaffolding to provide students to support inquiry without circumventing student ownership and construction of knowledge is a delicate balance with which each teacher must grapple during learning events (Roth, 2013), depending on what they see as the purpose of MADs-Prezis™.

16.4.2 Credibility, Relevance and Authority

Students generally found each Prezi™ produced to be a credible source of information. Credibility seemed to be closely linked to the type of data in the Prezi™, particularly scientific types of data. For example, when asked about what would make the Prezi™ more credible, students commonly expressed they would need to see "more detail and statistics" (John, survey), and "more primary research; by this I mean stats, data, and survey results to ensure the information is accurate" (Irtiqua, survey). Ciarah agreed with these sentiments, stating "More scientific research from different scientific organisations would benefit" (Survey). Credibility appeared to be dependent on the perceived 'expertise' or authority of the individuals from which the data came. For example, Sara stated, "I would like to see more quotes from cosmetologists on the issue, since they are experts on this topic and understand the harms of toxic chemicals" (Survey). Students appeared to have potentially problematic taken-for-granted trust in authority. Nisura, for example, advocated for "finding the quote of a reliable person, saying what you want to get across" (Survey), and Ned claimed "I don't ever doubt the information because it is used for educational purposes" (Survey).

One aspect of many Prezi™s about which students were critical was that they often presented issues theoretically, or from a perspective that was remote and that didn't establish personal relevance to them. When asked how these media might be made more relevant, comments such as "Add data about teens/kids who have cell phones and teens/kids who don't, to show how cell phones are taking over their lives" (Jon, survey), "… you could use a few more people from our everyday lives, instead of using popular individuals." (Darcy, survey), and "include more teenagers in high school struggling with a drug problem. I would then feel more engaged in the presentation" (Jaslan, survey) were made by students. Stephanie added she would "Talk about the opinions of different consumers, such as teenagers" (Survey).

The content of many Prezis™ often led students to view the opinion of scientists as aligned with those of activists. For example, Ciara stated "activists and scientists have a relationship showing marijuana is not that bad" (Survey), and Susan added

"If there were an interview between scientists and activists it would be clearer they were on the same sides" (Survey). Larry was concerned that these statements suggested students put scientists and science on similar moral and ethical pedestals, which may represent a naïve view of the Nature of Science (NOS) because scientific data is frequently wielded by powerful actants to support the consumption of consumer-based products. On the grand scale, this may be morally and ethically problematic because these practices of science may camouflage detrimental effects to communities and environments due to consumption of commodities.

16.4.3 Controversy, Relationships and Deduction

The controversial nature of topics and the clarity of relationships were rated somewhat lower than the other categories, and students expressed very simplistic and dualistic views of the relationships and controversies. For example, Ana stated after viewing the Prezi™ on drugs that she wanted "a clearer comparison and overview of *both sides* of the issue" (emphasis added) (Survey); Isabelle identified only "fast food nutrition VS healthy food" (Survey) as the controversy conveyed by the Fast and Manufactured Food Prezi™. Jaslan stated, "They did a good job at illustrating *both sides* of whether drugs are OK to do or not" (emphasis added) (Survey). These views appear to be preconceived, as many stakeholders (actants) were presented in each Prezi™ and connections were not drawn between actants purposely to allow students to draw their own conclusions (deductive) about the nature of relationships and controversies.

Students' comments suggest they wanted the controversies to be made clearer; essentially, to be clearly given the controversies. For example, Nisura commented, "Some of the changes I would make to better illustrate the controversial nature of this Prezi™ is by grouping the pros and the cons to put how controversial it is into perspective" (Survey). Ron said he would "Give an introduction to explain what the issue is" (Survey) and Jaslan stated "I would provide more real-life examples so the readers can be clearer about what the controversy is" (Survey). Students also wanted the relationships between the various stakeholders presented in the Prezi™ made clearer to them. For example, when asked to comment about stakeholder relationships, Michelle wrote "I would find a way to add how one stakeholder's opinion relates to the other stakeholder's opinions" (Survey), and Susan stated "Create arrows indicating how they are related to each other." (Survey).

16.4.4 Determination of Actor-Networks and Inquiry

Despite students' criticisms that many Prezi™ were not explicit enough in regards to controversies and stakeholders, data suggests students learned a great deal about the actor-networks associated with the SSI featured in the Prezi™ they viewed. For example, Ron made explicit reference to actor-networks in his comment:

> Cars benefit many people's lives but they have a large negative one as well. Cars exist in a network, which also includes there are jobs, roads, pollution, raw materials, industry, and massive business opportunities such as advertisement and marketing. This is all at the cost of our earth though as well as other social issues (Survey).

An apparent outcome of their self-determination of actor-networks was that many students recognised that actants frequently align to support a common message about commodities; for example, Derek stated:

> If you have a product and you are producing it in a Third World country, where you are not giving people the proper amount of pay and they are living in a low ['destitute'] place, then that society is not doing well – and that is partly due to your product ... (Survey).

This suggests the Prezi™ had some effect in inducing students' awareness of actor-networks, but at the same time stimulated inductive self-determination of relationships within actor-networks associated with SSIs. These appear to be important components of commitment to activism (Bencze & Krstovic, 2013a).

Identification of controversies and actor-networks proved to be very effective in generating further questions, as noted by Mirjan, who said "After watching a Prezi™, they (the students) had many questions about the issue, the various actors, and their own relationship to them" (Interview, July 2015). Some examples of questions student had include: "I would like to know what the government had to say about the why the FDA does not have to check what goes into those products" (Susan, survey), and, "I would like to investigate the 'what's inside your cell phone' slide because I would like to find out what's inside my phone" (Robert, survey). This was one of the desired outcomes of the Prezi™; such student generated questions can become the basis of research informed and negotiated action (Bencze & Carter, 2011).

16.5 Discussion and Conclusions

This study was exploratory in nature, meant to guide future development of Prezi™ as a vehicle for communicating socio-scientific issues, and research based on their use in classrooms. The results of the study highlighted several issues, focusing our attention on: i) the amount of content in a Prezi™; ii) the nature of the content in a Prezi™; iii) the location of Prezi™ in the STEPWISE learning cycle, and; iv) the pedagogy in which a Prezi™ is introduced and learned. These discussions have led to several insights into the creation and use of Prezi™ for SSIs education which we present here.

16.5.1 Amount of Content

The amount of content in Prezi™ was a concern of students, and the largest point of debate among members of the research team. Many students commented there was too much information in each Prezi™. Too much prescribed content can lead to issues of disengagement with the material, effecting students' interest and learning (Wood, 1998). On the other hand, students desire for greater clarity and more information about the controversy and relationships between stakeholders are in opposition to our original intention for using Prezi™. Their discomfort supports the notion that students may expect to be provided answers, knowledge products, from an authority, that they can acquire (Weinstein, 2012), rather than actively constructing knowledge themselves through critical thinking and inquiry.

The amount of content included on each Prezi™ was the most contested topic of discussion during group meetings. Several group members felt many of the Prezi™s that had been produced may be too information-laden. Darren, in particular, was concerned that the level of information may hinder inductive processes, such as knowledge construction, and affective orientations, such as feelings of ownership and commitment, that may be required to motivate further investigation and actions on SSIs. Regarding this possibility, Darren, stated "I'm concerned the abundance of information in the cells may actually cause students to not ask many questions, because many of the questions may be answered to the satisfaction of the students already" (Meeting 2). This phenomenon has been noted in other research on Prezi™ (Brock & Brohdal, 2013). Christina added to this that "I wonder if a less-informative Prezi™ might be better suited to student driven questioning and research?" (Meeting 2). Indeed, some evidence supporting this was collected. Most of the questions students asked, for example, appeared to be defined by the parameters of information presented on the Prezi™. In other words, some Prezi™ appeared to set up the boundaries of possible inquiry for students. While partially agreeing with this, Larry reminded the team that each Prezi™ was, initially, envisioned as a MAD, a vignette of a SSI, a teaching tool and form of scaffolding, along with other media, such as videos, other presentations, case studies, and science activities, to be used primarily during the *apprenticeship* phase of STEPWISE. In other words, a Prezi™ is not meant to be a stand-alone inductive experience that in and of itself results in RiNA. Larry noted, however, a less informative Prezi™ might make sense in the later stages of the STEWISE apprenticeship, as the teacher is modelling research informed activism, and students are gradually being released from dependence on the teacher to provide questions to ask and the resources to answer them (Meeting 2).

The amount of information that should be contained in each Prezi™ is the basis of ongoing discussion among the research group. Certainly, different levels of information may serve different pedagogical purposes. However, avoiding student misunderstandings of actor-networks that fail to de-punctualise them, may require explicit instruction and teaching. This is a strong argument for the inclusion of deductive teaching and learning in pedagogy surrounding the use of Prezi™ (Bencze & Alsop, 2008).

16.5.2 Nature of Content

The nature of the content in Prezi™ is a central consideration when deciding how it may best be used in education for SSIs. In particular, issues of interpretation, bias, and representations of NOS, are important considerations in selecting information. Several members of the group thought there was overrepresentation of criticism of the commodity in each Prezi™ that was made, and that this bias was a problem. Larry expressed the view that although both positive and negative perspectives might be represented, the use of Prezi™ represents an opportunity to emphasize the stakeholders (actants) involved in the production of commodities that are often hidden from view to consumers, and about which consumers might be critical, if they were apparent. From social justice perspectives, it may be necessary to give these stakeholders greater representation on Prezi™ presentations than those that support consumption of the commodity, which are already over-represented in society.

Further muddying the issue of bias were students' rather simplistic views of complex SSIs. These views may be due to the influence of Western worldviews, stemming from Judeo-Christian epistemological dualisms; i.e. right/wrong, good/bad, self/other (Cobern, 1991). In other words, these students may have developed mental schema from socialisation in Western society that causes them to interpret a Prezi™ in certain (perhaps unintended and detrimental) ways. One of the criticisms of inductive learning theories, such as constructivism is that these 'mental lenses' acquired through socialisation, which often unintentionally reproduce the status quo, may be used by learners to construct knowledge (Mathews, 2002). A simplistic, dualistic interpretation of the controversy may act to 'punctualise' the SSI, causing the viewer to see only two sides (e g., critical/supportive), hiding from view the complex network, and moral and ethical grey-scale, that is part and parcel of these issues. Clearly, this discursive feature of these networks is something SSIs education should illuminate. How to embed this complexity in Prezi™ should be an important consideration guiding its construction.

Students' comments also suggest they hold strong and perhaps unrealistically positive views about the trustworthiness of those with authority and power, particularly scientists. This is not necessarily surprising, as the non-democratic and authoritarian nature of school likely socialises students to respond to, and to an extent, need, authority figures (Weinstein, 2012). This is problematic in SSIs education, as acquiescence to authority appears to be antithetical to 'speaking truth to power' and other critical orientations that appear to be necessary for activism (Watts et al., 2003). Privileging science and other authorities can be problematic because products and processes of STEM, many of which may be detrimental to communities, are often supported by scientific research and others in positions of power (Bencze, 2008). How NOS is represented in Prezi™ should therefore likely be a central consideration in its construction. Minja, who created a large number of Prezi™ documentaries, suggested that it was often challenging to find scientific data that was supportive of the activist stance. This led the group toward discussion about how the NOS needed to be realistically represented in Prezi™. To accurately represent NOS, Larry suggested each Prezi™ might contain varying and even oppositional scientific data. Darren suggested scientific data might be absent altogether, which may put more responsibility on students to do secondary research on the issue, thus discovering variability in scientific data related to the commodity; this practice may provide students with more authentic NOS perspectives (Lederman et al., 2002).

A Prezi™ can be seen to represent a 'Sign' of the world (Roth, 2001), including the personal biases, subjective choices, and pre-existing beliefs relative to SSIs that make sense to the author. This makes the use of specific design criteria for Prezi™ important if it is to contain features conducive to learning about complex actor-networks associated with SSIs, and contain realistic representations of science. A negotiation or provision of criteria may be required if students are involved in constructing Prezi™. In addition, teaching about the complex nature of SSI actor-networks in general, and NOS, particularly how it is relevant in SSIs, may be helpful for students to develop realistic views of these networks and sciences' role in these issues.

16.5.3 Relevance and Location of Prezi™ in STEPWISE

Many group members thought Prezi™ could be used at various stages of STEPWISE. The stage in which Prezi™ is used, however, appears to create variation in its potential personal relevance to students. For example, Madj thought that a Prezi™ would work well in the 'students expressing ideas' stage. At this stage, a pre-produced Prezi™ appears to be required. Yet, Minja, Darren and others pointed out the challenge in making a Prezi™ that necessarily appeals to an unknown audience, which requires a degree of objectivity and generalization. This type of Prezi™ may be limited in its ability to present a SSI in a way that is personally relevant to students.

Recognising this, Darren and Christina suggested Prezi™ could be used throughout a complete STEPWISE project to document and communicate various steps of students' inquiry. Prezi™ cells in which students express their pre-existing ideas and knowledge relative to the issue, cells that communicate secondary research, and cells that communicate primary research could be part of this product. The Prezi™ itself could serve as the summative product of a RiNA experience.

Seeing personal relevance in issues is important to the development of socio-political orientations and actions (Watts et al., 2003). If an issue is relevant to students, it has more meaning, they are likely to be more engaged learning about the issue, and may feel a greater sense of ownership over what they learn (Krstovic, 2014). These are qualities of transformational learning that can motivate students toward taking actions (Bencze & Krstovic, 2013a).

It seems reasonable that the amount and nature of content contained in Prezi™ could be variable, depending on where and how it is used in various stages of STEWPISE learning experiences. When used in STEPWISE apprenticeship, it may be beneficial that each Prezi™ be highly informative, with provision of multiple sources of data. However, in stages where students become more responsible for knowledge production, there may be value in using a partially constructed or empty Prezi™, which students are required to complete. It is important to recognise again that Prezi™ was not meant to be a stand-alone experience; the pedagogy surrounding its use can and should draw connections between SSIs and students' lives (Krstovic, 2014).

16.6 Implications and Conclusions for Using Prezi™ in Learning About Critical SSIs

It became clear during research group meetings that the pedagogy surrounding the use of Prezi™ is critically important if it is to aid students in self-determining actor-networks associated with SSIs. Aside from being part of a larger context of SSIs education, the learning potential of Prezi™ itself may not be met without teacher providing scaffolding, such as questions meant to initiate inquiries, and other forms of support. Once again, this reinforces the notion that a Prezi™ is not meant to be, and likely would be ineffective, as a stand-alone learning event (Roth, 2013) resulting in activism.

One important component of effective education about SSI's may be teaching ANT concepts (Krstovic, 2014). Teaching ANT reinforces the complexity of SSIs, and fosters a rejection of simplistic and dualistic views of these problems that may reproduce punctualisation of SSIs (Latour, 2005). Some suggested topics for teaching about ANT include:

- Individual actants are *heterogeneous*, composed of influences from other actants;
- Types of actants include: *materials* (e.g. living & non-living things, inventions, inscriptions) and *semiotic messages*;
- Actants may co-affect each other, with effects that constantly change;
- Actants can align, particularly under influences from powerful actants, so that a common semiotic message is supported by all.
- Activism may involve introducing new actants and re-orienting existing ones so that dominant semiotic messages change.

Students might learn about ANT before examining a Prezi™; this may give them the perspectives needed to realise the complexity inherent in actor-networks associated with SSIs. Alternatively, Prezi™ might be used to teach students about

Pedagogy	Activity
General	• Ask students to provide their pre-conceived evaluations (+/-) of the commodity in question; • Teach students about the commodities' life cycles, using the Prezi and *The Story of Stuff* video[1]; • Teach students about ANT, using an actor network drawing related to the commodity; • Ask students to identify 'stakeholders' (e.g., companies, governments) relating to the commodity and then suggest possible SSIs relating to them; • Ask students to draw actor networks associated with the commodity — prior to and after their research and suggested actions to address issues.
RiNAApprenticeship *Students Expressing Ideas:*	• Present only one cell to the student. The selected cell can be chosen based on the extent to which students relate to it, or based on how close the content is to the unit being taught. Encourage students to express their preconceptions about the content of the cell by asking questions, such as: - What is the cell about? (Teachers should emphasize that it is not a right/wrong question, but a question to express opinions). - Do you agree/disagree with the content of the cell and why? - What do they (students) know about the commodity in the cell? - What are the problems/issues/controversies regarding this commodity? - Who/what would benefit from it, and how? - Who/what would be harmed, and how? - What might different stakeholders say about the commodity? - What might you do to learn more about this issue? - To what extent is this issue related to you?
RiNAApprenticeship *Students Learning Ideas*	• Ask students to explore the cells in the Prezi in any order they wish. Teachers can ask questions, such as: - What controversies are presented in the Prezi? - What significant relationship(s) are illustrated between the different stakeholders? - What other stakeholders can be added to this presentation, and why? • Teachers could then explain STSE, the controversies, possible forms of secondary and primary research, and possible actions related to different cells in the Prezi.
RiNAApprenticeship *Students Judging Ideas*	• Ask students to choose part of the Prezi they would like to investigate further, or what they would like to add. Students would conduct secondary and primary research and take actions and add them as cells to the original Prezi.
RiNA	• Students conduct their research regarding a commodity/issue of their choice and develop their own Prezi communicating this process. The Prezi might include cells that contain pre-conceived ideas about the SSI, resources (videos/websites) they used for secondary research, results of their secondary research, results of primary research, and examples of action.

Fig. 16.2 Pedagogy involving Prezi™ to teach about critical socioscientific issues

ANT. Some pedagogical strategies that might be used with Prezi™ are presented in Fig. 16.2.

Uses of Prezi™ may provide students with critical ontological insights about SSIs and rich, perhaps critical, contexts that can motivated them to construct knowledge about personally relevant relationships and controversies, which appear to be needed to facilitate deeper commitments to RINA on SSIs defined by students (Bencze, Sperling & Carter, 2012). Our ongoing research suggest, under certain conditions, students can benefit from use of MADs in Prezi™ to develop awareness of actor-networks of which they are a part, and are associated with SSIs. The result of this preliminary study are encouraging, and provide the foundation for future research to further evaluate how Prezi™ might be involved in education leading to student engagement with RiNA on SSIs.

Appendices

Appendix A: Individual Prezi™ Student Responses

Cosmetics Prezi feature (15)	Average score /10	SD	Variance
Controversial nature	7.86	2.13	4.55
Relevancy of Information	8.93	1.27	1.63
Clarity of relationship between stakeholders	8	2.56	6.57
Credibility/reliability of Sources	9.26	0.70	0.46
Comprehensibility of presentation	9.66	0.61	0.38
Attractiveness of presentation	9.8	0.56	0.31
Fast food Prezi feature (5)			
Controversial nature	7.4	2.51	6.3
Relevancy of Information	8.4	1.14	1.3
Clarity of relationship between stakeholders	6.1	3.47	12.05
Credibility/reliability of Sources	7.6	2.6	6.8
Comprehensibility of presentation	7.4	2.4	5.8
Attractiveness of presentation	8.4	1.14	1.3
Drugs Prezi feature (6)			
Controversial nature	7.5	1.37	1.9
Relevancy of Information	6.83	2.63	6.96
Clarity of relationship between stakeholders	5.5	2.42	5.9
Credibility/reliability of Sources	7.83	1.51	2.26
Comprehensibility of presentation	8.33	1.86	3.47
Attractiveness of presentation	8.12	0.98	0.97
Cell phone Prezi feature (2)			
Controversial nature	7	0	0
Relevancy of Information	7.5	.71	0.5
Clarity of relationship between stakeholders	6.5	.71	0.5
Credibility/reliability of Sources	6.5	.71	0.5
Comprehensibility of presentation		1.41	2.0
Attractiveness of presentation	6.5	2.12	4.5

Appendix B: Student Survey

Exploring The Nature of Effective Secondary Research

This activity is an investigation of the important components of secondary research. As highlighted in bold below, it is important that your secondary research: 1) includes a controversial subject. 2) is personally relevant to you/community. 3) illustrates a clear relationship between stakeholders. 4) uses credible/reliable sources. 5) includes an appealing mind map presentation.

The following is an example of a sample secondary research using Prezi. **First explore the Prezi cells in any order you choose, then answer the following questions**. After examining each cell. you can return to the main slide by clicking on the Home button (located on the middle right side of the screen).

You may choose to investigate any **ONE** of the following three Prezis:

Cosmetics: https://prezi.com/0hkdjb31tucw/cosmetics/

Fast & manufacture foods: https://prezi.com/th-fzepweg8w/fast-manufactured-foods/

Drugs: https://prezi.com/vbwha24etft3/drugs/

I. Controversial nature of the issue

1. (a) Explain what controversy or controversies is/are represented here.

   (b) On a scale of 1–10. overall, how controversial was this Prezi?
   (Not Controversial) 1 2 3 4 5 6 7 8 9 10 **(Very Controversial)**

   (c) Explain what changes you would make to illustrate the controversial nature of the Prezi.

II. Relevance of information

2. (a) On a scale of 1–10. how personally relevant is this presentation to you (or your community)?
   (Not Relevant) 1 2 3 4 5 6 7 8 9 10 **(Very Relevant)**

   (b) Explain what you will do to make this presentation more relevant to you.

III. Relationships between stakeholders

3. (a) Explain what significant relationship(s) are illustrated between the different stakeholders.

   (b) On a scale of 1–10. how obvious were these relationships were for you?
   (Not Obvious) 1 2 3 4 5 6 7 8 9 10 **(Very Obvious)**

   (c) Explain how you might make the relationships between different stakeholders more obvious.

   (d) What other stakeholders can be added to this presentation, and why?

IV. Credibility reliability of sources

4. (a) On a scale of 1–10. how 'credible/reliable' (close to the reality of the stakeholder's views) was the information in this presentation?
   (Not Controversial) 1 2 3 4 5 6 7 8 9 10 **(Very Controversial)**

   (b) Explain what you would do to make this presentation more credible reliable.

V. Presentation of information

5. On a scale of 1–10. how easy the contents of the Prezi were to read and understand?
   (Not Easy to Read/Understand) 1 2 3 4 5 6 7 8 9 10 **(Very Easy to Read/Understand)**

6. (a) On a scale of 1–10. how attractive is the presentation for you (and for students of your age)?
   (Not Attracting) 1 2 3 4 5 6 7 8 9 10 **(Very Attracting)**

   (b) Explain how you would make this presentation more attractive.

VI. Other considerations

7. What part of this presentation would you like to investigate further? Why?

8. Explain what other things you would suggest to make this presentation more useful to students your age.

References

Ahola, T. (2004). *Oppimistyylien tunnistamisen mahdollisuudet sää tö tekniikan opetuksen kehittämisessä*. Oulu: OAMK (in Finnish).

Bauman, Z. (2001). *The individualized society*. Cambridge, UK: Polity Press.

Bencze, J. L. (2008). Private profit, science and science education: Critical problems and possibilities for action. *Canadian Journal of Science, Mathematics & Technology Education, 8*(4), 297–312.

Bencze, & Alsop, S. (2008). A critical and creative inquiry into school science inquiry. In W.-M. Roth & K. Tobin (Eds.), *World of science education: North America*. Sense Publishers.

Bencze, & Alsop, S. (2014). *Activist science and technology education*. Dordrecht, The Netherlands: Springer.

Bencze, J. L., & Krstovic, M. (2013a, April 27–May 1). *Students' social studies influences on their socioscientific actions*. A presentation at the annual conference of the American Educational Research Association, San Francisco.

Bencze, J. L., & Krstovic, M. (2013b, April 6–9). *Epistemic contributions to students' autonomous socioscientific actions*. A presentation at the annual conference of the National Association for Research in Science Teaching, Puerto Rico: Rio Grande.

Bencze, L., & Carter, L. (2011). Globalizing students acting for the common good. *Journal of Research in Science Teaching, 48*(6), 648–669.

Bencze, L., Sperling, E., & Carter, L. (2012). Students' research-informed socioscientific activism: Re/Visions for a sustainable future. *Research in Science Education, 42*(1), 129–148.
Brock, S., & Brodhal, C. (2013). A tale of two cultures: Cross cultural comparison in learning the Prezi™ presentation software tool in the US and Norway. *Journal of Information Technology Education: Research, 12*, 95–119.
Brown, W. (2005). *Edgework: Critical essays on knowledge and politics*. Princeton, NJ: Princeton University Press.
Calabrese Barton, A. M. (2012). Citizen (s') science. *Democracy & Education, 20*(2). Article 12. Available online at http://democracyeducationjournal.org/home/vol20/iss1/12
Charmaz, K. (2014). *Constructing grounded theory* (2nd ed.). New York: Sage.
Cobern, W. W. (1991). *World view theory and science education research, NARST monograph no. 3*. Manhattan, KS: National Association for Research in Science Teaching.
Conboy, C., Fletcher, S., Russell, K., & Wilson, M. (2012). An evaluation of the potential and impact of Prezi™, the zooming editor software as a tool to facilitate learning in higher education. *Innovations in Practice, 7*, 31–45.
Cyphert, D. (2004). The problem of powerpoint: Visual aid or visual rhetoric? *Business Communication Quarterly, 67*, 80–84.
Fleming, N. D. (1995). I'm different; not dumb. Modes of presentation (VARK) in the tertiary classroom. In A. Zelmer (Ed.), *Research and Development in higher education, proceedings of the 1995 annual conference of the Higher Education and Research Development Society of Australasia (HERDSA)* (Vol. 18, pp. 308–313). Sydney, Australia: HERDSA.
Fransson, G., & Holmberg, J. (2012). Understanding the theoretical framework of technological pedagogical content knowledge: A collaborative self-study to understand teaching practice and aspects of knowledge. *Studying Teacher Education, 8*(2), 193–204.
Giroux, H. (2008). *Against the terror of Neoliberalism: Politics beyond the age of greed*. New York: Paradigm.
Gough, A. (2015). STEM policy and science education: Scientistic curriculum and sociopolitical silences. *Cultural Studies of Science Education, 10*(2), 445–458.
Guba, E. G., & Lincoln, Y. S. (1988). Naturalistic and rationalistic enquiry. In J. P. Keeves (Ed.), *Educational research, methodology and measurement: An international handbook* (pp. 81–85). London: Pergamon Press.
Hodson, D. (2009). *Teaching and learning about science: Language, theories, methods, history, traditions and values*. Rotterdam, The Netherlands: Sense Publishers.
Hoeg, D., & Bencze, J. L. (2014). Counter cultural hegemony: Student teachers' experiences implementing STSE-activism. In L. Bencze & S. Alsop (Eds.), *Activist science and technology education* (pp. 575–596). Dordrecht, The Netherlands: Springer.
Hoeg, D., & Bencze, L. (2015). Socio-political development of private school children mobilising for disadvantaged others. *Cultural Studies of Science Education., 2*(10). doi:10.1007/s11422-014-9611-2.
Kincheloe, J. L., Steinberg, S. R., & Tippins, D. J. (1992). *The stigma of genius: Einstein and beyond modern education*. Durango, CO: Hollowbrook.
Krstovic, M. (2014). Preparing students for self-directed research-informed actions on socioscientific issues. In L. Bencze & S. Alsop (Eds.), *Activist science and technology education* (pp. 399–417). Dordrecht, The Netherlands: Springer.
Kunkel, K. (2004). A research note assessing the benefit of PowerPoint software in different lecture courses. *Teaching Sociology, 32*, 188–196.
Latour, B. (2005). *Reassembling the social: An introduction to actor-network-theory*. Oxford, UK: Oxford University Press.
Lawson, A. E. (2005). What is the role of induction and deduction in reasoning and scientific inquiry? *Journal of Research in Science Teaching, 42*(6), 716–740.
Lederman, N. G., Abd-El-Khalick, F., Bell, R. L., & Swartz, R. S. (2002). Views of nature of science questionnaire: Toward valid and meaningful assessment of learners' conceptions of nature of science. *Journal of Research in Science Teaching, 39*(6), 497–521.

Levinson, R. (2010). Science education and democratic participation: An uneasy congruence? *Studies in Science Education, 46*(1), 69–119.
Marshall, T. H. (1998). Citizenship and social class. In G. Shafir (Ed.), *The citizenship debates: A reader* (pp. 6–93). Minneapolis, MN: University of Minnesota Press.
Matthews, M. R. (2002). Constructivism and science education: A further appraisal. *Journal of Science Education and Technology, 11*(2), 121–135.
Mueller, M., Tippins, D., & Bryan, L. (2012). The future of citizen science. *Democracy & Education, 20*(1), 1–12.
Ng, W. (2012). Can we teach digital natives digital literacy? *Computers & Education, 59*, 1065–1078.
Pedretti, E., & Nazir, J. (2011). Currents in STSE education: Mapping a complex field, 40 years on. *Science Education, 95*(4), 601–626.
Pierce, C. (2013). *Education in the age of biocapitalism: Optimizing educational life for a flat world*. New York: Palgrave MacMillan.
Prensky, M. (2001). Digital natives, digital Immigrants. On the Horizon, 9(5), Retrieved June 12.07.15, from http://www.marcprensky.com/writing/prensky%20%20digital%20natives%20 digital%20immigrants%20-%20part1.pdf. [link does not work].
Rennie, L., Venville, G., & Wallace, J. (Eds.). (2012). *Integrating science, technology, engineering and mathematics: Issues, reflections and ways forward*. New York: Routledge.
Roth, W.-M. (2001). Learning science through technological design. *Journal of Research in Science Teaching, 38*, 768–790.
Roth, W.-M. (2013). To event: Toward a post-constructivist of theorizing and researching the living curriculum as event-in-the-making. *Curriculum Inquiry, 43*(3), 388–417.
Tufte, E. (2003). PowerPoint is evil. *Wired, 11*. Retrieved from *Wired*, http://www.wired.com/wired/archive/11.09/ppt2.html
Turner, B. (2001). The erosion of citizenship. *British Journal of Sociology, 52*(2), 189–209.
Virtanen, P., Myllärniemi, J., & Wallander, H. (2012). Diversifying higher education: Innovative tools to facilitate different ways of learning. *Proceedings of the 12th International Conference on Information Communication Technologies in Education (ICICTE) 2012*, 105–116. Retrieved from http://www.icicte.org/Proceedings2012/Papers/03-2-Virtanen.pdf
Wasser, J. D., & Bresler, L. (1996). Working in the interpretive zone: Conceptualizing collaboration in qualitative research teams. *Educational Researcher, 25*(5), 5–15.
Watts, R., Williams, N. C., & Jagers, R. (2003). Sociopolitical Development. *American Journal of Community Psychology, 31*(1), 185–194.
Weinstein, M. (2012). Schools/Citizen Science. *Democracy & Education, 20*(1). Article 6. Available online at http://democracyeducationjournal.org/home/vol20/iss1/6
Wood, G. H. (1998). Democracy and the curriculum. In L. E. Beyer & M. W. Apple (Eds.), *The curriculum: Problems, politics and possibilities* (pp. 177–198). Albany, NY: SUNY Press.
Zeidler, D. L., Sadler, T. D., Simmons, M., & Howe, E. (2005). Beyond STS: A research-based framework for socioscientific issues education. *Science Education, 89*(3), 357–377.
Zuckerman, L. (1999, April 17). Words go right to the brain, but can they stir the heart? *New York Times*, A17–A19.

Chapter 17
Battle of the Bands: Toxic Dust, Active Citizenship and Science Education

Larry Bencze and Chantal Pouliot

17.1 Introduction

Many of us live in 'representative' democracies, in which we mostly depend on leaders like scientists and engineers, government officials (elected and unelected), judges and lawyers and others for major decisions. On the other hand, in this age of electronic social media, we also can become regularly-engaged in discussions, surveys and questionnaires about matters of public concern. For the most part, however, we are called on only periodically to vote for elected officials who ask for the right to represent our interests for about 4 years. Such systems can work well, allowing individuals to develop expertise in specific fields and provide the whole society with numerous benefits. In many societies around the world, however, leaders and experts in representative democracies do not appear to be serving interests/needs of all their members—and, moreover, their governance seems associated with considerable harms to local and global biotic and abiotic environments.

In our current age, a particular problem appears to be the immense power associated with vast and complex networks of actants (living, nonliving & symbolic) often linked by interests in wealth accumulation by relatively small fractions of societies. Income and wealth disparities are widening in many societies and we are experiencing and predicting increases in various environmental problems—including many personal, social and environmental harms linked to higher average global temperatures, various health problems associated with chemicals in many products and services and compromises to nutrient quality in manufactured foods. Many of these

---

L. Bencze (✉)
Ontario Institute for Studies in Education (OISE), University of Toronto, Toronto, ON, Canada
e-mail: larry.bencze@utoronto.ca

C. Pouliot
Faculté des sciences de l'éducation, Université Laval, Québec City, QC, Canada
e-mail: chantal.pouliot@fse.ulaval.ca

L. Bencze (ed.), *Science and Technology Education Promoting Wellbeing for Individuals, Societies and Environments*, Cultural Studies of Science Education 14,
DOI 10.1007/978-3-319-55505-8_17

problems appear to involve fields of professional science and engineering—which are key to development and management of production and consumption of goods and services. Consequently, it seems necessary for citizens to be continually vigilant about actions of politicians, other leaders and experts like scientists and engineers and—where problems appear to exist (or could emerge)—take informed actions to try to bring about a better world.

Given roles played by fields of science and technology (and engineering and mathematics, etc.) in many problems faced by societies, a logical context for encouraging and enabling citizens to become more critical and activist appears to be school science (and, perhaps, related subjects). For schools to educate students about citizen engagement in problems and actions, however, they need relatively authentic models of citizen participation. Accordingly, in this chapter, we provide a case study of citizen engagement in addressing an environmental problem (airborne metal dust accumulation) that they believe could cause significant health problems for humans and material damage. Through their actions, various entities have been rallied in support of reducing/eliminating hazardous dust contamination. In doing so, their movement also uncovered evidence that they suggest would have helped sustain, if more widely known, a relatively high level of acuteness of community concerns about the dust. This case, which continues to evolve, may help teacher educators and teachers in schools to enlighten students about relative roles for leaders and citizens in representative democracies and, perhaps, help them to consider societal shifts towards greater citizen participation in decision-making in democracies.

17.2 Theoretical Background

17.2.1 Benefits and Limits of Socioscientific Issues Education

Traditionally, science has been taught in isolation from other subjects, a priority partly fed by claims that its focus on abstract, decontextualized knowledge makes it fundamental to progress in related fields like technology and engineering (Ziman, 1984). For at least the last four decades, however, educational researchers, book publishers, school systems and others claim that it should be taught in relationships with many other disciplines (Rennie, Venville, & Wallace, 2012). There have been in this period, for example, numerous efforts around the world to integrate or interrelate fields of science and technology and societies and environments (STSE) (Pedretti & Nazir, 2011). While implementation of STSE education across many school science contexts has been difficult (Hodson, 2011), perhaps largely due to tendencies of school science systems to prioritize teaching/learning of 'products'—such as laws, theories and inventions—of science and technology (Bell, 2006), when it *has* been implemented, a strong tendency appears to be to emphasize *controversies* in STSE relationships. Known in many places in the world as *socioscientific issues* (SSIs) (Sadler 2011), students often are invited to debate conflicting positions held by various 'stakeholders'—including politicians, company executives, citizens, activists, scientists, and others—regarding possible harmful effects of products (e.g., biotechnologies) of science and technology on wellbeing of individuals, societies and environments (WISE). Approximately paralleling this movement have been foci in France and elsewhere on *socially-acute questions* (SAQs)—which often deal with broader social controversies that may have varying degrees of ties to fields of science and technology (Legardez & Simonneaux, 2006).

In dealing with controversies, it is apparent that students are encouraged to negotiate data and claims—frequently in social situations—and then develop highly logical *personal* positions on issues (Levinson, 2010). Zeidler et al. (2009), for example, who have had significant influences on the nature of SSI education, summarize the approach this way:

> Central to this approach is the concerted effort to provide opportunities for students to reflect on issues in order to evaluate claims, analyze evidence, and assess multiple viewpoints regarding ethical issues on scientific topics through social interaction and discourse (p. 75).

On the one hand, there appear to be clear benefits of such argumentation-based approaches, including: development of students' *socioscientific reasoning skills* (Sadler et al., 2007); learning of products of science (e.g., laws & theories) (Venville & Dawson, 2010) and, learning about the nature of science (Khishfe & Lederman, 2006). On the other hand, there also may be various problems associated with such individualistic emphases on controversies. Students often are, for instance, placed in the role of *receivers* of knowledge from experts and/or people with power—creating a certain *dependency* on them. In analyses of socioscientific issues education approaches, Ralph Levinson (2010), for example, concluded that most prioritized 'Deficit' (citizens needing to be informed) or 'Deliberative' (citizens engaged in discussions with fellow citizens) models of citizenship—both of which he claims place many citizens in deference to experts and/or people with power. Such learning outcomes seem reasonable in *representative* democracies, in which citizens only periodically influence governing—as they, for example, participate in elections of leaders every few years (Wood, 1998). However, this can be problematic if leaders and/or experts do not provide problematic information and advice.

Although financial and other elite have long had significant power over large fractions of societies, such influences have, apparently, dramatically escalated in about the last 50 years. According to McQuaig and Brooks (2010), after the second world war and subsequent re-building programmes that involved major increases in government spending and intervention in economies, it seems that shares of wealth enjoyed by the richest 1% of the population dropped from a pre-war level of about 24% to a post-recovery level of 10%. This seems to have led them to resurrect traditional pre-war policies of economic liberalism—which advocated individuals' and groups' (e.g., corporations) rights to engage in economic markets free (liberated) from government intervention. Instead of freedom from government intervention, however, neoliberalism appears to favour government activism to promote policies and practices that facilitate private sector desires. For example, new transnational organizations, like the World Bank, World Trade Organization, International Monetary Fund and various 'think tanks,' were formed to represent economic and social elite—free from obligations to any country or its people. Apparently in cooperation with governments and transnational organizations, agreements—such as tax laws and international trade agreements—have been arranged in ways to favour for-profit activities by individuals and groups (Ball, 2012).

Apparently key to neoliberal successes are fields of science and technology (and, likely, engineering and mathematics) (Ziman, 2000). On the one hand, people are grateful to such fields for helping to improve conditions for many living things and nonliving environments. Humans enjoy longer lifespans, for instance, through developments in medical and agricultural knowledge and innovation. Nevertheless, there appear to be many significant potential problems for wellbeing of individuals, societies and environments associated with decisions made by powerful people and groups influencing fields of science and technology. Particularly in the neoliberal era, governments and transnational entities appear to have provided investors with many and varied opportunities to enter into contractual arrangements with scientists and engineers in ways that often seem to compromise the integrity of aspects of topic choice, methods of investigation, conclusions and extent and nature of dissemination (Mirowski, 2011). Such 'business-science partnerships' often, in turn, generate commodities that are linked to many personal, social and/or environmental harms. People are concerned, for instance, about various commercial products and services, such as: genetically-modified foods, etc. (Kleinman, 2003); household cleaning and hygiene products (Leonard, 2010); pesticides (Hileman, 1998); tobacco (Barnes, Hammond & Glantz 2006); and, pharmaceuticals (Angell, 2004). Many people also expect serious personal, social and environmental problems associated with dramatic increases in average global temperatures that often are linked to excessive fossil fuel uses (Klein, 2014). People wanting to address such hazards may, however, find it difficult to get valid information on which to make judgements about them. Financiers and others with vested interests in commercial products and services sometimes (or often) have, apparently, taken steps to cast doubt on research findings and claims from fields of science and engineering that would perhaps expose problematic aspects of for-profit commodities like those listed above (Oreskes & Conway, 2010). This claim suggests that many controversies have their source, not so much in data or theory (core entities of science) but, rather, in *ideology*. Regarding potential and realized climate change, about which there now is considerable corroborating data, there are critics of claims such as: 'Human activity has greatly increased carbon dioxide composition of atmosphere,' 'Humans must develop new technologies to address inevitable climate change' and 'Businesses should pay for damages they cause through their 'greenhouse gas' emissions.' Klein (2014) suggests that critiques about such claims often depend on one's level of support for persistent economic growth and wealth creation as compared to negative side-effects of such perspectives and practices.

17.2.2 Towards More Participatory Forms of Democracy

In light of often-problematic influences of capitalist entities on governments and fields of science and technology, there appear to be significant needs for science and technology educators (among others) to help generate citizens who may deeply analyze and evaluate knowledge production and dissemination systems and, where significant problems are identified, develop and implement plans of action to address them (Hodson, 2011). Using Levinson's (2010) conceptions of citizenship, this implies that students may, through 'Praxis' (e.g., self-led research), develop 'Dissent' (e.g., becoming critical of fields of science and technology and their relationships with members of societies and environments) and develop and engage in 'Conflict' by, for example, developing and enacting socio-political actions that may disrupt power relations in ways that may lead to a better world. In George Wood's (1998) terms, students/citizens would be involved in activities associated with more *participatory* forms of democracy—in which power is more widely-distributed across populations. In such societies, citizens are highly vigilant and activist.

Despite apparent merits of more participatory forms of socioscientific issues education, school science systems seem to have been very slow to adopt them. Promotion of many kinds of actions—such as petitions to power-brokers—to address issues, regardless of perceived causes (e.g., capitalist influences on techno-scientists), are rare in school science, as educators often find it easier to facilitate student personal decision-making about controversies (Hodson, 2011). There are, perhaps, numerous reasons to explain such reticence to promote activism. Aforementioned preferences for instruction and assessment of widely-accepted claims ('products') of fields of science and technology may be a significant factor. Perhaps associated with this are tendencies for school science systems to portray professional fields of science and technology as, despite periodic disputes linked to socioscientific issues, relatively efficient and unproblematic in generating such products. It is apparent, for instance, that school science systems routinely avoid reference to potentially problematic 'business-science partnerships' discussed above (Carter, 2005).

In our current era, a significant agent apparently inhibiting more critical and activist forms of science education is 'STEM' education. To call this a 'movement' is, likely, a misnomer. There appear to be many forms of and claimed purposes for 'it.' At its base, however, proponents often tout its emphases on integration and/or interrelationships among the four STEM fields; that is, among fields of science, technology, engineering and mathematics—which is said to be common within work among these professional fields (Rennie, Venville & Wallace 2012). Such relationships are said, not only to be 'authentic,' but also *essential* for societies' successes—particularly in commercial terms. Indeed, there appear to be *salvationary* tones to many STEM initiatives (Pierce, 2013). It is common, for instance, to read about proponents' claims that, without a concerted effort by school systems to generate more STEM graduates and corresponding workers, jurisdictions (e.g., cities, states/provinces, nations, etc.) will suffer hardships in the face of international economic competitiveness (Zollman, 2012). These kinds of assertions seem, to a great extent, reminiscent of Klein's (2007) suggestion that neoliberal capitalists often either take advantage of or create 'disasters' (natural or otherwise) that frequently destabilize societies in ways that make them amenable to furtherance of neoliberal policies and practices. Claims by STEM education proponents that their initiatives will lead to many more jobs and economic prosperity, for instance, seem to be somewhat of a ruse. Perhaps more realistically, science/STEM education seems mainly focused on identifying and educating relatively few STEM professionals (scientists, engineers, etc.; but, also, accountants, lawyers, business managers, etc.) who may develop and manage production and distribution of innovative for-profit products and services. This priority, indeed, seems evident in the highly influential new US science education standards document:

> The primary driver of the future economy and concomitant creation of jobs will be innovation, largely derived from advances in science and engineering.…4 percent of the nation's workforce is composed of scientists and engineers; this group disproportionately creates jobs for the other 96 percent (NRC, 2011, p. 2).

Such a vision of social stratification is not new. Nevertheless, divisions between rich and poor around the world appear destined for rapid increases (Piketty, 2014). Perhaps as worrisome, however, is that campaigns prioritizing wealth concentration—perhaps like many STEM education initiatives—also appear to bring with them relative lack of attention to wellbeing of many individuals, societies and environments. Such an ethic, indeed, seems congruent with corporations' legal rights to *externalize* many of their costs. To maximize profit, in other words, corporations are allowed to reduce costs for such 'ingredients' in their commodities as materials, labour (including benefits for health and retirement) and, as well, costs to address negative side-effects of their commodities (e.g., increased cancer rates linked to chemicals in foods, etc.) (Bakan, 2004). In terms of STEM education initiatives, such 'externalization' may be evident in many ways. However, it is apparent that many such initiatives studiously avoid reference to matters pertaining to problematic aspects of political economy like those addressed above (Gough, 2015). Related to this, there seems to be a concerted effort to continue to portray fields of science and engineering as immune to influences from government-capitalist complexes in ways that prioritize private profit (Zeidler, 2016). Broadly, in other words, it is apparent that much less emphasis has been placed on *ecojustice* concerns—many of which problematize fundamentals of current capitalist systems, including foci on continuous growth, competitiveness, individualized entrepreneurialism and perhaps less concern for matters of social justice and environmental sustainability (Martusewicz, Edmundson & Lupinacci, 2015). Associated with such psychological and social conditioning, Clayton Pierce (2013) suggests that much of STEM education discourse is aimed at subjectifying citizens (as *biopolitical subjects*) into roles as individually-competitive entrepreneurs, ready to enthusiastically participate in various economic interactions and transactions, often with relatively minimal concerns about potential negative side-effects on other living and nonliving things.

Challenging STEM education movements is not likely to be easy. They appear omnipotent and omnipresent, successfully rallying many important actants on a worldwide scale—including, for example, corporations, think tanks, banks, trade organizations (e.g., the World Trade Organization), financiers, pro-capitalist governments and universities—all aligned for promoting pro-capitalist education (Ball, 2012). It seems to be, in other words, a resilient Foucauldian (2008) *dispositif* (refer below); that is, a powerful aggregate of actants resistant to significant change or replacement. Given its hegemonic character, it may be that, rather than replacing it, reform movements need to work towards providing societies with alternative models of STEM education dispositifs that focus societies on efforts to promote benefits for wellbeing of individuals, societies and environments in participatory democracies (Hardt & Negri, 2009).

17.3 Research Context and Methods

17.3.1 Research Contexts

A pro-ecojustice alternative to STEM education programmes has been developing for nearly a decade. As its acronym suggests, the 'STEPWISE' framework prioritizes 'science and technology education promoting wellbeing for individuals, societies and environments.' A major emphasis of this approach is to encourage and enable students/citizens to 'spend' some of their educational wealth (e.g., STEM literacy), not just on improving their own living conditions, but also on actions to bring about a better world for other living things, societies and environments (to learn more about STEPWISE, refer to www.stepwiser.ca). Although the original framework is arranged as a tetrahedron to acknowledge reciprocal relationships among five learning domains (e.g., STSE and Skills Education), work with teachers since its inception in 2006 suggests that they have preferred a more linear version of the STEPWISE framework (Bencze & Carter, 2011), as indicated in Fig. 17.1.

In this approach, students are first provided with one or more sets (cycles) of 'apprenticeship' lessons and activities to help them gain expertise, confidence and motivation for eventually self-directing research-informed and negotiated action (RiNA) projects to address socioscientific problems of their choice. These apprenticeships are based on basic *constructivist* learning theory (Osborne & Wittrock 1985), encouraging teachers to first ask students to explore and express their pre-instructional views about STSE relationships and actions perhaps needed to address perceived harms for individuals, societies and environments. Afterwards, teachers are likely to teach students about STSE relationships to which they may not have been exposed, along with research-informed and negotiated action projects that have been conducted by others. These would then be followed by small-scale RiNA projects to address students' concerns regarding STSE relationships, with teacher assistance where students deem necessary. As indicated in Fig. 17.1, teachers may

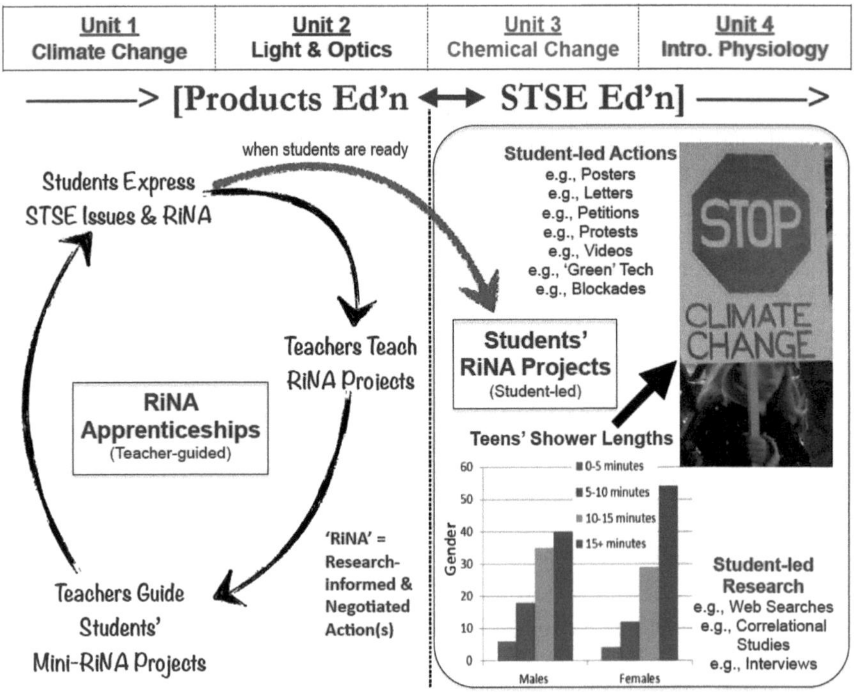

Fig. 17.1 STEPWISE Pedagogical framework

provide one or more such apprenticeship lessons and activities, depending on student needs, until the teacher believes that students likely have sufficient expertise, confidence and motivation to self-direct such projects.

Used in a variety of school, after-school and teacher education contexts since about January 2007, the approach illustrated in Fig. 17.1 seems to have enabled students (Krstovic, 2014), community youth (Sperling & Bencze, 2015) and student-teachers (Bencze & Sperling, 2012) to self-direct primary (e.g., correlational studies) and secondary (Internet searches) research as partial bases for developing and implementing personal and social actions to address socioscientific problems interesting them. It is, of course, impossible to accurately summarize the nature and extent of learners' projects. Nevertheless, we have generally found that, with sufficient support (e.g., apprenticeship and administrative & collegial consent) and motivation (e.g., grades or 'performances' [e.g., Parents' Night]), students have developed and often enacted a range of personal (e.g., more recycling, less use of bottled water), educational (e.g., posters, videos, Facebook™ posts) and political (e.g., letters to companies and government) actions—such as those illustrated in Fig. 17.2.

Particular successes promoting RiNA projects seemed realized when teachers helped students to use aspects of actor-network theory (Latour, 2005) to analyze SSIs and plan actions. Speaking for himself and on behalf of other students who had analyzed various commodities (e.g., deodorant, deicer, cologne, etc.), an

Sample Socio-political Actions on Socio-scientific Problems

A. Pamphlet summarizing secondary and primary research about risks/benefits of vaccination, distributed through a local community centre and the students' school library and main school office,

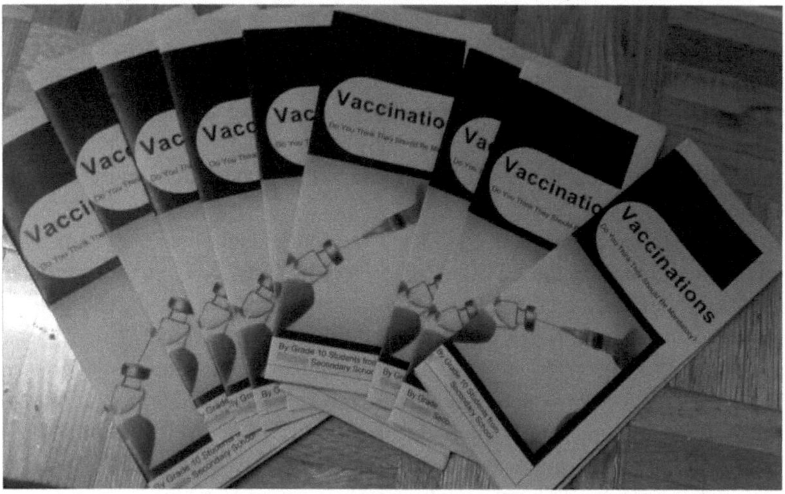

B. Letter sent to magazine read by teenaged girls, as a way of reaching girls in Catholic schools - where according to students' research, are not taught about birth control methods, despite using them.

Dear Vogue Magazine,

Our names are Marie LaMache and Kathy Leithwood; we are grade 10 students at Meadowlands Senior Public School in Canada. Our summative assignment for our science course consists of selecting a controversial issue that is related to human health, and developing a bias towards the issue. We must also conduct a study, (primary research), Learn more about our topic (secondary research) and propose an action plan to inform our community about what we have discovered.

The issue we chose is birth control, we wanted to choose a topic that is not typically talked about openly in a school environment so we could really learn about the issue and why it is avoided in schools. Also so we could inform ourselves and many other girls in difficult situations. After researching online we discovered that there are over 15 types of birth control, the four main types of birth control being: the pill, condoms (female and male), the patch, and vaginal rings. Since this was new and surprising information to us we based our study on surveying the general public about their knowledge and opinion of birth control. We went to a popular Mall in our area and surveyed 50 girls (13-19), 50 boys (13-19), 50 women (20 and up) and 50 men (20 and up), and asked the following questions: Gender?; Age?; Were you educated at a public school, catholic school or a different specialty school?; How many different types of birth control methods can you name?; Are you a supporter of Birth control methods (barrier, hormonal, emergency)?; Have you ever used any methods of birth control?; Have you been educated on different methods of birth control through school?; Are you religious? If yes, what religion?

After completing our survey and analyzing the data we collected, obvious correlations began to form, it was shocking to us that 74% of people that attended catholic schools did not learn about methods of Birth control in school. This surprising because if young women do not have a strong relationship with their parents and they do not learn this information in school, how are they preventing unwanted pregnancies and protecting themselves from STI's? We also learned that 93% of people support birth control; however, if the issue is so strongly supported why is it still an awkward topic to bring up? This is why we chose to write a letter to this magazine.

Vogue Magazine has the power to reach a much larger audience of young women then we ever could on our own. We think it would be a great article in your next issue to break down and simplify birth control in you health section, so girls can protect themselves and learn reliable information from a trusted source.
Sincerely,

Marie LaMache and Kathy Leithwood

Fig. 17.2 Common RiNA project action types

eleventh-grade student said: "[I]f you have a product and you are producing it in a Third World country, where you are not giving people the proper amount of pay and they are living in a low ['destitute'] place, then that society is not doing well – and that is partly due to your product" (Bencze & Krstovic, Chap. 10, this volume). Through development, partly based on Internet research, of actor-network maps about their respective commodities, students had *de-punctualized* them (where punctualization makes a network of actants seem like a single entity [Callon, 1991])—thus, exposing, for example, often-unacknowledged or hidden networks of problematic actants, such as poor workers labouring under deficient conditions in distant countries.

In reviewing RiNA projects conducted by youth/students and student-teachers since the beginnings of the STEPWISE project in early 2007, it became apparent that, although students had developed considerable expertise, confidence and motivation for self-directing RiNA projects to address problematic STSE relationships, projects tended to be relatively 'confined.' Frequently, for example, actions were aimed relatively locally—at fellow students, teachers/administrators, community members (broadly) and family members and friends. In that vein, although projects involving de-punctualization may have *anticipated* involvement of a breadth of actants, actions tended to be relatively narrow—aimed at a few of them at a time. Often, moreover, actions were relatively 'terminal'; that is, there often was little follow-up to determine effectiveness of actions. Such limitations may, for example, relate to confining characteristics of school. Teachers and students, for instance, often report that time to address large numbers of teaching/learning expectations in set amounts of time limit the extent to which students can develop and implement effective actions.

To educate students about other, perhaps more engaging and effective, forms of activism, we sought to analyze a 'real-life' (and ongoing) case of citizen participation in addressing potentially problematic dust accumulation in their community. Although some such cases—like citizen involvement in municipal water supply decisions (Roth, 2014)—have already been developed, citizen engagement in decisions involving fields of science and technology and their relationships with members of societies and environments often are highly complex, contextual and diverse and, therefore, it seems necessary to search for many and varied cases. A diversity of documentaries of such citizenship can benefit teachers and learners—providing them with more realistic choices, as seems congruent with democratic principles (Pierce, 2013).

In this chapter, we provide a documentary and analyses of an ongoing case involving deposits of grey-red dust in urban areas surrounding the Port of Québec City and citizen engagement in investigations into and actions regarding these deposits. The case outlined below began with actions of two citizens, Véronique Lalande and her partner, Louis Duchesne, who responded to a particular instance of dust accumulation in their neighbourhood, the borough of La Cité-Limoilou. Through their investigations, they concluded that the dust emanated from the city's port, where nickel ore is routinely received from mining areas and then shipped out to refineries. Examples of dust deposits in locations around the port are provided in

Fig. 17.3 Indications of dust accumulation in Québec City (Courtesy: www.vigilanceportdequebec.com)

Fig. 17.3. Limoilou is a 'downtown' neighbourhood of Québec City and is home to approximately 21% of the total population of the city. It ranks first among the city's boroughs for population density (Ville de Québec, 2015). Limoilou proper is, within this borough, the sector most affected by the dust in question. Comprising three neighbourhoods unto itself, Limoilou accounts for 2% (9.66 km^2) of the area of Québec City. According to statistics, 48% of inhabitants of the Limoilou sector are single-parent families and 46% of people aged 65 years and over live alone. Finally, the La Cité-Limoilou borough ranks third in the city for the number of immigrants. The Limoilou sector is semi-industrial, being bordered by the White Birch Paper™ mill, the city incinerator and the Port of Québec.

17.3.2 Data-Collection and Analyses Methods

As a study of a particular case (i.e., municipal dust deposits) with larger societal ramifications (e.g., citizen participation in socio-technical matters), our investigation resembles what Stake (2008) refers to as an *instrumental* case study. As such, it was necessary for us to collect broad and deep (meaningful) data to understand the case and, in doing so, to continually try to relate interpretations of them to situations, concepts and principles, etc. from larger contexts (e.g., global economic systems). In doing so, we believe this research should have *rationalistic* and *naturalistic* characteristics (Guba & Lincoln, 2011). From a rationalistic perspective, data were collected and analyzed pertaining, for example, to our contention that SSIs are composed of networks of living, nonliving and symbolic entities. More naturalistically, approaches common to ethnographic case studies were employed—with a view to attempting to react to unexpected phenomena as they arose. Accordingly, data types collected included:

- News media reports: Over 300 articles published in prominent Québec City and Montréal newspapers (*Le Soleil* and *Le Devoir*, among others) were collected and later analyzed. As well, about 20 extracts from televised news presentations about the dust situation—and possibly-related port expansion plans—in Québec City were analyzed.
- Activists' website contents: Text and graphics (videos and stills) appearing on the Initiative Citoyenne de Vigilance du Port de Québec website (www.vigilanceportdequebec.com), with material dating from July 2009 to present, were collected. This also was the source of photographs used in Fig. 17.3.
- Interviews of Citizens: Since 2013, citizens Véronique Lalande and her partner, Louis Duchesne, have been interviewed by Chantal Pouliot for a total of about 20 h of discussions. Lalande & Duchesne (L&D) are both well-educated. She studied law before switching to a Masters degree in Training Management. He has a Masters degree in ecology and directs forest ecology research for the provincial government. During conversations, L&D provided numerous insights into actions taken by them and other citizens regarding the dust situation, along with interpretations of actions and positions of various other stakeholders, including members of the port companies, port authority and city government. L&D also provided critical reviews of a book about the case written for citizens by Chantal Pouliot (2015).

From a naturalistic perspective, meanwhile, we also attempted to remain receptive to emergence of unexpected findings and conclusions. Indeed, citizen engagement in this situation continues and, so, new findings often emerge.

Regarding analyses, each of us coded data for categories and then developed encompassing themes—using constant comparative methods based on constructivist grounded theory (Charmaz, 2014). Categories and themes were then negotiated between us to achieve consensus (Wasser & Bresler, 1996). In doing so, as supporters of poststructural epistemological conceptions (e.g., Deleuze & Guattari, 1987), we were conscious that the case we construct may not perfectly represent the phenomena we studied—partly, for example, due to lack of comprehensive observation possibilities, varying and limited theoretical and ideological perspectives and, possibly, to limitations of human senses. Consequently, claims we make below must be seen as tentative, subject to ongoing study and debate.

17.4 Results and Discussion

17.4.1 A Dust-up

As with any report of material and semiotic phenomena, it is difficult to determine when problems and controversies surrounding accumulation of dust in Québec City neighbourhoods began. We only know that citizen concerns and government responses are ongoing. Nevertheless, as may become apparent from text below, a reasonable starting point for this documentary is October 2012. At that time, Véronique Lalande and her partner, Louis Duchesne, two citizens living in the Limoilou borough of Québec City, noticed significant accumulation of red-grey dust on their property and elsewhere in their neighbourhood (see Fig. 17.3). This prompted them to call local authorities to investigate the worrisome deposits. However, likely for complex reasons (e.g., not trusting governments), they sent samples of the dust to a private testing laboratory to determine its contents. Findings of these tests were alarming to L&D, suggesting—as indicated in Table 17.1—that air in their neighbourhood was a 'toxic soup,' comprised of potentially dangerous heavy metals, including: nickel, cobalt, arsenic, zinc, copper, barium, lead and cadmium. Moreover, the dust 'signature' (i.e., Nickel-Cobalt ratio) indicated it emanated from the city's port lands. At the same time, through secondary research, they determined that many other cities had lower metal content in air (Table 17.1) and that the nickel ore processed by the Québec City port emanated from a mine at Voisey's Bay (Newfoundland and Labrador) owned by Vale Corp. (www.vale.com) (Lalande & Duchesne, 2013a,b,c).

Armed with alarming data about dust accumulating in their environment, L&D proceeded to carry out several actions that they felt may lead to reductions in its dispersal or, ideally, its elimination. In the months following the beginning of the controversy, they produced a very detailed interactive website (www.vigilanceport-dequebec.com) (and associated page on Facebook™) that provides information (text, video and graphics) and commentary (e.g., in newsletters) about the case and opportunities for written feedback (blog posts). Over the last 3 years, they produced four analysis reports outlining their findings, concerns and recommendations (Lalande & Duchesne, 2013a,b,c, 2016). With increasing awareness of their data, concerns and actions, L&D also were invited to speak to students in various contexts, including: geography and chemistry classes at a local Cégep (post-secondary/pre-university institutions unique to Québec), university courses in science education and a student conference day in urbanism. They organized a major march (June 2013) (see Fig. 17.4) to remind politicians that they will not be satisfied until dust dispersal is eliminated, made public presentations about the situation, formulated two class action suits against Compagnie d'arrimage de Québec Ltée (the company processing the nickel ore) and the Québec Port administration (submitted in January 2013) asking, respectively, for compensation for initial damages from the October 2012 dust episode and asking for an injunction on dust emission from the port.

Table 17.1 Concentration en métaux lourds (mg/kg) dans le poussière de rue de Limoilou en comparaison avec diverses agglomérations urbaines

	Fer (Fe)	Arsenic (As)	Chrome (Cr)	Cobalt (Co)	Cuivre (Cu)	Étain (Sn)	Manganèse (Mn)	Nickel (Ni)	Plomb (Pb)	Vanadium (V)	Zinc (Zn)	Références
Limoilou, Canada 2012-10-26	170 000	9	50	26	644	38	333	474	118	49	998	
Ottawa, Canada	18 948	1	43	8	66	3	432	15	39	34	113	1
Hamilton, Canada					129				214		645	2
Sault Ste-Marie, Canada					81				64		193	3
New York, États-unis					355				2 583		1 811	4
Manchester, Angleterre	11 302				92		263		354		706	5
Birmingham, Angleterre					467			41	48		534	6
Londres, Angleterre	26				155				1 030		680	7
Oslo, Norvège	51			19	123		833	41	160		412	8
Madrid, Espagne	19			3	188		362	44	193	17	476	8
Hong Kong, Chine					173				181		1 450	9
Xi'an, Chine		11	167		95		687		230		424	10
Delhi, Inde			446		224			138	249		330	11
Buras, Turquie							521	121	486		121	12
Istanbul, Turquie					115		396	30	189		460	13
Seoul, Corée					101				245		296	14
Kuala Lumpur, Malaysie	1 790				36		153		2 460		344	15
Mutha, Jordanie	5 362				69		136	2	143		132	16
Karak, Jordanie	4 966		18	11	33		144	22	75		131	17
Concentrations naturelles		5	20	10	30	5		20	25	25	60	18

Fig. 17.4 Citizen march concerning Québec City dust pollution (http://ici.radio-canada.ca/regions/quebec/2013/06/02/003-marche-qualite-air-limoilou-citoyens-craintes-dimanche.shtml)

17.4.2 Rectifying Actions

Apparently in response to citizens' actions, city officials (including those governing the port and port company) took measures to address their concerns. These included purchase of water cannons that were used to dampen open piles of ore at the port, and washing of trucks leaving the port area. City streets also were more frequently cleaned to remove dust. Perhaps most significantly, the Québec provincial government and, more specifically, the Ministère du Développement Durable legislated changes to the allowable ambient nickel norm; that is, a *daily* (instead of yearly) quantity of nickel deposits permissible—a change in policy and practice that could reduce amounts of dust content in local air.

There are, likely, many ways to explain successes that Véronique Lalande and Louis Duchesne experienced in having authorities take actions to address what they perceived to be serious threats to citizens' health. One area of theory and research that has appeared to us to be useful in this context is that of *Socially Acute Questions* (SAQ; or, *Question Socialement Vive* [QSV] in French). These are public controversies that typically involve a broad range of actors (and actants)—often including experts and people with power and other citizens—that, due to the social 'acuteness,' become a subject of discussion and debate in educational contexts (Legardez & Simonneaux, 2006). In their investigations into the dust situation in their city, Lalande and Duchesne found that public consciousness and actions to address their concerns appeared to wax and wane over a several year period. From at least about 1970, many people in the city seemed to notice dust accumulating on various items (cars, windows, verandas). Which members of society were aware of the dust and had questions about it is difficult to fully know (e.g., from a poststructural position), but—as the 'story' (a construction) of the QC Dust case unfolds here, it is apparent that after some spikes in acuteness in the early 1980s, social consciousness seemed to subside until about 2009 when citizens of Beauport, a suburb of Québec City, complained to their city council about the dust (youtube.com/watch?v=9FjJJwj90m0) and, in 2011, when citizens of Cap Blanc, a borough of Québec City, organized and submitted a petition to the city asking for action to eliminate the dust. These citizen actions, however, seemed to generate little formal response from authorities. Actions initiated by L&D, on the other hand, seemed to have more substantive corrective responses (as noted above) from city officials. Again, while reasons for these differences in responses are somewhat uncertain, fluctuations in levels of 'social acuteness' seem pertinent. Perhaps, in light of needs for mobilization of awareness, understanding and concerns regarding the dust situation, we can turn to Michel Foucault's (2008) concept of *dispositif* (introduced above). It is apparent, in other words, that L&D's initiatives led to formation of a network of cooperating actants (living, nonliving & symbolic forms) that, together, may have increased collective public consciousness of the dust situation to the point that city officials found difficult to avoid. Our study of actions of Lalande and Duchesne and others note, for instance, significant media attention (e.g., refer to: www.vigilanceportdequebec.com) to their actions. Given its focus and goals, this activist dispositif may be called a 'heavy-metal band' (Interestingly, a Punk band sang about the situation, as heard here: youtube.com/watch?v=tzdl1eKwDdA). Using actor-network theory (Latour 2005), the 'loudness' of this band's messages may be understood in terms of a series of *translations*—in which ideas, perspectives, etc. about the dust situation move from one actant to another within the dispositif network, perhaps leading to somewhat common conceptions and motivations (e.g., need for actions to eliminate dust dispersal). One caveat with this idea is that, perhaps, due to gaps or inconsistencies in these translations (Roth, 2001; also refer below), conceptions and motivations may not be unanimous within the dispositif. Using Deleuze and Guattari's (1987) metaphor of the rhizome, translations often are unpredictable due, for example, to unforeseen events (e.g., citizen missing a news story). In this sense, we might relate formation of a dispositif similar to a *bricolage* (Roth, 2008); that is, arrangements that have more to do with contextual availability of actants than with purposeful arrangements – as we might expect if there was a 'conductor' (e.g., L&D) – of the many actants possibly involved.

17.4.3 Resistance from 'Developers'

Regardless of questions about the exact nature of any dispositif, further data from this study suggests that it is a useful concept. Although a dispositif appeared to be developed that raised the level of 'acuteness' (consciousness & understanding) of the nickel ore controversy that, in turn, led to some rectifying responses from city and private sector officials (as above), the extents of responses to the citizen dispositif were viewed in the eyes of activists as limited. In her letter, in which she resigned from the committee that included representatives of the City, the Public Health

Department, the Port and the company, Véronique Lalande wrote (translated from French) the following:

> [C]oncrete results in more than two years are slim [...] I will not repeat the observations made during the two annual reports. However, the recent announcement of a study on air quality, significantly far from requests made repeatedly by citizens and especially negotiated behind closed doors by three members of this committee will really sound the death knell regarding us. Our reading is severe and, as such, it seems important to put some facts in perspective. As a representative of the initiative, I am prepared to accept that all the members of this committee want to eventually solve the problem. Only for the most of them the air quality concepts, and atmospheric deposition of dust will remain theoretical concepts while they are our daily nightmare. [...] We come to the following conclusions:
>
>> Although known, simple and accessible, there is no will of the Port of Québec to engage freely in the only way that would solve the problem once and for all [...] For a variety of reasons, there [is] no political will, and therefore no will from the authorities to force the port to adopt these measures as essential to the actual development of a harmonious city-port relationship. Although they would like to see a settlement, the main motivation of the authorities and some agencies is first to protect their institution, not to expose even if it involves distortions of the truth, whether from assertions or more often from silence. We are announcing, you will have guessed by now, that Initiative Citoyenne de Vigilance du Port de Québec withdraws from the Vigilance des activités portuaires committee. We return from now, with serenity and I must say a huge relief, to our primary role: whistleblower. The very one that allowed an entire community out of ignorance and imagine an alternative (Nov. 20, 2015).

Indeed, in recent years, there has been considerable resistance from stakeholders with interests in promotion of handling of nickel ore at the Québec port—including the city government (e.g., politicians)—to more citizen involvement in decisions about the dust problem, often using various discursive strategies to depoliticize the issue and reduce citizen engagement. Indeed, the Mayor said the following to journalists:

> Québec City will totally support the Port of Québec in its projects of expansion of the port. I told him [the CEO of the Port …, "It is clear, we will not back down. I have confidence in the Port of Québec; I have confidence also to explain this to the population. We will fight any people wishing to decrease the value of the port or thwart its development efforts. We sincerely believe, sincerely, in the future of the maritime industry in Québec City, and the Mayor of Québec and the city of Québec will be behind you, count on us (November 20th, 2015).

Despite a relatively long history of dispersal from the port through the air and adherence to objects, it appears that public consciousness about the dust has waxed and waned. Although some progress has been made—largely, it seems, as a result of citizen activism and related emergence of influential data—in reducing dust emissions and augmenting quality standards regarding them, apparent resistance of those in support of port development to dramatic reductions in metal dust dispersal suggests that there may be—and may have been for some years—a concerted effort to minimize attention to the matter. Evidence for this claim is circumstantial. Nevertheless, official documents uncovered by Lalande and Duchesne are supportive of this possibility. In 2015, their research revealed existence of two official documents (from Pluram and Roche Consultants, in 1981 and 1983, respectively) that

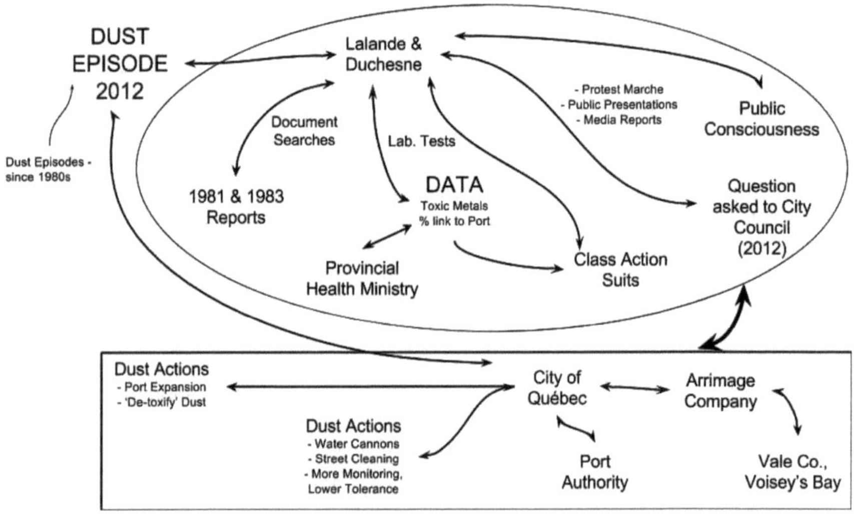

Fig. 17.5 Battle of the bands

indicated that the city and port authorities knew as early as 1981 about the dust emanation problem, but failed to act on it. This finding suggests to us that, perhaps, pro-development stakeholders could have, in effect, acted to align actants in ways that formed a *quieting dispositif*; that is, an aggregate of actants that served to more or less minimize—apparently between at least 1981 and 2009—the acuteness of the dust controversy. This pro-development dispositif may have exposed many citizens to unhealthy environments for the sake of profit. If such a dispositif exists, it suggests that, as depicted in Fig. 17.5, it may be in a kind of 'cultural clash' ('battle of the bands') with the activist dispositif—which, to a great extent, seems to have acted in support of ecojustice goals (refer above).

17.5 Summary and Conclusions

17.5.1 Educational Benefits of the Québec City Dust Controversy

The case study of ongoing dust deposits surrounding the Port of Québec City reported here appears to have excellent potential for those interested in educating students about realistic contexts of research-informed and negotiated actions to address perceived problems for wellbeing of individuals, societies and/or environments associated with influences of powerful people and/groups on fields of science and technology (and, likely, engineering and mathematics). The case indicates, for instance, that citizens can, particularly with assistance from professional laboratory services, generate data and findings that can help rally a range of actants to form an activist *dispositif* (Foucault, 2008) that can function collectively to achieve particular ends. Accordingly, it seems students should be encouraged to analyze STSE relationships in terms of actor-network theory and to network their actions; that is, to work to align several cooperating actants towards their cause. We can imagine some students, for example, after conducting secondary and primary research about fast foods, developing activist materials (e.g., posters, petitions, videos, websites, Twitter™ feeds, Facebook™ groups) and, to help mobilize facts and perspectives in them, interacting with various fast food 'stakeholders'—including people in their schools responsible for food supplies, local media, parent groups, managers of sports and recreational centres, teenager groups, local politicians, etc. With enough appropriate actants supporting their causes, they may initiate—like Lalande & Duchesne—a dispositif that realizes results they desire. In the Québec City dust case, it was apparent that two citizens took actions that resulted in alignment of a great range of actants—including: data concerning dust content and sources, many news reports, an interactive website, a citizen march, and class action suits—all, more or less, calling for reductions, if not full elimination, of dust dispersal from the city's port lands. In the sense that their actions were, largely, aimed at asking powerful members of their community to take rectifying actions, they were 'speaking [what they perceived to be] truth to power' (AFSC, 1955).

Based on experiences of activists in Québec City, students may also learn that an activist dispositif can be somewhat successful; in this case, with city and port officials choosing, for example, to: spray dust piles with water, clean streets more frequently and monitor the air for dust particles on a more regular basis. Such actions, moreover, can embolden citizens to gain a sense of agency in power relations. Véronique Lalande, who seemed so central to development of the activist dispositif in Québec City with regards to the apparent heavy metal dust contamination, made various statements indicating her increased vigilant-activist stance:

> My balcony hasn't changed since October 26. It's still covered with dust. But I have changed. I am no longer a passive bystander (May 6, 2013, Porter, 2013).

> We're fighting furiously for this – particularly to give people back the feeling of having some control over things by showing that science can be used from the citizen's perspective, without feeling intimidated (Sept. 29, 2013, personal email).

> I urge you to consider the message that rewarding a delinquent industry with more powers would send to thousands of men, women and children who live close to port facilities and lack the industry's resources to plead their case (Nov. 23, 2014, Deposition to Canadian House of Commons).

As Ms. Lalande's May 6, 2013 statement (above) indicates, however, she and other citizens of their community were not fully satisfied with officials' rectifying actions. There were still significant amounts of (potentially toxic metal) dust circulating in the Québec City air and, ultimately, the piles of ore at the port (see Fig. 17.3, above) were still open to the air. Moreover, city and port officials seemed intent on *expanding* transshipment capacities. However, perhaps most significantly, L&D uncovered existence of two reports about the dust situation commissioned by the city in the 1980s that, if more widely circulated and broadcast, may have led to more public consciousness and actions. That these reports were not made more public and part of community discourse suggests to us existence of a dispositif aimed at continued or, even, expanded port developments—and perhaps limited and/or oppositional acts affecting dust dispersal across the city. In other words, 'discovery' of the 1980s reports seemed to serve as an agent of *de-punctualization* (Callon, Lascoumes, & Barthe, 2001) of a phenomenon (dust), revealing a network of actants (perhaps a 'pro-development' dispositif) that could be serving to maintain a relatively low level of public consciousness of potential health problems linked to the dust dispersal. With further de-punctualization, citizens also may gain insights into existence of a larger network of actants linked to the dust, such as: port workers, miners, shipping workers, mine and ocean environments, transnational agreements pertaining to mining and manufacturing, etc.

As discussed earlier, possible existence of a pro-development dispositif appears to align with claims in Oreskes and Conway's (2010) book, *Merchants of Doubt*, that powerful people and groups may sometimes take actions to cast doubt on potentially incriminating science data and claims for the sake of profit. Also aligned with some (or many) capitalist activities, it suggests an ethic of *externalization* of costs—in this case, perhaps in the sense of health costs borne by citizens in relation to illnesses stemming from an airborne toxic stew of heavy metals. In the context of a STEPWISE-informed programme in science and/or technology education, the Québec City dust case would fit very well within the "Teachers Teach RiNA Projects" phase of the apprenticeship component of the pedagogical framework (Fig. 17.1). As in this article, particular attention should be paid in the apprenticeship to apparent struggles between two opposing dispositifs. Apparently driving alignment of actants to form 'activist' and 'pro-development' dispositifs were a different set of ideological perspectives—with, broadly, the former erring on the side of caution with regards to development and the latter enthusiastically supporting it, perhaps willing to sacrifice possible personal, social and/or environmental harms for the sake of economic growth and benefits associated with it, such as jobs and capital gains. In teaching about such contrasting dispositifs, teachers might focus on characteristics of research-informed and negotiated action (RiNA) projects evident in the project; that is, as given in Fig. 17.6 (Bencze, Carter & Krstovic, in press). Regarding effectiveness of translations between 'World' (e.g., dust accumulation) and 'Signs' (statements about nickel ore), students could learn about *ontological* and *ideological* gaps. The former kinds of gaps are inefficiencies in translations due to ontological differences between each entity. For example, it is apparent that geographical features (landforms in a country), which are aspects of the 'World,' cannot fully be represented by maps of them (Signs). This suggests that there always will be inconsistencies in such translations. In terms of *ideological* gaps, however, inefficiencies in translations may be—to varying degrees—*intentional*. Advertisers, for instance, often purposely idealize for-profit products and services (i.e., produce *idealized* Signs for commodities [World]) (Bencze et al., in

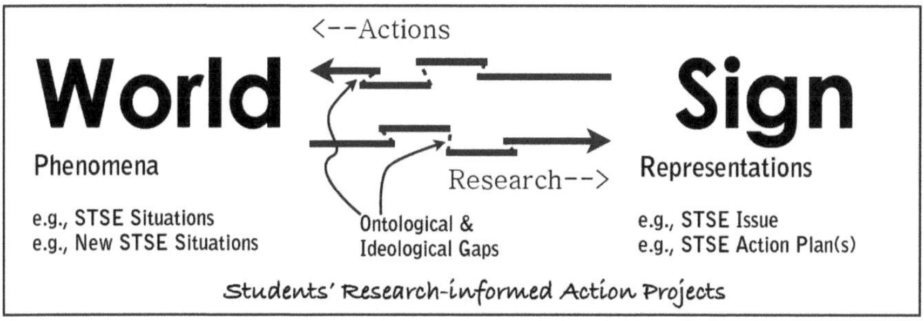

Fig. 17.6 Variations in ideological gaps in RiNA projects

press). Regarding the Québec City dust case described here, students could be given examples of how the two different dispositifs might engage in such translations on different ideological bases. For example, for World → Sign translations, activists might label dust to be 'toxic heavy metals'; whereas, pro-development individuals/groups may say that 'the dust is relatively harmless iron oxide.' Meanwhile, for Sign → World translations, activists might recommend costly installation of ore pile covers; whereas, pro-development individuals/groups may advocate for less-expensive periodic spraying of open-air ore piles with water cannons.

In the eyes of activists, individuals/groups promoting economic growth with less than desirable attention to wellbeing of many individuals, societies and/or environments may be considered *oppressors* (Freire, 1997). Educating students about potentially problematic power relations, as may be the case in pro-development dispositifs described here, may, therefore, represent a kind of *conscientization*—a critical consciousness about a (and/or one's own) social milieu (Freire, 1997). At the same time, educators in democracies may not want to be guilty of oppression, in the sense of providing students with mistranslations of 'real-world' documentaries like the one here—presenting pro-development individuals/groups in an unrealistically-bad light. It seems that no educator can avoid ontological gaps and, likely, ideological gaps. Accordingly, Paulo Freire (1997) recommended that, to be free of potential oppressors (including teachers), learners need to be given full control over 'praxis'; that is, critical, reflective, practice. Levinson (2010) echoes this call in his discussion of possible citizenship roles in the context of socioscientific issues education. This recommendation is, indeed, built into the STEPWISE framework—when students are encouraged to engage in student-directed and open-ended (when conclusions are not predetermined; but, rather, determined by learners in the context of experiences and their existing theory, etc.) RiNA projects (see Fig. 17.1). Moreover, there appears to be evidence to suggest that students may become relatively free of teacher influence in their RiNA projects, after having been provided with nurturing apprenticeship lessons and activities beforehand (Bencze & Alsop, 2014).

Acknowledgements We would like to thank Véronique Lalande and Louis Duchesne for their continued feedback on versions of the case study reported here.

References

American Friends Service Committee [AFSC]. (1955). *Speak truth to power: A Quaker search for an alternative to violence*. Philadelphia, PA: American Friends Service Committee.

Angell, M. (2004). *The truth about the drug companies: How they deceive us and what to do about it*. New York: Random House.

Bakan, J. (2004). *The corporation: The pathological pursuit of profit and power*. Toronto, ON: Viking.

Ball, S. J. (2012). *Global education inc.: New policy networks and the neo-liberal imaginary*. Abingdon, UK: Routledge.

Bell, R. L. (2006). Perusing Pandora's box: Exploring the what, when, and how of nature of science instruction. In L. B. Flick & N. G. Lederman (Eds.), *Scientific inquiry and nature of science: Implications for teaching, learning, and teacher education* (pp. 427–446). Dordrecht, The Netherlands: Springer.

Barnes, R. L., Hammond, S. K., & Glantz, S. A. (2006). The tobacco industry's role in the 16 Cities Study of Secondhand Tobacco Smoke: Do the data support the stated conclusions? *Environmental Health Perspectives, 114*(12), 1890–1897.

Bencze, J. L., & Alsop, S. (Eds.). (2014). *Activist science & technology education*. Dordrecht, The Netherlands: Springer.

Bencze, L., & Carter, L. (2011). Globalizing students acting for the common good. *Journal of Research in Science Teaching, 48*(6), 648–669.

Bencze, L., Carter, L., & Krstovic, M. (in press). Youth uses of actor-network theory undermining societal consumerism. In G. Reis, M. P. Mueller, R. A. Luther, L. Siveres, & R. Oliveira (Eds.), *Sociocultural perspectives on youth ethical consumerism* (pp. xx–xx). Dordrecht, The Netherlands: Springer.

Bencze, J. L., & Sperling, E. R. (2012). Student-teachers as advocates for student-led research-informed socioscientific activism. *Canadian Journal of Science, Mathematics, and Technology Education, 12*(1), 62–85.

Callon, M. (1991). Techno-economic networks and irreversibility. In J. Law (Ed.), *A sociology of monsters: Essays on power, technology and domination* (pp. 132–161). London: Routledge.

Callon, M., Lascoumes, P., & Barthe, Y. (2001). *Agir dans un monde incertain: Essai sur la démocratie technique*. Paris: Éditions du Seuil.

Carter, L. (2005). Globalisation and science education: Rethinking science education reforms. *Journal of Research in Science Teaching, 42*(5), 561–580.

Charmaz, K. (2014). *Constructing grounded theory* (2nd ed.). New York: Sage.

Deleuze, G., & Guattari, F. (1987). *A thousand plateaus: Capitalism and schizophrenia* (trans. Brian Massumi). Minneapolis: University of Minnesota Press.

Foucault, M. (2008). *The birth of biopolitics: Lectures at the Collége de France, 1978–1979* (Ed., M. Senellart). New York: Palgrave Macmillan.

Freire, P. (1997). *Pedagogy of the oppressed* (New Revised 20th-Anniversary ed.). New York: Continuum.

Gough, A. (2015). STEM policy and science education: Scientistic curriculum and sociopolitical silences. *Cultural Studies of Science Education, 10*(2), 445–458.

Guba, E. G., & Lincoln, Y. S. (2011). Competing paradigms in qualitative research. In N. K. Denzin & Y. S. Lincoln (Eds.), *The SAGE handbook of qualitative research* (pp. 105–117). Thousand Oaks, CA: Sage.

Hardt, M., & Negri, A. (2009). *Commonwealth*. Cambridge, MA: Belknap Press of Harvard University Press.
Hileman, B. (1998). Industry's privacy rights: Is science shortchanged? *Chemical & Engineering News, 76*(17), 36.
Hodson, D. (2011). *Looking to the future: Building a curriculum for social activism*. Rotterdam, The Netherlands: Sense.
Khishfe, R., & Lederman, N. G. (2006). Teaching nature of science within a controversial topic: Integrated versus nonintegrated. *Journal of Research in Science Teaching, 43*(4), 395–418.
Klein, N. (2007). *The shock doctrine: The rise of disaster capitalism*. New York: Henry Holt.
Klein, N. (2014). *This changes everything: Capitalism vs. the climate*. Toronto, ON: Simon & Schuster.
Kleinman, D. L. (2003). *Impure cultures: University biology and the world of commerce*. Madison, WI: University of Wisconsin Press.
Krstovic, M. (2014). Preparing students for self-directed research-informed actions on socioscientific issues. In L. Bencze & S. Alsop (Eds.), *Activist science & technology education* (pp. 399–417). Dordrecht, The Netherlands: Springer.
Lalande, V., & Duchesne, L. (2013a). *Analyse des concentrations de métaux dans l'air ambiant du territoire de Limoilou*. Initiative citoyenne de vigilance du Port de Québec, 20 pages. (disponible en ligne: vigilanceportdequebec.com).
Lalande, V., & Duchesne, L. (2013b). *Analyse de la provenance du nickel dans l'air ambiant du territoire de Limoilou en fonction de la direction des rafales de vent*. Initiative citoyenne de vigilance du Port de Québec, 4 pages (accessible en ligne : vigilanceportdequebec.com).
Lalande, V., & Duchesne, L. (2013c). *Analyse des concentrations de nickel dans l'air ambiant des quartiers Limoilou et Vanier (d'avril 2012 à mai 2013)*. Initiative citoyenne de vigilance du Port de Québec, 10 pages (accessible en ligne : vigilanceportdequebec.com).
Lalande, V., & Duchesne, L. (2016). *Collection de témoignages de citoyens victimes d'empoussièrement dans les quartiers centraux de la ville de Québec*. Initiative citoyenne de vigilance du Port de Québec, 26 pages (accessible en ligne : vigilanceportdequebec.com).
Latour, B. (2005). *Reassembling the social: An introduction to actor-network-theory*. Oxford, UK: Oxford University Press.
Legardez, A., & Simonneaux, L. (2006). L'école à l'épreuve de l'actualité – *Enseigner les questions vives*. ESF, Issy-les-Moulineaux.
Leonard, A. (2010). *The Story of Stuff: How our obsession with stuff is trashing the planet, our communities, and our health - and a vision for change*. New York: Free Press.
Levinson, R. (2010). Science education and democratic participation: An uneasy congruence? *Studies in Science Education, 46*(1), 69–119.
Martusewicz, R., Edmundson, J., & Lupinacci, J. (2015). *Ecojustice education: Toward diverse, democratic, and sustainable communities* (2nd ed.). New York: Routledge.
McQuaig, L., & Brooks, N. (2010). *The trouble with billionaires*. Toronto, ON: Viking.
Mirowski, P. (2011). *Science-mart: Privatizing American science*. Cambridge, MA: Harvard University Press.
National Research Council [NRC]. (2011). *Successful STEM education: Identifying effective approaches in science, technology, engineering, and mathematics*. Washington, DC: National Academies Press.
Oreskes, N., & Conway, E. (2010). *Merchants of doubt*. London: Bloomsbury Press.
Osborne, R., & Wittrock, M. (1985). The Generative Learning Model and its implications for science education. *Studies in Science Education, 12*, 59–87.
Pedretti, E., & Nazir, J. (2011). Currents in STSE education: Mapping a complex field, 40 years on. *Science Education, 95*(4), 601–626.
Pierce, C. (2013). *Education in the age of biocapitalism: Optimizing educational life for a flat world*. New York: Palgrave MacMillan.
Piketty, T. (2014). *Capital in the twenty-first century*. Cambridge, MA: Cambridge University Press.

Porter, I. (2013, May 6). Point chaud: L'Erin Brockovich de Limoilou. Le Devoir. Retrieved from http://www.ledevoir.com/politique/ville-de-quebec/377506/l-erin-brockovich-de-limoilou
Pouliot, C. (2015). *Quand les citoyen.ne.s soulèvent la poussière*. Montréal, QC: Carte blanche.
Rennie, L., Venville, G., & Wallace, J. (Eds.). (2012). *Integrating science, technology, engineering and mathematics: Issues, reflections and ways forward*. New York: Routledge.
Roth, W.-M. (2001). Learning science through technological design. *Journal of Research in Science Teaching, 38*(7), 768–790.
Roth, W. M. (2008). Bricolage, métissage, hybridity, heterogeneity, diaspora: Concepts for thinking science education in the 21st century. *Cultural Studies of Science Education, 3*(4), 891–916.
Roth, W.-M. (2014). From-within-the-event: A post-constructivist perspective on activism, ethics, and science education. In L. Bencze & S. Alsop (Eds.), *Activist science and technology education* (pp. 237–254). Dordrecht, The Netherlands: Springer.
Sadler, T. (Ed.). (2011). *Socio-scientific issues in the classroom: Teaching, learning and trends*. Dordrecht, The Netherlands: Springer.
Sadler, T. D., Barab, S. A., & Scott, B. (2007). What do students gain by engaging in socioscientific inquiry? *Research in Science Education, 37*(4), 371–391.
Sperling, E., & Bencze, L. (2015). Reimagining non-formal science education: A case of ecojustice-oriented citizenship education. *Canadian Journal of Science, Mathematics, and Technology Education, 15*(3), 261–275.
Stake, R. E. (2008). Qualitative case studies. In N. K. Denzin & Y. S. Lincoln (Eds.), *Strategies of qualitative inquiry* (pp. 119–149). Los Angeles: Sage.
Venville, G. J., & Dawson, V. M. (2010). The impact of an argumentation intervention on Grade 10 students' conceptual understanding of genetics. *Journal of Research in Science Teaching, 48*(8), 952–977.
Ville de Québec. (2015). La Cité-Limoilou. Retrieved from http://www.ville.quebec.qc.ca/apropos/portrait/arrondissements/lacitelimoilou/index.aspx
Wasser, J. D., & Bresler, L. (1996). Working in the interpretive zone: Conceptualizing collaboration in qualitative research teams. *Educational Researcher, 25*(5), 5–15.
Wood, G. H. (1998). Democracy and the curriculum. In L. E. Beyer & M. W. Apple (Eds.), *The curriculum: Problems, politics and possibilities* (pp. 177–198). Albany, NY: SUNY Press.
Zeidler, D. L. (2016). STEM education: A deficit framework for the twenty first century? A sociocultural socioscientific response. *Cultural Studies of Science Education, 11*(1), 11–26.
Zeidler, D. L., Sadler, T. D., Applebaum, S., & Callahan, B. E. (2009). Advancing reflective judgement through socioscientific issues. *Journal of Research in Science Teaching, 46*(1), 74–101.
Ziman, J. (1984). *An introduction to science studies: The philosophical and social aspects of science and technology*. Cambridge, MA: Cambridge University Press.
Ziman, J. (2000). *Real science: What it is, and what it means*. Cambridge, MA: Cambridge University Press.
Zollman, A. (2012). Learning for STEM literacy: STEM literacy for learning. *School Science and Mathematics, 112*(1), 12–19.

Chapter 18
Supporting Pre-service Teachers to Teach for Citizenship in the Context of STSE Issues

Sarah El Halwany, Majd Zouda, Chantal Pouliot, and Larry Bencze

18.1 Introduction

Science education for citizenship appears to underlie current conceptualizations of scientific literacy (Roth & Désautels 2004). Science education for citizenship implies a shift from the traditional role of school science concerned with preparing "insiders who identify with the subculture of science" (Aikenhead, 2006, p. 3) — that is, scientists and engineers — to a view of science education concerned with forming *competent outsiders* (Feinstein, 2011) who engage in science given its usefulness to their everyday life.

In a similar vein, Roberts (2011) outlined two visions for science education. Vision I is targeted towards acquiring science knowledge and skills to facilitate students' entry into science-related fields while vision II defines scientific literacy in terms of its relevance to students/citizens. For instance, a vision I perspective would teach students about the circulatory system while a vision II would teach them how to identify and act in case of a hemorrhage (Fourrez, 1997). Recently, Jesper Sjostrom, Ingo Eilks and Vania Zuin (2016) proposed a vision III oriented towards critical perspectives on technoscience fields coupled with activist dispositions for social and environmental justice. One main difference between vision II and vision III seems to be levels of politicization of students' education. Vision II may be more interested in forming competent outsiders for personal and humanist reasons while

S. El Halwany (✉) • M. Zouda • L. Bencze
Ontario Institute for Studies in Education (OISE), University of Toronto, Toronto, ON, Canada
e-mail: sarah.elhalwany@mail.utoronto.ca; majd.zouda@mail.utoronto.ca; larry.bencze@utoronto.ca

C. Pouliot
Faculté des sciences de l'éducation, Université Laval, Québec City, QC, Canada
e-mail: chantal.pouliot@fse.ulaval.ca

L. Bencze (ed.), *Science and Technology Education Promoting Wellbeing for Individuals, Societies and Environments*, Cultural Studies of Science Education 14,
DOI 10.1007/978-3-319-55505-8_18

vision III is, perhaps, more concerned with forming competent outsiders willing to engage in socio-political actions.

Understandably, for students to engage in socio-political activism on issues related to Science, Technology, Society and the Environment (STSE) — such as processed foods, pharmaceutical drugs — they need to acquire informed science bases, along with skills to critically reflect on STSE issues. Students able to critically reflect on STSE issues are likely to exhibit aspects of socio-scientific reasoning (Sadler, Barab, & Scott, 2007) that consists of: "recognizing complexity of SSI issues, examining issues from multiple perspectives, appreciating that SSI are subject to ongoing inquiry and exhibiting skepticism when presented with potentially biased information" (p. 374).

However, merely developing students' socio-scientific reasoning and allowing them to form their own personal positions on STSE issues "will not bring about social justice and will not save the planet" (Hodson, 2010, p. 201). Rather, classroom practices need to be reworked to nurture more critical and activist dispositions on STSE issues, in line with a vision III of science education (Santos, 2009).

To prepare students for participatory forms of citizenship, science teachers need to be equally supported in reworking their pedagogical practices. In this chapter, we explore extents to which a teacher training course — which was structured around the STEPWISE framework — influenced pre-service teachers' pedagogical dispositions to teach for socio-political activism regarding STSE issues.

18.2 STSE Education and Democratic Participation

STSE issues may be considered controversies or problems stemming from interactions among fields of science and technology and societies and environments. In the Canadian educational context, socio-scientific issues are discussed as 'STSE' issues (Pedretti & Nazir, 2011). Traditionally, science and technology fields have been perceived as fundamentally different and unrelated (Gardner, 1999). It was believed that, while science observes phenomena of the world and generates representations of them (e.g. falling object → conceptions of gravity), technology translates representations into phenomena of the world (e.g. conceptions of gravity → airplane). More recently, it has been suggested that science and technology fields involve cyclical translations between phenomena and representations of them and that, rather than seeing them as separate, students need to see them as mutually constituted (Roth, 2001). The term *technoscience* (Sismondo, 2008) has been coined to characterize this hybridity.

Describing relationships among science, technology, societies and environments, Ziman (1987) noted two variations for depicting those relationships that he termed internal and external sociology of science. An internal sociology of science refers to workings of science taking place independent of the society and where scientific knowledge is generated for "its own sake without any thought for its possible applications" (Ziman, p. 4). An external sociology of science, meanwhile, deals with technological effects of scientific knowledge where theoretical knowledge from the 'pure' sciences is used by technologists and engineers to produce 'inventions,' which are then used by society. Scientific processes remain hidden from public scrutiny and the focus is rather on the 'instrumental' capabilities of science (e.g. for military, economic reasons). Internal and external sociologies of science may reinforce a linear depiction of relationships among science, technology and society. This linear, unidirectional, relationship is often transmitted through school science, subtly positioning science as authoritative body of knowledge and de-valuing technology as a field of 'applied' science (Gardner, 1999). When scientists are represented as epistemologically superior, citizens become excluded from participating in knowledge production and managing STSE issues, limiting their roles to mere consumers of scientific and technological products (Rudolph, 2005). In contrast, transmitting views that scientists' work is heavily influenced by their own subjectivities, assumptions, and even vested interests of stakeholders (Venturini, 2010) would 'humanize' the field, possibly paving the way for students/citizens to see their potential role as knowledge producers and decision-makers on STSE issues.

Inviting students, as present political subjects (Alsop & Bencze, 2014), to take socio-political actions on STSE issues is believed to increase their political efficacy as citizens of participatory democracies (Hodson, 1999). Wood (1998) defines participatory democracy as a form of governance whereby citizens take a self-governing role in managing public issues (e.g. STSE issues). Consistent with the words of Dewey (cited in Wood, 1998), this democracy "is more than a form of government; it is primarily a mode of associated living, of conjoint communicated experiences" (p. 87) in which decisions are made by those who will be directly affected by the decision. On the other hand, in representative democracies, citizens' participation in their own governance is limited to electing representatives who take decisions *for* them.

Participatory citizenship through science education might be achieved through models such as "science education as praxis" and "science education for dissent and conflict" (Levinson, 2010). Both of those models foreground learning that is: relevant to students and their larger community, embedded in an eco-reflexive stance towards STSE issues (Sjostrom et al., 2016) and culminating in collective actions for social and environmental justice (Santos, 2009). The teacher ceases to be "epistemologically privileged" (Levinson, 2010, p. 83) and scientific knowledge, among other bodies of knowledge, is continuously sought in the situated activities of learners (Roth & Barton, 2004).

If schools were to become sites of knowledge production (praxis) for social change (dissent and conflict) (Freire, 1970), roles of teachers need to be re-conceptualised from teacher-technician, concerned with implementing prescriptive 'teacher-proof' curricula, to teachers as political agents (Carlone, Haun-Frank & Kimmel, 2010). Rooted in traditions of participatory democracies, politicized science teachers are those who make salient, through their practice, issues of power and social inequities and act to politicize their students as knowledge producers and change agents on STSE issues (Alsop & Bencze, 2014). Amidst a fairly 'apolitical' science education climate that privileges content learning (Bell, 2006) over equally important knowledge domains of scientific literacy, namely nature of science, authentic inquiry and sociopolitical activism (Hodson, 2011), science teachers as political agents are those who teach *against the grain* (Cochran-Smith, 1991) to "promote a certain type of citizenship and civic responsibility of which transformation, agency and emancipation are key features" (Pedretti & Nazir, 2011, p. 617).

18.3 Teaching for Democratic Participation: Some Challenges

Science teachers who are conscious of their roles in preparing critical thinkers and active citizens are likely to detract from controlling students' learning and from teaching them exclusively for the test. Those teachers are likely to favour student-directed, open-ended science inquiries followed by greater citizen involvement in issues related to wellbeing of individuals, societies and environments (Krstovic, 2014).

Incorporating student-centered and activist pedagogies into one's repertoire of teaching is likely, however, to be faced with personal as well as structural barriers. Some teachers may find comfort in the predictability of familiar teaching strategies, leading them to avoid unknown and risky pedagogical approaches (Sutton & Wheatley, 2003). Moreover, parents and society in general may ascribe to conservative views of science education — expecting classrooms to be a place where students accumulate scientific 'facts' (Lakin & Wellington, 1994). Students themselves might be overwhelmed by demands of student-centered learning approaches, perhaps, as a result of low exposure to open-ended and student-led investigations (Reis, 2014). Also, overcrowded curricula would deter teachers from implementing student-centered activities which could take away from 'instruction time' (Reis, 2014). Those challenges and others are likely to exacerbate the task for pre-service teachers who might resist teaching *against the grain*, especially in the early years of their careers (Carlone et al., 2010). Not only are novice teachers unwilling to position themselves as outsiders to their communities of practice, their initial experiences teaching in formal classroom settings might further constrain their agency as teachers who might, otherwise, teach in non-standard ways and/or for uncommon purposes (Hodson, 2009). For instance, Darren Hoeg and Larry Bencze (2014) found that while pre-service teachers' pre-practicum orientations and beliefs were conducive to implementation of socio-political activism on STSE issues, their post-practicum orientations demonstrated more skepticism in their ability and desire to implement such atypical practices. The authors discussed this misalignment in pre- and post-practicum experiences in relation to possible influences of a hegemonic school culture embodied by the associate teacher. This apparently hegemonic culture, which prioritises content teaching, is believed to be co-extensive with neoliberal structures that position science knowledge as a commodity to be consumed by students (Bencze & Carter, 2011). In light of those challenges facing pre-serving teachers to incorporate sociopolitical activism in their future teaching, it would be more feasible to prepare beginner teachers to be able to teach *within and against the system* (Campano, 2007). Indeed, reforms in teacher education programmes are not to be viewed as static, resulting in 'teachers as finished products' (Brickhouse & Bodner, 1992); but, rather, as 'ongoing accomplishments steeped in both historical traditions and innovation' (Roth as cited in Carlone et al., 2010). In other words, if teacher training programmes were to prepare novice teachers to teach against the grain, in this case preparing teachers for sociopolitical activism, teacher education courses may need to support student-teachers to navigate tensions between institutional/political structures and agency (Carlone et al., 2010).

18.4 The Context of the Study

In this chapter, we explore extents to which a teacher education course, entitled *Science for Democracy*, which was structured around the STEPWISE pedagogical framework (see Fig. 18.1), might prepare student-teachers to infuse socio-political activism in their future practice. The general aim of STEPWISE is to "encourage and enable students to apply their science and technology education, including their

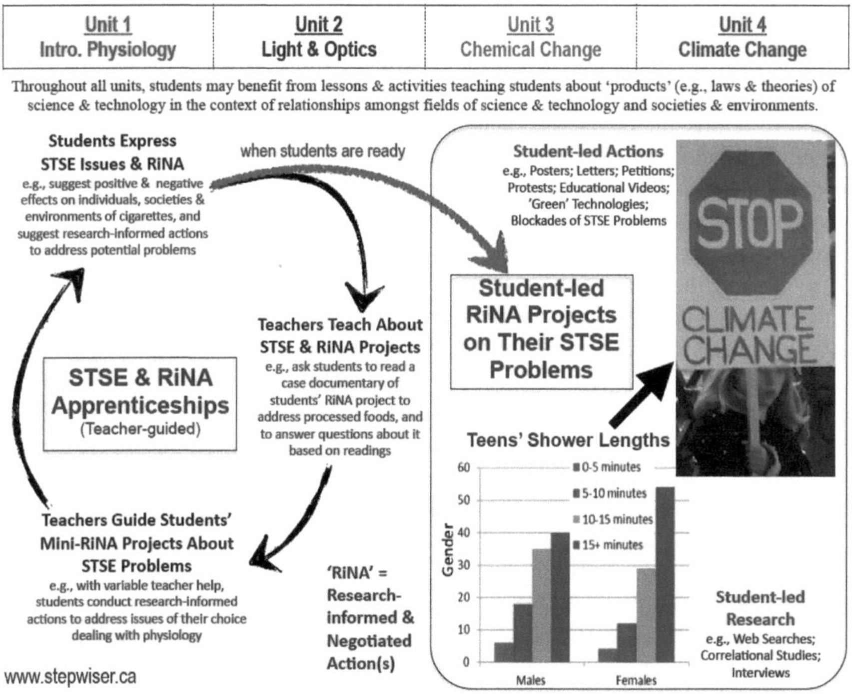

Fig. 18.1 STEPWISE pedagogical framework

primary and secondary research findings, to take socio-political actions to address STSE issues" (www.stepwiser.ca). Actions could include developing and distributing informational posters, petitions, letters and/or videos targeted to powerful people and groups that may assist in reducing environmental and social problems due to techno-science applications.

The STEPWISE framework can be operationalized through a pedagogical approach that engages students in Research-informed and Negotiated Action (RiNA) projects on STSE issues. In the context of this pre-service teacher education course, student-teachers led their own RiNA projects in conjunction with pedagogical learning about RiNA. Prior to the RiNA project, the course typically started with an apprenticeship phase, which is mostly teacher-directed. In our study, the course instructor (fourth author) and his TA (second author) initiated the RiNA apprenticeship by allowing student-teachers to: 1) *reflect on and express their preconceptions* about an STSE issue, 2) *learn about* RiNA to stimulate discussions on powerful stakeholders related to an STSE issue and 3) *engage in mini*-RiNA projects. The apprenticeship is intended for participants to consider various types of actions to address potential harms to wellbeing of individuals, societies and environments (WISE) due to business-government-technoscience partnerships (Venturini, 2010). As its name suggests, RiNA stresses needs for research to inform such actions. To model RiNA projects, the course instructor showed a video (www.youtube.com/watch?v=WhN6PS1GT9c) discussing an STSE issue surrounding makeup and liquid foundation. This video was developed by high school students as one possible form of action to raise awareness on problems associated with this industry. Showing student-teachers successful examples of student/citizen activism would give some sort of visibility that might encourage them to see activism as a possibility in their lives. In return, the course instructor was pointing to pedagogical significance of using such videos to sensitize students to research-informed activism. The instructor also made available to student-teachers in this course an array of similar downloadable resources to support them in implementing STSE activism in their prospective teaching.

Following the apprenticeship phase, student-teachers self-led secondary and primary research projects (in the form of correlational studies) working in groups and planned appropriate actions related to an STSE issue of their choice.

This course was offered weekly (4 h/week) during the fall semester (Sep.-Dec., 2014). The objectives of the course were twofold: 1) To develop student-teachers' critical awareness about potential problems linked to decisions about science and technology made by powerful individuals and groups, followed by designing RiNA projects to address problems of their choice and 2) To enable and support student-teachers in implementing research-informed and negotiated activism on STSE issues in their future practice.

With regards to the first objective, we believe that allowing student-teachers to experience a self-led RiNA project first-hand, in contrast to merely 'telling' them about RiNA, might be a powerful way to support them as active citizens and as politicized teachers. Derek Hodson (1999) appears to concur when saying that, "if teachers are to politicize students, prepare them for and engage them in sociopolitical

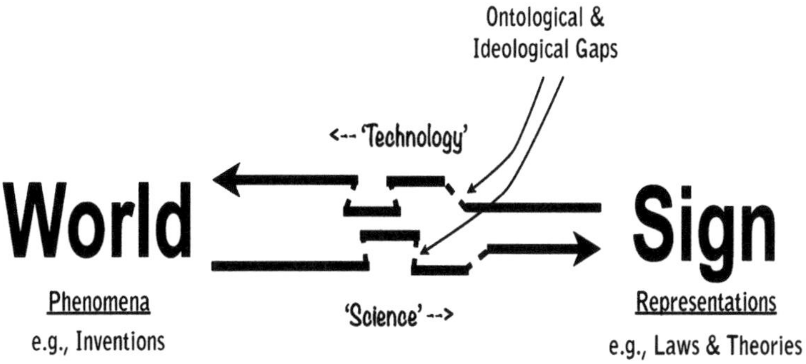

Fig. 18.2 Science and technology interrelated model (Roth, 2001)

action (…), they, too, must be politicized. They also must have been prepared for, and have engaged in, sociopolitical action" (p. 793).

Based on Roth's (2001) science and technology interrelated model (Fig. 18.2), the more individuals have control over dialectical relationships between the World (e.g. technological inventions) and representations (Signs) of them (e.g., science laws and theories), the deeper and more committed may be their learning. If we were to adapt Roth's model to teacher education programmes and, more particularly, to the objectives of this course, developing teachers' pedagogical expertise and commitment to teaching for socio-political activism might be limited if the course instructor solely presented future teachers with pedagogical models and theories about 'activist' teaching (Signs) without equally emphasizing reciprocal relationships (Sign ←→ World). In the case of the course *Science for Democracy*, student-teachers had opportunities to experience a double layering of Sign ←→ World transitions. For the first layer, by conducting their own RiNA projects, student-teachers transformed their own developed representations (e.g., their primary research) into the world (as actions). For the second layer of Sign-World translation, the pre-service course required them to design their own lesson plans about RiNA (as a Sign) and implement it in an actual classroom setting (World). Because the course was designed to support parallel dialectical Sign←→ World translations, it becomes relevant to explore extents to which student-teachers might have developed deeper and more meaningful connection with RiNA on STSE issues. Indeed, previous research suggests that when students-teachers are given opportunities to self-direct their own RiNAs, they develop greater emotional attachment and commitment to implement RiNA in their future practice (Bencze & Sperling, 2012). Our study attempts to further characterize relative development of those pedagogical dispositions.

Consistent with the course's second objective, we define pedagogical dispositions in terms of student-teachers' level of confidence, motivation and expertise to implement RiNA in their future teaching. The course might have further developed student-teachers' pedagogical dispositions to adopt RiNA (second objective of the course) by enabling them to develop their own pedagogical resources based on their

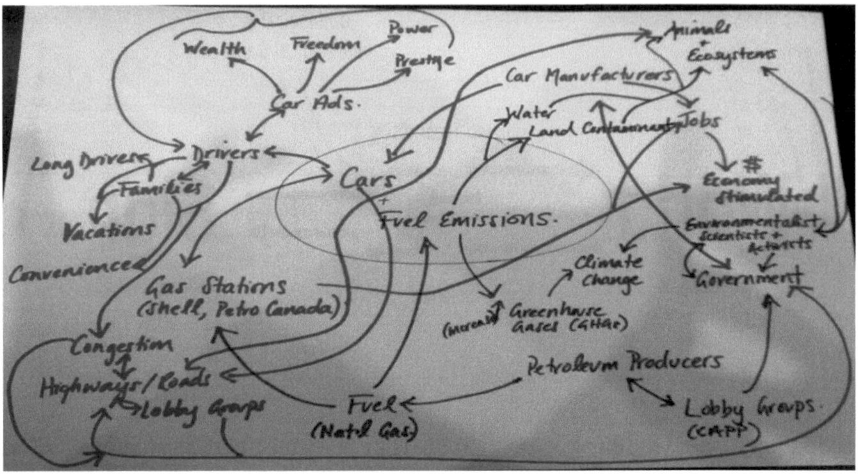

Fig. 18.3 An actor-network map (Venturini (2010); Drawn by student-teachers on the STSE issue related to cars)

own RiNA projects, possibly resulting in greater self-identification with their practice. Moreover, and towards the end of the course, student-teachers were given the opportunity to practise teaching 'against the grain'. Working in pairs and assessed by their peers, student-teachers introduced their RiNA project to a group of grade 9 students at a public high school in Toronto. This short episode of teaching (approximately 30 min), henceforth referred to as microteaching, consisted — to varying degrees — of student-teachers following three pedagogical components: 1) eliciting grade 9 students' existing conceptions about an STSE issue of their choice; 2) introducing Actor-Network Theory (ANT) (Latour, 2005), whereby actants (both semiotic and material) related to an STSE issue are interconnected in complex and dynamic ways (see Fig. 18.3 for an example of an ANT drawn by student-teachers investigating the STSE issue of cars); and, 3) making use of their own RiNA project to incite grade 9 students to consider sociopolitical activism as a rightful/democratic option for citizen involvement in knowledge production and decision-making on STSE issues.

Engaging student-teachers in this episode of microteaching, albeit short in duration, was planned to help boost their confidence, motivation and expertise to implement RiNA in the future by reinforcing translations from Sign to World (Fig. 18.2), as described above. Moreover, terms such as undemocratic education (used in the context of this course to refer to an education that does not convey to students an accurate/complete representation of STSE relations), citizen participation in knowledge production and decision-making were repeatedly used by the course instructor and his assistant hoping to extend student-teachers' orientation towards not merely teaching content knowledge but also teaching for social justice and democratic participation.

Since, this chapter is mainly concerned with research regarding support for student-teachers in implementing RiNA in their prospective teaching (second objective of the course), discussions of findings related to the course's first objective are only highlighted whenever they were found to contribute to increasing student-teachers' pedagogical confidence, motivation and expertise to implement RiNA on STSE issues.

18.5 Research Methodology and Methods

To account for how the pre-service course might have motivated and supported student-teachers to use RiNA in their future classrooms, we begin by exploring their perspectives on whether school science should prepare students to engage in sociopolitical activism on STSE issues. Then, we describe extents to which student-teachers perceived that the course helped them to develop their confidence, motivation and expertise to teach for sociopolitical activism on STSE issues.

Fourteen participants (50% of class size) consented to participate in this study. All participants were pre-service teachers studying at a university in Toronto, Canada. They were enrolled in the pre-service course '*Science for Democracy*,' which is an elective course, open to student-teachers from various educational backgrounds and teaching qualifications. Given space limitation and for more in-depth analyses, we focused our study on differential learning outcomes that emerged among three main clusters of teacher candidates. Aligned with constructivist leaning theory, individual student-teachers came into this elective course with differing educational backgrounds and research experiences that appeared to have an influence on how they would interact with the knowledge presented to them during the course. With regards to participants' level of involvement in previous research and their educational degrees, three main clusters emerged: (1) Science degree and research background, (2) Science degree/limited research background, (3) Non-science degree/no research background. Those with science and research backgrounds included some with an advanced university science degree or science-related field and some research experience in *fully* self-directing research (i.e., full control over design, data collection, analysis and conclusions). Student-teachers with a science background and limited research experience, included those who have a university science (or science-related) degree and those who had *partial* control over a previous research (either collecting or analyzing data). Finally, our last cluster of student-teachers comprised those with no science degree nor significant post-secondary research background. It is important to note here that those categorizations are not discontinuous and variations within-clusters are evident in our data. However, we chose to organize findings according to those three clusters because patterns in student-teachers' overall learning in this course emerged in relation to participants' research experience and science background.

Table 18.1 represents a summary of the three main clusters of student-teachers and of the data sources used to analyze their levels of pedagogical confidence,

Table 18.1 Description of the study sample on the three clusters

Cluster 1: Science and research background	Highest degree earned	Research experience	Data sources
Max	PhD (2nd year, population dynamics)	Theoretical ecology	STP+ 3 interviews + Teaching RiNA and views on citizenship survey + Reflections on micro-teaching
Luke	PhD (cancer research)	Cancer research	STP+ 3 interviews + Teaching RiNA and views on citizenship survey + Reflections on micro-teaching
Ben	PhD (Immunology)	Immunology	1 interview
Mary	PhD (evolutionary biology)	Fungal biology	STP+ 1 interview + Teaching RiNA and views on citizenship survey + Reflections on micro-teaching
Cheryl	B.S and M.A education	Social studies	STP+ Teaching RiNA and views on citizenship survey + Reflections on micro-teaching
Alan	B.S Neuroscience	Gender studies	1 interview
Cluster 2: Science degree/ limited research background			
Blake	B.S Physics	Limited	STP+ 2 interviews + Teaching RiNA and views on citizenship survey + Reflections on micro-teaching
Ava	Bachelor in Physical and health education	Limited	STP+ 3 interviews + Teaching RiNA and views on citizenship survey + Reflections on micro-teaching
Layla	M.S Biochemistry	Limited	STP + Reflections on micro-teaching
Hannah	Bachelor in Mechanical engineering	Limited	STP+ 2 interviews + Teaching RiNA and views on citizenship survey + Reflections on micro-teaching
Leo	B.S chemistry and biology	Limited	STP+ 1 interview + Reflections on micro-teaching
Julie	B.S Biology	Limited	1 interview
Cluster 3: Non-science degree/no research background			
Alexander	Hospitality services	None	STP+ Teaching RiNA and views on citizenship survey + Reflections on micro-teaching
Lewis	Hospitality services	None	STP+ 3 interviews + Teaching RiNA and views on citizenship survey + Reflections on micro-teaching

All names are pseudonyms

motivation and expertise to use RiNA. Note that data sources for all fourteen participants are not the same given their various degrees of consent to participate in aspects of this study.

We analyzed participants' learning experiences based on the following five data sources:

1) Cathleen Loving's (1991) *Scientific Theory Profile* (STP) was used with student-teachers before and after the course to determine whether their views about knowledge trustworthiness and knowledge production changed over the period of the course. The STP is composed of a grid with the x-axis referring to methods for judging theory. Philosophers on the Rationalist side of the x axis view that science is a rational process and that scientists make decisions about their work in systematic and logical ways (despite that they might acknowledge outside influences, those philosophers say that scientists are able to reduce outside influences to work in objective ways). Philosophers on the Naturalist side view that science theories are influenced by inside/outside factors (gender, social, economic, cultural, etc.). The y axis relates to the truth value of scientific theories (or the extent to which scientists are able to achieve the truth). On the Realist side are those philosophers who say that scientists develop claims that match phenomena while the Anti-realists think that scientists can never reach absolute truth and that there will always be uncertainties and doubts about any claim. Between the Rationalist and the Anti-realists are *Instrumentalists* who say that scientists can develop laws and theories that serve as tools or instruments that work for the time being but do not necessarily represent the truth. The course provided pre-service teachers with many opportunities to reflect on nature of science, including but not limited to: designing and conducting correlational studies (primary research), classroom discussions surrounding critical STSE issues and playing a NOS card exchange game (Cobern & Loving, 2002). In this game, student-teachers convinced their peers to trade cards containing single statements about which they least agreed (e.g. 'scientific knowledge corresponds directly with reality') with statements/cards that most resonated with their personal views (e.g. 'scientific knowledge is our understanding of reality').
2) Interviews: The first author conducted interviews with participants before and towards the end of the course (without the instructor knowing the identity of the interviewees). The interview instruments included items from the VOSTS (Views on Science-Technology-Society) questionnaire (Aikenhead & Ryan, 1992), follow-up and probing questions derived from students' written artefacts (namely the "Teaching RiNA and views on citizenship survey", see below for description). The interview questions also were designed to allow student-teachers to express the extent to which they feel that their confidence, motivation and expertise to implement RiNA have developed. After the course grades were submitted, the fourth author (course instructor) met with some pre-service teachers, who consented to a follow up interview, eliciting their views on the course in general, on the STP, and their level of commitment and expertise to prepare

student-activists. All four authors met regularly to discuss and revise all interview instruments.
3) Teaching RiNA and views on citizenship survey: This survey was developed collaboratively by all four authors. Almost all student-teachers in this course completed this survey as part of class work assignments. Participants rated their views AND justified their opinions on the extent to which they feel that it is important for science teachers to promote science activism on STSE issues and the extent to which they feel able to promote RiNA on STSE issues, among other items.
4) Reflections on micro-teaching: All student-teachers were required to submit a 1–2 page(s) reflection following their microteaching. Those reflections typically included a general overview of the lesson and their opinions about its strengths and its weaknesses. Student-teachers also incorporated in those reflections their peer assessment and some even used those assessments as bases for further reflections on their practice.
5) Researcher's field notes: The first author observed almost all classes (6 out of 7) and took field notes from classroom interactions and from 30-min microteaching sessions (per pair of student-teachers). Also all four authors had access to classroom lesson plans through an online forum that student-teachers used to post assignments and access resources related to RiNA.

To determine participants' views on science education's role in preparing citizen-activists and on whether the course supported them to incorporate RiNA in their teaching repertoire, we coded data from all the five sources, listed above, for recurring themes using constant comparative methods (Charmaz, 2014). Themes were continuously negotiated between all four authors to ensure further validity. Some of the themes seemed to be consistent on the three previously mentioned clusters of student-teachers. Other themes materialized as a result of within-cluster, as well as across-cluster, analyses.

18.6 Findings

To account for extents to which the course politicized student-teachers, able to help their future students become citizen-activists, we begin by exploring whether the course influenced participants' conceptions of STSE relationships and nature of science. Next, we report their opinions on whether school science should encourage students to design and conduct RiNA projects on STSE issues. Later, we examine their perceptions on the extent to which they considered the course to have had an influence on their dispositions (in terms of confidence, motivation and expertise) to implement RiNA apprenticeships and student-led projects.

Views about Science: Towards a Critical STSE Education Science teachers' understanding of nature of science (NOS) might influence their students' NOS views (Abd-El-Khalick & Lederman, 2000). Teaching explicitly about NOS was

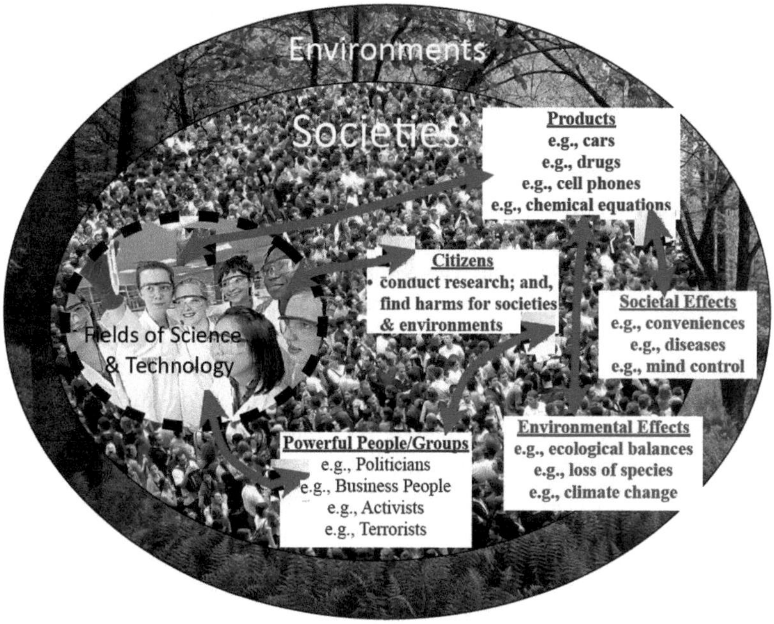

Fig. 18.4 STSE relationship model

not, however, an end by itself but rather a vehicle towards critically questioning STSE relationships for increased democratic participation. Given that the STEPWISE pedagogical framework prioritizes greater citizen involvement in knowledge production and decision-making on STSE issues, this pre-service course was tacitly rejecting a linear view positing science as prior to and essential for technology design for societal uses (Ziman, 1987) and, rather, promoting more fluid and reciprocal relationships, as depicted in Fig. 18.4. At the end of the course, some participants seemed to espouse this, rather dynamic, view of STSE — in which Science and Technology, as interactive fields (Roth, 2001), are permeable (as indicated by the dotted lines) to outside influences. While we acknowledge that such influences may be considered positive or negative, a critical understanding of STSE relationships warrants the view that "technological and scientific developments are inextricably linked with questions about the distribution of wealth and power and that problems of environmental degradation are rooted in societal practices and the values and interests that sustain and legitimate them" (Hodson, 2010, p. 200).

Such interactive and reciprocal relationships among fields of science and technology and societies and environments implies a Naturalist view of nature of science (Loving, 1991), whereby scientists' work is perceived to be subject to many influences by powerful people and groups (see examples in Fig. 18.4 that are based on participants' aggregated responses). We even argue here that promoting Naturalist views about science may be conducive to greater citizen engagement in decision-making on techno-scientific issues. In contrast, Rationalist-Realist views (Loving,

1991), whereby scientists are thought to work in systematic and disinterested manners to discover 'the truth,' might be conveying the message to students that scientists with superior skills, attitudes and knowledge (Allchin, 2003) are better positioned to take decisions for them.

Student-teachers who had an advanced science degree (PhD) and who spent some time doing empirical research in their fields (cluster 1) identified themselves as 'scientists,' given their formal training; e.g., "I don't think of myself as a social scientist but rather as a scientist teaching" (Mary). They felt that the course merely reinforced their previously-held views that: 1) science leans towards Naturalist-Anti-realist side (Loving, 1991) and that 2) science and technology fields are embedded in political, economic and social structures (Fig. 18.4). With regards to their views about nature of science, one participant within cluster 1 revealed domain-specific views on STSE issues, perceiving physics to be more rationalist than biology since, according to him, less funding goes into physics, which would in return makes it less susceptible to outside influences.

> Physicists are more rationalist. Science in average is slightly to the naturalist side of things for reasons that biology is much more naturalist, more funding goes into biological research, more funding (and biotech) anything that is much more readily applicable has more potential for being influenced… Geologists reaching the past could be more rationalist than geologists predicting the future (Max).

As the course progressed, some of the participants in clusters 2 and 3 shifted their views about nature of science towards the Naturalist end on Loving's (1991) spectrum.

> Science is on the Naturalist side (…) before the course, I might have been closer to the middle (…) doing the card exchange game (nature of science) and listening to everyone's inputs in class discussion and why they stand where they do had a great influence on my views (Ava).

> Since I took this course it changed a bit, so now it's more towards naturalist (…) From a Naturalist perspective, just doing the correlation study, we had some perspectives, we were looking for some types of information (…) Also, seeing the projects in class that kids did, the one on makeup, it had a great impact on me, it was great to see how the kids push back against this industry (Lewis).

Engaging with the video developed by high school students on makeup and liquid foundation, constructing actor-network maps, designing and conducting the correlational study, and playing the card exchange game related to nature of science are some of the major factors that student-teachers in clusters 2 and 3 considered to have had an influence on their views. There is, however, not enough evidence to suggest that the course changed student-teachers' views on the truth value of science (Realist vs. Anti-realist).

Other participants, namely from the second and third cluster, felt that the course enlightened them to roles of various 'hidden' actants (Latour, 2005) who might influence work of scientists and engineers; e.g., "I never paid much attention to how government and society might affect decision-making and policies on STSE and how I was blind to those external factors" (Ava). Those participants valued how the course primarily empowered them to reconsider their role as present political subjects. The course might have somewhat developed their political interest:

> Thank you for opening my eyes wider to STSE issues (…) I was rarely involved in STSE issues before this course. As a citizen, I always had the thought that someone else is handling these problems. Some may view these STSE issues as government problems and feel they are not capable of changing the issues or do not know the avenue on how to become part of the solution. The non-trust of citizens on joining STSE groups is based on corrupt organizations who ask for funding to solve problems of society, but in turn use these profits for personal gains such as CEO's of large 'non-profit organizations'. Perhaps this is a social justice issue in itself (Alexander).

School Science's Role in Research-Informed Activism on STSE Issues When asked whether school science should prepare students to do their own research AND take actions on STSE issues, almost all participants agreed that developing students' critical engagement with knowledge and increasing their sense of social responsibility should be promoted through school science.

> I think it is absolutely critical that students learn how to think for themselves when it comes to science and technology-based issues (…) I also think that students are ready to tackle issues head on, and to feel engaged through more meaningful mediums like problem solving around real-world issues, leading campaigns of change, developing relations with local, national and international partners, finding a sense of motivation through their own community-oriented work, and I think that RiNA projects have the potential to offer this type of learning atmosphere for students (Cheryl).

> Having students go through the process of researching a problem, identifying key actants within a system and formulating a plan of action will help develop their skills in critical thinking. Ultimately, I believe that teaching science can be considered "gateway activism", whereupon students acquire the basic skills to research issues, whether they pertain to science or not, and come up with their own views and plans of action (Ben).

However, others believed that it is mainly "the scientists' job" (Ava) to produce knowledge, since citizens lack technical equipment, are not formally trained as scientists and have other commitments. Student-teachers seemed to rationalize the relative exclusion of laypeople from processes of knowledge production with relation to division of labour (Michael, 1996). Still, they believed that it's important for students/citizens to critically evaluate knowledge from secondary sources in Ava's saying: "I think we should teach students to question the information they are getting, it would be more realistic to prepare them to be critical than do primary research".

There is some evidence that suggests that some student-teachers (mainly from cluster 2) may not have recognized possibilities that by teaching students specific skills they will be simultaneously teaching citizens. Asking them the question, 'Should we be encouraging the society to do their own research?', Julie replied: "I never thought about it in that sense (…) I think that primary research is important to teach to students but to citizens, I never thought of that..." Some of those pre-service teachers from cluster 2 seemed to view their role in preparing students to identify with the subculture of science as insiders rather than outsiders (Aikenhead, 2006), such as Leo who said: "I think it's a good thing to teach students how to do their own research because maybe you will hit those who are interested in becoming scientists". Many student-teachers mentioned that correlational studies, as approaches to knowledge production, are more accessible to citizens than experiments and that citizens can nowadays more readily engage in data collection facilitated through social media, online surveys and other technological tools. Lewis particularly expressed this latter view: "I think already the world is fascinated with surveys, everything you see on Facebook… People are always investigating what they eat in their daily routines".

Pedagogical Dispositions (Confidence, Motivation, Expertise) to Implement RiNA In an attempt to ease student-teachers' anxiety and possibly increase their willingness to implement RiNA in their prospective teaching, the course instructor made sure to highlight that STSE activism doesn't necessarily oppose the mandated curriculum but, rather, reorients its purpose:

> Everything we are doing is part of the Ontario curriculum, the only thing we are adding is for students to actually test their plan of action. The Ontario curriculum stops at developing a plan of action without testing it (Larry, Field notes, class 5).

Such a strategic comment could be considered a way to prepare student-teachers to work within and against the system. Although actions are a supplement to traditional STSE teaching, emphasizing sociopolitical actions underlines critically examining the status quo of societies and roles of citizens in representative democracies. Thus, 'adding' an action component to STSE teaching is not to be understood as a superficial revision but, rather, as a basis to question roles of schooling in a democracy for a democracy (Carr, 1998).

Student-teachers' level of confidence and motivation to implement RiNA in their teaching varied. Some seemed to be committed to adapt RiNA into their teaching practices.

> I am already considering methods to incorporate a study like this into a hospitality program. It is a great way to inform and motivate students to seek answers to challenges the world faces. Students need to see that they can impact the world. This project (RiNA) empowers students to consider activism as a possible method for dealing with STSE issues (Lewis).

Moreover, those who considered themselves 'activists' in their everyday lives seemed to strongly endorse preparing active and informed citizens through science education — such as is indicated in a comment by Mary:

> I consider myself a citizen activist and spend a lot of time involved in public education programs, and lobbying and advocating for policy reform (…) I will always make this type of education (research-informed activism) at the forefront.

Most participants mentioned that while they believed that it is important to promote attitudes of critical thinking and social responsibility among students, they noted various challenges that might hinder those efforts: (1) students might not be motivated enough to take actions mentioning apathy as a symptom of individualistic societies (2) primary research is hard to implement, (3) structural challenges (e.g., school/department cooperation) and (4) RiNA takes a lot of time since it needs to be consolidated over various loops of apprenticeships across various curricular units (refer to Fig. 18.1), a bit overwhelming for beginner teachers.

> In my practicum, I thought how would I be able to do RiNA if I have all this material to be covered, especially if you are new teachers (Hannah).

> RiNA is hard for new teachers…you need lots of expertise in teaching students the basic stuff and then to teach them how to think critically….I think I will teach the content first (Leo).

Leo was very explicit saying that as a beginner teacher, he might prioritize teaching science content over critical thinking.

Some participants in the first cluster (science and research background) mentioned that their advanced science background, coupled with taking this course, gave them increased confidence and expertise to teach about RiNA:

> In schools now there is no connection between primary, secondary research and action. It is important in the future to make connections between them. Having gone through this course and seeing how they connect to each other, I can now construct a yearlong project that incorporate all of these. It takes time to do and it takes different levels of knowledge and skills to do each step. Maybe other teachers are not prepared to do this but having taken this course and having a science background, I feel better able to take students through all those steps (Luke).

> I came into this course with a wealth of information about misappropriations of science, the course gave me a framework to work with those ideas with students (Alan).

Some other participants, regardless of their educational background and research experience, revealed some pedagogical dispositions to implement RiNA as a result of doing their own RiNA projects but also as a result of the microteaching:

> I felt more confident teaching RiNA after the microteaching after seeing how students were excited about the things that we did (Hannah).

> I think I could really work the RiNA framework into my own curricular design. Now that I have been through my own project, understood the pitfalls and my own misunderstandings, as well as having some experience working with the framework in a high school classroom (i.e. micro-teaching), I can build on this knowledge to better inform my pedagogy in the sciences (Cheryl).

> We have more expertise in our project to use it as pedagogical resource and also possibly more emotional/personal investment in it, we know the challenges (like coming up with questions) we have expertise (…) to be able to talk about our successes and failures and tell students what to expect (Lewis).

Furthermore, some expressed views that using their own RiNA project to inform their pedagogy increased their emotional attachment to what they were teaching and, as a result, they felt they could motivate their students to lead similar projects.

> By using your own RiNA as a pedagogical tool, you take ownership, it motivates the students in that they will be doing something that you did yourself and you can tell them about the limits, weaknesses as you have experienced them, and you can relate more personally to the topic (Blake).

> I feel having done it, especially if I were to teach RiNA….if I can do it, you can do it….it's a doable project and it made me more passionate teaching about it (Hannah).

However, with additional prompting, most participants revealed that they might not have developed enough pedagogical expertise to implement such projects in their classrooms especially that the microteaching was limited to implementing the first phases of the STEPWISE framework.

> I would have been less prepared to engage students in RiNA if I didn't take this course but I don't see myself very well prepared (Max)

> The course was so focused on the RiNA project at the expense of the pedagogy (Lewis)

> I don't feel quite prepared. I need more practice, the micro-teaching was not enough (Ava)

While their pedagogical dispositions to implement RiNA might not have fully developed, participants demonstrated better pedagogical dispositions to teach about complex relations that might exist between science, technology, society and the environment. Almost all participants emphasized components of ANT teaching in their microteaching lesson plans:

> There was evidence that students had positioned themselves in the material, when they later drew themselves into the ANT Map, and also began comparing their own perceptions on the STSE issue, with what they thought their parents would think. We purposely highlighted hidden actants that students would find surprising (Cheryl).

Since almost all participants emphasized ANT teaching in their microteaching lesson plans, this might serve as an indication that they have developed some pedagogical commitment to using a 'networked' approach when teaching about STSE issues. Specifically, Layla developed a lesson plan addressing the problem of plastic bag pollution as a global issue based on local problems associated with plastic bag pollution in her own home country, Somalia. Pairing with Luke, she started her lesson eliciting grade 9 students' preconceptions about plastic bag pollution as an STSE issue. Layla gave students the following cards: landfill, delegates, plastic bags, pirates, manufacturers, and asked them to arrange them as they find appropriate. Students initially arranged the actors "in a linear, one-way fashion" (field notes, Layla's reflection). She then introduced ANT and the case of plastic bag pollution in Somalia. Below is an excerpt from her reflections indicating that she might have been successful teaching about ANT:

> For the second activity we have given students the same picture cards and asked them to arrange them again (…) we were surprised how quickly they understood how these are interconnected and how these connections are not linear.

18.7 Discussion

Our findings are grounded in participants' subjective reflections on their own learning experiences in this course. Mirroring a democratic dialogue, this chapter attempts to put in conversation participants' views of their learning (and of themselves) and our understanding of those views and experiences. While student-teachers were primarily implementing the STEPWISE framework, they were invited to evaluate this 'activist' pedagogy to inform their own practice, as evidenced in Max's saying:

> Most of the learning goals associated with the RiNA project should be incorporated in the curriculum, and important for students to know. I would rather incorporate those in smaller lessons. I don't think RiNA is the only way of reaching those goals.

Max's reflection on RiNA allowed him to take ownership over his future practice by envisioning himself teaching for socio-political activism beyond the STEPWISE framework.

A major finding of our study seems to be that student-teachers' educational background and research experience influenced the extent to which the course had an influence on their views about STSE and NOS, as well as on their dispositions to teach for socio-political activism. The course allowed participants, mainly in clusters 2 and 3, to begin to think critically about STSE issues while simultaneously shifting their views about how scientific knowledge is generated. Since we have limited data regarding students' views about truth value of knowledge from the sciences, future direction for this course might be to make more explicit the notion that democratic participation on STSE issues involves an appreciation of the open-endedness and interpretive nature of scientific knowledge.

Participants in cluster 1 (Science and research backgrounds) were slightly more assertive of their willingness and ability to engage their students in research-informed and negotiated activism compared to participants from clusters 2 and 3. This finding partly resonates with previous research that suggests that the likelihood of pre-service teachers using inquiry in their classrooms is increased for those with significant research background and advanced science degrees (Windshitl, 2004). Some participants (regardless of their belonging to a specific cluster) mentioned that doing the RiNA project followed by the episode of microteaching contributed to developing, to a certain extent, their pedagogical dispositions to implement RiNA in their future practice. There seem to be at least two possible explanations for this. First, as a result of having control over the reciprocal translations from Sign to World and World to Sign (Roth, 2001), student-teachers might have developed deeper and more meaningful connections with RiNA on STSE issues (Bencze & Sperling, 2012). This personal identification with their own self-led project might have resulted in greater pedagogical commitment and expertise to implement RiNA in their teaching. A second explanation could be that student-teachers were deliberately making use of their metacognitive skills to reflect on how their own experiences doing the RiNA project might help them address future concerns that might come up with their students.

Although the course was intended for pre-service teachers to reconsider their roles in teaching citizens and not only students, some participants in cluster 2 revealed that they might still ascribe to a traditional model of schooling that primarily values content teaching over preparing critical and active citizenry. On the other hand, the course seemed to have influenced student-teachers' pedagogical dispositions to teach about Actor-Network Theory (ANT). In contrast to an isolationist portrayal of science that might alienate citizens from participating in decision-making (Rudolph, 2005), allowing students to see complex relations among actants regarding STSE issues seems to be an important step to institute change (Pierce,

2013) and to design effective action plans that target specific stakeholders. Layla's ability to present complex relationships surrounding plastic bag pollution in a familiar context (Somalia) might be further indication of her pedagogical expertise to teach about ANT. As a foreign teacher in a Canadian context, Layla's personal investigation and emotional attachment to this STSE issue (Bencze & Sperling, 2012) being both a global and local concern afforded a space to connect with her students by bridging geographical and cultural boundaries. As a result of this pre-service course, she might have become empowered as a 'glocalized' citizen (Luke 1994 cited in Vasquez, 2013) infusing the local with the global and the global with the local (Local ←→ Global) to act and push students to act on environmental problems that ultimately affect everyone.

Given its open-endedness and its emphasis on self-directed learning, the STEPWISE framework is aimed at increasing students' political efficacy for well-being of individuals, societies and environments, while simultaneously valuing the cultural backgrounds and experiences of students and perhaps, as evidenced here, those of teachers. Going back to Roth's (2001) dialectical model of World-Sign, Layla's ability to critically read her World (deconstructing the local issue of plastic bag pollution in Somalia) allowed her to read her *word* (Freire, 1987). This is evident in her ability to design a personally relevant lesson plan (Sign) that somehow informed her creative approach when teaching about ANT (World).

Contributing to student-teachers' limited pedagogical dispositions to teach for socio-political activism are some perceived challenges in the school system that mainly stresses science content teaching. Although the course instructor underlined that research-informed and negotiated activism does not necessarily oppose the curriculum, the course needs to develop further strategies to support pre-service teachers as both insiders and outsiders to the hegemonic culture of teaching and learning (Carlone et al., 2010).

18.8 Implications

This pre-service course could be viewed as a model to prepare science teachers to reconsider their roles in democratizing their classrooms as sites of authentic inquiry and action on STSE issues. Since student-teachers' backgrounds were found to influence their learning outcomes in this course, it might be important to tailor teacher-training programs to individual student-teachers' experiences and levels of education. The course could have further supported pre-service teachers to use research-informed and negotiated activism in their practice if it engaged them in multiple translations from World to Sign and Sign to World (Roth, 2001). For instance, student-teachers might have better consolidated their knowledge, attitudes and skills to conduct research-informed and negotiated actions if the course was designed over various loops of apprenticeships as part of a spiral curriculum. Also, the course needed to ideally balance engaging pre-service teachers in designing and conducting RiNA projects and teaching about research-informed and negotiated activism in actual classroom settings. Lastly, student-teachers' pedagogical dispositions and commitment to use socio-political activism in their prospective teaching might have been improved if they had multiple opportunities to practice teaching for participatory citizenship, beyond the microteaching.

References

Abd-El-Khalick, F., & Lederman, N. G. (2000). Improving science teachers' conceptions of nature of science: A critical review of the literature. *International Journal of Science Education, 22*(7), 665–701.

Aikenhead, G. S. (2006). *Science education for everyday life: Evidence-based practice*. New York: Teachers College Press.

Aikenhead, G. S., & Ryan, A. G. (1992). The development of a new instrument: 'Views on Science—Technology—Society' (VOSTS). *Science Education, 76*(5), 477–491.

Allchin, D. (2003). Scientific myth-conceptions. *Science Education, 87*(3), 329–351.

Alsop, S., & Bencze, J. L. (2014). Activism! Toward a more radical science and technology education. In J. L. Bencze & S. Alsop (Eds.), *Activist science & technology education* (pp. 575–596). Dordrecht, The Netherlands: Springer.

Bell, R. L. (2006). Perusing Pandora's box: Exploring the what, when, and how of nature of science instruction. In L. B. Flick & N. G. Lederman (Eds.), *Scientific inquiry and nature of science: Implications for teaching, learning, and teacher education* (pp. 427–446). Dordrecht, The Netherlands: Springer.

Bencze, J. L., & Sperling, E. R. (2012). Student teachers as advocates for student-led research-informed socioscientific activism. *Canadian Journal of Science, Mathematics, and Technology Education, 12*(1), 62–85.

Bencze, L., & Carter, L. (2011). Globalizing students acting for the common good. *Journal of Research in Science Teaching, 48*(6), 648–669.

Brickhouse, N., & Bodner, G. M. (1992). The beginning science teacher: Classroom narratives of convictions and constraints. *Journal of Research in Science Teaching, 29*(5), 471–485.

Campano, G. (2007). *Immigrant students and literacy: Reading, writing and remembering*. New York: Teachers College Press.

Carlone, H. B., Haun-Frank, J., & Kimmel, S. C. (2010). Tempered radicals: Elementary teachers' narratives of teaching science within and against prevailing meanings of schooling. *Cultural Studies of Science Education, 5*(4), 941–965.

Carr, W. (1998). The curriculum in and for a democratic society. *Curriculum Studies, 6*(3), 323–340.

Charmaz, K. (2014). *Constructing grounded theory* (2nd ed.). New York: Sage.

Cobern, W. W., & Loving, C. C. (2002). The card exchange: Introducing the philosophy of science. In *The nature of science in science education* (pp. 73–82). Dordrecht, The Netherlands: Springer.

Cochran-Smith, M. (1991). Learning to teach against the grain. *Harvard Educational Review, 61*(3), 279–311.

Feinstein, N. (2011). Salvaging science literacy. *Science Education, 95*(1), 168–185.

Fourez, G. (1997). Scientific and technological literacy as a social practice. *Social Studies of Science, 27*(6), 903–936.

Freire, P. (1970). *Pedagogy of the oppressed*. New York: Continuum.

Freire, P. (1987). The importance of the act of reading. In P. Freire & D. Macedo (Eds.), *Literacy: Reading the word and the world*. South Hadley, MA: Bergin and Garvey.

Gardner, P. L. (1999). The representation of science-technology relationships in Canadian physics textbooks. *International Journal of Science Education, 21*(3), 329–347.

Hodson, D. (1999). Going beyond cultural pluralism: Science education for sociopolitical action. *Science Education, 83*(6), 775–796.
Hodson, D. (2009). *Teaching and learning about science: Language, theories, methods, history, traditions and values*. Rotterdam, The Netherlands: Sense Publishers.
Hodson, D. (2010). Science education as a call to action. *Canadian Journal of Science, Mathematics, and Technology Education, 10*(3), 197–206.
Hodson, D. (2011). *Looking to the Future*. Rotterdam, The Netherlands: Springer Science & Business Media.
Hoeg, D., & Bencze, L. (2014). Counter cultural hegemony: Student-teachers' experiences implementing STSE activism. In J. L. Bencze & S. Alsop (Eds.), *Activist science & technology education* (pp. 575–596). Dordrecht, The Netherlands: Springer.
Krstovic, M. (2014). Preparing students for self-directed research-informed actions on socioscientific issues. In J. L. Bencze & S. Alsop (Eds.), *Activist science & technology education* (pp. 399–417). Dordrecht, The Netherlands: Springer.
Latour, B. (2005). *Reassembling the social: An introduction to actor-network-theory*. Oxford, UK: Oxford University Press.
Lakin, S., & Wellington, J. (1994). Who will teach the 'nature of science'? Teachers' views of science and their implications for science education. *International Journal of Science Education, 16*(2), 175–190.
Levinson, R. (2010). Science education and democratic participation: An uneasy congruence? *Studies in Science Education, 46*(1), 69–119.
Loving, C. C. (1991). The scientific theory profile: A philosophy of science model for science teachers. *Journal of Research in Science Teaching, 28*(9), 823–838.
Michael, M. (1996). Ignoring science: Discourse of ignorance in the public understanding of science. In A. Irwin & B. Wynne (Eds.), *Misunderstanding science?* (pp. 107–125). Cambridge, MA: Cambridge University Press.
Pedretti, E., & Nazir, J. (2011). Currents in STSE education: Mapping a complex field, 40 years on. *Science Education, 95*(4), 601–626.
Pierce, C. (2013). *Education in the age of biocapitalism: Optimizing educational life for a flat world*. New York: Palgrave MacMillan.
Reis, P. (2014). Promoting students' collective socio-scientific activism: Teachers' perspectives. In J. L. Bencze & S. Alsop (Eds.), *Activist science and technology education* (pp. 547–574). Dordrecht, The Netherlands: Springer.
Roberts, D. A. (2011). Competing visions of scientific literacy. In C. Linder, L. Ostman, D. A. Roberts, P. Wickman, G. Ericksen, & A. Mackinnon (Eds.), *Exploring the landscape of scientific literacy* (pp. 11–27). New York: Routledge.
Roth, W.-M. (2001). Learning science through technological design. *Journal of Research in Science Teaching, 38*(7), 768–790.
Roth, W. M., & Barton, A. C. (2004). *Rethinking scientific literacy*. London: Psychology Press.
Roth, W.-M., & Désautels, J. (2004). Educating for citizenship: Reappraising the role of science education. *Canadian Journal of Math, Science & Technology Education, 4*(2), 149–168.
Rudolph, J. (2005). Inquiry, instrumentalism, and the public understanding of science. *Science Education, 89*(5), 803–821.
Sadler, T. D., Barab, S. A., & Scott, B. (2007). What do students gain by engaging in socioscientific inquiry? *Research in Science Education, 37*(4), 371–391.
Santos, W. L. D. (2009). Scientific literacy: A Freirean perspective as a radical view of humanistic science education. *Science Education, 93*(2), 361–382.
Sismondo, S. (2008). Science and technology studies and an engaged program. In E. J. Hackett, O. Amsterdamska, M. Lynch, & J. Wajcman (Eds.), *The handbook of science and technology studies* (3rd ed., pp. 13–31). Cambridge, MA: The MIT Press.
Sjöström, J., Eilks, I., & Zuin, V. G. (2016). Towards eco-reflexive science education. *Science & Education, 25*(3–4), 321–341.

Sutton, R. E., & Wheatley, K. F. (2003). Teachers' emotions and teaching: A review of the literature and directions for future research. *Educational Psychology Review, 15*(4), 327–358.
Vasquez, V. (2013). Living and learning critical literacy in the university classroom. In C. Kosnik, J. Rowsell, P. Williamson, R. Simon, & C. Beck (Eds.), *Literacy teacher educators: Preparing teachers for a changing world*. Rotterdam, The Netherlands: Sense Publishers.
Venturini, T. (2010). Diving in magma: How to explore controversies with actor-network theory. *Public Understanding of Science, 19*(3), 258–273.
Windschitl, M. (2004). Folk theories of "inquiry": How preservice teachers reproduce the discourse and practices of an atheoretical scientific method. *Journal of Research in Science Teaching, 41*(5), 481–512.
Wood, G. H. (1998). Democracy and the curriculum. In L. E. Beyer & M. W. Apple (Eds.), *The curriculum: Problems, politics and possibilities* (pp. 177–198). Albany, NY: SUNY Press.
Ziman, J. M. (1987). *An introduction to science studies: The philosophical and social aspects of science and technology*. Cambridge, MA: Cambridge University Press.

Chapter 19
Teaching Girls to Fish?: A Case of a Co-Curricular Food Justice Education Program for Youth

Erin Sperling and Larry Bencze

19.1 Introduction

Sustainability education requires understanding of complex systems and skills to investigate and actively address challenges within systems in the short term with an eye on long range impacts (Tilbury & Wortman, 2004). Often involving controversial aspects of fields of science, such challenges are sometimes called *socio-scientific issues* (e.g., Sadler et al., 2005). Issues such as climate change, food insecurity, access to fresh water and natural resources are some examples of how sustainable, systems-oriented citizen actions are required. Thus, within the field of sustainability education, there is impetus to engage learners with knowledge and skills of scientific, social and economic systems and their interrelations. This impetus is a foundation from which citizens can begin, as agents of change, to address challenges that have arisen in opposition, either intentionally or not, to the wellbeing of individuals, societies and environments. Sustainability education (among many types/approaches to environmental education) offers a particular and more holistic view of the human as a part of ecosystems (Shiva, 2005). Such a view may be considered an integral, but not necessarily central, piece of the 'science, technology, society and environment' (STSE) education paradigm that is promoted in some parts of the world (Pedretti & Nazir, 2011). Some STSE relationships may involve controversy and, in that sense, be comparable to socio-scientific issues. A holistic view, in partnership with an action orientation and engagement, can offer hope for wellbeing from a sustainability standpoint, and may include opportunities to consider injustices perpetrated through social inequities, such as classism, racism and sexism (Martusewicz, Edmundson, & Lupinacci, 2011). In previous studies that engaged students through a research-based, action-oriented curriculum (i.e., Bencze, Sperling, & Carter,

E. Sperling (✉) • L. Bencze
Ontario Institute for Studies in Education (OISE), University of Toronto, Toronto, ON, Canada
e-mail: erin.sperling@mail.utoronto.ca; larry.bencze@utoronto.ca

L. Bencze (ed.), *Science and Technology Education Promoting Wellbeing for Individuals, Societies and Environments*, Cultural Studies of Science Education 14,
DOI 10.1007/978-3-319-55505-8_19

2012), it was concluded that there are certain constraints in formal, institutional schooling that inhibit youth from full participation in civic engagement with sustainability issues. Among these, limitations of time in combination with an official science curriculum that does not explicitly require student action around issues, at least in the Ontario context, caused disconnection between best intentions of sustainability education and reality. Accordingly, this chapter considers that non-formal educational encounters can be rich sites of citizenship development around STSE issues, and can work in supporting development of strong citizenship attributes in youth that emerge through STSE issues.

There are multiple ways that youth are engaged in non-formal educational experiences. Broadly, non-formal education has historically provided ‘real-life’ experiences for youth, by meeting them in their communities, often in a form of place-based education (Gruenewald & Smith, 2008). Educational experiences that, either intentionally or not, connect to realities of youth (or any learners) have been noted as being very successful ways of engaging them in learning and collaboration (Rahm, 2010). There is a sense of freedom found in non-formal education that allows for localized, youth-driven curriculum that may inherently work toward sustainability by (learning to) knowing one’s own environment, physically, emotionally and socially (Barton, 2003).

This chapter explores a case of non-formal education that provides several opportunities for youth to explore many outcomes for sustainability education—intersecting science and citizenship education, in partnership with the university-based research team using the ‘STEPWISE’ (Science & Technology Education Promoting Wellbeing for Individuals, Societies & Environments) framework (www.stepwiser.ca). In this case, as defined by us in reviewing objectives of local social justice organizations (e.g., via its website and advertising materials), we contacted an organization offering community-based programming that: provides access and exposure to conflict and structural power imbalances (that is perhaps more obvious than in the formal system); organically creates spaces of culture (as opposed to forcibly); works at creating strength through localized, community-based scientific and other knowledge forms (uncited to protect anonymity for ethical purposes). This ideology works in concert with the STEPWISE program, which, as we see in other chapters of this and other publications, is a framework for organizing instruction in science & technology education that encourages and enables students/citizens to use their literacy to try to bring about a better world. The framework features relationships among common learning goals, including learning ‘products’ (e.g., laws & theories) of science and technology and developing science inquiry skills, and self-directed student research-informed and negotiated action (RiNA) projects to address potential problems related to decisions made by powerful people/groups about fields of science and technology, all working toward the wellbeing of individuals, societies and environments. How and why youth become engaged in these processes and to what degree they are able to create and use science knowledge through a sustainability action project were analysed from data gathered through this ethnographic case study.

There is need for attention to socioscientific issues (SSIs) associated with problems stemming from relationships among fields of science and technology and societies and environments (STSE), such as physical and mental health, impacts of manufactured foods, power allotted to mega-corporate pharmaceuticals, and radiation from electronic devices. Often because of associated threats to the wellbeing of individuals, societies and environments, many scholars and citizens urge educational systems to actively address SSIs. As such, school science needs improvement in many ways related to SSIs (MOE, 2009). Too often, it is oriented towards content instruction (e.g., facts, laws, theories) that can compromise education in other important learning domains, such as skills and attitudes (Bell, 2006). Through repeated experiences of working with teachers in formal education systems, despite some success (ie. Sperlng, & Bencze, 2010), it has become apparent that there are many challenges to pedagogies that can support and enhance student learning and action around SSIs. Particularly difficult to promote in such environments are student-led research-informed and negotiated action projects (Bencze & Sperling, 2012). The outcomes of this project may contribute to the body of knowledge surrounding the growing need for youth to become engaged as citizens addressing issues of local and global import.

19.2 Theoretical Background

There is a great body of research literature suggesting that each of us is in dynamic relationships, to varying degrees, with other living and non-living entities on earth (Latour, 1999). From this perspective, many scholars strongly suggest that educators encourage and enable students to accept more collectivist epistemological stances and take actions that may contribute to improvements to wellbeing of individuals, societies and environments (Hodson, 2010). One approach to understanding and implementing educational experiences that lead to collectivist ideology of wellbeing has been identified through use of socioscientific issues-based (SSI) education. SSI education exposes students to ideas, information and knowledge through issues that are ideally, but not always, connected in some way to the lived realities of the students. The contextualization of content, in this way, provides more authentic connections to experiences of learning and doing science, as well as focusing on empowering students to consider how issues reflect and engage with their own lives, as well as the physical and social world around them (Zeidler, Sadler, Simmons, & Howes, 2005). When students are exposed to and have begun to create knowledge about issues, they may also formulate action plans to undertake in response to the information that has been gathered (Bencze & Alsop, 2009).

Student actions can take different forms, which may result in or be directed by different targets and goals. Actions they might take include: educating others (e.g., via presentations, posters and pamphlets), lobbying power-brokers (e.g., via petitions and letter-writing campaigns), imagining and developing potentially-improved products and systems (e.g., organizing a school food program) and/or making personal improvements (e.g., using a travel mug), as well as others (Bencze & Alsop,

2009). In previous research, we have noted success in student self-efficacy and outcome expectation around their involvement in actions (Bencze, Sperling, & Carter, 2012).

In several cases, learners have shown some benefit from student-directed research, engaging in various primary and secondary research practices (see other chapters in this book), and using findings to inform personal and social actions to address perceived problems. Knowledge that is produced by learners, based on their own set of inquiries, helps to support their understandings of other knowledges, as well as their impetus for taking action around SSIs. Students benefit from RiNA when they have more control over both directions of RiNA; that is, representing phenomena and using negotiated representations to self-direct actions on the world. For example, we encountered students who investigated bottled water as an STSE issue within their science curriculum, and found that tap water was the best option based on an analysis of ten different bottled and tap samples. These results led them to champion a campaign to drink from the tap using reusable bottles (Bencze et al., 2012). That being said, despite some successes, in many cases, this type of student-directed research-informed and negotiated activism programming has been difficult in school-based contexts. This seems linked to economic forces on science education systems, which seems to favour selection and education of small numbers of students who may become scientists and engineers and other producers and managers of knowledge (refer to Bencze, 2010, for more about this). Outcomes related to self-efficacy and agency of the youth participants were based on knowledge duality theory (Wenger, 1998), which suggests that deep attachments to ideas and actions arise when people have personal involvement in dialectical relations between phenomena (e.g., food-related choices) and representations (e.g., drawings and graphs) of them. In the study outlined below, we attempted to provide youth with opportunities to self-direct primary and secondary research that may significantly inform, motivate and direct actions taken to address SSIs related to food justice and security.

Formal and Non-formal Learning There are potentially different goals, structures and possibilities through non-formal education settings that can lend space and flexibility to youth-led concerns and initiatives. We imagined that there may be fewer constraints in community group programming, due to fewer requirements to implement standardized evaluation and, therefore, more focus on community well-being, which is more participant- and issue-oriented than a traditional, formal school-based curriculum. In particular, non-formal education can, as Ann Higgins-D'Allessandro (2010) suggests, provide:

> learning programmes to immerse youth in community problem solving [that] promotes the development of youth cognitive and social capabilities as well as enhancing civic understanding and the willingness to work for social justice. Working for social justice enfranchised youth (p. 572).

With these understandings in mind, we turned to a local CBO as a pilot study in a non-formal education setting. The partner CBO is not-for-profit, multifaceted (educational and provisional), and is issues-based in its form and function around issues of food security. To paraphrase from its website, it *attempts to provide more access to healthy food while addressing dignity, community building and dismantling of inequality*. Because of the situated nature of socio-scientific issues, we consider that place-based pedagogy offers an opportunity to engage with complex realities that people are facing in tangible ways, from local impacts of industry on a watershed, or of climate change, whether fruit crops are pollinated in a given year or what food choices and costs are available in a particular neighbourhood. This way of approaching education also acknowledges social, economic, physical and cultural realities and dynamics of youth as a particular segment of society. This is often different from the dominant, or hegemonic, cultural discourses, and thus creates tensions that may impact on learning and outcomes, and desire of these. However, place-based education, rather than de/constructing identities through difference, "deconstructs and reconstructs identities through the construct of *relationship*" (Gruenewald & Smith, 2008, p. 147). Our sense of place and attachment to it bring about change in the world through processes of decolonization and reinhabitation (Gruenewald, 2003). These seemingly esoteric notions can become very tangible in thinking through neocolonial issues and injustices at a local level. For example, discussing and addressing the spatial colonization of the grocery shelves by mega-corporations calls for a decolonization of our food spaces, away from those who use the power of money to take ownership of our visual inputs, and lead a charge toward real choices beyond product placements, to support our health, farmers and local food systems (Winson, 2013). Using place-based pedagogy may prove to be the most integrative way of understanding and enacting the intersection between citizenship empowerment and a transformative model in science education as it exists in food.

Food as Pedagogy The notion of food as a particular site of cultural, social, environmental concern and intersection brings an additional richness to questions and understandings of this project. Popular writer Michael Pollan (2008) wrote that food is "about pleasure, about community, about family and spirituality, about our relationship to the natural world, and about expressing our identity...culture as [well as] about biology" (p. 8). We all have relationships to food and we are all impacted by it; to what degree and with what degree of agency in food-related SSIs are both interesting concerns to consider within the context of sociopolitical markers, oppression(s) and knowledge sources. Food provides a platform to address numerous social and environmental inequities and to provide tangible opportunities for transformative learning. As Jennifer Sumner (2008) stated, in relation to her work with adult learners, food "goes beyond formal learning by highlighting various forms of learning in both the home and the community...it politicizes transformative learning to deliberately deal with questions of power" (p. 37). There is a gap around the questions of how youth engage in food as pedagogy, within a pedagogy of place, toward action on socioscientific issues and impacts on individuals' sense of civic identity. With all of this in mind, it is time to consider how this research project comes to the table.

19.3 Assembling the Ingredients

After reflecting on difficulties that educators have had in encouraging and enabling students in school contexts to address socio-scientific issues, we explored such activism in the context of after-school youth groups. Students' performances in such optional clubs are not evaluated to nearly the same extent as in schools, and expectations and outcomes are more often student- or community-centred in such settings. For similar reasons, Derek Hodson (2010) also has recommended such an approach.

This qualitative, ethnographic, action-research project took place at one main site (CBO community kitchen) and three secondary sites (CBO community garden, a local high school and an urban university campus). The program took place for 2 h after school each week, from October to May, and included an intensive four-day spring break camp. The primary researcher took on the role of researcher-facilitator, working closely with the CBO program facilitator to provide curriculum and facilitation support as requested, as well as ongoing data collection. As a researcher-participant, working with the facilitator through an ongoing negotiation process, our goal from the research perspective was to determine ways and means of infusing — in negotiation with CBO personnel — student-directed research-based activism into the program. The research emerged through both naturalistic and rationalistic perspectives, in the sense that we had initial understandings of expected outcomes, based on discussions with the CBO facilitators and directors, and then as the program unfolded several unexpected outcomes occurred, which are discussed in detail in the findings.

The CBO offers a diverse range of programming for community members. These range from a drop-in for those in need to come in days a week for healthy hot meals, to community advocacy groups to bring issues to the attention of powerbrokers. There are pre- and neo-natal nutrition programs, a cooking group for Indigenous peoples and an afterschool program for elementary-aged kids. The particular program for youth with which we worked was founded 2 years prior to our involvement. The program founder described that it.

> … is aimed at high school girls who are from the community who can benefit from opportunities to learn about cooking, food issues and gain self-confidence to address issues for themselves and their families. ... It is meant to be a welcoming space where youth could learn about and develop agency around food issues …where girls could have the necessary social support to develop … skills [and] gain confidence to think of themselves as leaders on food issues within their social group and beyond (uncited for confidentiality purposes (if this makes sense)).

In each of the 2 years of study, 2010–11 and 2011–12, eight female youth participants ranging in age from 14 to 17 were from three different high schools in the community. They were recruited from local public and Catholic board high schools early in the school year through awareness activities, such as organized lunch time presentations or information tables in the main hallway set up by the program facilitator at the school. Ideally, the participants were not involved in many other activities which would inhibit their participation in the food program. There were eight to fifteen participants throughout the year, with eight attending semi-consistently, who gave informed consent to participate in the research.

Multiple data sources included observational data, audio-recorded discussions and interviews, artefacts of participants' work, multimedia material, and public documents produced by the CBO about or referring to the youth food program. Data were collected from facilitators and participants with three group interviews, recordings and transcriptions, field notes, ten interviews with staff/facilitators, photos taken by participants and researchers, documents created by the CBO and by the youth, including paper-based and multimedia documents. Four action-oriented videos were produced by participants each year. All names of people and organizations have been changed to mask their identities as much as possible. We have used constant-comparative methods in order to find some themes and codes of analysis of the rich and deep amount of data, including over 100 h of audio and video recording (Charmaz, 2014) as well as hundreds of photos taken by the participants. Ultimately we were trying to understand the ways that youth take self-directed action around STSE issues, and what are some of the factors that influence their outcomes. All names have been changed to protect anonymity.

19.4 Placing the Meal on the Table

19.4.1 Outcomes

We have observed, documented and uncovered a number of outcomes as a result of participation in the STEPWISE-CBO program relationship. Both entities are plainly interested in educating youth about socio-scientific issues related to food security. We have noted several examples of guided activism and we observed increased self-efficacy as well as leadership among youth. We also encountered some challenges to the expected changes in youth, apparently including lower levels of outcome expectancy among them.

19.4.1.1 Guided Activism

There were several examples of guided activism. One student, Alison, commented about her experience of the program: "Knowing what's really in our food and where it comes from or what it was made with has changed my opinion about eating in a lot of different ways". This also shows a certain degree of understanding about the complexity of food systems. She chose to have this quote published in the zine at the program end that the group collaboratively produced. Guided activism projects included the following: video-making to share with the community and the public; presenting new knowledge and experiences to the school groups; making presentations to their parents and other community members; a student-written article in

CBO end-of-year report; and, the production of a 'zine.' We can evaluate in some capacity these activism outcomes through Hodson (2003) who addresses the political nature of issues-based scientific literacy in a four-level framework. Each level moves through a degree of STSE sophistication of student attainment: Level 1, appreciating the societal impact on scientific change; Level 2, recognition of stakeholders in scientific decisions and the link to wealth and power; Level 3, development of individual views and establishing valuepositions; and Level 4, preparation for and taking of action. For these participants, knowledge of STSE relationships was displayed most prominently through levels 1–3, for example by acknowledging the carbon footprint of industrial farming, and having repeated discussions on the impact on farmers and migrant workers, and who uses food banks.

An important climax and highlight of the guided activism aspect of the program was when they created activist videos, which involved both taking political action based on a position and accepting a leadership role. These were created in the context of apprenticeship activities provided by us. In doing so, we encouraged them to conduct secondary and primary research to learn more about food issues and use their findings to develop activist videos that made sense to them. The youth chose, as a group, to investigate food based on a theme they developed; that is, "What is the true cost of food?" As shown in Fig. 19.1, after learning about materials and energy costs of generating beef, one group conducted a study of males' and females' uses of local fast food and fresh food restaurants (M = male; F = female and Fr = fresh food; Mc = fast food) and based on their findings (from second and primary research), developed a video to educate people about relative 'costs' of manufactured vs. whole foods. This group, along with the others, then used their videos to tell other students and their families about food-related STSE issues. Many of these youth also expressed desire for making change in their own choices and for their peers and community. 'Karole,' for example, said that before the spring camp, "I never thought I would have been able to make a video to show other people, but now I totally think I could do my own and I will!" She also confidently spoke to other students from her school at a large assembly, and explained both the video development process and its message to over 100 of her peers.

19.4.1.2 Self-Efficacy

Increased self-efficacy is an important marker for youth programming. One participant during our first year with the program wrote that "We have all gained confidence and are not scared of changing something in a recipe to make it ours". Cathy, a program alumna, was enlisted to recruit new participants for the following year's program. At the outset, she was very shy and reserved, often having to be asked by a facilitator to contribute to group discussions. One year later, and only months after the program ended, she spoke with great enthusiasm about her participation in it and her current aspirations to be a chef. This was a direct change from when she began the program a year previous and was very hesitant and nervous about being in the kitchen, talking to others about food and nutrition and had little to no knowledge of

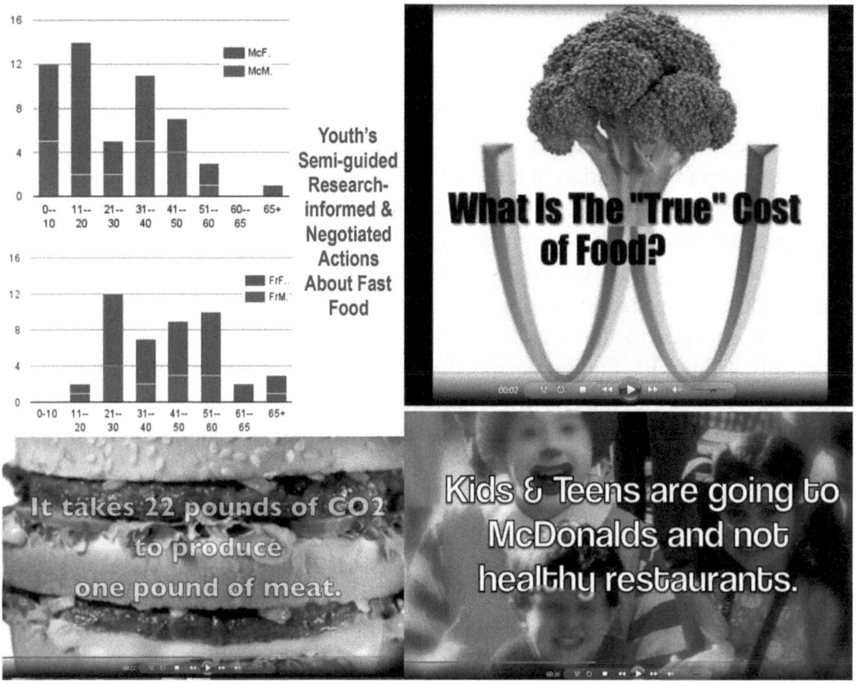

Fig. 19.1 Screenshots from an activist video about foods

food systems. She talked to potential participants at her school about her view of the importance of learning about our food systems, with passionately delivered statements, such as "Corn is in everything! Did you know that?!" This reference to corn was part of the learning about the industrial food system and processed foods. Another student, Patti, wrote in her program journal about effects of her involvement:

> On our big day [presenting at their school] we were nervous because we had more than 100 grade nine students staring at us, expecting us to say something amazing.... Through this presentation and the other activities we did, [the program] has made us test our ability to lead. To become a leader you have to really know yourself before you can guide others towards positive goals. We are definitely on our journey towards healthy food leadership.

This was reiterated in an interview with her at the end of the program. "I saw myself as a leader before, in some ways, but not in food. Now I do". This exhibits a degree of positive outcome expectancy [this is self-efficacy] that increased as a result of participation in the food justice program.

Karole also wrote in her journal and said in a group meeting that "The videos we made could seriously change people's minds about the food they eat. It's changed my perception about food for sure. This makes me feel good about teaching others something new." Based on these many positive outcomes, where we see that the girls took up positions around food justice and STSE issues, and then felt confident to speak to and educate others through actions about the issues, such as with the videos they made, the guided action could be called a success. There was repeated acknowledgement by the participants that they felt their actions with the videos could impact their peers and the choices they might make, however, several participants also expressed caution in terms of addressing sites of power.

Patti, a student who exhibited a great deal of quiet leadership throughout the program, and Karole, who was initially quite shy but became much more outspoken, both spoke on separate occasions about their sense that their school's principal would not be interested in hearing from students about food (or other) issues, such as recent changes to the cafeteria that eliminated healthy choices for students. Ironically, from a facilitation perspective, their school principal was very engaged and supportive of the students' involvement and outcomes in the program! But there was a common sentiment among the participants of "Why would they listen to me, I am just a kid."

19.4.1.3 Change and Challenges

Generally, we observed quite a lot of change and engagement with the youth participants over the program. However, our desired engagement with the group was based on an end goal of *student-led* action on STSE issues. And this did not completely come into being. We could describe this with the euphemism that "Girls were taught to fish but not encouraged to fish." Indeed, the participants were not *presented* with opportunities for entirely student-led actions. While it seemed feasible in development, with the right trajectory, unfortunately there were many challenges that inhibited the full experience/project from the perspective of the STEPWISE team. While moving along the trajectory, it was a much slower process than we imagined. Perhaps we as researchers were judging it through the lens of school-based practices where there are certain control practices over students; with evaluation and mandatory attendance, culturally and systemically enforced structure, and relatively more contact time.

Basically, while we observed guided activism, a large component of the STEPWISE model, there was little room for youth-led activism to take place. While some youth-led, community-based activism took place, such as one of the participants volunteering with seniors, and recruiting her friends, in a community garden, we are working toward a longer term review of the impact of the program on participants. At this point, we are only one to 2 years past the youth participation in the program; thus, some ideas may take longer to settle into one's consciousness, and may be taken up at a later date (Freire, 1970).

19.4.2 Factors Influencing Outcomes

Three themes emerged from the data and experience of participating in the project as a framework for understanding factors that influenced the outcomes. These are (1) the paradigmatic differences between the CBO and the research group, (2) the contextual or situated conditions, and (3) the nature of the youth. Each of these themes is explored through different categories.

19.4.2.1 Theme 1: Paradigmatic Discontinuity Between the CBO and the Research Group

The first theme that emerged is seen as the paradigmatic differences in approach to education between the CBO and the post-secondary institution-based research group. This became apparent through two main categories: pedagogical approaches and understandings of curriculum, and orientation to science.

Pedagogy and Curriculum

Differing approaches to pedagogy emerged as a challenge to ensuring mutual attainment of learning goals for the youth in the CBO program. From our perspective as STEPWISE researchers, we had attempted to work with the CBO to elicit more participant, versus facilitator, control over their experiences, which was more prevalent in the weekly delivery of the program. Through implementation of the STEPWISE framework, we have found that learners take greater ownership of their learning, and also their actions when they are engaged in cycles of student-directed, open-ended research informed and negotiated action projects (Bencze et al., 2011), as opposed to teacher-directed and -delivered learning and activism projects. We observed a strong preference for knowledge transmission versus community-based research, which we also noted as a focus on content more so than skills, which can be seen as problematic to the overall outcome expectation of student ownership as well (Krstovic & Bencze, 2012). We believe that this was the case partly because the CBO has built up so many resources and information that are connected to their mission that is very important to them to share and disseminate as directly as they can, with the short time they have with the students. However, this was in contrast to one of the tacit goals of the program. Because it was voluntary to attend, the facilitator repeatedly expressed that to keep the youth returning, "the programming is based on the interest of the students", and thus could not be altered to a great degree to push a particular agenda if they did not express interest in it. This created considerable tension in the planning and delivery of the program, in attempting to base it on interests of the youth that attended most regularly.

An additional challenge in the form of pedagogy was through the lens of technology. Specifically, while there had been an initial and ongoing discussion about the youth using video cameras and computers to document issues raised through their experiences in the program and in their homes, there was some reluctance to give the participants the cameras to take home. The facilitator also expressed resistance to using the technology herself, often stating that she did not know how to use the video cameras or software. She also voiced concerns surrounding possible losses or damages to the cameras. Perhaps this influenced decisions to provide less time and space for the youth to create their own projects. This seemed to highlight a resource imbalance between the CBO and the research team, that a resistance to the technology as a pedagogical and action tool was not considered but rather emerged through the program.

Orientation to Science

Within the culture of the CBO, there was a certain degree of discrepancy between seeing oneself as a 'scientist'/researcher or 'user of science' which impacted the openness to having students do their own community-based research and activism. There was a tendency towards accessing others' research (secondary research) versus conducting one's own research, relying on documentary films and other products of the professional food justice community to access information. Science as a body of information or a discipline was positioned outside or as a marginal set of information in relation to the food justice curriculum. Supporting this claim, the facilitator said "I am not going to get into the carbon cycle here" when presenting the idea of the overabundance of corn in processed foods. Science and scientific research was often presented as outside the scope of what the CBO was trying to achieve, thus the youth were not encouraged to draw connections between food and science, despite what we as researchers perceive to be strong and important connections.

As a result of these discontinuities, there was at times a sense of incommensurability of different educational cultures. For the first author, this meant that in order to build and maintain the research relationship, she acted as a culture broker, attendant to various demands and ownerships of such a position (Aikenhead & Jegede, 1999). While Glen Aikenhead's work is mainly in reference to indigenous and non-indigenous cultural boundaries, we use this idea in acknowledging the cultural differences of the university space and the community space, particularly with regard to youth research. These factors highlighted the difference of conceptions despite goals, to bridge the cultures of two spaces which we tried to accommodate, using the first author as a boundary agent in the role of as participant-researcher. There was a lot of pressure for Erin to align with the needs and desires of the CBO program and to release academic goals, which was a challenging position.

19.4.2.2 Theme 2: Contextual – Situated Conditions

The particular context of this research presented issues, strengths and challenges to the outcomes. These manifested through the negotiation of the context, in food as a thematic issue and in the nature of the youth participants.

Negotiation of the Context

Collaboration between the research and CBO group began, after a pilot year in 2010–11, as a more formal program in early October of 2011. We did not begin our research relationship until late October, once the program was running with participants and already had a loose curriculum planned, with room for expansion. The STEPWISE project has its own history and traditions. It is likely that because we met once the program was running with our offer of research and program support, in our excitement we did not put enough effort into negotiating details of the arrangement. Discussing ideas with the organizers, such as student-directed research and video-making for action-taking, early in the program calendar, was mainly an abstract concept that we found was harder to put into practice in reality later in the year.

Additional challenges to the context included the schedule of weekly, voluntary meetings for only 2 h that were often in competition with other demands in students' lives, such as Leadership camp and sports teams, as well as homework and evaluations for the older participants who were more aware of the impact of their school performance. A positive impact in this non-formal context was that many of the youth were able to get volunteer hours toward their school record (a mandatory requirement for the region of their formal schooling) for the hours they spent doing service with the program, including the activist videos they created and shared with their peers. This also created a stronger link between the non-formal CBO program and the various schools that the participants attended. This meant that of the four feeder schools, those with teachers or administrators with stronger connections to the program facilitator seemed to have greater participation by their students, and were more open to the participants sharing their work back in the school, which reinforced the leadership experience for the students.

Food as a Context

The pedagogy of food became a powerful emerging theme in this research. Warren Belasco (2007) describes a rich narrative when he wrote that "Food is an 'edible dynamic' binding present and past, individual and society, private household and world economy, palate and power" (p. 5). Enacting this edible dynamic means putting the community in control and at stake of food sovereignty, in that the local is the site of food power, thus it is changeable from site to site — that not every community has the same needs, resources, or capabilities. "Power is exercised through everyday mundane activities and processes, what Foucault calls technologies', [which are] hybrid assemblages of diverse forms of knowledge such as advice, techniques, judgments, experts, texts, and sanctions" (Flowers & Swan, 2012, p. 535). These sites of power are units of change in the food movement and seeing them as part of the food system means we must also teach about them, and not marginalize or undermine them. Thus, from a pedagogical perspective, the edible dynamic means adaptability, recognizing socially constructed contexts, and teaching skill sets instead of disembodied content. In other words, the pedagogy of food means critical thinking, collaboration, participatory democracy and accessing resources. Each of these is apparent at moments of the food program that was observed and participated in, with the greatest focus on critical thinking and collaboration.

Within the food program, the edible dynamic offers particular opportunities that are unique. Systemic or abstract issues can become tangible to participants, such as understanding the complexity of inputs and energy required to produce an end product of a granola bar or fast food hamburger. An example of this occurred when the youth first created their own food stories posters, which they then placed in a network of themes with each other. Themes of family, garden, heritage and friendship were identified by the participants. This activity gave participants a context for moving between the specific to the abstract, through their memories and histories of food that were shared in the group. This also provided a context for deep community building within the group.

19.4.2.3 Theme 3: Nature of the Youth

Regarding our goal of promoting student-directed open-ended research-informed and negotiated actions (RiNA), there were strengths and challenges in terms of the nature of the youth in the program. Students are seen as 'digital natives' with ease of use in computing and other digital technologies, embedded in the phones they use daily. They had an overall comfort-level and not a lot of help was needed. For example, Karole commented: "I have used iMovie™ before to make movies" ... "but not to show other people". Using Actor-Network Theory (ANT), we can understand how technology was a productive mediating tool for the outcomes, while being a detrimental factor to the facilitators (van Eijck, 2010).

The youth expressed self-confidence and expertise in food-related issues and actions, which was supported by their apprenticeship — guided step-by-step at first in doing primary research and then offered some freedom in how they would proceed with their findings. This also was enhanced in their taking leadership roles to teach each other. Using ANT, we can theorize that these positive changes were associated with who is facilitating, role models, the type of apprenticeship, location at and status of the university, the modelling of other videos like watching Food Inc., and overall, the relationship building, between the facilitators and the students, and among the students. The specific culture of apprenticeship was in many ways a spiral curriculum — thinking and rethinking of food in new ways, and considering what do we have and what can we do (Bruner, 1966). The participants became more conscious of different types of actants (commercialization, corporatization, industrialization) connected to the foods they prepare/consume which may have impacted their impetus to act on behalf of the issues. For example, one group of participants became very interested in why food at McDonald's™ was so attractive to teenagers. They were especially concerned after learning about some of the contents of the products and observing that many more teens visit the fast food chain over a healthier option in the same neighbourhood. This resulted in several awareness videos being produced in the group that included the statement "What is the REAL cost of food?", showing a deeper or more systemic understanding of food. From an ANT perspective, we consider this moving from *punctualization* (Callon, 1991), where people perceive of an actant as a single, independant, actant — e.g., a chocolate bar — unaware of contents, manufacturers, labourers, etc. toward *depunctualization*, through greater exposure to and opportunities to deconstruct the actant to recognize the multiple associated components. This curriculum in food justice can be seen as constructed to understand individual and community relationships to and knowledge of food; it became unpacked, like a picnic basket, building the youth confidence from early steps of basic cooking to the greater, more complex systemic issues. Cathy was a participant who went from "I don't know how to cook anything" to applying to go to chef school, as well as being very shy at the beginning to being very vocal about food issues with her friends and peers, such as health and commercialization. This was also the case with primary research for all participants, which was not a comfortable tool initially, but some strides were made along the way. This was cut short however due to the challenges we mentioned earlier. In addition, this program was more of a survey of what the organization does and it was claimed by several staff members much later in the program that it was not meant to offer experiences in activism, as we were told "there are other people who do that" at the CBO. This was unexpected and challenging to the researchers.

19.5 Summary and Conclusions

There were many important moments of growth for both the members of the CBO as well as the research group in relation to the outcomes achieved by the youth participants. Both the program facilitators and the researchers felt at different times confined and free within the paradigm of an extra-curricular, non-formal education setting. We have been able to identify some key influencing factors, such as the situated context of food as the curriculum, discrepancies in pedagogical approaches that were and were not negotiated, and the nature and commitment of the youth themselves. Upon reflection that programs such as those offered by the CBO have emerged out of perceived gaps in the schooling opportunities based on the desired curriculum of the CBO, it is worth considering whether the curriculum of community organizations are challenged by similar constraints to the formal schooling experience, such as in terms of time and access to authentic resources. Further, the CBO seemed constrained by inherent institutional challenges — such as funding, accountability to funders and space. Some theorization points to a deficit of availability and overall comfort with research-informed and negotiated action programming on the part of the facilitator and the CBO. It is worth considering that there are deficits due to both a lack of possible scaffolding from the formal school space, as well as from the CBO funding accountability and expectations to be met to other stakeholders. There is a need to co-develop, from the beginning, programs with the CBO, with a reflection of Habermas' (1971) emancipatory, versus practical and technical interests.

The outcomes of data analyses suggest a hesitation by the CBO to support youth-led research, apparently associated with a focus on didactic pedagogies, despite outward social justice intentions. Greater opportunities for informal research-informed activism would in fact support locally-focused CBO work. Proponents of participatory action research in community settings would find similarities and benefits of RiNA from a social justice perspective, which is ultimately in line with CBO objectives – with regard to the development of participant ownership and empowerment of and by data, leading to locally focused actions (see for example Torre, Fine, Stoudt, & Fox, 2012). Given challenges of enacting the full breadth of research-informed and negotiated actions in both formal and non-formal sites, we see a need to encourage this sort of pedagogy and have it become normalized, rather than marginalized. To this end, we are beginning to formulate possible strategies for future collaborative endeavors to create a space for SD-OE research-informed and negotiated actions based on SSIs. Our initial findings indicate a more pronounced need to identify core goals and methods, early on in the collaborative relationship and, in fact, recruit and nurture a partnership based on core goals and methods. Finally, a substantial possibility is to form one's own community group in response to the complexities and pressures of collaboration. It is yet to be determined, in this case, if the benefits of the collaboration outweigh the challenges. This case is explored further in another chapter of this publication. In recognition of the growing social and ecological challenges our world is facing, it is perhaps more prudent to collaborate, share and merge in partnership than to split our resources.

Acknowledgements This project was funded through a generous grant from the Social Sciences and Humanities Research Council (Canada).

References

Aikenhead, G. S., & Jegede, O. J. (1999). Cross-cultural science education: A cognitive explanation of a cultural phenomenon. *Journal of Research in Science Teaching, 36*(3), 269–287.

Barton, A. (2003). *Teaching science for social justice*. New York: Teachers College Press.

Belasco, W. (2007). *Appetite for change: How the counterculture took on the food industry*. Ithaca: Cornell University Press.

Bell, R. L. (2006). Perusing Pandora's box: Exploring the what, when, and how of nature of science instruction. In L. B. Flick & N. G. Lederman (Eds.), *Scientific inquiry and nature of science: Implications for teaching, learning, and teacher education* (pp. 427–446). Dordrecht: Springer.

Bencze, J. L. (2010). Exposing and deposing hyper-economized school science. *Cultural Studies of Science Education, 5*(2), 293–303.

Bencze, J. L., & Alsop, S. (2009). A critical and creative inquiry into school science inquiry. In W.-M. Roth & K. Tobin (Eds.), *The world of science education: North America* (pp. 27–47). Rotterdam: Sense.

Bencze, J. L., & Sperling, E. (2012). Student-teachers as advocates for student-led research-informed socioscientific activism. *Canadian Journal of Science, Mathematics and Technology Education, 12*(1), 62–85.

Bencze, J. L., Sperling, E., & Carter, L. (2011). Students' research-informed socioscientific activism: Re/Visions for a sustainable future. *Research in Science Education, 42*(1), 129–148. doi:10.1007/s11165-011-9260-3.

Bruner, J. S. (1966). *Toward a theory of instruction*. Cambridge, MA: Belkapp Press.
Callon, M. (1991). Techno-economic networks and irreversibility. In J. Law (Ed.), *A sociology of monsters: Essays on power, technology and domination* (pp. 132–161). London: Routledge.
Charmaz, K. (2014). *Constructing grounded theory* (2nd ed.). New York: Sage.
Flowers, R., & Swan, E. (2012). Pedagogies of doing good: Problematisations, authorities, technologies and teleologies in food activism. *Australian Journal of Adult Learning, 52*(3), 532–572.
Gruenewald, D. (2003). The best of both worlds: A critical pedagogy of place. *Educational Researcher, 32*(4), 3–12.
Gruenewald, D., & Smith, G. (Eds.). (2008). *Place-based education in the global age: Local diversity*. New York: Lawrence Erlbaum.
Habermas, J. (1971). *Knowledge and human interests*. Boston: Beacon.
Higgins-D'Allessandro, A. (2010). The transdisciplinary nature of citizenship and civic/political engagement evaluation. In L. Sherrod, J. Torney-Purta, & C. Flanagan (Eds.), *Handbook of research on civic engagement in youth* (pp. 559–592). New York: Wiley. doi:10.1002/9780470767603. ch21.
Hodson, D. (2003). Time for action: Science education for an alternative future. *International Journal of Science Education, 25*(6), 645–670.
Hodson, D. (2010). Science education as a call to action. *Canadian Journal of Science, Mathematics and Technology Education, 10*(3), 197–206.
Krstovic, M., & Bencze, J. L. (2012). The use of correlational studies to encourage activism in science. *Crucible, 43*(4). Online at: http://www.stao.ca.
Latour, B. (1999). *Pandora's hope: Essays on the reality of science studies*. Cambridge, MA: Harvard University Press.
Martusewicz, R., Edmundson, J., & Lupinacii, J. (2011). *Ecojustice education: Towards diverse, democratic and sustainable communities*. New York: Routledge.
Ministry of Education [MoE] (2009). Acting today, shaping tomorrow: A policy framework for environmental education in Ontario schools. Toronto: Queen's Printer for Ontario. Retrieved May 30, 2011: http://www.edu.gov.on.ca/eng/teachers/enviroed/ShapeTomorrow.pdf.
Pedretti, E., & Nazir, J. (2011). Currents in STSE education: Mapping a complex field, 40 years on. *Science Education, 95*(4), 601–626.
Pollan, M. (2008). *In defense of food: An eater's manifesto*. New York: Penguin.
Rahm, J. (2010). *Science in the making in the margin*. Rotterdam: Sense Publishers.
Shiva, V. (2005). *Earth democracy: Justice, sustainability, and peace*. Boston: South End Press.
Sumner, J. (2008). Eating as a pedagogical act: Food as a catalyst for adult education for sustainability. *Kursiv – Journal fuer politishe Bildung, 4*, 23–37.
Sperling, E., & Bencze, J. L. (2010). 'More than particle theory': Citizenship through school science. *Canadian Journal of Science, Mathematics and Technology Education, 10*(3), 255–266.
Tilbury, D., & Wortman, D. (2004). *Engaging people in sustainability*. Gland, Switzerland: IUCN.
Torre, M., Fine, M., Stoudt, B., & Fox, M. (2012). Critical participatory action research as public science. In H. Cooper, P. Camic, R. Gonzalez, D. Long, A. Panter, & K. Sher (Eds.), *APA handbook of research methods in psychology* (pp. 171–184). Washington, DC: American Psychological Association.
van Eijck, M. (2010). Addressing the dynamics of science in curricular reform for scientific literacy: The case of genomics. *International Journal of Science Education, 32*(18), 2429–2449.
Wenger, E. (1998). *Communities of practice*. Cambridge, MA: Cambridge University Press.
Winson, A. (2013). Spatial colonization of food environments by pseudo-food companies: Precursors of a health crisis. In M. Koç, J. Sumner, & A. Winson (Eds.), *Critical perspectives in food studies* (pp. 186–207). Don Mills, ON: Oxford University Press.
Zeidler, D. L., Sadler, T. D., Simmons, M. L., & Howes, E. V. (2005). Beyond STS: A research-based framework for socioscientific issues education. *Science Education, 89*(3), 357–377. doi:10.1002/sce.20048.

Part III
Commentaries

Overview

The chapters in this section provide theoretical and practical analyses and evaluations of STEPWISE frameworks. Authors in this section have published educational reports in various contexts and work in various parts of the world, including: Australia (3); Brazil (1); Canada (19); France (2); United Kingdom (2); and, United States of America (5). In some cases, authors report examples of educational experiences they have provided students that align well with STEPWISE principles and practices. Others discuss, in broad terms, how principles and practices can be analyzed and evaluated with regards to particular theoretical perspectives, such as biopolitics and poststructuralism. Among such discussions are suggestions for ways in which aspects of STEPWISE perhaps need to be revised. Many such suggestions may be useful for ongoing development of educational programmes that encourage critical and activist forms of citizenship.

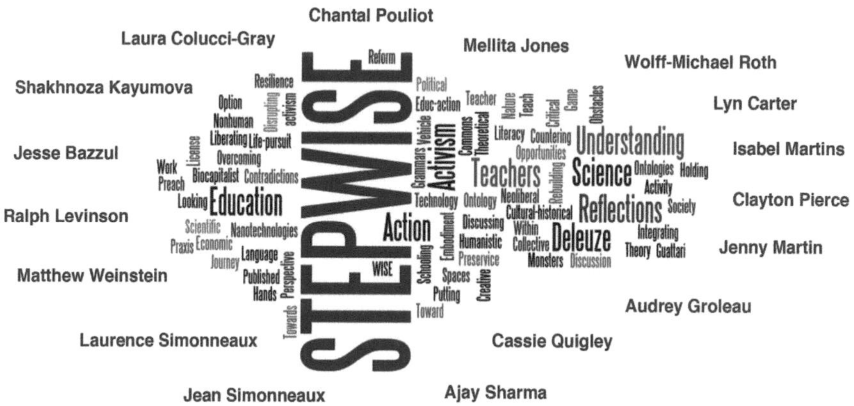

The Chapters

20	**Actor Network Theory and STEPWISE: A Case Study on Learning About Food Justice with Plants**
	Clayton Pierce
21	**Rebuilding Community Spaces: Integrating Resilience into STEPWISE**
	Cassie F. Quigley
22	**Socio-scientific Issue-Based Learning: Taking Off from STEPWISE**
	Ralph Levinson and The PARRISE Consortium
23	**"Preach or Teach?": An Ongoing Journey to Becoming STEPWISE**
	Mellita Jones
24	**"I Had to Take Action Straight Away." Preservice Teachers' Accounts of Pro-environmental Action**
	Lyn Carter and Jenny Martin
25	**Science Education for a Better World? Reflections on Embodiment, Language and Sensitive Action**
	Laura Colucci-Gray
26	**WISE Preservice Teachers Discussing Social and Economic Disparities During a Discussion Game Dealing with Nanotechnologies**
	Audrey Groleau and Chantal Pouliot
27	**STEPWISE as a Vehicle for Scientific and Political *Educ-action*?**
	Laurence Simonneaux and Jean Simonneaux
28	**Understanding Opportunities and Contradictions in the Grammars of Activism and Schooling**
	Matthew Weinstein
29	**In Which Ways Can (Science) Education Promote the Well-Being of Individuals, Societies and Environments?**
	Isabel Martins
30	**'STEPPING' Toward a Critical-Activist Science Education: Dialoguing Subjectivity, Social Ontology and Multiplicities**
	Jessie Bazzul and Shakhnoza Kayumova
31	**Countering the Neoliberal Ontology of Nature: The STEPWISE Option**
	Ajay Sharma
32	**STEPWISE: A Societal-Historical Activity (Activism) Theoretical Perspective**
	Wolff-Michael Roth

Chapter 20
Actor Network Theory and STEPWISE: A Case Study on Learning About Food Justice with Plants

Clayton Pierce

20.1 Introduction: Learning Alternatives to the Corporate Management of SSI's

Over the past year we have witnessed two contradictory milestones in the modern history of science and society. On the one hand, the International Panel on Climate Change (IPCC), comprised of the world's top climate scientists, provided the public grim data showing how the earth's CO_2 threshold will be broken by 2030 if nations (especially the U.S. and China) continue to produce CO_2 emissions at the present rate (IPCC, 2014). On the other hand, with a Republican majority now in place in the U.S. congress and the election of Donald Trump, the governments of Canada and the United States are continuing to pressure policymakers to push through the building of the Keystone XL Pipeline that will carry millions of tons of raw tar sands across North America and a multitude of ecological regions, First Nations and American Indian land, and countless towns and cities in the U.S. and Canada. This latest petroleum industry mega project is important because the latest study on the impact of burning the fossil fuel carried by the Keystone XL Pipeline is estimated to increase the world's total output of carbon dioxide emissions by 121 million tons—compared to the 36 billion tons of carbon dioxide which is currently the world's total emissions output (Erickson & Lazarus, 2014).

The Keystone XL pipeline controversy is emblematic of a socioscientific issue (SSI) that pits science (expert produced evidence) against society (government/corporate groups and deceptive media campaigns that shape public knowledge around SSI's). It also perfectly illustrates how two sets of experts are largely determining the outcome of a controversy that, in this case, has catastrophic planetary

C. Pierce (✉)
Fairhaven College of Interdisciplinary Studies, Western Washington University, Bellingham, WA, USA
e-mail: clayton.pierce@wwu.edu

L. Bencze (ed.), *Science and Technology Education Promoting Wellbeing for Individuals, Societies and Environments*, Cultural Studies of Science Education 14,
DOI 10.1007/978-3-319-55505-8_20

ramifications; one group provides the data (though among this group are also scientists hired by the petro-chemical industry to provide its own version of the facts) the other disregards or selectively utilizes these facts depending on how they fit within the economic development plans of corporations that drive public policy in the U.S. and across the globe (Bencze, 2008). What I focus on this chapter is how using an actor network approach to teaching science and technology within the STEPWISE framework can move away from the gridlock created through the expert scientific and corporate governance model to one that promotes democratic public engagement with urgent SSI's.

As creator of STEPWISE, Lawrence Bencze has defined the science education framework as "an educational and research project aimed at encouraging students to apply their science and technology education, including their primary and secondary research findings, to take socio-political actions to address 'STSE' ('socioscientific') issues—such as controversies surrounding merits of nuclear reactors".[1] In this chapter, I look at how using pedagogical spaces such as community gardens can lead to STSE actions rooted in students and communities' own practice of research and knowledge production into relevant public problems as a way to highlight the use of ANT within STEPWISE approaches to science education. Just as the Keystone XL pipeline, thought of as an actor within a complex set of socio-ecological relationships, could be studied as a way to map a whole host of interrelated controversies from species extinction to well-water poisoning to atmospheric overload of pollution, I use community school gardens as an example of a STEPWISE laboratory that can connect existing social movements against neoliberal governance of humans and nonhumans across the globe. Grounding research and learning in school community gardens where food justice issues frame inquiry, I argue in this chapter, leads to a much more democratic approach to science education as well as deeper learning about a reality riddled with complex problems and unequal power relations that determine, among other things, how we eat and learn. In this sense, both STEPWISE and ANT offer a non-deterministic and more democratically open learning approach to conducting student and community led research around pressing SSI's.

20.2 Actor Network Theory and STEPWISE

One of the biggest challenges facing the STEPWISE framework's broader adoption across science and technology education contexts, as Lawrence Bencze points out in the introduction to this book (Chap. 2, this volume), is that it fundamentally challenges the dominant STEM education reform movement in North America that is deeply entrenched within the economic development plans of nations focused on free

[1] See Lawrence Bencze's website that provides an overview and examples of how to apply the STEPWISE approach in classrooms available here: www.stepwiser.ca/ Also please see Chap. 2 in this book for a more detailed description of STEPWISE as a science education approach.

market expansion. The field of science and technology education, in other words, needs to be contexualized within the 'flat world' economic policy framework of states that frame STEM education as a national security problem and economic recovery tool (Pierce, 2012). There is not an advanced industrialized nation that does not promote rapid development of more STEM workers as part of its economic recovery plan where a high tech workforce is seen as the competitive field where success will be measured against other competing nations. Yet just as the example of the Keystone XL pipeline and countless other socioscientific issues (SSI) highlight, teaching and learning science and technology now more than ever needs to focus on the wellbeing of individuals, societies, and environments (WISE) if we are going to have a chance of changing the path of planetary catastrophe we are currently headed down. In this sense STEPWISE as an approach to science and technology pedagogy might be viewed as an example of John Holloway's suggestion for individuals and communities to work within the 'cracks' of capitalism to create alternatives to the exploitative and destructive values capitalist society requires people to embody and reproduce (Holloway, 2010).

In this chapter, I use actor network theory (ANT) to offer an example of a science and technology learning context that allows the STEPWISE framework to work within the 'cracks' of neoliberal society. Actor network theory (ANT), developed in the work of science studies theorists Bruno Latour (1987), Michel Callon, and John Law (1992) is predicated on the idea that in order for science to enter into democratic relation with the public (or communities), human and nonhuman agency must be accounted for in politic life and not treated as objects to be interpreted and controlled by scientists/experts (Latour, 1993) . Moreover, as Latour (2005) points out, ANT can be understood as a "political epistemology" in that gaining better knowledge of controversies and public debates that center around science/society problems (such as genetically engineered food) requires that communities develop ways to map complex arrangements of actors (both human and nonhuman) that create controversies in the first place (Latour, 2005).

Over the past decade STSE educators have been drawing on actor network theory as a way to disrupt dominant approaches to teaching science and technology that fail to take into account the agency of things (called nonhumans) like oil pipelines, genes, plastic bottles, or plants. Mathew Weinstein (2008), for example, has argued that teaching science in schools enrolls students into the modern scientific project in ways that reproduce the nature-culture binary and thus anti-democratic learning about science in society. Others (Bencze & Carter, 2011) have also argued that science education in the service of consumer society fails to teach students how scientific networks in neoliberal society are unsustainable. Yet others (Bencze & Alsop, 2014) have suggested that it is precisely the corporate actants shaping how science is practiced in society that science education should be focused. Pierce (2007) has also argued that ANT can be used to broaden the epistemic horizon of science education to take into account nonhumans as democratic actants.

The public controversy around the engineering and building of the XL pipeline is a good example of how an ANT approach to science education is both richer and more democratically attuned. For instance, an ANT perspective would involve a mapping of the entirety of actors that create the political power necessary to assemble such a massive engineering and energy project. In the case of the XL pipeline, prominent actors that would need to be mapped through an actor network approach include the TransCanada corporation, various agencies in the U.S. and Canadian government, tar sands, environmental scientists, climate scientists, petro-chemical scientists, engineers, a multitude of investors, the oil refinery industry in Texas, all of the impacted ecological zones the pipeline runs through, CO_2, and so on. However more than just a listing of actors involved in bringing together one of the largest fossil fuel extractive projects in history, an actor network mapping of the XL pipeline provides a more comprehensive and accurate understanding of how the controversy is managed by separating nature from culture—the way scientists and policy experts in other words speak for ecological regions or the earth's atmosphere thereby maintaining a political monopoly over how the public is to understand *and* respond to the controversy. By focusing on the actors themselves (say tar sands) without the mediating power of corporate and scientific experts, communities can come to their own conclusions and decisions regarding the mobilization of tar sands into the energy lives of people across the continent and globe. Mapping actor networks such as the ones that comprise the XL pipeline helps citizens and communities better see and understand the type of power relations involved in orchestration of controversies and ways in which powerful corporate and governmental actors have constructed a networks capable of realizing untold amounts of wealth for the fossil fuel industry and its investors while also speaking for all the nonhumans involved. Yet, in addition to highlighting the constructed actor network and power relations maintained through the corporate/governmental actors organizing the XL pipeline, an actor network approach can also help identify the 'cracks' in actor networks from which to destabilize the necessary power relations holding the assemblage together.

In the following section I provide an example of an actor network approach from my own research in 7th and 8th grade science education that I put into conversation with the STEPWISE model of STSE education. Specifically, I use the example of a school community garden where I have been doing work with students, teachers, and community members over the past 5 years as a case study of sorts for applying ANT within the STEPWISE model. In doing so I take each phase of the STEPWISE model for teaching STSE education, skills education, products education, students' research, and STSE actions and discuss how each phase occurs during research on a community problem (or a SSI) through the learning context of a community garden. I argue that one of the pedagogical strengths of using an ANT approach within the STEPWISE model is how it encourages learning critical literacies capable of understanding how corporate and state governing strategies are constructed and maintained in society in ways that limit and regulate public debate and action around SSI's. Community concerns around food justice and community health problems such as type II diabetes and access to healthy and affordable food, using ANT to map and generate the scaffolding for critical food literacies in 'food deserts', it is the hope of our research, resituates where scientific knowledge and action is produced—empowering communities to name both the SSI problem *and* appropriate (and just) responsive actions. Put differently, one important potential of using ANT as a learning theory within the STEPWISE model is that students, teachers, and communities can begin to identify 'cracks' within the governing structures that organize, for instance, how communities and students access food and sites of its production that are more just, sustainable and healthier than the ones administered by state/corporate food actor networks. Let us now look at each phase of the school garden research framed by ANT through the STEPWISE framework of STSE.

20.3 Food Justice and Community Health as STSE Education

Centering a community problem or controversies (SSI) as the point of departure for science and technology education is an approach developed by a number of critical science educators over the past two decades (Hodson, 1994, 2003). One of the unique features of an STSE approach is that it situates knowledge production (or how we understand phenomenon or problems in the world) within student learning and their connection to place and community (Mueller & Zeidler, 2010). Put differently, it is an educational practice that places development of scientific literacy as something that moves from the ground up—expert knowledge around a community problem or controversy is decentered as supreme power (Roth & Barton, 2004). STSE education embraces the fact that student and community produced knowledge is equally if not more important than expert scientific knowledge for understanding problems created by the fields of science and technology in society (Bencze & Sperling, 2012). Problems facing communities in the STSE model, in other words, are liberated from highly controlled and many times ideologically distorted scientific enterprises (i.e. XL Pipeline) and instead set into horizontal relationships of scientific research with students and communities. In the context of my research with communities and students using a school garden as a site of STSE education, the SSI (or controversy/community problem) chosen to guide research was food access/production in the community and school and the related health concern of type II diabetes. Here there is also an important intersection with work in the field that has bridged science education with ecojustice education (Mueller & Zeidler, 2010) models where science education moves away from the traditional 'gatekeepers' of science literacy (work force development for example) and instead is set within the context of environmental justice problems facing communities (Mueller, 2011). The choice of food and community health as an SSI to guide STSE research with 7th and 8th grade students was identified by members of the community garden comprised of largely Latina mothers who also work in the school's community center. For this group of community members, who also played a central role in the construction of the school garden, having access to healthy, affordable food and a place for their children (of which many are students at the school) to learn and contribute to growing healthy food is an important problem facing the community.

In our role as university researchers and teachers in a classroom, myself and two graduate students (one of which is a teacher in the classroom in which we are conducting the research) have designed a critical food literacy curriculum to implement with 7th and 8th grade students over a semester that builds out from the identified community SSI—access to healthy food and related community health concerns. In our design of the critical food literacy curriculum we framed the project within a food justice model that adopts the position that

> the food justice movement is an analysis that recognizes the food system itself as a racial project and problematizes the influence of race and class on the production, distribution, and consumption of food. Communities of color and poor communities have time and again been denied access to the means of food production, and, due to both price and store location, often cannot access the diet advocated by the food movement. Through food justice activism, low-income communities and communities of color seek to create local food systems that meet their own food needs (Alkon & Agyeman, 2011).

By situating the critical food literacy curriculum around a food justice framework our hope is that, as co-researchers with students and community members, the question of food access and community health can be connected to other local movement networks also working toward greater food justice in underserved communities in the area. In this sense, the particular SSI chosen to animate STSE education in our case has the potential to learn from and work with other community run groups who are creating their own sets of knowledge and practices.

This broader goal of the critical food literacy curriculum, however, is also one that will not be artificially constructed but instead organically take place through research conducted by the students themselves. The primary goal of the critical food literacy curriculum is to start with students' current food literacy practices they use in their daily lives that then can be used as a baseline from which to engage in research geared toward developing critical food literacies.[2] The critical food literacy curriculum is comprised of three phases or research focus areas. The first focuses on issues of food pathways in the students' community/school. In the beginning of the semester students will be asked to create food maps and answer questions about, for example, where food comes from that students tend to eat on a daily basis; what or who determines the kinds of foods that are distributed within their communities/school; how such a food network impacts food decisions students make and factors that go into this decision making process; How does food access in your community compare to more affluent-white communities across the city? The second focus area

[2] In the design of the critical food literacy curriculum project we as researchers are not starting from a rigid or clearly defined definition of what a critical food literacy will look like for each student. Instead, following ANT, critical food literacy may mean different things to different students depending on the actors and specific type of inquiry they follow as well as the local context from which the research is being done. However this is not to say that there will not be generalizable aspects to what a critical food literacy should include. For example, as our framing of the curriculum itself reflects, emphasis on research within a food justice framework that uses ANT as a method of analysis gives great importance to the actor networks that govern students and communities' food lives such as high-end grocery corporation's decision to not build stores in working class and communities of color.

of the curriculum organizes student research around food related community health problems. During this phase of research students will be asked to identify existing community health problems related to diet and evaluate scientific data on such problems as type II diabetes, obesity, and long term heart and cardiovascular health. The third and final research phase focuses on inquiry around food production—what is the dominant way food is produced in the students' community; do the students have experience producing their own food; and what alternative food production models exist currently in the community/school?

From these three general research areas that comprise the critical food literacy curriculum and the inquiry it promotes, greater specificity or fine-grained analysis will be provided through a focus on specific actors. For example, in asking students to research food pathways they will be prompted to map out their usual food consumption habits, how they access their favorite foods, and what people/actors are involved in shaping their food choices. If, for instance, soda is identified as a food type that is consumed regularly by a student, then soda will become a key actor from which students will conduct further research into how it is a prominent food choice in their daily lives by mapping out the dominant pathway it takes as a food commodity. Questions such as who or what produces soda, what ingredients are used in making it, what type of manufacturing or growing practices are associated with soda production are all possible question avenues students can use to chart the food pathway of their food example (or actor). Here emphasis will be on having students map out systems of food access and distribution in their communities and who controls the sites of access and distribution within their communities. Moreover, local grocery stores and the school cafeteria will be sites where students can research what the construction of food pathways look like in their everyday lives and a number of actors who are involved in co-constructing the network at their final distribution point.

Building out from the food pathway map of particular food actors, students will be asked to connect their analysis with health problems in the community and specifically if their food example is associated with any of the identified top health concerns. If soda is the food actor they are tracing, then students will be asked to find and analyze links made by science between diseases such as type II diabetes and obesity and roles their food actor may or may not play in the health issue. Finally, students will be asked to include in their maps of specific actors where and how they are produced. The focus on food production of their food actor will eventually be compared to alternative food production sites in the community (specifically the community garden at the school) as a way to evaluate the different interests and values comprising the two food production models (corporate/Big Agra and local, community produced food). Below, as a way to further draw out how ANT is a particularly useful pragmatic learning approach for engaging SSI's in STSE contexts, I work through each area of the STEPWISE framework as it relates to the critical food literacy curriculum project described above. I think it is useful to consider what an ANT emphasis in each phase of the STEPWISE framework looks like because it can illustrate the strength of such a pairing in STSE contexts and the complex socio-ecological settings being fundamentally transformed in a biocapitalist society. Put differently, now more than every STSE approaches need to be thinking about and experimenting with ways to pedagogically engage with problems that stem from products science and technology have created by changing nature into new commodities such as GE food.

20.4 Students' Research: Follow the Food Product

One of the main reasons ANT is an excellent method to use for STSE education is because it is suited particularly well for the complex socio-ecological problems that exist in the biocapitalist societies in which we live (Bencze, 2014). In other words, within the current phase of capitalism known as neoliberalism, all manner of forms of life have become a part of the production and exchange of commodities in society. Genetically engineered food, for example, created through rDNA manufacturing in genetic laboratories (the spliced gene sequences used to recode plant and animal organisms into new commodities) have created entirely new challenges that STSE should be at the forefront of confronting. In biocapitalist society creating cognitive maps with students around SSI's is a powerful educational tool that can help students and communities better understand ways in which an actor (say a GE corn plant) is not a singular object existing in nature (Pierce, 2015). Rather ANT ignores the modern idea of unitary objects existing in nature that can only be understood by the scientific expert (who is the gatekeeper of knowledge for the public's understanding of objects) and instead lets inquiry into an actor dictate the way we understand and think about its social and political agency Moving away from modern scientific explanations of things like GE food and toward more democratic ones involves tracing networks of power relations that constitute the lives of things like GE corn in society. So, thought of from an ANT epistemological standpoint, actors are best understood when they are set into relation to the constellation of other actors that shape our thinking about and experiences with things in the world—largely in ways that preclude democratic political relationships from forming between human and nonhuman actors (Latour, 2004). In other words, human agency is dialectically constructed through corporate and expert understandings of public problems in ways that promote an engagement with nonhuman actors such as air pollution particulates in undemocratic relations. How the public understands something like air pollution is almost always shaped through channels of power (pollution scientists-state bureaucracy-extractive industries) that deny the public coming to their own understandings *and* types of democratic practices around the problem of air pollution particulates because they are not the experts nor would it be beneficial to one of the most powerful industries in the world to allow knowledge of public problems to be generated from below (Harding, 2008).

Schools, as Matthew Weinstein (2008, 2009) has pointed out, are a particularly important site where dominant modern scientific literacies are maintained and thus non-democratic relations with actors and the controversies they are involved in are guarded through scientific/government/corporate gatekeeping mechanism—such as the way we learn about scientific enterprises like GE food. In this sense, as Weinstein points out, schools "are part of the networks that constitute science: takeaway schools and the population of bodies that are interested in science (in the sociological sense, i.e., as future scientists or as boosters, funders, etc.) become curtailed. In Latour's language, schools have been enrolled in the general enterprise of science (for a general history of this enrollment, see Montgomery, 1994)" (Weinstein, 2008, p. 400). Weinstein's ANT analysis of the way schools enroll students and the public in general into the modern scientific enterprise points to the importance of developing ANT practices within schools, teaching and learning about science where learning about something like food literacy using ANT can be seen as "shift[ing] the intentions from science as a deferred practice/authority (good or bad) to materially present and life-shaping set of social structures that constitute being a student, working in uneven ways both inside and outside the space-time of 'science education'"(Weinstein, 2008, p. 400). Weinstein's characterization of schools as incubators of non-democratic relationships between the public and the enterprise of science points to the importance of using ANT as mode of student research, a way to learn about the complex actor networks that governing things in our everyday lives such as the food we eat. It can become a way where, through student designed research around a food actor in this case, alternative political relationships can be established within the practice of STSE around SSI's such as food justice in a community. Let me give an example within the critical food literacy curriculum project of how ANT moves away from modern, anti-democratic ways of knowing and understanding controversies and the actors involved in their construction.

One of the design features of the critical food literacy curriculum project that uses ANT as a pragmatic learning approach and method of inquiry focuses on how particular food actors' agency has changed within different socioculture contexts and over time. For instance, if student researchers are studying corn (as a primary ingredient to soda) they are asked as part of the mapping research to think of the food actor in three ways: in its bioregional origin and the dominant relations it had with humans in that place; how the food actor moved from this context to other food systems (i.e. through migration, industrialization, etc.); and finally to describe and characterize the agency of the food actor in its dominant form today (such as in soda, condiments, bread, etc.). The point of emphasizing these three phases in the history of a food actor is that it encourages student researchers to understand how the agency of actors (in this case corn) change within different sociocultural and political contexts. In other words, one important lesson students can learn from researching the history of food actors is that actants such as corn are tied to local contexts differently; food desert communities, for instance, are saturated with industrial food products loaded with corn and corn byproducts such as high fructose corn syrup. As food actors such as corn get wrapped up and integrated into larger, more complex systems where science and technology play a bigger role in their social existence, students learn, the further removed people and communities are from determining what the relationship between the nonhuman actor and themselves looks like. To even ask the question of whether or not GE food is healthy and poses great risks to humans, communities are ushered into a maze of scientific debates and arguments where both their agency and that of the food actor are effectively neutralized by the terms established by the governing powers of food system networks. Here we can see, in terms of how students' research fits within the STEPWISE framework, that inquiry into SSI's such as food actors allows for a deeper level of literacy to develop that challenges the modern scientific assumption that objects in nature (corn in this case) have always been inert things waiting for scientists to come along and discover and explain their secrets (as well as their risks) to the unknowledgeable public. By following the plant, and more specifically different moments of its agency within sociocultural contexts where disagreements have emerged between the public and science (such as whether GMO food products should be required by law to be labeled), student research within the STEPWISE model builds in its practice an immunity (or least developing immunity) to modern scientific views of nonhuman actors such as corn that stymie more democratic relations between humans and nonhumans.

Here it is important to point out however that ANT and the STEPWISE framework both emphasize and encourage students to conduct research on controversies or disagreements around SSI's in a way that is not deterministic or a 'one size fits all' approach. That is, a pedagogical strength of ANT and the STEPWISE framework as learning tools is that it teaches students, teachers, and communities how to construct their own knowledge around a problem or disagreement that very well may be unique due to the differences particular actor networks have upon certain local communities. Consider again how industrial food products disproportionately affect food desert communities—researching the problem of the actant corn, in other words, looks different in food desert communities than it does in more affluent communities with greater access to healthier and fresher food.

20.5 Skills and Products Education: Digging into the Practices that Shape Food Actors

One of the important emphases of the STEPWISE framework is a focus on how to encourage students to develop skills (skills, attitudes, and ideas) and products (i.e. laws, theories, and inventions) in a variety of ways from research conducted on an SSI such as GMO food actors (Bencze, 2014). Within the particular context of our most recent work with 7th and 8th grade students, one of the primary skills the research project focuses on is developing critical food literacy relevant to the particular community from which the students are a part. I refer to 'food literacy' here as both a skill and type of knowledge; learning about food systems in one's community for instance translates to food habits and decisions made by individuals. Our research project and lesson plans define critical food literacy in the following way: an ability to critically assess and evaluate how food decisions in student lives are made and what actors and sets of food governance policy shape these decisions. Thought of in this way, critical food literacy, as we are operationalizing it in the study and through research with students, asks students to begin to create an actor map of three areas of impact on their own food lives.

The first area (or phase of research) framing student inquiry and the development of a critical food literacy within the broader framework of the study is a food desert analysis of their community. The term 'food desert' is one that has been used by food justice activists and researchers as well as governmental agencies such as the United States Department of Agriculture (USDA). As Nathan McClintock (2011) points out, the USDA defines food deserts as areas "with limited access to affordable and nutritious food, particularly such an area composed of predominantly lower income neighborhoods and communities" (p. 89). Yet, as McClintock also points out, this institutional/governmental definition is seriously lacking in that it does not take into account structural ways in which capitalist and racist economic and social policies create the very conditions of a food desert in the first place. Adopting McClintock's more comprehensive definition of food deserts our study has students research *both* access to affordable and nutritious foods (i.e. fresh foods that do not contain high amounts of fats and processed sugars) and what actors play the most influential role in shaping food policies for low income neighborhoods such as the one in which the school is located. In other words, the way we are framing the food desert analysis conducted by the students on their own community is interested in both understanding access and distribution of quality fresh/healthy foods and corporate and governmental policies that have manufactured and profited from these food landscapes.

The skills accompanying or rather that are developed during this phase of the critical food literacy research are the ability to (a) engage in a geographic mapping and assessment of food distribution sites in the surrounding community of the school and (b) a cognitive map focused on key corporate and governmental actors that construct and maintain food desert conditions (or what Larry Bencze has called 'mind maps' in his STSE action project with science students charting actors involved in cell phone actor networks) (Bencze, STEPWISE webpage). It is important to note that the skill developed within this phase of the critical food literacy is generated from an ANT approach that, again, allows for a more comprehensive way in which to understand a socioecological problem such as food justice and equity in this case. Students, in mapping the food geography of their community alongside the cognitive mapping of actors that put into motion and enforce food landscapes within communities we are hoping that students cultivate a level of critical thinking that penetrates the neoliberal rationality that frames food justice issues as a matter of personal responsibility and sound economic decision making. Food justice in communities, put differently, has more to do with levels of investments and deficit views of working class communities of color by government and corporate actors than it does with students working hard in school so they can obtain a high enough paying job to shop at Whole Foods. A food desert analysis helps students develop skills *and* critical thinking abilities that rejects such neoliberal assertions about relationships between access to and production of healthy affordable foods in underserved and disinvested communities.

The second phase of research guiding the critical food literacy project zooms in focus to food actors themselves. If the scope of the food desert analysis discussed above is on broader spatial and policy food systems, this second phase uses particular food actors to guide inquiry. Specifically, as the curriculum relates to the development of a critical food literacy, students will choose a food product they identified through an initial survey on current food eating habits to study from an ANT perspective. Here the results of the survey show that most of the foods identified as part of students' daily eating habits contain processed sugars (such as high fructose corn syrup) and saturated fats, trans-fats, preservatives, salts, and so on. From the point of departure of students' daily eating habits we are hoping that by following and creating an ANT map of key ingredients to the identified food products such as soda and pizza students can see how these actors are connected to important community health problems such as type II diabetes, obesity, and heart disease. In this sense the students' ANT research connects to aspects of the food desert analysis conducted in phase one. Food products that make up a large portion of the industrial food system of the US are high in sugars and saturated fats (among other unhealthy ingredients) and are also overrepresented in urban areas. For example, a recent study of Detroit's food desert regions showed that "over half a million Detroit residents live in areas that have an imbalance of healthy food options. They are statistically more likely to suffer or die prematurely from a diet-related disease, holding other key factors constant" (Mari Gallagher Research & Consulting Group 2007, p. 5).

Set within the context of health-related illnesses caused by food desert environments, students' ANT mapping of everyday food products such as soda become connected to larger community health problems. What's more, through fine grain foci on food products such as soda, students are also able to make relevant connections between 'hidden' ingredients such as high fructose corn syrup and type II diabetes that is a leading cause of shortened life spans in food desert communities. What an ANT mapping of a food actor might look like (and what has been mapped in previous work I've done with science students) in the classroom could be as follows. Starting with a can of soda, students are asked to identify the second most abundant ingredient after water. Once high fructose corn syrup is identified as the prime actor, students are asked to investigate these research lines of inquiry: where and how is corn grown in the US; who owns and controls how corn is grown in the US; and finally what ecological and social problems are known to be attributable to the massive amounts of corn that exists as a base food product within the industrial food system of the US?

One of the strengths of using an ANT mapping approach within the STEPWISE framework to study a food actor is that, as a learning approach, local community health issues such as type II diabetes can be connected to larger actor networks. As multiple students take on researching these guiding inquiry questions a comprehensive actor map comes into being from a collective research approach that is highly relevant to a community disproportionately affected by the industrial food system. That is, investigating a food actor such as high fructose corn syrup involves a complex assemblage of actors that would be hard to capture through one student's construction of a map—instead through a collective research process where multiple students may be focused on the same food actor there stands a much higher chance that key actors that bring high fructose corn syrup to the food lives of students will be traced and analyzed. Because the food actor of high fructose corn syrup is connected to highly complex systems such as industrial farming practices and policy, GMO corporations, the creation of huge ecological 'dead zones' in the Gulf of Mexico from fertilizer runoff that seeps into the Mississippi river, community health problems, industrial livestock farming, the exploitation of migrant farmworkers, and a host of other actor networks, a collective research format provides a much needed comprehensive view (and knowledge) to complicated actor networks. In short, creation of ANT maps by students on food actors such as high fructose corn syrup create an epistemic landscape that is both a complex and highly accurate manner in which to understand how the food lives of communities are involved (and are constructed by) complicated yet comprehensible power relations. Within this phase of the critical food literacy project, one knowledge and related skill that hopefully emerges is an epistemic standpoint from which to participate in more meaningfully and effectively in public debates around community health issues for instance. This is an important knowledge and skill, what we might call actor network critical thinking, because it provides students (and citizens of communities) with knowledge and vision in which to reject and counter simple and misleading corporate/governmental explanations of food justice isuses/controversies such as 'its your own responsibility to manage your eating habits and buy healthier foods'. Armed with an ANT analysis of food actors, students and citizens would immediately know that the issue of food justice is much more complex than they might think. Students would, instead, know that it is framed and constructed by powerful actors (who many times take on the role of expert) who have other driving interests that are in direct conflict with that of the health and wellbeing of a community and the ecosystems in which they are nested.

The third and final research phase of our critical food literacy project focuses on questions and knowledge around the school garden itself. In particular, in this final phase of student research that incorporates findings from the food desert analysis and an ANT study of a particular food product, we are asking students to think about food production in their community and from where knowledge about food production might come. In the context of this phase of critical food literacy research students will conduct a survey of the accessible land where food can be produced legally within the school boundary of the district. Here, again, how land and green spaces in the community will be analyzed is through an ANT approach. Now let me explain the two primary goals of this phase of research in order to make the ANT approach clearer.

The first goal, as indicated above, is to have students create a working map of the immediate area around and adjacent to the school—focusing on land and space where food could be grown. Once students identify what land/space is suitable for growing food (framed by questions such as: is it paved land, is it polluted in any way, is it private property?) they will then be asked to research whether or not there are any legal or governmental policies that would prevent the land from being used to grow food. The key point to this aspect of research is for students to determine how land/space is governed by different, interlocking governmental and private entities and how difficult it would be to gain access to this land for growing food for the community. The hope in framing this line of inquiry around land/space mapping is for students to make connections between privatization (both of public land and private land) and how this system of land/space governance ties into the industrial food system of the US that the students analyzed in the first two phases of research. One important question we want to raise, in other words, is how control of land and space in communities limits what can or cannot be their food lives. Utilizing an ANT approach, in this case, is a highly useful tool because it helps students construct for themselves a working map of the land/space in their community that leads to a focus on what power relations regulate potential growing space. Land/space, put differently, set within the context of this phase of research cannot be thought about in isolation but rather in relation to policies and laws that serve interests of industrial food distribution within communities.

The second connected goal in this phase is to have students identify alternative models of food production within their community. Here, the school garden will be the central focus of research; but, so will local ethnic markets, farmers markets, and other food distribution sites where fresh, healthy, and affordable food is made available. Important to this line of inquiry is for students to identify what types of knowledge are being used in the garden and by whom. For example, in the school garden where our food literacy research is taking place there is a strong and united group of committed community members who have run, organized, planted and harvested, and generally taken care of the garden for the past 3 years. From this organizational context, where a committed group of Latina mothers have been working and running the garden for 3 years, we will ask the students to determine what types of knowledge are being used in the garden to grow food. To determine what kinds of knowledge is being used in the garden students will interview the parents that run the school garden and ask questions such as "how do you decide what type of plants to grow in the garden?"; "where did you learn how to grow food and care for plants?"; "what is the difference between food from the garden and food from the grocery store?". Once students have collected their survey data from the parents who run the garden, students will then be asked to compare these answers with similar queries about dominant sites of food distribution in the community such as a grocery store. That is, are these simple questions about food production even ones that make sense within the dominant industrial food system? The hope here is that students will be able to see and contrast what types of knowledge and agency can be used and enacted in community gardens versus what the dominant industrial food paradigm requires: a passive, consumer model where cultural knowledge such as the Latina mothers at the school garden are not valued and in fact are barred from participation.

From a skills and products vantage point, this third phase of research offers multiple lines of development. For the land analysis students will be asked to measure and quantify how much land is needed to produce enough produce for a family of four. Students will also be doing soil analyses (checking the pH level for example) of soil taken from different areas on the school grounds including in the garden.

Particular scientific laws/concepts included here will be defining and setting into the context of the school garden decomposers, producer, and consumer. Lastly, guiding the research in this phase will be an emphasis on what constitutes good data collection and how to present in visual terms data to audiences such as the maps students will create in this phase.

20.6 STSE Actions: Growing Alternatives to Industrial Food Practices

One of the important and distinctive qualities of the STEPWISE framework is its emphasis on integrating the practice of science education with civic engagement around identified community problems. In this sense, STEPWISE is a model of science education that is an alternative to hegemonic models of science education that focus on developing human capital skills as part of the broader neoliberal economic development project (Pierce, 2013). It is a model of science education that is inherently democratic in this sense; it allows the learning about and production of scientific knowledge in the service of communities as opposed to dominant top-down models of science education that comprise most approaches in schools in the U.S. and other advanced industrialized nations. STEPWISE allows for a practice of science education that looks a lot like what, in the context of Europe, has been called the science shops movement which has focused on providing access to research in institutions of higher education in ways that serve community needs instead of profit growth for corporations (www.livingknowledge.org/livingknowledge/). In the example I have offered here of a school garden and the connected critical food literacy research project, there are a few STSE actions that I think are important to highlight and that are made allowable by the STEPWISE framework for science education.

Taking a problem such as food injustice in a community is a complex one that is difficult to research in one semester of class. Moreover, the way the industrial food system has been normalized over the past century in the US makes it challenging to get students (as well as teachers and community members) to question how and from where we get our daily food and in what ways this system threatens our health and wellbeing while simultaneously creating windfall profits for some of the most powerful corporations in country. Yet this is precisely the type of question the STEPWISE framework is designed to deal with as a science education model. In particular, through the phases of research described above, students and community members will participate in research that hopefully has a direct impact on their daily food lives while also generating alternatives to the industrial food system that surrounds their community. One important STSE action that is built into the critical food literacy research project is to have students present and discuss their findings to the community garden board rooted in the family center at the school.

Specifically, by the end of the semester of research students will have created a comprehensive actor network map that incorporates each phase of research through use of Mental Modeler software (www.mentalmodeler.org/). The MentalModeler software was designed by environmental science and political science researchers interested in developing a better way to conceptually map socioecological problems in a way that included community stakeholders within the research process. In other words, one of the most important virtues of the software is that it lends itself to collective decision making based on data collected by researchers and community stakeholders—thus breaking down the usual expert driven research model practiced in the academy and the sciences more broadly (which also agrees with ANT). Our goal in using the MentalModeler software as a visual method for presenting the students' research is to provide a collective platform from which to make evidence based actions in the community around food justice concerns. For instance, if students' research during the critical food literacy project suggests that expanding the size of the existing school garden so more fresh food could be offered in the school cafeteria as a way to offset the daily consumption of industrialized, highly processed food then we could run this scenario through the MentalModeler software to see what would be needed in terms of land use and production levels. One of the strengths of the MentalModeler software is that it can be used as a tool to show, through visual graphs, positive, negative, or neutral impacts on different proposed scenarios such as increasing the yield of food grown in the school garden to replace highly unhealthy food (negative impact) from the school's cafeteria. In this sense, students not only get to drive the research around questions of food justice in their community and school but they also are given the opportunity to present in a compelling and accurate way the data collected throughout the semester to community stakeholders who are interested in enlarging the school garden's growing area. The principal and the school garden board have already doubled the size of the garden recently and hopes to do the same in the near future—research produced by students who are studying food justice in their community, it is hoped, will provide compelling evidence to realize garden expansion.

Finally, in addition to using MentalModeler software to build conceptual maps where the impact of increasing the size of the garden can be visualized and measured, students will also be working in the garden itself, learning about different plants and growing practices from the community members who run the garden. So, in this sense, the garden, at the end of the final phase of research, will become a learning laboratory in which to develop alternative food literacies at the level of food production habit formation. Production of food, cultural knowledge about how to grow different types of plants and specific practices (such as how to plant corn seeds in small mounds of soil), and what the nutritional value of the food compared to the food actors students already studied are all part of the learning experience students will engage at the end of the research project. Because the particular class of 7th and 8th graders with whom we are working with are enrolled in a service/civic engagement class focused on the Latino/a community in the area, students are going to be asked what role the school garden can play in providing an important resource for the community. Here we hope to generate dialogue with students about the different production model of food practiced in the school garden and that of the food distribution sites they studied in their community. Who and what, in other words, have direct relations with food production in each model and which is more democratic in its organization?

Within the context of the school garden, then, STSE action is focused on students' interaction with the garden, community members, and the productive relations that construct this learning space. I think it is also important to emphasize this experimental learning space where students can learn growing and eating habits within the garden because it demonstrates another dimension to the STEPWISE model and in particular what STSE action can look like science education contexts: it also can include and foster the development of alternative habits and values to heavily schooled spaces (think highly standardized, industrial schooling) that are more in line with healthy communities and ecosystems. STSE actions, put differently, can provide the pedagogical space where alternative value systems to industrial models of school and food can be developed, one that Dilafruz Williams and Jonathan Brown's work on learning gardens have pointed out (2012). Using ANT within the STEPWISE framework thus allows for not only the epistemic (how we create knowledge) room to learn about complex food actor networks and the ways they maintain food injustice but also educational spaces for learning alternative habits to industrial schools and food systems. Freeing both our communities' body and minds from these industrial delivery systems of food and education needs to begin from somewhere. STEPWISE is one such exit point for us to learn our way out problems that now require complex thinking and empowered communities especially when those problems are guarded by the scientific experts of the education and food industries.

References

Alkon, A. H., & Agyeman, J. (2011). The food movement as polyculture. In A. H. Alkon & J. Agyeman (Eds.), *Cultivating food justice: Race, class, and sustainability* (pp. 1–20). Cambridge, MA: MIT Press.

Bencze, J. L. (2008). Private profit, science, and science education: Critical problems and possibilities for action. *Canadian Journal of Science, Mathematics, and Technology Education, 8*(4), 297–312.

Bencze, J.L. (2014). *'STEPWISE': Science & technology education promoting wellbeing for individuals, societies, & environments: A summary for science teachers*. Available http://webspace.oise.utoronto.ca/~benczela/STEPWISE_Sumry4Tchrs.pdf

Bencze, J. L., & Alsop, S. (2014). Afterword: Toward technoscience education for healthier networks of being. *Cultural Studies of Science Education, 9*, 607–622.

Bencze, J. L., & Carter, L. (2011). Globalizing students acting for the common good. *Journal of Research in Science Teaching, 48*(6), 648–669.

Bencze, J. L., & Sperling, E. R. (2012). Student teachers as advocates for student-led research-informed socioscientific activism. *Canadian Journal of Science, Mathematics, and Technology Education, 12*(1), 62–85.

Erickson, P., & Lazarus, M. (2014). Impact of the Keystone XL pipeline on global oil markets and greenhouse gas emissions. *Nature Climate Change, 4*, 778–781.

Harding, S. (2008). *Sciences from below: Feminisms, postcolonialities, and modernities*. Durham, NC: Duke University Press.
Hodson, D. (1994). Seeking directions for change: The personalisation and politicisation of science education. *Curriculum Studies, 2*(1), 71–98.
Hodson, D. (2003). Time for action: Science education for an alternative future. *International Journal of Science Education, 25*(6), 645–670.
Holloway, J. (2010). *Crack capitalism*. New York: Pluto Press.
IPCC. (2014). Summary for policymakers. In C. B. Field, V. R. Barros, D. J. Dokken, K. J. Mach, M. D. Mastrandrea, T. E. Bilir, M. Chatterjee, K. L. Ebi, Y. O. Estrada, R. C. Genova, B. Girma, E. S. Kissel, A. N. Levy, S. MacCracken, P. R. Mastrandrea, & L. L. White (Eds.), *Climate change 2014: Impacts, adaptation, and vulnerability. Part A: Global and sectoral aspects. Contribution of working group II to the fifth assessment report of the intergovernmental panel on climate change* (pp. 1–32). Cambridge, UK/New York: Cambridge University Press.
Latour, B. (1987). *Science in action: How to follow scientists and engineers through society*. Cambridge, MA: Harvard University Press.
Latour, B. (1993). *We have never been modern*. (C. Porter, Trans.). Cambridge, MA: Harvard University Press.
Latour, B. (2004). *Politics of nature: How to bring the sciences into democracy*. (C. Porter, Trans.). Cambridge, MA: Harvard University Press.
Latour, B. (2005). *Reassembling the social: An introduction to actor-network-theory*. New York: Oxford University Press.
Law, J. (1992). Notes on the theory of the actor network: Ordering, strategy, and heterogeneity. *Systems Practice, 5*(4), 379–393.
Mari Gallagher Research & Consulting Group. (2007). *Examining the impact of food deserts on public health in detroit*. Available from: http://www.yaleruddcenter.org/resources/upload/docs/what/policy/detroitfooddesertreport.pdf
McClintock, N. (2011). From industrial garden to food desert: Demarcated devaluation in the flatlands of Oakland, California. In A. H. Alkon & J. Agyeman (Eds.), *Cultivating food justice: Race, class, and sustainability* (pp. 89–120). Cambridge, MA: The MIT Press.
Montgomery, S. L. (1994). *Minds for the making: The role of science in American education, 1750–1990*. New York: Guilford.
Mueller, M. P. (2011). Ecojustice in science education: leaving the classroom. *Cultural Studies of Science Education, 6*, 351–360.
Mueller, M. P., & Zeidler, D. L. (2010). Moral–ethical character and science education: Ecojustice ethics through socioscientific issues (SSI). In *Cultural studies and environmentalism* (pp. 105–128). Dordrecht, The Netherlands: Springer.
Pierce, C. (2007). Designing intelligent knowledge: Epistemological faith and the democratization of science. *Educational Theory, 57*(2), 123–140.
Pierce, C. (2012). The promissory futures of education: Rethinking scientific literacy in the era of biocapitalism. *Educational Theory and Philosophy, 44*(7), 721–745.
Pierce, C. (2013). Learning about a fish from an ANT: Actor network theory and postgenomic science education. *Cultural Studies of Science Education, 10*(1), 83–107.
Pierce, C. (2015). Against neoliberal pedagogies of plants and people: Mapping actor networks of biocapital in learning gardens. *Environmental Education Research, 21*(1), 460–477.
Roth, W.-M., & Barton, A. C. (2004). *Rethinking scientific literacy*. New York: Routledge.
Weinstein, M. (2008). Finding science in the school body: Reflections on transgressing the boundaries of science education and the social studies of science. *Science Education, 92*, 389–403.
Weinstein, M. (2009). *Bodies out of control: Rethinking science texts*. New York: Peter Lang.
Williams, D. R., & Brown, J. D. (2012). *Learning gardens and sustainability education: Bringing life to schools and schools to life*. New York: Routledge.

Chapter 21
Rebuilding Community Spaces: Integrating Resilience into STEPWISE

Cassie F. Quigley

Communities across the globe are encouraging progressive, citizen-driven environmental practices. In his book, *Blessed Unrest*, Paul Hawken (2007) discovered over two million organizations work toward ecological sustainability and social justice worldwide—most occurring on a very small scale in classrooms, homes, and backyards. He calls this movement, 'the largest social movement in history that no one saw coming' and carefully connects it with an ideology focused on improving life through social, economic, and environmental justice—similar principles to the wellbeing of individuals, societies, and environment (WISE). That said, I concur with Neil Adger and his colleagues (2001), who say that environmental changes at the local level are largely obscured because of larger dominant discourses from policy makers and government officials. This rhetoric is almost entirely subjugated by impressions that environmental issues are only resolved through globally-coordinated action(s)—as documented during the Stockholm Conference in 1972, Club of Rome in 1973, the World Conservation Strategy in 1980, Rio in 1992, Johannesburg in 2002, and the present-day Earth Summits. Motives for this rhetoric are varied, and include economic and cultural globalization—meaning those in control of the global 'solutions' will advance financially and influence culture worldwide often without consideration of local knowledge (Quigley, Dogbey, Che, & Hallo, 2014). Dismantling dominant rhetoric is difficult. It requires forces that reject top-down government approaches enforcing specific actions that are counter to values of local communities. That said, we honor the work in schools and teacher education programs that are infusing STEPWISE-informed approaches. 'STEPWISE' (Science & Technology Education Promoting Wellbeing for Individuals, Societies & Environments; see: www.stepwiser.ca) is a curricular and pedagogical framework that conceives educational domains, such as skills and

C.F. Quigley (✉)
College of Education, Clemson University, Clamson, SC, USA
e-mail: Cassieq@g.clemson.edu

L. Bencze (ed.), *Science and Technology Education Promoting Wellbeing for Individuals, Societies and Environments*, Cultural Studies of Science Education 14,
DOI 10.1007/978-3-319-55505-8_21

knowledge, as co-dependent—while arranging such domains in ways that promote personal and social actions to try to rectify perceived problems associated with fields of science and technology.

As a high school, environmental science teacher, I incorporated a STEPWISE-informed approach while teaching at an international school in Southeast Asia. The students voiced concerns about residue left on their desks from frequent mosquito spraying that occurred in the school; they were worried about dangers of the chemicals. Coincidently, we were beginning individual inquiry projects, and the students asked if they could research the mosquito spraying practices in order to examine if there would be alternative strategies to spraying. As a teacher who readily incorporated place-based/project-based instruction, I jumped at the opportunity to have a project that was not only student-driven, but had potential to impact the health of the school as well as the local environment. The students invited the pest control company, Pest Busters™, to class and asked them to present the types of chemicals used in order to learn more about their pest control practices. Importantly, the students created a list of pre-questions surrounding content necessary to hold a meaningful discussion with Pest Busters employees. As discussed in Chap. 2 of this book, this step, *Students' Research*, was critical to inform any actions they could suggest in place of mosquito spraying. Some questions included, "What is the life cycle of a mosquito?" and "What diseases are spread by mosquitos?" and "What is the anatomy of a mosquito?" After explorations regarding *Students' Research*, we held a discussion with employees who presented the types of chemicals used (i.e., cypermethrin, thiamenthoxam, and chloropyrifos). The students discovered most of the chemicals were being used in drains despite warnings to not use in any water source for potentially contaminating water sources. Additionally, almost all of the chemical containers warned against use in areas where children were present. During our discussion, Pest Busters™ admitted they felt the chemicals were being overused and had warned the school board about risks of chemical resistance in mosquitos. Despite these concerns, the chemicals were being sprayed, misted, or spread in crystal form, twice a week.

As the discussion came to a close, the students began to conduct *Students' Research,* to understand the toxicity levels by the World Health Organization, and alternative uses for chemicals. Additionally, using various sampling techniques, the class determined the relative densities of mosquitos (both larval and adult) in different habitat types around the school. The focus was on the Aedes mosquito, as this is the primary way Dengue fever is spread, a major health concern for the community. Therefore, a portion of the *Students' Research* was on identification of Aedes and learning structures of the insect. The students became interested in the transfer of other diseases through insects, such as Lyme disease through ticks; and, therefore, a subset of the class explored this topic. After obtaining initial data of the relative densities, the students plotted the largest densities on a map to look for patterns. Then, they began the activist portion of study, which was to work to reduce the numbers of mosquitos by covering drains and creating systems for drying up pooled water. They then created a list of high-risk zones, areas that were still problematic and intervention was needed. The students recommended landscaping and draining to prevent further pooling of water—a breeding place for Aedes.

After a 3-month investigation, the students analyzed the data and found powerful results. They were able to reduce the number of mosquitos in a target area by almost 30%, which was higher than the reduction rate of Pest Busters™. In addition, because the students were also preventing breeding of mosquitos instead of reducing the number of live adults, they predicted the numbers to continually decrease. However, the exact prediction could not be calculated precisely given our small sample.

Next, the students presented their findings and alternative solutions at the school board meeting. The school board listened as the students presented and then asked questions about Dengue fever and the ability to target the Aedes mosquito. The students responded with evidentiary support about the ability to remove breeding grounds by reducing the amounts of standing water. The school board persisted and said, "We need a guarantee the Aedes mosquitos would be killed." Then the school board president stated, "In the 8 years we've used Pest Busters™, we've never had a reported case of Dengue on campus." At the end of the conversation, as their science teacher, I was proud. The students effectively and clearly communicated their results and responded to the questions with scientific evidence. Despite this success, I realized I had not prepared my students for one element of activism—rejection. We were so thrilled by our results that I forgot our audience was a group of businessmen and women who were running a profitable school. As we walked back to the classroom, I could see the disappointment on my students' faces and realized that STEPWISE-informed practices required something else of students—*resilience*. As an environmental scientist, I understand the resistance towards change, but also understand needs for persistent activists. So, a question becomes, 'How do we evoke persistence of activism in our science classrooms?'

Connecting scientific knowledge to activism begins and is strengthened in our classrooms. Even in the standards-driven United States curricula, activism remains a part of the goals of science education (National Research Council, 2012). Social activism involves taking personal accountability and actions in solving societal problems, but also influencing actions of others—including parents, friends, neighbors, local businesses, and in our case, the school board (Lester, Ma, Lee, & Lambert, 2006). Louise Chawla (1999) studied motivations of environmental action and found one of the most influential aspects was education especially when it involved a sense of agency, student voice, and belonging.

As I examined the STEPWISE-informed practices in my classroom, I noticed student voice, belonging, agency was prevalent. For example, during the mosquito project, the students had voice in topic of study, voice in the way in which they conducted the study, and because of the legitimacy of situation. This was a real situation that affected the students. Social justice was present in the form of student agency—the act of influence and effect in a specific circumstance—in opportunities to be agents of change in their school by working to improve conditions in their school. Dana Mitra and Stephanie Serriere (2012) found students, particularly middle schoolers, valued their schooling the most when teachers privileged their voices.

The concept of belonging includes developing relationships marked by supportive, positive experiences with teachers and peers. This includes opportunities to learn from one another (Mitra, Serriere & Stoicovy, 2012). Belonging increases student attachment to peers and teachers but also their broader community. This is critical to STEPWISE-informed practices, as they are situated in local, relevant socio-scientific or 'STSE' (Science, Technology, Society, Environment) issues. This sense of belonging helps to motivate students and provides a support system when faced with challenges in their activism (Mitra & Serriere, 2012).

Two days later, when I walked into school, I witnessed this persistence. Posted on almost all of the classroom doors were handmade signs that read, "No Pesticide Zone. Children Learning" and "No Spraying in Our Classroom." Without my knowledge, the students had talked to all the classroom teachers, presented information they discovered about the toxicity of the chemicals used in the spraying and risks of long-term exposure to high-risk groups such as children. One by one, the teachers were convinced and posted signs asking the company not to spray. The school board ignored the resistance but the company listened. The company stated it did not feel comfortable spraying in areas where employees were asking them not to do so. As a result, the school board issued a statement, "Teachers have the authority to not have his/her classroom sprayed and the pest company will not spray on any classroom with a sign. The teachers do not have authority beyond their classroom walls. Thus, any shared space, such as offices, cafeterias, etc. will be sprayed according to the guidelines established by the pest company." When discussing this victory with the students, they felt vindicated but understood decisions the school board made were not guided by scientific evidence. This frustrated them but also motivated them. They talked about a sense of belonging to a problem that was situated squarely in their school and, armed with their voice, knowledge and a sense of obligation to continue, the students persisted despite the school board's decision.

When students are activists, they often face barriers. But during these challenges, leadership can provide pathways to persist. Mitra and Serriere (2012) explored student voice in an elementary school, highlighting the case of "Salad Girls," three young students who evoked changes in the food choices at their school cafeteria. By conducting an inquiry that involved gathering data school-wide and communicating results to the decision-makers at the school, the girls were ultimately able to galvanize changes in the school menu. Similar to my students, the Salad Girls faced standoffs. However, unlike our situation, the Salad Girls had the support of the principal, who encouraged them to meet with district-wide officials. So, why did my students continue to fight without a clear conduit towards change?

As I reflect on what motivated them to continue their quest, I realized the STEPWISE-informed approach, included opportunities for students to learn about their ecological home—their school. The investigation into pesticide use and mosquito behavior armed them with knowledge and gave them agency to meet with the school board. Even though the school board refused to listen to the evidence, they felt they had a responsibility to improve their community. They were resilient. Environmental education acting alone, in the absence of opportunities for youth to gather knowledge about their community, cannot be expected to create youth that adapt to changes around them and to make better uses of their resources. Without these opportunities, we cannot expect resilience needed to persist when faced with resistance.

Reflecting on the first 20 years of the United States' National Environmental Education Act (NEEA), it succeeded in introducing thousands of people to environmental issues in their communities. Nevertheless, the public's awareness is not accompanied by a commensurate understanding to solve these environmental problems (Tidball & Krasny, 2011). This is despite considerable transformations in ways in which we teach environmental education (Hungerford, 2009). These changes include development of STEPWISE-informed approaches that include both activism and scientific content. Unfortunately, there is one component of environmental education that tends to detach and disengage youth. Overwhelmingly, environmental education continually emphasizes that humans negatively impact their environment, which implies that humans are separate from the environment (Tidball & Krasny, 2011).

Tim Ingold (2000) examined risks of viewing humans as separate from environments and offers differences between them. He contends the world can only be 'nature' without humans inhabiting it, while 'environment' is nature with relation to humans. In this view, 'environment' is always viewed from the human perspective. By delineating nature from environment, we avoid seeing ourselves as beyond the world and, thus, somehow able to intervene on its processes. Environments are continually changing through activities of humans and, therefore, 'environment' is not static. As science educators, we can reframe dichotomies of human and environment to view them as interconnected and interdependent.

This is an area where I think the process could be more transparent to students during implementation of STEPWISE and a place where I fell short during instruction. In my environmental science class, even with the infusion of STEPWISE approaches, the students repeatedly focused on humans' negative role in the environment. The students were caught in an anthropocentric view of the environment—that humans were in control of environments and their actions, while influencing their view of the environment, seemed to be in control of environmental forces. I saw this as a roadblock. For me, it was difficult to isolate scientific principles I felt were necessary for students to understand while placing it in a context that was student-driven and relevant. During our inquiry into mosquito control, there was an overwhelming focus on human's control of the mosquitos and little investigation into other factors that could control the insects (i.e., temperature, predators). Even when predators were discussed, it was how humans could introduce bats into the environment versus those naturally occurring predators. However, with each introduction of factors, solutions and scientific concepts got muddled. As a teacher, I avoided the chaos by looking at a few human-induced factors instead of a larger systems view. In this way, one challenge of STEPWISE is to reframe this anthropocentric view and embrace the messiness that characterize the environment.

STEPWISE has another opportunity to impact EE curricula in meaningful ways. Numerous experts in the field agree that there are negative perceptions of implementation of EE worldwide. There are several reasons for such perceptions:

integration difficulties that create curricular disorganization resulting in a patchwork of a variety of subjects without meaning; puzzling connections between social and natural sciences leading to over-reliance on specific fields, such as geology, biology, or geography; and lack of professional learning for teachers, which leads to difficulties selecting effective pedagogical methods to implement EE (Almeida & Vasconcelos, 2013). These deficiencies often lead to schools' one-off activities (such as trash pick-up in a park or planting trees in school grounds) or 'shock doctrine' (Mueller & Bentley, 2007) to scare children into participating in 'green' activities. While these activities provide external benefits (parks are cleaner), they fail to provide students with connections between their actions and environmental knowledge. They also fail to produce long-term effects on environments. Worse yet, these activities result in the alienation from nature, as youth feel the issues are too extensive for them to unravel (Louv, 2005). Another considerable issue with short-term approaches to EE is that they exclusively reflect an anthropocentric view of the human–nature relationship, as if this view was the only one possible (Almeida & Vasconcelos, 2013). When examining policies that are most often enacted in schools, there is an over-reliance on natural resource management to ensure the needs of humans (Feygina, 2013). In contradiction, it is exactly these views—that the value of nature is derived from the human use—that are at root of the current environmental state.

Certain theories, such as those that describe social-ecological systems, view humans as integral components of ecosystems. These theories help to link learning at the individual level with changes at the community and ecosystem level. Integrated social-ecological systems (SES) provide a lens to examine environmental education as a part of a larger system (Walker et al., 2006). SES can help make learning and education transformational at different levels (e.g., individual, community and ecosystem).

Before turning to a discussion on how SES can play a more substantial role in STEPWISE, I explore learning as an interactive process. Whereas environmental scholars have assumed cognitive views of learning (Jordan, Messner, & Becker, 2009), other scholars have focused more on interactive social contexts of learning (Alexander, Schallert, & Reynolds, 2009). Scholars who adhere to interactive learning theories suggest that, during learning, there are changing levels of participation in authentic *communities of practice* (Lave & Wegner, 1991). For example, learning occurs when someone interested in improving water quality in their community moves from being an observer of others doing a cleanup of a local lake (water quality community of practice), to peripheral participation (someone who participates but has not yet mastered the practice), to a full participant who is able to monitor and clean up the lake. However, other scholars suggest learning is more complex than novices becoming active community participants. Instead, a more recent understanding of learning describes multiple interactions of learning systems and a learning process that is continual but not linear (Alexander et al., 2009). These scholars consider learning as more dynamic and interactional, exerting a "reciprocal effect on the learner's surroundings" (p. 178).

A SES view of learning expands on this effect on the learner's surroundings and suggests that the learner has access to many tools, concepts, strategies, people and contexts that 'afford' learning. These are referred to as affordance networks (Greeno, 1998). For example, when participants clean up a local watershed, which creates a fishing spot wherein people interact with the watershed, the fish, and the environment, this creates a new affordance network. These ideas of learning are often found in ecology literature and assume an iterative feedback between the learner and their environment. In short, interactive and ecological views of learning define successful participation and increasing possibilities for action in a social-ecological system (Barab & Roth, 2006). Thus, learning is composed of individuals interacting with each other and the physical environment.

Building on this SES view of learning, environmental education is defined as a program or activities in which students interact with social, physical and biological environments, under the guidance of more experienced individuals (Tidball & Krasny, 2011). These programs have two goals: to change individual behaviors and impact social and ecological processes that foster social-ecological system wellbeing. There are many theoretical frameworks used to understand the interaction between environmental education and social-ecological systems (Jordan et al., 2013) and resilience theory is one of the theories.

Resilience theory (Folke, 2006) readily integrates social and physical environments that are critical to environmental education. Resilience in social-ecological systems examines the following: (1). The amount of change a system can undergo and still maintain its identity (such as a city maintaining a green space that is used for a children's playground even though a large overflow parking lot is built through it). (2). The ability of a system that has degraded but can still rebuild (i.e., the previously described children's playground becoming a vacant area used for dumping but then is rebuilt by the community through civic action). As I work to promote resilience in activism in my classrooms through STEPWISE, I found it important to demonstrate the resilience of our physical environments. I noticed that if students could witness the resilience of the environments, they were able to mimic this resilience to evoke change. It was as if witnessing the resilience of the environment strengthened their resolve.

SES is helpful for considering ways to increase environmental literacy, while understanding that involving students in civic action, particularly those that are community-based, do not always translate into positive outcomes. Thus, an important component of applying resilience theory is the concept of virtuous and vicious cycles (VVC) (Powell, Selman, & Wragg, 2002). These cycles represent interactions that are typically self-sustaining and reinforce one another. If the influence is negative, they are considered vicious cycles. If the influence is positive, they are considered virtuous cycles. Understanding these virtuous and vicious cycles can provide a way to understand how environmental education programs might interact to create a desirable social-ecological system.

When working with STEPWISE-informed practices with students in South Carolina, our class considered dredging of a local watershed, the Savannah River. Dredging compromises freshwater wetlands leading to health issues by affecting drinking water for millions of South Carolinians. Dredging also results in economic loss, as the Jasper port, a local port located on the Savannah, would be forced to close. In our research, we learned 25% of proposed dredging was located on the port site. The students worked with the community that lived in the neighborhood adjacent to the port. Together, the students and the community self-organized to prevent dredging and encouraged opening of the Jasper port by lobbying government officials. This work transpired over the course of 2 years and demonstrates needs for students and communities to be resilient. What I, as the teacher, noticed is that community and the students leaned on each other, offering support, feedback, and solutions as they needed to problem-solve through government agencies, webs of conflicting scientific data, and unfamiliar economic models. However, through STEPWISE, which included scientific research, community efforts and activism, the site was transformed into social and economic capital through jobs at the port, salvaged the wetlands, and protected the water source for drinking. This type of civic action ‘flipped’ the vicious cycle to a virtuous cycle of renewal and demonstrated *resilience*—of the students and of the physical environment.

Using this theoretical framework with STEPWISE-informed practices bolsters current narratives of traditional environmental education, which tend to expose youth to green spaces (e.g., hiking, camping) instead of providing them opportunities for action on spaces that are emerging from vicious cycles. In this way, the students do not act in isolation and can be a part of efforts that are ongoing and are already moving from vicious to virtuous. Environmental education without context-based, real-life situations for students to discuss their perspectives about environmental sustainability cannot realistically engage civic action and bring about change. Moreover, as stated earlier, much environmental education literature focuses on humans as the destructing force. As with STEPWISE, the goal of using this theoretical framework of resilience is to teach students not to be that destructing force by advancing environmental literacy and civic action.

Overwhelmingly, through STEPWISE-informed experiences, my students developed a sense of belonging in the community, had voice in the process, and developed agency through the activist component of the practice. All of these components are well documented in civic engagement and efficacy literature and are critical to creating an informed populous (Mitra & Serriere, 2012). However, the needs and the abilities to be resilient should be presented as an asset to promoting change in the community. Future research should consider whether specific ways in which STEPWISE-informed practices can support/encourage/promote resilience in our youth. The question remains how do we create contexts for our youth to engage with STEPWISE-informed practices in multiple ways. Within schools, potential STEPWISE-informed practices at the curricular level such as science or social studies classrooms. If STEPWISE-informed practices are readily infused in schools, youth could have meaningful experiences that could create a sense of belonging, agency, voice, and resilience.

References

Adger, N., Benjaminsen, T., Brown, K., & Svarstad, H. (2001). Advancing a political ecology of global environmental discourses. *Development and Change*, 32(4), 681–715. doi: http://dx.doi.org/10.1111/1467-7660.00222

Alexander, P. A., Schallert, D. L., & Reynolds, R. E. (2009). What is learning anyway? A topographical perspective considered. *Educational Psychologist, 44*(3), 176–192.

Almeida, A., & Vasconcelos, C. (2013). Teachers' perspectives on the human-nature relationship: Implications for environmental education. *Research in Science Education, 43*(1), 299–316.

Barab, S. A., & Roth, W. (2006). Curriculum-based ecosystems: Supporting knowing from an ecological perspective. *Educational Researcher, 35*(5), 3–13.

Chawla, L. (1999). Life paths into effective environmental action. *The Journal of Environmental Education, 31*(1), 15–26.

Feygina, I. (2013). Social justice and the Human–Environment relationship: Common systemic, ideological, and psychological roots and processes. *Social Justice Research, 26*(3), 363–381.

Folke, C. (2006). Resilience: The emergence of a perspective for social–ecological systems analyses. *Global Environmental Change*, 16(3), 253–267. doi: http://dx.doi.org/10.1016/j.gloenvcha.2006.04.002

Greeno, J. G. (1998). The situativity of knowing, learning, and research. *American Psychologist, 53*(1), 5.

Hawken, P. (2007). *Blessed unrest: How the largest social movement in the world came into being and why no one saw it coming*. New York: Penguin Books.

Hungerford, H. R. (2009). Environmental education (EE) for the 21st century: Where have we been? where are we now? where are we headed? *The Journal of Environmental Education, 41*(1), 1–6.

Ingold, T. (2000). *The perceptions of the environment*. London: Routledge.

Jordan, R.C., DeLisi, J.R., Brooks, W.R., Gray, S.A., Alvarado, A., & Berkowitz, A.R. (2013). *A collaborative model of science teacher professional development*. Paper presented at the International Journal of Modern Education Forum, 2(2).

Jordan, S., Messner, M., & Becker, A. (2009). Reflection and mindfulness in organizations: Rationales and possibilities for integration. *Management Learning, 40*(4), 465–473.

Lave, J., & Wegner, E. (1991). *Situated learning: Legitimate peripheral participation*. New York: Cambridge University Press.

Lester, B. T., Ma, L., Lee, O., & Lambert, J. (2006). Social activism in elementary science education: A science, technology, and society approach to teach global warming. *International Journal of Science Education, 28*(4), 315–339.

Louv, R. (2005). *Last child in the woods: Saving our children from nature-deficit disorder*. New York: Algonquin Books.

Mitra, D. L., & Serriere, S. C. (2012). Student voice in elementary school reform examining youth development in fifth graders. *American Educational Research Journal, 49*(4), 743–774.

Mitra, D. L., Serriere, S. C., & Stoicovy, D. (2012). The role of leaders in enabling student voice. *Management in Education, 26*(3), 104–112.

Mueller, M. P., & Bentley, M. L. (2007). Beyond the "decorated landscapes" of educational reform: Toward landscapes of pluralism in science education. *Science Education, 91*, 321–338.

National Research Council. (2012). *A framework for K-12 science education: Practices, crosscutting concepts, and core ideas*. Washington, DC: The National Academies Press.

Powell, J., Selman, P., & Wragg, A. (2002). Protected areas: Reinforcing the virtuous circle. *Planning Practice and Research, 17*(3), 279–295.

Quigley, C. F., Dogbey, J., Che, M., & Hallo, J. (2014). Investigating local sustainable environmental perspectives of Kenyan community members and teachers. *Cultural Studies of Science Education, 10*(3), 551–580. doi:10.1007/s11422-014-9584-1.
Tidball, K. G., & Krasny, M. E. (2011). Toward an ecology of environmental education and learning. *Ecosphere, 2*(2), 1–17. , art21.
Walker, B., Gunderson, L., Kinzig, A., Folke, C., Carpenter, S., & Schultz, L. (2006). A handful of heuristics and some propositions for understanding resilience in social-ecological systems. *Ecology and Society, 11*(1), 13.

Chapter 22
Socio-scientific Inquiry-Based Learning: Taking off from STEPWISE

Ralph Levinson and The PARRISE Consortium

22.1 Introduction

In the contemporary post-industrial world, where effects of neoliberalism and globalisation on education policy are becoming increasingly insidious (Ball, 2013), it is encouraging to see a resource such as STEPWISE,[1] which explicitly challenges underpinning social values and epistemologies of school education encouraged by a market adapted for extreme consumerism. STEPWISE provides not only encouragement for socio-political engagement but a theoretical framework that justifies the strategies adopted. For reasons discussed in Chap. 2, it is challenging for the STEPWISE philosophy to gain leverage in science school curricula. Those situations where STEPWISE has gained momentum perhaps emphasize the exceptional characteristics of pedagogy and democratic school structures where such practice is enabled. Enacting the STEPWISE philosophy cannot separate itself from the broader educational and social context. Science curricula in most post-industrial countries have outcomes based on meeting certain defined targets and a teacher

Based on: Levinson, R. & The PARRISE consortium (2014). *Initial SSIBL framework*, D1.2 PARRISE, co-funded by the European Union under the 7th Framework Programme, Freudenthal Institute for Science and Mathematics Education, Utrecht, The Netherlands/University College London – Institute of Education, United Kingdom.

[1] STEPWISE' is the acronym for Science & Technology Education Promoting Wellbeing for Individuals, Societies & Environments. It is a theoretical and practical framework that organizes teaching/learning goals in ways that encourage and enable students to self-direct research-informed and negotiated actions to address personal, social and environmental problems linked to fields of science and technology. To learn more about this framework, refer to Chap. 2 in this book (and: www.stepwise.ca).

R. Levinson (✉)
University College London, Institute of Education, London, UK
e-mail: r.levinson@ucl.ac.uk

L. Bencze (ed.), *Science and Technology Education Promoting Wellbeing for Individuals, Societies and Environments*, Cultural Studies of Science Education 14,
DOI 10.1007/978-3-319-55505-8_22

culture that is underpinned by 'presentism' (Hargreaves & Shirley, 2009), a persistent focus on short-term measurable outcomes.

22.2 The Political Climate

Schools and their curricula do not exist in a socio-political vacuum (Apple, 2004). Science curricula, and their associated STEM agendas, have been particularly susceptible to political influences (Pierce, 2015). This can be seen, for example, in the 2012 curriculum reforms by the UK conservative government, which imposed greater prescription and focus on content (Vasagar, 2012). Added to this are political pressures generated from international comparisons through the PISA results where England was seen to do badly compared with the Asian tiger economies. '"England needs a rocket under them to improve their PISA scores", says Minister for Education'" (Baird et al., 2011, p. 140).

One of the reasons why there is such panic about achievement in STEMsubjects can be seen in resources that privilege science, technology and engineering in higher education as well as in rhetoric that accompanies pushes towards the STEM agenda. Such rhetoric is couched in language of national economic competitiveness, human capital resource (Thomasian, 2011), supply and demand, high-end technologies and added values — together with 'softer' language of climate change and sustainability (Ravetz, 2005). The new science-society formulation of the EU, however, is 'RRI' (Research, Responsibility and Innovation): emphases being on science *for* society and *with* society (Owen et al., 2009). Advancements in science and technology need not be detrimental to the planet or to human communality; on the contrary, with public participation and goodwill, technoscientific progress in a market-driven economy could, according to RRI philosophy, go hand-in-hand with technologies that can remediate some of the more harmful effects to environments. Such proposals need to be treated with caution, however, particularly in light of dismantling of welfare state policies in Europe, rise of free marketism and entrepreneurship, as well as the complexity of relations of technical expertise and lay knowledge and concerns (Jasanoff, 2003).

No one except an extreme Luddite (in fact there is a lot we can learn about collective bargaining from the Luddite movement and the political organisation in response to the introduction of labour-saving technologies) would gainsay that technologies can be enhanced for the public good. For those of us, particularly those born between the end of WWII and the 1960s, who live in the post-industrial world in relative affluence and employment, benefits of a highly developed science and technology base are manifest in vastly improved health, longevity, mobility, educational possibilities compared with our grandparents. But the problem is not that science and social egalitarianism are mutually contradictory; i.e., that science and technology are associated with markets and free enterprise, as opposed to fair distribution of goods. Despite attempts to exploit possibilities of digital technologies, for example, to address social exclusion and enhance social mobility little progress has been made (Selwyn et al., 2001). One of the problems is the nature of the consumer-led and driven market that creates goods which harm social life, and a hyper-reality (Baudrillard, 1994) that is self-referencing. Social and material inequalities reduce social trust and drive consumerism; there is a correlation, for example, between a country's income inequality and low levels of waste recycling and high carbon emissions (Wilkinson & Pickett, 2010). When tracing back materials that give us so much value (e.g., superconductors, gems, rare metals) to their source (a problem addressed by STEPWISE), benefits of material progress for the affluent in rich countries need to be measured against material and social devastation caused to those in producer countries (Shiva, 2000).

Over the last thirty years in the UK and much of the industrialised world, there have been shifts through the 'Third Way' (Giddens, 2008) towards more overt neoliberal discourses; the recent TTIP (Transatlantic Trade and Investment Partnership) agreements reflect the extent to which the EU treads carefully with global corporate giants. New information and social media technologies have accompanied these changes, which have also enhanced possibilities for, and economics of, globalisation. In terms of science, there have also been concomitant changes in the nature of citizenship, from one which was dependent on the goodwill of the state and scientific expertise to one which has become sceptical of expertise, and organising itself in new ways (Novas, 2006). STEPWISE's response to this problem is to enhance social empowerment through school science education (Bencze & Carter, 2011).

22.3 Socio-scientific Inquiry-Based Learning (SSIBL)

In this article, I discuss a framework for a European project (EU) that is influenced by philosophy associated with STEPWISE, namely an aspiration to social justice through authentic action, but which builds up from inquiry and citizenship through the EU formulation of RRI.

SSIBL The European Union has a broad commitment to Inquiry Based Science Education (IBSE) (Rocard, 2007). Inquiry-based methods have been shown to increase 'both children's interest and teachers' willingness to teach sciences' (p. 12). Much of IBSE funded by the EU to date has focused on developing scientific knowledge and procedures (in STEPWISE terms, Products and Skills Education) and has been broadly inductive. Pierce (2015) has described this separation of science from social and cultural concerns as 'purification,' which stems from a broader Enlightenment problematic.

Socio-scientific inquiry is challenging for teachers because it takes students to unexpected and unanticipated areas of knowledge. Some of the inquiries discussed below are similar to the kinds of activities proposed by STEPWISE. I am part of a consortium of science teacher educators in universities in Europe, with the acronym PARRISE (Promoting Attainment of Responsible Research and Innovation in Science Education) funded by FP7 ('FP7' stands for the '7th *Framework Programme for Research and Technological Development*' and is designed to respond to

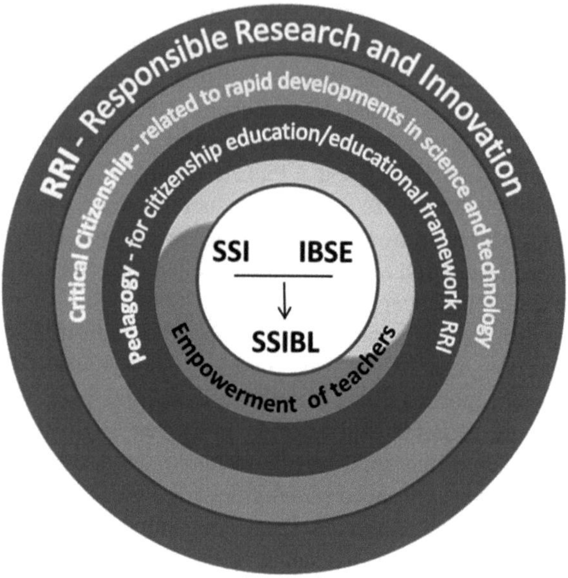

Fig. 22.1 The components of SSIBL

Europe's needs in terms of jobs and competitiveness, as well as enhancing the global knowledge economy), developing inquiry activities in the context of Research, Responsibility and Innovation (RRI), Citizenship Education (CE) and Socio-Scientific Issues (SSI) (see Fig. 22.1).

Our project acknowledges importance of social participation: scientific research and production should be carried out *with* and *for* society (Owen et al., 2009). How this can be achieved presents political and structural challenges (von Schomberg, 2013) through 'anchor points' that are ethically acceptable, sustainable and socially desirable. Influences of political literacy; i.e. critical citizenship education on inquiry-based activities frame science inquiry within contexts of social and political questions, what we have termed SSIBL (Socio-Scientific Inquiry Based Learning). At the heart of SSIBL is researching a question aimed at improving local and/or global conditions, producing realisable outcomes through democratic processes, and drawing on scientific knowledge that may be recontextualised as part of this process. The inquiries should stem from the concerns and pre-occupations of the young participants, although scholars such as Laurent Humbel et al. (2012) recognise that social inquiries stimulated by controversy need to incorporate a pedagogical triggering mechanism, an 'element declancher'. Hence, part of the SSIBL programme at the scaffolding stage has much in common with apprenticeship activities in STEPWISE. In the next section, we describe some examples that reflect the spirit of SSIBL.

What SSIBL Activities Might Look Like Before depicting the SSIBL framework, below are a few examples to illustrate its philosophy.

22.3.1 Campaigning Against the School's Sugary Drink Dispenser

This account was given by a college principal and formed part of the evidence for the Valuable Lessons (Levinson & Turner, 2001) research study.

Senior management in a college for students in the 16–19 age range installed a drinks dispenser to raise money for extra-curricular activities. Noting the problem of a dispenser of high-sugared drinks in their college, a small group of students decided that the action by the school authorities was detrimental to the students' interests. Such drinks were deemed to be unhealthy and to inhibit concentration. They approached the Principal, asking for the dispenser to be withdrawn. The Principal refused, arguing that money raised by use of the dispenser helped to fund out-of-school activities and was used by many students.

The group then decided to collect as much secondary evidence as they could to buttress their argument and to campaign within the college for its removal. They brought their argument to the College Council, a representative student body, which decided that the dispenser was not in the college's best interests but to also form a group to find alternative ways to make good any losses incurred by the removal of the dispenser. The college management agreed to implement their decision and to work with them to find alternative ways to raise money.

22.3.2 Assisted Reproduction

This activity is based upon challenging representations of assisted reproduction (AR) often promoted by private clinics (Fig. 22.2).

Questions about AR are commonly-discussed in older age groups in secondary schools, aged 15+. Young adults are reaching an age where having children becomes realisable and many values of family and status are related to having children. AR is a medical resource that can enable a couple to have children, but problems as presented in schools are often conceptualised as medical ones: the biology of the couple's reproductive systems, and associated psychological problems (Reis, 2015). But these raise other 'hidden' questions.

1. Should AR be publically funded through the health service. What is health? Does AR come under the category of health? In some countries, AR may be positively encouraged through public health services for political reasons; e.g., population growth. In others, it may be very difficult to gain access to them.

Fig. 22.2 Use of images to promote raising questions (Taken from http://www.dailymail.co.uk/news/article-2143184/Fertility-firm-appeals-Cambridge-University--girls-egg-donations.html)

2. If AR is sought privately, this raises questions of social justice — a balancing of rights ('I have the right to spend my money to follow options open to me') as against egalitarianism and social cohesion ('why should advantages of embodied nature be available for some and not others?'). In a society where it is seen as desirable, or having status, to have children accessibility to this technology is, therefore, a political question. What about political, cultural and religious issues in extending these rights to same-sex couples?
3. Regulation of the fertility industry. Fertility clinics regularly advertise through media (see Fig. 22.2). How they represent themselves is open to critique. But how are they regulated against malpractice?
4. That technologies are available makes it important to understand risks involved. What information would we need to assess these risks?
5. What about ethical questions implicit in AR? Such a technology effectively makes selection of particular attributes possible. The most common is sex selection but also selection against or sometimes for particular disabilities; e.g. there has been a debate in the deaf community about selecting *for* deaf children (Mand et al., 2009).
6. Poorer countries have become suppliers of cheap womb labour and services; e.g., egg provision. Globalised economies, reproductive tourism (www.eggsploitation.com).

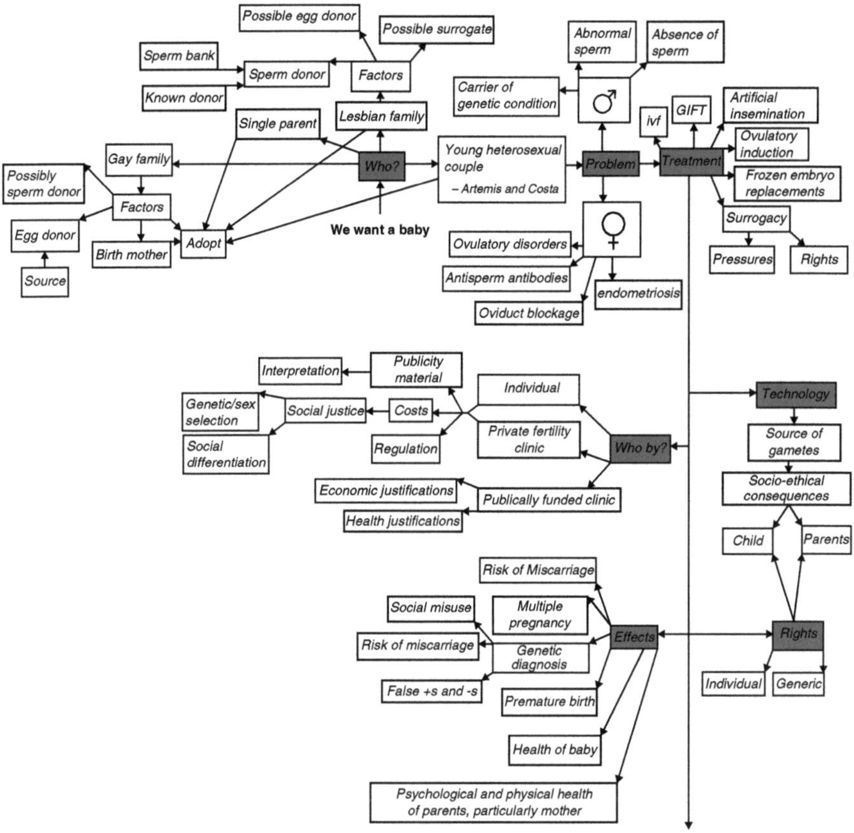

Fig. 22.3 Concept map – Assisted Reproduction

Hence, this raises questions at personal, social and global levels:

- What does this mean for me and my family? How do I feel about the possibility of AR?
- What do these questions mean for the society I live in?
- What are the global issues connected with this? What are the practices now and how do they promote diversity and inclusivity, compassionate justice and renewal of life?

Figure 22.3 represents the inter-connected issues that arise from the above questions about AR and link the social and political questions to the scientific context. An outcome of inquiry into this issue might be a leaflet produced by students which raises some of these questions, and suggested stimuli for discussion in science lessons.

Table 22.1 Constituents of SSIBL activities

Activity/Aspect	Interest for students	Outcome	Science knowledge
Animal house	Animal welfare and fuel costs	Model selected for a more fuel efficient animal house	Transfer of energy Homeostasis
Sugary drink dispenser	Health aspects of high-sugar drinks	Alternative means of raising money through healthier drinks; increased political participation	Diabetes; effects of sugar on body
Assisted Reproduction	Interest in new reproduction technologies	New teaching resource	Reproduction; ivf

22.3.3 School Animal House

Studying heat transfer, the teacher might use a number of examples, including the school animal house. Pupils in a school might be aware that the animal house is quite old, and can become over-warm in summer and too cold in winter. As a result, in winter, the heating system has to be kept on to keep the animals warm and, in summer, the fan has often to be kept running to cool the animals. This extra use of electricity impacts on the school's electricity bills and pupils can relate this to excessive and unnecessary use of fossil fuels at a global level. They could draw on their knowledge of heat transfer to solve the problem and, as a class, generate the question: 'What is the best design for the school animal house?' It has to maintain a steady temperature (knowledge of small warm-blooded mammals) under different weather conditions. The planning stage might involve different groups testing different materials and designing small-scale models to check their predictions. Each group designs their own model and tests how well they maintain a steady temperature in different ambient temperatures (high and low). Some time is allowed to change designs, if necessary, and then each group presents their findings to the year group. The best design is selected and the pupils build the animal house (or employ a company to build the house according to their plans). Tests are carried out once the animal house is built to check that it is working properly, and they also assess changes in fuel bills as a result of their design. The details of their inquiry are presented at the local teachers' science education meeting.

Table 22.1 outlines the age range, the scientific knowledge that needs to be recontextualised and applied for each inquiry activity and possible action points.

22.4 Explaining the Framework

SSIBL is comprised of an overarching context, RRI, and three interconnected pillars: CE, SSI, IBSE underpinned by an engaged pedagogy (Fig. 22.1). I discuss each of these, in turn, below.

22.4.1 RRI

Technological developments, inspired by research and innovation, both have an impact on, and are influenced by, social values and social change. Owen et al. (2009) identify three underpinning features of RRI:

(i). Science *for* Society (SfS),
(ii). Science *with* Society (SwS), and
(iii). coupling of research and innovation with responsibility (R&R).

Science *for* society focuses on public values, i.e. normative motivations; science *with* society on dialogue and deliberation, i.e., substantive motivations; and, coupling of research and innovation with responsibility as a recognition of practices of science, uncertainties and risks associated with development of any technology and how these might be anticipated and managed (Ravetz, 2005).

Science *with* society is participative. This acknowledges that those affected by the technology, as well as scientists, can influence decisions both at the upstream stage (that is, when the scientific ideas are initiated and possible consequences anticipated) as well as downstream at the point of production, application and distribution. Participation and dialogue in research assume knowledge and understanding of the underlying science, as well as critical appreciation of processes of the research both in its scientific and social components. Participative R&D is, therefore, a multi-agency approach to research and innovation because knowledge is differentiated and distributed in form (i.e., from academic knowledge, including different disciplines, professional knowledge, knowledge-for-living) (Layton et al., 1993). These foreshadow interactions between formal (curricular) and *informal* (non-curricular) knowledges.

In discussing values, people are identifying not only norms by which societies cohere but also those that are desirable. These can reflect a conflict between market-driven economies and needs for ethical relationships between people within a sustainable society. For example, drives for economic growth can potentially stimulate development of alternative technologies that support sustainability and zero carbon outputs. However, it can also endanger desirable outcomes because economic growth drives increasing levels of consumption; hence, the need for critical approaches that identify, problematise and raise questions about underpinning *values*. SfS is the process where science takes into consideration 'the values, needs and expectations of society.' (ec.europa.eu/programmes/horizon2020/sites/horizon2020/files/FactSheet_Science_with_and_for_Society.pdf).

Underpinning a curriculum and pedagogy that aims at enhancing human capacities within socio-cultural realities in which people live, Roger Simon (1992) derives three principles: *securing diversity, compassionate justice, renewal of life* (Fig. 22.4).

Securing diversity assumes differences between people (and non-human species) from classrooms to the whole planet entailing an ethic of respect. This implies opening participation in classrooms to young people who are often prevented from fully-engaging, and understanding that needs, interests and voices of people and communities across the world are mediated by power relations and have unequal status. Recognising diversity means inclusivity in terms of impacts science and technology have on a whole range of stakeholders, including those who cannot

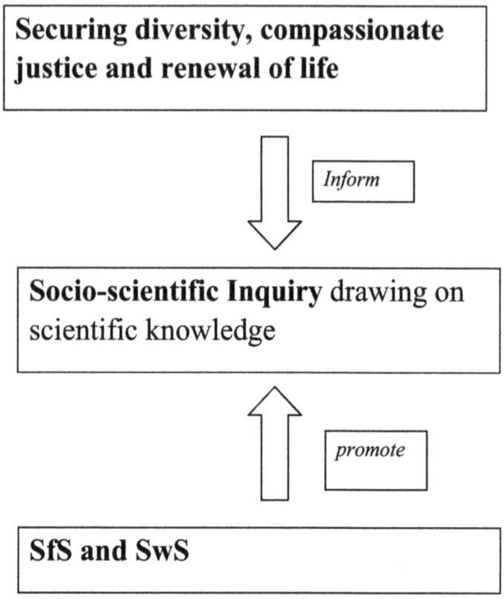

Fig. 22.4 Relationships between RRI and socio-scientific inquiry

claim a stake for themselves but who are affected by impacts of the technology. It is an opportunity for disadvantaged groups to gain and use relevant knowledge.

Disadvantage and marginalization can be problematic to recognize and address, particularly where school systems are not adapted to such needs, and where there are deep-rooted social, cultural and economic factors. These can take different forms and kinds of solutions. Where there is purposeful liaison with a particular community; e.g., the Roma communities in different parts of Europe, and recognition of legitimate identities, there can be clear gains in ways in which students meet their legitimate aspirations (Nistor et al., 2014). Where schools can seem threatening and oppressive to some groups, arrangements can be made to carry out inquiries in other arenas outside of them (Ellsworth, 1989).

While recognising diversity implies openness (willingness to listen to others, respect what others have to say, and change one's mind if convinced by better reasons), it does not imply agreement. What drives dialogue is difference and controversy (Hess, 2009).

Compassionate justice, minimisation of suffering, is a driving factor within RRI. Science *for* society, means that fruits of technology are distributed fairly, and that we have a mutual obligation to fellow inhabitants of the planet and a sensitivity to power relations which often distort those obligations. Renewal of life can be expressed as recognising "the interdependence of human life within a living planet as a source of both constraint and indeterminacy of human plans" (Simon, 1992, p. 27). Taking sustainability seriously entails respect for responsibilities we have towards each other.

22.4.2 Citizenship Education (CE)

CE can be seen as a continuum from knowing what is entailed by citizenship to having a more active concern for seeking justice. The term *critical* CE can also be interpreted in different ways from one which focuses on critical thinking to an emphasis on praxis; i.e., reflection and action as well as constructive dissent (Levinson, 2010). A useful framework, in the light of activities discussed, to represent the dimensions of critical CE is adapted from Johnson and Morris (2010) (see Table 22.2).

The horizontal row: politics, social, self and praxis, represent the component elements of critical citizenship education while the vertical column represents the necessary attributes. Each cell describes how each attribute exemplifies each element with the brackets indicating how they might be manifested in the context of SSIBL within the classroom.

Deliberative dialogue is at the heart of the democratic process that incorporates the substantive meaning of dialogue as communication between participants but also the appropriate dispositions, such as listening, equality, respect and openness (Rice & Burbules, 1992) that presuppose constructive dialogue. In addition, this dialogue incorporates criticality, an ability to identify and respond to logical inconsistencies and unsupported assertions. Reasonable people hold their views open to criticism and are prepared to justify them or revise their views in the light of more compelling arguments. Deliberation goes beyond dialogue in that, in the democratic context, it involves free and equal citizens giving reasons to settle socially urgent questions (Simonneaux, 2014) on which they have divergent views (Enslin & White, 2003). In the context of schools, deliberative dialogue has much in common with Neil Mercer and Karen Littleton's (2007) construct of group exploratory talk, in which students share relevant information about a problem, listen actively, and where everyone contributes, helping to build up on ideas to reach agreement. This drive towards consensus is one of the historic features of democratic deliberation (Habermas, 1984). While these features need to be aspired to in the democratic classroom, this is not always the case because dialogue is always mediated by power, which can be through positions of status (teacher and student), differential access to knowledge (scientist and layperson) and inequalities in social and cultural capital (Gamarnikow & Green, 2000). Effectively, this means that what might be seen as a normative view by most students in a classroom might be seen very differently by one or two others who might feel disinclined to make their views known, again a case of inclusivity. For example, a teacher and the class might start off from the proposition that global warming is an important issue to address while a small few may feel differently, perhaps because they feel far more pressing concerns or that people close to them have very different views from the rest of the group.

Democratic deliberation needs to be fostered in the classroom and cannot be assumed. It also presupposes an environment where students trust each other as well as the teacher and where questioning habits have been encouraged. The SSIBL process itself can encourage such an environment but encouraging constructive dialogue in the classroom might need to be built up and nurtured over a period of time. How conditions for constructing democratic deliberation in the classroom are facilitated depends on the teaching and learning context. In an environment where teachers and students are used to arguing and discussing in an open and respectful manner, attain-

Table 22.2 Dimensions of critical citizenship education which incorporates principles of compassionate justice, securing diversity and renewal of life

	Politics (ideology)	Social (collective)	Self (subjectivity)	Praxis (engagement)
Knowledge	Knowledge and understanding of political systems and power structures. (understands where authority lies, e.g. that school student council, governing body and Principal, will need to be influenced to effect change)	Knowledge of interconnections between culture, power and transformations; knowledge of non-dominant as well as dominant discourses. (appreciates that there are a variety of opinions and to look out for marginal voices, e.g. 'silent' third world egg donors)	Sense of identity (understands how they are positioned in relation to a particular issue, e.g. right to buy fertility treatment)	Knowledge of how to collectively effect change for social justice. (knows how to garner support to effect change, e.g. campaigning against, sugary drink dispenser)
Skills	Critical political analysis. (Understands relationships between power, culture and knowledge; hence ideas of status of knowledge – the relationship between expert, anecdotal and communal knowledge)	Capacity to engage in dialogue and deliberation. (e.g. take part constructively in classroom discussions, both face to face and online)	Reflect on own status in society. (can place themselves in others' shoes while aware of their own position)	Imagining a better world; active participation in acting collectively to change status quo. (Articulates a vision of a better world and how to implement that aspiration practically)
Values	Commitment to values opposing injustice and oppression. (Advances an understanding of causes of injustice and how it relates to their own value system)	Inclusive dialogical relationship with others; ability to reflect others' values and commitments. (Can articulate viewpoints of others even where there is disagreement)	Consideration of self-worth. (Expresses why they have a particular perspective and its meaning to them)	Informed responsible, reflective ethical action. (Action taken is thoughtful and reflects underpinning values)

(continued)

Table 22.2 (continued)

	Politics (ideology)	Social (collective)	Self (subjectivity)	Praxis (engagement)
Dispositions	Actively questioning social injustice and oppression. (Raises critical questions about acts of injustice which can then generate questions for enquiry)	Responsible towards self and others. (keeps social responsibility foremost in thinking)	Autonomous and critical (Can listen to others' perspectives but maintains their own view, albeit self-critically)	Commitment and motivation to change society responsibly. (Communicates reasons for actions to others)

Adapted from Johnson and Morris (2010)

ing SSIBL will be relatively unproblematic. However, in more authoritarian learning environments many adjustments will need to be made, hence the structural and political positioning of schools within a broader social domain. A more gradual approach is necessary where students could be taught procedures for group talk then go on to develop their own procedures.

Critical citizenship education also incorporates a knowledge of political and moral concepts such as rights and equality. These are not necessarily mutually supportive concepts: ensuring equality might mean restricting rights. While rights — with responsibility — and equality are desirable, they can only be discussed in relation to their limitations. For example, if parents have the right to pay to choose the sex of their baby, this will have implications for equality, ethical values and natural justice. Interdisciplinary arrangements in school may be needed to foster these components, for example, planning for SSIBL with science, history and citizenship teachers collaborating.

22.4.3 Socio-scientific Issues (SSI)

Socio-scientific issues comprise conflicting opinions about a course, or courses, of action that have a scientific content and impact upon communities or society. They are controversial when good reasons can be given for conflicting opinions and/or courses of action (Dearden, 1981). There can be different levels of controversy. At one level, a controversy might be solved upon the production of relevant evidence, e.g. differences about the best material for lagging an animal house can be tested based on experiments to measure temperature difference. On the other hand, there may be core differences of values which are less easily settled, such as whether it is right or wrong to abort a foetus under certain conditions (Levinson, 2006). Where there is controversy, particularly as they impinge on core values, strong *emotions* may be aroused. It is important that such matters are dealt with sensitively, that participants are listened to with critical respect, and encouraged to be open and honest (Hodson, 2014). This is not an easy situation to achieve and will depend on the culture and the nature of collaboration within the group. One of the skills underpinning teaching SSIs and also SSIBL is to help create an atmosphere of mutual respect in the classroom, attempting to understand what is in the mind of the 'other'.

Learner competencies in SSIs include employing 'scientific ideas and processes, understandings about science and social knowledge (e.g. ideas about economic and ethical influences) to issues and problems that affect their lives' (Sadler, 2009, p. 13). Goals for student participation in SSIs vary. Some see the main goal as being legitimate participants in social dialogues that are science-related (Sadler, 2009) while others maintain that *socio-political action* is a more urgent outcome (Bencze & Alsop, 2014). Socio-political action implies asking questions about 'how research priorities in science are determined' (Hodson, 2014, p. 68), whose interests are considered in formulating policy, and how action can influence policy decisions. It also implies commitment to reflective change, while Wolff-Michael Roth and Angela Calabrese Barton (2004) propose that such action is necessarily collective (see Table 22.1). In SSIBL, we encompass goals emphasizing participation and socio-political action, indeed the second presupposes the former. While RRI presupposes participative dialogue, inquiry into SSIs is *non-trivial*, i.e. it involves students as critical citizens who learn how to enact goals which reflect aspects of social justice. (By non-trivial we draw a distinction between activities which involve simulation, i.e. writing a letter to a political leader as an exercise where the letter will never be sent, as compared with actions which are enacted, and *realised*, through the process of social and political participation. However, an action might involve deciding not to change if, for example, such a change risks too much harm.)

SSIs, and hence SSIBL, present particular challenges for organising learning and assessment, precisely because they are transdisciplinary and context-dependent. Approaching an issue depends to a large extent on our personal history, our social situation, our intentions, needs and wants, and our knowledge and experience of the issue. Stein-Dankert Kolstø (2001) offers a framework for examining the science dimensions of SSIs which have potential for contributing towards an assessment framework. These are:

(i). Science-in-the-making and the role of consensus — how is scientific knowledge made and how do its claims come to be validated?

(ii). Science as one of several social domains that contribute towards decision-making. This is central to SSIBL because there are a number of issues which are ostensibly based on science but where science knowledge may not be the main factor in decision-making, see for example Chris Dawson (2000). In these formal school science might be redundant (Ryder, 2001) and expert knowledge might itself be contested (Layton et al., 1993).

(iii). Distinguishing between descriptive and normative statements.
(iv). Demands for underpinning evidence. In some cases evidence may be unambiguous but in complex SSIs this is rarely the case. It also involves matters of trust about whose evidence is more convincing, and how that evidence was amassed.
(v). Scientific models as context-bound which raises questions as to how scientific models are applied to complex situations involving a range of social and political factors.
(vi). Values; the way in which values influence our thinking and responses to an SSI.
(vii). The relationship between scientific evidence, i.e. that which comes from experts and anecdotal evidence, which comes from a range of lay sources.
(viii). Suspension of belief, being sceptical about the relationship between evidence and the conclusions which can be drawn.
(ix). A critical attitude, learning to ask the questions which are able to scrutinise knowledge claims.

All these dimensions have relevance in SSIBLs and values and critical attitudes apply to transdisciplinary inquiries generally. While these dimensions are unlikely to feature simultaneously in SSIBL they are, nonetheless, helpful as pedagogical resources to support decision-making and argumentation.

Rosemary Hipkins et al. (2014) use the term 'wicked problems,' which illustrate well the kinds of controversies to which SSIBL aspires. These are serious social challenges that span multiple domains (social, economic, moral, aesthetic, political) and link closely with other problems. There are no clear solutions and different groups of people believe they have answers which often contradict one another. 'Wicked problems' do not have finite or unambiguous answers but in dealing with them, other interesting questions emerge.

22.4.4 Inquiry Based Science Education (IBSE)

At the core of the SSIBL framework is inquiry-based learning. Inquiry in the U.S. was promulgated by the philosopher and educationalist, John Dewey (1916). Dewey saw its democratic potential as a means for citizens to participate through solving problems of mutual concern and developing habits of mind of curiosity and communality. IBSE has been influential in science education policy, both through the National Research Council (2000) and the EU (Rocard, 2007) who conceive of science practice as question-driven and open-ended. The fundamental features of inquiry based learning are consistent with the proposed SSIBL framework: purposeful research-driven learning through collaboration, critical examination of evidence and experience. The main distinguishing point of SSIBL is that it involves an authentic open-ended question or hypothesis formulated by students, teachers or other interested parties, and taking action. Since student interest, research,

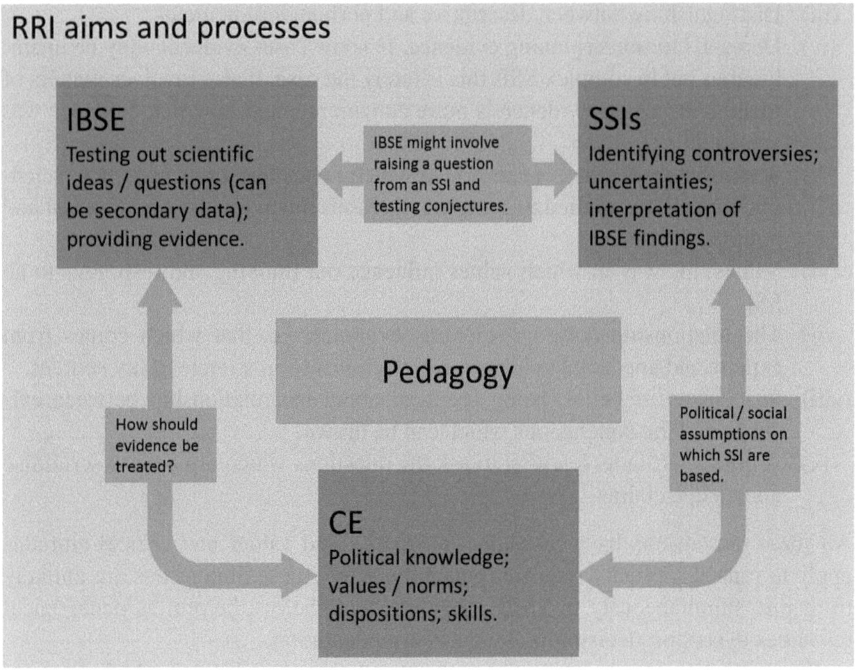

Fig. 22.5 SSIBL framework. Interconnections of the main pillars in SSIBL with RRI as the overarching context

questioning and the collection and interpretation of evidence underpin inquiry, they are components of the role of inquiry-based science learning in the SSIBL framework (Fig. 22.5). However, there are no specific design stages to inquiry-based learning in SSIBL. Student interest is not always spontaneous and, in most cases, will involve teacher preparation and competence in nurturing student interest. Inquiry in the context of SSIBL has features that are quite distinct from those normally attributed to inquiry based learning in science education.

22.4.5 *Scaffolding Inquiry Teaching and Learning*

Since our characterisation of inquiry is seeking knowledge through evidence to answer authentic questions, inquiries need to be based on student interests. Through the introduction of inquiry-based learning, students should feel empowered to direct their own learning through collaboration within a community of learners. Students might find it difficult to generate researchable questions if inquiry-based learning is new to them. One strategy for reaching the stage of genuinely open inquiry is first through a structured approach, then through guidance with teacher support and then open inquiry, similar to a STEPWISE apprenticeship approach, although teacher judgment here is crucial. Too great a dependence on structured inquiry could impede moves to open inquiry.

Scaffolding is the process whereby learners are given appropriate support to help them learn something which they could not achieve on their own. It is a central pivot to social constructivist learning because it presupposes support can be given at a stage when the student is ready for it and can then be phased out when the learner has acquired the required competence. The precise nature of the support depends on a range of factors, what needs to be learned, the knowledge and skills the learner already has, the experience they have of the context of learning, the complexity of the concepts and skills to be learned, the knowledge and skills of the facilitator.

Time is also a factor. Short term inquiries would have outcomes that could be completed in one or two lessons or sessions and carried out mainly within school. Long term SSIBLs would go beyond this time and often include external agencies. Examples of short term SSIBLs are;

- Situating a feeder for nesting birds;
- Designing a poster to reduce school energy consumption;
- Organising a system for building the school compost heap;
- Bringing in plants for a community garden;
- Producing a leaflet to show how to estimate maximum salt intakes.

Such short term projects can meet the framework for SSIBLs. As well as different time spans for SSIBL these can also be structured from inquiries which are mainly closed and directed mainly by the teacher to those which are more open. Structured inquiries will help make explicit to students the knowledge and procedures necessary to carry out an inquiry.

22.4.6 Authenticity

Questions generated through inquiry are deemed to be 'authentic'. However, 'authentic' risks being a catch-all term with multiple, sometimes contradictory, meanings. In the context of SSIBL, authentic questions can be the kinds of questions that scientists raise, although the discourse between scientists in a research project, often influenced by political, cultural and economic factors, will be very different from that of school science (Quigley, 2014). In NRC terms, authentic practice is linked to student ownership of the learning process, although that raises difficulties when students encounter learning experiences that are genuinely challenging and need guidance. The Galileo Educational Network (Galileo.org) conceives of authenticity as focusing on problems and issues relevant to students in the 'real world'.

For Anne Hume and Richard Coll (2010), authentic problems are those that are ill-defined, have no obvious solution, where data has not been collected and there are no established goals and methods, a condition which is unlikely to be attained in the vast majority of school-based inquiries in science. Rather than attempt an overarching definition for authentic practice in SSIBL, it would be more helpful to identify its main components:

(i). Proceeds from questions that interest and engage students and through which they express a wish, and choose, to find answers;
(ii). A mutually-agreed purpose of all participants (i.e. a social authenticity);
(iii). What is relevant and has value and meaning (i.e. personal authenticity); and,
(iv). Where scientific ideas are a resource and can be activated to help find a solution to the problem.

There are, therefore, implications. A mutually-agreed purpose may go beyond the bounds of the school walls for participants, particularly where in finding the answers to questions, students might work with scientists, or other people with expertise. SSIBL might involve interaction either in informal education contexts and/or working with agencies outside the school. An example of this is a collaboration between scientists and students in a school in London with a high proportion of students of Bangladeshi origin. The collaboration stems from an inquiry into the pattern of diabetes in the family histories of the Bangladeshi community in east London. Students at the school, using their background socio-cultural knowledge, work with university scientists, health practitioners and the local political authority in devising a questionnaire. In the university laboratories under the guidance of scientists, students learn sophisticated analytical techniques on DNA found in affected families.

To ascertain what is relevant and has meaning is made real through participation and democratic dialogue where participants become agents of change transforming a reality that can be improved. Finally, activating scientific ideas as a resource might not be straightforward. It might involve distributed knowledge where different parties can contribute through their own experiences and expertise (Roth & Lee, 2002). But it also encompasses questions of scientific uncertainty. Consider, for example, a project that involves testing the pH of potentially polluted waters in a stream. Most school students when they encounter pH measurements use a pH meter or pH papers in ideal conditions. However, measuring the pH of a stream means taking flow, turbidity and temperature into account, thinking about sampling techniques, and being able to assess error. When students begin to work in non-ideal situations the limitations and uncertainties of scientific practice become clearer.

22.5 Developing the Framework for SSIBL

Based on the account above in elaborating the principal features of SSIBL, Fig. 22.5 models the possibilities for SSIBL incorporating the components from Fig. 22.1 and listed in the previous sections.

There is no set format for the order in which the pillars of an inquiry might be arranged. Inquiries might start with a question, followed by planning, perhaps reframing the question after planning, data collection and interpretation, and subsequently communication and taking action based on findings. These stages are likely to be iterated at various points, however. Alternatively, students might be exploring data, and research questions might emerge from the data. Data might involve carrying out surveys or using ethnographic methods.

Figure 22.5 is *a framework to be aspired to*. It is recognized that teachers will go through different routes in building up to SSIBL.

A ubiquitous question, particularly from younger students, is 'What are we doing this for?' For some activities, such as early-stage reading, the answer is long-term, complex and a straightforward answer might demean the purpose. But in the case of SSIBL, it is quite a legitimate question, and the socio-scientific purpose needs to be clear if the process is to have meaning.

Actions are linked to authentic practices, the aim is to change affairs from being unsatisfactory to more desirable ones. For example, there is a difference between students discussing the most efficient ways to conserve fuel use, or answering a set question on this topic, and those who design and build the school animal house based on an inquiry into the best way to cut down electricity bills. This action component is, arguably, a distinctive feature of SSIBL, and models that of STEPWISE. Hence authentic action components are oriented in the students' educational and social settings, and they play a role in transforming the materiality of students' lives, in however small a way.

Ideas or questions or hypotheses for SSIBL should aspire towards the following attributes:

(i). Openness (i.e. no pre-set answer)
(ii). Authenticity
(iii). Comprise different and conflicting perspectives (i.e. controversy)
(iv). Links between personal and social relevance.
(v). Participatory (i.e. all students should be able to take part and co-operate in addressing the question)
(vi). It should be researchable (i.e. either primary or secondary data can be gathered and interpreted to answer the question)
(vii). Focus (i.e. it should be narrow enough so the relevant data is containable)
(viii). Feasibility (i.e. it should be possible within time and curriculum constraints to answer the question)
(ix). Epistemologically appropriate (i.e. it should draw on science knowledge which students have or can be taught, and/or support the building of relevant knowledge).

When studying a topic, students can have a space in which to formulate their own questions. There are a variety of ways in which this can be opened up. Students could brainstorm where they suggest various 'raw' questions, there is a follow up time to choose questions which students prioritise, followed by group work in which they frame the questions with the properties above (Table 22.3).

Table 22.3 Examples of SSIBL questions

Question	Level	Scientific knowledge	Other knowledge	Personal to social
Sun-tanning parlours should be banned for young people under the age of 18	Upper secondary	Radiation	Risk and uncertainty;	Relevant for young people in colder climates but raises broader questions about freedom of choice, global warming and damage to the ozone layer.
		Structure of the skin	Human Rights	
Selection against certain disabling genetic conditions is the thin edge of the wedge for wholesale genetic selection.	Upper secondary	Genetics;	Social justice;	Personal questions about attitudes to disability but broader questions about access to fertility treatment.
		Ivf techniques	Rights;	
			Culture	
Is it possible to avoid resistance to antibiotics?	Upper secondary	Bacteria	Risk;	How do different legislation scenarios across the world influence individual decision-making about the rational use of antibiotics?
		Antibiotics	Legislation;	
		Infectious diseases	Personal and social decision-making	
		Evolution		
		Selection pressure		
Does recycling paper do more harm than good?	Lower secondary	Manufacture of paper;	Process of paper production and recycling;	Personal attitudes to waste as against economic interests of those who pulp wood and produce paper.
			Cost-benefit analysis;	
		Chemical structure of paper;	Local recycling legislation;	
		Solvent chemistry	Interest groups	
What's the best way to feed small birds?	Primary	Bird nutrition;	Conservation	Local aesthetic pleasures of birdlife in the context of broader species competition and interaction.
		Food webs;		
		Sampling; techniques		
How can we reduce car pollution outside our school?	Primary and Lower	Fuel combustion;	Use of secondary data;	Local concerns about pollution related to global use of fossil fuels and alternatives
	Secondary	Sampling;	Pros and cons of car use	
		Measuring particulates		

22.6 Assessment

Assessment of students in SSIBL depends on the purpose of the assessment and the nature of the assessment — whether it is diagnostic, formative or summative.

Laurence Simonneaux (2014) identifies four didactic strategies, slightly adapted, that can also reflect assessment purposes. These are:

- A doctrinal strategy that aims at the acceptance of authoritative scientific concepts.
- A problematising strategy that focuses on students' reasoning through SSIBL.
- A critical strategy that aims to develop capabilities in scrutinizing claims, to be questioning of expertise and to appreciate the uncertain nature of science and its applications and that the development and production of technology carries risks.
- A pragmatic strategy to engage students and to promote student action.

The framework of SSIBL assessment is adapted from Table 22.2 and incorporates:

- Knowledge about an issue (both scientific and transdisciplinary).
- Skills in organizing and operationalising a socio-scientific based inquiry.
- Values that reflect issues of social justice and wellbeing.
- Dispositions that include recognition of inclusivity and democratic deliberation.

Table 22.4 is a grid that can be adapted depending on the context of SSIBL (Fig. 22.6).

22.7 Considerations in Relation to STEPWISE

In conclusion, I want to emphasise three distinctive aspects of SSIBL.

1. Inquiry in SSIBL is not formulaic and might be quite different from inductive-based inquiry. It involves asking authentic questions where the solutions are diverse, politically-constituted and complex, and involves drawing on domains of knowledge beyond science. In that sense, they have much in common with the Socially Acute Questions approach (Simonneaux, 2014).
2. Inquiries should stem from students' own interests and motivations. This might not always be possible and is an aspiration. An important aspect of skilful pedagogy in SSIBL is helping to stimulate questions that promote a genuine sense of inquiry in students.
3. Actions of SSIBL are *non-trivial*. They involve informed actions which make a difference to individual and social wellbeing.
4. Actions are collaborative and enmeshed within a web of interested human and non-human relationships. Their realisation is therefore uncertain and the processes of achievement of aspirations based on social justice are risky. That leads to the production of knowledge-in-action, reflecting and acting on the interrelationships between knowing the world and the vagaries of action, rather like disturbing a network of human and non-human actants (Hoeg & Bencze, 2014).

Table 22.4 Example of assessment framework using school animal house project as examplar

Didactic approach	Knowledge	Skills	Values	Dispositions
Authoritative (mainly structured)	Focus on learning substantive theoretical scientific knowledge and inquiry skills. (learn principles of heat transfer and experiments to measure them)	Follow through procedures as taught and relate findings to science knowledge. Explicate learned procedures. (Investigates conductivity of different materials using given experimental procedures and makes conclusions based on evidence.)	Learning that the applications of science are not value-free. (Principles of heat transfer can be applied and associated with improvement of wellbeing at personal and social level).	Discusses problem in groups.
Problematising (mainly guided)	Applying scientific and transdisciplinary knowledge into new contexts (explain how principles of heat transfer apply to the construction of an animal house, and reflect on human responsibilities to non-human species)	Devises procedures as taught for carrying out inquiry. This can be done with some guidance, if appropriate. Demonstrates awareness of uncertainty in considering empirical and second hand data. (Devises a method for investigating appropriate materials to build animal house. Takes into account concepts of accuracy, validity, precision and so forth in making and interpreting data).	Articulating that an inquiry is based on a range of value judgments. (Can articulate values associated with inquiry such as importance of fuel conservation, its local and global implications, and non-human wellbeing).	Recognises when to collaborate in groups and when to be autonomous. Participates fully.
Critical and pragmatic (mainly open)	Recontextualising and scrutinizing relevant scientific knowledge and research in the light of contingencies of specific contexts. (Recontextualise and adapt knowledge of heat transfer in the light of constraints of constructing an animal house, e.g. use of composite materials, ventilation factors, humidity. Research on habits of small mammals to reflect design constraints. Reflect on ethical relations between humans and non-humans in terms of concepts such as rights and responsibilities, and process of fuel conservation).	Suggests and collaborates as group to generate questions or hypotheses for building animal house using research material, scientific and other knowledge, reports and surveys. Follows through inquiry. Adapts knowledge and understanding depending on circumstances. (Develops inquiry and outcome, eg working model of animal house, and evaluates process and product).	Inquiry driven by value considerations recognizing various aspects such as inter-relationships between personal and social values. Recognises interconnectedness. Identifies political nature of action where necessary. (Construction of animal house driven by awareness of need and knows how to take action to meet that need.)	Can operate between full autonomy and collaboration. Recognises importance of inclusivity and how to negotiate consensus.

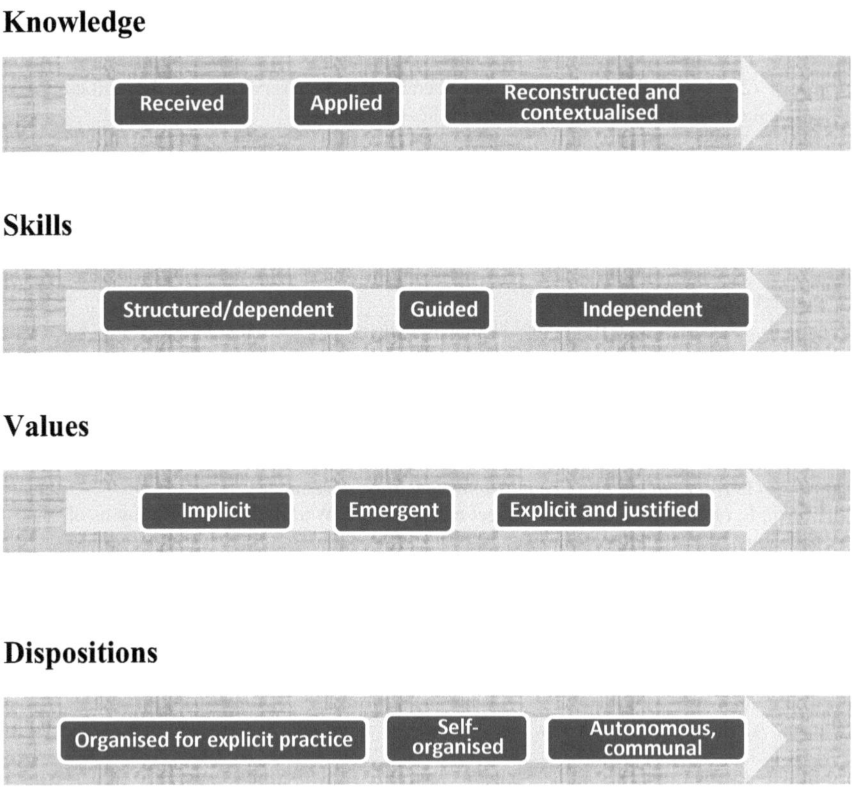

Fig. 22.6 Summarises assessment for progression through the four dimensions of knowledge, skills, values and dispositions

The last point does raise the question of what is meant by action. Just as the social applications of technoscience carry accompanying hazards, risk and uncertainties (Ravetz, 2005), so the intentions of actions are carried out in a sea of uncertainties. An example is an incident based on the AR activity described above where a discussion among students resulted in homophobic sentiments being expressed, which were then challenged. For any action in the social world to succeed, it must rely on collaboration and an element of reliance on others (Arendt, 1998). Participation and trust are crucial in a diverse and plural society where values might vary enormously. So, the achievement of a particular outcome is only a partial measure of success; the importance of negotiation and participation based on shared knowledge, and the understanding of what is possible in sometimes unpromising circumstances, is a core part of the learning process.

References

Apple, M. (2004). Creating differences: Neo-liberalism, neo-conservatism and the politics of educational reform. *Educational Policy, 18*(1), 12–44.
Arendt, H. (1998). *The human condition* (2nd ed.). Chicago: University of Chicago Press.
Baird, J., Isaacs, T., Johnson, S., Stobart, G., Yu, G., Sprague, T., & Daugherty, R. (2011). *Policy effects of PISA*. Oxford, UK: Oxford Centre for Educational Assessment.
Ball, S. J. (2013). *The education debate* (2nd ed.). Bristol, UK: The Policy Press.
Baudrillard, J. (1994). *Simulacra and simulation*. Ann Arbor, MI: University of Michigan Press.
Bencze, L., & Alsop, S. (Eds.). (2014). *Activist science & technology education*. Dordrecht, The Netherlands: Springer.
Bencze, L., & Carter, L. (2011). Globalizing students acting for the common good. *Journal of Research in Science Teaching, 48*(6), 648–669.
Dawson, C. (2000). Selling snake oil: Must science educators continue to promise what they can't deliver? *Melbourne Studies in Education, 41*(2), 121–132.
Dearden, R. F. (1981). Controversial issues in the curriculum. *Journal of Curriculum Studies, 13*(1), 37–54.
Dewey, J. (1916). *Democracy and education*. New York: Macmillan.
Ellsworth, E. (1989). Why doesn't this feel empowering?: Working through the myths of critical pedagogy. *Harvard Educational Review, 59*(3), 297–325.
Enslin, P., & White, P. (2003). Democratic citizenship. In N. Blake, P. Smeyers, R. Smith, & P. Standish (Eds.), *The Blackwell guide to the philosophy of education* (pp. 110–125). Oxford, UK: Blackwell Publishing.
Gamarnikow, E., & Green, A. (2000). Citizenship, education and social capital. In D. Lawton, J. Cairns, & R. Gardner (Eds.), *Education for citizenship* (pp. 93–113). London/New York: Continuum.
Giddens, A. (2008). *The third way: The renewal of social democracy*. Cambridge, UK: Polity Press.
Habermas, J. (1984). *The theory of communicative action*. Boston: Beacon Press.
Hargreaves, A., & Shirley, D. (2009). The persistence of presentism. *Teachers College Record, 111*(11), 2505–2534.
Hess, D. (2009). *Controversy in the classroom*. New York: Routledge.
Hipkins, R., Bolstad, R., Boyd, S., & McDowall, S. (2014). *Key competencies for the future*. Wellington, NZ: New Zealand Council for Educational Research.
Hodson, D. (2014). Becoming part of the solution: Learning about activism, learning through activism, learning from activism. In L. Bencze & S. Alsop (Eds.), *Activist science and technology education* (pp. 67–98). Dordrecht, The Netherlands: Springer.
Hoeg, D., & Bencze, L. (2014). Counter cultural hegemony: Student teachers' experiences implementing STSE activities. In L. Bencze & S. Alsop (Eds.), *Activist science and technology education* (pp. 575–596). Dordrecht, The Netherlands: Springer.
Humbel, L., Jolliet, F., & Varcher, P. (2012). *3 key competencies in ESD for learners and teachers: Make a deconstruction, operate a reconstruction and ask critical questions*. A case study in college classrooms about some SAQ (Socially acute questions) concerning 'le fait religieux' (unpublished paper).
Hume, A., & Coll, C. (2010). Authentic student inquiry: The mismatch between the intended curriculum and the student-experienced curriculum. *Research in Science and Technological Education, 28*(1), 43–62.
Jasanoff, S. (2003). Technologies of humility: Citizen participation in governing science. *Minerva, 41*, 223–244.
Johnson, L., & Morris, P. (2010). Towards a framework for critical citizenship education. *The Curriculum Journal, 21*(1), 77–96.

Kolstø, S. (2001). Scientific literacy for citizenship: Tools for dealing with the science dimension of controversial socioscientific issues. *Science Education, 85*(3), 291–310.
Layton, D., Jenkins, E., Macgill, S., & Davey, A. (1993). *Inarticulate science?* Driffield, UK: Studies in Education Ltd..
Levinson, R. (2006). Towards a theoretical framework for teaching controversial socio-scientific issues. *International Journal of Science Education, 28*(10), 1201–1224.
Levinson, R. (2010). Science education and democratic participation: An uneasy congruence? *Studies in Science Education, 46*(1), 69–119.
Levinson, R., & Turner, S. (2001). *Valuable lessons.* London: The Wellcome Trust.
Mand, C., Duncan, R. E., Gillam, L., Collins, V., & Delatycki, M. V. (2009). Genetic selection for deafness: The views of hearing children of deaf adults. *Journal of Medical Ethics, 35*(12), 722–728.
Mercer, N., & Littleton, K. (2007). *Dialogue and the development of children's thinking: A socio-cultural approach.* Abingdon, UK: Routledge.
National Research Council. (2000). *Inquiry and the National Science Education Standards: A guide for teaching and learning.* Washington DC: The National Research Council.
Nistor, N., Stanciu, D., Vanea, C., Sasu, V. M., & Dragota, M. (2014). Situated learning in young Romanian Roma successful learning biographies. *European Education Research Journal, 13*(3), 311–324.
Novas, C. (2006). The political economy of hope: Patients' organisations, science and biovalue. *BioSocieties, 1*(3), 289–305.
Owen, R., MacNaghten, P., & Stilgoe, J. (2009). Responsible research and innovation: From science in society to science for society, with society. *Science and Public Policy, 39*, 751–760.
Pierce, C. (2015). Learning about a fish from an ANT: Actor network theory and science education in the postgenomic era. *Cultural Studies of Science Education, 10*(1), 83–107.
Quigley, C. (2014). Expanding our view of authentic learning: Bridging in and out-of-school experiences. *Cultural Studies of Science Education, 9*(1), 115–122.
Ravetz, J. (2005). *The no-nonsense guide to science.* Oxford, UK: New Internationalist/Verso.
Reis, G. (2015). Social dimensions of scientific knowledge: The '$5 bottle of hope'. *School Science Review, 96*(356), 27–30.
Rice, S., & Burbules, N. (1992). Communicative virtues and educational relations. *Philosophy of Education, 1992*, 34–44.
Rocard, M. (2007). *Science education now: a renewed pedagogy for the future of Europe.* Brussels, Belgium: European Commission.
Roth, W.-M., & Calabrese Barton, A. (2004). *Rethinking scientific literacy.* New York/London: RoutledgeFalmer.
Roth, W.-M., & Lee, S. (2002). Scientific literacy as collective praxis. *Public Understanding of Science, 11*, 33–56.
Ryder, J. (2001). Identifying science understanding for functional scientific literacy. *Studies in Science Education, 36*, 1–44.
Sadler, T. (2009). Situated learning in science education: Socio-scientific issues as contexts for practice. *Studies in Science Education, 45*(1), 1–42.
Selwyn, N., Gorard, S., & Williams, S. (2001). Digital divide or digital opportunity?: The role of technology in overcoming social exclusion in U.S. education. *Educational Policy, 15*(2), 258–277.
Shiva, V. (2000). *Stolen harvest.* London: Zed Books.
Simon, R. (1992). *Teaching against the grain: Texts for a pedagogy of possibility.* New York: Bergin & Garvey.
Simonneaux, L. (2014). Questions socialement vives and socioscientific issues: New trends of research to meet the training needs of post-modern society. In C. Bruguiere, A. Tiberghien, & P. Clement (Eds.), *9th ESERA conference selected contributions. Topics and trends in current science education.* Dordrecht, The Netherlands: Springer.

Thomasian, J. (2011). *Building a science, technology, engineering and math education agenda: An update of state actions*. Washington, DC: National Governors Association.
Vasagar, J. (2012, June 12). Michael Gove's curriculum attacked by expert who advised him. *The Guardian*. Retrieved from https://www.theguardian.com/education/2012/jun/12/michael-gove-curriculum-attacked-adviser
Von Schomberg, R. (2013). A vision of responsible innovation. In R. Owen, M. Heintz, & J. Bessant (Eds.), *Responsible innovation*. London: Wiley.
Wilkinson, R., & Pickett, K. (2010). *The spirit level: Why equality is better for everyone*. London: Penguin books.

Chapter 23
"Preach or Teach?": An Ongoing Journey to Becoming STEPWISE

Mellita Jones

23.1 Introduction

> Our world contains grave evils, which can be remedied if men [sic] wish to remedy them. Those who are aware of these evils fight against them are likely, it is true, to have less everyday happiness than those who acquiesce in the status quo. But in place of everyday happiness they will have something which I, for my part, value more highly, both for myself and for my children. They will have the sense of doing what lies in their power to make the world less painful…They will have the knowledge that they are amongst those who prevent the human race from sinking into stagnation or despair. This is something better than slothful contentment. (Russell, 1932/2009, pp. 67–68)

This chapter outlines my ongoing journey as a science teacher educator towards a more activist approach to science teacher education. Science teacher education curriculum is typically framed within the same neoliberal influences that governments prescribe in school science curriculum documentation. The science teacher educator is thus expected to prepare emerging generations of science teachers to know and be able to teach this prescribed curriculum. Subsequently, the status quo citizenship demanded by a neoliberal, industrial society and supported by school systems is maintained. Perpetuating the status quo like this can be useful and even necessary in order to establish and maintain a well-functioning society where a productive life can be enjoyed. However, the neoliberal climate that is currently evident in societies of the "developed" (or global North/minority) world advantages the privileged few at the expense of the less privileged many, creating an unjust world of increasing inequity. This inequity has led to some (e.g., Ayers, Quinn, & Stovall, 2009; Clover, 2002; Hodson, 2010) to call for an activist approach to education in order to create a 'better' world, one in which the wellbeing of individuals, societies, animals and the broader environment, is promoted.

M. Jones (✉)
Faculty of Education and Arts, Australian Catholic University, Ballarat, VIC, Australia
e-mail: Mellita.Jones@acu.edu.au

L. Bencze (ed.), *Science and Technology Education Promoting Wellbeing for Individuals, Societies and Environments*, Cultural Studies of Science Education 14,
DOI 10.1007/978-3-319-55505-8_23

Calls for education to inspire a better world are not recent. Some 70 years ago, social theorist and educational philosopher, Bertrand Russell (1932/2009) was calling for education to address social inequity to secure a better future for all. In spite of this, a truly democratic education is yet to emerge as a prevailing global paradigm. In science education there are a few dedicated voices (e.g., Aikenhead, 2006; Bencze & Alsop, 2009; Hodson, 2003, 2010; Roth & Désautels, 2002) calling for science to critically examine relationships between science and society to enhance the wellbeing of all peoples and the world in which we live. Their messages reflect the ideas of Russell, showing how his work is still of relevance today. Accordingly, Russell's work has been heavily drawn on in the shaping of this paper.

One framework that appears to embed principles outlined by Russell (1932/2009) is that of Science and Technology Education Promoting the Wellbeing of Individuals, Societies and Environments [STEPWISE] (Bencze & Alsop, 2009). STEPWISE aspires to contribute to a transformational education for a better world by challenging the nature of science education and working towards wellbeing for all. STEPWISE encourages teachers to move beyond traditional approaches to teaching science that focus on conceptual and procedural knowledge, which ultimately reinforce neoliberal agendas. Instead, STEPWISE links these traditional forms of knowledge in a framework that centers on action and makes explicit associations among scientific ideas and their important social and political implications.

The activist education approach central to STEPWISE aspires to challenge the status quo of science education so that citizenship becomes one of *active* concern for societal wellbeing and, thus, reflect the type of citizen that Russell (1932/2009) describes in the opening quote of this chapter—one who fights to make the world less painful. This contrasts with the construction of citizenship in neoliberal societies, where the privileged excel at the expense of both others and nature. Fostering a more equitable, global citizenship and sustainable future presents a challenge for science teacher educators. First, they must recognise and, second, adopt a curriculum that inspires socio-scientific activism in both citizenship and teaching. In this paper, socio-scientific education for activism refers to the sorts of world issues that Derek Hodson (2010) terms as "civic scientific literacy" which "comprises the knowledge, skills, attitudes and values necessary for making decisions on matters such as energy policy, use of natural resources, environmental protection, and moral-ethical issues raised by technological innovations" (p. 197). Adopting a curriculum that is focused on such societal and ecological wellbeing may, in turn, encourage pre-service teachers to adopt similar practices once they enter the profession as well as in their personal lives.

One tension that can arise from this is that *teaching* for activism could be viewed as *preaching* about activism and, thus, a framework that inspires socio-scientific activism must be carefully constructed and applied. One theory that could inform such a framework is Pierre Bourdieu's (1984) notion of social and cultural capital, and his conception of socialised norms or 'habitus.' His general framework is relevant to education because of ways in which educational institutions create particular social spaces in which social and cultural relations exist individually and institutionally. Such a space, or 'field' (Bourdieu, 1984), involves establishment of class groups "of which some assume dominant positions and others find themselves subordinate" (Fenge, 2011, p. 378)—reflective indeed of the neoliberal scene.

A particular field has its characteristic features, structures and conventions that guide thinking and behavior—both consciously and sub-consciously. Simon During (2007) provides a useful example to explain the effect of the field on its members: "[I]f you are a writer you can't write anything you like, you find yourself positioned in a field which structures your possibilities" (p. 88). A similar set of invisible boundaries is established through the traditional rules and processes characterising science education. These boundaries act to form the habitus of thinking and behaviour of which we are often unaware so much are they embedded in 'normality' of daily life. Normality formed by the habitus of the field in this way provides legitimacy to its products and outcomes, including the inequities and injustices in the world.

Fenge (2011), reporting on Weick (1995), tells us that habitus is "grounded in both individual and social activity" (p. 379). In science, and in science education, the field's structures and conventions involve laws and theories that govern different disciplines of science, which tend to maintain traditional foci on what Bencze and Alsop (2009) refer to as 'products education' within fields of science learning—that is, a focus that supports products associated with ongoing industrialization and consumption in society. Dispositions and behaviours emerging from fields of science education—i.e. the habitus—invoke science processes of inquiry and investigation (skills education) that involve 'fair tests' and evidence-based theorising.

These forms of knowledge and practice drive the nature of science education and influence ways in which science is used in society. Habitus is linked to reproduction and change in society (Rawolle & Lingard, 2013). It also sets particular curriculum and pedagogical dispositions that Rawolle and Lingard (2013) recount, leading to reproduction of "class structure, class codes and class relations through schooling" (p. 121). The western view of science creates a habitus that attempts to be objective and value free. Thus, habitus helps to explain how science education as an institution leads to the reinforcement of the neoliberal state that dominates western culture in the global north minority (developed) world and increasingly, in the global South majority (developing) world.

Fortunately, habitus is not a fixed, permanent state. Navarro (2006) suggests that it can be altered through encountering different contexts and environments that create tensions and, upon reflection, challenge ways in which we think and act. Such reflection can assist in recognising what may have previously been invisible due to the manner in which the field can normalise certain ways of thinking and behaving. Indeed, Bourdieu (1984) reports that habitus is formed from both past experiences as well as current events that can alter our perceptions. Education is one particular institution that can re-shape habitus by challenging the status quo and building in experiences of critical reflection to help ensure there is not an unconscious acceptance of the social/cultural field and, subsequently, a legitimization of social, cultural and ecological inequity.

23.2 Purposes of Education

Education is an institutional concept. It has evolved from its early purposes to induct the rich into educated and privileged society to that of 'training' individuals to work and operate in the type of citizenship seen as desirable by the State (Russell, 1932/2009). Bourdieu referred to education as "a sorting institution that functioned to divide groups primarily through the valuing of cultural capital" (Rawolle & Lingard, 2013, p. 120). Today, still, education is viewed as a major vehicle through which one achieves social mobility: success, affluence, and (supposedly) wellbeing (Kalantzis & Cope, 2008). Education is also seen as key to growing a 'knowledge economy' and, perhaps in conflict with this notion, also a precursor to resolving significant inequities in the world.

Different stakeholders can view purposes of schooling differently. In current neoliberal contexts, governments tend to view education as a process for developing emerging knowledge economies of the world and, thus, become/remain competitive in a global economy. To this end, despite rhetoric claiming its power to resolve inequities in the world, education for neoliberalism is one in which the prevailing habitus values individual power and wealth above equity and access to resources for everyone and maintenance of functioning ecosystems. This neoliberal view of education creates an individualistic sense of purpose—that of preparing individuals for productive, working lives, through which they contribute to building of the economy (and by unexamined implication, the society) of their respective countries. However, as Kalantzis and Cope (2008) attest, education is central in this shaping of "certain types of citizens" (p. 71). With such centrality, careful consideration is needed to determine what content, skills, values and attitudes should be included in education programmes, and whether present foci on individualistic approaches are, in fact, suitable for emergent global citizenship required for wellbeing in the twenty-first century and beyond.

It has been argued that education focused on producing 'good individuals' should naturally foster a society of 'good citizens' (Russell, 1932/2009). The notion of a 'good citizen' can, however, mean very different things to different people. For example, some view 'good citizens' as those who do achieve individual success (usually measured in terms of monetary wealth), and subsequently contribute to larger society through their services, taxes, and/or philanthropic ventures. In this model, it is quite likely that 'success' comes at the expense of others; for, in the competitive neoliberal archetype, success is based on competition and personal gain deriving from a capitalist political model. Alternatively, good citizenship can be viewed in a more egalitarian manner; as equity and working toward the betterment of all. The variability in how the notion of good citizenship can be perceived demands that the ways in which it is characterised receives careful attention if it is to be a focus of educational outcomes.

In contrast to the dominant neoliberal discourse, Kalantzis and Cope (2008) report that "many political and community leaders present education as a mechanism for ensuring social equity" (p. 6). It is difficult, however, to see equality as the product of an education that is so focused on individuals' aspirations and achievements (Kalantzis & Cope, 2008). In fact, Mirra and Morrell (2011) report that, in the United States (US), the neoliberal agenda has led to entrenchment of educational inequality. They also allege that this leads to a "mechanistic purpose for teaching" and promotes the "capitalist purpose for education" (p. 409). Such a view attributes both success as well as any lack of success to the individual (Kalantzis & Cope, 2008). This further promotes an education system that is geared towards individual performance and outcomes, and individual accountability and blame when outcomes are not achieved. This attribution to individual effort often comes with little consideration of one's position, or the impact of one's actions and outcomes on nature, or on others in the local or global community. Bourdieu recognised that education geared this way leads to *reproduction* of cultural and social inequities (Rawolle & Lingard, 2013), rather than resolving them.

This often-unconscious disregard contributes to a number of socio-scientific transgressions, including over-consumption of resources; excessive waste; unsustainable population growth; food security risks; loss of biodiversity and detrimental climate change. It is also linked to exploitation of peoples from disadvantaged communities who may be recruited into slavery types of roles, such as child slavery for cocoa and coffee production; primitive and often dangerous working conditions for production of clothing in sweat shops; and generally, threats to the livelihood and wellbeing of current and future generations.

These and other injustices in the world have resulted in a number of activist groups mobilising against disparity and inequity in an effort to achieve greater parity in the world and to promote a more sustainable way of living. Campaigning for a better and more just world, however, should not just be the concern of activists, but rather, of all citizens of the world, and education plays a central role in addressing needs of humankind (Hopkins, 2013). Through education, there are greater chances of reframing the field, fostering dispositions that engender knowledge, values, attitudes and desires to take actions required to secure a more sustainable, equitable and ethical citizenship. "It is only through the will and through the exercise of power that the individual … becomes an effective member of the community" (Russell, 1932/2009, p. 3). Science education, in particular, offers a natural conduit for such a citizenship-focused education, due to its link with many of the most significant issues threatening the world (e.g. climate change; food security, land and water usage, biodiversity, unethical development/use of technology).

23.3 Scientific Literacy and School Science

The relationship between science and social, ecological and cultural world issues has led to many within science education communities rethinking purposes of science education (e.g., Aikenhead, 2006; Bencze & Carter, 2011; Hodson, 2003; Roberts, 2007; Roth & Lee, 2004). Traditionally, science education has been conceived as serving to foster scientific literacy (De Boer, 2000). Generally, this has meant a science education focused on developing knowledge and skills for evidence-based thinking and argumentation associated within the long-established science disciplines (e.g., physics, chemistry, biology). In recent times, the usefulness of this dichotomisation has been questioned and there have been claims that science education is in a state of crisis in most post-industrialised countries (Tytler, 2007).

While scientific literacy is commonly considered to be the overarching purpose of science education, historically there has been little consensus on its definition (De Boer, 2000; Fensham, 2004). In the past decade, however, there has been an increasing shift away from the traditional products-based approach to one more sympathetic to the socio-scientific issues that plague the contemporary world. Many (e.g., Bencze & Carter, 2011; Roth & Lee, 2004) argue that science education needs to respond to these societal issues by embedding them in a more active and overt manner in the curriculum. Such deliberate focus on the social implications of modern day science should help the wider public to participate in effective, informed decision-making about "personal and public science-based issues" (Tytler, 2007, p. 4).

Recent discourse around "re-imagining" (Tytler, 2007) science education has reinforced existence of two main themes about the purpose of science education, which Roberts (2007) tells us are competing for precedence. The first theme aligns with his 'Vision I' view of processes and products of science, in which students are essentially prepared for an expert science career path—a "propaedeutic" approach (Roth & Lee, 2004, p. 275) and something seen as important for "carry[ing] the nation into a technologically driven future" (Tytler, 2007, p. 1). The second theme recognises needs for accessibility and engagement in science by all citizens to ensure "lifelong participation in and learning of science-related issues" (Roth & Lee, 2004, p. 263). Roberts (2007) views this 'Vision II' form of science as being concerned with ways in which students are likely to encounter science in every day life.

The first of these themes tends to encourage content-based approaches to teaching that deal with products and processes of science and often manifests in the delivery of abstract concepts (Aikenhead, 2006). This approach is representative of what Roth and Lee (2004) describe as the "competitive and individualistic nature [of science] and its claims to objectivity, value-free inquiry, and being an isolated enterprise" (p. 265)—that is, a neoliberal approach geared towards business-as-usual capitalist outcomes. Alas, it is this approach that often discourages ongoing participation in science education (Roth & Lee, 2004) due to its lack of relevance to contemporary life and the perpetuation of the image of science as being for the elite.

The second theme of science education encourages a contextualized approach, providing a means for education about ideas and ethics related to "fundamental societal conditions" (Tytler, 2007, p. 2). This second approach also relates to issues encountered in everyday life, making school science more relevant to most students, rather than just the relatively small proportion who follow it into further study and careers (Aikenhead, 2006). Approaching science in this way involves a shift from the traditional objective, value-free, view of science that produces 'answers' to the questions of the world. There is no doubt that the sense of certainty provided by traditional science is no longer enjoyed. The current milieu in which scientific progress has enabled science to be used in highly controversial ways: the atomic bomb; unraveling the human genome and subsequent designer babies; genetic engineering in foods, among others, has seen science become less certain in providing answers to important, value-laden, and subjective concerns of the world. This places natural and timely socio-cultural emphases on science that should be reflected in education if the world is to move to a more equitable and sustainable future.

Given the problematic nature of science-related issues in the modern world, and the uncertainty of science in providing definite solutions/resolutions to these problems, it seems appropriate to view scientific literacy in the same way that do Wolff-Michael Roth and Stuart Lee (2004)—as a social practice; and a more authentic and relevant school science as "citizen science". This change in focus generates what I see as a third theme for science education, that of *socio-eco-activism* in which ideas of science related to society and nature are not just explored in cognitive ways, but their ethical implications are debated and authentic action is incorporated through an activist science education. Education framed around citizen science in this way would better address significant ethical problems and questions of the world. Indeed, During (2007) recognises that "Science has…become of more interest to cultural studies in response to the increasing technologicalization of nature and the human body as well as in response to global warming" (p. 23), which further strengthens its association with a more social, cultural, ecological and citizen-based definition. Such a view is consistent with other discourses, not only within science education, but also about education more generally (Mirra & Morrell, 2011). It aligns with what Kalantzis and Cope (2008) portray as a "new learning", which they describe as being "about action as well as cognition…about the capacity to be productive in the world as well as knowing about the world" (p. 9).

New learning requires a significant shift in thinking about organisation and delivery of curriculum (Kalantzis & Cope, 2008). Typically, school curriculum programmes and, indeed, teacher education programmes, are fragmented into subjects and disciplinary areas, such as English, mathematics, and science. Citizen-based education programmes require more holistic approaches, and these require reform of both school and teacher education programmes to ensure that teachers have adequate skills and knowledge to challenge, change and implement more relevant, citizen-based education. With gross inequity and significant science-based issues prevailing in the world, today's version of citizen-based education is going to require a further critical element of educating for activism. This chapter goes on to consider this transformation from the perspective of science teacher education, where the discourse and relevance of citizenship that is tied to science-related ideas and issues is paramount.

23.4 Teacher Education

Roles of teacher education are ill-defined beyond their general purposes in preparing teachers for the profession. There are a number of differing views about what such preparation should involve, as is reflected by what Louden (2008) denotes as the "101 damnations of initial teacher education", referring to the large number of reviews into teacher education in Australia and how it should be conducted. For example, it has been argued that educators should be "agenda-setters and change-makers" (Kalantzis & Cope, 2008, p. 33), whereby current practices and approaches to teaching, learning and school organisation are challenged and reformed. This would require teacher education to inspire pre-service teachers to challenge the status quo, and focus their education and subsequent teaching practices on new ideas, approaches and structures. Others view roles for teacher education in preparation of pre-service teachers to learn about characteristics of curricula that they will be charged with delivering upon entering the profession. Such preparation supports schools in satisfying requirements of teacher accreditation bodies, government agencies, and other governing bodies. However, the sometimes complementary and sometimes competing demands of stakeholders driving education leaves little room for reform in most schools—especially when government school funding is often tied to student outcomes and/or adherence to government initiatives such as accountability measures and national testing. Moreover, Mirra and Morrell (2011) suggest that quality of teachers and teaching are often based on measures of these "unexamined assumptions about what constitutes desirable student learning outcomes" (p. 408). These ties to important resources and measures of accountability represent yet more mechanisms for advancing the neoliberal agenda.

The same unexamined outcomes for school student learning create pressure on faculties of education to deliver teacher education in ways that support the system in place. However, if teacher education acts merely as a prop for the status quo, how does change occur—in the school or classroom, let alone in the world? Transformational education (Mezirow, 1991) towards a focus on active citizenship means a more authentic curriculum is needed in which students and teachers can connect knowledge and skills with key issues in the local and global community and actively participate in measures to improve conditions for the greater good. Situations need to be utilised to engage students in "participatory modes" (Roth & Lee, 2004, p. 267) where they can make their own decisions and pursue their own interests in authentic situations that not merely *reflect* daily life, but rather, are *embedded* in daily life. Tytler, Symington, Kirkwood and Malcolm (2008) refer to such an approach as "knowledge 'in action' and 'in context'" (p. 17). Aikenhead (2007) introduces this knowledge in action as a Vision III expansion of Roberts' (2007) visions I/II of scientific literacy.

Kalanzis and Cope (2008) do warn, however, that transformational education requires more than just authenticity. They say:

> [w]e have the power to transform our classrooms and our schools. As we embark on these transformations, we also make our own contribution to the transformation of broader

> society. Better learners will better contribute to the making of a better society … This means more than being "authentic". Being authentic may produce a better fit between education and society, but leaves society fundamentally the way it is. It sets out to reflect the realities of the world more than to change them. (p. 33)

This is particularly evident in science education and, subsequently, science teacher education. Much curriculum tends to be written quite conservatively and, thus, supports propaedeutic approaches described by Roth and Lee (2004). In this curriculum, it is 'safe' to stick to abstract ideas and science inquiry skills. Authenticity may be attempted by contextualising content and skills within thematic topics. Even within contextualised themes, however, science education is primarily about acquisition of a body of knowledge—knowledge of science and knowledge to follow general scientific processes—the habitus of the field. The risk with this is that science is represented as an objective, value-free discipline; yet, as mentioned earlier, in the twenty-first century science is inherently entwined with ethical and value-based issues that plague the world. In an activist approach, science education would frame curricula to reflect knowledge and obligations to use knowledge in socially and ecologically responsible ways that strive for equity and justice for all. However, rarely does learning in science demand, or sometimes even discuss, notions about students as agents of change within communities to benefit the world as a whole.

23.5 Activist Science Teacher Education

Gallavan and Webster-Smith (2012) claim that "[t]eacher education is a powerful mechanism for helping teachers to understand the importance of agency" (p. 55) and that this occurs through rich opportunities and reflection. Thus, even though knowledge and action "are ultimately entwined" (Alsop & Bencze, 2010, p. 178), it is not likely to be sufficient to merely impart knowledge of concepts and ideas about socio-scientific issues with hopes that, with such knowledge, action will result. One need only reflect on the still-widespread inaction on climate change in some quarters, despite overwhelming scientific evidence to recognise this. Instead, education that is research and community-based is required, which has potential to lead to lifelong learning and action where "the collective praxis of the community takes precedence over the individual" (Roth & Lee, 2004, p. 284). This is the sort of science education that has potential to create more global forms of citizenship. Increasing participation in community issues is more likely to result in an education that moves beyond the acquisition of knowledge to one that encourages "discovery and action" (Mirra & Morrell, 2011, p. 412).

This socially responsible activist outcome of science education is reflected in STEPWISE (Science and Technology Education Promoting Wellbeing for Individuals, Societies and Environments). The four vertices of the STEPWISE tetrahedral framework: Science Technology Society and Environment (STSE) Education; Skills Education; Students' Research, and Products Education, are already, to some extent, reasonably-well embedded in both school and science teacher education. For example, socio-scientific issues are prevalent through topic areas such as genetic engineering, climate change, biodiversity, sustainability, and are explicit in most formal curriculum documents (e.g., ACARA, 2013; Ministry of Education (Ontario), 2011; NCCA, 1999; UNESCO, 2009). Science inquiry skills and processes have also had an increasing profile in curriculum documents over past decades and students are regularly involved in research projects, some about socio-scientific issues, and others on more traditional science-related topics (e.g., famous scientists, inventions, diseases etc.).

In spite of the prevalence of individual aspects of STEPWISE already present in schools, it appears that there has been limited uptake of the framework across education sectors, and pre- and in-service teachers find the framework difficult to implement (Bencze & Carter, 2011). There may be a number of reasons for this. For one, there appears to be a gap between how different aspects of STSE, Science Inquiry skills and Students' Research are approached in schools—often in a disconnect from one another. This is exemplified through traditional modes of science teaching where content is often delivered through transmission approaches; recipe-style practical work is conducted, in which some inquiry skills are privileged over others (such as collecting and analysing data to form a conclusion, but rarely posing a question and designing the processes to collect evidence); and student research is completed in isolation from these two arms as well. This creates particular social fields (Bourdieu, 1984), in which the habitus underpinning practice in fields perpetuate objective, value-free, views of western science.

What STEPWISE does that is perhaps unique is to encourage traditionally-separate aspects of science education to be brought together such that they work in harmony: student-led research about a science inquiry where students contemplate, pose, process and attempt to answer a question and relate to existing ideas within the science field. This does not so much *change* the social field, but rather *expands* it. This expansion helps to ensure that critical inquiry and societal concerns underpin development of habitus within the field. Such an approach reflects Hodson's (2010) urging for alignment of issues-based learning with traditional subject-based curriculum; not as an 'add-on' but, rather, as an inter-related activity. Students' research then embeds science inquiry skills and contexts of inquiry are selected from STSE content areas. This marriage between aspects of science learning provides a more holistic learning framework (Bencze & Carter, 2011). It allows science research, content and theory to become relevant and useful rather than abstract and disconnected.

Critically, STEPWISE requires yet a further aspect of education to emerge—the vital step of *action* towards enhancing societal and ecological wellbeing. Supporting Hodson (2003), Bencze (2014) points out that this aspect of scientific literacy is relatively rare in science learning experiences but is central to STEPWISE and central for a citizenship education. Teaching for activism is emerging as an increasingly urgent requirement if the survival of earth's ecosystems is to prevail into the future. In spite of the increasing urgency of the global situation and the recognition of similar arguments dating back to the 1930s (e.g., Russell, 1932/2009), there remains limited uptake of activist approaches to education. Such resistance to activism, or even learning about socio-scientific issues, is exemplified in the current United Kingdom (UK) curriculum (see UK Department of Education, 2013), in which the terms 'sustainability' and 'climate change', perhaps two of the most significant issues in the modern world, are conspicuously missing—purportedly due to "reservations about the inappropriate use and, indeed, over-use of the term 'sustainability'" (UK Environmental Audit Committee, 2005, p. 3). The ongoing omission of these significant areas of global concern from the UK curriculum is reflective of general neoliberal positions of the wider global North minority world.

Bourdieu argues that the dominating field of power can be responsible for both the social production and the social consumption that occurs within the field (Rawolle & Lingard, 2013). This reinforcing structure makes the nature of the science education field one of importance, given its power to influence how society is produced, or reproduced, and how those within it behave. With the current neoliberal approach, reinforcing powers encourage social consumerism that is individualistic and inequitable. This perpetuates construction of this model of society at the expense of others. If a new field of influence is to emerge, where a more socially aware and equitable habitus is to ascend, the present dispositions shaping production and consumerism of the social condition must be transformed. Activist education offers the possibility of such a transformation. 'Real' action on issues like climate change, sustainability, and general resource inequity in the world, exposes students to underpinning issues and *involves* them in affirmative action. Bearing witness to the results of such action can be empowering (Stevenson & Robottom, 2013) and, thus, small shifts may begin in the disposition and ultimately the social field in which science education manifests. This would, however, require a significant shift away from the capitalist drivers that underpin neoliberal philosophy and current education systems that support it.

A further factor that may be exacerbating limited uptake of activist science teaching may stem from confusion, or sense of ethical responsibility, that teachers may have about what and how they present this type of learning—which, by its nature, can be quite controversial in wider society. Teachers are accountable to a range of stakeholders—students, parents, their colleagues, school managers, system agencies, as well as to their own sense of ethics—as to content and pedagogies they adopt in teaching for activism. Hodson (2010) alludes to this, stating importance of care to be "taken to ensure both the appropriateness of a set of actions for the particular students involved and the communities in which the actions will be situated" (p. 203).

23.6 A Journey Towards Activist Teaching

Certainly, my own journey towards a more activist approach to science teacher education has been hesitant. I have struggled with a sense of competing pressures; between a profound sense of responsibility and capacity to 'do some good' and a deep concern for the potential misuse of my position and power. Continually, I have (and still do) question whether my teaching could be construed as preaching, or perhaps what Gramsci (2007) describes as "moralistic sermons" (p. 47). Russell (1932/2009) expresses such a tension as a type of burden, stating that "[i]n this world of flux men [sic] *bear* their part as causes of change, and in the consciousness of themselves as causes they exercise will and become aware of power" (p. 3, emphasis added). Commitment to educating for social justice and equity in the world has been compromised by my uncertainty about content and approaches I have used to teach about it and consequently, has at times felt like a burden of conscience.

When I examine the STEPWISE framework now, I can see clearly that my early engagement in teaching about socio-scientific issues reflected the typical, disconnected approach—that is, concerned with content, products, skills and, to a limited extent, research about particular ideas and issues. To this end, I believe I delivered a reasonably good Products Education—the government mandated curriculum frameworks helped ensure this; I was reasonably good at incorporating Skills Education, mainly because I enjoyed teaching when learning was focused on science skills and processes rather than just conceptual knowledge. I was also reasonably effective in addressing Science Technology, Society and Environment (STSE) education, as I already had a passion for learning and teaching about socio-scientific issues in the world and their associated ethics. I did not, however, have an explicit awareness or understanding of links between these aspects of science education.

I can also see that I did not have a very good understanding of the impact my teaching had on my students; believing, without ever really questioning, that by equipping them with knowledge and informing them about issues, they would feel inspired to take action in their own lives. With time, it has become obvious that these conventional forms of STSE education are, as Hodson (2010) describes, "inadequate to meet the needs and interests of students faced with the demands, issues and problems of contemporary life" (p. 197). Using Hodson's (1994) levels of sophistication, I was operating at the lowest level (Level 1): "appreciating the societal impact of scientific and technological change and recognising that science and technology are, in substantial measure, culturally determined" (p. 85).

In preparing for lectures and tutorials in the core science education courses I taught in a Bachelor of Education (Primary) program, I began to learn more about particular injustices in the world. I also gained some experience working in the Pacific country of Solomon Islands, both in a capacity-building role, co-planning and presenting teacher professional development alongside local teacher leaders; and supervising a small group of pre-service teachers on a four-week teaching practicum. Solomon Islands is one of the poorest and least developed countries in the world (United Nations Conference on Trade and Development [UNCTAD], 2011), and education is difficult for most people to access. My work, over a seven-year period (and which is ongoing), provided direct observation of impacts of social inequity. I was able to witness first-hand impacts of global issues like climate change, and see how those with the least power to effect any change were the ones already suffering consequences imposed on them by the global North minority's way of life.

These encounters and experiences expanded my own social field and, with reflection, altered dispositions informing my personal habitus. My sense of responsibility to others and my sense of being able to contribute in some way influenced the amount and the manner in which I taught about and increasingly, *for*, socio-scientific issues. In an effort to minimise what I feared was 'preaching' about these issues, I introduced a range of student-led research and presentations. One particular strategy involved running debates about particular genetically modified foods/crops (e.g., Flavr Savr tomato (no longer available); Bovine Growth Hormone, Roundup Ready, etc.). Here, I believed (perhaps somewhat naïvely) that students would uncover information for themselves and, in trying to form an argument using their research-evidence, would come to see risks, injustices and unethical behaviours of self-interested corporate and capitalist organisations. To some extent, this did occur. Disappointingly, however, often the 'winners' of debates (as selected by the remaining peer group) were those who exhibited good debating skills, rather than the information alone acting as bases for decisions. The power and danger of charisma, and ability to present an argument with confidence, was far more convincing than even some of the most frightening of statistics and information presented.

I was quite confronted, and somewhat disappointed by these experiences, although they did enable me to better see how people can be swayed into particular actions and ways of thinking by the power of the person or media used to inform them. The self-interest of certain groups (government and for-profit organisations) became increasingly obvious to me and I began to advance through Hodson's levels of sophistication. I became better able to recognise links between the presentation and uptake of certain commodities and interests of associated funding bodies/capitalist organisations. I began to realize how this power and influence had potential for conglomerate control (Level 2); and consequently, my own values, attitudes and actions; i.e., habitus, began to transform (Level 3). I also recognised that lack of direct involvement with an issue led to a disconnect between the issue and the real-life impact it had on people. My students may have learnt about the topic, but the task lacked authenticity to really engage students' values about the examples and their use in the world.

The transformation occurring in my own thinking and actions, in turn, further impacted my teaching, which also began to better reflect Hodson's (2010) levels of sophistication. I designed tasks, questions and information to reflect, far more explicitly, levels of power that those with privilege, money and position can influence on those without (Level 2). I engaged students in critical reflection on their own actions that might be supporting or challenging the outcomes of the organisations wielding such power (Level 3). Yet, I still hesitated to challenge students to participate in their own socio-scientific action (Level 4). I was still in a false state of belief that with knowledge and reflection, my students would, as I had, change their values, attitudes and subsequently, their own actions.

In transitioning between educating about socio-scientific issues to teaching to challenge students' attitudes and beliefs for action, I again became self-conscious that my teaching might be construed as preaching. Placing explicit expectations on them for action created a conflict in my mind that I was misusing my position.

However, my teaching did tend towards providing explicit and, by the nature of my delivery, implicit learning in a one-sided manner. Many students' unit (course) evaluations began to demonstrate the value of such teaching, with comments like, "Learning about the state of the world/sustainability issues enlightened us and gave us information we can use" and "I valued the range of sustainability topics… It made me see how science is related to everyday life and gave me an understanding of human impact". However, there would also be an occasional criticism with comments urging me to "keep [my] personal political views out of the classroom." Although these criticisms were by no means in the majority, these latter types of comments made me question whether I was misusing my position and authority as a means of propaganda.

23.7 Propaganda in Education

Russell (1932/2009) tells us that propaganda in education occurs when educators present information that, whilst may possibly be completely accurate, tends to be selective in its content and given at the exclusion of other existing, contrary, information. Russell also asserts, however, that it is impossible for educators to avoid propaganda, as attempting to do so would be an unnatural suppression of their personality. Foley (2004) also recognises the personal effect of the teacher, stating that: "anything educators do should be grounded in their values based on the deepest possible understanding of their work" (Foley, 2004, p. 10). Teaching and learning cannot be value-free; particularly in a world where all actions can have direct and indirect effects on others, and especially when these effects are often detrimental. As Foley states:

> Every technique you use, every theory you employ, has moral and political effects … Critical educational theory alerts us to the moral and political implications of educational interventions. It does so with an emancipatory intent. It is interested in learning and education that frees people from exploitation and oppression, and helps them develop their capacities and take control of their lives. It focuses on collective educational efforts in community and worker organisations, social action and social movements. (p. 16)

Foley's view helps us to see that propaganda, *per se,* is not the issue; but rather, as Russell (1932/2009) also recognises, that educators need to ensure they present more than just one side of any topic of controversy. This allows for transparency leading to critical reflection and personal decision-making. This decision-making and reflection should be couched not so much in science/technology content, which is useful for informing the decision, but rather, in ethical uses of science/technology so as to preserve interests of the wider public (Bencze & Carter, 2011). A further challenge for the educator is to move this sort of learning beyond what is known as the *interpretivist* approach to education. Interpretivism recognises social and cultural contexts associated with individual ideas and values; it does not critically examine the manner in which particular ideas and values are presented, how they are shaped, or to whose benefit or detriment they affect. As Hodson (2010) recognises, these affects can be quite profound. Foley (2004) reports that:

> [a]ccording to Habermas and other critical theorists, the limitation of the interpretive framework is that it over emphasises the subjective dimension of knowing and learning, and pays insufficient attention to the ways in which our understandings are shaped by the structure and culture of the institutions in which we live and work. (p. 14)

Critical theory recognises ways in which understandings are shaped. It acknowledges inter-relationship between theory, ideology and power (Foley, 2004) and, thus, allows for emancipatory intent of more democratic education. This is similar to how Bourdieu (1984) describes formation and re-shaping of habitus. It is inter-relationships between social and cultural dimensions that influence dispositions of habitus within a given social-cultural field and critical reflection on these influences and dispositions that can re-shape it. Such emancipatory approaches to education also align with conceptions of *new learning*. New learning also aspires for emancipation, which Kalantzis and Cope (2008) say is about making the world a better place, rather than settling for encouraging students to do their best, as if somehow they are bound by prevailing social conditions in which they live. An education aspiring to such an outcome requires promotion of voice, choice and ownership (Gallavan & Webster-Smith, 2012). This means giving each person opportunities to express themselves and be heard; the right to select from a range of possible outcomes, so long as rights of others are safeguarded in any selection; ownership and responsibility for what is expressed, choices that are made, and subsequent actions that follow (Gallavan & Webster-Smith, 2012).

23.8 An Application of Activist Teaching

In an attempt to develop my teaching to better reflect an activist framework, I used my growing knowledge and awareness in the design of a unit (course) for futures learning. In this unit, I worked with two colleagues to implement Mezirow's (1991) transformative learning. This was my first experience of planning for intentional teaching of activism. It was also the basis of a small research grant through which we collected data about types of triggers and actions to which students reported committing, or intending to commit, as a result of their learning. Results of this study were limited (see Carter, Castano & Jones, 2014) in that there were some personal actions taken up as a result of the unit (e.g., greater awareness and commitment to buying free-range chicken/eggs); but, overall, impacts appeared minimal. A second iteration of the study in 2013, demonstrated more substantial commitments to personal action, including two students who reported becoming vegetarian as a result of information about (1) the impact of meat-eating on the environment and/or (2) because of animal cruelty. In fact, I also became vegetarian as a result of this work and, two years after that, vegan. Three years later, the lasting affects of this work have led to my ongoing commitment to a vegan lifestyle.

Interestingly, my personal action to become vegetarian and later, vegan, and taking other personal actions based on my developing ethics and values, has once again made me self-conscious as to how my teaching might be interpreted. I am fearful that my students will see my choice of content and strategies as a judgment on them and/or a form of recruitment into a lifestyle similar to my own. As a result, I communicate to my students less than I normally would about my own personal stance on particular issues or, if I feel that I am expressing an opinion that tends one way over another, I try to be explicit that this is what I am doing. Otherwise, I now teach passionately and overtly ideas for and against a range of issues, particularly those concerned with sustainability and animal cruelty. Outside of the two-cycle experience of teaching explicitly for action using Mezirow's (1991) transformative learning, however, I have still not developed a strong emphasis of Hodson's (1994) Level 4 of sophistication: Preparing for and taking action on socio-scientific and environmental issues in my individual teaching.

Recently I collected some feedback from students participating in my third year core science unit in a Bachelor of Early Childhood and Primary Education; and my second year Bachelor of Education (Primary) course. The 12-week unit focused on energy for a sustainable future and other sustainability issues, such as food security, access to fresh water, waste production and climate change. When asked whether learning in the unit had inspired them to take, or consider taking, any action in regard to any of the socio-scientific issues covered in the unit, a range of small to medium personal actions were reported. These included types of actions that are easily implemented, like conserving energy through reduced electricity use; not over-filling the kettle when boiling water; buying Australian-made products where possible to reduce embodied energy; and buying locally grown/farmed foods where possible to reduce food miles. The embodied energy of products appeared to have been one of the most significant topics, with many students indicating never having thought about/realised this issue before. They felt, generally, that their everyday consumer choices were something on which they could easily improve. Some of the statistics around food wastage were also quite powerful and many indicated that they had taken measures to reduce their food waste and, for the food waste they did generate, they had started composting, with a few reporting that they had established worm farms.

The more substantial actions mentioned included two students who decided to become vegetarian and one, who was already vegetarian, deciding to become vegan. A further five claimed to have reduced the amount of meat in their diet from daily consumption to between two and four times per week. These sorts of actions were not necessarily sought-after, and I am not advocating that these are in any way coveted forms of action. They are, however, substantial in that they can be more challenging to adopt and are generally less likely to emerge as an outcome, especially when unsolicited. They demonstrate that raising awareness and encouraging personal reflection in general teaching can influence changes in the dispositions of some people in quite substantial ways. However, it is worth noting that each of the people involved (including myself) were already part way through the thinking process about making such a change. Collectively, the personal actions reported reflected that there was a range of more informed, effective measures and behaviours towards a more sustainable lifestyle. Some also mentioned that they had begun taking part in signing activist petitions and trying to influence their family and friends to take similar actions to themselves.

23.9 Conclusion and Future Implications

Overall, there was a strong sense from the feedback provided that there had been considerable increases in knowledge and awareness reflecting Hodson's (1994) Levels 1 and 3 stages of sophistication, and some evidence of Level 4 action; and, even though this had not been an explicit requirement of the unit, it did appear to come about as a consequence of the learning. This is perhaps an outcome of what Russell (1932/2009) terms "herd education" where: "every collection of human beings in close proximity develops a herd feeling, which is shown in a certain instinctive uniformity of behavior" (p. 60). There was a general consensus of attitudes and opinions voiced throughout the semester in a variety of forms (lectures, discussions, student presentations, online blogs) that could be representative of this "herding".

These results have further inspired my thinking about my teaching. They demonstrate the beginnings of a shift in the disposition of many of the students involved. Thus, explicit teaching and raising awareness, not just of STSE issues, but also of power and influence of those with vested interests in particular issues/actions, appears to have some effect on habitus. It suggests also, that science teacher education framed in certain ways can create particular social fields that, in turn, affect dispositions and habitus of students involved towards a socio-eco-activist field. It is not just what was presented that was highly valued, but also how it was presented and approached that was important.

Results here have engendered in me both a sense of relief from my fear that my teaching would be viewed as preaching and a sense of inspiration to further this style of teaching. At the end of the semester, I mentioned to my students that I was concerned about 'preaching' rather than teaching, and I would be interested in any comments that they wanted to add to their unit evaluations about the matter. A number did comment and indicated that my teaching was inspiring and not at all commensurate with preaching. This has certainly assisted my confidence and intensified my commitment to pursue a teaching agenda that does require more explicit forms of socio-eco-activist education. STEPWISE appears to serve as a useful framework to pursue such an agenda.

Implications of the journey to date are that I am more committed and more courageous about delivering an activist science teacher education. Critical analyses of my practice, undertaken through self-reflection, student feedback and reviewing literature in the field, has also been influential in my thinking about how I might undertake activist science teacher education in future. It seems likely from the journey to date, that I have potential to affect dispositions/habitus of my students by creating a social field that focuses on equity, justice and wellbeing for all people and for nature. The tetrahedral framework of STEPWISE will also be useful in helping me place action at the centre of my teaching, although I will be interested to compare impacts of more indirect approaches to activism I have recently taken with more direct and explicit activist approaches I am planning and that is promoted through STEPWISE. I wonder if explicit identification of my hesitation and concern about 'preaching' is perhaps beneficial in creating a sense of concern that it not be construed this way that subsequently engenders the opposite effect. It is also important to avoid propagandist approaches described by Russell (1932/2009) by ensuring both sides of arguments are exposed and explored. The student research aspect of STEPWISE can certainly allow for this, as does the opportunity to critically reflect on results of this research. As Gallavan and Webster-Smith (2012) recognise:

> By giving teacher candidates rich opportunities to reflect within themselves and to design the tapestries that tell their stories, each teacher can find her or his personal power by discovering voice to choose the option that yields choice, ownership and action for societal change. The change may be global or local, public or private, loud or quiet, grandiose or humble. All the same, their actions can overcome situational boundaries and not allow conditions to thwart their positions. (p. 55)

Investigating these wonderings will form bases of my ongoing journey towards a new vision for science teacher education or, perhaps, a Vision IV for scientific literacy, one that is concerned with education as socio-eco-activism.

Acknowledgement My sincere thanks to my dear friend and colleague Dr. Caroline Smith who has been an inspiration from the very beginning of my career in teacher education, and who read multiple drafts of this chapter and provided invaluable feedback.

References

ACARA. (2013). *Australian curriculum*. Retrieved from http://www.australiancurriculum.edu.au/

Aikenhead, G. (2006). *Science education for everyday life: Evidence-based practice*. New York: Teachers College Press.

Aikenhead, G. (2007). Expanding the research agenda for scientific literacy. In C. Linder, L. Östman, & P. Wickman (Eds.), *Promoting scientific literacy: Science education research in transaction*. Proceedings of the Linnaeus Tercentenary Symposium, Uppsala University, Sweden, May 28–29, 2007. Retrieved from http://eprints.qut.edu.au/12883/1/Linnaeus_Tercentenary_Symposium.pdf#page=70

Alsop, S., & Bencze, J. L. (2010). Introduction to the special issue on activism: SMT education in the claws of hegemon. *Canadian Journal of Science, Mathematics, and Technology Education, 10*(3), 177–196.

Ayers, W., Quinn, T., & Stovall, D. (2009). *Handbook of social justice in education*. New York: Routledge.

Bencze, J.L. (2014). *'STEPWISE': Science & technology education promoting wellbeing for individuals, societies and environments: A summary for science teachers*. Retrieved from http://webspace.oise.utoronto.ca/~benczela/STEPWISE_Sumry4Tchrs.pdf

Bencze, J. L., & Alsop, S. (2009). A critical and creative inquiry into school science inquiry. In W.-M. Roth & K. Tobin (Eds.), *The world of science education: North America* (pp. 27–47). Rotterdam, The Netherlands: Sense.
Bencze, J. L., & Carter, L. (2011). Globalizing students acting for the common good. *Journal of Research in Science Teaching, 48*(6), 648–669.
Bourdieu, P. (1984). *Distinction: A social critique of the judgment of taste*. London: Routledge.
Carter, L., Castano, C., & Jones, M. (2014). Transformative learning in science education: Investigating pedagogy for action. In L. Bencze & S. Alsop (Eds.), *Activist science & technology education* (pp. 531–545). Dordrecht, The Netherlands: Springer.
Clover, D. (2002). Traversing the gap: Concientizacion, educative-activism in environmental adult education. *Environmental Education Research, 8*(3), 315–323.
De Boer, G. (2000). Scientific literacy: another look at its historical and contemporary meanings and its relationship to science education reform. *Journal of Research in Science Teaching, 37*(6), 582–601.
During, S. (Ed.). (2007). *The cultural studies reader* (3rd ed.). New York: Routledge.
Fenge, L. (2011). 'A second chance at learning but it's not quite higher education': Experience of a foundation degree. *Journal of Further and Higher Education, 35*(3), 375–390.
Fensham, P. (2004). *Engagement with science: An international issue that goes beyond knowledge*. Paper presented at the science and mathematics education conference, Dublin, September 23–24, 2004. Centre for the Advancement of Science Teaching and Learning.
Foley, G. (Ed.). (2004). *Dimensions of adult learning: Adult education and training in a global era*. Berkshire, UK: Open University Press.
Gallavan, N., & Webster-Smith, A. (2012). Commentary. In R. Flessner, G. Miller, K. Patrizo, & J. Horwitz (Eds.), *Agency through teacher education: Reflection, community, and learning*. Plymouth, UK: Rowman & Littlefield Education.
Gramsci, A. (2007). The organization of education and culture. In S. During (Ed.), *The cultural studies reader* (3rd ed.). New York: Routledge.
Hodson, D. (1994). Seeking directions for change: The personalization and politicisation of science education. *Curriculum Studies, 2*(1), 71–98.
Hodson, D. (2003). Time for action: Science education for an alternative future. *International Journal of Science Education, 25*(6), 645–670.
Hodson, D. (2010). Science education as a call to action. *Canadian Journal of Science, Mathematics, and Technology Education, 10*(3), 197–206.
Hopkins, C. (2013). Educating for sustainability: An emerging purpose of education. *Kappa Delta Pi, 49*(3), 122–125.
Kalantzis, M., & Cope, B. (2008). *New learning: Elements of a science of education*. New York: Cambridge University Press.
Louden, W. (2008). 101 Damnations: The persistence of criticism and the absence of evidence about teacher education in Australia. *Teachers and Teaching: Theory and Practice, 14*(4), 357–368.
Mezirow, J. (1991). *Transformative dimensions of adult learning*. San Fransisco: Jossey-Bass.
Ministry of Education (MOE). (2011). *The Ontario curriculum grades 9–12: Environmental education: Scope and sequence of expectations* (2011 ed.). Toronto: Queen's Printer for Ontario. Retrieved from http://www.edu.gov.on.ca/eng/curriculum/secondary/environment.html
Mirra, N., & Morrell, E. (2011). Teachers as civic agents: Toward a critical democratic theory of urban teacher development. *Journal of Teacher Education, 62*(4), 408–420.
Navarro, Z. (2006). In search of cultural interpretation of power. *IDS Bulletin, 37*(6), 11–22.
NCCA. (1999). *National council for curriculum and assessment: Curriculum online*. Retrieved from http://www.curriculumonline.ie/
Rawolle, S., & Lingard, B. (2013). Bourdieu and educational research: Thinking tools, relational thinking beyond epistemological innocence. In M. Murphy (Ed.), *Social theory and educational research: Understanding Foucault, Habermas, Bourdieu and Derrida* (pp. 117–137). London/New York: Routledge.

Roberts, D. (2007). Scientific literacy/science literacy. In S. Abell & N. Lederman (Eds.), *Handbook of research on science education*. Mahwah, NJ: Lawrence Erlbaum Associates Inc..

Roth, W.-M., & Désautels, J. (Eds.). (2002). *Science education as/for sociopolitical action. Counterpoints: Studies in the postmodern theory of education*. New York: Peter Lang Publishing, Inc., 275 7th Avenue, 28th Floor, New York, NY 10001.

Roth, W. M., & Lee, S. (2004). Science education as/for participation in the community. *Science Education, 88*, 263–291.

Russell, B. (1932/2009). *Education and the social order.* New York: Taylor & Francis.

Stevenson, R., & Robottom, I. (2013). Critical action research and environmental education: Conceptual congruences and imperatives in practice. In R. Stevenson, M. Brody, J. Dillon, & A. Wals (Eds.), *International handbook of research on environmental education* (pp. 469–479). New York: American Educational Research Association/Routledge.

Tytler, R. (2007). *Re-imaging science education: Engaging students in science for Australia's future, Australian Education Review No. 51*. Camberwell, Australia: Australian Council for Educational Research/ACER Press.

Tytler, R., Symington, D., Kirkwood, V., & Malcolm, C. (2008). Engaging students in authentic science through school-community links: Learning from the rural experience. *Teaching Science, 54*(3), 13–18.

UK Department for Education. (2013). *National curriculum in England: Science programmes of study*. Retrieved from https://www.gov.uk/government/publications/national-curriculum-in-england-science-programmes-of-study/national-curriculum-in-england-science-programmes-of-study

UK Secretary for Education and Skills. (2005). *Government response to the EAC report on education for sustainable development*. Norwich, UK: The Stationary Office Ltd. Retrieved from https://www.gov.uk/government/publications/government-response-to-the-eac-report-on-education-for-sustainable-development

UNESCO. (2009). *Education for sustainable development (ESD)*. United Nations Educational, Scientific and Cultural Organisation. Retrieved from http://www.unesco.org/new/en/education/themes/leading-the-international-agenda/education-for-sustainable-development/education-for-sustainable-development/

United Nations Conference on Trade and Development (UNCTAD). (2011). *The least developed countries report*. Retrieved September 19, 2012, from http://www.unctad.org/en/docs/ldc2011_en.pdf

Chapter 24
"I Had to Take Action Straight Away." Preservice Teachers' Accounts of Pro-environmental Action

Lyn Carter and Jenny Martin

24.1 Looking into Action

'STEPWISE' (*Science and Technology Education Promoting Wellbeing for Individuals, Societies and Environments*) is a framework for science education that encourages students to work towards a better world utilising, in part, their understanding of science and technology. It aims to inform decisions about actions they could take to address socio-scientific issues (SSIs) or issues in science, technology, society and environment (STSE). As we can see from many of the chapters in this volume, Bencze diagrammatically represents STEPWISE as a tetrahedron with four learning domains at its peripheral points and 'STSE Actions' suspended in the centre. The pivotal positioning of STSE Actions, suggests Larry Bencze (Chap. 2, this volume), emphasises the altruistic nature of STEPWISE, fostering in students belief "that they can—and perhaps should—'spend' some of their literacy, not just on themselves, but also on efforts to improve the wellbeing of other individuals, societies and environments (WISE)."

In this chapter, we focus on the 'suspended centre'; that is, the STSE *actions* fundamental to the framework. Bencze describes what he means by action and the conditions under which it may be facilitated in several of his writings. Influenced by Derek Hodson (2003) and others calling for science education reform that prioritises sociopolitical actions, Bencze (e.g., Bencze, in press) sees actions as including one or more of: "educating others (e.g., via posters and pamphlets), lobbying power-brokers (e.g., via petitions and letters to politicians), developing potentially-improved products and systems (e.g., a cell phone with recyclable components) and/or making personal improvements (e.g., using a travel mug)." Moreover, having

L. Carter (✉) • J. Martin
Faculty of Education and Arts, Australian Catholic University, Melbourne, VIC, Australia
e-mail: Lyn.Carter@acu.edu.au; jenny.martin@acu.edu.au

L. Bencze (ed.), *Science and Technology Education Promoting Wellbeing for Individuals, Societies and Environments*, Cultural Studies of Science Education 14,
DOI 10.1007/978-3-319-55505-8_24

investigated conditions under which students may be motivated to take action, Larry Bencze (2014, p. 5) suggests:

> (i)t is possible that, after hearing about issues from a teacher, students may not be particularly motivated to take action(s) to address SSIs (e.g., merits of trans-fats in manufactured foods). In our research, we have found that students are more likely to be motivated to act on SSIs if they have *developed* some ideas, conclusions, etc. as a result of their *own* research. … (T)he more activities are 'student-directed' and conclusions are 'open-ended' (with many possible conclusions from data and theory) the more personally-engaged students will be in the activities. If students have self-directed open-ended research about SSIs, their findings (e.g., that road salt inhibits seedling germination) can motivate them to take action(s) to address the SSIs they investigated.

Students '*own* research' in these instances, Bencze (2014) notes, would likely be a combination of 'primary' and 'secondary' research where the former refers to investigations in which students generate original knowledge through their own activity, while the latter involves them in gathering knowledge discourses from available references sources.

In and of itself, STSE action suspended and cloistered within the tetrahedron raises some interesting questions for us. For example, action (and also altruism) like many others, are constructs without a compass and limits that don't inevitably circumscribe the critical attitude seeking less repression that is assumed by the STEPWISE framework. Recent studies of the Tea Party in the United States for instance, show clearly that all forms of socio-political action like 'lobbying power brokers' can be used just as much in the service of socially progressive goals, as against them (Burris, 2001). Within France to take another example, there has been strong action by those groups both supportive and opposed to gay marriage. Furthermore, action is culturally located and needs to be considered as such. Some of these and other significant questions are canvassed within the various chapters of this volume and have been discussed by us elsewhere (see for example, Carter, Castano & Jones, in press).

While undoubtedly important considerations, in this chapter, though, we venture a different approach. Here we consider 'action', which lies at the heart of STEPWISE and other SSI/STSE approaches, from a discursive psychological perspective. For Rom Harré (1984), the discursive or cultural turn in psychology was a reaction to cognitive psychological assumptions that separate an inner psychological realm of the human mind from outer social contexts. Thus, in discursive psychology, the meaning of any word such as 'action,' can only be understood in its context of use. As such, discursive psychology takes an alternative view to cognitive psychology that underpins much science education research in the area enabling a somewhat unique contribution. It is worth noting that we have not analysed STEPWISE for its cognitive psychological assumptions or viewed it through a discursive lens. That is a task for another day. Instead we offer our experience in a project similar to STEPWISE in the hope that STEPWISE may take something of value in reflecting on the notion of action.

Hence, to illustrate our discursive psychological perspective, we describe part of a study undertaken with our first-year science course within a Bachelor of Education programme for preservice elementary or primary teachers at an Australian University (around 400 students). The course's topic area of environmental sustainability and pro-environmental engagement, its focus on action, and a self-directed and open inquiry pedagogy involving both primary and secondary research ensures our course is empathetic with the tenets of STEPWISE. Our course aimed to empower preservice teachers to become science educators in their own right, and to position themselves as pro-environmentally active, both in their lives and in their teaching. In our research, we are concerned with ways in which preservice teachers develop responsibility for pro environmental action.

In this chapter then, we report on the section of the larger study where the preservice teachers completed an Eco Challenge[1] that required them to conduct an action-focused and evidence-based appraisal of their current sustainability practices. They were then required to redesign and implement practices (action) to reduce their ecological footprint in terms of food and energy consumption and their production of waste, evaluating and explaining their success or otherwise. The preservice teachers recorded their progress in an open-ended, reflective journal format of their own choosing (e.g., a paper notebook or on-line blog) kept across the 12-week semester of the course. The preservice teachers provided evidence which included statistics related to their adopted practices, such as quantities and frequencies showing reduced consumption, photographs, reports of conversations with significant others as well as reflections upon their own or others' attitudes, and information they researched that would elaborate their positions. Formal classes during the semester were designed to support their understanding of environmental sustainability, which in STEPWISE terms, was largely through the study of secondary research. The journals could be regarded as a mix of STEPWISE's primary and secondary research. While the journals were not part of the preservice teachers' assessment for the course, they could have been referred to in their final reports as evidence of their action and changed perspectives.

We move onto the next sections and outline aspects of discursive psychology relevant for our study as well the stance we have taken on pro-environmental engagement as applied scientific literacy. We then report some of our findings before concluding with a discussion on action that may be insightful for the STEPWISE framework given the unique lens of discursive psychology.

24.2 A Discursive Psychological Approach

Discursive psychology (Davies & Harré, 1990; Harré, 1984) is an approach to psychology developed to overcome Cartesian dualism that is a foundational assumption in cognitive psychology. Unlike cognitive psychology, the discursive approach acknowledges cultural and relational aspects of any action in the social world and makes no distinction is made between social and psychological phenomena (Harré,

[1] Dr. Caroline Smith originally designed the Eco Challenge in 2008. It has since been adapted.

1984). Cognitive psychological approaches to science education research, by contrast, typically privilege generalised inner mental states when looking at (in our case here) preservice teachers' stated intentions (for example, Basu, 2008), or the attribution of intention to students (see Sharma, 2007) as central to the operationalisation of action. In other words, students expressing their intention to take action when interviewed about action contribute to that particular practice with some understanding of expected responses. Rather than looking to what students say (or do) in a social setting as representing general psychological states (e.g. wanting), a discursive psychological approach, looks to the *function* of students' *sayings or doings in their context of use* (Wood & Kroger, 2000). Discourse-based approaches in science education research explore action in science as a complex social activity suggesting that action cannot be explained as individual intentions (as in cognitive psychology), rather it needs to be understood as the social meaning achieved in the intention, and in terms of resultant social practices (see for example, work from Anderson & Zuiker, 2010; Arnold, 2012; Arnold & Clarke, 2014; Seah, Clarke, & Hart, 2011; Yerrick & Gilbert 2011). We believe this approach, less usual in science education research, can avoid limitations posed by much of the cognitive psychology at large in our field.

In discursive psychology, social psychological phenomena are manifest as social acts. A social act is defined as the relatively determinate social meaning of action, and is the smallest unit of analysis. The analysis of social acts involves three mutually interdependent features of a conversation (discourse). These are the actual doings and sayings ('action'), the conversational 'storylines,' and the 'positioning' of actors (Harré & van Langenhove, 1999). A 'position' is a person's psychological location in an ongoing 'conversation'—or where they locate themselves. An example of a function of language use is the 'positioning' of persons in interaction or realisation of social identities (Davies & Harré, 1990) and social practices. Personal development occurs as individuals 'position' themselves in 'conversation' and develop social identities in relation to their obligations and *responsibilities* in a local moral order (Arnold, 2010; Harré, 1984; see also Martin & Carter, 2015). Pronouns, modality and tense become then, the language devices used to index 'position' or responsibility to persons (see Muhlhäusler & Harré, 1990 for a discussion of functions of pronoun use). Jenny Martin (writing as Arnold, 2012) has referred to these grammatical devices as the Grammar of Agency.

In our analysis of preservice teachers' reflective accounts of their Eco Challenge, we analyse their accounts as social acts. The journal entries provided a space to record and construct (story) the preservice teachers' own thoughts and ideas, narratives, choices of life aspects to which to attend, and how they position themselves in relation to action for environmental conservation. The discourses appropriated or transformed by preservice student teachers in 'storying' their own pro-environmental engagement included course readings, peer conversations, generalised discussion and discourses circulating more widely in society through the media and online sources of information. The section of our overall study reported in this chapter is concerned with the preservice teachers' use of the concept of 'action' in their own journal accounts, thus responding to relevant questions like:

- How do the preservice teachers employ the language of 'action' in their accounts?
- How do preservice teachers 'story' their own pro-environmental action? How do they position themselves and and their responsibility to take action?
- In what ways do these discourses promote or limit the preservice teachers' pro-environmental action and their positioning as responsible for action regarding EfS?

24.3 Pro-environmental Action as Applied Scientific Literacy

Beyond STEPWISE, several other research strands also inform our work. Clearly, environmental sustainability education is a key influence, as the highly significant 1992 United Nations Conference on Environment and Development blueprint, *Agenda 21*, suggests:

> Education is critical for promoting sustainable development and improving the capacity of the people to address environment and development issues (Chapter 36, Para 36.3).

Much has been written about environmental sustainability education since that landmark report including the 2009 *Climate Change and Sustainable development: The Response from Education.* It is one of an increasing number of reports, policy documents and research publications that identify characteristics of sustainability education able to promote sustainable agendas. Education for environmental sustainability encompasses a complex agenda around the natural, the social, the cultural and the economic and shares with STEPWISE, issues that include health, poverty and redistributive justice. Importantly, *The Response from Education* report found that one of the key obstacles to successful implementation of sustainability education is teacher knowledge, understanding and action due to the lack of sufficient and high quality pre- and in-service teacher education. Clearly, if we wish educational settings to be proactive in implementing sustainability education either within their science programmes or elsewhere, to promote pro-environmental engagement and action, then sustainability education needs to become a core element in both in-service and pre-service teacher education. That we have taken this call seriously is evidenced by the fact that one of our three undergraduate science courses is devoted solely to EfS and pro-environmental engagement.

Preservice teachers' actions can also be regarded as a subset of the research into youth and adult pro-environmental engagement. Louise Chawla (2006) and Joya Palmer, Suggate, Robottom, & Hart, (1999) have found that opportunities for affirmative experiences of nature as well as the shaping by family or other role models can facilitate pro-environmental engagement. Anja Kollmuss and Julian Agyeman (2002) identified pro-environmental knowledge, awareness, values, attitudes, emotion, and locus of control, responsibilities and priorities as all impacting pro-environmental behaviour. Similarly, Beth Robelia, Christine Greenhow and Lisa Burton (2011) found when investigating on-line communities, that being part of a like-minded group spurred involvement in many participants and encouraged them to learn more about climate change and change their behaviour and take action. These and other studies have all been useful in offering insight into the important factors facilitating preservice teachers' engagement and action, and helping us prepare positive curricula and other educational experiences.

Moreover, Lyn Carter (2012) identified eleven preferences to learning that may engage preservice teachers with environmental sustainability. These preferences find some resonance with the currents and trends Lucie Sauvé (2005) identified as readily apparent with the environmental education field. However, Carter (2012) suggested that preservice teacher motivation for engagement and action was more complex, as the preservice teachers' developing sense of their own professional responsibilities intersects with competing discourses of individualism in Western society and social justice, for example. Lou Preston (2011) also focused on preservice teachers and EfS in her study, showing that 'discourses of ecological crisis' limited preservice teachers' views of pro-environmental action to absolving personal guilt and to individual acts of green consumerism which they saw by the way, as beyond their financial means. In ecological crises discourses, individual responsibility is uncritically taken as the cause of environmental problems rather than them as part of the broader complex sociopolitical networks with powerful interests at work. Clearly a multiple phenomenon, our study using a discursive psychological approach, builds upon Carter's (2012) and Preston's (2011) efforts to explore how preservice teachers engage with prevailing discourses, including discourses of professional responsibility (DEWHA, 2010), and position themselves as responsible or obligated for engaging in pro-environmental action, and how they account for their action in terms of structural contingencies or affordances.

Through the Eco Challenge and coterminous with STEPWISE, our first year science course takes an action (applied) orientation to scientific literacy similar to that of Crowell and Schunn (2014). Amanda Crowell and Christian Schunn's applied view of scientific literacy shifts the preoccupation of a more generalised understanding of scientific literacy as knowledge acquisition to an emphasis on engaging with practices of science. More importantly, it encodes possibilities for taking actions (the 'social act' in discursive psychology), deemed essential as any "problem is unlikely to improve if subsequent actions do not condense with … understanding" (Crowell & Schunn, 2014, p. 719). Crowell and Schunn argue that applied dimensions of scientific literacy have received less attention in science education research than other types of scientific literacy, describing it as a level of public understanding of science that encourages decision-making in concert with scientific consensus. Pro-environmental engagement and action is an example of applied scientific literacy or scientifically literate action related to environmental sustainability.

24.4 Reporting Some of the Findings from Our Study

For this section of our study, we randomly chose fifty individual journals from the large cohort of preservice teachers. We coded the participating preservice teachers' journal entries according to their use of the language of action, and the Grammar of Agency—pronouns, modality and tense—(Arnold, 2012; Arnold & Clarke, 2014) and analysed how the preservice teachers used the concept of action in their accounts. After elaborating our coding of the preservice teachers' accounts, we present four themes that emerged from the data.

24.4.1 Coding the Journal Entries

Muhlhäusler and Harré (1990) refer to an indexical progression by which persons can locate responsibility for action to themselves to varying degrees, ranging from the unmarked first person to the use of first person pronouns and epistemic verbs such as 'to think' or 'to know', which can strongly index responsibility to an individual (individual agency). The use of the first person plural can index responsibility for action to an individual as a member of a group (collective agency). The use of third person pronouns can deflect responsibility from an individual and index it to public personae. Sentence modality, like the use of the first person pronoun can index responsibility for action to a speaker or writer. Usually, the present tense will be used when responsibility is indexed to persons within an ongoing interaction. The use of the past tense can indicate reported action, the social meaning of which cannot be reliably gauged (Harré & van Langenhove, 1999). However, past and future accounts can be analysed for their function in the immediate context.

Our coding using the language of action was an iterative process. For example, we included not only derivatives of the word 'act,' such as 'actor' 'acting' and 'action', but also words that the participants used synonymously as 'action' and its derivatives in defining 'the language of action' for our study. The preservice teachers used the words, 'practice', 'change', 'changing', 'choosing', 'trying' and 'doing' synonymously with 'action' as illustrated in Excerpts 1 and 2. The bold font in these excerpts has been added to highlight the preservice teachers' language of action.

Excerpt 1 "*The first lecture for this class I found overwhelming. Although I have heard of the concept of sustainability discussed many times, I don't think I ever thought about it on any meaningful level. Similarly, although I* ***attempt to make*** *environmentally sustainable* ***choices*** *(e.g.* ***choosing*** *recycled paper and energy efficient light bulbs) I feel as though these* ***choices*** *are made without much consideration or understanding of what I'm* ***doing****. I am hoping that this subject will give me a more thorough understanding of the impact that* ***making different choices*** *can make.*"

Excerpt 2 "*The video we watched in the lecture really reinforced this idea that small* ***changes*** *made on an individual level can make a large difference to the world's environmental sustainability. It made me feel a little guilty as I know it is something I should be more conscious of.*"

In Excerpts 1 and 2, the preservice teachers use the idea of consciousness and making choices in doing certain things as synonymous with pro-environmental action. For these preservice teachers, taking action is equated with knowledge of impacts doings have on environments and *consciously choosing* and *doing* things that have less impact. The idea of acting with volition was found to be thematic in the preservice teachers' accounts of pro-environmental action, elaborated below as Theme 2.

24.4.2 *Emergent Themes*

In our analysis of the language of action, four themes emerged from the preservice teachers' accounts:

1. The language of action was employed by the preservice teachers in reference to individual action.
2. The psychological category of intentionality was employed by the preservice teachers in their explanations of pro-environmental action.
3. Action was linked by the preservice teachers with their own becoming, or their identities.
4. The preserve teachers evoked the concept of inaction and accounted for inaction in various ways.

These four themes are elaborated in the following four sections. The Excerpts are from different preservice teachers' journals and have been chosen as representative of the themes. Preservice teacher language use has been highlighted using bold font as related to the themes.

24.4.2.1 Theme 1 – Individual Action

The preservice teachers employ the language of action in reference to individual action. This is illustrated in Excerpts 2, 3, 4 and 5. In Excerpt 2, the idea of individual action is evoked using the phrase, "on an individual level", and the use of the pronoun, 'I', indexes the preservice teacher's sense of personal responsibility for action. Excerpt 3 evokes the idea of individual action when the preservice teacher describes herself as "only a single person" and uses "everyone" to refer to individuals in the plural sense.

Excerpt 3 "*It can feel like* ***I am only a single person*** *and how will* ***I*** *make a difference. However, through my research I am beginning to understand that* ***everyone*** *makes a difference no matter how small their actions are!*"

The preservice teachers described individual action as making "small changes" (Excerpt 4) and "doing small things" (Excerpt 5).

Excerpt 4 "*Although* ***I*** *have made* ***small changes*** *that have led to my ecological footprint getting smaller it makes it clear to me that small changes can make a difference. I do not believe that the issue of sustainability is being made clear enough. I feel as though the* *issue has been put on the back burner. Yes, there are government bodies working on ways and listing ways people can be more sustainable, but maybe someone should start saying* ***we*** *NEED to do this rather than* ***you*** *can do this." [capitalisation n the original]*

The preservice teachers associate individual action to psychological categories, such as motivation and efficacy. In Excerpt 5, for example, the preservice teacher links pro-environmental practices with a sense of personal empowerment.

Excerpt 5 "*After my ecological footprint has reduced by 0.4 planets, though it seems to be small,* ***I have realised doing small things*** *around the house and in the garden can make a difference whether it be small or large. Creating a veggie patch, taking more public transport or reducing the consumption of animal products can have a great impact, and in doing this project* ***I have realised*** *that I don't really need animal products that much and that* ***I can*** *produce my own edible food. I think* ***I will*** *continue with these practices because not only is it healthy, I feel like* ***I am*** *giving back to the environment and* ***its empowering****. Though it may be little progress and may have had a few setbacks, I am happy with what I have done.*"

The concept of individual action employed by the preservice teachers includes the idea of intentional non-action, for example, pro-environmental action includes 'refusing' (Excerpt 6) and 'going without' (Excerpt 7).

Excerpt 6 "*Instead of buying what I wanted I began to think more about whether I really needed the product before purchasing it. On several occasions,* ***I refused*** *to 'impulse buy' and instead with home without them.*"

Excerpt 7 "*In my plan, I mentioned that* ***I would*** *attempt a shopping detox i.e.* ***go*** *for two weeks* ***without*** *buying unnecessary items such as clothing and make up. At first this seemed overwhelming as I rarely go a week without buying a new item of clothing to wear out on the weekend.*"

In summary, the preservice teachers storied themselves as developing a sense of efficacy in their individual actions using statements like "I now realise that small changes can make a difference". This extended to the anticipated cumulative effect of "small things" and a sense that every individual has a responsibility to do "small things". The preservice teachers storied individual action as the solution to the challenge of world environmental sustainability. However, the focus on the individual limits the scope of the concept, "pro-environmental action" to absolving guilt (see Preston, 2011).

24.4.2.2 Theme 2 – Action and Intentionality

The preservice teachers drew upon the psychological category of intentionality in identifying pro-environmental action. They differentiated acting with knowledge and volition from "instinctive" action, action undertaken without consciousness of its effect on the environment. In Excerpt 1, the preservice teacher refers to acting without "consideration or understanding" and, in Excerpt 2, the preservice teacher states that acting towards the world's environmental sustainability is "something I should be more conscious of". In this way, the preservice teachers make a distinction between acting in habitual ways and acting with an intention towards limiting ones impact, or consciousness and knowledge about ecological sustainability.

Excerpts 8–12 provide examples of the language the preservice teachers used to refer to lack of consciousness or volition:

Excerpt 8 "*Even though I do such things as have a recycling bin, and I am always turning off lights and I have solar panels on my house, I am* **not actively** *taking part in trying to reduce my ecological footprint.*"

Excerpt 9 "*In my daily life* ***I don't generally take account*** *of the damage I am doing to the planet*"

Excerpt 10 "*When searching the sustainable living tips, I realised that* **we already do** *several of the suggested tips* ***without even thinking***"

Excerpt 11 "Growing up during one of the longest droughts in Australian history has made me more aware of my water usage… **instinctively I take** shorter showers"

As differentiated from acting instinctively or without thinking, the preservice teachers described action stemming from new knowledge (illustrated in Excerpts 12 and 13).

Excerpt 12 "*These figures [electricity, gas and water bills] came as a shock to me as I truly had no idea that these simple services cost so much! Looking at these figures* ***provides me with motivation*** *to try and reduce the costs*"

Excerpt 13 "*I really love shopping but* ***I need to*** *reduce how much clothes I buy and also because I buy online which means the transport for the package to get here also impacts the environment*"

In Excerpt 12 and 13, the preservice teachers link new knowledge with the "motivation" (Excerpt 12) or "need" (Excerpt 13) to act. In Excerpt 14, the preservice teacher makes an explicit reference to a distinction between acting with or without volition.

Excerpt 14 "*On a weekly basis households are expected to push out their recycle bins and rubbish bins to collected and thus getting rid of unwanted waste appropriately. For most families and mine, this is as a far as it goes for making* ***a conscious effort*** *to apply sustainable practices. This continuous act is* ***not a voluntary decision*** *made by every person in Australia, but rather a societal rule and law of the government to rid waste… Sustainable practices need to become a way of living rather than rule obeying.*"

The duality of acting with, or without, intention alluded to in the written reflections by the preservice teachers is extended by many to an imperative to act once consciousness or knowledge is established. The idea of imperative action is illustrated in Excerpt 4 and also in Excerpt 15. In Excerpt 4, the preservice teacher refers to individuals taking action in plural ("we NEED to do this") and, in Excerpt 15, the preservice teachers refers to the imperative to take action in terms of a personal sense of responsibility ("I had to take action").

Excerpt 15 "*6.4 global hectares are needed to support by lifestyle. As soon as I completed my ecological footprint I realised* ***I had to take action*** *straight away. I was seeing friends that day so I rode my bike rather than driving*

The analysis shows that the preservice teachers employ psychological language (intentionality, consciousness) in their understanding of what pro-environmental action entails. In other words, they understand pro-environmental action as necessarily involving conscious decision making by autonomous individuals. In their use of the language of intentionality, the preservice teachers situate pro-environmental action as an individual responsibility.

24.4.2.3 Theme 3 – Action and Identity

In their reflective accounts of pro-environmental action, the preservice teachers employed grammars of identity and personal agency (Arnold, 2012). In Excerpt 16, the preservice teacher uses the first person pronoun and psychological categories ("intrigued", "angry" and "interested") and positions herself as agentic and personally responsible for pro-environmental action.

Excerpt 16 *"This week after the tutorial* ***I*** *was really* ***intrigued*** *about the variety of energy sources that there are.* ***I*** *even* ***got angry*** *that we are still using coal and petrol.* ***I*** *am really* ***interested*** *in different water energy sources, especially to do with salt water because* ***we*** *have an abundance of that.* ***I*** *have gone around the house and* ***I*** *have taken out appliances that are plugged in that don't need to be and* ***I*** *have checked the washing machine to see the star rating".*

The preservice teacher's use of 'we' in Excerpt 16 positions her as a member of a collective, in this instance, as a citizen of the State, Victoria. She takes collective responsibility in this way for "still using coal and petrol", and collective ownership of a resource (salt water). Despite the acknowledgement of collective responsibility, she takes personal responsibility for action, and her pro-environmental action is limited to the individual response of reducing energy consumption by unplugging appliances, for example.

In Excerpt 17, the preservice teacher uses first person pronouns to identify herself as a person with particular capacities and dispositions and, importantly, links changes in her identity ("someone who") to the development of a capacity to act ("implement strategies" and "influence others").

Excerpt 17 *"The most important aspect about this challenge for* ***me*** *was* ***my own personal shift in values****. As previously mentioned, I feel as though* ***I have moved****, and will continue to move, from* ***someone who*** *passively accepts sustainability as something that 'someone else will deal with', to* ***someone who*** *can actively implement personal strategies to benefit the environment, as well as helping to educate others on how they can do so as well.*

In Excerpt 18, the preservice teacher links her developing identity as an educator to acting. She emphasises being a person who acts as a necessary aspect of her professional identity.

Excerpt 18 *"I instantly thought,* ***I am*** *an educator and if* ***I am*** *going to teach children about caring for our world* ***I better get to it and act*** *in order that* ***I am*** *not hypercritical to what I teach."*

In summary, action and identity are linked by the preservice teachers in the following, related ways: (1) Action is accounted for as personally motivated, (2) Action is accompanied by a shift in personal identity, and (3) An identity as "someone who acts" is a professional imperative.

24.4.2.4 Theme 4 – Accounting for Inaction

The notion of inaction was evoked by the preservice teachers. In their reflective accounts, the preservice teachers offered explanations for 'inaction'. The preservice teachers used the idea of inaction as action that was not directed towards ecological sustainability (Excerpts 19–21) and action that was not intentionally directed towards ecological sustainability (Excerpt 14).

Excerpt 19 *"The reason* ***we don't do enough*** *to help at the present moment is due to having an* ***attitude*** *of "nothing I can do will help so I will not bother doing it"*

Excerpt 20 "one of the most difficult things about implementing all these practices has been the initial **financial outlay** and even more so the **time** factor. It takes **so much longer** to do my groceries now that I go to several local suppliers for produce instead of simply purchasing everything from the supermarket… I think **time** has been the biggest factor to why **we haven't done anything** before now."

Excerpt 21 *"Sustainable living, reducing our individual footprint for me is one of those important issues that* ***feels too big for me*** *to be able to make an impact on. Of course, in our family we do the token things, recycle plastic, turn power points off and attempt to have shorter showers. Are these little things enough?"*

Inaction was accounted for by the preservice teachers using psychological categories, such as "attitude" (Excerpt 19) and "feel" (Excerpt 20), and socio-material realities, such as "time" "money", and availability of products (Excerpt 19).

The preservice teachers accounted for inaction in individual terms and did not refer to action beyond the responsibility of an individual, such as collective action or political action. They did not critically reflect on the kind of action required solving some of the bigger issues, such as equitable access to the earth's resources and communities' transitions to energy sources other than petroleum, even though the preservice teachers named Victoria's (the state within Australia where they are located) continuing use of coal and correspondingly high eco footprint as a problem.

24.5 Discussion

Our approach utilised the theoretical lens of discursive psychology where the social meaning of language is understood to be relatively determinate, to investigate how the concept of action is used by our preservice teachers in their reflective journal accounts. We found that our preservice teachers linked taking action to individual intentionality and responsibility (Themes 1–4). In this way, our research supports claims made by Preston (2011) that young people tend to adopt ecological crises discourses. As noted above, in ecological crises discourses, individual responsibility is uncritically taken as the cause of environmental problems. In Preston's study, participants attributed blame to individuals due to laziness, failing to understand the broader complex power-vested interests. Rather than taking action, Preston reported that ideals of green consumerism were drawn upon by preservice teachers in interviews to help absolve their sense of individual responsibility or guilt. Like Preston's study, many of our preservice teachers linked feelings of guilt and shame with their formal learning experiences, specifically the calculation of their Ecological Footprints as a requirement of the course. However, in taking action, and in contrast to Preston's findings, our preservice teachers 'storied' their action as positive, claiming that small things done by individuals can make a difference. The experience of taking action and course requirement of evaluating the action using data was drawn upon as resources by our preservice teachers in their accounts to story their own sense of individual responsibility in a positive way.

Preston's (2011) study was concerned with understanding individual attitudes, a common approach within a cognitive psychological framework. Unlike Preston's study, we took a discursive psychological approach and were concerned with preservice teachers' self-positioning. Our analysis was not limited to individual attitudes and knowledge, but provided a scope for preservice teachers to 'position' themselves as collectively responsible. We found that despite this, the majority of our preservice teachers still utilised individualistic ontologies. In their accounts, our preservice teachers' pro-environmental engagement was limited to individual action, or "small things" (Theme 1, Excerpts 4 & 5). They also differentiated between instinctive and deliberate action (Theme 2). Rather than looking to action as meaningful in context, or how they were positioned in social contexts salient in their everyday lives (e.g. as unempowered citizens or family members), the preservice teachers themselves became entrapped in a loop of deliberation over the attribution of individual intentionality, induced by the their adoption of cognitive psychological constructs like 'motivation' and 'intention'. Discursive psychology provided an alternative view and we looked to the function of psychological categories in participants' discourse, rather than treating these constructs as inner realities. From this perspective, it is clear that the uncritical adoption of storylines involving individual responsibility for environmental problems can limit action to 'everyday' activism, possibly at the expense of developing other forms of activism such as 'conventional political' activism, which rely more upon students developing a sense of collective agency (Martin & Carter, 2015). Carter et al. (in press) have described 'everyday' activism as acts by individuals usually performed privately or on a small scale, while 'conventional political' activism typically involves organisational or institutional structures such as political parties, religions, organised community groups and other more amorphous 'social movements' that may also coalesce temporarily around a particular goal.

Moreover, drawing on Sauvé (2005), our findings suggest that the Eco Challenge elaborated a "conservationist/resourcist" approach to environmental education or pro-environmental engagement in our terms, limiting our preservice teachers' critical engagement. Our way forward is to explore "socially critical" and/or "values-centred" (Sauvé, 2005) approaches and build into the Eco Challenge the opportunity for our preservice teachers to reposition themselves as collectively responsible in communities salient to them (Martin & Carter, 2015).

The research report here informs our work in promoting pro-environmental engagement with preservice teachers and suggests that opportunities for the preservice teachers to position themselves as knowing members of collectives in relation to current local and global sociopolitical contexts could be a way forward to broaden their concepts of action. Identifying that need as STEPWISE has already done and facilitating its occurrence are of course, different issues. How successful we will be in moving preservice teachers into understanding and acting upon broader social responsibility remains to be seen.

In summary, our study shows that discursive psychological perspectives have much to add and should be fully explored for researching applied dimensions of scientific literacy as promoted by programs such as STEPWISE. Critiques of theory and practice in areas that education for sustainability, STEPWISE and other socio-political and STSE projects should involve an examination of cognitive psychological assumptions that depict individual minds and knowledge as separated from their social realisation, which we have found can limit pro-environmental engagement to individual action. We hope that these results offer some fresh insight into how STEPWISE conceives of it action at the heart of its tetrahedron.

References

Anderson, K., & Zuiker, S. (2010). Performative identity as a resource for classroom participation: Scientific Shane vs. Jimmy Neutron. *Journal of Language, Identity & Education, 9*(5), 291–309.

Arnold, J. (2010, July). *Identity projects and practice in a science classroom: A case study of sudents' reflexive positioning*. Paper presented at Australasian Science Education Research Association annual conference, Port Stephens, Queensland.

Arnold, J. (2012). Science students' classroom discourse: Tasha's umwelt. *Research in Science Education, 42*(2), 233–259.

Arnold, J., & Clarke, D. (2014). What is 'agency'?: Perspectives in science education Research. *International Journal of Science Education, 36*(5), 735–754.

Australian Government Department of the Environment, Water, Heritage and the Arts (DEWHA) (2010). *Sustainability curriculum framework*. Commonwealth of Australia.

Basu, S. (2008). How students design and enact physics lessons: Five immigrant Caribbean youth and the cultivation of student voice. *Journal of Research in Science Teaching, 45*(8), 881–899.

Bencze, L. (2014). *'STEPWISE': Science & technology education promoting wellbeing for individuals, societies & environments. A summary for science teachers*. Toronto, ON: University of Toronto. Available at: http://webspace.oise.utoronto.ca/~benczela/STEPWISE.html.

Bencze, L. (in press). Student-led learning for 'altruistic' socio-political actions. In B. Shapiro (Ed.), *Actions of their own to learn* (pp. xx–xx). Rotterdam, The Netherlands: Sense.

Burris, V. (2001). Small business, status politics, and the social base of new Christian right activism. *Critical Sociology, 27*(1), 29–55.

Carter, L. (2012). *Investigating preservice science teachers' preferences in developing pro-environmental engagement*. Paper presented at the XVth International Organisation for Science and Technology Education symposium Hammamet, Tunisia, 29th October–3rd November, 2012.

Carter, L., Castano, C. & Jones, M. (in press). Sociopolitical activism and transformative learning: Expanding the discourse about what counts in science education. In L. Bryan, & K. Tobin (Eds.), *Science education as a political issue: What's missing in the public conversation about science education?* (pp. xx–xx). New York: Peter Lang.

Chawla, L. (2006). Research methods to investigate significant life experiences: Review and recommendations. *Environmental Education Research, 12*(3), 359–374.

Crowell, A., & Schunn, C. (2014). Scientifically literate action: Key barriers and facilitators across context and content. *Public Understanding of Science, 23*(6), 718–733.

Davies, B., & Harré, R. (1990). Positioning: The discursive production of selves. *Journal for the Theory of Social Behaviour, 20*, 43–63.

Harré, R. (1984). *Personal being: A theory for individual psychology*. Oxford, UK: Basil Blackwell Publishers Limited.

Harré, R., & van Langenhove, L. (1999). The dynamics of social episodes. In R. Harré, & L. V. Langenhove (Eds.), *Positioning theory* (pp. 1–13). Oxford: Blackwell.

Hodson, D. (2003). Time for action: Science education for an alternative future. *International Journal of Science Education, 25*(6), 645–670.

Kollmuss, A., & Agyeman, J. (2002). Mind the gap: Why do people act environmentally and what are the barriers to pro-environmental behavior? *Environmental Education Research, 8*(3), 23–260.

Martin, J., & Carter, L. (2015). Preservice teacher agency concerning education for sustainability (EfS): A discursive psychological approach. *Journal of Research in Science Teaching, 52*(4), 560–573.

Muhlhaüsler, P., & Harré, R. (1990). *Pronouns and people*. Oxford: Blackwell.

Palmer, J., Suggate, J., Robottom, I., & Hart, P. (1999). Significant life experiences and formative influences on the development of adults' environmental awareness in the UK, Australia and Canada. *Environmental Education Research, 5*(2), 181–200.

Preston, L. (2011). Green pedagogy – guidance and doubt in teaching outdoor and environmental education. *Asia-Pacific Journal of Teacher Education, 39*(4), 367–380.

Robelia, B. A., Greenhow, C., & Burton, L. (2011). Environmental learning in online social networks: adopting environmentally responsible behaviors. *Environmental Education Research, 17*(4), 553–575.

Sauvé, L. (2005). Currents in environmental education – Mapping a complex and evolving pedagogical field. *The Canadian Journal of Environmental Education, 10*, 11–37.

Sharma, A. (2007). Making (electrical) connections: Exploring student agency in a school in India. *Science Education, 92*(2), 297–319.

Seah, L., Clarke, D., & Hart, C. (2011). Understanding students' language use about expansion through analyzing their lexicogrammatical resources. *Science Education, 95*(5), 852–876.

Wood, L., & Kroger, R. (2000). *Doing discourse analysis: Methods for studying action in talk and text*. Thousand Oaks, CA: Sage Publications.

Yerrick, R. K., & Gilbert, A. (2011). Constraining the discourse community: How science discourse perpetuates marginalization of underrepresented students. *Journal of Multicultural Discourses, 6*(1), 67–91.

Chapter 25
Science Education for a Better World? Reflections on Embodiment, Language and Sensitive Action

Laura Colucci-Gray

> *We have a responsibility*
> *to think of life in terms of both limits and crossroads, where*
> *new intersections of technology, interpersonal relations,*
> *desire, and imagination can sometimes, against all odds,*
> *propel unexpected futures.*
>
> *(Biehl & Locke, 2010, p. 318)*

25.1 Introduction: Touching the Earth on an Autumn Day

University campus, Aberdeen, city: a day in September in the northern hemisphere. Situated at the crossroads between a busy road and a social housing estate owned by the local council, the campus is a mixture of old and newness; the original buildings dating back to 1495 sit alongside a range of modern constructions, amidst landscaped grassy areas, paved paths and several car parks. A group of first year students, prospective primary teachers, were invited to go for a walk across campus to the botanic gardens and to engage with sensory exploration of their surroundings (Gray, 2012). When they returned inside, they wrote down their immediate sensations from the walk: "*Refreshing and colourful autumn season*"; "*content*", "*calm and content*". The sensorial practice was noticeably both fulfilling and restorative for the self as students experienced 'being in touch' with their surroundings: "*overwhelmed with the connection to the environment*"; "*revitalised*".

It might appear preposterous, if not dangerously relaxed an approach to teaching to send students 'out for a walk' and away from indoor classrooms, the places they

L. Colucci-Gray (✉)
School of Education, University of Aberdeen, Aberdeen, UK

Interdisciplinary Research Institute on Sustainability, Turin, Italy
e-mail: L.t.gray@abdn.ac.uk; http://www.iris.unito.it

L. Bencze (ed.), *Science and Technology Education Promoting Wellbeing for Individuals, Societies and Environments*, Cultural Studies of Science Education 14,
DOI 10.1007/978-3-319-55505-8_25

had come to be taught. This kind of thinking is common and follows a familiar logic in Western thought. Formal education is expected to emancipate from trivial business of everyday life; override family folk knowledge and personal beliefs with powerful and consolidated descriptions of the world. Taking this logic to its full expression, what we first experienced as the world, since our time in the womb and throughout our childhood—is 'inverted', and turned in upon itself: textual descriptions and numerical accounts overrule the primal feeling of connection to a warm body. Yet, the Earth as the 'oikos' is indeed our family home, the warm body which nourishes us and that we 'inhabit', in a mesh of genetic, historical, and ecological threads. How could we resolve this tension? How could the trend be reversed?

25.2 Step Wisely in a World in Transformation

Profound transformations affecting social and ecological systems globally are calling for a revision of the epistemological foundations of knowledge, the sciences included. Complexity and uncertainty necessitate new forms of inquiry, open to a multiplicity of voices, disciplines, and methods (Colucci-Gray, Perazzone, Dodman, & Camino, 2013). In this context, 'Science & Technology Education Promoting Wellbeing for Individuals, Societies & Environments', with its witty acronym 'STEPWISE', sets the aims of science education neatly within the broader sphere of teaching science for citizenship (Bencze, Chap. 2, this volume). This educational approach invites students to act as informed citizens, who can draw connections between local events and global issues and are prepared to take action.

Following the muffled sound of the students' steps into the gardens, I will aim to introduce a further educational dimension. I will start by exploring a 'change of posture' for knowing and learning, one stemming from deeper recognition of intimate relationships between humanity and nature. 'Stepping wise' is conceived of as a way of being, 'in touch with' the world and unfolding … within the boundaries of the Biosphere. Ecological awareness can only be cultivated in one's own body; it is foregrounded here as a basis for ethical action.

25.3 Living in the Anthropocene: The Material and Cultural Impacts of Science and Technology

> Our globalization is powered by around thirteen terawatts (TW) while the flux of energy from the centre of the Earth is around forty TW. Yes, we now measure up with plate tectonics (…). And if all humans were to be powered at the level of North Americans we would operate at a hundred TW, that is, with twice the muscle of plate tectonics. That's quite a feat. "Is it a plane? Is it Nature? No, it's Superman! (Latour, 2011).

In this recent essay, Bruno Latour concisely described the status of humanity on the Planet: the ability to transform and impact on the natural systems has become immense. The last 200 years of human history have seen the growth of a powerful infrastructure connecting scientific and technological knowledge with financial powers, bringing humanity to become one of the greatest evolutionary forces affecting *Life on the Planet* (Steffen, Crutzen, & McNeill, 2007). In this scenario, it becomes paramount to inquire into the socio-cultural infrastructure from which scientific and technological knowledge are being developed, and raise awareness of their impacts, on society and supporting ecosystems.

Whilst the word 'science' is often used to refer to knowledge in singular form, the scientific enterprise comprises of a multiplicity of stakeholders; it is transnational, branching out into a multiplicity of sub-disciplines, each one with its own purposes and methods for the investigation of the natural world. From nuclear physics to biotechnology, scientific research reached deeply into the texture and workings of Nature, with narrowing research questions, specialised modelling tools and techniques. A profoundly diverse array of approaches is adopted to interrogate, interpret and intervene in the world (Kellert, Longino, & Waters, 2006), with each discipline presenting a discrete way of seeing and perceiving a problem and its solutions.

Meanwhile we are also aware of the increasing number of unforeseen consequences derived from scientific and technological interventions (Harris & Sarewitz, 2012). Scientific research and all its applications are no longer confined to laboratories but they take place in environments, involving local economies, societies and natural systems:

> Now that science has come out of the lab into the worlds of people, economies and nature, its effects can no longer be contained or predicted. Hence, we now live in an age of awareness of 'unanticipated consequences', of 'unknown unknowns', when we simply cannot afford to do business or science as usual (Ravetz, 2006, p. 4).

From such recognition, a number of important considerations can follow:

> On the laboratory Earth, the effects of science and technological applications are felt by all, each person, group or community experiencing the effects differently, at different timescales and different understandings.

Meantime, processes of globalization have brought into view the biophysical limits of our Planet (Rockstrom et al., 2009) with the related problems of ecology and equity (Sachs & Santarius, 2007). It is apparent that the current scenario challenges established systems of knowledge. It is a time when science and scientists and educational institutions alike have the responsibility to question if current models of development can effectively respond to people's needs. Such inquiry however is inevitably rooted into different ways of viewing humanity on the Earth: classical environmental literature has depicted human communities as 'passengers' on the spaceship Earth, with increasingly more limited resources; this view calls for urgent and radical solutions. Conversely, consideration of the complex ways in which society and ecological systems are coupled together and interact re-positions humans as 'inhabitants' of a textured Earth, a métissage of material and biological unfolding... Depending on the view that is being adopted, different concerns, actions and educational possibilities can follow.

25.3.1 Combining Power and Knowledge: Overcoming Limits Through Innovation

In the context of global socio-environmental change outlined above important fractures and contradictions exist: despite the extraordinary amount of knowledge at our disposal, future expectations are challenged by complex, and apparently intractable problems. For instance, meeting food demands of a growing global population seems to be irreconcilable with environmental preservation. As reported by Julia Lefevre in her editorial for *Science,* it is difficult to strike a balance between conservation and development needs, and it appears that we will be short of meeting the targets for biodiversity by 2020 (Lefevre, 2014).

Under such conditions, local governments around the world and international organisations are devoting large investments to support 'the innovation challenge': through deliberate political sponsoring, techno-scientific research is deployed to make 'visible impacts' on the social and ecological systems that are the recipients of innovation (Kiers et al., 2008). Proposals such geo-engineering the planet provide are some of the examples of this kind of research. The 2013 report from the Intergovernmental panel on climate change (IPCC) pointed out that, according to some climate models, geo-engineering may even be necessary to reduce temperature rises to pre-industrial levels. However, there is limited evidence of impacts of both Solar Radiation Management (SRM) and Carbon Dioxide Removal (CDR) on the climate system. In order to pursue pressing and immediate goals, the long-term impacts of the new, and ever more powerful technologies are not included in the analytical frame.

So, techno-scientific innovation is expression of a scientific enterprise entrusted with solving problems and taking care of the future: power over nature, control and urgency are its propelling narratives (Benessia et al., 2012). By appealing to 'powerful innovations', some societies hold up the conviction that 'innovation challenges' will help humanity *to step out* of the current ecological and socio-economical impasse, securing progress and better opportunities for human life. This belief is confirmed by recognition that humans have demonstrated great creativity in ways in which they have used technological, social and cultural tools to overcome resource limitations (Matthew & Boltz, 2012). Science is one of the greatest cultural innovations of human history and, as such, it can provide attitudes of mind and infrastructure to overcome what may be insurmountable constraints.

The expectations placed upon science and technology, however, are recalibrated by the realisation that the gap between ecology and equity, which should be filled, supposedly, by innovative interventions, is not only widening, but it may actually be accelerating (Rockstrom et al., 2009). In a detailed analysis of impacts of genetically modified crops to alleviate world hunger, David Quist, Heinemann, Myhr, Aslaken and Funtowicz (2013) highlighted the long, winding path that leads from a stated solution to spread of unexpected outcomes. Political decisions over internal and external subsidies, issues of environmental quality and consumers' choices all affected ways in which a new technology interplays with any specific context and its rules, customs and structures. In addition, a techno-scientific enterprise of large proportions is required to deliver results to its commissioners. Agribusiness industries hold the economic power to sponsor research which may inform decision-making on quality and safety of agricultural products and practices, as well as to lobby policy-making to favour corporate goals (Pollan, 2011). As a result, scientific research that serves corporate interest, rather than public good, has become the norm. Such complexity poses serious challenges to scientifically-driven development policies that are applied at a global scale. Not only is science enmeshed with political and economic interests but there is a growing recognition that the effects of a pervasive intervention, from GM crops to the construction of a large dam or airport will extend into the very fabric of society and workings of democratic institutions. Key questions arise about modes of investigations and framing of problems, as well as the values and purpose of science and technology. There is a necessity to clarify if powerful techno-science is in fact deployed to the services of oppressive regimes.

25.3.2 Deploying Powerful Knowledge: Managing Complexity

Confronted with the dilemmas and limitations of techno-science operating in a complex world, the field of 'sustainability science'—as a relatively new branch of epistemological reflection—proposes a revision of conventional, disciplinary–based scientific inquiry to respond more aptly to global system change. Biermann et al. (2012), for example, argued for more efficient ways to emulate nature in uses of resources and management of waste in order to implement more responsible stewardship. Solutions based on modelling and collection of large amounts of quantitative data on complex, socio-environmental systems are sought to foster major changes in production processes and individual life-styles. It is within the field of sustainability science that we find a growing field of studies related to sustainability indicators (Bell & Morse, 2008) aimed at providing policy-makers and citizens with tools for taking decisions and making informed choices in the face of socio-environmental challenges.

The field of sustainability science recognises the critical interplay between human culture and the current state of the Planet (Folke & Gunderson, 2012). It promotes research devoted to new, social and economic practices for sustainable living. Such goals are pursued through affirmation of predictive aspects of science: it is an approach based on collecting 'solid' data and on measurement, designed to preclude subjective readings of a contextual reality in order to produce objective and reliable knowledge. However, there are challenges coming from the variety of 'sciences' to which to resort, with their different methods, languages and research questions (Sarewitz, 2004). Different types of science offer different solutions! A comprehensive view of the world as it might be derived from the abstract schemas built in our minds and in our disciplines may not be achieved by putting together of isolated, objective pieces.

25.3.3 Revisiting Assumptions About Power and Knowledge

The problematic role of science and technology in the face of contemporary global issues is to do with the key question of whether such knowledge is 'apt' or fit to address global environmental change and whether it could guide people to take appropriate actions. Indeed, the gap that exists between the ability to penetrate the secret workings of nature and the impossibility to foresee system changes is such that some philosophers of science have elucidated a fundamental contradiction or paradox in our knowledge systems. Such contradiction is inherent to the cultural frame of modernity, which assigned to science the power to predict and to control the natural systems, waiting for a single, epistemically exhaustive, pristine, expert definition to rely upon for our future (Benessia et al., 2012). Science's predictive abilities and deployment of remedial technology were privileged as the primary tool for shaping policies and actions, while eroding the fundamental commitment between humanity and nature: "*this means giving up our agency as members of civil society and most importantly, it implies a continuous procrastination, because the future is in fact, irreducibly indeterminate and intrinsically complex*" (Benessia et al., 2012, p.77). Besides, the paradox of modernity is made more acute by the realisation that future developments will inevitably escape determination precisely because of our greater power to act and transform.

So, the conventional idea of science is entering a 'crisis' and from such realisation we can derive interesting ways to redefine it and contextualise it. In order to begin to see features of a new science let's go back for a moment to the image presented at the beginning of this chapter; it was the picture of a human being who is moving through a world of abstract representations: he/she is taking steps but where is he/she heading to? What is the direction of their errands? To err, as in the translation of the English word, contains the double meaning of 'moving from point to point' but also of getting into error/ make a mistake. So we might be running an errand to get to a set place or target, such is the goal of powerful innovation, but we may also be 'wandering', in an errand from alley to alley, exposed to multiplicity, creativity and the possibility of going wrong… All such options may coexist, in rhetorical plurality. Acknowledging such complexity however requires a change of speed, a change of posture and a change of attitude: we are turning our attention from describing and calculating to sensing and being sensitive to the nuances of contexts around us.

25.4 Embracing Complexity and Uncertainty: A New Science for Earthlings?

Starting from very different realms of inquiry—from philosophy to law, anthropology, ecology, linguistics and so on—many scholars are contributing to building a new idea of science. Insights from complexity theory contributed to a shift from

Newtonian mechanics towards quantum physics, highlighting unpredictability, emergence and self-organisation in biological systems. The Universe was now conceived as an open system, whereby notions of change, fluidity and heterogeneity overtook ideas of solidity, constancy and stability brought on by earlier ontologies (Holland, 2013). In this context, the notion of a 'tentative science', which progresses by 'rules of thumb', improvisation and with whatever resources are available at hand appeared more 'fitting' than the science of pre-designed plans. Deleuze and Guattari (1976) referred to this type of science as 'nomadic' to distinguish it from the classical, State science. If the latter is concerned with laws and description, nomadic science is concerned with experience, process and engagement with materials, places and context. There are in effect.

> two formally different conceptions of science" the authors explain, "competing... royal science continually appropriates the contents of nomad science while nomad science continually cuts the content of royal science loose (Deleuze & Guattari, 1976, p. 367).

The philosophical contribution offered by Deleuze and Guattari resonates with the analysis offered by Alice Benessia et al. (2012) of ways in which scientific knowledge has shaped our understanding of sustainability in contradictory, problematic ways. The authors highlighted the extent to which a linear approach to problems, permeated as it were by the modern framework of royal science, set humanity on a path of over-reliance on statistics for risk assessment and cost-benefit analysis, continuously chasing a schema for reproducing the world as a fixed and predictable entity. With the intention of casting light on such contradictions, the authors uncover "*the inherently modern divides such as the ones between facts and values, reason and passion and between knowledge and experience*" (p. 75). The proposition is to implement a variety of approaches to knowing that could help overcome the modern predicament by bringing forth new kinds of "*hybrid* knowledge and practice... through extended participatory processes" (p. 76).

Arguably, such a proposition is profoundly associated with experiences of everyday living in this age of globalisation. Following Bruno Latour (2007), all 'matters of fact' or *objects* have become 'matters of concern' or *things*, that is, "*issues, gatherings, assemblies of some sort*" (p. 6). In this view, he continues, what matters is to understand and recognise the nature of the relationships that brings such assemblages together, legally, scientifically, religiously or artistically: "*politics is not a domain, is a type of relation*" (p. 8). Indeed, as Latour continues, the epoch in which we are living is the epoch of total interdependence and transformation of humans and nature: "*every product, every biological species, every packaging... along with every river, every glaciers and every bug...*" (p. 9). In this view, the Earthlings as inhabitants of the Earth have the power to bring forth the politics of the Earth by virtue of relationships and interdependencies: in a world of assemblages, no hard distinction can be made between issues pertaining to different stakeholders. What counts is the multifaceted human experience which brings people together around a shared concern. Issues arise because of our inherent complicity and by the same token, they can be addressed and transformed through sharing of experiences and reciprocal empathy (Colucci-Gray et al., 2013).

Such considerations challenge the idea of science *speaking truth* to policy. However, by rejecting the conventional idea of science as fixed and reliable knowledge, we may find ourselves left empty-handed, caught in between the impossibility to achieve a comprehensive view of the world and the prospect of relativism, where anything goes. The task is, thus, to proceed with clarifying further what other dimensions of knowing, beyond the cognitive and the factual, may be involved, when trying to step wisely, into an ever changing world.

25.4.1 The Primacy of Movement

In light of contributions provided by philosophy and sociology of science, we can see quite clearly the troubles of humanity in the Anthropocene. Canonical knowledge often does not match with experience on the ground, and this recognition would apply to the realm of sustainability issues as much as it does to the realm of school curriculum more generally and science education in particular.[1]

Overcoming what was outlined earlier as the dualism of *either* holding consolidated knowledge *or* falling into chaos requires overcoming what a range of authors and thinkers—from philosophy, to anthropology, linguistics and more recently neurosciences—have referred to as the split between the mind and the body, knowledge and experience, objective and subjective. Returning to the argument set out at the start, the case for a re-inversion is being introduced: the man with the head inside a schema is stretching and turning outwards…head and body together, into the world….

It may appear ironical to make a case for retuning the head to the body as if they were separate parts. Yet this condition of 'separation' (or 'inversion') is not strange to mankind. In an essay focussing on 'walking' as a dimension of human development, traceable across cultures at different times and in different environments, the anthropologist Tim Ingold points to the anatomical changes inscribed first in the bodies of humans as compared to other apes, and secondly in the bodies of different human populations, as a result of differences in the ways each species and then cultures *inhabited* their own environment (Ingold, 2011). Reflecting on Charles Darwin's description of the human figure, Ingold emphasizes Darwin's remarks on the relative and hierarchical position of the hands and the feet: man could not have attained his dominant position in the world without the use of his hands now freed from the task of locomotion. And as the story goes, "*Marching head over heels – half in nature, half out – the human biped figures as a constitutionally divided creature*" (Ingold, 2011, p. 35).

[1] The problem of *relevance* of science content for everyday life has been addressed by many scholars in science education grappling with conflicting aims for science education and curriculum design (see Aikenhead & Mitchell, 2011; Hodson, 2011; Priestley & Biesta, 2013; Reiss & White, 2014).

At the time of Darwin, already the differences between the anatomies of the Western man, a city dweller, walking with booted feet on paved surfaces were known to anthropologists and biologists alike. Indigenous populations, for example; had not lost their abilities to use their feet more skilfully than simply as a standing support. No real hierarchy or superiority was established between feet and hands, hands and brain. The point Ingold wishes to make is to help us recognise that connections exist between the world of experience and the realm of cognition: it is the direct, physical interaction with the world that gives rise to our mental schemata. But there is more to this.

We often resort to the idea of walking to express other experiences that are abstract, difficult to put into words: walking can be used as a metaphor for our life often compared to a journey, and how we grow and learn as a result of living and being in the world. Hence far than simple locomotion, walking can be equated to the extended and life-long process of experience and learning. In this sense, in the world of booted feet inhabited by city dwellers, living amongst landscaped gardens and square buildings, walking equates to striding across the streets to reach a destination. "*Between them, the boot and the chair establish a technological foundation for the separation of thoughts from action and mindfrom the body – that is the fundamental groundlessness so characteristic of modern metropolitan dwelling*" (Ingold, 2011, p. 39). It is perhaps not so difficult to see similarities with the world of formal education, combining 'learning on the chair' with utilitarian values, such achieving goals and destinations… even if they may prove disconnected from real life necessities. It is also now possible to build a perspective on the experience of the students mentioned in the opening of this chapter, who were exploring new ways of learning and being while walking in the familiar, urban environment.

Delving a little deeper into 'walking' as a form of knowing we can see how the world of bare feet, contrasted by the world of the booted foot, is the world of light stepping, of ongoing direct experience: of finding a way, sensing and adjusting to the terrain, a process of learning akin to Deleuze and Guattari's notion of nomadic science. A body that walks, slowly, is also a body that encounters the world, and in so doing, learns, interacts and produces a view of what the world is like. Drawing on insights offered by phenomenology, Ingold details aspects of the importance of direct, physical experience—mediated by a range of tools—in shaping our perception and, thus, what becomes part of our cognitive repertoire. So in order to 'step wisely,' we must understand the technology of how we walk; thus, how we learn, and how we approach the world; dextrously and lightly was the option followed by our ancestors and which propelled our evolution on the Earth surface. I now turn to provide more detail about the significance of such an approach for cognition and learning.

25.4.2 Experiencing the World Through Language

Drawing upon the philosophical branch of phenomenology, Bruno Latour extends the notion of experiential knowing (and learning) by returning to the writings of William James (1996): the 'physical, or bodily experience is conceived of as a proper and different form of empiricism which is understood much more broadly than what traditional, reductionist science had confined to the definition of objects, without relations. As James summarised:

> Propositions, copulas and conjunctions, 'is' 'isn't', 'then', 'before', 'in', 'on', 'beside', 'between', 'next', 'like', 'unlike', 'as', 'but', flower out of the stream of pure experience, the stream of concretes or the sensational stream, as naturally as nouns and adjectives do, and they melt into it as fluidly when apply them to a new portion of the stream' (James, cited by Latour, 2007, p. 95).

What brings such disparate fields together is the notion of *knowing in interaction* with nature, and in which nature is not a fixed entity but a system of relationships, exchanges and ongoing, mutual transformations. Understanding a fully, contextual experience of one's being *in relation* with the many facets of the cosmos is the fundamental step of knowing as 'inhabitants' of the Earth or 'Earthling': conjunction, prepositions provide the material out of which our experience is woven. An interesting feature of William James' empiricist account of world perception is the aesthetic quality associated with knowing and learning. Walking or journeying both physically and metaphorically is a *material experience*—that is—a moment of encounter with what is important, or salient to us. Current understandings of human evolution, for example, describe human behaviours as responses that do not come simply from external stimuli, but are rooted into early physical experiences, which have a role in the development of memory and language. So the words we produce are tools through which we express what the body perceives and such words retain in themselves the germs of an earlier experience in a place of relations, both human and non-human. The linguistic dimension is often at the core of debates amongst scholars who draw upon the philosophical insights of agential realism (Barad, 2007) and post-structuralist, feminist epistemology (see Ahmed, 2008). I am not entering into details about the nuances of the debate. It is important to specify, however, that language is not treated here as a tool for representing or mirroring cognition, but as a tool that is integral to the process of perception and associated meaning-making. Thus, language constructs our understanding of matter and it is through language that matter becomes salient to the self.

Besides, by means of the multiplicity of such experiences and the fluidity of neural connections, words are only 'fixed' to the extent to which we wish them to define them within particular parameters; more realistically, we find that words and gestures to express particular concepts are often unstable due to the many different types of experience that may be at the roots of their creation. We make full use of such faculty in the production of metaphors but also in our usage of analogies, verbal organisation of ideas and non-verbal gestures and movements. We might argue that by means of this ability of transferring and mapping a domain of experience onto another domain we are able to give meaning to new situations: thus we are able to respond to novelty and produce new knowledge, drawing on metaphors retained in language, including the language of science (Aikenhead & Mitchell, 2011; Brown, 2003).

Most crucially, this discussion suggests that scientific knowledge is not simply 'given' but it is continuously produced in the act of the subject being 'introduced' to the world; knowledge is not 'something' to be acquired; knowledge equates to 'knowing' itself, unfolding at the point of contact with the world; for such reason knowing is always contingent and inherently temporary. Equally, the act of knowing can be qualified through the relationships that are established between the knower and the world. Far from being simple spectators, we can become participants in the world with a role in its making and with a degree of responsibility for how and what we know.

25.4.3 From Describing the World to 'Leaving a Trail'...

The extensive body of research that is currently available on pre-eminence of body in cognition and language points to the primacy of movement versus vision and perception. Such a distinction—or 'inversion'—is critical to this argument. The world of experience is a world in perpetual flux and actions of an organism are more akin to movement of improvisation rather than controlled, logical, algorithmic steps of a machine. Indeed in order to understand how the organism orients itself into the world we need once again to look more attentively to the body. Important connections are found between physical perception (rather than simple vision), language and emotion (Damasio, 2012). For example, emotions such as 'interest' and 'attention' are powerful means for directing the self towards particular encounters. Equally, emotions play a role in the retrieval of past experiences from the memory. So the emotional quality of an event which comes to be perceived is paramount to the person's knowledge of the world.

Recognising the role of emotion in perception contains some important lessons for education and science education in particular. In the first instance, the idea that our learning can be 'biased' by an emotional state in which we may find ourselves is well founded. The common response to this state of affairs is 'normally' to remove emotions in order to try and process information objectively. From a knowledge point of view however, emotional states can be generated and accounted for to enable the individual self to explore, to become 'interested' in the world and thus develop a connection that will lead to knowing. Deleuze and Guattari recover the word 'affect' to express more powerfully that emotional states are in fact affective states, that is, they are connected to the fundamental ability of the body to act.

At a basic level, experiences of teaching practice may confirm how the introduction of fun elements can aid to a better climate for learning, positive dispositions and even recall of information. At another level, generating an emotional response may challenge the conventional opposition between subjectivity and objectivity, suggesting that rather than obstructing rational thought, emotional states enable reasoning to occur in the first place through affective, inter-personal connection. Indeed, it appears that an organism can recognise an optimal range of operation: "*optimal ranges express themselves in the conscious mind as pleasurable feelings; dangerous ranges as not so pleasant or even dangerous feelings*" (Damasio, 2012, p. 55). In addition, what is at stake in triggering emotional states is not simply recall of an immediate feeling but possibilities for raising consciousness about ourselves and the *quality of relationships* we establish with other people, places or events at a particular time. If metaphors retain an element of an earlier experience, emotional expressions can provide a means to understand the essence of that experience.

Ostensibly one of the greatest challenges when looking at the world of science education through lenses of movement is recognition that an organism that is 'in the world' will be exposed to consolidated knowledge encapsulated in nouns that may no longer have any connection with the original experience. Science educators have recognised the extent to which metaphors produced by teachers or scientists often contradict experiences of the child, cause confusion, and do not make sense to the learner. Hence, there is an important role for science educators and researchers alike to promote direct encounters with the natural world to which we can respond in inter-action as well as in *intra-action* (Barad, 2007), by letting the experience resonate emotionally inside us for no separation exists between ourselves and the world. The world is unfolding not only before us but 'within us' and 'with us inside it', and it is a world of verbs and agents, colours, smells and emotional responses. What is involved in the production of a conscious mind is a series of *images: "the self comes to mind in the form of images, relentlessly telling a story of such engagements"* (Damasio, 2012, p. 203).

At this point, we can fully appreciate effects of making a case for 'inversion'. The human self has changed from its closed-in figure, separated against the background, to a layered or threaded process, with each thread telling something of places and encounters, which are genetic, sexual, linguistic, social, and which are experienced and re-told by each and every gesture. Identities are fluid, co-constructed in a field of materiality on which we depend and to which we become 'sensitive'. From such a shift, other significant moves occur in the way we look at science literacy and learning. Very differently from the idea of accumulation of factual information, learning science can be more akin to a form of literacy and story-telling, an explication and re-telling of our activities in the world. Learning leaves a physical trail on ourselves, as well as on the path upon which we travel… what is the story that we tell? What relationships have we established? Could this be the beginning of ethical development?

25.5 Living on the Earth: Interrupting Given Schemas and Re-connecting with Experience

The insights derived from the range of scholarly contributions illustrated earlier point to a view of knowing and learning that is profoundly subjective on the one hand and cultural on the other. Drawing on contributions derived from embodied cognition, we can observe acts of knowing as moments of encounters between the subject and the world 'out there'; yet we are conscious that such encounters have no clearly-defined boundary—as our subjectivities and perceptions are continuously affected by flows of neural and material connections which make for our continuous 'coming and becoming' into the world (Biehl & Locke, 2010). In this view, our movements and actions into the world are just one amongst many paths and possibilities for interaction and for exchanges. Underscoring Deleuze's notion of "becoming", Biehl and Locke repeatedly refer to 'action' and 'learning' as movement across emotional, social, and experiential registers.

Arguably, however, there are trajectories of material experience that bring us progressively further away from the fabric of nature's time and nature's flows, thus reinforcing dualist schemas and view of nature as fixed and 'out there'. A number of authors have pointed to the phenomenon of 'psychic numbness' to refer to inabilities of people living in the Westernised, urban world to perceive with the senses. Robert Michael Pyle (1993) described the state of personal alienation from nature due to lack of exposure and physical contact. Yet, there is also a perceptual difficulty due to the inability to focus on detailed observations at a slow pace as we are continuously over-stimulated by fast imaging and engage with technologically-mediated experiences (for a more detailed explanation of the process of overstimulation and attention fatigue see Kaplan & Kaplan, 1989). As Bill McKibben (2005) points out, we are faced with a curious paradox. In the course of a couple of generations, our species has managed to powerfully raise the temperature of an entire planet but, as odd as it may seem, it has not bared within us: "*We can register what is happening with satellites and scientific instruments, but can we register it in our imaginations, the most sensitive of all our devices?*"

Now, let's go back to the image of the students in the gardens. Within the slowness of life unfolding in the gardens, some students '*saw everything within the first ten minutes'*, detached, and moving as 'spectators'/onlookers in the world. It is stark the contrast with the ones who trailed across the cold air in the autumn sun and were aware of themselves and their perception "*the weather affects what I see*", and through the physical encounter felt the "*excitement of experiencing the unknown*".

In light of the discussion conducted so far, it is crucial to recognise that learning is not simply describing or acquiring nominal knowledge about 'something'. Rather, learning is more akin to a process of 'becoming' as we trail through the mesh of everyday interactions with natural, social and materials flows. Under such conditions, scholars working in the field of environmental education have stressed the importance of re-connecting with nature to restore ability to be sensitised to nature, develop an intimacy with it and learn not only about nature but to register the presence of nature as a constitutional element of ourselves.

The notion of a narrative self, as outlined so far, that is always on the cusp of becoming, makes it possible to reinterpret ideas of civic 'participation': it is not only the act of political consensus or dissent, or the practice of democratic ideals, but it is the moment of tuning in with other beings by means of the emotional, physical and material stories that connect ourselves with other people. To participate in social life equates to being conscious of how 'our stories' are made possible or are prevented by the stories and the lives of others…

This idea resonates with the insightful notion of 'pedagogy of dislocation' (Edwards & Usher, 1998) whereby "location" is used as a metaphor to foreground the notion of (re)positioning and being (re)positioned. Edwards and Usher point to "moving" quality of any location and consider the part globalisation has played in recasting the significance of location in the contemporary condition. In this sense, as human creatures we are embedded in a planetary web of material and energy flows that connect our everyday living with global material cycles and the lives of people both near and distant from us. I, thus, suggest that there is a greater role for education to interrogate the nexus of material assemblages that satisfy our demands on the one hand and cause depletion and disparities on the other.

Taking wise steps in the face of uncertainty is more than just 'protecting' nature or an act of 'conservation' made by some altruistic or enlightened minds. Education and science education in particular must create opportunities for exploring the capabilities of the narrative self across a range of material relationships afforded by the digital, social and the natural world. This might well be another form of resistance to the neoliberal forces and the enforced trajectories of material production and consumption; find spaces whereby such trajectories are being interrupted and disclose new imaginations for living, within the boundaries of the Earth.

25.6 Final Remarks

The reformulation of thinking in science toward a more holistic view of the world and an awareness of intimate relationships between humanity and nature is allowing for a re-composition of knowledge to include cognitive, emotional, linguistic and ethical aspects and, arguably, also spiritual ones. Experiences of direct engagement with a world more than human are at the basis of the cultural heritage of human civilisations across the world: ancient stories and myths recount the good and the evil of mankind as a shared, universal experience. In this sense, informed or 'intelligent' action is not simply the technical outcome of being 'informed' by facts or evidence. It is an interpretive action that is intuitive, cultural and connected to earlier emotional encounters with Nature; such experiences can support and nurture the ability for self-identification with other life forms capable of emotions.

Science inquiry, in particular, offers opportunities to tap into a multitude of interpretive schemes, each one shaped by the physical, material and temporal experience of the knower. In this way, the type of education we are advocating is not dissimilar from what Bencze (Chap. 2, this volume) refers to 'students/citizens that are capable of engaging with 'praxis' (e.g. primary research), dissent and conflict. However it is also radically different from an approach to research which excludes the self and its internal world. Critical, ecological awareness, I argue, is deeply embedded in the multiplicity of embodied experiences that account for the possibility of human existence, shape humans' ability to self-restore but, most significantly, explore ways for living *from within* the boundaries of the Biosphere (Colucci-Gray et al., 2013).

A sustainable approach to life on the Planet, therefore, requires a fundamental revision of established views of science knowledge as a body of truths to recover the profound connections between the body and the Earth, the life-giving soil and life-swelling ground.

In this view, it is possible to appreciate the contribution given almost a century ago by thinkers such as Gandhi and his followers, Vinoba and Kumarappa, who marched to re-claim the land for the people, seeking to integrate science, economics and ethics (Colucci & Camino, 2016; Ninan, 2009). More than a century ago, Gandhi summarises his view of development as "*living simply, to allow other people to simply live*". Science and awareness of oneself should always progress hand in hand.

References

Ahmed, S. (2008). Open forum imaginary prohibitions: some preliminary remarks on the founding gestures of the 'New Materialism'. *European Journal of Women's studies, 15*(1), 23–39.

Aikenhead, G., & Mitchell, H. (2011). *Bridging cultures: Indigenous and scientific ways of knowing nature*. London: Pearson.

Barad, K. (2007). *Meeting the universe halfway: Quantum physics and the entanglement of matter and meaning*. Durham, NC: Duke University Press.

Bell, S., & Morse, S. (2008). *Sustainability indicators: Measuring the immeasurable* (2nd ed.). London: Earthscan.

Benessia, A., Funtowicz, S., Bradshaw, G., Ferri, F., Raez-Luna, E., & Medina, C. P. (2012). Hybridizing sustainability: Towards a new praxis for the present human predicament. *Sustainability Science, 7*, 75–89.

Biehl, J., & Locke, P. (2010). Deleuze and the anthropology of becoming. *Current Anthropology, 51*(3), 317–351.

Biermann, F., Abbott, K., Andresen, S., Backstrand, K., Bernstein, S., Betsill, M. M., et al. (2012). Navigating the Anthropocene. Improving earth system governance. *Science, 335*, 1306–1307.

Brown, T. L. (2003). *Making truth:Metaphor in science*. Chicago: University of Illinois Press.

Colucci, L. & Camino, E. (2016). Looking back and moving sideways: Following the Gandhian approach as the underlying thread for a sustainable science and education. *Visions for Sustainability, 6*(2), 23–44.

Colucci-Gray, L., Perazzone, A., Dodman, M., & Camino, E. (2013). Science education for sustainability, epistemological reflections and educational practices: From natural sciences to trans-disciplinarity. *Cultural Studies of Science Education, 8*(1), 127–183.

Damasio, A. (2012). *The self comes to mind: Constructing the conscious brain*. London: Vintage.

Deleuze, G., & Guattari, F. (1976). *Rhizome: Introduction*. Paris: Editions de Minuit.

Edwards, R., & Usher, R. (1998). "Moving" experiences: globalisation, pedagogy and experiential learning. *Studies in Continuing Education, 20*(2), 159–174.
Folke, C., & Gunderson, L. (2012). Reconnecting to the biosphere: A socio-ecological renaissance. *Ecology and Society, 17*(4), 55.
Gray, D. S. (2012). Walking in the mindfield. *International Journal of Holistic Education, 1*(1), 1–10.
Harris, P., & Sarewitz, D. (2012). Destructive creation and the new world disorder. *Current History, 111*(741), 29–33.
Hodson, D. (2011). *Looking to the future: Building a curriculum for social activism*. Rotterdam, The Netherlands: Sense Publisher.
Holland, E. W. (2013). *Deleuze and Guattari's a thousand plateaus*. London: Bloomsbury.
Ingold, T. (2011). *Being alive: Essays on movement, knowledge and description*. London: Routledge.
James, W. (1996). *Essays in radical empiricism*. London: University of Nebraska Press . [1907].
Kaplan, R., & Kaplan, S. (1989). *The experience of nature: A psychological perspective*. New York: Cambridge University Press.
Kellert, S., Longino, H., & Waters, K. (2006). *Scientific pluralism*. Minneapolis, MN: University of Minnesota Press.
Kiers, E. T., Leakey, R. R. B., Izac, A.-M., Heinemann, J. A., Rosenhal, E., Nathan, D., et al. (2008). Agriculture at a crossroads. *Science, 320*(5874), 320–321.
Latour, B. (2007). *A plea for earthly sciences*. Keynote lecture for the annual meeting of the British Sociological Association, East London, April 2007. To be published in J. Burnett, S. Jeffers, & G. Thomas (Eds.) *New Social Connections: sociology's subjects and objects,* London: Palgrave MacMillian.
Latour, B. (2011). *Waiting for Gaia. Composing the common world through art and politics*. A lecture at the French Institute for the Launching of SPEAP in London, November 2011. To be published in Albena Yaneva (Ed) What is Cosmopolitical Design? Farnham: Ashgate, 2015.
Lefevre, J. M. (2014). Planet at the crossroads. Editorial. *Science, 346*(6209), 525.
Matthew, J. H., & Boltz, F. (2012). The shifting boundaries of sustainability science: Are we doomed yet? *PLoS Biology, 10*(6), e1001344. doi:10.1371/journal.pbio.1001344.
McKibben, B. (2005, April). Imagine that: What the warming world needs now is art, sweet art. *Grist Magazine*. Online: www.grist.org/comments/soapbox/2005/04/21/mckibben-imagine
Ninan, A. S. (2009). Gandhi's technoscience: Sustainability and technology as themes of politics. *Sustainable Development, 17*(3), 183–196.
Pollan, M. (2011). *The omnivore's dilemma: The search for a perfect meal in a fast-food world*. London: Bloomsbury.
Priestley, M., & Biesta, G. (2013). *Reinventing the curriculum: New trends in curriculum policy and practice*. London: Bloomsbury.
Pyle, R. M. (1993). *The thunder tree: Lessons from and urban wildland*. Boston: Houghton Mifflin.
Quist, D. A., Heinemann, J., Myhr, A., Aslaken, I., & Funtowicz, S. (2013). Hungry for innovation: pathways from GM crops to agroecology. In E. Miljøagentur (Ed.), *Late lessons from early warnings: Science, precaution, innovation* (pp. 458–485). Bruxelles, Belgium: European Environmental Agency.
Ravetz, J. (2006). Towards a non-violent discourse in science. In B. Klein Goldewijk & G. Freks (Eds.), *New challenges to human security: Empowering alternative discourses*. Wageningen, The Netherlands: Wageningen Academic Publishers.
Reiss, M., & White, J. (2014). An aims-based curriculum illustrated by the teaching of science in schools. *The Curriculum Journal, 25*, 76–89.
Rockström, J., Steffen, W., Noone, K., Persson, A., Stuart Chapin, F., Lambin, E. F., et al. (2009). A safe operating space for humanity. *Nature, 461*, 472–475.
Sachs, W., & Santarius, T. (2007). *Fair future: Resource conflicts, security and global justice*. London: Zed Books.
Sarewitz, D. (2004). How science makes environmental controversies worse. *Environmental Science & Policy, 7*(5), 385–403.
Steffen, W., Crutzen, P., & McNeill, J. R. (2007). The anthropocene: Are humans now overwhelming the great forces of nature? *Ambio, 36*(8), 614–621.

Chapter 26
WISE Preservice Teachers Discussing Social and Economic Disparities During a Discussion Game Dealing with Nanotechnologies

Audrey Groleau and Chantal Pouliot

Economics isa social science; complex economic phenomena can seldom be understood if presented in a vacuum, removed from their sociological, political, and historical contexts. To properly discuss economic policy, students should understand the broader social impacts and moral implications of economic decisions

(International Student Initiative for Pluralism in Economics, 2014).

From economics…

In September 2013, a book entitled *Le capital au XXIe siècle* (Piketty, 2013) was published in France. An English translation under the title *Capital in the Twenty-First Century* came out approximately six months later. By June 2014, some 150,000 French copies and 400,000 English copies had been sold (Jaxel-Truer, 2014). This book presents a historical, empirical and comparative study of the rise in economic inequality. In addition to documenting the unequal distribution of wealth, Piketty aimed to identify modes of social organization, institutions and public policies that could lead to fairer democratic societies. Of particular interest for the purposes of this chapter, Piketty presents the dual notion that: 1) underlying the growing economic inequality is a history of the distribution of wealth which "has always been deeply political, and… cannot be reduced to purely economic mechanisms." (p. 20); and, 2)

Audrey Groleau and Chantal Pouliot have contributed equally to this chapter.

A. Groleau (✉)
Département des sciences de l'éducation, Université du Québec à Trois-Rivières, Trois-Rivières, QC, Canada
e-mail: audrey.groleau@uqtr.ca

C. Pouliot
Faculté des sciences de l'éducation, Université Laval, Québec City, QC, Canada
e-mail: chantal.pouliot@fse.ulaval.ca

L. Bencze (ed.), *Science and Technology Education Promoting Wellbeing for Individuals, Societies and Environments*, Cultural Studies of Science Education 14,
DOI 10.1007/978-3-319-55505-8_26

the dynamics of inequality are "shaped by the way economic, social, and political actors view what is just and what is not, as well as by the relative power of those actors and the collective choices that result" (p. 20).

On May 5, 2014, one month after the publication of the English translation of Thomas Piketty's book, twenty-two[1] associations of economics students signed an open letter calling for a renewal in ways in which economics is taught at university. The signatories, supported by over eighty economics researchers, professors and practitioners (including Thomas Piketty), denounced narrowing of curricula and called for greater pluralism in theoretical perspectives taught. The latter, they stated, should include neoclassically-based approaches as well as the post-Keynesian, institutional, ecological, feminist, Marxist and Austrian traditions, among others, and should address "the multidimensional challenges of the 21st century – from financial stability to food security and climate change" (International Student Initiative for Pluralism in Economics, 2014).

… To science education

Science education must not be blind to economic dimensions to which it contributes and by which it is influenced. This is the position expressed by the authors of the collective work entitled *Activist Science and Technology Education* (Bencze & Alsop, 2014), which points to needs for deep-rooted change, "tak[ing] more seriously wider social, political, economic and environmental contexts in which our practices reside and also seek to resist and influence" (Alsop & Bencze, 2014, p. 2).[2]

In this chapter, it will be seen that, in discussing development and commercialization of nanotechnologies in the context of a game aimed at encouraging sociopolitical discussion in the area of techno-science, the student participants brought up some of the economic aspects of the subject under discussion. More specifically, the participants referred to limited access to benefits of nanotechnologies, unequal distribution of their costs and benefits, risks of offshoring factories and exploiting child labour and development of medical treatments for profit. We begin the chapter with a presentation of the group discussion game, Decide, illustrating that it shares several democratic values with STEPWISE. We then briefly describe the sociopolitical context of the study. Next, we illustrate, through excerpts of the participants' conversations, how these students expressed their views regarding some economic aspects of the controversy surrounding development and commercialization of nanotechnologies. We conclude by discussing contributions that Decide may bring to STEPWISE.

---

[1] As of January, 2015, this number had risen to over 65 associations.

[2] Similarly, the authors of *Risky Business*, published in June, 2014, provide arguments for those who maintain that if science education aims to promote a nuanced and deeper understanding of socio-scientific problems, it cannot disregard the economic and political issues involved (Risky Business Project, 2014). In this report, influential authors Michael R. Bloomberg, Henry M. Paulson and Thomas F. Steyer strongly urge business leaders and investors to play an active role in public discussion (p. 47) and support an aggressive push to bring down carbon emissions.

26.1 Decide: A Game that Prompts Discussion on Inequality and Legitimizes These Discussions

Decide is a group discussion game that broadly shares STEPWISE orientations. First, Decide is distributed under a Creative Commons License (Attribution—Share Alike 3.0 Unported), which means that it is highly accessible and shared free of charge in an altruistic spirit. The user is granted several rights: the game can be copied and adapted as needed, provided that it is attributed to the author or licensor. It can also be distributed (in its original or adapted version), but only under the same license terms. Second, the game's instructions give the players a great deal of leeway, both in terms of the form the discussion will take and its content. For example, the players can focus on any particular aspect of the controversy that they consider to be essential, or exclude any aspect that they deem to be less important. Moreover, the game cards invite the players to consider, in the course of their reflection and discussions, the well-being of individuals, societies and environments. Lastly, not only does Decide invite players to discuss various issues that often are overlooked in science and technology education, but it also legitimizes these discussions. For example, there are game cards that explicitly ask questions relating to the uncertainties involved in the controversy under discussion, respect for human rights, pertinence of public engagement in these debates and in socio-political decision-making processes, costs and benefits associated with the development of new technologies, and different types of inequality (inter-generational; between rich and poor countries; between humans and non-humans).

The fact that it ends upstream of any concrete action is a criticism that can be levelled at uses of Decide as a pedagogical tool in the science classroom. In this sense, although it can fit into a STEPWISE approach and contribute to the achievement of STEPWISE goals, it does not, in itself, constitute such an approach because it does not aim to lead to action. In order for this pedagogical tool to draw closer to STEPWISE goals, it could, for example, be used as a starting point to help students identify significant issues underlying a controversy, pinpoint those that interest them in particular, and form an informed opinion about them.[3] The students could then be asked to pursue their own investigations and engage in social action. It would also be possible for students who are already familiar with Decide to put together a game kit on a current or local socio-technical controversy that interests them and put it online, or organize sessions of the game with members of their community. In this case, the social

[3] It was based on this perspective that Romain Martiny developed a Decide game kit on the controversy surrounding the presence of metal dust in the central neighbourhoods of Quebec City (Martiny, 2015), which he then used in the chemistry classes that he taught in a pre-university college program. In addition to appropriating the controversy and learning about the socio-political and economic contexts surrounding it (Pouliot, 2015), the students were asked to give their opinion on actions that could be taken by the actors concerned. They were also invited to watch an excerpt from a television show addressing the issue of metal dust and take a stand on the nature of citizen expertise as well as the ins and outs of the economic arguments put forward by the Port of Québec.

action taken by the students would be to create a Decide game kit, make it available, and use it as tool to raise awareness of and foster citizen involvement in the controversy in question.

26.1.1 Goals and Rules of the Game

Decide is a group discussion game that aims to help players become more fully acquainted with current socio-technical controversies. Distributed free of charge by the FUND organization and accessible online in PDF format at www.play-decide.eu, Decide is based on the Democs group discussion game.[4] However, it is not played online. It must be printed on paper or cardboard. While the recommended number of players is four to eight, we observed that sessions involving three or four players usually turned out to be the most productive and the most agreeable. Several versions of the game are available—in several different languages (e.g., French, English, Italian and Portuguese) and exploring various socio-technical controversies (e.g. orphan drugs, biomedical tests or climate change). At the time of writing, there were 32 kits available in English (see Appendix for a list).

Each game session involves four phases: a preparation phase and three in-game phases (information, discussion, and shared group response). The **preparation** phase involves preparing the material (printing up the kit and cutting out the cards) and consulting the rules of the game, which are simple and quite flexible. The first phase of the game itself (the **information** phase) lasts approximately 30 min. Essentially, the players learn about the controversy by reading four possible policy positions on the controversy, as well as cards explaining some of the issues involved. One set of cards, called the Story Cards, present the point of view or story of fictitious individuals, bringing out different aspects of the socio-technical controversy under discussion. These fictitious individuals might be business owners, researchers, religious leaders, etc. There are also Info Cards presenting definitions, statistics, current or future applications of the technology, etc. Lastly, the Issue Cards invite the players to think about various issues surrounding the controversy. These cards present thought-provoking questions, quotes and various points of view. The players read several of each of the types of cards and select those they consider to be the most significant, which they then summarize for their co-players. The second phase of the game invites the players to **discuss** the controversy (for approximately 30 min), either taking turns or choosing an open discussion format. If they wish, they can refer to the Story Cards, Info Cards and Issue Cards to back up their arguments. During the third and last phase of the game, the players try to **formulate a shared group response** (this phase lasts approximately 20 min). The players reread the four policy positions presented during the information phase and can add others as they see fit.

[4] Democs was created by the New Economics Foundation (NEF) and aims to foster discussion on public policies.

They then vote individually on all four policies. Lastly, they negotiate and attempt to find some common ground, without necessarily having to reach a consensus. It is possible to upload the results of the game session on the Decide web site.

26.2 Socio-political Context of the Study

The discursive interactions presented below were produced as part of Audrey Groleau's doctoral research, conducted at Université Laval under the supervision of Chantal Pouliot. These interactions were recorded during two play sessions of the group discussion game Decide, focusing on controversies surrounding development and commercialization of nanotechnologies. Each session took place in French and involved three to four participants,[5] all of whom were planning to become teachers, were in their last term of a pre-university college program and were enrolled in a sociology of science course.

The empirical component of this study was undertaken in spring 2012 during a major student strike protesting an increase in tuition fees that had been announced by the provincial government. The strike mainly called for a more equitable distribution of wealth among individuals and between generations. In other words, as pointed out by André Drainville (2013) and André Frappier, Richard Poulin and Bernard Rioux (2012), it represented a resistance movement against the neoliberal economic system in place. These events coloured discussions of one of the teams of participants, who referred to this situation when backing up their opinions during the game.[6]

26.3 The Participants Discussed the Controversy Surrounding Nanotechnologies in Economic Terms

Using Decide in the classroom creates an opportunity to discuss, among other possible subjects, the controversy surrounding development and commercialization of nanotechnologies and to explore the various issues involved, including economic issues. In the play sessions discussed here, the students brought up the disparities between the rich and poor. In particular, they shared their views on the offshoring of factories, the exploitation of child labour, the priority given the well-being of Western societies over the common good of all, and profitability.[7]

---

[5] For a total of 7 participants: 6 female students, 1 male student; 5 participants planned to become elementary school teachers, 1 a phys. ed. teacher, and 1 did not specify the teaching level or discipline.

[6] The excerpts in question are not presented in this chapter. It should, however, be noted that they referred to the ability or inability of citizens to become involved in controversies that concern them.

[7] It should be noted that some of these issues are not addressed in the game kit.

26.3.1 Limited Access to Products and Services Allowing Individuals to Benefit from Nanotechnologies

In the following excerpt, the members of the team expressed their views on the possible application of nanotechnologies aimed at reversing, or at least slowing down, the aging process among humans. They predicted that only wealthy individuals would likely benefit from access to this application. They associated this privilege with a form of discrimination that would lead to a widening in the wealth gap between individuals and between the populations of different socio-political regions.

Rosalie[8]*:* *Yeah, well, that [the nanotechnology application making it possible to slow down the aging process] is going to lead to discrimination. Because, like it said in here [in my cards], it's going to be really expensive.*
Charlotte: *True.*
Rosalie: *It's not going to cost ten dollars to make yourself younger. Plastic surgery is already really expensive.*
Alice: *It's going to cost a [inaudible].*
Charlotte: *Just imagine! Having yourself made younger, it's going to cost an arm and a leg!*
Rosalie: *Yeah, only rich people will be able to afford it.*
Charlotte: *Super rich and powerful men, they're the ones who're going to have themselves made younger.*[9]
Alice: *And that's going to widen the gap between the rich and poor and between countries too.*
Rosalie: *Right.*
Alice: *Not every country is going to have ...It's going to be more common in the West. In the East, you won't see much of that.*
Charlotte: *True.*

26.3.2 Unequal Distribution of the Costs and Benefits

Later on in the discussion, the members of the same team referred again to the idea that inequalities between the populations of various socio-political regions[10] could become more pronounced as a result of development and commercialization of nanotechnologies. Xavier mentioned the unequal distribution of the social, economic and medical costs and benefits associated with development of these technologies. In other words, the following excerpt expresses the view that the development and commercialization of nanotechnologies will take place to the detriment of the populations in Southern countries.

---

[8] These excerpts have been modified slightly for readability. The names of the participants have been changed to preserve their anonymity.

[9] Although it was not the aim of this chapter to address the way the participants described the individuals that would have access to the benefits of nanotechnologies, Charlotte's comment appears significant as she associates wealth, power and masculinity with the elite who would benefit from nanotechnologies.

[10] Here, it is a question of opposition between the countries of the North and South rather than between the East and West.

Xavier: *Nanotechnologies are going to be exclusively available to a certain population.*
Alice: *Yeah.*
X avier: *Probably the people in the North, you know, the people in the South aren't going to benefit from them. In fact, they'll probably be the ones to pay for them – they'll pay with their labour and also, I don't know, maybe it's going to be discovered that, to produce nanoparticles, you need a particular mineral that's really rare.*
Alice: *Yeah.*
Xavier: *And this mineral will only be found in mines in theSouth. So you'll have firms that go there specifically to exploit platinum or whatever, and it might be really rare. We don't really know. But I think there's a risk that the gap between the rich and poor will just get wider.*
Rosalie: *For sure.*
Xavier: *...in a really big way, I mean, in terms of who gets priority and who gets the rights*
Alice: *Yeah.*
Xavier: *Some people will get the rights. So there will be more than just a wealth gap, there will be a gap in terms of who gets the exclusive rights.*
Alice: *That's true.*
Xavier: *So**this person**here will be able to live well, you know, have access to a particular treatment –**this**person will benefit from it, but**that**person won't.*
Alice: *This person will benefit but others won't.*

26.3.3 The Risk of Offshoring Factories and Exploiting Child Labour

Decide consists of cards containing questions that aim to launch a conversation about specific issues related to a given controversy. The next excerpt corresponds to the following questions: "Could nanotechnology widen the poverty gap? Might strict regulations in the West cause manufacturers to move to poorer countries, forcing people there to deal with hazards that are prohibited here?" The members of the second team argued that development and commercialization of nanotechnologies could lead to the offshoring of factories. They drew parallels between nanotechnologies and the textile industry, which has set up factories in Asia. They referred to the fact that the cost of labour is lower in Asia and also denounced the fact that children are made to work there.

Florence: *For sure, it's always like that. I'm a little pessimistic, that's just how I am, but I think humans are sort of screwed up. I'm sure that if regulations are put in place here that aren't put in place in other countries...*
Olivia: *They're going to go...*
Florence: *...well, it's going to lead to the same thing, they're going to leave, like they do now, say, if they aren't allowed to exploit, you know, when they make jeans, for example.*
Olivia: *Yeah, they're not allowed.*
Florence: *So they go into companies, say, in India, and it costs them...*
Emma: *Yeah.*
Florence: *...almost nothing in labour. They go there to exploit young children and then they come here and sell us their jeans at crazy prices and line their pockets with the profit.*
Emma: *True.*

26.3.4 Development of Medical Treatments for Profit

Emma, Olivia and Florence referred twice to the idea that economic profitability is one of the main criteria when it comes to choosing which medical treatments will be developed by the pharmaceutical industry. In one of these excerpts,[11] the participants compared two different diseases, one that mainly afflicts rich populations, namely cancer, and another, AIDS, which mainly afflicts less economically privileged populations. They predicted that, while considerable effort will be made to find a cure for cancer, the same will not be true when it comes to finding a cure for AIDS.

Olivia: *Well, it's true that we'll find a cure for cancer, but we won't find a cure for AIDS. Because that won't give* ***us*** *anything. It's the children in Africa who have AIDS and we could care less about them.*[12] *We want to benefit. We want to make a profit so we're going to find a cure for cancer. The people who can afford it will pay, so…*

Florence: *Yeah, exactly.*

26.4 Concluding Remarks

In this chapter, we set out to demonstrate that Decide is coherent with an approach based on STEPWISE as it provides the opportunity to address the well-being of individuals, societies and environments. To this end, we examined excerpts of conversations on socio-economic themes in which the participants were critical of certain issues relating to the development and commercialization of nanotechnologies. In particular, the participants discussed the limited access to products and services allowing individuals to benefit from nanotechnologies, the unequal distribution of the costs and benefits of these technologies, the risk of offshoring factories and exploiting child labour, and the development of medical treatments for profit. One of the pertinent contributions of Decide with regard to the philosophical and pedagogical aims of STEPWISE certainly lies in opportunities it provides participants to discuss development of techno-science while considering, in the words of Larry Bencze and Lyn Carter (2011), that "[w]ealth and wellbeing are funneled towards traditional elites, typically at the expense of the vast majority of other people and to the detriment of living and non-living environments" (p. 650).[13] Because it allows

[11] The other excerpt is not presented in this chapter.

[12] It was observed that Olivia takes on the voice of the people who will benefit financially from the production of these treatments (Potter (1996, p. 160–162) refers to this process as "Active voicing").

[13] The participants discussed the effects of the development of nanotechnologies on humans and non-humans. For example, one participant said, "*Who benefits from the use [of nanotechnologies]? I think it should be everyone, animals as well as people. I don't think it should be restricted to any one group in particular. [...] Yeah, if we want to avoid the situation where some species go extinct because of them, well, I think everyone should be able to benefit" (Emma).* The fact that the participants dis-

de-punctualization of nanotechnologies (Callon, 1991), namely the identification of actor-networks that interact, Decide can be mobilized during the Teacher Teaches phase of the STEPWISE apprenticeship.

It is not common practice to investigate socio-technical controversies in science classrooms by examining economic systems of which they are part. However, in the current period, which Pierce (2013) refers to as the postgenomic era (p. 111), marked by the unequal distribution of wealth (Piketty, 2013) and social and environmental effects of neoliberalism, addressing these concerns in science education has become imperative.

Acknowledgements We would like to thank the participants of this study and their teacher. We would also like to express our gratitude to the Social Sciences and Humanities Research Council of Canada, which funded this study.

Appendix: List of Themes Addressed in the Decide Game Kits Available in English

1. Ambient assisted living
2. Animal testing in biomedical research
3. Blood pressure
4. Climate change (3 versions)
5. Cross border health care
6. Diagnosis, information to the patient, genetic counselling
7. Digital world (2 versions)
8. Energy and sustainability
9. Environmental ethics
10. eTRIKS: The value of medical research data and its reuse
11. Global migrations
12. Health technologies: scoping the ‘value of innovation’
13. Healthy diet and lifestyle
14. HIV/AIDS and legal responsibility
15. Human enhancement
16. Integrating community care and medical care
17. Malaria
18. Nanotechnology
19. Neonatal screening
20. Neuro-Enhancement
21. Neuroscience – “brain enhancements”

---

cussed these effects of nanotechnologies is coherent with Pierce’s view (p. 112) that science education should lead to more democratic relations between humans and non-humans. This conversation also illustrates that Decide can lead participants to consider not only the well-being of individuals and societies but also that of environments (animals or biodiversity).

22. Orphan drugs
23. Patient-team relationships
24. Preimplantation genetic diagnosis (PGD)
25. Science camps
26. Smart cities
27. Stem cells
28. Structuring of healthcare among regions
29. Sustainable use of forests
30. Tuberculosis in Moldova and Romania
31. Xenotransplantation
32. Young people and the media

References

Alsop, S., & Bencze, L. (2014). Activism! Toward a more radical science and technology education. In L. Bencze & S. Alsop (Eds.), *Activist science and technology education* (pp. 1–19). Dordrecht: Springer.

Bencze, L., & Alsop, S. (Eds.). (2014). *Activist science and technology education*. Dordrecht: Springer.

Bencze, L., & Carter, L. (2011). Globalizing students acting for the common good. *Journal of Research in Science Teaching, 48*(6), 648–669. doi:10.1002/tea.20419.

Callon, M. (1991). Techno-economic networks and irreversibility. In J. Law (Ed.), *A sociology of monsters: Essays on power, technology and domination*. London: Routledge.

Drainville, A. C. (2013). Present in the world economy: Quebec's students in the springtime. *Globalizations, 10*(6), 785–801. doi:10.1080/14747731.2013.814454.

Frappier, A., Poulin, R., & Rioux, B. (2012). *Le printemps des carrés rouges: lutte étudiante, crise sociale, loi liberticide et démocratie de la rue* (p. 159). Mont-Royal: M éditeur.

International Student Initiative for Pluralism in Economics. (2014). *An international student call for pluralism in economics*. Retrieved from http://www.isipe.net/open-letter/

Jaxel-Truer, P. (2014, June 28). Thomas Piketty: Pourquoi ses courbes affolent la planète? M. *Le Magazine Du Monde*, 27–33.

Martiny, R. (2015). *Création d'un jeu de discussion à propos de la controverse autour des impacts sanitaires et environnementaux des activités portuaires de transbordements de minerais*. Unpublished Master's dissertation, Université Laval, Québec.

Pierce, C. (2013). *Education in the age of biocapitalism: Optimizing educational life for a flat world*. New York: Palgrave MacMillan.

Piketty, T. (2013). *Le capital au XXIe siècle*. Paris: Éditions du Seuil.

Potter, J. (1996). *Representing reality: Discourse, rhetoric and social construction*. London/Thousand Oaks: Sage.

Pouliot, C. (2015). *Quand les citoyen.ne.s soulèvent la poussière: La controverse autour de la poussière métallique à Limoilou*. Montréal: Carte blanche.

Risky Business Project. (2014). *Risky business: The economic risks of climate change in the United States*. Retrieved from http://riskybusiness.org/uploads/files/RiskyBusiness_PrintedReport_FINAL_WEB_OPTIMIZED.pdf

Chapter 27
STEPWISE as a Vehicle for Scientific and Political *Educ-action*?

Laurence Simonneaux and Jean Simonneaux

27.1 Introduction

In contemporary society, interactions between science, technology, technoscience and society are pervasive. Socio-Scientific Issues (SSI) education was developed in response to this situation and has become one of the main contemporary trends in science education. But it is possible to distinguish different variations related to SSI education; among others, variation in educational stakes that can 'cool down' or 'heat up' these issues (Simonneaux, 2013). Science and society are now seen as mutually interdependent in an educational context. The orientation of European programmes such as U FP7 and Science & Society programme and Horizon 2020 illustrate this. One of the goals of science education is to help students develop their understanding of how society and science are mutually dependent. This is the educational school of thought known as 'Science-Technology-Society' (STS) and, for several decades, the study of socio-scientific issues education has developed along these lines. The origin of the STS movement can be traced back to the 1930s and was led by scientists into the field of science education. It immediately fell in line with the citizenship education trend (Hogben, 1942). In Great Britain, after the Second World War, two movements had an influence on the promotion and development of STS education: the first was initiated by scientists who felt a sense of responsibility towards the public in view of the environmental impacts of scientific and technological developments, such as nuclear weapons and pesticides; the second movement was much weaker and aimed to break down barriers between the 'two cultures', the arts and science (Ratcliffe, 2001). Encouraging individuals to take a personal position was a major challenge for STS education: « In traditional

L. Simonneaux (✉) • J. Simonneaux
UMR EFTS, ENSFEA, University of Toulouse, Toulouse, France
e-mail: laurence.simonneaux@educagri.fr; jean.simonneaux@educagri.fr

L. Bencze (ed.), *Science and Technology Education Promoting Wellbeing for Individuals, Societies and Environments*, Cultural Studies of Science Education 14,
DOI 10.1007/978-3-319-55505-8_27

science instruction personal opinion is not involved and may be actively avoided. STS instruction, on the other hand, seeks out exchanges between students to help them arrive at personal positions that combine scientific knowledge with moral responsibility » (Solomon, 1981, p. 78). The STS movement was revisited by Derek Hodson (2003) who integrated the environmental dimension and became a campaigner for the development of STSE education in order to incite students to engage in decision making and action.

Alain Legardez and Laurence Simonneaux coined the term *'Questions Socialement Vives'* – in English: 'Socially Acute Questions' (SAQs). These questions are 'acute' in society, in research and professional fields, in classrooms and are often discussed in the media. The field of SAQs represents a French orientation to the teaching of SSIs. But SAQ instruction is part of the educational movement which advocates the study of the interactions between Sciences-Technologies-Societies-Environments (STSE). This study supposes to acknowledge the links that exist between sciences, politics and business.

Liora Salter (1988) uses the term mandated science, John Ziman (1996) post-academic science, Sheila Slaughter and Larry Leslie (1997) academic capitalism. The sciences have "*entered into a polygamous union with the economy, politics and ethics*" (Beck, 2001, p. 53). This raises the question of the moral responsibility for uses of scientific applications. Society? Scientists? Technologists? State? Jerome Ravetz (1975) raises this issue in his own way: "*Scientists take credit for penicillin, but Society takes the blame for the Bomb*" (p. 46). Knowledge and nature itself found themselves as 'goods.' That is to say, turned into salable and purchasable things. Parallel to development of science and the technosciences, in 1994, in the context of the 4th EU Framework Programme, ELSA was introduced as a label for developing and funding research integrating the Ethical, Legal and Social Aspects of emerging sciences and technologies. Currently, particularly in the context of EU funding initiatives, such as Horizon 2020, a new label has been forged, namely Responsible Research and Innovation (RRI). We are not going to summarize here the analysis carried out by Hub Zwart, Laurens Landeweerd, and Arjan van Rooij (2014) about this semantic shift. They consider that "*the newness of RRI does not reside in its interactive and anticipatory orientation, as is suggested by authors who introduced the term, but rather in its emphases on social-economic impacts (valorisation, employment and competitiveness)*" (p. 1). These labels (ELSA, RRI) indicate that the political sphere has understood the need to take into account social and societal aspects of the development of the technosciences in order to avoid a rejection by society, as was the case in France with GMOs. This is typically what has been done as soon as the first concerns about nanotechnologies were expressed. The goal is to prevent public controversy to hinder innovation. ELSA or RRI labels reveal the great importance of humanities and social sciences in the (Techno)Sciences – Societies interactions.

In science education, the notion of SSI has been introduced as a way of describing social dilemmas impinging on scientific fields (Zeidler, Walker, Ackett, & Simmons, 2002). Within SAQ education, the educational challenge is to enable students to develop informed opinions on these issues, to be able to debate such issues, to be capable of making choices with respect to preventive measures and intelligent use of new techno-sciences (Simonneaux, 2006). In order to solve most problems arising in contemporary society, scientific solutions alone are not enough and considerations must be given to the social implications of decisions relating to scientific investigations (Sadler, Chambers, & Zeidler, 2004).

SSI/SAQ education contributes to the 'educations for': (scientific) citizenship education, sexuality education, education for health, education for safety, education for the environment and for sustainable development. 'Educations for' focus on complex issues involving uncertainties that inextricably associate questions of a scientific, and social nature along with values and ethics. SAQ education raises the problem of teaching and learning in an uncertain world influenced by development of techno-sciences and environmental and health crises. These questions situate social and scientific controversy, complexity, building of expertise, assessment of evidence, and uncertainty and risk at the very heart of teaching-learning processes. It is not only experts who make decisions on SAQs; all citizens are involved (consumers, voters, legislators) (Simonneaux, 2006). Not only is it not possible to make just one valid and rational decision, but conflicting interests may lead to divergent decisions.

An SSI-oriented approach provides the motivation that students often do not find in traditional science education. This approach gives meaning to learning, makes operational the abstract concepts of science and promotes students to make connections between concepts. Nevertheless, it has been shown that teachers 'heat up' or 'cool down' SSIs, according to the questions that are under discussion, according to the educational risk that they are prepared to accept and according to the rationality to which they subscribe.

At the 'cold end', teaching about SSIs is used to motivate students to learn science, or even to convince them of merits of technosciences. At the 'hot end' of the continuum, teaching foci go beyond the purpose of developing science conceptual and procedural knowledge to the nurturing of activist commitments amongst learners. Pioneers of the 'activist' movement have developed a framework called STEPWISE (Science and Technology Education Promoting Wellbeing for Individuals, Societies and Environments) for organizing teaching and learning in science and technology[1]. The STEPWISE program aims to promote social and environmental justice and tries to foster a desire for change as well as a sense of responsibility (Bencze, Sperling, & Carter, 2012). Bencze (2000) suggests that students work on student-directed and open-ended research projects. This involves getting students to work on projects based on their own research to provide information on socio-scientific issues and encouraging them to make their results public by way of socio-political action (for example, organizing demonstrations and exhibitions, posting militant videos on YouTube™).

Between these two ends, there is a continuum of educational stakes. These range from learning stabilized scientific concepts underlying the issues addressed, developing a capacity for critical thinking and decision-making, learning about the nature of scientific knowledge and taking part in high-level cognitive procedures (identifying the conflicting interests of stakeholders, evaluating risks and uncertainties, generating debate and pinpointing fallacies, cultivating socio-scientific reasoning, identifying the actors' values, assessing evidence and critically analyzing research

[1] http://www.stepwiser.ca

methodology). These procedures contribute to development of critical thinking. When critical thinking occurs, foci move towards the 'hot end.' In the current field of French education, where the educational stakes are high, it is asserted that SAQs can develop high-level thinking, decision making and critical thinking with a focus on promoting an engaged citizenship.

Erminia Pedretti and Joanne Nazir (2011) identify and explore six currents in STSE education: application/design, historical, logical reasoning, value-centered, sociocultural, and socio-ecojustice currents. They consider that the latter four can be associated with SSI education. Most of the SSI-based instructions aiming at high level thinking abilities may be referred to the logical reasoning current. According to Pedretti and Nazir, "*the aim of science education in this current reflects a citizenship and civic responsibility emphasis through the transaction of ideas. As such, the dominant approaches are cognitive and reflexive*" (p. 612). Nevertheless, socioscientific reasoning may also be connected to the value-centered current. "Activities within this current tend to target students' moral and emotional identities to stimulate cognitive and moral development. As such, the dominant approaches are affective, moral, logical, and critical" (p. 614). And, of course, the STEPWISE program may be related to the socio-ecojustice current. According to Pedretti and Nazir, "the dominant pedagogical approaches in this current are creative, affective, reflexive, critical, place based, and experiential" (p. 617). The ecojustice current is also a political education (Lowenstein, Marusewicz, & Voelker, 2010).

SAQs are not only encountered in the (more or less stabilized) 'hard sciences' and invariably in the disciplines within the field of humanities and social sciences, but also in the area of social and vocational knowledge. We consider that many different actors take part in knowledge production. These include scientists, citizens, philosophers, professionals and, even, whistleblowers. The epistemological exploration preceding any didactic undertaking thus takes on a particular form because it cannot exclude interactions among actors and the diversity of reasoning involved in economic, political or scientific fields. It is undoubtedly a primary epistemological position to consider that scientific production processes are oriented and are results of interests of the different stakeholders. Consequently, Jean Simonneaux (2011) asserts that the knowledge involved in SAQs can be conceived of as plural (polyparadigmatic) and/or engaged (analyzing controversies, uncertainties and risks) and/or contextualized (observing empirical data within a given context), and/or distributed (constructed by different knowledge producers).

Decisions taken on SAQs cannot be based solely upon scientific knowledge (be it knowledge from the area of the social sciences or the hard sciences) but must also take into account social implications, ideologies and values. Unlike the work on SAQs, the SSI approach is mainly based on the didactics of the 'hard sciences.' Although complexity and uncertainty are recognized in SSIs, the role of interdisciplinarity is rarely studied, nor are concepts of the humanities and social sciences or those of social or vocational knowledge seriously taken into consideration. If we take the example of the controversial question of pesticide use, we can see that over and above the interdisciplinary aspect, it is the production of distributed situated knowledge that enables us to comprehend the issue. Farmers are not merely receivers of innovations designed upstream, but rather the producers and holders of knowledge distinct from agronomists (Goulet, 2013). There is no more ONE chemical solution designed upstream. "Singularity, and idiosyncrasy would thus be required in the modes of knowledge and management at the expense of "recipes" established by an experimental science that criticized the farmers involved in these movements" (Goulet, 2013, p. 439). Recognition of farmers as producers of knowledge displays a political dimension.

Consequently, the STEPWISE and SAQ approaches may not only contribute to scientific literacy, but they also can develop students' political literacy by including such topics as risk analysis, analysis of patterns of political and economic governance as well as decision making and action. Even though Dana Zeidler et al. (2005) have provided evidence that SSI education is a better way than the STS movement to integrate the Nature of Science, arguments, values and moral judgements, Derek Hodson (2011) has critiqued both of these approaches and asserts that STS and SSI education have given too low a priority to the promotion of critical thinking. He asserts that neither STSE nor SSI-oriented teaching go far enough.

27.2 Modernity/Reflexive Modernization and Education

A way to consider science-society relations and their connection to science education is to situate them in terms of historical sociology.

This amounts to situating education within pre-modernity, modernity, post-modernity patterns. Does the ternary pattern pre-modernity, modernity, post-modernity reflect the progressive emancipation of the individual in society? It is difficult to situate the temporal boundaries of the pre-modernity period: from antiquity to medieval times or up until the Age of Enlightenment. It is based on tradition and/or religion. The beginnings of modernity are sometimes associated with the end of the Byzantine Empire in the fifteenth century or the discovery of America and sometimes linked to the literary dispute between the Ancients and the Moderns in France in the seventeeth century. Be that as it may, what characterizes modernity is the pursuit of the ideal developed by Enlightenment philosophers, that is to say uses of reason to fight against the arbitrariness of the authorities, against prejudice and against the contingencies of tradition. The authorities and tradition are replaced by reason and science which will foster progress based on true and objective knowledge. Max Weber talks of instrumental rationality. Thanks to modern science, Man (sic) should dominate nature. A new mode of production and consumption, capitalism, is established supported by technological innovation. Modernity is associated with an increase in individualization. Education should liberate the individual thanks to rational knowledge. Scientific knowledge is glorified and transmitted via a top-down process. Scientists get a privileged position and replaced the priests of the pre-modern time. There is an unproblematic link between scientific reasoning and social, moral, ethical reasoning.

Modernity is an « ideal type » as defined by Weber, that is to say a theoretical construction that seeks to match with an empirical historic reality. We observe that the following period is more difficult to define, that the authors have proposed different ideal types (post-modernity, late-modernity, reflexive modernization, advanced modernity, second modernity, etc.). In the twentieth century, philosophers from the Frankfurt School consider that modernity as a project for social emancipation, has not kept its promises. In the process of dominating nature, Man has made himself the slave particularly because of the development of the technosciences. According to Jürgen Habermas, modernity is an unfinished project that humanity should defend and reclaim in order not to lose its humanity.

Others believe that we have entered the period of postmodernity which will finally allow us to fulfill the project of emancipating the individual by freeing us from the last transcendental figures of modernity. Hope for progress is shattered by the excesses of the technosciences (nuclear weapons, pollution, health problems). Hope in the future is replaced by a cult of the present, but also by an anxiety for the future due to concern over the adverse effects of the capitalist model, especially on environments. Links between science and society are problematic and complex. That scientific research, cultural norms, socio-political contexts, applications influence each other is a recognized fact. The optimism of modernity is replaced by skepticism or even pessimism. Relativism develops alongside a recognition of true and objective knowledge. Traditional images of academic science have changed. Trends in sciences are now criticised as being more and more determined by economic interests.

Bruno Latour (1991) seeks to show that the project of modernity cannot ever be realized. It is a project built on two parallel contracts, i.e. the 'social contract' which is the ground for social order and the 'nature contract' which is the ground for modern science. Together these contracts should make it possible to draw a solid line that separates the society from nature. Such a project, however, is caught by contradictions that become evident as soon as we face such 'hybrids' as AIDS and the ozone hole, which are neither nature nor culture, but something in between. Bruno Latour considers that all cultures have produced hybrids. The specificity of this epoch is based on two things: (i) the scale and the threatening nature of our hybrids, (ii) their existence calls into question modern attempts to draw a solid line between nature and culture. According to him, the time line of modernity and its by-product, progress, is not straight. We have never been modern because we have never progressed towards increased efficiency and profitability. "The old idea of progress, the one we recently abandoned, let us stop being wary, let us throw caution to the wind. The new concept appears rather to oblige us to be cautious, to make selective choices, to meticulously consider all the possibilities" (Latour, *Le Monde*, 24 août 1996).

Ulrich Beck refuses the post-modernist approach; he considers we have entered a new modernity, but we are still within modernity. We have moved from an industrial modernity to a reflexive modernity. He calls this era the 'risk society.' Beck (1986/2001) suggests that these days we are emotionally aware of man-made hazards: society is concerned about the risks posed by techno responses to past problems. The production of new scientific knowledge is to resolve the multiple impacts (waste, pollution, new diseases) that have been generated by technoscience. One could sum up this late-modernity as an epoch during which individuals have become aware to risks, uncertainty, complexity, disorder, distrust of social and scientific institutions and traditional authorities.

Beck postulates that institutions, including science, are struggling with effects of what they have created, and even though they have begun to change. It is necessary that research anticipates the consequences, uncertainties and risks of scientific advances. Using Beck's analysis, in our late-modern society, scientific rationality would not be sufficient to justify any technoscience and would need to be accompanied by reflexive criticism of its impact. The status of experts in science and technology is criticized and the political nature of technological choices is revealed. Beck believes that, faced with the risk society, with crises, with the uncertainty of knowledge, individuals will develop a reflexive modernization, that alternative rationalities will surface and new social movements, a 'sub-politic' may emerge in the interstices of the official society. According to Anna Olofsson and Susanna Öhman (2007), to be defined as 'reflexive, ' people have to show both awareness and some kind of active strategy to handle new risks. But awareness does not implicate always action, far from it, people can remain in a fatalist position.

The risk society approach tends to adopt a critical realist (Bhaskar, 1975) approach, contending that the real social and natural world exists apart from and is independent of human perception and understanding. Thus, human knowledge of reality is fallible and incomplete and is historically, socially, culturally and politically situated. Experts' judgments of risk cannot be objective and neutral.

Beck's works are sometimes criticized as being strictly theoretical, unsubstantiated by empirical work. Mette Jensen and Anders Blok (2008) did a test, in the form of a case study, on the pesticides issue as perceived in Denmark. Their aim was to study whether or not we live in a risk society. Beck is sometimes accused of exaggerating, especially by Arthur Mol and Gert Spaargaren (1993), who advocate an alternative paradigm, called 'ecological modernization,' in which green lobbies are seen to guarantee environmental interests. Therefore risk society does not exist because of ecological progress. In this case, the technoeconomic progress of modernity will happen under the control of ecological progress. During their study, Jensen and Blok (2008) observed that lay respondents had different 'risk habitus' (p. 765); in particular, they were less anxious when they trusted in a form of ecological modernization to guarantee control. « While a majority of lay-people (and a minority of counter-experts) may be said to broadly inhabit a 'risk' society, a majority of experts (and a minority of lay-people) rather inhabit an 'ecological modern' one". These authors consider then that "as a societal narrative, 'risk society' is hence clearly contested" (p. 773).

Anthony Giddens (1994) also rejected the concept of post-modernity. He calls the current era advanced modernity. For him, no knowledge is ever stabilized forever; progress is a myth. For Göran Therborn (2003), 'multiple modernities' coexist; i.e., people from different lives (traditional, modern, late modern) share the same society. This is similar to the position of Mary Douglas (1985) who emphasizes cultural impacts on judgments about risks. She considers that, within a same culture, different groups can have different conceptions of risks. For her, risk judgments are political, moral and even aesthetic. Her theory reflects social biases that influence a person's perception of risk.

According to Douglas (1992), the theoretical construction of any social organization is based on two key dimensions: an internal structure characterising social groups that gives them a definite place, and a hierarchy that delimits the boundaries of each group compared with other groups. Douglas has focussed on very significant relationships between the organizational form of a cultural group and its values. She has identified four organizational types that occur in modes of social participation and cultural principles. These four types correspond to different perceptions of knowledge, nature and risk. She distinguished these types as: the bureaucrat, the individualist, the egalitarian and the fatalist.

Bureaucracy is an archetype of hierarchy. Within this structure of organised social groups people are attached to values such as order, decency and laws. Individuals within this type agree with the authorities and the scientific knowledge of the scientists who advise them. Nature is regarded as robust and adaptable to human disturbance, but there are lines that should not be crossed. This group perceives that, beyond these limits, irreparable damage can occur and the ecological balance may be irreversibly disrupted. This hierarchical type shows a very pronounced risk aversion but when risks are identified this group expects that the State and the experts will find a good solution.

The primacy of the ego is dominant in the individualist type. Among representatives of this type, shared values are those of an enterprising spirit, free competition and individual achievement. Scientists, innovators and entrepreneurs are respected and followed. Nature is regarded as very robust with a self-balancing system that allows it to cope with all situations. The environment is viewed as a homeostatic system that seeks to recover its original state when a disturbance unbalances it. Balance is the steady state of nature and any imbalance is only temporary. This group believes that the benefits of science and technology always outweigh the harm. Such a conception of nature encourages this group to support boldness and innovation in all fields of technology. The individualistic type is rather risk friendly as they see risk is an opportunity to seize so that they can assert themselves and control the future.

Egalitarians promote the primacy of the group. This type of social organization refers primarily to small groups that are formed around an ideal, an ideology or a fight that they think is legitimate. Within this group are found whistle blowers. Values that have the most weight among egalitarians are equality, fairness and justice. These individuals are wary of academics and they call on knowledge produced by the group itself. Nature is seen as fragile and in a very unstable equilibrium. The action of humanity is regarded as harmful to nature and any imbalance is felt as irreversible. Egalitarians accuse participants within a hierarchical structure and especially those with the individualistic structure, of systematically plundering natural resources and threatening the ecological balance and the common good, including future generations. They have an aversion to technological risk.

Groups belonging to the fatalist type are not integrated into society and are without means for organising and developing a structured group identity. It appears that their internal disorganization and subordination to other social groups plunges them into a kind of fatalism. They do not really think much about knowledge, but express a general mistrust of it. Patrick Peretti-Watel (2001) defines this group as having poorly determined values and that they display fatalism about their condition and the situations they encounter. They have a view of nature as capricious and unpredictable. They perceive ecosystems as changing randomly that are impossible to predict and control. Chaos theory is, according to Peretti-Watel, the best example of this group's view of risk. For them, risk is inevitable and they have to cope with it. Maybe we can consider that fatalists rely on pre-modern notions of fate and lack of control, while egalitarian refer to late-modern notion of reflexive control over risk.

As science experts are mistrusted, everyone has to make his/her own decision. "We have no choice but to choose how to be and how to act" (Giddens, 1994, p. 75). Hence, there are needs for schools to train for action or activism.

According to Gilles Lipovetsky and Jean Charles (2004), a hypermodern society has emerged which is replacing the postmodern society because of an anxiety associated with awareness of serious issues linked to socio-economic, health and environmental deregulations.

SAQs lies within the field of Post Normal Science (PNS), as defined by Silvio Funtowicz and Jerome Ravetz (1993), as a science with strong links to human needs, thereby leading to large uncertainties, major issues, values, and requiring urgent decisions. According to Ravetz (1997), the question 'what if?' justifies strong consideration 'to extended facts'; that is to say, data from sources outside the orthodox research. These authors emphasize that decision processes on the PNS should include open dialogue with everyone concerned. They introduced the concept of 'extended peer community.' It is important to train students to participate within the 'peer extended community.'

In the perspective of reflexive modernization, SAQs and STEPWISE also question foundations of science and rationalist utopias according to which reason and truth emerge from confrontation of ideas. Thus, for Beck (1986/2001), we must go beyond the "*successive attempts to rescue the 'underlying rationality' of scientific knowledge*" (p. 360) implemented whenever science is confronted with failure or adverse effects. In the research cited above, Jensen and Blok (2008) conclude that the real value of the work of Beck might be its 'performative' dimension with reference to Latour (2003). It is in this vein that we consider STEPWISE to be of interest because reflexivity on modernization is not self-evident. STEPWISE advocates the vital importance of raising awareness of this reflexivity through 'educ-action,' in order to ensure that citizens remain vigilant, do not off-load their responsibility by trusting the government to exercise ecological control. How far should reflexivity be developed? Should education promote the exercise of reflexivity about expert knowledge or empower students to generate their own risk knowledges?

Educ-action aims to encourage not only the involvement of students and teachers but also their commitment to individual and collective action, what Beck calls sub-political engagement. In this sense, both the STEPWISE and SAQ movements defend a humanistic, scientific, political and economic education (Table 27.1).

Table 27.1 From modernity to late-modernity

	Time	Main ideas	Science education policy
Pre-modernity	Ancient and medieval thought	Search for patterns in nature. Hierarchical view of society	Elitist. Scholastic
Modernity	17th to early 20th or even until today	Overarching idea of Enlightenment, Science as rationalist. Rationality is superior to other ways of thinking. Logical positivism, Karl Popper	Lay people need to know more science to appreciate and support good policy. Aim to think scientifically. Understand science first then apply to society. There is an unproblematic link between scientific reasoning and social- moral-ethical reasoning
		Empiricism. Mertonian sense of important values of science such as search for truth, objectivity, impartiality, etc.	
Late-modernity	Since the middle of the 20th	Science seen as imbued with power relationships. Link to society is problematic and complex. Science has a role but meshed in economic, political and cultural dynamics. Ideologies, values recognized	Contextual and situated education
		Critical Realism (Bhaskar). Postnormal science (Funtowicz & Ravetz) even relativism. Society of Risk (Beck)	Consideration for complexity and uncertainty
			Socio-scientific reasoning, moral reasoning Controversial SSI Education for Sustainability SAQ Scientific AND political education

Table elaborated with the contribution of Levinson

27.3 Educ-action and Activism

The development of educ-action is not a new trend, we can refer back, for example, to Freire, but it is evolving with the emergence of a late–modernity. Educ-action meets with resistance on the part of teachers particularly because of its ideological and political dimension. Furthermore, this educ-action implies varying forms of commitment that we need to examine.

27.3.1 Teachers Involvement/Resistance and Rationality

"The need for the inclusion of socio-scientific issues (SSI) into science curricula has been generally accepted, but relatively few science teachers have incorporated SSI into their courses. Most science teachers feel that their most important task by far is to teach the principles of science, and any substantive pedagogical changes represent a burden" (Lee & Witz, 2009, p. 931). There is a perception amongst many science teachers that science education is about the delivery of facts, and that science is value-free (Levinson & Turner 2001). However, some teachers address SSIs out of their own personal initiative and heat up the issues; that is, they 'teach against the grain' (Cochran-Smith, 1991). Some examples of these practices were presented in a symposium at the last ESERA conference (Levinson & Martins, 2013).

One difficult problem is of the neutrality of teachers leading the debates. Thomas Kelly (1986), one of the first researchers who considered using debates for classroom study of controversial issues, postulated four positions that teachers might adopt: exclusive neutrality, exclusive partiality, neutral impartiality and committed impartiality. Those in favour of exclusive neutrality believe that teachers should not broach controversial themes and that scientific discoveries are value-free truths. They subscribe to a positivistic approach that has been widely criticized. There are two main arguments against their position: first, teachers always convey values, if only through the examples they choose; secondly, the task assigned to schools in a democratic society is to train citizens who are capable of debating controversial scientific issues, which means that the school must stay in touch with real life. Exclusive partiality is characterized by the deliberate intention to bring students to adopt a specific point of view on a controversial issue. In this case, teachers ignore contradictory positions or brush them aside as insignificant. They believe that their mission is to provide students with intellectual certainties. Those in favour of neutral impartiality believe that students should debate controversial issues as part of their education to become citizens and that teachers should remain neutral and not reveal their points of view. For some supporters of this position, teachers should remain silent and neutral so as to maintain their authority and should not reveal their uncertainty or ignorance, while others believe they should remain neutral in order not to influence students' argumentation. This position, which is nevertheless quite appealing, has been criticized. It is important that students have the opportunity of comparing their points of view to those of a 'role model' adult such as the teacher. Moreover, as we have said previously, teachers always convey their values, albeit unconsciously and neutrality is an illusion.

Concerning the latter position, an apparently paradoxical position, teachers gave their points of view while encouraging analysis of competing points of view on the controversial issues. This was the position recommended by Kelly. And Hodson (2011) believes that "it is incumbent on teachers to share their views on these matters with students and to make explicit the ways in which they have arrived at their particular position. It is also incumbent on teachers to adopt the same stance of critical reflection and open-mindedness that they demand of their students, and to be willing to change or modify their views in the light of new evidence, a new way of interpreting evidence, a reappraisal of underlying values, or whatever" (p. 61). He believes this is a way to explicitly develop their own critical thinking vis-à-vis their previous positions.

Research was carried out on commitments to climate change teaching declared by teachers of different disciplines. It was observed that, depending on their discipline, these teachers engaged in three types of pedagogical models (positivist, interventionist and critical). These models ranged from educating students in accordance with their own opinions to teaching students how to make their own choices (Urgelli, Simonneaux, & Le Marec, 2010). In the case of the positivist model, the teachers focused on teaching the reference knowledge of the discipline presented as non-controversial and presumed that this approach would enable the students to make choices as responsible « informed » citizens. In the interventionist model, the teacher intended to question the environmental consequences of human development in relation to the urgency of the climatic issue, or to scientific and technical progress. The declared objective was to stress the need to change behavior and consumption patterns in the face of the rising demand for energy. In the case of the critical model, the teacher declared s/he planned to get the students to take a global view of ways in which expertise on the climate is portrayed in the media—underlining, in particular, that the complexity of the issue is inconsistent with a consensual scaremongering approach by the media to climate-related risks.

The diversity of these engagements can be explained by the ecological convictions and/or epistemological doubts the teacher holds. Epistemological doubt, that is to say the acknowledgement that these questions are controversial and fraught with uncertainties, may be crucial to the way these questions are taught. If the teacher accepts the doubt, he/she may choose a critical approach to the question. Sometimes, in spite of her/his personal doubt, the teacher chooses not to engage students in a critical approach for fear of influencing them on account of his/her institutional position. On the question of climate change, the ecological convictions of the teachers studied by Urgelli (2009) justified an interventionist approach. In the case of issues related to health (gene therapy, the use of embryonic stem cells), we assume that ethical convictions can determine ways with which these questions are dealt.

The nature of the teachers' rationality has an influence on their choice of teaching strategies, depending on whether they adopt a techno-scientific rationality (the techno-sciences will resolve the problems raised by current technosciences) or a critical rationality which implies reflexivity towards the techno-sciences. The teacher's rationality can vary according to the issue.

A study on teachers in agricultural education in France has been conducted. The study focused on SAQs related to animal husbandry (the evolution in meat consumption, the contribution of animal breeding to the greenhouse gas effect, animal welfare). We wanted to discover whether they approached these SAQs on the basis of their ecological or ethical convictions and called breeding practices into question and/or a critical analysis of animal husbandry knowledge. This group tended towards a techno-scientific rationality (Simonneaux, 2012). Faced with these SAQs, the teachers took sides with the breeders above all else. They empathized with the farmers who were angered by the criticism fired against them and by the measures they were required, by law, to take. These teachers believed techno-science would resolve the SAQs. They would like to see more targeted research associated with the development of the techno-sciences in breeding.

The majority of those teachers took a positivist approach to the environmental issues offered up for debate. They assimilated sustainability rhetoric as long as it is associated with productivity. They were confident that techno-scientific progress would resolve the SSIs linked to the environment. But, fundamentally, they minimized the responsibility of animal husbandry and the part it played in the issues raised (climate change, the food crisis). They were also reticent about the regulations on animal welfare. However, another group of teachers revealed their critical rationality when dealing with the question of pesticides by denouncing the environmental problems and to a lesser extent the problems linking the health of consumers and farmers to pesticide use (Simonneaux & Simonneaux, 2013).

For many authors, such as Agnieszka Jeziorski and Alain Legardez (2013), sustainable development is an SAQ. They have tried to identify what representations future secondary school teachers have of sustainable development and educating for sustainable development, and to analyze the results in terms of what fosters and what hinders a critical education focusing on socially acute questions. Consequently, data were collected using two complementary tools: firstly, a questionnaire was administered on one hand to 223 French Canadian trainee teachers in science and technology and social sciences, and on the other to future teachers of French, Earth and life sciences, history and geography; and, secondly, a semi-directive interview was conducted with 12 respondents to the questionnaire.

From the point of view of socially acute questions, Jeziorski and Legardez consider that ESD is in line with a transformative, participatory approach to education, as referred to by Bob Jickling and Arien Wals (2013). According to them, the position practitioners and academics adopt towards ESD depends on their conceptions of education and the people being educated. They distinguish two conceptions of education: transmissive and transformative. The aim of transmissive education is to unilaterally convey ideas defined by a limited number of external experts. Its goal is efficiency and social reproduction. Transformative education is in complete contrast to transmissive education in that, in the latter, knowledge is co-created within a given context. Thus, the creation of new knowledge is influenced by prior knowledge and different cultural perspectives. The aim is to provide an education for critical citizenship which trains students to question the world in which they live to empower them to create their own world. In general, citizens are educated to conform with a view to social reproduction, that is to say they are trained to accept the role traditionally assigned to them in the work society. In a transformative approach to education, citizens participate in decision-making. Figure 27.1 illustrates the different ways to engage in ESD depending on the representation of education on one hand and the citizens being educated on the other. The vertical axis represents the conceptions of education and the horizontal axis the conceptions of the people being educated. ESD in terms of SAQ would be in quadrant IV.

This research shows that "the positions adopted by the trainee teachers questioned, fluctuate between transmissive education and socio-constructivist transformative education. Both positions may coexist in the same person and come into conflict when it comes to choosing a didactic strategy. The socio-constructivist transformative approach thus limits itself mainly to exposing different points of view (most of the time concerning the implementation of sustainable development, without really discussing it) and providing the students with a context (territorialized education, project-based teaching). The importance of reflexive and interdisciplinary activities and debate on the subject of sustainable development are expressed, but their implementation seems to run counter to the positivist school paradigm of which is still dominant" (p. 31).

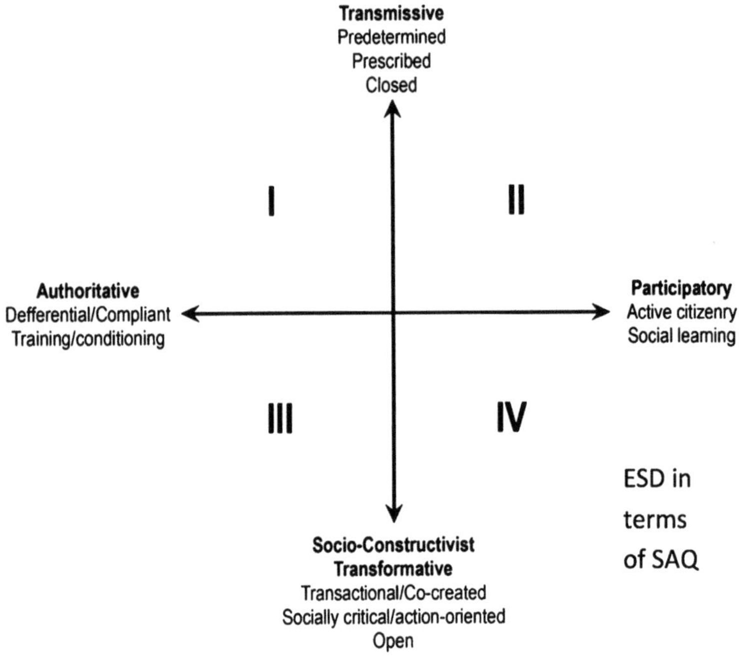

Fig. 27.1 ESD in terms of socially acute questions based on the conception of education and on the conception of the educated person (Adapted by Jeziorski and Legardez from Jickling and Wals, 2013)

The actual work of teachers is not only a personal interpretation of the curriculum prescribed according to the constraints related to the context of the teaching-learning situation. The difference between the task prescribed to teachers and what they actually do can be explained in terms of 'professional genre.' "A collective professional genre retains the transpersonal memory of a work environment. It preserves and transmits social history" (Clot, 2008, p. 77). The results of numerous research projects conducted within the context of 'learning to produce differently' (a program set up by the French Ministry of Agriculture to foster more sustainable forms of production which are more respectful to living organisms and the environment) show that the main obstacle to the so-called agroecological transition is the professional genre to which teachers adhere (Vidal & Simonneaux, 2013). Teachers identify primarily with the professional genre of conventional farmers which is focused on productivity and not with the professional genre of teachers who promote agroecological innovation. They seek legitimacy in the field in order to be accepted by their students who they deem resistant to the agroecological transition, or unaware of the health hazards of pesticide use, for example. We have shown in our research that students are well aware of the issues, but the risk in teaching them as perceived by teachers, leads them to cool down the issues. Jensen and Blok (2008) refer to this as "risk habitus" (p. 765).

27.3.2 From Involvement to Activism via Commitment or Promoting a Sub Political Engagement

This question of commitment and action is becoming increasingly significant to the didactics of SAQ and 'educations for.' It has been apparent in the STEPWISE program for quite some time. The transformative aim is an essential marker but it is necessary to substantiate the possible forms that transformation and change may take within the education system.

For Freire, the activity of teaching has an influence on the world and it cannot be neutral. Freire's perception is set in a specific context (postwar Brazil), where the predominant social challenge is how to tackle poverty in a context of populism and military dictatorship. Logically, emancipation seemed to be the overriding issue and was to lead to protests against injustice in order to let people imagine how to fight against oppression of individuals and communities (Zanchetta, Kolawole-Salami, Perrault, & Leite, 2012). The emancipation of thought remains the factor that allows education to take root in social reality, even if the social challenges are different today (Santos & Mortimer, 2002). In the case of Freire's approach, over and above the steps taken by individuals, it is important to insist upon roles played by the community. Paulo Freire (1972, 1974) affirms that education is a human activity that is inserted in human reality; therefore, its task is to transform the human world. The humanistic education he advocates goes beyond teaching contents without social meanings. It focuses on the human condition and in its transformation. According to Freire, "those that believe that the teacher has to be 'apolitical' are unintentionally and naively supporting the dominant ideology imposed by the technological systems. They reinforce it when they do not discuss it with their students" (Santos & Mortimer, 2002 p. 647). These references to Freire really echo positions of SAQ education and the STEPWISE approach.

We examine here how this educational form targeting commitment to action can actually take shape.

If we consider learning as a process of change and/or empowerment the extent of the changes taking place in the students may be measured in different ways, may be more or less specific and sometimes ambiguous. We first propose to make a distinction between motivation, involvement, commitment, empowerment and activism. We are not claiming that the definitions given here are definitive or that they represent a consensus. Motivation may be seen as the willingness on the part of the students to participate in the educational activity. Understanding instructions, the difficulty of the task, the extent of the challenge or competition and the pleasure factor are all elements which can explain motivation (Simonneaux, Leboucher, & Magne, 2014).

Motivation is certainly a criterion which is conducive to the educational process but does not in any way measure the effectiveness or the extent of the changes occurring in the students.

Involvement may be considered as the students' capacity to become active in the collective training process. The degree of involvement helps us examine the intensity with which the individuals mobilize their attention, their interest and their enthusiasm in carrying out the learning tasks (Cheffers, Brunelle, & Von Kelsch, 1980). Here, again, the intensity of the involvement, whilst certainly an indicator of the success of the educational process, does not allow us to measure the extent of the changes taking place.

Commitment represents an individual's capacity to take a stance on issues, to undertake action and /or to comply with a more or less pre-determined form of behavior.

Empowerment applies to individuals' ability to make decisions and take control over their lives. Nina Wallerstein and Edward Bernstein (1988) refer to 'individual, social, collective empowerment.'

Activism involves learning about and experiencing participation (Linhares & Reis, 2014), and can go as far as convincing other people to influence the decision makers and to develop actions with a view to improve the well-being of individuals, of societies and of the environment (Bencze, Alsop, & Bowen, 2009). Of course, not everyone agrees with this definition. In a French context, activism may be perceived as a synonym for militancy, sometimes suspected of scaling-up actions without giving them proper thought or may even be associated with violent behavior.

We consider that activism applies to three key elements: awareness, reflexivity and the implementation of actions which are assessed and modified according to what is at stake.

In the case of commitment, empowerment or activism, the indicators correspond to the students' stances regarding what is taught or the educational goals and not simply a form of behavior that is expected in class showing motivation and implication. It is necessary to make this distinction from the outset, but it needs to be developed and clarified. In particular, it raises a methodological problem of observation. There are two ways of interpreting the attitudes of learners who show a predisposition for action, either we look at the components in their language assuming that they will provide an insight into what an individual thinks and that these thoughts determine his behavior, or we base our analysis on directly observable behavior.

Social psychology has identified different action models or theories including the following:

- Involvement is the intensity with which the individuals undertake in terms of attention, interest and enthusiasm in the tasks required by the teachers (Cheffers et al., 1980).
- The theory of planned behavior focuses on an individual's intentions to explain his/her behaviors, which can be understood by his/her attitudes, perceptions of norms, and behavioral controls (Ajzen, 1991).
- Pierre Bourdieu also developed a theory of action around the concept of habitus. This theory seeks to demonstrate that social agents develop strategies based on a small number of *dispositions* acquired through socialization. The identification of these dispositions allows us to determine the potential commitment of individuals to the action.
- The dispositions for action, initially put forward by Bourdieu (1998) are considered by Ria (2012) to be a set of perceptive, interpretative, cognitive, emotional, intentional and actional components mobilized in the same type of situation.

- The commitment theory predicts effects that influence the behavior of another person not by resorting to persuasion but by stimulating a previously minimal behavior which subsequently leads to greater commitment (Kiesler, 1971; Joule & Beauvois, 1987). It is not simply a question of a person being committed or not but rather a question of the extent of the commitment
- Habermas (1987) distinguishes communicative, strategic, normatively-regulated and dramaturgical action. According to him, communicative action presents itself as an interactive activity moving towards agreement and whose function it is to coordinate the actions between participants.
- Neil Mercer (1995) distinguishes the following types of discourse: disputational, cumulative and exploratory talks. The latter are supposedly dominant in collaborative approaches and reveal the collective commitment of the actors.
- The common operational referent is defined as a process shared by a team in order to carry out an action on the basis of each member's skills (De Terssac et Chabaud, 1990).
- A community of practice is a group of people who work together in a situated context. Their objective is to increase their skills in a given practice (Lave & Wenger, 1991) (Table 27.2).

Involvement, individual and collective commitment can be identified through language components (answers to questionnaires and interviews, interactions) and behaviors observable in context.

In Table 27.3, we consider that the minor eco-gestures correspond to the aspects of commitment in the lighter shaded boxes, militancy to the aspects in the grey boxes and finally activism to all the aspects in level 4 in the dark grey boxes.

The 'minor gestures' have often been highlighted in ESD in the form of eco-gestures encouraged by teachers (Jeziorski & Ludwig-Legardez, 2013). However useful they may be, these minor gestures have been called into question by many actors. They do not make it possible to construct and understand a project for society or a community in all its complexity. They may even hinder the understanding of global issues by letting us think that environmental questions can be resolved by these civic eco-gestures. This said, these eco-gestures may, however, be a first step.

A future citizen is not only responsible for his own actions, he must also be able to participate in public decisions, to commit himself to the development of

Table 27.2 Psycho-sociologic models about implication and engagement

Implication	Voluntary accomplishment of learning tasks (Cheffers et al.)
Individual commitment	Theory of planned behavior (Ajzen)
	Dispositions (Bourdieu)
	Theory of commitment (Kiesler), Voluntary submission (Beauvois & Joule)
Collective commitment	Discourses in collaborative practices (Mercer)
	Communicative action (Habermas)
	Common operational referent (De Terssac & Chabaud)
	Community of practice (Lave & Wenger)

Table 27.3 The scope of commitment in educational activities

Criteria for analyzing commitment	Level 1	Level 2	Level 3	Level 4
The intention to act	*Claims an action is relevant*	*Claims that an action should be taken*	*Claims his support for such an action*	*Claims he will commit to such an action*
The individual's role in the action	*Participates occasionally in the action*	*Participates regularly in the action*	*Makes suggestions to assess / improve the action*	*Gets others involved in the action ; is a driving force*
The impact of the action	*A set of minor gestures or individual behaviors*	*Collective assessment of the tasks/gestures*	*Inclusion, assessment of actions in a Long Term plan*	*The action is publicized in the media outside school*
The collective dimension of the action	*Action shared by a group of students*	*Institutionalized action (evaluated) in school*	*Action taken in conjunction with external partners*	*Action taken outside school*
The critical perspective	*Identifies complimentary factors or variations to the action taken*	*Identifies the limits of the action*	*Identifies the controversies and risks*	*Compares the different positions and argues his personal point of view*

a 'sustainable' society and, to do this, he needs to acquire knowledge, values and an ability to live in a community. This conception of the future citizen means we have to define educational goals that are more ambitious than these simple "minor green gestures". The socio-political action which is developed upholds a critical perspective particularly in reference to controversial issues (Linhares & Reis, 2014). It involves fostering commitment AND reflexivity.

Activism is sometimes interpreted as engaging in action without giving it much thought. This is in no way what is meant by the pioneers of the 'activist' movement in schools who have developed a framework called STEPWISE which, as we have already indicated, aims at social and environmental justice and attempts to foster a desire for change and a sense of responsibility among individuals (Bencze, Sperling, & Carter, 2012). These different angles for analyzing and / or fostering action may be seen as a graduation of the goals of educ-actions, ranging from simple adhesion to a project and the development of expected behavior, through adapting behaviors, deciding and reasoning behavioral changes and to societal transformation. Beyond these goals, another focus for the analysis could be the range of actors concerned by these actions. At one end of the scale we may find the student concerned as an individual actor and at the other end this action may concern a wider community outside of school. The degree of autonomy in the learning community, the time-scale (short or long term) and scope (local or global) of the action may constitute other lines of analysis. Over and above these indicators (level of commitment, the actors concerned, the time-scale and scope) and the extent of the action, we must also examine the purposes and methods used which are pursued according to a given context in all its complexity. This is vital if we are to understand the dynamics involved in an educational project.

27.4 Conclusion: STEPWISE for a Committed Educ-action

The educational perspective of STEPWISE implies interaction between schools and society, between scientific processes and sharing knowledge, between individual and collective processes, between reflexivity and actions ...The socio-political issues and the question of commitment stimulate the sought after critical perspective. STEPWISE promotes the concept of the engaged school and research which contributes to the emergence of critical education. This, to us, seems to be an essential step towards the development of the emancipated eco-citizen. Schools must be transformed in keeping with this critical education. We consider this to be a vital step towards dealing with the challenges facing society today and in the future. This transformative goal for schools may take on different forms: critical education, socio-political education, and activism. We can see huge similarities between the SAQs approach and the STEPWISE program in their aims for scientific, social, political and economic education but there is also a similarity with the humanistic science education sought after by Freire. "The Humanistic Science Education is a slogan that tries to contribute to changing the context of the modern society through educational processes (…) Science Education has a potential to contribute for the transformation of modern society through helping make visible the pitfalls of the system and make people aware of their role as citizen and consumer in this society" (Santos & Mortimer, 2002, p. 641). According to his dialogic action theory, action started in dialogue, word is a transformable praxis, which acts on the world. It supposes a collective action in which subjects meet in cooperation to transform the world. Wildson Santos and Eduardo Mortimer add a humanistic argument to STS education. "This argument brings to discussion to the need of transforming scientific and technological modern society through human values, preparing the students for a society in which sustainable knowledge and responsible action are the norms. This is not a movement anti-technology, but a movement against a particular model of economic development and technological practice" (p. 646). The inclusion of SSI or SAQ in education is necessary but it must integrate not only science contents but also "the understanding of environmental risks; the power of domination that the technological system impinges in culture; the difference between human needs and market needs; and the developing of attitudes and values consistent with a sustainable development" (p. 647).

Yves Chevallard (2010, 2014) a pioneer in the didactics of mathematics in France, who developed the concept of didactic transposition, challenges, in his later writings, what he calls the paradigm of visiting Works; that is to say, a form of schooling based on the transmission of knowledge that is disconnected from the issues that led to its production. This is what he refers to as the *old school paradigm* which aims to create differences, to select an elite by venerating knowledge presented in a monumentalist, frontal fashion with an authoritarian relationship to truth (the teacher proclaims). He contrasts this with *the paradigm of questioning the world;* that is to say, the pedagogy of inquiry for a democratic school which creates citizenship where knowledge is alive and is an instrument for improving community life and taking control of the world. He considers that the didactitian should not withdraw into his/her discipline but should become "gyrovague" [2]; in our opinion, this means he should be open to interdisciplinarity and the integration of lay-knowledge.

In the European project PARRISE (Promoting Attainment of Responsible Research and Innovation in Science Education) within the framework of the 7th European Science and Society program in which several authors of the present chapter participated, an investigative approach to SSI is modelled « SocioScientific Inquiry Based Learning » (SSIBL)[3]. This approach should lead students to set up actions. It remains to be seen whether these actions will be 'cooled down' or 'heated up' from an activist point of view.

References

Ajzen, I. (1991). The theory of planned behavior. *Organizational Behavior and Human Decision Processes, 50*(2), 179–211.

Beck, U. (2001). *La société du risque, sur la voie d'une autre modernité*. Paris: Flammarion.

Bencze, J. L. (2000). Procedural apprenticeship in school science: Constructivist enabling of connoisseurship. *Science Education, 84*(6), 727–739.

Bencze, J. L., Alsop, S. J., & Bowen, G. M. (2009). Student-teachers' inquiry-based actions to address socioscientifc issues. *Journal for Activist Science & Technology Education, 1*(2), 68–102.

Bencze, L., Sperling, E., & Carter, L. (2012). Students' research-informed socio-scientific activism: Re/Visions for a sustainable future. *Research in Science Education, 42*(1), 129–148.

Bhaskar, R. (1975/1997). *A realist theory of science* (2nd ed.). London: Verso.

Bourdieu, E. (1998). *Savoir faire. Contributions à une théorie dispositionnelle de l'action*. Paris: Seuil.

Cheffers, J., Brunelle, J., & Von Kelsh, R. (1980). Measuring student involvement. In G. Schilling & W. Baur (Eds.), *Moyens audio-visuels dans le sport* (pp. 216–229). Basel, Switzerland: Library of Congress.

Chevallard, A. (2010). La didactique, dites-vous? *Education et Didactique, 4*(1), 136–143.

Chevallard, A. (2014). Des didactiques des disciplines scolaires à la didactique comme science anthropologique sur un obstacle épistémologique, psychologique et institutionnel. *Education et Didactique, 8*(1), 35–44.

Clot, Y. (2008). *Travail et pourvoir d'agir*. Paris: PUF.

Cochran-Smith, M. (1991). Learning to teach against the grain. *Harvard Educational Review, 61*(3), 279–310.

[2] The word *gyrovague*, (from Late Latin *gyrovagus*, from Latin *gyro-* gyr- + *vagus* wandering) originally refers to a wandering and usually dissolute monk of the early church. Chevallard, 2014, p. 40.

[3] Also see Chap. 22, this volume, for more information about the SSIBL framework.

De Terssac, G., & Chabaud, C. (1990). Référentiel opératif commun et fiabilité. In J. Leplat & G. de Terssac (Eds.), *Les facteurs humains de la fiabilité dans les systèmes complexes* (pp. 110–139). Toulouse, France: Octarès Editions.
dos Santos, W. L. P., & Mortimer, E. F. (2002). Humanistic science education from Paulo Freire's 'Education as the practice of freedom' perspective. In *X International Organization for Science and Technology Education (IOSTE) symposium – PR*, Foz do Iguaçu, 2002. Proceedings, Vol. 2, pp. 641–649.
Douglas, M. (1985). *Risk acceptability according to the social sciences*. New York: Russell Sage Foundation.
Douglas, M. (1992). *Risk and blame: Essays in cultural theory*. London: Routledge.
Freire, P. (1972). *Pedagogy of the oppressed*. Harmondsworth, UK: Penguin Books Ltd..
Freire, P. (1974). *Education for critical consciousness*. London: Sheed and Ward.
Funtowicz, S. O., & Ravetz, J. R. (1993). Science for the post-normal age. *Futures, 25*(7), 739–755.
Giddens, A. (1994). Living in a post-traditional society. In U. Beck, A. Giddens, & S. Lash (Eds.), *Reflexive modernization: Politics, tradition and aesthetics in the modern social order* (pp. 56–109). Cambridge, UK: Polity Press.
Goulet, F. (2013). Narratives of experience and production of knowledge within farmers' groups. *Journal of Rural Studies, 32*, 439–447.
Habermas, J. (1987). *Théorie de l'agir communicationnel, Tome 1: Rationalité de l'agir et rationalisation de la société*. Paris: Fayard.
Hodson, D. (2003). Time for action: Science education for an alternative future. *International Journal of Science Education, 42*(6), 645–670.
Hodson, D. (2011). *Looking to the future – Building a curriculum for social activism*. Rotterdam, The Netherlands: Sense.
Hogben, L. (1942). Biological instruction and training for citizenship. *School Science Review, 23*(91), 263–281.
Jensen, M., & Blok, A. (2008). Pesticides in the risk society: The view from everyday life. *Current Sociology, 56*(5), 757–778.
Jeziorski, A., & Ludwig-Legardez, A. (2013). Éducation au développement durable: La difficulté de concevoir une action éducative interdisciplinaire. *Revue francophone du développement durable, 1*, 59–74.
Jickling, B., & Wals, A. E. J. (2013). Probing normative research in environmental education. Ideas about education and ethics. In R. B. Stevenson, M. Brody, J. Dillon, & A. E. J. Wals (Eds.), *International handbook of research on environmental education* (pp. 74–86). New York: Routledge Publishers.
Joule, R.V. et Beauvois, J.L. (1987), Petit traité de manipulation à l'usage des honnêtes gens. Grenoble, France: *Presses Universitaires de Grenoble,*.
Kelly, T. (1986). Discussing controversial issues: Four perspectives on the teacher's role. *Theory and Research in Social Education, 14*, 113–138.
Kiesler, C. A. (1971). *The psychology of commitment: Experiments liking behavior to belief*. New York: Academic Press.
Latour, B. (1991). *Nous n'avons jamais été modernes: Essai d'anthropologie symétrique*. Paris: La Découverte.
Latour, B. (2003). Is re-modernisation occurring – And if so, how to prove it? A commentary on Ulrich Beck, theory. *Culture and Society, 20*(2), 35–48.
Lave, J., & Wenger, E. (1991). *Situated learning: Legitimate peripheral participation*. Cambridge, UK: University Press.
Lee, H., & Witz, K. G. (2009). Science teachers' inspiration for teaching socio-scientific issues: Disconnection with reform efforts. *International Journal of Science Education, 31*(7), 931–960.
Levinson, R., & Martins, I. (2013). Political literacy as a component of science teacher practice: Teaching against the grain. In *A Symposium presentation at the 2013 ESERA conference*, Cyprus.
Levinson, R., & Turner, S. (2001). *Valuable lessons: Engaging with the social context of science in schools*. London: The Wellcome Trust.

Linhares, E., & Reis, P. (2014). La promotion de l´activisme en futurs enseignants partant de la discussion de problèmes socio-environnementaux et socio-scientifiques. *Revue francophone du Développement durable, 4*, 80–93.
Lipovetsky, G., & Charles, S. (2004). *Les temps hypermodernes*. Paris: Grasset.
Lowenstein, E., Marusewicz, R., & Voelker, L. (2010). Developing teachers' capacity for ecojustice education and community-based learning. *Teacher Education Quarterly, 37*(4), 99–118.
Mercer, N. (1995). *The guided construction of knowledge: Talk amongst teachers and learners*. Philadelphia, PA: Multilingual Matters LTD.
Mol, A. P. J., & Spaargaren, G. (1993). Environment, modernity and the risk-society: The apocalyptic horizon of environmental reform. *International Sociology, 8*(4), 431–459.
Olofsson, A., & Öhman, S. (2007). Views of risk in Sweden: Global fatalism and local control—An empirical investigation of Ulrich Beck's theory of new risks. *Journal of Risk Research, 10*(2), 177–196.
Pedretti, E., & Nazir, J. (2011). Currents in STSE education: Mapping a complex field, 40 years on. *Science Education, 95*(3), 1–26.
Peretti-Watel, P. (2001). *La société du risque*. Paris: La Découverte & Syros. Coll. Repères.
Ratcliffe, M. (2001). Science, technology and society in school science education. *School Science Review, 82*(300), 83–92.
Ravetz, J. R. (1975). …et augebitur scientia. In R. Harré (Ed.), *Problems of Scientific Revolution. Progress and Obstacles to Progress in the Sciences* (pp. 42–75). London: Oxford University Press.
Ravetz, J. R. (1997). Simple scientific truths and uncertain policy realities. *Studies in Science Education, 30*(1), 5–18.
Ria, L. (2012). Variation des dispositions a agir des enseignants débutants du secondaire : entre croyances et compromis provisoires. In D. P. Guibert & P. Périer (Eds.), *La socialisation professionnelle des enseignants du secondaire* (pp. 107–125). Rennes, France: Presses Universitaires de Rennes.
Sadler, T. D., Chambers, F. W., & Zeidler, D. L. (2004). Student conceptualisations of the nature of science in response to a socio-scientific issue. *International Journal of Science Education, 26*(4), 387–410.
Salomon, J. (1981). STS for schoolchildren. *New Scientist, 89*(235), 77–78.
Salter, L. (1988). *Mandated science: Science and scientists in the making of standards*. Boston: Kluwer.
Simonneaux, J. (2011). *Les configurations didactiques des questions socialement vives économiques et sociales*. Habilitation à diriger des recherches. Université de Provence.
Simonneaux, J., Leboucher, F., & Magne, M-A. (2014). *Using a serious game to encourage the design of innovative environmentally friendly agricultural systems*. ERIDOB, Haifa, juin 2014.
Simonneaux, L. (2006). Quel enjeu éducatif pour les questions biotechnologiques ? sous la direction de. In A. Legardez & L. Simonneaux (Eds.), *L'école à l'épreuve de l'actualité – Enseigner les questions vives* (pp. 33–61). Issy-les-Moulineaux, France: ESF.
Simonneaux, L. (2012). Les rationalités d'enseignants en productions animales face aux questions socialement vives en élevage. *Revue de Didactique des Sciences et Techniques, 5*, 9–46.
Simonneaux, L. (2013). Questions socialement vives and socioscientific issues: New trends of research to meet the training needs of post-modern society. In C. Bruguière, A. Tiberghien, & P. Clément (Eds.), *9th ESERA conference selected contributions. Topics and trends in current science education* (pp. 37–54). Dordrecht, The Netherlands: Springer.
Simonneaux, L., & Simonneaux, J. (2013). *The reasoning of teachers in the French agricultural education system on issues linked to pesticide use*. Conference of the European Science Education Research Association September 2–7, 2013, Electronic conference proceedings, Nicosia, Cyprus.
Slaughter, S., & Leslie, L. L. (1997). *Academic capitalism: Politics, policies, and the entrepreneurial university*. Baltimore: The Johns Hopkins University Press.
Therborn, G. (2003). Entangled Modernities. *European Journal of Social Theory, 6*(3), 293–305.
Urgelli, B. (2009). Les logiques d'engagement d'enseignants face à une question socioscientifique médiatisée, le cas du réchauffement climatique. Thèse de doctorat de l'Ecole normale supérieure de Lettres et sciences humaines Sciences de l'éducation – Sciences de l'information et de la communication.

Urgelli, B., Simonneaux, L., & Le Marec, J. (2010). Réchauffement anthropique et développement durable -Quelle(s) éthique(s) pour une éducation scientifique citoyenne. Education au développement durable et à la biodiversité : concepts, questions vives, outils et pratiques. Digne les bains.

Vidal, M., & Simonneaux, L. (2013). *Les enseignants refroidissent la question socialement vive du bien-être animal*. Penser l'éducation, Hors Série, pp. 431–446.

Wallerstein, N., & Bernstein, E. (1988). Empowerment education: Freire's ideas adapted to health education. *Health Education & Behavior, 15*, 379–394.

Zanchetta, M. S., Kolawole-Salami, B., Perrault, M., & Leite, L. C. (2012). Scientific and popular health knowledge in the education work of community health agents in Brazilian shantytowns. *Health Education Research, 27*, 608–623.

Zeidler, D. L., Sadler, T. D., Simmons, M. L., & Howes, E. V. (2005). Beyond STS: A research-based framework for socio-scientific issues education. *Science Education, 89*, 357–377.

Zeidler, D. L., Walker, K., Ackett, W., & Simmons, M. (2002). *Tangled up in views: Beliefs in the nature of science and responses to socio-scientific dilemmas. Science Education, 27*, 771–783.

Ziman, J. (1996). Post-academic science: Constructing knowledge with networks and norms. *Science Studies, 9*(1), 67–80.

Zwart, H., Landeweerd, L., & van Rooij, A. (2014). Adapt or perish?: Assessing the recent shift in the European research funding arena from 'ELSA' to 'RRI'. *Life Sciences, Society and Policy, 10*(11), 1–19.

Chapter 28
Understanding Opportunities and Contradictions in the Grammars of Activism and Schooling

Matthew Weinstein

28.1 Introduction

In this chapter, I wish to celebrate the STEPWISE model, described in Chap. 2,[1] but also probe its boundaries — with the intent of pondering how to make it even more faithful to its own ends. My context for thinking is the post-financial crisis in the U.S.A., where neoliberalism has taken a particular form of austerity logic playing on nationalist tropes of independence, self-sufficiency, and racial privilege. It is in this U.S. context that I ask what is the nexus of social activism and science education? This is the heart of the problem that STEPWISE as a model of educational praxis seeks to clarify. To help me grapple with the possibilities and limitations of this model, I want to consider some concrete instances of what we call in the ed-biz 'socio-scientific issues' (SSIs). In posing these concrete instances, I am drawing on Donna Haraway's (1997) use of 'figures' as means of critical interrogation. I have three such figures that haunt me at the moment of this writing, and each implodes science and society in particular ways that have implications for teachers, curriculum workers, and scholars:

- Development and use of new sensor technologies, combined with drones that intensify the surveillance state, magnify the will to conflict (no skin off of our noses) as demonstrated in Yemen, Somalia, and elsewhere. Drones are perhaps the perfect weapons of the neoliberal era: small, integrated into mass

[1] 'STEPWISE' is the acronym for Science & Technology Education Promoting Wellbeing for Individuals, Societies & Environments. It is a theoretical and practical curriculum/teaching framework encourages and enables students to engage in self-directed research and social actions to address personal, social and environmental problems linked to fields of science and technology. To learn more about this framework, refer to Chap. 2 in this book and visit: www.stepwiser.ca

M. Weinstein (✉)
School of Education, University of Washington, Tacoma, WA, USA
e-mail: mattheww@uw.edu

L. Bencze (ed.), *Science and Technology Education Promoting Wellbeing for Individuals, Societies and Environments*, Cultural Studies of Science Education 14,
DOI 10.1007/978-3-319-55505-8_28

communication systems, at the edge of global power struggles, and highly prone to entrepreneurial repurposing for either spectacular visualization or commercial exploitation. For me, Amazon's proposal to use drones to deliver packages seems like a reincarnation of an earlier search for peaceful uses of fission immediately after the bombing of Hiroshima and Nagasaki.

- If drones are one figure of the neoliberal moment combining computer science, technology, and global management and warfare, the trickle down of such technologies to police represents a second implosion of science and technology. In the current 'state of emergency' (Fassin & Pandolfi, 2010) in the U.S., local police have integrated military technique and technology. Images of tanks on the street of Ferguson, Missouri, to quell protests of the shooting of Michael Brown, an unarmed Black teen, have become emblematic of a change in which citizenry, especially if marked by race, become 'terrorists'; protest becomes 'assault'; and, a plea for help becomes an occasion for 'state murder.' The killing of Michael Brown is just one of many police murders of unarmed Black men, and not by far the most horrific. Georgia Ferrell, rising football star, was shot by police seeking help by knocking on neighbors' doors after a car accident. As Ann Morning has shown, race as biological difference is not dead 60 years after the end of World War II and the supposed death of eugenics (Morning, 2011). Many (most?) scientists very much embrace racial essentialism, which acts, as Michael Hetzfeld (1992) has shown, as the foundation of social production of indifference ('they' are inherently different than 'us'). Not just one type of indifference, but multiple occasions for uncaring: indifference to dead bodies, indifference to asymmetric applications of laws and, thus, justice. Black power has shifted to the plea 'don't shoot' in Ferguson, Missouri. In this new racial economy, science and technologies (genetics and tanks) shape particular forms of social order and disorder. These order/disorder systems are of this moment in which neoliberal political geography so easily divides what Naomi Klein (2004) has called, *disaster capitalism*, following the maps of Iraq after the U.S. invasion, red and green zones: areas in which the civil society functions and those in which the disenfranchised are left to a social Darwinian logic often accompanied by hyper-policing (not to help but to contain populations seen as disposable).
- And what is that military, with so much 'surplus equipment,' is pursuing around the world? Energy, even as we extract record amounts of oil from shale sands and natural gas from fracking in the U.S. and increasingly elsewhere, e.g., England. The nature of fracking is by decree ambiguous. The chemicals that are forced underground and likely into the drinking water are protected by trade secrets. If that were not enough, additional laws have been passed to gag any attempts to divine, analyze or otherwise inform the public of what is involved chemically or hydrogologically in the fracking process. The tar sands have been producing a more toxic oil than that pumped earlier, and has been at the root of a series of exploding train cars from Quebec to Louisiana. Meanwhile, the wide spread use of fracking that has had the amazing effect of producing earthquakes caused by the subcutaneous extraction of gas are common occurrences far from tectonic plates. Oklahoma is now experiencing more earthquakes than California, though

these are human-created ones. Energy demands and earth science have always been soul mates—mining, etc.

In each of these figures, science, technology and engineering are completely interwoven with grotesque struggles for power and profit. It is into the world in which these problems (along with desertification, ocean acidification, global climate change, and antibiotic resistant tuberculosis, not to mention Ebola, diarrhea, and other problems for the world's poor) are norms. These are figures peculiar to neoliberalism in as much as neoliberalism has (1) bound science more tightly to corporate interests than in the past and (2) taken advantage of market logics to undermine science as space for critique (Mirowski, 2011). In education, schools and whole countries are pushed into STEM (Science, Technology, Engineering and Mathematics). To understand how this reflects the neoliberal moment and stages, the figures bulleted above, one has to read this as what Stuart Hall described as an 'articulation', a connection between certain topics in the pursuit of certain political interest and thus a disassociation between others (Hall, 1986). These technical subjects are reduced to *techne* or skills (Grundy, 1987) by their mutual association. In other words, what counts in STEM is what the acronym leaves out: the social, the political, the historical, and the ethical. While some blocs within the coalition pushing STEM seek exclusion of those pesky SSI/STS/STSE topics and analyses, which seek to revive these excluded analyses (social, political, etc.), interestingly, the U.S.'s new science standards, the 'NGSS' (Next Generation Science Standards), include, but marginalize, the issues (burying them within what the standards call crosscutting ideas), reflecting, I believe, an ambivalence about the neoliberal (or corporate) intentionality of the STEM movement (converting biology to bioengineering as it were).

Into this pedagogical economy comes STEPWISE, which has to be understood as a resistant counter-articulation of science to STEM. At its core is a relational analysis. As Michael Apple (1990) explains,

> [a relational analysis] involves seeing social activity as tied to the larger arrangement of institutions which apportion resources so that particular groups and classes have historically been helped while others have been less adequately treated... Thus to understand, say, the notions of science and the individual, as we employ them in education especially we need to see them as primarily ideological and economic categories that are essential to both the production of agents to fill existing economic roles and the reproduction of dispositions and meanings in these agents that will 'cause' them to accept these alienating roles without too much questioning (p. 10).

The relational analysis Larry Bencze crafts explicitly draws on Bruno Latour's (1987) and Michel Callon's (1999) Actor Network Theories (Rouse, 1993, hereafter ANT) to articulate the relational nature of technoscientific knowledge—Latour's name for the contiguity of the STEM fields, though properly it's *socio-techno-scientific* knowledge. It is not just that living and non-living are embedded in the same network, but that science and technology mask sociotechnical networks needed to stabilize theories and products.

STEPWISE takes this relational logic farther than Latour's vision, which was frustratingly neutral in its politics, Latour infamously adopted a quasi-objectivist point of view (Rouse, 1993) and instead embraces an engaged pedagogy. STEPWISE draws on Marxist traditions to problematize consumption and ecological crisis within capitalist relations. STEPWISE even in its initial (non-simplified) formulation places student action at the center of the curriculum and teaching, scaffolding for students the relationship of science to challenging and changing the world.

As noted, my intention here is to celebrate and push STEPWISE. The celebration is deserved as the model creates a space that allows students to 'refigure' my exemplars above. Larry Bencze and I (and many others) share a passionate concern about capitalism sans alternative, ecological catastrophe, and the related alienation from 'the world,' by which I mean a living in relation and obligation to others. Technoscience in all its STEMish variants certainly is central to these imbricated catastrophes, in as much as it enables mastery over nature and transformation of the world (including us) into commodities, primarily through framing nature, problems, and solutions exclusively within the realm of the technical. But to what extent can schools, which are themselves part of these same technoscientific networks (as I will discuss), become capable of teaching against the neoliberal grain? To what extent can teachers misbehave—meaning behave differently than the sorting, selecting, standardizing instruments that schools have become under neoliberal stewardship, at least here in the U.S., home of the Chicago School of Economics that spread neoliberalism globally? This is the question that I seek to explore in this chapter. Again, my context is the U.S. where neoliberal reforms shape every level of the school organization from standardized testing of students, to value added measures of teachers, to data driven management of districts.

28.2 STEPWISE and the Grammar of Schools

STEPWISE builds on the century long pneuma of social reconstruction. This vision of social reconstruction partially founds modern schooling in the U.S. With roots in pragmatism and early twentieth century radicalism (Lasch, 1965), schools have, since their modernization in the U.S.'s post Committee of Ten reconceptualization, partially reoriented themselves towards visions of social change. This radical thread is as essential to schools as the business/factory driven format of the school house (bells, schedules, hierarchical authority, etc.) according to historian Herbert Kliebard (1986). For Kliebard, curriculum—and I would add pedagogy—is the negotiation between this radical and business intentionalities as well as those of child-centered growth logics and Deweyan pragmatism. Schools are means, according to Larry Cuban and David Tyack (1995), of 'tinkering towards utopia' (the title of their book). Schools are sites of both conservative and progressive politics to capture futures and bring about social visions through their actions on the young. Teachers, of course, are active in letting in some elements of reform and filtering out others. Cuban and Tyack's main focus is, in fact, on the 'ways schools change reforms rather than the usual reverse question, which sees the school room (including both teachers and students) as the passive object of policy ('How do reforms change schools?'), Cuban and Tyack both grant teachers agency and schools an institutional culture, which reforms have to contend with to succeed and, if they succeed, they are likely changed by the interaction with the wills of teachers. At the heart of the filtering of reform is what Tyack and Cuban call 'the grammar of school,' a deep material and cultural structure that sorts reforms into useable and unusable parts. Certainly, part of that structural filter revolves around a preservation of know-how and prior labor of both teachers and students. Attending to the ways reforms interact with know-how helps to account for the price of reform for students, teachers, and reformers.

The grammar of schooling also involves pedagogical forms and pedagogical knowledge. Basil Bernstein (1990) has analyzed how discourses, when detached from their fields of origin (e.g., professional biology) and relocated to the classroom (e.g., grade 10 biology), are transformed in the classroom: simplified, stripped of controversy, reduced to transmittable facts, reduced to the testable items. In these transformations, the grammar of schooling becomes visible. The transformation of STEPWISE from its tetrahedral form to its 'simplified,' 'linear' form is an exemplar of the ways 'schools change reforms' or more specifically, the ways that reforms are changed to fit the grammar of schooling. In this grammatical transformation, purpose is changed into the general coins of 'schooling': scores, grades, rankings, and teacher approval. Activities within school are always at risk of being co-opted, by even the best and most sincere teachers, into being for schools themselves, what I call the 'school effect.' By the very fact of an activity happening within the grammar of schooling, it is reinterpreted by students as merely being for the sort of capital schools parcel out (grades etc.). For example, as a student-teacher, I watched a colleague introduce her students to *The Autobiography of Malcolm X* as an attempt at critical pedagogy. The students, however, seeing words on the board, 'read' the lesson as one on vocabulary. Her lesson got 'schooled.' Note that, here, it is students not the teacher who enforced the grammar of schooling. Students are highly invested in the economy of school, and critical pedagogy can have the ironic effect of deskilling them in the immediate pursuit of grades.

With this in mind, it is worth thinking about the multiple grammars involved in the STEPWISE vision. STEPWISE both is a resistance to 'school science' (science transformed into the grammar of schooling) and a pedagogical approach within 'school science.' However, in its design, there is clearly a portion of STEPWISE that is organized in the grammar of schooling: 'product' (i.e., traditional reified content), skills education and students' research are hardly alien to science education. On the other hand, STSE while long pushed for as focus for science education, turns out to be highly resistant to the grammars of schooling: its interdisciplanarity, its refusal to accept science as the metadiscourse, its relationality all push against the grammar of schools. Nidaa Makki (2008) has explored ways that teachers give lip service to STS but rarely implement it. There is a reason for this; STS replaces science as a Truth, i.e., as a metadiscourse on nature, with sociology or cybernetics (in the case of Actor Network Theory) as a metadiscourse on science. To study something that is supposed to transcend discourse as a discourse is ultimately to take a critical stance on it. Furthermore, the very authority of the teacher rests, in part, on science as metadiscourse, i.e., as a higher level of truth. Pedagogy is supposed to be 'evidence based', after all. So understanding (aka questioning) science risks undermining the standing of the pedagogue him/herself. Like schools teaching Illich's *Deschooling* (1971), which I do, by the way, becoming critical or even sociological about science risks unravelling the whole ball of yarn: By what authority other than the game of schooling can I have my students read *Deschooling* and then question schooling? Ultimately, Bencze's central point of activity plays a similar role in crossing out of the comfort zone of schooling. It is not that service learning or community projects are uncommon, but the STEPWISE notion of activity is specifically about activity that questions authority, power, economy, and established institutions.

Two grammars are combining here. First the grammar of schooling embodied, as noted, in the products, skills, and research vertices of the tetrahedron. But this is being fused into a pedagogical pidgin with a grammar with clear ties to Freire, informal education, and the long history of activist education (e.g. Highlander school (Adams & Horton, 1975)). Many critical theorists, including Freire, turn to 'informal education' meaning out of the eye of the state, education to enact their radical pedagogies. Relocating education 'off school grounds' helps mitigate 'the school effect.'

Informal education, of course, has its own grammar, including an expectation of egalitarian spirit and dialogue. Experience often matters more than credentials for the teacher, in these settings. This experience has to be conveyed (not being materialized in certificates) through narrative and density of material contextualized in sites of practice. The 'product' of this education is the transformation of practice, the activity that STEPWISE promotes, for instance, rather than marks. Discipline, classroom management, etc., are transformed by the voluntary nature of informal education. It should be of little surprise that many critical pedagogues prefer or focus on the informal rather than the formal, Angela Calabrese Barton's work on GET CITY for instance (Calabrese Barton, Tan, Turner, & Gutiâerrez, 2012). By developing pedagogies for the informal, the ecology removes points and grades, it stresses voluntary participation, critical pedagogues can keep foregrounded the larger purpose of such education. Similarly, using 'informal educational grammar' means that inter-disciplinarity can be employed without challenges from boundary police (other teachers who want to claim a monopoly on their subject matter) who might complain about such border crossings (Giroux, 1992) or third spaces (Soja, 1996).

It is precisely for these reasons that when I wanted to study how radicals adapted science for social justice ends. I turned to the informal education of street medics, i.e., providers of medicine to protesters in situations fraught with risk, due usually to uses of physical and chemical weapons (tear gas or pepper spray) by police or national guards, rather than classrooms. Trainings were led by established medics who shared stories of dealing with epileptic seizures, baton wounds, tear gas, and broken limbs. We were also given direct notes of protocols we should use for both medicine and for establishing rapport with patients. Finally, we were to practice these protocols in increasingly complex skits or scenarios. In this informal setting, no hedging was necessary. Police brutality, the complicity of counter protesters and the media, and danger to medics could all be spelled out. In classrooms, by contrast, where, whatever the politics of the teacher, he or she is pressured by numerous obligations that can sideline their own pedagogical interests: standards, administrators, parents at political or philosophical odds with themselves, content demands by other teachers (e.g., concerns that students will not be 'ready' for their class).

STEPWISE very much depends on the grammar of informal education. This is evidenced in its multidisciplinarity (science, sociology, political theory, rhetoric) as well as its ethos of questioning authority both in the abstract through STSE education and in the concrete, e.g., in promoting boycotts as potential actions. Even in its revised (linear) format, it uses the model of apprenticeships that explicitly evokes informal educational settings. The very shift from knowledge students must learn to identifying questions students 'want' to pursue is a move towards the informal. And, yet, the formal co-exists with this informal grammar throughout the STEPWISE model. STEPWISE is explicitly a model for formal education, because, and I strongly share this same conviction, the formal is in fact a commons. So, how does this pidgin play out? When does one grammar supersede the other? What can be said and what not said within the rules of these grammars? For instance, does STEPWISE provide entrees for my three SSI figures?

STEPWISE in its formulation is backed by a complex theoretical framework drawing on Marxism, ANT, and semiotics. In its practice, what hegemonic limits does it confront?

28.3 Boundaries of Practice

While there clearly is a dialectical relationship between the grammars of informal and formal education, they are not mutually exclusive realms. The formal and informal work together, borrow from each other. People operating in both contexts selectively poach on the practices developed in the other. It is important not to overstate the distinction between the two. In this section, I focus a little more closely on how the tangle of grammars shapes and limits the content of STEPWISE. This includes shaping what can be addressed as a socio-scientific issue and the kind of scaffolding schools seem able and not able to provide students.

While two grammars are present in STEPWISE, the grammar of schools dominates in as much as formal schooling is the material condition of STEPWISE teaching, which means that classroom points, teacher approval, grades, etc. are at play in student thinking. The informal is rolled in, to trouble 'school science,' but it is a school science classroom. To understand what the consequences of this grammatical transformation to STEPWISE looks like, I want to consider, in general, the projects published in the *Journal for Activist Scienceand Technology Education* (JASTE, 5(1) [[goo.gl/N00b3s]), which are student reports from a teacher following the

STEPWISE model (Krstovic, 2014). This is not STEPWISE the theory; this is, following Latour (1987), reports from STEPWISE-in-action (SWiA). This is STEPWISE adjusted by the time frames and professional logics in which teachers work. These were reports drafted in the context of a science education class, so they are do not reflect the long-term consequences of the curriculum. Does the political consciousness provoked here remain or is it washed away in subsequent 'product' oriented classes?

To me, the most notable aspect of the JASTE work in the special issue featuring student reports is that it primarily focuses on consumption. Of the nine published projects, six emphasize and take action around either use (e.g., of automobiles) or purchase (e.g., of lotions). This may be an effect of the framing of the assignment, again, not a consequence of the abstract model, per se. This focus on consumption comes about through, what from a pedagogical point of view is a natural decision point, students contributing their own ideas and concerns at the outset of each unit. It is also sometimes promoted explicitly, as in the case of another published STEPWISE example (Bencze, Carter, & Krstovic, in press) in which the teacher explicitly asks his students to 'pick one personal hygiene product.' This launch point has the clear advantages of (1) starting with the students' personal consumption habits, (2) emphasizing objects familiar to students, and (3) offering a relatively simple set of political actions that students might take: alternative consumption choices and sharing information with peers or others. Starting with the concrete (items close at hand) in this way makes sense pedagogically and developmentally. The focus on consumption, however, is made complex (a good thing) by the approach's use of Actor Network Theory to examine reification in an STSE framework. Students investigate and create 'graphs' (concept map like flow diagrams) to trace circulation of people, corporations, and materials, etc. This helps move beyond the reification of consumption itself (as some sort of autonomous activity) and demonstrably leads students to questions of extraction, production, and marketing. This stood out most clearly in one of the chapters focused on the role of Coltan (columbite–tantalite) in cell phones, leading students to examine the processes of production and the production of warfare and instability to support their digital habits (it should be noted that these students did get a bit distracted in analyzing their cell phones by a simultaneously examining internet privacy). One of the clear successes of SWiA is that students develop a relational understanding of their consumption choices. The ANT approach (or a guided network approach, at any rate) has one other critical advantage, it helps students identify social movements with which they can ally and, in this way, moves (albeit slightly, as the connection to collective action/organizations seems very abbreviated) beyond mere individualized consumption activism. Students working on gene patenting, for instance, wrote letters of support with BCA, an anti-patent organization; students studying cars and consumption promoted a collective day of not driving. This is a critical step beyond neoliberalagency or good deeds as noblesse oblige.

However, in starting the investigation in this way, some issues are eliminated even while others are included. While beginning with the personal makes developmental sense, it immediately renders, at least initially, out of bounds two of my initial socio-scientific issues: drones as remote control warfare, and militarized police in the wake of increasing states of emergency/wars on 'terror.' While themes of consumption are at play in these issues, the consumption is done by the state, and citizens are merely the target of ruthless control and surveillance, surveillance that fuses Foucault's sovereign and disciplinary power, powers often invisible until attempts are made at resistance—the militarization of the police in Ferguson, MO only became visible when people tried to protest the killing of Michael Brown.

It is easy for many SSIs to become invisible because the news does not report them, or because they involve bodies that are not supposed to matter, to use Judith Butler's expression (1993). Henry Giroux (2009) notes (drawing on the Italian philosopher Giorgio Agamben) that racialized populations are positioned as disposable in racist capitalism. Racism, in essence, becomes a way of fragmenting the oppressed, through biologized forms of capital (Roediger, 1991/1999). Those with insufficient capital are deemed, in Agamben's phrasing, *Homo sacer*, outlaw or disposable humans (Agamben, 1998). This suggests a whole set of political relations that involve SSI that are not easily addressed if the terrain of activity is defined by consumption and choice. Beyond consumerism, neoliberal capitalism has produced new populations of slavery (acknowledged however briefly by the STEPWISE project on cell phones), new forms of economic exchange such as high-speed stock trading (Lewis, 2014), new roles and functions of the state, and new forms of social organization and resistance. STEM threads through all of these emergent forms and relations. Sometimes they are rendered visible in STEPWISE as students map out networks, but they are often too distant, too behind the scenes, or perpetrated on populations with less access to power than even those in our classrooms. Kim Fortun, drawing on post-coloniality, has charged that such blindness to the issues of the marginal, is inherent in Latour's cosmology (2014). For Fortun, the question of networks and the accounting of them must follow political economy, post-colonial critique, and racial analysis, not the mere 'following' of scientists, as Latour describes. While STEPWISE does premise political economy, the question remains what actants are revealed and obscured in this pedagogy? How might teachers shift the questions and scaffolding, i.e., the apprenticeship, so that these other SSIs, ones that are the occasions of many nascient social movements, become curricular objects?

These junctures of science and power are not outside of STEPWISE except that, once it is enacted in real schools with a forced scarcity of time, they are marginalized. The SWiA logic of starting with student consumption makes perfect sense, but to deepen the pedagogy STEPWISE needs to be freed from the linearity of the curriculum and spiraled so that more remote, more challenging and complex problems can be posed in subsequent turns of the curricular wheel. The idea is to move from consumption-at-hand to problems that impact people 'out of favor,' vulnerable, and less obviously entangled in the actor networks defining students' lives. While this is a small tweak in terms of theory, it is an enormous struggle in terms of the politics of schools, but before considering that politics I want to reframe the question of schools and ANT.

28.4 Pushing the Bus We Are Riding On

In the last section, I argued that certain SSIs that are powerfully significant yet not immediately or easily connected to chains of student consumption risk marginalization or even invisibility in the consumer exchange focus of SWiA. However, some SSIs or nodes in the actor-network of science that are even close at hand may be missed due to our apperception of science.

As I have elsewhere argued, schooling itself has to be seen as a technoscientific node, part of the networks that the STEPWISE students are drawing (Weinstein, 2008). Schools do not merely represent science, they also embody social understandings of science in both pedagogy and curriculum: from child-centered pedagogies to 'evidence based' methods, science is internal to the logic of modern schools. I have indexed this in my writing through the signifier science/education (the slash is a nod and a wink to slash fiction writers, i.e., fans of television, books, and movies who write their own fiction extensions of those texts and indicate readers the romantic entanglements involved through listing character-one/character-two in the summary of their stories). Science/education is the topos of all locations that combine scientific discourses and processes of education. Schools are organic-mechanical instruments of subject production. The neoliberal turn in schools parallels the neoliberal turn in laboratory science with a non-coincidental introduction of value-added measures of teaching, STEM-driven curriculum, and pharmaceutically managed children is beautifully captured in Clayton Pierce's *Education in the Age of Biocapitalism* (2013). These are hallmarks of neoliberal scientificity. Pierce has been able to connect value-added, STEM, ADHD medication, and logics of human capital that drive school policy and demonstrate how they emerge together from a particular habitus and interest in particular world-orders.

The 'new' neoliberal scientificity of schools has not been without resistance. Resistance to these reforms in the U.S. comes from groups like the Badass Teachers Association (BATS), Diane Ravich's Network for Public Education, and United Opt Out. Elsewhere (Sao Paulo, Brazil; Madrid, Spain; Santiago, Chile), the resistance has taken the form of mass protests and strikes by students and teachers against austerity politics. In the U.S., the discourse of these resistant groups is varied and includes a call for professionalism (via a nostalgia for former times), parent autonomy (often born out of conservative Christian ideology), and critique of overreach of either states or plutocrats-cum-corporations. In reframing questions of schools as questions of science/education, my purpose is to highlight that struggles, of which clearly STEPWISE is one, over purposes and loyalties of 'reliable' knowledge production, aka, science, also need to include consideration of the institution of schools themselves since schools turn out, when you draw the Actor Network Map, to be essential to the reproduction of the network itself. Science education cannot be a meta-discourse on science, since it is a discourse of science. The anthropologist of science, Emily Martin, borrowing from Berger and Luckmann, captured this problem when she described the anthropology of science as akin to 'pushing the bus we are riding on' (1998, p. 25)—though M.C. Escher's hands drawing each other also captures the self-referentiality of this work. Furthermore, this pushing, I would argue, even as it identifies corporate interest, research bias, rhetorics of science-in-action, and cultural doxas that found the enterprise of science, must struggle to hold onto reliable knowledge of the world because science allows a certain sort of sharing of frameworks, challenging of authority, and grounding for future action (Haraway, 1991). So the problem is how to put the very process or context of reflection, which is institutionally situated, into the maps that those reflections generate?

28.5 A Better Politics

I think the way out of this conundrum is simple: understand that representations are not depictions of reality but tools for action, or as Ian Hacking has phrased it, ‘we represent in order to intervene’ (1983, p. 31). By taking a pragmatist turn, the problem of metadiscourse is ‘evaded’ (pragmatism evades most of the foundational paradoxes of European philosophy (West, 1989)). The trick is to think outside of the usual dichotomies that frame schools as outside of the discourses they study. The entire school/real world dichotomy has to be replaced by one that sees struggles over schools as contiguous and part of struggles over commodities, capitalism, colonialism, etc. For example, a great deal of effort has been expended to integrate schools and communities. This includes opening schools for different uses after hours, and promoting ‘authentic instruction’ (Newmann, 1996) during the school day. One implication of my argument about science/education is that, in addition to these steps to integrate schools into the communities they serve, schools themselves must also be the site of politics over democratic participation. Schools are not inauthentic places, they are real places where students and teachers live. If STEPWISE is a curriculum designed to inflict insight into the relational webs of science and capitalism and to spur activism, that activism should have a special focus on its very setting. This means a renewed focus on not just the questions of democracy and participation in the classroom (though it certainly means that), but also over the nature of schools and their ever emergent expressions of capitalism.

What this leads to regarding SWiA is a second level of apprenticeship (again drawing on the SWiA linear model). This is a modeling of resistance. Teachers need to demonstrate what activism looks like through their own struggles over education neoliberalization. Modeling struggle involves sharing with students a wide variety of considerations: democratic participation, collective action, strategy and tactics, and ethics and respect. Neoliberal infection of education takes a wide variety of forms and is contested by numerous local movements, often not covered in the press, and I would not dare suggest particular alliances or tactics, as they vary as much within countries as between them. In my own work as a teacher educator, that has meant promoting the work of the local Badass Teachers Association (www.badassteacher.org) and local teachers’ unions to educate students about neoliberalism and its resistance. In doing so, I am trying to pave paths for continuing resistance through their professional lives. I have spoken up when colleagues try to promote neoliberal accountability programs that I believe research has shown to be invalid. I have also modeled for students my own small and large steps resisting 'performativity' (neoliberal accountability theater) and talked about what I have chosen to agree to within these regimes and how I have tried to create alternate channels while doing so. As science educators, we are good at modeling science ratiocination and experiments; we are less comfortable modeling rebellion and resistance necessary to make education more just (the goal of STEPWISE) and equitable.

My purpose in pushing STEPWISE here, as in my earlier critique of the way that SWiA can easily miss SSIs that are not immediately connected to students' actor networks, is not to deny the power of this model. The evidence is that SWiA has allowed students to (1) demonstrate relational thinking about materiality in their lives, (2) identify individual actions that promote change, and, from my critical theoretical point of view, more importantly, (3) act in concert with and in support of existing social movements to move towards change. At some point, SW becomes SWiA as it meshes and grinds against the rhythm (grammar) of schools, however. That mis-meshing is visible in Fig. 2.3 of Chap. 2 (this volume) with the tension between the circular and organic STSE education in the lower half and the coverage oriented curriculum at the top of the illustration organized as traditional units. Such impetus to coverage—critiqued even by such curricular conservatives as McTighe and Wiggins (2004)—is an essential product of the current ecology of schools (in which tests, habit, parental and peer demands combine to insist on more-and-shallow rather than less-and-deep).

In pointing to horizons, my argument is not that we accept them. Instead, I am arguing that neoliberalism and its limitations are also the pedagogical moment in which teachers can push back in the site of their own labor. In other words, the aforementioned ecology has to be shifted, through teacher and student activism within the school and that this activity is within the logics of STEPWISE as it resists neoliberal science/education. This is a pedagogy in the making, or to borrow from the early protesters of neoliberalism at the meetings of the World Trade Organization, this has to be an *autre-didactique* as their vision of democracy was *autre-mondialisation*; not against globalization, just as *autre-didactique* cannot be against science, accountability, or teacher authority; but each of those terms has to be radically redefined and taken back within a vision of social justice: very much STEPWISE's wise (WISE) vision.

References

Adams, F., & Horton, M. (1975). *Unearthing seeds of fire: The idea of highlander*. Winston-Salem, NC: J. F. Blair.

Agamben, G. (1998). *Homo Sacer: Sovereign power and bare life*. Stanford, CA: Stanford University Press.

Apple, M. W. (1990). *Ideology and curriculum* (2nd ed.). New York: Routledge.

Bencze, L., Carter, L., & Krstovic, M. (in press). Youth uses of actor network theory undermining societal consumerism. In G. Reis, M. P. Mueller, R. A. Luther, L. Siveres, & R. Oliveira (Eds.), *Sociocultural perspectives on youth ethical consumerism* (pp. xx–xx). Dordrecht, The Netherlands: Springer.

Bernstein, B. (1990). *The structuring of pedagogic discourse: Class, codes and control* (Vol. 4). London: Routledge.

Butler, J. (1993). *Bodies that matter: On the discursive limits of "sex"*. New York: Routledge.

Calabrese Barton, A., Tan, E., Turner, E., & Gutiâerrez, M. V. (2012). *Empowering science and mathematics education in urban schools* (pp. 1 online resource (215 p.)). Retrieved from http://JHU.eblib.com/patron/FullRecord.aspx?p=977911

Callon, M. (1999). Some elements of a sociology of translation: Domestication of the scallops and the fishermen of St. Brieuc Bay. In M. Biagioli (Ed.), *The science studies reader*. New York: Routledge.

Fassin, D., & Pandolfi, M. (2010). *Contemporary states of emergency: The politics of military and humanitarian interventions*. New York: Zone Books.

Fortun, K. (2014). From Latour to late industrialism. *Hau: Journal of Ethnographic Theory, 4*(1), 309–329.

Giroux, H. A. (1992). *Border crossings: Cultural workers and the politics of education*. New York/London: Routledge.

Giroux, H. A. (2009). *Youth in a suspect society: Democracy or disposability?* (1st ed.). New York: Palgrave MacMillan.

Grundy, S. (1987). *Curriculum: Product or praxis*. New York: The Falmer Press.

Hacking, I. (1983). *Representing and intervening: Introductory topics in the philosophy of natural science*. Cambridge, UK: Cambridge University Press.

Hall, S. (1986). On postmodernism and articulation: An interview with Stuart hall (Edited by Lawrence Grossberg). *Journal of Communication Inquiry, 10*(2), 45–60.

Haraway, D. J. (1991). Situated knowledges: The science question in feminism and the privilege of partial perspective. In D. J. Haraway (Ed.), *Simians, cyborgs, and women: The reinvention of nature* (pp. 183–201). New York: Routledge.

Haraway, D. J. (1997). Modest_Witness@Second_Millennium.FemaleMan©_Meets_OncoMouse™. New York: Routledge

Herzfeld, M. (1992). *The social production of indifference: Exploring the symbolic roots of Western bureaucracy*. New York: Berg: Distributed exclusively in the U.S. and Canada by St. Martin's Press.

Illich, I. (1971). *Deschooling society* (1st ed.). New York: Harper & Row.

Klein, N. (2004, May 2). *The rise of disaster capitalism*. Retrieved July 27, 2008, from http://www.thenation.com/doc/20050502/klein

Kliebard, H. M. (1986). *The struggle for the American curriculum: 1893–1958*. New York: Routledge.

Krstovic, M. (Ed.). (2014). Editorial: Student-led research informed action projects on socio-scientific issues. *Journal for Activist Science & Technology Education, 5*(1): 63. Available at: goo.gl/N00b3s

Lasch, C. (1965). *The new radicalism in America (1889–1963): The intellectual as a social type*. New York: Alfred A. Knopf.

Latour, B. (1987). *Science in action*. Cambridge, MA: Harvard University Press.

Lewis, M. (2014). *Flash boys: A wall street revolt* (1st ed.). New York: W. W. Norton.

Makki, N. (2008). *A naturalistic inquiry into preservice teachers' experiences with science, technology, and society (STS) curricular approaches*. Ph.D., Kent State University, Kent, OH.

Martin, E. (1998). Anthropology and the cultural study of science. *Science, Technology, & Human Values, 23*(1), 24–44.

McTighe, J., & Wiggins, G. P. (2004). *Understanding by design: Professional development workbook*. Alexandria, VA: Association for Supervision and Curriculum Development.

Mirowski, P. (2011). *Science-mart: Privatizing American science*. Cambridge, MA: Harvard University Press.
Morning, A. J. (2011). *The nature of race: How scientists think and teach about human difference*. Berkeley, CA: University of California Press.
Newmann, F. M. (1996). *Authentic achievement: Restructuring schools for intellectual quality* (1st ed.). San Francisco: Jossey-Bass.
Pierce, C. (2013). *Education in the age of biocapitalism: Optimizing educational life for a flat world* (1st ed.). New York: Palgrave MacMillan.
Roediger, D. R. (1991/1999). *The wages of whiteness*. New York: Verso.
Rouse, J. (1993). What are cultural studies of scientific knowledge. *Configurations, 1*(1), 1–22.
Soja, E. W. (1996). *Thirdspace: Journeys to Los Angeles and other real-and-imagined places*. Cambridge, MA: Blackwell.
Tyack, D. B., & Cuban, L. (1995). *Tinkering toward utopia: A century of public school reform*. Cambridge, MA: Harvard University Press.
Weinstein, M. (2008). Finding science in the school body: Reflections on transgressing the boundaries of science education and science studies. *Science Education, 92*(3), 389–403.
West, C. (1989). *The American evasion of philosophy: A genealogy of pragmatism*. Madison, WI: University of Wisconsin Press.

Chapter 29
In Which Ways Can (Science) Education Promote the Well-Being of Individuals, Societies and Environments?

Isabel Martins

29.1 Well-Being and Science Education

Conceiving of education as a threefold process, which involves students' humanization, socialization and enculturation (Charlot, 2006), calls for an understanding of how these dimensions can be articulated both at individual and collective levels. In STEPWISE, this conception is formulated in terms of clear aims concerning the promotion of social justice and environmental sustainability (Bencze & Carter, 2011) and is expanded so as to include a discussion of how education can inform socio-politicalactivism. Similarly to STSE and SSI related approaches, STEPWISE addresses relationships between science, technology, societies and environments as well as controversies by problematizing their effects in human life in at least three ways: Firstly, it by contextualising socio-scientific issues in both history and political economy. Secondly, through the consideration of axiological dimensions, in addition to the customary epistemological considerations, that characterise the discussion of the nature of science knowledge, scientific activity and its role in society. Thirdly by problematizing semiotic and rhetorical dimensions of sign making as a major component of learning. In doing so, it lays important foundations to explore students' construction of their subjectivity with respect to science in ways that foreground democratic participation and the betterment of life conditions.

STEPWISE has at its core the idea that education should promote the well-being of individuals, societies and environments. However, despite its centrality, the concept of well-being has not been extensively problematised in the original STEPWISE framework. The issue becomes relevant if we consider that different views of well-being mobilise aspects, as diverse as quality of life, welfare, satisfaction of desires and wealth. How can we then discriminate between such different views? To what

I. Martins (✉)
Universidade Federal do Rio de Janeiro, Rio de Janeiro, Brazil
e-mail: isabelmartins@ufrj.br

L. Bencze (ed.), *Science and Technology Education Promoting Wellbeing for Individuals, Societies and Environments*, Cultural Studies of Science Education 14,
DOI 10.1007/978-3-319-55505-8_29

extent is it possible, or even desirable, to define universal parameters in order to establish adequate threshold levels for each one of them? How would they apply to the individual, the societal and the environmental domains? And what would be the values to inform and constitute an account of well-being? From this perspective, in the next sub-sections, I explore some of these different meanings of well-being. The difficulty of arriving at a unique definition will be evident. However, some account of well-being is, nonetheless, necessary if not to distinguish its relationships with cognate concepts and how they constitute educational discourses.

29.2 The Polysemy Around the Concept of Well-Being

According to Mark McGillivray (2007), early conceptualisations of well-being were too narrowly focused and nowadays treating it as a multi-dimensional concept is not only more common but also more appropriate. It is not unusual to find the concept of well-being in connection with terms as diverse as satisfaction, quality of life, welfare, human/social development as well as with others, such as utility, desires, needs and capabilities. Such diversity not only points at relevant domains where well-being has been an object of theorising and debate but also to the different corresponding ways of characterising or, even, "measuring" it. For instance, views which equate well-being and development are usually related to an Economics framing. They are typically found in economic organisations' reports and are likely to be translated into statistics, data concerning people's access to goods and commodities, public health indicators and others such as Human Development Index (HDI) and Gross National Product (GDP). On the other hand, considerations that relate well-being to desire, utility, needs and capabilities usually draw upon (moral) Philosophy and are commonly expressed through logical analytical argumentation about actual or hypothetical dilemmas involving individuals and social groups. No less important are the psychological accounts of well-being, often linked with both physical and mental health aspects, as well as with subjective perceptions of happiness and fulfilment. The task of conceptualising well-being becomes even more problematic when one considers that all statistics, arguments and perceptions are likely to be traversed by cultural, political and historical aspects.

A number of theories were developed throughout history in attempting to answer such matters. Some of them are briefly sketched below.

29.3 Theories of Well-Being

In his account of theories of well-being, Roger Crisp (2016) explains that hedonist theories have their roots in ancient Greece and characterise well-being as an issue of greater pleasure over pain. However not all pleasure, or pain, is equivalent. Therefore, measuring this balance is not a simple matter and even Bentham's proposition that intensity and duration of experiences would constitute adequate parameters can be defied. In addition, as pointed out by John Stuart Mills, the quality of experiences, that is, the issue that some pleasures may be more valuable than others may represent a challenge to hedonist perspectives.

Such difficulties are also present in accounts of well-being in terms of desire (or preferences) satisfaction. Desire satisfaction is a complex matter because of the inherent complexity of desire, a concept that has bearings on issues that relate pleasure, good, reason, reward or a combination of these (Schroeder, 2015). In addition, the fulfilment of desires may happen at the level of experiences that are situated in time and space, like satisfying one's cravings for sweets, or in contexts that give respect to broader and long-lasting aspects in a person's life, like keeping within a healthy body weight range. This leads to the conclusion that desires may be ill-informed, a fact that is explained by Mozaffar Qizilbash (1998) through the hypothetical case of a person who enjoys smoking and is unaware of its harmful effects. There also are other problems with the desire satisfaction view, such as:

> (1) actual desires may have no bearing on the person's well-being (quality of life, etc.), but primarily on the lives of others […] (3) actual desires may not match with the agent's values, which may be reflected in 'second-order' desires; and (4) actual desires may be adaptive (Qizilbash, 1998).

This example, and indeed many other contexts concerning health education, illustrates the impact that knowledge about the object of desire may have on desiring such objects in the context of desire satisfaction theories of well-being. However, as we know, having (all) relevant information about the benefits of healthy eating and exercising does not prevent people being overweight. In other words, it may not always be possible to decide what is best for oneself, because being human also means facing and dealing with limitations, ambivalences and contradictions. However, the issue becomes more complex when our choices involve, both in their origins and consequences, harmful circumstances to other people. For instance, buying a mobile phone may be the end point of a chain of exploitation that begins with children working under degrading conditions. Being aware of that may not be sufficient to stop us buying these consumer goods as they also carry on them the appeal and, sometimes, the actual possibility of social inclusion and participation and organization in communities and in wider networks. This ambivalence is typical of contemporaneity and summons us all up to be critically-active about our role in the complex network of interests and desires and individual and collective interest should be problematised (especially when they clash).

Thus, criticisms of desire satisfaction approaches can help inform our analyses of the promotion of well-being as an educational goal. One aspect that may make it easier to address is the dependency between desires and knowledge of some sort. This way, they point to the connection between learning and well-being, here understood as a state attainable by individuals once knowledge—conceptual, processual or attitudinal—is acquired. Another aspect is the need to consider the impact that aspects which are not strictly rational, such as affect, emotions and may have on decisions concerning one's well-being. Yet another relevant aspect is the relative and changeable character of an individual's desires and the extent to which they depend on other individuals' desires, an issue that bears direct relationship to considerations about the pseudo neutrality of educational activities. Moreover, the idea that actual desires may be adaptive calls our attention to the mutually constitutive character between education and, in this case, the satisfaction of desires. Knowledge would then influence desires and, in this case, (the promotion of) well-being would be best seen as a process, as opposed to a state. In addition, the relationship between knowledge and desire is itself problematic as it is not possible to assert that we have all relevant information about objects of desire and, even if we did, our calculative abilities could not be sufficient to work out what is best (Qizilbash, 1998, pp. 59, 60).

An alternative view to the desire account for well-being, known as the capabilities account, is mostly identified with Amartya Sen's (Sen, 1993) and Martha Nussbaum's work (Nussbaum, 2000). For Sen, well-being, or quality of life, concerns a person's capabilities to achieve adequate functionings in society and not his or her access to goods or resources. This approach, which can be translated into an opposition between intrinsic and instrumental values, is problematic if the moral dimension entailed in the ways through which functionings can be achieved is not taken into account as, for instance, in the context of Qizilbash's example of a starving man who expands his capability of thieving to feed himself (1998, p. 54). In his theory, Sen widens the scope of agent as someone who is capable of bringing about change in a way that is coherent with one's values but does not offer a list of capabilities. Nussbaum's approach addresses this issue by calling upon Aristotelian views to pinpoint a list of ten capabilities that would make good human life and by separating capabilities themselves from the external conditions that would enable them. Despite the list's breadth of scope, such a view may still be in contradiction with pluralism; that is, the idea that there may be "a diversity of forms of good and that there is no one form of life that is best" (Qizilbash, 1998, p. 57).

The desire satisfaction account on well-being is also rejected on the grounds of the possibility of determining objectively valuable goods. Different authors have suggested different lists of values. A possible list, as suggested by Qizilbash's would include:

> a version of Griffin's view of well-being, with an extended list of prudential values, [...]: (A) minimum levels of nutrition, health, sanitation, shelter, rest and security; (B) certain basic intellectual and physical capacities and literacy; (C) certain levels of self-respect and aspiration; (D) enjoyment; (E) autonomy or self-determination ('positive freedom'); (F) liberty ('negative freedom'); (G) understanding; (H) significant relations with others and some participation in social life; and (I) accomplishment (p. 67).

Comprehensive though it may be, such a list is still open to the objection that its definition can be thought of as either subjective (relativistic and arbitrary) or objective (authoritative and paternalistic). The emphasis on autonomy may not be sufficient to resolve this matter and the argument that *"we must share some values that make any distinctively human life go well"* (Qizilbash, 1998, p. 71) must be brought to bear on the debate if we are to defend a list of prudential values.

29.4 Well-Being and (Social) Justice

The discussion made so far, although sketchy, suggests relevant links between theories of well-being and theories of justice. This is important, as both concepts are strongly linked in STEPWISE framework.

Both hedonist and desire accounts of well-being can be thought of as directly related to the concepts of 'utility', especially those held by classical utilitarians, such as Jeremy Bentham and John Stuart Mills. Utilitarian views are consistent with an idea of well-being as related to achieving a distribution of the greatest amount of good to the greatest number of people. Together with 'welfare', utility became the one of the most commonly-adopted foundations to think about well-being amongst both economists and utilitarian moral philosophers (Qizilbash, 1998, p. 52). In fact, utilitarians provide us with a simple clear-cut universal criterion—(the maximisation of) human welfare on which to base decision making. For instance, assessing the consequences of action in terms of its consequences for the majority can be regarded as a powerful logics to sustain not only individual action but also public policy. Take the example of large-scale vaccination. Despite the costs involved and arguments that vaccine administration in situations where there is no significant presence of pathogenic agents may put people at risk of side effects, in several countries BCG (Bacillus Calmette–Guérin) and MMR (measles, mumps & rubella) vaccines are either compulsory or recommended by health authorities. The cost-benefit balance tilts to the consequent benefits in order to justify the act. However, there are other cases in which utilitarianism may be challenged in terms of the dissociation between the moral and consequential dimensions of the act, as in the extreme example provided by Walter Sinott-Armstrong (2014) of transplanting organs of a healthy individual to save the lives of many critically-ill patients. Thus, both the appeal and rejection of utilitarian arguments may be seen as derived from their consequentialist dimension. Other sources of criticism of utilitarianism can be found in context of debates concerning access to and distribution of wealth amongst people as, for instance, in arguments against higher taxation of fortunes in order to fund social policies of poverty alleviation. Another version of utilitarianism, which relates to the interplay between freedom and equity, and has strong consequences for thinking about educational opportunities, is meritocratic utilitarianism. According to this interpretation,

> (1) educational policies are to be evaluated on the bases of their effects on economic productivity; and (2) educational opportunities are to be distributed (and designed) on the basis of economically valuable skill" (Howe, 2001, p. 205).

The consideration of a wider social historical political conjuncture of such discourses is necessary to signify them since they appear to be ubiquitous in curriculum policy documents throughout the world and can be linked to the commodification of education, understood as a feature of discursive changes brought about in late modernity (Fairclough, 1993).

Thus, the consequences of linking well-being indicators to attainment of targets and benchmarks related to educational standards must be examined in a critical way. Stronger links between performance and investments, as favoured by neoliberal policies, help create a cycle where educational indicators became goals to be reached as well as tools for the evaluation of public policies. Usually defined by international agencies, parameters such as universal healthcare, access to primary education and reduction of populations living below poverty threshold reflect, on the one hand, the current startling levels of inequality amongst nations and, on the other, a trap for countries that are yet more penalised when they fail to reach established levels. This reality can be more dramatic in developing countries, which suffer the pressure to aspire to be players in the current geopolitical scenario, but have to follow guidelines of multilateral institutions as part of loan agreements.

Other accounts of well-being and its relationships to educational aims, by contrast, are based upon humanistic perspectives and put forward the argument that "schools should be mainly about equipping people to lead a fulfilling life" (White, 2011, p. 1). In his book "Exploring well-being in schools", John White starts by questioning the bases of our beliefs about well-being and by relating them to religious and secular views as well as to issues concerning morality and reason. Throughout his book, he re-claims positive aspects related to different accounts of well-being in order to establish connections between, for example, pleasure and enjoyment to learn, or between achieving success and being an autonomous agent. He does that while problematizing the kind of places schools turned out to be in contemporary times, that is, environments where practices such as the learning of disciplinary knowledge, examinations and rankings, and preparation for the employment market have acquired exaggerated importance. These considerations call our attention to the direct ways through which a number of opposing values such as consumerism and wealth, on the one hand, and moral goodness, pleasure and worthwhile activities have been shaping our understanding of educational goals and aims.

Back to examining relationships between well-being and justice, it is worth remembering John Rawls' strong objections to utilitarianism (1971). In an account that attempted to conciliate freedom and equity, the author sustains that justice is the first virtue of social institutions. Rawls' claim for a universal theory of justice is rejected by those who believe that culture, traditions and belonging to social groups are important elements in the characterisation of pluralism in society and in shaping conceptions of and standards to assess justice. It is possible to place those who criticize the liberal premise that "politics should not be concerned solely with securing the conditions for individuals to exercise their powers of autonomous choice, as we also need to sustain and promote the social attachments crucial to our sense of well-being and respect" (Bell, 2016) under the label of communitarians.

The arguments between utilitarians and communitarians are also key to inform the complex debate concerning relationships between multiculturalism and citizenship (Kymlicka, 1995).

29.5 Measuring Well-Being

The different conceptualisations of well-being may account for the diversity of ways to measure it. Whilst some studies favour individual subjective perceptions about one's feelings of happiness and achievements as adequate indicators, other studies tend to objectify parameters through which to characterise both individual and collective well-being, such as income and GDP. Qualitative or quantitative, subjective or objective, all approaches to measurement have nonetheless limitations and flaws. This has led to multi-dimensional approaches to measuring well-being so as to encompass aspects as diverse as knowledge, health, wealth and freedom. One example of attempts to capture multiple relevant dimensions to well-being is the Human Development Index that combines parameters such as life expectancy, number of years of schooling and gross national income per capita. However, according to John Finnis (1980*apud* Akire, 2002), although the dimensions are "self-evident, in that they are potentially recognizable by anyone" they can also be seen as "incommensurable, in the sense that all of the desirable qualities of one are not present in the other; irreducible, as there is no one denominator to which they can be totally reduced; and non-hierarchical, since at any point in time anyone dimension can seem to be the most important".

29.6 Well-Being, Participation and Decision Making

The STEPWISE agenda enhances and complexifies targets related to the promotion of well-being in so far as it seeks to encourage and support critical actions that are identified with aspects of political life such as participation and decision making.

The issue of participation has been problematized in terms of relationships between science education and democracy in frameworks that describe participation in schools (Levinson, 2010) in terms of social, epistemic and pedagogical dimensions. At the core of the 'science education as praxis' framework, of which STEPWISE is an example, we find references to communal interests as well as to the framing and re-framing of contextualised problems. However, the sheer consideration of the diversity of communities and of the range of features that may bind them together warns us that cross-community dialogue is not a simple straightforward goal to achieve.

Views of well-being that address relationships between individuals and social groups can be productively used to tease out the aspects involved in the discussion of the role of education in the empowerment of individuals and the emancipation of social groups for political action. For instance, there may be profitable dialogues between communitarianism and Paulo Freire's critical humanistic pedagogy (1970), in which the awareness about social elements that constitute our lives is seen as a key element for the argument of education as liberating praxis. Likewise, perspectives that draw upon the Foucaultian concept of biopolitics have eloquently criticised the ways through which power relations frame and shape people's identities and perceptions of needs. If, on the one hand, the desire account entails an objective character, by equating well-being with achievement of success of some sort, on the other hand, it does not problematize the quality of this success, in particular, whether it should be best thought of as depending on intrinsic values, like freedom and satisfaction, or on instrumental values, such as goods and material resources. Apart from that, identifying the origins of needs and the nature of choices are crucial issues if we are to identify degree of autonomy we exercise in making our choices in (political) life, especially those concerning issues related to domains that are highly valued and legitimised in society, like science and technology.

The question of autonomy also reverberates in schools in terms of demands for changes that are placed by STEPWISE and related approaches. They can be located both at the level of the micro-politics of the classroom as well as between other individuals and macro-social groups as diverse as principals, educational researchers and policy makers. The former level can be exemplified in curriculum terms, for instance, with reference to ways through which contextualisation must be effected in science learning. Acknowledging historical, discursive and cultural aspects of contexts goes beyond superficial references to students' daily lives and need-to-know pedagogies. It involves problematizing the nature and dynamics of the relationships between teachers and students and the establishment of a didactic contract regarding aims, expectations and needs for knowing. The latter calls upon rethinking aspects concerning curricula, systemic evaluation and classroom assessment practices. It is important to investigate the extent to which such practices end up configuring curricula and assessment as instruments of coercion and not of emancipation. Such tension has been increasingly challenging, for instance, for Brazilian school teachers who must follow guidelines of local Educational Authorities with respect to curriculum planning that regulate the selection of contents and methodological approaches, choices of educational materials and the adoption of compulsory assessment practices. Together they provide the basis for periodic evaluations of students' performance, the results of which determine school funding and teachers' salary bonuses. In such a context, it is not easy to compromise between autonomy and respect to diversity, on the one hand, and obedience to rules that seem necessary to meet targets and standards, on the other.

29.7 Implications for STEPWISE Research and Practice Agendas: Well-Being as a Dialectic Generative Process

The discussion made so far questions the possibility of promoting well-being outside practises defined by dialogue and co-responsibility. One reason would be the risk of dismissing diversity and eliminating pluralism as an important element in the construction of identities and as source of reflection and alterity. Another reason would be the reinforcement of the asymmetry between academic/disciplinary and social/cultural contexts. This is not to say that any view is as good as any other. It is not an argument for relativism but, instead, a call for deeper understanding of processes through which some views become hegemonic and acquire differentiated degrees of social legitimation.

Analyses of challenges faced by science education (and science educators) in contemporaneity demand an analysis and a tentative conceptualisation of contemporaneity itself. This is especially difficult because contemporaneity is an inherently complex and mutable set of force relations/power struggles, which has expressions in different fields of societal life (economic, political, social etc.). Another problem is the fact that it is impossible to adopt a bird's eye view on contemporaneity, since we are fully immersed and implicated in the processes of production, enactment and consumption of contemporary discourses. Nonetheless, an analysis of key features of contemporaneity can potentially illuminate relevant aspects of science education goals (finalities), policies and strategies in different levels. Critical discourse analysis may be an apt framework for such an analysis, in so far as it explores dialectical pairs, such as structure/agency, colonisation/appropriation, reflexivity/ideology (Chouliaraki & Fairclough, 1998) in promoting articulations between local experiences and socio-historical dimensions. The idea would be, therefore, to generate an overview of the ways through which relevant aspects of contemporaneity (e.g. individualism, efficiency, competitiveness, space-time 'compression', technologisation of social life etc.) are represented in science education.

Critical analysis of dialectical relationships in discourse may also help break with asymmetric views. Those who seek to promote well-being are impacted not just by reflexive dimensions of their actions but also by the reconfigurations in the social relations that can result of emancipation and of extended possibilities of participation and knowledge production of targeted groups. Promoting well-being can be, in this sense, a dialectical generative process of both discourse change and social change.

Another point that can be made in the light of the discussion the need for adopting a political philosophical outlook on well-being in order to problematize relationships between education and different models of socio-economic development in contemporary society (neoliberalism is one example).

Throughout the chapter, I have sketched relationships between well-being and moral and political arguments and argued for the relevance of Moral Philosophy as a relevant foundation upon which STEPWISE can be grounded, expanded and evaluated. The connections between well-being, social justice, power and wealth that are suggested in the original STEPWISE framework can be further elaborated in terms of contemporary critique on theories of justice.

Furthermore, the articulation of social theory and discourse, as present in critical discourse perspectives, were suggested as a powerful analytical tool to examine such connections as present in educational literature, multilateral documents and curriculum materials. The irreducible nature of the relationships between discourse and society may also help foreground nuances in global accounts of ways through which education, in general, and science education, in particular, has been recruited as a major component of a hegemonic project of society based on the values of capital.

References

Bell, D. "Communitarianism", The Stanford Encyclopedia of Philosophy (Summer 2016 Edition), Edward N. Zalta (ed.), URL = https://plato.stanford.edu/archives/sum2016/entries/communitarianism/

Bencze, J. L., & Carter, L. (2011). Globalizing students acting for the common good. *Journal of Research in Science Teaching, 48*(6), 648–669.

Charlot, B. (2006). A pesquisa educacional entre conhecimentos, políticas e práticas: especificidades e desafios de uma área de saber. *Revista Brasileira de Educação, 11*(31), 7–18. doi:10.1590/S1413-24782006000100002.

Chouliaraki, L., & Fairclough, N. (1998). *Discourse in late modernity*. Edinburgh, UK: Edinburgh University Press.

Crisp, R. "Well-Being", The Stanford Encyclopedia of Philosophy (Summer 2016 Edition), Edward N. Zalta (ed.), URL = https://plato.stanford.edu/archives/sum2016/entries/well-being/

Fairclough, N. (1993). Critical discourse analysis and the marketization of public education: The case of universities. *Discourse and Society, 4*(2), 133–168. doi:10.1177/0957926593004002002.

Finnis, J. (1980). Natural law and natural rights. Oxford, UK: Clarendon Press.

Freire, P. (1970). *Pedagogy of the oppressed*. New York: Herder and Herder.

Howe, K. R. (2001). In defense of outcomes-based conceptions of equal educational opportunity. In P. H. Hirst & P. White (Eds.), *Philosophy of education. Major themes in the analytic tradition. Volume III Society and Education*. London: Routledge.

Kymlicka, W. (1995). *Multicultural citizenship*. Oxford, UK: Clarendon Press.

Levinson, R. (2010). Science education and democratic participation: An uneasy congruence? *International Journal of Science Education, 46*(1), 69–119.

McGillivray, M. (2007). *Human well-being*. New York: Palgrave MacMillan.

Nussbaum, M. (2000). *Women and human development: The capabilities approach*. Cambridge, UK/New York: Cambridge University Press. ISBN 9780521003858.

Qizilbash, M. (1998). The concept of well-being. *Economics and Philosophy, 14*, 51–73. doi:10.1017/S0266267100004934.

Rawls, J. (1971). *A theory of justice*. Cambridge, MA: Harvard University Press.

Schroeder, T. "Desire", The Stanford Encyclopedia of Philosophy (Summer 2015 Edition), Edward N. Zalta (ed.), URL = https://plato.stanford.edu/archives/sum2015/entries/desire/

Sen, A. K. (1993). Capability and well-being. In M. C. Nussbaum & A. K. Sen (Eds.), *The quality of life* (pp. 30–55). Oxford, UK: Clarendon Press.

Sinnott-Armstrong, W. (2014). *Consequentialism. The Stanford Encyclopedia of philosophy* (Edward N. Zalta, Ed., Spring 2014 edn.). Retrieved October 13, 2015 from http://plato.stanford.edu/archives/spr2014/entries/consequentialism/

White, J. P. (2011). *Exploring well being in schools. A guide to making children's lives more fulfilling*. London: Routledge.

Chapter 30
'STEPPING' Toward a Critical-Activist Science Education: Dialoguing Subjectivity, Social Ontology and Multiplicities

Jesse Bazzul and Shakhnoza Kayumova

Jesse: Shakhnoza Kayumova and I met in the late summer of 2014 as overwhelmed new assistant professors at the University of Massachusetts, Dartmouth. We were shocked to find we shared an interest in critical theory and science education (yes, what are the odds!)—specifically how science education is involved in subjectification practices that enable oppressive social orders. The following metalogue (Roth & Tobin, 2004) represents our ongoing discussion about what critical activist science education might look like. We use Larry Bencze's STEPWISE framework as a springboard into these nebulous, yet productive, considerations. In short, we conclude that the ontological framework set out by Bencze (Bencze & Carter, 2011) allows for a subversion of current science education practices that shape subjectivities and ways in which students approach science and the world. We begin our metalogue with subjectivity and then discuss social ontologies, multiplicities, and what these could mean for a critical activist science education. As shall be seen, this dialogue has also provided a space to exchange and deepen our understanding of the work of Michel Foucault, Gilles Deleuze and Felix Guattari. We have chosen to write this chapter as a metalogue, which can be thought of as dialogue that reflects upon theory, methodology, or research data, not just to retain our individual voices concerning critical scholarship, but to synthesize our ideas and thoughts together. Writing in metalogue allows us to come out the other side changed, enhanced, and challenged by each other. Metalogue also challenges traditional forms of writing

J. Bazzul (✉)
Faculty of Education, University of Regina, Regina, SK, Canada
e-mail: Jesse.Bazzul@uregina.ca

S. Kayumova
Department of STEM Education & Teacher Development, University of Massachusetts-Dartmouth, Dartmouth, MA, USA
e-mail: skayumova@umassd.edu

L. Bencze (ed.), *Science and Technology Education Promoting Wellbeing for Individuals, Societies and Environments*, Cultural Studies of Science Education 14,
DOI 10.1007/978-3-319-55505-8_30

where an abstract unity, or third-person, speaks with an objective, and often authoritative, voice; thereby allowing us more agency to move beyond traditional representations of writing and scholarship.

Shakhnoza: Bencze's framework is based on five principal components of engagement with learning science, two of which are based on conceptual understanding of science, while the other three are on development of skills, practices, and actions. Conceptually, students begin with an understanding of social and environmental issues (STSE Education), and move towards learning about products of science and technology (e.g. knowledge, concepts). In STEPWISE, students are first positioned as *producers* of that knowledge through sociopolitical actions and locally-situated research, and in doing so incorporate pre-existing scientific knowledge into their learning, developing skills, attitudes, and practices of science as they move along. What is useful about this approach is that it enables students to inquire into their lived realities and use science as a space to take actions towards social, environmental, and political change. One question I have about this framework is: what kinds of subjectivities are made possible for students? I think this is where we should focus our metalogue.

Jesse: It's probably a good idea to outline what we are thinking about when we say "subjectivity." Beginning with a basic conception is helpful for opening a larger conversation. Subjectivities can be thought of as the various, ever-changing, values, beliefs, outlooks, emotions, convictions, and ideologies held by a 'subject' or individual. In terms of Bencze's STEPWISE framework, I feel we are speaking about affecting a shift in the subjectivities of students of science. Traditionally speaking, the outlooks, beliefs, have been relatively free of attitudes like caring, civic-mindedness, etc.: STEPWISE moves these subjectivities toward a political consciousness. From here, scholars and researchers can ask all kinds of important questions in education, including what kinds of outlooks, beliefs, and identities institutional practices and discourses of schooling constitute (Bazzul, 2012). This question focuses on how practices and discourses shape subjects and subjectivities (Bazzul, 2014a). However, questions of subjectivity can also begin with outlooks of students. For example, a phenomenological approach may ask how students experience science or how science learning involves synthesis with the life worlds of students (Roth, 2014; Bazzul, 2014b). Studies that deal with worldviews (e.g. 'scientific') or teacher beliefs, culturally relevant pedagogies, and epistemological debates are also studies into subjectivity. Scholarship into how subjectivities are constituted has a lot to say about the social, political, and cultural contexts of science education practice and research. I feel science education can be a place for fostering an activist, sociopolitically engaged subjectivity—what can be called a political subjectivity (see Rancière, & Corcoran, 2010; Bazzul, 2015).

Shakhnoza: My understanding of subjectivity is couched in feminist readings of poststructural theory (Butler, 1995; Weedon, 1997). Contrary to humanist assumptions, poststructuralism decenters the subject as the main source of meaning making. Many feminist scholars use Foucauldian theory of the subject to argue that the subject "constitutes him/herself through practices that are basically related to power and knowledge" (Foucault 1988, p. 10). To make it clear, the notion of subjectivity in poststructuralism is not similar to traditional positivist binaries of subjective vs. objective. Instead, the concept suggests that the subject is a part of socio-cultural, historical, and discursive production. It is more in line with notions of identity, without implications of inherent traits. Implications of this view on the activist and social justice oriented science education research and practice include that subjects, whether teachers, students, and/or researchers, are not the sole actors behind their act (practice). In other words, as subjects we are continually produced and formed within practices and knowledge systems—social, discursive, cultural and economic grids of intelligibility—of which we are a part (Butler, 1995). Processes of formation by which we become intelligible to even ourselves and others are contingent on our social, cultural, economic, gendered, and even raced positions. Although not fixed, these locations determine both how we read the world and how we are read by others (Weedon, 1996).

Jesse: In a similar way, my interest in subjectivity began with poststructural feminist readings of Foucault's work from scholars like Bronwyn Davies (2006) and Judith Butler (1997). One important consideration that comes from structuralism and poststructuralism, and one I think science educators have not considered enough, is how a subject emerges (is produced) through technologies of power and official discourses of science and education. This subject is repeatedly produced through subjectifying practices and discourses that distribute effects of power. But as Foucault (1997) argues, power is relational and, therefore, resistance is always possible. A major critical project for science education involves tracing the limits of subjectivities produced through discourses and practices, with the intention of ways of relating to each other, the world, and ourselves. This latter aspect is where a 'political subjectivity' comes into play, something that remains a perceived limitation of poststructuralist approaches to subjectivity and politics.

Shakhnoza: Jesse, you touched upon a critical point. Some critics argue that in poststructural thought agency is ripped off of humans. But, as you said, what it does is to make visible how power is always part of practice. I agree that science education is one of the fields that can take advantage of poststructural theory and readings of feminist poststructuralist scholars to understand effects of subjectification practices and discourses on teachers, students, and on us, as researchers. And I cannot agree more with what you said that, "[a] major critical project for science education involves tracing the limits of the subjectivities produced through discourses and practices with the intent of reimaging ways of relating to each other, the world, and ourselves." This call also goes along with an argument that for teachers, students, and even researchers to "problematize' what they are, what they do, and the world in which they live…to transform themselves…to change themselves in their singular being…into arts of existence," (Foucault, 1985, p. 10) we need "a robust new theory [or theories] of the subject as a multi-layered entity that is not unitary and still capable of ethical and political accountability (Braidotti, 2006, p. 144). For instance, in my own research in science education, feminist poststructuralist readings of the subject have been helpful for me to (de)center teachers and students as the sole agents of action/practice and enactments. It helps me to make visible discourses and intricate power relations on teachers', students', and our own, subjectivities. Foucault argues that locally, culturally, socio-historically, and materially situated conditions can enable and/or constrain individuals through ways they are brought to think, speak, write, measure, and know. When we think about teaching and learning science, we must understand that what is given falls onto 'grids of intelligibility' that are historically, politically, and culturally situated. What STEPWISE does differently is that it does not introduce (subjectify) students to knowledge and practices that are 'already there', instead it starts with situated contexts where students work to create their own knowledges and practices. So, then a task of researchers becomes not tracing the progression, but to locate those instances when practices, such as STEPWISE, become enabling or constraining of certain modes of action, and what are the implications of those instances on the subjectivities of students. When we take this view into account, we always need to question notions of the critical subject, critical agency, and critical activism. What kind of subjectivities are these notions bringing along, what is made possible in this discourse and impossible, what kind of subjects are enabled and/or disabled?

Jesse: One way of looking at discourses of science education is in terms of how they work to constitute or give importance to particular subjectivities. This involves investigating day to day science education practices regarding the nature of power relations in science education, including how these power relations are embodied or come to operate on bodies. Foucault (1977, 1982) makes the point, implicitly in *Discipline and Punish,* and explicitly in the *Subject and Power*, that 'modern' subjects are not just subject to technologies/discourses of power, but are produced by (are the very effect of) these technologies of power. An implication is that whatever subject positions exist, they are both subject to technologies of power (subjectification) and simultaneously a product of these technologies—which does not mean subject positions constituted by technologies of power may not also be productive and/or valuable. A historical example of this 'double-bind' that relates to science is how 'homosexual/heterosexual' identities were produced through discourses and practices of hygiene, biology, and health in the nineteenth century (Foucault, 1980). These categories emerged out of normalizing discourses, yet were taken up in ways unintended by these discourses. My research so far has looked at 'official discourses' in curriculum and policy (Bazzul, 2014a). Judith Butler (1997) casts the modern rational subject as a site of political contestation. For Butler, the subject is produced again and again through repetitive subjectification practices that constitute the subjects' relation to themselves, others, and the world. Resistance is found at the *site of repetition*; that is, where these practices are repeated. This is the value of STEPWISE; it disrupts depoliticized science lab activities at the site of repetition and replaces them with community-oriented science work. An activist science education must continually maintain a critical stance on how subjects are produced through practices and discourses of science education. Some questions have yet to be engaged adequately, however. How does viewing the subject as a site of contestation allow students and teachers to challenge oppressive discourses and practices? How do teachers and students go about reshaping subjectivities?

Shakhnoza: Your questions in relation to STEPWISE remind me of our recent work (Bazzul & Kayumova, 2016) titled "Toward a Social Ontology for Science Education: Introducing Deleuze and Guattari's assemblages". Working with Deleuze and Guattari's concept of arborescent and rhizomatic structures in relation to Foucault's notions of the subjects allowed us to analyze how science teachers, students, and researchers are phenomenon of already established science practices and classroom discourses, what Deleuze and Guattari might call arborescent structures. An issue with arborescent structures, such as official and prescribed curricular practices, is that they channel things in one direction. And the example of STEPWISE as a community-oriented science is a good example *of a rhizomatic,* multiple, dynamic, and open-ended assemblage with possibilities to disrupt dominant and unidirectional science practices by replacing them with socially engaged activities. At the same time, I was wondering if you can give some examples how precisely Bencze's STEPWISE program works to 'disrupt', what you named as, "depoliticized science" lab activities, at the different sites of repetition. And how does precisely the community oriented science and activist science education engage the question of subjectivity. Wouldn't you say that the very notion of producing subjects for certain purposes, whether it is for critical science activism, and/or traditional science, both are inherently aimed at structuring individuals rationalities in one direction?

Jesse: The disruption is quite literal and not unlike what many science educators attempt to do in their daily practice. Bencze's framework advocates re-centering science investigations around community problems or issues that affect students and, in that sense, disrupt unidirectional aspects of traditional state curricula and other technologies of control. Some examples include investigating issues of sustainability, risk, and social inequality through project based learning (Bencze, Sperling & Carter, 2012). Along with repeated practices, mastery of skills is vital to production of subjects (See Butler, 1997; Althusser, 1998). Science education subjectifies students through cookbook labs that do not engage social/community interests and having them master a particular set of science skills unrelated to community, social contexts. While we might imagine that science students become "more free" as they adopt these skills from science activities, this acquisition also furthers their own subjection. Bencze's framework disrupts these subjectification practices by literally replacing depoliticized activities with activities that *begin* with taking social action, thereby having students master sociopolitically relevant skills as well as those typically related with science, inference-making, observation skills, etc. These sociopolitically relevant skills better fit a *politicized subjectivity*, what I think is a productive extension of the critical, structural/poststructural approach to subjectivity. Your question about the production of subjectivity is, I think, one of those standpoint questions that is difficult to answer. One way to answer it is to say that subjectification practices are ongoing – so the question on ethical and political terms will always be, 'How do we wish to live in relation to others and ourselves?'

As we engage these questions at home, at work, in spiritual communities, or in schools, we are constituting/challenging subjectivities with particular aims. I do not believe this is ever a 'neutral' or totally 'free' terrain. However it does bring up a fissure in what constitutes resistance—is it finding an 'outside' or does it involve radically altering apparatuses, or both? Working for both emancipation and creative, rhizomatic expression is important, and I believe institutions have some role to play. If the subject is a site of political contestation, it means it can be/is being reworked. All in all, the reworking of subjectivities towards goals of social justice and equality must be an active project.

Shakhnoza: These are all very interesting and fascinating ideas. But at the same time I must admit that Bencze's framework for critical social activism sounds more of an activist than a critical approach. Or if I re-frame what I have said, it seems to be situated in a more progressivist camp than a critical camp (by the way I am not implying a dichotomy, just the existence of some theoretical tensions). Therefore, I want to caution us not to approach Bencze's work as given, and start with questioning assumptions of this work. And as we work with theory, it is important not to use poststructural subjectivity as another framework to replace and/or to add on to existing framework. For instance, Deleuze, in his dialogue with Foucault, defined a theory "exactly like a box of tools. It has nothing to do with the signifier. It must be useful. It must function. A theory does not totalise; it is an instrument for multiplication and it also multiplies itself" (Foucault & Deleuze, 1977, p. 208). So, using insights of postructuralist subjectivity in Bencze's work is to be able to apply the theory not for totalizing purposes, but for the theory itself to gain multiplicities in its application to practice. So working with a theory means to unpack taken for granted assumptions, subjectivities, and discourses of a practice, in order to make room for more multiplicities.

Jesse: Gathering some assumptions from Bencze's writing (see Bencze & Carter, 2011), relations of power that seem most pressing to him relate to global capitalism, neoliberalism and neo-conservatism. Your separation of the terms *critical* and *activist*, I loosely think of as the 'negative' and 'positive' aspects of a social justice, community oriented science education. Foucault (2003a) is careful to stress that critique exists only in relation to 'something else'—and so it is the reconstitution of 'something else' that could be the focus of activist work. Critique works to tear down hierarchies and subvert taken for granted practices, material distributions, and discourses; and politicized activism works to build different relationships, distributions, networks, structures, etc. The entwining of critique and activism can result in the 'gaining of multiplicities', to use the language of Deleuze and Guattari's rhizomes and lines of flight. So how would you describe a critical activist project along these lines?

Shakhnoza: Well, for me, to contemplate this question, it is important to go back to what Deleuze said; that is, that "[a] theorizing intellectual, for us, is no longer a subject, a representing or representative consciousness. Those who act and struggle are no longer represented, either by a group or a union that appropriates the right to stand as their conscience. Who speaks and acts? It is always a multiplicity, even within the person who speaks and acts. All of us are 'groupuscules.' Representation no longer exists; there's only action-theoretical action and practical action which serve as relays and form networks" (Foucault & Deleuze, 1977, p. 207). This quote reminds me that people who are supposed to be a part of this discourse—teachers, parents, students, and other minority groups—are already the ones who are usually excluded from these critical conversations. And, if there is a certain level of representation from different social groups in these conversations, it is usually at the very miniscule level. As Deleuze (1994) contends, "[r]epresentation fails to capture the affirmed word of difference. Representation has only a single center, a unique and receding perspective and in consequence a false depth. It mediates everything, but mobilizes and moves nothing" (p. 67). Taking into consideration notions of Deleuzian representation, and asking what is included and excluded in critical science activist work, means we need to be aware of our own privileged positions and make room in our conversation, not only for our own "voices" and "actions" to be visible and represented, but also of those who already do critical and activist work in their day to day practices. This is precisely what STEPWISE is doing when it allows students to have a role in knowledge production. To say otherwise, the critical and activist science is already there. However, those activities and practices are not privileged, not visible, and often times subjugated. I argue when we make those *already there* critical and social work visible, that's when critical science activism can become a "movement, for its part, implies a plurality of centers…a tangle of points of view, coexistence of moments…" (Deleuze, 1994, p. 67).

Jesse: Yes, it involves making those who do not count or were not visible count—and it involves a co-existence of movements and struggles as Deleuze says. Gaining multiplicities seems similar to gaining possibilities and I think Deleuze and Guattari (1987) are saying that the ethical-political move involves differentiating oneself or allowing for this differentiation. However, I think they would also say that, just because multiplicities (and lines of flight) exist, it doesn't mean they are necessarily equitable, good, or safe. It reminds me again of the criticism of poststructuralism's political apathy and or lack of normative political grounds. Foucault claims he does not let politics inform his analysis and instead asks what politics has to say about problems it faced (Foucault, 2003b)—which is what makes his critical methods so powerful! However, the powerful critical toolbox Foucault and other poststructuralists provide, I feel, needs to be with active (activist) political projects. I find Hardt and Negri's (2000) work useful in how it outlines revolutionary potential of hybrid subjectivities through linking-together of social struggles (in sexuality, labour, against white supremacy, etc.). According to them, this involves 'listening to' subjectivities that find it difficult to live within the sociopolitical order as they have a better understanding of how power is exercised.

Shakhnoza: I can see how poststructuralism could be accused of having no normative grounds for politics. But, to me, that does not mean that a poststructural perspective is against sociopolitical action. Instead, a poststructural stance reminds us critical scholars that we should always be wary of our own practice. Regardless of how well-intended is the social activism, we always need to be cautious of replacing one normative discourse with another normative discourse. To remember norms can easily become essentializing, totalizing, and alienating overtime and re-produce power structures. I am not sure how much of what is available to us as a site of struggle is not another mechanism of power. And, you are actually giving a similar argument by using the example of multiplicity. Multiplicity seems to be everywhere, and in all of our discourses, and yet it is not achieved at the practical level. I do not think that we "know" yet enough what is "multiplicity" and/or "equality," in order for us to be able to fight for it. To me, we can set up ourselves of another slippery slope of what Freire called "oppressor vs. oppression," and or re-producing new subjectivities, and/or new discourses, without adequately challenging power relations inherent in our day to day practices of which I would argue each of us is already a part. As Sara Ahmed (2007) says, even the "project that aims to dismantle or challenge the categories that are made invisible through privilege is bound to participate in the object of its critique. [Therefore,] we might even expect such projects to fail, and be prepared to witness this failure as productive" (p. 149).

Jesse: I think Ahmed may also be saying that critique and subsequent redirection of power is simultaneously participation in original problematics. I agree that a critical attitude means being skeptical of all things 'necessary'. I think this is why political struggles need to be seen as a horizon—similar to the immanent horizon of Deleuze and Guattari to which you refer in your work. As critical activists, we should think of 'building'—but we can be reflexive in how we organize and build the kind of future we want. We can use temporal concepts, structures, and even 'non-place' universals—all of which are best seen as tentative, because time and again we find new exclusions—and demand that these abstract, temporal universals, institutions and structures include them! (see Butler, Laclau, & Žižek, 2000, p. 39). But you are right, it is a very slippery slope! There is also need for a material, ontological dimension to this critical project. Taking some of the themes from our discussion so far, let's move the conversation to how ontological, material questions need to be a part of a critical activist project.

Shakhnoza: The approach to critical science activism from an ontological perspective is intimately related to multiplicities. As you and I said, multiplicities are ontologically *already there*. So, to me, an aim of critical science activism is not, necessarily, to produce something new but, instead, to make visible things that are already there. It is a political project because, through subjectification practices, our multiplicities are continuously subverted. The repetitive practices are acting as centrifugal forces to achieve the sameness and normativity. Anything outside of the established norm and repetition becomes pathological, not good, and problematic. Making visible existing ontologies is about legitimizing already present materiality and resources and to recognize their affective dimensions. However, this recognition is not to benefit the capitalist or neoliberal base It is not another sort of capitalization on the ingenious power; instead, it is providing space for differences to be empowered. I find feminist and postcolonial scholars to be exemplary in this sense. They have already been openly political about gender and colonialism issues, without playing an 'objective' and/or 'relative' card; yet, they opened up spaces to deconstruct normative discourses through listening, including, and dialoguing with others and being aware of their own shifting power relations.

Jesse: What I also find inherent to postcolonial and feminist work is the power to remake. Multiplicities are the substrate of social life, and Deleuze and Guattari (1987) show very beautifully how these exist irrespective of humans. For me, the critical project is more than an uncovering of multiplicity but rather to actively change power relations that also simultaneously produce multiplicities. I also do not think there is an priori multiplicity that can be separated from subjectification. Deleuze and Guattari emphasize a difference in 'actual' and 'virtual' ontologies and multiplicities encouraging us to see the latter as immanent possibilities. Bencze's STEPWISE framework does bring in ontological questions but does not have a clear theory about how an activist science education would engage on a materialist or ontological level. Possibilities for reconfiguration of material realities can be found in Deleuze and Guattari's concept of *the assemblage*, socially imbedded ontologies complete with their discursive and material components (in Deleuze and Guattari's terms, assemblages of enunciation and machinic assemblages). That is, seeing human groups as co-extensive and interconnected with material entities such as bee populations and bodies of water (Delanda, 2006; Bazzul & Kayumova, 2016). These assemblages have *emergent properties* that arise from connections between components. In this way, we can create emergent possibilities for social and environmental change in a merging of worlds, biotic, abiotic, non-human, etc.

Shakhnoza: I completely agree with this argument. Similar to Latour's Actor-Network-Theory (ANT), the space is a part of discursive, bodily, temporal and material network of entities. And not only discourse but also material entities are important parts of a science practice. When different entities come together during a science practice, there is already a potential for divergences and multiplicities. But questions remain: What are those practices that we value and legitimize?, Why those and not others?, What are the roles of objects in our practices? and What is the role of a space and time?

Jesse: Creative ontological dimensions provide a 'way out' of the structures/practices/discourses where subjects are made, and made to be functional. Lines of flight with their *n-dimensions* and their rhizomatic forms are not exactly in the same ontological dimension as the structures and technologies of power that give rise to them (see Bazzul & Kayumova, 2016). What may be more important are processes of reterritorialization and deterritorialization, rather than ontological beginning points. That is, becoming is what's important since it is hard to say at any moment what is truly arborescent and what is rhizomatic, in the same way it is difficult to say what practices, discourses, and modes of being can be attributed to a subject. A question I have is how a focus on affective networks and embodiment achieves a materialist, ontological approach to critical activist science education. Perhaps our conversation should conclude around Deleuze's notion of imminence?

Shakhnoza: It is helpful to remember that Deleuze and Guattari's work on immanence (which is actually stemmed from the Deleuze's earlier philosophical work) combines Spinoza's notion of immanence and Bergson's idea of multiplicities as inventive and creative evolutions. Building on Bergson's work, Deleuze problematizes multiplicity as manifested in the modern projects. He argues that modern multiplicity is made possible mathematically. Modern multiplicity is an entity that can be observed, codified, quantified, and put into patterns. When applied to social fields—the subjects are continuously identified, categorized and simultaneously produced as gendered, classed, and raced subjects through the rationalization of their identities, experiences, social positions, and roles in the society. Through the repeated practices and prescribed modes of engagement identities, categories, and multiplicities become immanent part of the reality. These immanent multiplicities are further quantified, calculated, and illustrated in the national and international data sets. Based on my reading of Deleuze and Guattari, to subvert these mechanisms, which operate through common-sense, observable, and rationalized discourses, is to make visible non-quantifiable multiplicities that already existent in the human nature, such as affect, feeling, and attitude. These are lived in a moment, qualities in relation to one each other, immanent lines of flight in the spatial and temporal moments of interaction. These are not-predictable, dynamic, and non-quantifiable multiplicities, continuous variances in scientific term, which re-make the assemblages of our lives continuously and creatively. In a way, these are affective moments of rupture and lines of flight in the repeated practices, when subject can no longer survive within given discourse or practice. Thomas Kuhn's idea of paradigm change can be an example of a rupture in the history of science. I see the value of these concepts for an a critical activist science education here, because for multiplicities to come together in their immanence require a creative connection in which lines of flight for the desire of something different is re-united. I see Bencze's project as an immanent space, in which you and myself, for critical science activism, are coming together in multiplicities of desires, and this work is another example of a line of flight we are taking in our collective endeavor.

Jesse: Truth! We began this metalogue by concluding that Bencze's STEPWISE framework can be a means to subvert subjectification practices that may lead to depoliticized, disengaged 'subjects of science education'. However, we feel that an activist, science education needs to be simultaneously critical to subvert multiple technologies of power and forms of oppressions. This critical project, in our opinion, needs to engage with critical theory, creative ontologies, and multiplicities. Much still needs to be done for science education to be critical and activist. With an intensification of social inequality and environmental destruction a response by the science and education communities may be imminent! I would like to conclude with two short lists: (1) What we've outlined in this discussion that is key to a science education that is both critical and activist; (2) What needs more discussion.

Critical activist science education should engage:

- How student and teacher subjectivities are constituted through practices and discourses
- Multiple critical perspectives such as feminist, post-colonial, anti-capitalist, queer

- Inclusion of material, non-human components in sociocultural, political frameworks
- The 'reality' of virtual and actual multiplicities as well as ontological entities

What requires further discussion:

- Goals of a critical, activist science education
- The terrain of social and political struggles in science education
- Oppressions associated with science education
- What it means to be critical and activist in science and education
- How power operates in the field of science education
- The importance of immanence in modernity, biology, time (Bergson), subjects (Spinoza)
- What constitutes effective resistance to forms of power
- How the work of Foucault, Butler, Deleuze can guide a critical activist science education
- How creative ontologies can fit with critical activist work
- What does it mean to take a materialist approach to critical work

Paulo Freire differentiates between dialogue and conversation by attributing the search for truth or theoretical clarity to dialogue. Shakhnoza, as you stated above, we are very grateful to Larry Bencze for providing this dialogic space. In joining others working for a critical activist science education we can follow Freire's (1993) call to dialogue where none of us can speak a true word alone nor can we speak for others! We call on critical activist educators to move us in a more just direction by speaking, acting together.

References

Ahmed, S. (2007). A phenomenology of whiteness. *Feminist theory, 8*(2), 149–168.

Althusser, L. (1998). Ideology and ideological state apparatuses. In J. Rivkin & M. Ryan (Eds.), *Literary theory, an anthology* (pp. 294–304). Malden, MA: Blackwell.

Bazzul, J. (2012). Neoliberal ideology, global capitalism, and science education: Engaging the question of subjectivity. *Cultural Studies of Science Education, 7*(4), 1001–1020. doi:10.1007/s11422–012–9413-3.

Bazzul, J. (2014a). Critical discourse analysis and science education texts: Employing Foucauldian notions of discourse and subjectivity. *Review of Education, Pedagogy, and Cultural Studies, 36*(5), 422–437.

Bazzul, J. (2014b). The sociopolitical importance of genetic, phenomenological approaches to science teaching and learning. *Cultural Studies of Science Education, 10*(2), 495–503. doi:10.1007/s11422-014-9605-0.

Bazzul, J. (2015). Towards a politicized notion of citizenship for science education: Engaging the social through dissensus. *Canadian Journal of Science, Mathematics and Technology Education, 15*(3), 221–233.

Bazzul, J., & Kayumova, S. (2016). Toward a social ontology for science education: Introducing Deleuze and Guattari's assemblages. *Educational Philosophy and Theory, 48*(3), 284–299.

Bencze, J. L., & Carter, L. (2011). Globalizing students acting for the common good. *Journal of Research in Science Teaching, 48*, 648–669.

Bencze, L., Sperling, E., & Carter, L. (2012). Students' research-informed socio-scientific activism: Re/visions for a sustainable future. *Research in Science Education, 42*(1), 129–148.
Braidotti, R. (2006). *Transpositions: On nomadic ethics*. Cambridge, UK: Polity Press.
Butler, J. (1995). Burning acts: Injurious speech. In A. Parker & E. K. Sedgwick (Eds.), *Performativity and performance* (pp. 197–227). New York: Routledge.
Butler, J. (1997). *The psychic life of power: Theories in subjection*. Stanford University Press.
Butler, J., Laclau, E., & Žižek, S. (2000). *Contingency, hegemony, universality: Contemporary dialogues on the left*. New York: Verso.
Davies, B. (2006). Subjectification: The relevance of Butler's analysis for education. *British Journal of Sociology of Education, 27*(4), 425–438.
DeLanda, M. (2006). *A new philosophy of society: Assemblage theory and social complexity*. London: Bloomsbury Publishing.
Deleuze, G. (1994). *Difference and repetition*. Columbia University Press.
Deleuze, G., & Guattari, F. (1987). *A thousand plateaus: Capitalism and schizophrenia*. London: Bloomsbury Publishing.
Foucault, M. (1977). *Discipline and punish: The birth of the prison*. New York: Vintage.
Foucault, M. (1980). *The history of sexuality. Vol. 1: An introduction*. New York: Vintage Books.
Foucault, M. (1982). The subject and power. In H. L. Dreyfus & P. Rabinow (Eds.), *Michel Foucault: Beyond structuralism and hermeneutics* (pp. 208–226). Chicago: University of Chicago Press.
Foucault, M. (1985). The use of pleasure: Volume 2 of the history of sexuality (R. Hurley, Trans.). New York: Vintage.
Foucault, M. (1988). The ethic of care for the self as a practice of freedom. [interview]. In J. B. D. Rasmussen (Ed.), *The final foucault*. Cambridge, MA: MIT Press.
Foucault, M. (1997). Sex, power and the politics of identity. In R. Hurley, (Trans.) & P. Rabinow, (Eds.) *Ethics: Subjectivity and truth* (pp. 163–173). New York: The New York Press.
Foucault, M. (2003a). What is critique? In P. Rabinow & N. Rose (Eds.), *The essential Foucault: Selections from essential works of Foucault, 1954–1984* (pp. 263–278). New York: New Press.
Foucault, M. (2003b). Polemics, politics, and Problematizations: An interview with Michel Foucault. In P. Rabinow & N. Rose (Eds.), *The essential Foucault: Selections from essential works of Foucault, 1954–1984* (pp. 18–24). New York: New Press.
Foucault, M., & Deleuze, G. (1977). Intellectuals and power. In D. F. Bouchard (Ed.), *Language, counter-memory, practice: Selected essays and interviews* (pp. 205–217). Ithaca, NY: Cornell University Press.
Freire, P. (1993). *Pedagogy of the oppressed* (Rev. ed.). New York: Continuum.
Hardt, M., & Negri, A. (2000). *Empire*. Cambridge, UK: Harvard University Press.
Ranciére, J., & Corcoran, S. (2010). *Dissensus: On politics and aesthetics*. London: Continuum.
Roth, W.-M. (2014). Enracinement or the earth, the originary ark, does not move: on the phenomenological (historical and ontogenetic) origin of common and scientific sense and the genetic method of teaching (for) understanding. *Cultural Studies of Science Education, 10*, 1–26.
Roth, W.-M., & Tobin, K. (2004, September). Co-generative dialoguing and metaloguing: Reflexivity of processes and genres. *Forum Qualitative Sozialforschung/Forum: Qualitative Social Research, 5*(3), 1–12.
Weedon, C. (1996). *Feminist practice and poststructuralist theory*. Oxford, UK: Basil Blackwell.
Weedon, C. (1997). *Feminist politics and poststructuralist theory*. Oxford: Blackwell.

Chapter 31
Countering the Neoliberal Ontology of Nature: The STEPWISE Option

Ajay Sharma

31.1 Introduction

Educators have long positioned schools as engines of social change. For instance, over a century ago, John Dewey, Hickman, and Alexander (1998) had averred that school was "the primary and most effective instrument of social progress and reform" (p. 234). This tradition of educationalizing social issues has been critiqued by some scholars, such as David Bridges (2008), Lynn Fendler (2008) and David Labaree (2008), for various reasons ranging from ineffectiveness to aiding in cultivation and governance of a pliable population. However, unwilling to let go of this powerful instrument of change, I remain of the view that given the role of schools in influencing our mental dispositions (*habitus*), social, economic and cultural resources (*capital*) and the space in which we live our social lives (*field*), schools and school-based efforts can indeed be catalyzing nuclei for and/or be an integral component of a wider, sustainable movement for social change (Bridges, 2008). It is with this standpoint that I view and understand the 'STEPWISE'[1] framework for activist science and technology education. If science curricula and instruction at the K-12 level can serve as vehicles of social reproduction, then it also stands to reason to assume that they can be repurposed to serve progressive goals. And that is indeed what the STEPWISE project has shown to us in its years of existence.

[1] 'STEPWISE' is the acronym for Science & Technology Education Promoting Wellbeing for Individuals, Societies & Environments. It is a theoretical and practical framework that organizes teaching/learning goals in ways that encourage and enable students to self-direct research-informed and negotiated actions to address personal, social and environmental problems linked to fields of science and technology. To learn more about this framework, refer to Chap. 2 in this book.

A. Sharma (✉)
College of Education, University of Georgia, Athens, GA, USA
e-mail: ajay@uga.edu

L. Bencze (ed.), *Science and Technology Education Promoting Wellbeing for Individuals, Societies and Environments*, Cultural Studies of Science Education 14,
DOI 10.1007/978-3-319-55505-8_31

However, much of the lessons and insights that STEPWISE offers may remain inaccessible unless read diffractively by critique and propelled through dialogue. It is in this spirit that this chapter critically examines the STEPWISE framework in light of its potential to offer the theoretical space and resources needed for evidence based, scientifically grounded understanding and collective democratic environmental and social justiceactivism. In particular, I compare and contrast the neoliberal ontology of the natural world as embedded in the mainstream, official science curricula in the United States with the ontology undergirding the STEPWISE framework to showcase the strengths and limitations of STEPWISE oriented science instruction for environmental action. I also discuss ways in which the STEPWISE framework can be further strengthened to better serve the goal of preparing students that can act as sharing, cooperative and public good maximizing *homo reciprocans* to create a more ecologically and socially just world.

31.2 Neoliberal Ontology of Nature

Let us begin with a brief analysis of how we come to understand nature. 'Nature' is a much-used phrase in both every day and academic language. However, as Raymond Williams (1980) famously declared, nature may also be "perhaps the most complex word in the language" (p. 219). In keeping with the focus of this chapter, when I speak of nature I am referring to the external material world that may or may not include human beings. It is possible that this particular description may come across as a fairly unproblematic and objective signification of the word nature. We tend to perceive nature as a directly perceived and objectively verifiable part of our world. It is out there, we just have to open our eyes to it. However, taking inspiration from Karl Marx's thesis of unity of nature with society as produced in practice through human labor, critical human geographers from David Harvey (1982) and Neil Smith (2008) onwards have argued that 'Nature' as we ordinarily understand it is not an objective component of our realities but instead is a social production (Castree, 2005). Kevin Archer (2010) explains this perspective on nature as follows:

> Human material activities produce both what we consider to be nature and the surrounding landscapes that are considered natural. An implication of this perspective is that because what humans consider to be nature and natural are only artifacts of their own material activities, they are fully historical and open to change as human material activities change over time. Nature is always social nature, in this respect, not something outside social reproduction to be merely observed, protected, saved, restored, conserved, or otherwise managed by somehow extranatural humans. Nature is what results from the various cultural and historical ways in which humans socially (re)produce both their sciences and their societies (p. 2560).

However, we tend to perceive nature as an objective reality, and find it is difficult to see how nature could be socially produced. According to Bruno Latour (2012), this is because since the age of enlightenment the *modern constitution* — the overall philosophical framework of modernity — has been so constructed that, ontologically speaking, humans were consistently consigned to a totally separate domain from nonhumans. That is, as Braun (2006) explained, "Society is often depicted solely as 'humans among themselves', an autonomous realm that obeys its own historical dynamics, while non- humans enter the story only as fetishized commodities or fixed capital" (p. 192). Thus, the *modern constitution* created a dualist ontology that split the world in two: a 'social world' and a 'natural world' that interacted and impacted each other but remained apart. This distancing enabled humans to objectify, dominate and exploit the 'natural world' for their purposes, besides also presenting it as an objective reality of their world.

It is clear then that, as Bruce Braun (2006) opined, "to speak of nature is to presuppose an ontology" (p. 193). Of course, our ontological perspectives are not formed in isolation. They come to us embedded in discourses that constitute us and are used by us to make sense of our existence and navigate our daily lives (Chouliaraki & Fairclough, 1999). One such discourse - neoliberalism — has had an enormous influence on us, both within schools and without (Davies & Bansel, 2007). Admittedly, neoliberalism is not the only discourse at work in schools and beyond. In different societies, one will find different mixes of discourses constituting people and their lives. But it can be well argued that forces of globalization have allowed or encouraged neoliberalism to circulate globally more successfully than others (Harvey, 2005). It is quite likely then that we, in the West, may have become much influenced by this global and hegemonic discourse in terms of how we interact with and understand nature (Heynen, 2007).

Neoliberalism, both as a concept and a global phenomenon, has gained widespread circulation in scholarly literature as well as public media. However, because of its perceived complexities and diverse avatars, it is mapped onto multiple and even inconsistent significations. So much so that it has even been labeled as "a rascal concept — promiscuously pervasive, yet inconsistently defined, empirically imprecise and frequently contested" (Brenner, Peck & Theodore 2010, p. 1). Broadly speaking, however, one can say that it has become a "general descriptor for the dominant ideological sensibilities and imperatives of the post-Keynesian globalization era" (James, 2013, p. 31). Based on Michel Foucault's (2010) critique of neoliberalism in his lectures at Collège de France in 1979, I see neoliberalism as a discourse of governmentality that seeks to elevate individualized, market-based competition and exchange as the central and most desirable governing principle for organizing human action and social life, even in areas hitherto considered non-economic. One such 'non-economic' area has indeed been nature (as in natural systems).

When extended to natural systems, neoliberalism has profound implications for our ontology of the world. First, because of its roots in classical liberalism and hence in the *modern constitution*, neoliberalism endorses and further reifies the modern dualist ontology mentioned earlier that mereologically divides our planet in two separate and distinct 'worlds' — a social world and a natural world. But, it doesn't just stop there. Neoliberalism takes another critical step of embedding both worlds within an economic world governed by a market-based rationality. As a result, ontologically-speaking, the world gets constituted in terms of only those entities and relations that are intelligible under an economic grid of intelligibility. For instance, Niyamgiri Mountain in the state of Chattisgarh in India, though rich in bauxite ore, is only seen as a sacred site by the Kondh tribe that lives in that region. However, significations like sacred or profane are not intelligible in a market-based economic discourse. Thus, we find that for the Vedanta mining company that wanted to acquire mining rights there, Niyamgiri Mountain could only occupy the ontological status of a mining site.

Second, because nature is brought within the purview of an economic logic, each entity and its relations are first individuated either as a resource (of goods or services) or a sink (of waste matter or energy) by separating them from their context through legal and material boundaries. For instance, a forest ecosystem is likely to be decontextualized from the local socioecological context and positioned as a resource for timber or as a sink for atmospheric carbon. Third, once atomized, entities and relations in nature get signified and valued primarily in monetary terms for their exchange value. Carbon cap and trade approaches to pollution and climate change are, for example, based on this ontology of nature, as they seek to create a market where carbon pollution and carbon sinks can be monetarily valued and traded accordingly. Fourth, individuation and monetary valuation of entities and relations makes them abstract in nature in the sense that now entities existing in different place and time can be sorted, classified, compared and treated as equivalent or different based on their monetary value and other decontextualized attributes. For instance, under the payment for environmental services (PES) program, such as the United States' Conservation Reserve Program, environments in different space and time can be compared and treated as equivalent to enable any corporation damaging or polluting one location to 'offset' its actions by preserving some other location. Fifth, because monetary valuation is based on linear, static models of neoclassical economics, neoliberal ontology reduces poorly understood complex, nonlinear and nondeterministic relations between entities in natural systems to simple, linear and deterministic relations.

Such on ontology of nature transmutes entities and relations of the natural world into commodities that can be extracted, transported and traded in a global market. Commodification of natural resources is certainly not a new phenomenon. The resignification of nonhuman entities of the world as 'fictitious commodities' (Polanyi, 1957) has always been an integral component of capitalist modes of production and consumption (Harvey, 2005). What is relatively new with neoliberalism is the internalization of natural resources within the economic regime as commodities that are to be conserved and sustainably exploited in the conservation-as-development paradigm of green or natural capitalism (Büscher, Dressler, & Fletcher 2014). Finally, humans find a place in the neoliberal ontology of nature as environmentally responsible subjects that view environmental action more as a personal virtue than as a matter of collective action (Treanor, 2010).

Despite its proclaimed intentions of resolving "the paradoxical idea that capitalist markets are the answer to their own ecological contradictions" (Büscher, 2012, p. 30), evidence suggests that neoliberal conservation and green or natural capitalism has neither saved environments nor helped the vast majority of the people (Castree, 2008). In fact, the state of our planet has worsened since neoliberalism came to define capitalism and advanced capitalist societies. It is not surprising then that economist Nicholas Stern (2009), while admitting the culpability of neoliberal capitalism in damaging our planet, acknowledged that climate change is "the greatest and widest-ranging market-failure the world has seen" (p. 11). Thus, we find ourselves not only facing in the form of climate change "the biggest problem our civilization has ever had to face up to in its 12,000 years" (Clery, 2007), but also appear to be in the early stages of the Earth's sixth mass extinction (Barnosky et al., 2011). So much so that it has now become moot if a collapse of global civilization can even be avoided (Ehrlich & Ehrlich, 2013).

A critical part of coming to know the world relates to labeling the world, categorizing entities thus labeled in different categories, linking them with meanings or interpretations through chains of signification, and ascribing relationships among them. Highlighting the importance of the emerging ontology of children, Howard Gardner (2011) averred that "the way in which children come to think of classes of entities affect the kinds of theories they develop about these classes and the kinds of inferences they are prepared to draw" (p. 94). For instance, if a student lumps all ocean dwelling creatures as belonging to the class of fishes, she will naturally tend to see a whale as a fish, ascribe all fish-like behavior to it and relate to them accordingly. Although naïve ontologies of nature developed outside school have been proved to be remarkably robust, science education remains one of the chief socially-sanctioned ways through which the society seeks to influence how children come to understand the world and figure their place in it (Driver, Asoko, Leach, Mortimer, & Scott, 1994). Thus, officially-sanctioned ontologies as reflected in official discourse of school science come to acquire an important role in shaping students' views of the world. In the next section, I present ways in which the STEPWISE framework centered science instruction at K-12 level presents an appealing alternative approach that counters the now dominant neoliberal ontology of nature.

31.3 The STEPWISE Framework as a Counter to the Neoliberal Ontology of Nature

Research on representations of nature in school science, such as that by Graça Carvalho, Tracana, Skujiene, and Turcinaviciene (2011), Kostas Korfiatis, Anastasia Stamou and Stephanos Paraskevopoulos (2004) and Ajay Sharma and Cory Buxton (2015), indicates that official discourse tends to offer representations of natural systems' relationships with social systems, and roles of human agency in these relationships in ways that align well with neoliberal ontology of nature discussed earlier. For instance, science textbook and science standards tend to represent natural systems as stable and space-time independent systems that are functionally and ecologically distinct from social systems. This dualist ontology of nature, while largely erasing human presence from nature, reifies human dominance over nature by categorizing it as a separate and independent resource for material ends. Thus, nature while externalized from the social is nevertheless brought within the overall fold of the economy and made to serve ever increasing demands for profit and capital accumulation. Further, this substantivist ontology privileges entities over relations by: (a) showing the natural world as populated by stable entities, both living and nonliving, that possess some intrinsic properties on the basis by which they interact with the world; and (b) 'thingifying' the relations. (Barad, 2003). It can be well argued that, once internalized, such an ontology may lead students to accept commodification of nature as a natural corollary.

The STEPWISE framework has much potential to enable teaching practices and learning experiences that counter substantivist and neoliberalism-aligned ontology of nature. I will begin with STEPWISE's focus on STSE (science, technology, society and environment) issues, with particular attention on those that relate to local and/or global political economy. My own study with middle school students (Sharma, 2013), as well other studies, such as by William Cobern (2000) and Kai Nielsen (2012), indicate that students' understanding of nature, while originating from diverse sources, does not deviate much from school science representations of nature in at least one crucial aspect: seeing nature as distinct and separate from society. Students associate nature with pristine wilderness of wildlife parks, coral reefs and polar ice caps — places where human presence or influence is assumed to be minimal or non-existent. Further, in my study, I found that students tended to think that, while natural world may be under threat and some animal and plant species may be endangered because of our actions, they themselves are safe and far removed from such dangers. To me, such views are strong evidence of how ever-widening *metabolic rifts* in advanced capitalist societies (Foster 1999) — spatial and temporal space between humans and their sources of material sustenance — has occluded our abilities to understand and protect the nonhuman world. For example, most of us get our bananas from plantations in Amazon rainforests — locations that are far removed from us. Thus, even though most of us may care for the earth (Kempton, Boster & Hartley, 1995), it is difficult to have intimate and valid knowledge or abilities to do much about environmental degradation that banana and other agricultural plantations are causing to Amazon rainforests (Vandermeer & Perfecto, 2005). John Foster (1999) developed the notion of *metabolic rift* based on Karl Marx and Friedrich Engels' (1967) insight about how capitalist production "disturbs the metabolic interaction between man and earth, i.e. it prevents the return to the soil of its constituent elements consumed by man in the form of food and clothing; hence it hinders the operation of the eternal natural condition for the lasting fertility of the soil…" (pp. 504–505). As Foster (1999), Bruno Moore (2011) and several other social scientists have argued, global capitalism is the single most important reason for the ever-widening metabolic rift in advanced capitalist societies. Advanced capitalist societies need to figure ways to reduce metabolic rift, not just because it reduces our agency to understand and take successful action against socioecological problems; but, more importantly, because metabolic rift creates and aggravates these problems. For instance, metabolic rift has been implicated in the worsening conditions for Climate Change (Clark & York 2005).

Now, it is my understanding that school science usually does not do a good job at helping students understand the nature and causes of metabolic rift (Sharma, 2012). However, that can change if the STEPWISE framework is used to teach science. In the hands of a capable and committed science teacher, this framework can set up problem contexts, offer learning spaces and resources that enable students to use their *STSE Education, SkillsEducation, Products Education* and *Students' Research* to undertake critical investigations on the nature and causes of *metabolic rift* within their own communities. For instance, an investigation on production and consumption chains of processed food industry, as mentioned in chapter two of this book by Larry Bencze would be well-suited for helping students understand how globalized production, processing, consumption and removal of agricultural and food wastes sustain and exacerbate metabolic rifts at both local and global levels. Similarly, a well-designed inquiry unit on climate change can help students appreciate the global fluxes, pools and feedback loops of carbon cycle that connect spatio-temporally separate ecosystems despite the metabolic rift that exists at a local level.

Thus, the STEPWISE framework appears well-suited to counter the nature-society dualism integral to the neoliberal ontology of nature. Investigations conducted jointly by students and the teacher under this framework can help students understand that, even though modern discourses, such as neoliberalism, encourage us to view nature as distinct from the society, from scientific and ontological standpoints this nature-society dualism is not only outdated and incorrect but also complicit in creating and aggravating ecological crises all over the world. The STEPWISE framework can provide students learning experiences that let them understand that, "there is no 'nature in general' any more than there is a 'society' which exists as a unified totality — there are only hybrid networks composed of *specific* human and non-human actants, that are of greater or shorter length, are more or less dense, and 'hold together' for longer or shorter periods of time" (Braun, 2006, p. 202). It can help them perceive the world as comprising of "specific assemblages of humans and non-humans," in which each entity or actant is constituted by relationships in which it is embedded.

For instance, a STEPWISE investigation on domestic cats can reveal how the nature-social boundary becomes blurred in an assemblage that include pet-owners, their neighbors, pet-food industry and local undomesticated flora and fauna. In the United States, cats and their owners emerge as important actants in this assemblage, as domestic cats kill a median of 2.4 billion birds and 12.3 billion mammals a year (Loss, Will, & Marra, 2013). Similarly, investigations on local rivers and forest areas can reveal the extent to which their natural pristineness owes to the careful management by humans. In this way, students can learn to substitute subtantivist neoliberal ontologies of nature with relational ontologies that privilege relations over entities and acknowledge that social and ecological entities and phenomena are best understood as expressions of socio-ecological relationships that operate at different spatial-temporal scales (Castree, 2005). Mainstream school science and neoliberalism-dominated political economy tend to present relations within and across socioecological systems as linear, simple and deterministic and in support of a ‘balance of nature’ standpoint on natural phenomena. Though it may not be easy, I would love to see science educators come up with ideas for science investigations that allow students to appreciate complex, nonlinear and nondeterministic characteristics of such relationships. There already exist some good exemplars that can be used as spring boards for other investigations. For instance, a science inquiry activity developed by Candace Dunlap of TERC™ and EarthLabs™ project (serc.carleton.edu/earthlabs/carbon/lab_5.html) allows students to investigate relationships between microbial soil respiration and decomposition with carbon cycle and consider roles of carbon cycle feedback loops in climate change. Further, through STSE investigations, students can come to understand that, contrary to their perceptions, not only are they not immune from environmental threat but, more importantly, the poor and marginalized amidst their communities face many of the same environmental problems that distress people in less developed societies. For instance, based on the current understanding about the existence of a ‘climate gap’ between rich and poor in the cities of the US in terms of vulnerabilities to adverse effects of climate change (Morello-Frosch, Pastor, & Shonkoff, 2009), students can undertake STEPWISE investigative science projects that explore leading indicators of ‘climate gap’ in their own communities.

I am also much enthused by the fact that STEPWISE’s framework focuses on controversial issues in its pedagogy. This is because such issues often relate to movement of matter and energy across different ecosystems that clearly indicate how deeply embedded nature is in global capitalist economic regimes. Secondly, controversial issues are good at revealing to students conflicting interpretations, interests and motivations of different stakeholders in STSE issues and, thus, better prepare them to reconcile differences and employ evidence-based arguments to engage in student-led activism. Finally, and most importantly, I am of the view that controversial issues can be critical in revealing agency of specific actants in creating and aggravating socio-ecological problems. Systemic-functional linguistic analyses of school science texts reveal that the grammar of such texts is shaped to “exclude and suppress human agency in natural — social relationships or to attribute that agency to some anonymous, amorphous and non-individuated group, labeled simply as ‘people’ or ‘humans’” (Sharma & Buxton, 2015, p. 276). As a result, students either do not get to understand specifics of human involvement in (re)creating nature and its attendant crises or come to see ordinary people as largely responsible for causing socioecological problems through excessive consumption and other ecologically damaging actions, such as transporting invasive species. This positioning of actions of ordinary individuals as the leading cause of our environmental woes suits capitalist systems just fine, as it does a nice job of obfuscating larger and more significant sociocultural and politico-economic factors behind environmental stress and degradation (Hempel, 1996). For instance, contrary to what school science often implies, production activities, such as farming, industry and mining, rather than the end-usage by people, are indeed major sources of chemicals-based pollution (McKinney, Schoch, & Yonavjak, 2012). A focus on controversial issues, I therefore believe, can do much to put the spotlight on private corporations’ role in recasting nature as a commodity to be 'sustainably' exploited for ever-increasing profits and capital accumulation – a mission that invariably leads to ecological ruin and social injustices.

My support of the STEPWISE framework as a bulwark against the influence of neoliberalism in science education does not assume a facile translation from potential to realization. A framework can, at best, offer broad parameters, orientation and syntax in which to embed locally relevant issues and activities that would achieve the desired educational goals. The open-endedness of the STEPWISE framework allows science teachers much flexibility and creative space to adopt and adapt ideas developed elsewhere — a sort of globalization of grass-root ideas from below — that counters top-down globalization of science education imposed by policy elite much enamored of neoliberal ideas. However, this open-endedness also encumbers the STEPWISE framework with some risks and limitations that in contexts marked by limited resources and preparation can potentially lead to outcomes that do little to counter the neoliberal ontology of nature. The next section, thus, presents some recommendations that, in my opinion, will make the STEPWISE an even better program for prioritizing altruistic actions to address socioscientific issues in science education.

31.4 STEPWISE: Some Recommendations

A key feature of STEPWISE is the culmination of student projects in sociopolitical action. Student participation in such actions is, indeed, critical to the laudable goal of preparing students as agentive, critically minded citizens who know how to gather and use scientific evidence on STSE issues to improve the wellbeing of other individuals, societies and environment (WISE). As Bencze opines in the Chap. 2, "because of the seriousness of the problems for the WISE linked to decisions about science and technology …" we certainly need citizens who are "ready to critique decisions of the powerful and take actions to bring about a better world" (personal communication with the editor). STEPWISE is open-ended about the nature of action students may take to bring about change on the issue related to their investigative project. That is good, because students and the teacher, as local actors, are best placed to decide the nature and objectives of their research-informed and negotiated action (RiNA). However, STEPWISE framework offers little articulation on the nature of actions needed for bringing about a sustainable and noticeable improvement in the wellbeing of other individuals, societies and environment. As a result, a perusal of possible kinds of STSE actions undertaken by students as given in Fig. 2.8 in Chap. 2 by Larry Bencze shows that, in most cases, RiNA takes the form of actions undertaken individually by students. Individual actions, such as writing letters to members of the government and business leaders or educating others on STSE issues (refer Fig. 2.8, Chap. 2, this volume), can certainly play an important role in bringing about change and also are generally easier and safer to undertake for students. However, I am not sure to what extent such actions match the ecological and sociological scales at which socioecological problems express themselves. For instance, most RiNAs take the form of local actions. But the fact is that most STSE issues exist on local as well as larger regional, national and international scales — both socially and ecologically. Local action here is necessary, but rarely sufficient in solving such problems (Ostrom, Burger, Field, Norgaard, & Policansky, 1999).

Further, individual responses to larger socioecological problems cohere well with the mainstream environmentalist discourse that tends to present environmental action as personal virtue (Treanor, 2010). They also correspond well with the neoliberal discourse, in which it is assumed that an individual is an autonomous agent in sole possession of her agency, and her actions as a citizen reflects her freely chosen lifestyle. In light of research findings, see for instance Irene Lorenzoni et al. (2007), that show that there exist significant social barriers in modern societies that inhibit even knowledgeable and environmentally conscious people from acting in environmentally responsible ways, such assumptions are not only highly questionable, but can also lead to actions that may feel good but contribute little to resolving the issue. In fact, such RiNAs may even reinforce the impression that ordinary people are indeed responsible for the sorry state of the planet and, if we all just made good environmentally friendly 'green' lifestyle choices, our planet would be saved. Such a perspective absolves capitalism of its role in jeopardizing life on Earth. It also runs counter to current wisdom amongst ecologists and environmental sociologists, in which it is generally agreed that individual environmental actions can only be effective when they take place in a context of just and democratic governance of ecological resources (Dietz, Ostrom, & Stern, 2003). That is, individual actions are effective only when they form a part of wider collective action. From my distant understanding of the STEPWISE framework, it appears to me that this entire effort would benefit from a more robust articulation of needs of and strategies for RiNAs to be part of or nucleus for some wider and sustained collective action.

STSE issues, by definition, are as much social as they are scientific. Further, as discussed earlier in the current Anthropocene era, "to talk about nature is to talk about humans and vice versa" (Sharma, 2012, p. 47). The STEPWISE framework, however, is focused exclusively on science education. This limitation is understandable on pragmatic grounds as a viable strategy to gain a foothold for STSE issues based education within the mainstream education system. However, in the longer term, it would be advisable to integrate school science more closely with social studies within the STEPWISE framework. Such an integration, I believe, is critical, not only for understanding typical STSE issues in science education, such as climate change and pollution of rivers, but also for social studies topics, such as the genocide in Rwanda or the conquest of Americas by Spanish conquistadors. As Lyn Carter (2008) recommends, perhaps we can bring about this integration by learning from the example of *Sustainability Science* — a fast emerging disciplinary field that integrates research and knowledge from diverse disciplines of natural and social sciences for fostering sustainability of human life on our planet.

Another way in which the social is under-articulated within the STEPWISE framework relates to the role of local community in helping students appropriate the four learning domains of the framework — STSE Education, Skills Education, Students' Research and Products Education. I am of the view that STSE focused education and student-led RiNAs should also involve the local community in all the four learning domains. Cornel West (1999), while talking about Gramscian ideas about roles of philosophy in social activism, mentioned in an interview that, "... the aim of philosophy is not to become worldly by imposing its elite intellectual views upon people, but to become part of a social movement by nourishing and being nourished by the philosophical views of oppressed people themselves for the aims of social change and personal meaning" (p. 173). For STEPWISE projects, this Gramscian advice would translate as a suggestion that students should not be led to think that they or their teachers are the 'experts' who are best placed to both understand local STSE problems and come up with solutions that locals need to adopt if they are serious about resolving their issues. Local communities are the best repositories of expertise in most learning domains, especially as they pertain to local STSE issues. There is considerable scope within the STEPWISE framework for community involvement in all aspects of STSE focused and sociopolitical action oriented education. It is possible that some RIAs that students have so far done are indeed based on the premise of community involvement from start to finish. However, it is my hope that future iterations of STEPWISE framework will be founded on the assumption of the engagement and equal partnership of local community members as experts in student learning and action on STSE issues.

Lastly, it would be very helpful for students as well as local communities if the STEPWISE framework can elaborate in much greater detail its own distinct ethical stance on living in the Anthropocene era, and also equip students to develop one for themselves. Such an ethical stance, I imagine, would be critical of the ontological implications of neoliberalism. But, more importantly, it should also offer staging grounds for constructive suggestions that offer alternatives to neoliberal commodification of nature. As Nancy Fraser (1995) argued in the context of feminist struggles, I believe that efforts to counter ill-effects of capitalism on our planet need "both deconstruction and reconstruction, destabilization of meaning *and* projection of utopian hope" (p. 71). One possible repository of positive visions of the future could be local knowledge and practices, as these often embody different and diverse interpretations of the actants and relations found within and across local human and non-human assemblages. Unfortunately, these interpretations have been long subjugated and supplanted by the one favored by neoliberalism in which everything gets valued, sorted and classified as a commodity. Frameworks like STEPWISE are well-placed to articulate an ethical stance that enables students and local communities to recuperate and re-establish the local and alternative meanings so that we all can (re)imagine alternative democratic and ecologically sustainable futures.

31.5 Conclusions

Michel Foucault (1980) once said that,

> [t]he essential political problem for the intellectual is not to criticize the ideological contents supposedly linked to science, or to ensure that his own scientific practice is accompanied by a correct ideology, but that of ascertaining the possibility of constituting a new politics of truth. The problem is not changing people's consciousnesses - or what's in their heads - but the political, economic, institutional regime of the production of truth (p. 133).

The ability of global capitalism to commodify nature rests upon its hegemony over regimes of truth that naturalize neoliberal perspectives on nature for elite and masses alike. Science education is an important regime of truth that tells students what the world is like. As a result, it is also one of the leading discursive pathways for instilling a neoliberal perception of the world on future citizens. The STEPWISE framework can help science teachers and students alike to delink the power of truth from the hegemony of neoliberal thought and, thus, denaturalize neoliberal ontologies of nature by delegitimizing them and by recuperating alternate ontologies through their RiNAs. As I argued in this chapter, the STEPWISE framework offers discursive space and resources needed for this task within the context of science education. Equally importantly, it eases students into a performativity that allows them to occupy subjectpositions of citizen-activists — first as apprentices and then as mature practitioners. Of course, as indicated in this chapter, there is some scope for further elaboration of core ideas and development of overall framework that should help STEPWISE achieve its goals.

New creative solutions to big problems often emerge in "interstitial locations — nooks and crannies in and around the dominant institutions" (Brecher, Costello & Smith, 2000 p. 24). The STEPWISE project is indeed one such solution. However, from my distant vantage point, it still appears fragile and in need of sustained support and collaboration from like-minded efforts in other regions. In recent decades, we have seen numerous attempts by such local efforts to link-up transversally with similar efforts to outflank dominant forces and change the status quo (Galvan, 2005). This is akin to globalization-from-below that depends neither on the power of the state nor on the market to effect large-scale social change. I end this chapter with the fond hope that, in near future, the STEPWISE project will indeed become an integral part of wider collaborative, democratic efforts that attempt to renegotiate theory and praxis of mainstream science education for a social and ecologically just, sustainable future.

References

Archer, K. (2010). Social construction of nature. In B. Warf (Ed.), *Encyclopedia of geography* (pp. 2558–2561). Thousand Oaks, CA: Sage Publications.

Barad, K. (2003). Posthumanist performativity: Toward an understanding of how matter comes to matter. *Signs, 28*(3), 801–831.

Barnosky, A. D., Matzke, N., Tomiya, S., Wogan, G. O. U., Swartz, B., Quental, T. B., et al. (2011). Has the earth's sixth mass extinction already arrived? [10.1038/nature09678]. *Nature, 471*(7336), 51–57.

Braun, B. (2006). Towards a new earth and a new humanity: Nature, ontology, politics. In N. Castree & D. Gregory (Eds.), *David Harvey: A critical reader* (pp. 191–222). Malden, MA: Blackwell Publishing.

Brecher, J., Costello, T., & Smith, B. (2000). *Globalization from below: The power of solidarity*. Cambridge, MA: South End Press.

Brenner, N., Peck, J., & Theodore, N. I. K. (2010). Variegated neoliberalization: Geographies, modalities, pathways. *Global Networks, 10*(2), 182–222.

Bridges, D. (2008). Educationalization: On the appropriateness of asking educational institutions to solve social and economic problems. *Educational Theory, 58*(4), 461–474.

Büscher, B. (2012). Payments for ecosystem services as neoliberal conservation: (Reinterpreting) evidence from the Maloti-Drakensberg, South Africa. *Conservation and Society, 10*(1), 29.

Büscher, B., Dressler, W., & Fletcher, R. (2014). Nature inc.: Environmental conservation in the neoliberal age. Tucson, AZ: University of Arizona Press.

Carter, L. (2008). Sociocultural influences on science education: Innovation for contemporary times. *Science Education, 92*(1), 165–181.

Carvalho, G. S., Tracana, R. B., Skujiene, G., & Turcinaviciene, J. (2011). Trends in environmental education images of textbooks from Western and Eastern European countries and non-European countries. *International Journal of Science Education, 33*(18), 2587–2610.

Castree, N. (2005). *Nature*. London/New York: Routledge.

Castree, N. (2008). Neoliberalising nature: Processes, effects, and evaluations. *Environment and Planning A, 40*(1), 153.

Chouliaraki, L., & Fairclough, N. (1999). *Discourse in late modernity* (Vol. 2). Edinburgh, Scotland: Edinburgh University press.

Clark, B., & York, R. (2005). Carbon metabolism: Global capitalism, climate change, and the biospheric rift. *Theory and Society, 34*(4), 391–428.

Clery, D. (2007). U.K. Science Adviser offers some parting shots. *Science, 318*(5858), 1862–1863.

Cobern, W. W. (2000). *Everyday thoughts about nature: A worldview investigation of important concepts students use to make sense of nature with specific attention to science*. Dordrecht, The Netherlands: Springer.

Davies, B., & Bansel, P. (2007). Neoliberalism and education. *International Journal of Qualitative Studies in Education (QSE), 20*(3), 247.

Dewey, J., Hickman, L., & Alexander, T. M. (1998). *The essential Dewey: Pragmatism, education, democracy*. Bloomington, IN: Indiana University Press.

Dietz, T., Ostrom, E., & Stern, P. C. (2003). The struggle to govern the commons. *Science, 302*(5652), 1907–1912.

Driver, R., Asoko, H., Leach, J., Mortimer, E., & Scott, P. (1994). Constructing scientific knowledge in the classroom. *Educational Researcher, 23*(7), 5–12.

Ehrlich, P. R., & Ehrlich, A. H. (2013). Can a collapse of global civilization be avoided? *Proceedings of the Royal Society B: Biological Sciences, 280*(1754), 20122845.

Fendler, L. (2008). New and improved educationalising: Faster, more powerful and longer lasting. *Ethics and Education, 3*(1), 15–26.

Foucault, M. (1980). Truth and power. In C. Gordon (Ed.), *Power/knowledge: Selected interviews and other writings 1972–1977* (pp. 109–133). New York: Pantheon Books.

Foucault, M. (2010). *The birth of biopolitics: Lectures at the Collège de France, 1978–1979*. New York: Picador.

Foster, J. B. (1999). Marx's theory of metabolic rift: Classical foundations for environmental sociology. *American Journal of Sociology, 105*(2), 366.

Fraser, N. (1995). False antitheses. In S. Benhabib, J. Butler, D. Cornell, & N. Fraser (Eds.), *Feminist contentions: A philosophical exchange* (pp. 59–74). New York: Routledge.

Galvan, R. T. (2005). Transnational communities en La Lucha: Campesinas and grassroots organizations "globalizing from below". *Journal of Latinos & Education, 4*(1), 3–20.

Gardner, H. (2011). *The unschooled mind: How children think and how schools should teach.* New York: Basic Books.
Harvey, D. (1982). *The limits to capital.* Oxford, UK: Basil Blackwell.
Harvey, D. (2005). *A brief history of neoliberalism.* New York: Oxford University Press.
Hempel, L. C. (1996). *Environmental governance: The global challenge.* Washington, DC: Island Press.
Heynen, N. (2007). *Neoliberal environments: False promises and unnatural consequences.* New York: Routledge.
James, M. (2013). Neoliberal heritage redress. In J. A. Henderson & P. Wakeham (Eds.), *Reconciling Canada: Critical perspectives on the culture of redress* (pp. 31–62). Toronto, ON: University of Toronto Press.
Kempton, W., Boster, J. S., & Hartley, J. A. (1995). *Environmental values in American culture.* Cambridge, MA: MIT Press.
Korfiatis, K. J., Stamou, A. G., & Paraskevopoulos, S. (2004). Images of nature in greek primary school textbooks. *Science Education, 88*(1), 72–89.
Labaree, D. F. (2008). The winning ways of a losing strategy: Educationalizing social problems in the United States. *Educational Theory, 58*(4), 447–460.
Latour, B. (2012). *We have never been modern.* Cambridge, MA: Harvard University Press.
Lorenzoni, I., Nicholson-Cole, S., & Whitmarsh, L. (2007). Barriers perceived to engaging with climate change among the UK public and their policy implications. *Global Environmental Change Part A: Human & Policy Dimensions, 17*(3/4), 445–459.
Loss, S. R., Will, T., & Marra, P. P. (2013). The impact of free-ranging domestic cats on wildlife of the United States. *Nature Communications, 4*, 1396.
Marx, K., & Engels, F. (1967). *Capital: A critique of political economy.* New York: International Publishers Company, Incorporated.
McKinney, M. L., Schoch, R. M., & Yonavjak, L. (2012). *Environmental science: Systems and solutions.* Burlington, MA: Jones & Bartlett Learning, LLC.
Moore, J. W. (2011). Transcending the metabolic rift: A theory of crises in the capitalist world-ecology. *The Journal of Peasant Studies, 38*(1), 1–46.
Morello-Frosch, R., Pastor, M., & Shonkoff, S. B. (2009). *The climate gap.* Los Angeles: USC Center for sustainable cities, University of Southern California.
Nielsen, J. A. (2012). Arguing from Nature: The role of 'nature' in students' argumentations on a socio-scientific issue. *International Journal of Science Education, 34*(5), 723–744.
Ostrom, E., Burger, J., Field, C. B., Norgaard, R. B., & Policansky, D. (1999). Revisiting the commons: Local lessons, global challenges. *Science, 284*(5412), 278–282.
Polanyi, K. (1957). *The great transformation.* Boston: Beacon Press.
Sharma, A., & Buxton, C. A. (2015). Human–nature relationships in school science: A critical discourse analysis of a middle-grade science textbook. *Science Education, 99*(2), 260–281.
Sharma, A. (2013, April). Where are the people? Understanding representations of society-nature relationships in a middle grades science classroom. In *Paper presented at the Annual Meeting of American Educational Research Association,* San Francisco.
Sharma, A. (2012). Global climate change: What has science education got to do with it? *Science & Education, 21*(1), 33–53.
Smith, N. (2008). *Uneven development: Nature, capital, and the production of space.* Athens, GA: University of Georgia Press.
Stern, N. (2009). *The global deal: Climate change and the creation of a new era of progress and prosperity.* New York: Public Affairs.
Treanor, B. (2010). Environmentalism and public virtue. *Journal of Agricultural and Environmental Ethics, 23*(1), 9–28.
Vandermeer, J. H., & Perfecto, I. (2005). *Breakfast of biodiversity: The political ecology of rain forest destruction.* Oakland, CA/New York: Food First Books.
West, C. R. (1999). *The Cornel West reader.* New York: Basic Civitas Books.
Williams, R. (1980). *Ideas of nature: Problems in materialism and culture* (pp. 67–85). London: Verso.

Chapter 32
STEPWISE: A Societal-Historical Activity (Activism) Theoretical Perspective

Wolff-Michael Roth

> Philosophers have but differently interpreted the world, the point is to change it. (Marx & Engels, 1978, p. 7)

In the introductory quotation, Karl Marx and Friedrich Engels point out that philosophers—today they might have added the pure (natural, social sciences, humanities)—have only interpreted the world in the effort to understand it when the real point is to change it: for the well-being of *all*. The STEPWISE[1] approach is described as using science education and technology education promoting well-being for individual, societies, and environments as a unit (category) that cannot be further broken down into elements (e.g. Bencze & Carter, 2011; Bencze & Krstovic, 2013). Any one of its (five) parts—STSE education, Skills education, STSE actions, Students' research, and Product education—can be understood only in its relation to all the other parts and to the STEPWISE approach as a whole. In other words, any part is like the proverbial raindrop that reflects its universe. If we were to isolate any one part for the purpose of analysis, taking the STEPWISE framework, we would find (or want) all the other parts reflected in it. In this approach, science and technology education is all over and integral to every part. That is, science and technology education is the quintessence of the approach, its driver and its outcome.

The authors and developers of STEPWISE state an underlying goal, which is laudable and with which I cannot but agree: the well-being of individuals, society, and environment. Moreover, students *are asked to take* a communitarian view.

[1] STEPWISE' is the acronym for Science & Technology Education Promoting Well-being for Individuals, Societies & Environments. It is a theoretical and practical framework that organizes teaching/learning goals in ways that encourage and enable students to self-direct research-informed and negotiated actions to address personal, social and environmental problems linked to fields of science and technology. To learn more about this framework, refer to Chap. 2 in this book.

W.-M. Roth (✉)
Faculty of Education, University of Victoria, Victoria, BC, Canada
e-mail: mroth@uvic.ca

L. Bencze (ed.), *Science and Technology Education Promoting Wellbeing for Individuals, Societies and Environments*, Cultural Studies of Science Education 14,
DOI 10.1007/978-3-319-55505-8_32

However, *why* they should take such a view is not so clear. It is particularly important to think through this relation between individual and collective, because some members of society—and many others less intentionally—do exploit others for the express purpose of enriching themselves. This is a form of thought that is kept alive in/as societal relation and, therefore, will not be eliminated by teaching—especially not by teaching that thinks the collective on the base of the individual—e.g., altruism is simply the counterpart to ego(t)ism. Negation does not eliminate that which it negates; negation reinforces, and is built on, that which it negates. In the context of schooling, which is still oriented to the reproduction of society, STEPWISE becomes a contradiction: it aims at the well-being (of all) but may in fact support the differentiation typical for an unjust society.

In this chapter, I formulate an approach for theoretically grounding something like promoting well-being of individuals, society, and environment (WISE) in ways that take into account society and dialectical relations with its members. That is, whereas the WISE approach has been formulated from within science (and technology) education—in the way other approaches have been offered, i.e., as "new drivers" (e.g., Fensham, 2002)—the approach I describe takes as its starting point history of society and division (of labor) of society into activities that serve generalized satisfaction of needs (Marx & Engels, 1983).

32.1 Understanding *Activity*: A Societal-Historical Approach

STEPWISE is a framework for thinking about curriculum design. It is not a theory in the psychological, sociological, or natural science sense, where we would have latent and overt factors that make a system with determinate or mediating effects upon each other. As an individual who has had a life-long interest in theory for philosophical reasons, I would want a theory, especially I would want a theory that links what students do to society at large. Societal-historical activity theory is one such theory that allows me to understand the STEPWISE approach not only as currently used in schools but also when it becomes the guiding idea for organizing informal learning settings, such as summer camps, after-school settings, or museums.

In the societal-historical approach, theorists are concerned with categories and holistic analytic units (Leont'ev, 1983). The problem of traditional psychology is that it reduces human characteristics to elements (factors) from which the phenomenon is constructed. On the other hand, a unit is "an analytic result that—in contrast to elements—*contains all the fundamental properties that are characteristic of the whole*, and which represents the irreducible living parts of the unitary whole" (Vygotskij, 2005, pp. 672–673). When we investigate properties of phenomena characteristically *human*, the smallest unit turns out to be productive activity (*dejatel'nost'*) (Leont'ev, 1983)—farming, manufacturing, trading, and educating constituting some typical examples. The English word *activity* is actually not clean, because it confuses two forms of events that are separate in the original formulations of activity theory: societally motivated activity and (vital) activity (*aktivnost'*). Student assignments, experiments, and tasks are not activities in the societal-historical approach but are part of schooling, the unit of analysis required for understanding what students do and why. The latter in fact retains characteristically human properties, such as language, practices, and (*societal* rather than *social*) relations. It has been noted, for example, that human relations within schooling are typical of the middle and upper classes, and atypical of human relations within working classes (Eckert, 1989). For this reason, schooling inherently is biased against working class students, who either give up their relations or fail because of a mismatch between what they practice and what schools enforce; this leads to the fact that working class kids become working class adults (Willis, 1977). Thus, as long as STEPWISE is part of schooling, I would theorize it in terms of associated characteristics. It is only when the subjects of interest are participants in activism that the characteristics will fully reflect those of activism.

The different activities that constitute society have evolved historically through increasing division of labor as existing capacities—e.g., division of labor and tool production already exist among chimpanzees—come to be generalized and coordinated. Thus, at the beginning of anthropomorphosis, collective hunt existed in the division of labor between beaters and killers. Tools were fashioned by those using them (Leont'ev, 1959). The first generalized division of labor occurred when some members of the group (society) stayed in the camp fashioning the knives and spears with which others hunted. In *exchange*, those fashioning tools would receive products of the hunt. In this way, provision of needs was shifted from the individual to the society and, thereby, it became generalized. As long as the individual contributed to the *general provision of needs* (control over condition), s/he was enabled to meet *individual needs* (e.g., through exchange processes). Historically, the increasing division of labor led to a patchwork where formal education became one of the forms of activities that make a society.[2]

Once I adopt activity as the unit, then I no longer need to introduce contextualization so important to other approaches. Activity includes the (individual or collective) *subjects* of activity, their *objects*, *tools*, *communities*, *rules/laws*, *division of labor*, and the final *products*. Each of these aspects of an activity reflects all of the other aspects and the activity as a whole (Roth, 2014c). Participating as subject in activism, I (Wolff-Michael Roth) am different from the subject I am in the university (activity = tertiary schooling) or in the supermarket (activity = exchange relations). The differences arise from the differences in the societal activity that constitutes the ways in which we make sense both as participants and as researchers looking at the participating subjects. Some tractor is a very different tool when part of farming than part of the repair industry or of tractor pulling and tractor stock racing events. These differences are reflected in the different forms of consciousness between the

[2] On an absolute time scale, different societies evolve formal schooling at different time points. Moreover, schooling may initially exist on a voluntary level and for those who can afford it and later become obligatory for all youth. Some societies did not have formal schooling right into the twentieth century.

farmer and the tractor stock racer. The rules of engagement and relations differ, and so does the division of labor. The products of each activity also are different. Thus, STEPWISE that is part of realizing schooling activity is different than if STEPWISE were part of an after-school or museum (leisure) activity.

Societal activity is a unit that contains within it a full production cycle, beginning with raw materials to the final product (Roth, 2014d). Such changes include those at the biological level: even thinking uses energy supplied by food; and even the most boring job leads to incremental changes in my capacities to do it (Y.-J. Lee & Roth, 2005). Time and change, thereby, are inherent in the analytic unit and category of thought. I do not need to theorize learning separately. Participation in activity is equiprimordial with change: in the individual, society, and environment. That is, participation in activism, when it is in the *general* interest common to all, *inherently is for well-being*. Anything said to be stable, such as knowledge, comes to be problematic, because in a unit of change everything changes as well. Only dead things do not change.

Societal activity also frames the goal-directed actions that compose it: actions realize the activity, but activity motivates actions. In the context of STEPWISE, this would be equivalent to saying that in pursuing WISE issues, some students will find that their range of actions increases if they learn to graph. Some action, in a different activity, would be a different action. Goal-directed conscious actions consist of sequences of conditioned operations—like signing with one's name is composed of writing individual letters. However, we are not conscious of writing individual letters but of signing. Learning to graph inherently means labeling graphs and writing units—we do not have to think about these parts of the graph but they come from our pens or computer actions in the pursuit of producing graphs. Some of the research within the societal-historical approach has shown that this determination of smaller and smaller parts of the whole goes right down to the neural level (Luria, 1973).

Activities are also the most useful frames for understanding the person. Within an activity, the person undergoes a process of *subjectification* (Roth, 2013). That is, as subjects of activity they contribute to producing the conditions within activity to which they are subject and subjected. Individuals, therefore, not only are agents but also patients within activity, which they undergo as much as assist in bringing about. Our subjectivities are shaped in and through participation, both determining and being determined by the societal relations we entertain with others (Vygotskij, 2005). The second important category is *personality* (Leont'ev, 1983). Because we participate in multiple activities, all of which are connected into a patchwork that makes up society, personality is the ensemble of the societal relations we entertain in these different activities—which not only include schooling but also all those other ones that make our day, such as life in the family and community, exchange (shopping), gardening, apiculture, or leisure. Any person therefore is thought in terms of a patchwork that reflects the patchwork of society, but the personal hierarchies of these activities are different, which is expressed in the form of different personalities. Thus, a person will exhibit very different characteristics when her most important activity is that of activism than if activism is lower in the hierarchy.

I am a very different kind of activist than the leader of the activist group that had an unhealthy Hagan Creek / ḰENES as its object and its revitalized state as its motive.

32.2 Theorizing the Four Principles of STEPWISE

STEPWISE is organized around four principles: constructivism, holism, contextualization, and communitarianism. In the opening parts of the preceding section, I suggest that STEPWISE is not a theory in the traditional sense. In this section, I comment on how to integrate these principles into a theory of activity.

32.2.1 Constructivism

More than 25 years ago, I became "a constructivist" after having done a PhD in the neo-Piagetian paradigm, which combined developmental theory with short-term memory cognitive theories. Initially, radical constructivism appealed to me. But I abandoned the theory when returning to the classroom where I became aware of the tremendous social dimensions of learning. My classroom research contributed to the establishment of the social constructivist approach in science education (e.g., Roth & Roychoudhury, 1992) and to the notion of authentic science in science classrooms (Roth, 1995). Since then, having conducted many studies especially in the workplace and other informal learning settings, I have come up with a lot of evidence in support of the conclusion that constructivism is a wrong theory (e.g. Roth, 2011). This is so in particular because *to construct* is a transitive verb and needs an object: we construct something (with the materials and tools at hand). But, as pragmatic philosophers suggest, those who are in the process of evolving something new—which readers may articulate in their favorite theoretical discourse as evolving a new language, a new discourse, or new knowledge—cannot know what they are learning and creating (Rorty, 1989). Students are frustrated and ask teachers questions like "Am I right so far?" precisely because they do not know what the endpoint of their actions lies (i.e., the new knowledge, discourse) and, therefore, also cannot be metacognitive about what they are doing (Roth & Radford, 2011).

Abandoning the constructivist principle does not hurt the STEPWISE approach. Indeed, I would think that the agency | passivity approach is in fact providing it with a new tool. If we— like scientists, are subject to radical uncertainty with our actions (Roth, 2009)—never know what our situated actions are until after these have happened (e.g. Suchman, 2007), then the precautionary principle is de rigueur. The precautionary principle, however, is *the* one principle that environmental activists have elevated to dominate all discourse not only about the actions of industrialists and global market actors but also about the actions scientists, especially those who work for Monsanto and similar companies that have become the epitome of evil for environmentalists. If STEPWISE were to reproduce a culture of devil's apprentices by training GMO-happy scientists, then certainly it would not act for the general (common) good.

Constructivism is problematic even in its social version because it is built on the idea of the individual as the source of its knowledge, identity, actions, motives, or beliefs. It therefore fits into an individualist worldview. On the other hand, societal-historical theorists note that humans distinguish themselves from other animals in terms of their particular society. I therefore had the insight of an appropriate aphorism: *mind is in society*, as the title of one book with translations of Vygotsky texts suggests, *because society is in the mind*. This certainly is the underlying message in the diction that "every higher-order psychological function was external, that is, was societal; before it could at all be a function it was a societal relation between two people" (Vygotskij, 2005, p. 1021). Any idea I may have, any thought, even any invention, inherently is a societal phenomenon or is nothing at all—because the word in which I can express the idea always is a word for two or, as Vygotsky points out, is not a word at all.

32.2.2 Holism

Integral to STEPWISE is the contention that it is a holistic framework. Thus, "learning in one domain (e.g., Products) is and should be related to learning in other domains (e.g., Skills Education)" (STEPWISE, n.d.). However, if it is holistic, then there is no choice about the relation between domains: In a truly holistic approach, the relation always already *is* so that there is no choice whether it should or should not be. In the societal-historical approach, whether it focuses on activity or activism (Roth, 2010), it makes no sense to separate operations (i.e., skills), which are conditioned parts that make actions, independent of the societal activity. It makes no sense to say that teachers should be "addressing elements around the periphery of the framework (e.g., Products Education)" (STEPWISE, n.d.). Operations are integral parts and functions of the societal activity; and when the activity changes, its constitutive actions change, and so do the operations (Leont'ev, 1983). This explains why engineers and scientists make very different use of the same mathematical equation (Brown, Collins, & Duguid, 1989).

Thus, if the framework is holistic then it does not make sense in my world to claim the framework to "impl[y] that teaching and learning can begin (and end) with any domain" (Bencze & Carter, 2011, p. 659), for every domain is a reflection of, tied to, and active in every other domain and in the framework as a whole. In my own teaching, therefore, the societal activity was taken as the minimal unit. In experimenting, students were developing everything that is required to experiment. If operations ("skills") are a function of the activity in a holistic approach, there is no way that they can be taught independently. Rather than suggesting that teachers and researchers can start with any one of the five domains—STSE Education, Skills Education, Students' Research, Products Education, and STSE Actions—my activist theory includes all of these domains (Roth, 2010). It is of the same fractal nature as the societal-historical theory of *production* (Marx & Engels, 1983). Thus, production entails (is for the purpose of) *consumption*, *exchange*, and *distribution*. Each of these theoretical terms is again thought in terms of productive activity. For example, there are subjects, objects, tools, division of labor, rules/laws, and community in (market) exchange activity. Thus, in my approach to activism, students participate from the beginning in activism that already exists as possibility at the societal level. Within this participation, all the other domains have their place. But because the division of labor is integral to my thinking about activity, I do not require every student to engage in the typical science process skills. Instead, I think about scientific and technological literacy as *collective praxis* (Roth & Lee, 2002). Thus, we need these forms of literacy to emerge at the collective level. Even if some students were to focus solely on the legal, ethical, or indigenous aspects of their actions—and therefore not do science experiments—they would still contribute to making scientific and technological literacy emerge collectively. But because of the dialectical relation between collective control and individual control, these students and their actions would still be constitutive of the collective praxis. Forcing every student to do the same experiment or do the same skills education is hegemonic and undemocratic.

Some readers may suggest that we harm students if we do not make them develop skills. But we do not have to make them learn skills—if the students see that learning something inherently increases their action possibilities, their room to maneuver, and therefore their control over conditions, then not learning not only would be illogical through their own eyes but also detrimental (Holzkamp, 1993). This was beautifully illustrated in the work of Paolo Freire with peasants, who, once they saw how much their action possibilities would increase if they knew how to read, they really wanted to learn how to read. Why would they not engage in something that increases their control? In another example that turned out to be an interesting experiment in informal science learning in a museum, students were invited to construct airplanes (Leont'ev, 1983). In the first part of the experiment, the young museum visitors showed no interest in learning the physics of flying. They saw no need for it given that they were pursuing the goal of making nice planes. Then the motive of the construction activity was changed to making the planes cross a certain distance. At this point students began to use the physics resources by an order of magnitude more following the early crashes of the planes and following their realization that with some physics they could gain control over the performance of their planes. We found the same to occur in our activist research, where students, seeing what others were doing and how they employed particular tools, began to use and learn the use thereof without having to be coaxed or required (Roth, 1998a; Roth & Barton, 2004).

32.2.3 Contextualization

The founders of the approach suggest that "STEPWISE education *should be* set within personally relevant and practical 'real-world' contexts" (STEPWISE, n.d.). Again, from the societal-historical perspective, every activity is contextual. There are

Fig. 32.1 A student reports her findings—from a project including photographing and describing a local creek and interviewing the mayor, community members, and indigenous elders

some important distinctions to be made. If STEPWISE occurs as part of schooling, *this* constitutes the context. Schooling, however, one of the many society-constituting activities, has its own motive: the reproduction of society—not only in terms of its cultural knowledge and practices but also in terms of social structure (class). STEPWISE would be just another "driver" of science education subject to the production and reproduction of an inequitable society (Roth & McGinn, 1998). Little to nothing would have been gained, for students engage in the STEPWISE curriculum in the way they are engaging in all curriculum: defensive doing/learning to avoid negative consequences and harm (low marks/grades for most middle-class students) (Holzkamp, 1993). That is, even in my own former teaching, where students contributed to community activism and whose final evaluation was their participation in an activist-organized open house event in the community (Fig. 32.1), the students still participated within schooling, which shaped the division of labor (teachers and chaperones versus students), the rules, or community of practice (school science). I came to the conclusion that to change science learning we have to deinstutionalize science and technology education (Roth & McGinn, 1997), making provision for participating in activism as it already occurs in and through societally motivated activity rather than using programs such as STEPWISE to organize *school curriculum* in yet another way. If STEPWISE is used in school systems intending to graduate students with differentiating report cards, then "What'd I get?" will continue to be the most important question students ask; and (middle and upper class) parents will continue to be concerned with grades and report cards that become the key to entering the colleges and universities of their choice.

Contextualization should not only refer to situating student inquiry in the world they know, which is the very foundation for anything scientific even when the scientific overturns their everyday intuitions. Instead, contextualization also should refer to students' interests, inclinations, and preferences. In the way some people like rock over classical music, we all have preferences for the different fields of human endeavor. In this context, not all students take to science; and forcing them to do science and technology oriented tasks is just as harmful a context for individual development as is presenting scientific and technological book knowledge without reference to the settings where such knowledge might come in handy. One of the seventh-grade students I worked with, Michelle, experienced little interest in doing the scientific experiments that some of her peers did; and she showed little interest in the naturalist observations and leaf printing that other girls were doing. Her contribution to the *overall project* of well-being of individuals, society (individualized in our municipality), and environment by means of a journalistic endeavor: reporting, verbally and photographically, and interviewing people for the purpose of informing her community (Fig. 32.1). Forcing her to do a correlational study, we would certainly have lost her interest in contributing to WISE. Moreover, the selections of academics and educational bureaucrats concerning what is relevant knowledge to be implemented in formal education are not used in the field by those taking courses. Our research in maritime education showed that the seafarers taking college courses for upgrading their status could not use what they knew from working onboard; and what they were taught in the courses turned out to be useless onboard (Emad & Roth, 2008). The electrician apprentices that we followed over a four-year period had to learn mathematical approaches (trigonometry) that were inferior to far more fail-proof, tool-centered approaches characteristic of electricians' practices (Roth, 2014b). All the contextualization of trigonometry in the college classroom failed to make these apprentices better apprentices.

32.2.4 Communitarianism

An important aspect of STEPWISE framework is communitarianism. This aspect is defined by stating that "the ultimate goal of all STEPWISE-related lessons and student activities should be to encourage and enable them to take action(s) to address WISE Problems" (STEPWISE, n.d.). The issue could be approached differently at a conceptual level—focusing on the constitutive relation between society and its individual members. It has been noted that an appropriate determination of individual subjectivity in its relation to societal subjectivity requires us to realize that *individual* life conditions always already are individually relevant *societal* life conditions (Holzkamp, 1979). As individuals we are exposed to the conditions and relatively powerless. The excesses of globalization and neoliberal trends against which STEPWISE are directed (Bencze & Carter, 2011) are but one area in which the individual is powerless in the face of superhuman forces. We overcome such powerlessness only in expanding our individual action possibilities (room to maneuver) by participating in the (collective) control over general, collective conditions (Holzkamp-Osterkamp, 1978). The potentiation an individual experiences is proportional to the extent to which the conditions are relevant to the individual. This is so "because the higher order of societal interconnections necessitates a higher order of collective force display to control those conditions" (Holzkamp, 1979, p. 11). The formulation pertains directly to the STEPWISE framework, which, through student engagement intends to enable addressing well-being of individual, society, and environment. We may observe this to occur not through students' individual actions but through their participation in the *collective* control of which their individual actions are a constitutive part. That is, the collective control is the motive and individuals increase the control over their own conditions through participation in the (collective) control over generalized collective conditions, which in turn are concretized in individual conditions.

Most of us do not realize this connection between individual and collective and, thereby, come to act against our own interests. For example, buying something that is cheaper tends to be against the interests of well-being of our own communities because cheaper products tend to come from nations where the workforce is exploited (China, Bangladesh, India, Philippines). At the same time, buying foreign products—e.g., fruit and vegetables—means not contributing to the maintenance of a local workforce (farmers, farm hands), whose own spending would in turn put money back into local employment. Most people do not appear to realize that they are acting against their own well-being in purchasing cheap products from abroad rather than supporting employment in their own community, where the wages have to be sufficiently high to allow workers to survive.[3] Moreover, buying foreign increases the ecological footprint, pollution levels through the burning of fossil fuels, and pollution through the use of pesticides and herbicides. Thus, I live in an area of Canada where there is the capacity to produce almost 90% of all the vegetables we consume in the area—but there are only few of us who are willing to shift (a large part of) their consumption to less than a 100-mile and even 0-mile diet and to contribute to producing the food locally.

The STEPWISE approach focuses on the *common good*, which is thought in terms of "altruism" (Bencze & Carter, 2011, p. 659). I, too, had been thinking about the other as integral part of theorizing science education using the term *solidarity* (Roth, 2007b). To think this part of the STEPWISE approach requires theorizing the relationship between the individual and the collective, between the individual control over the conditions for the satisfaction of basic and extended needs and the collective, generalized control over the generalized conditions—which occurs at the societal level. *Altruism* is a rather new word, created during the early 1900s by A. Comte—sometimes considered to be the first philosopher of science in the modern sense—and his followers. The word was constructed on the model of its opposition, *egoism*, a term referring to the belief that takes the personal mind as the center of being and to the self-centered interests. The Latin *alter*, other, is the basis of the idea of altruism, the selfless concern for the well-being of others.

---

[3] In Brisbane (Queensland, Australia) bananas, which are grown just outside of the city, are three times as expensive as in Vancouver (British Columbia, Canada) where the bananas have come from Central America. Brisbanites support Australian farmworkers; Vancouverites contribute to the exploitation of farmers and workers in Central America.

Inherently, therefore, the other is thought from the position of the self (ego), with all the theoretical and philosophical weaknesses that come from doing so despite the good intentions underlying the concern for the other.

Instead of emphasizing the individual, which then leads to the need to introduce the notion of altruism, we may want to focus our attention on *general, common* interests as opposed to special interests of certain groups. Special interests remain special interests and cannot be overcome by juxtaposing it by other special interests—as K. Marx did with his call to a battle between social classes. Unjust working conditions cannot be overcome by the special interests of labor unions. The special interests might be those gaining from the globalization of markets, which are confronted by the special interests of those opposed to such trends. All we get from such opposition is confrontation of special interests. On the other hand, common interests are general in the sense that these are mine and those of everyone else. I do not need to have a special orientation towards altruism if I act in the common interest, because I act both in the communal and personal interest. General interests therefore are those of the other and acting in the general interest is acting in the interest of self and others. We no longer need the altruism; and self-interest (ego(t)ism) will be the same as other-interests (altruism).

32.3 If I Were a (Science) Teacher Again Adopting WISE as a Motto …

The creators and developers of the STEPWISE approach invite and work with teachers to reorient the ways in which they teach science. I have been a science teacher and also directed a science department in the province of Ontario where STEPWISE is implemented and where a special version of it is developed (STEPWISE, n.d.). As a teacher, I am not a technician but develop ways of thinking and implementing curriculum. I also have considerable experience in teaching science and technology, especially by means of independent, student-centered inquiry and in design contexts where students contribute through activism (e.g. Roth, 2001b; Roth & Lee, 2004). Most importantly, I also taught personal health and development, physical education, and the fine arts. These, too, are dimensions of human experience that cannot be disconnected from those that are said to be quintessential to the sciences and technology. Concerned with the well-being of my students, whether or not they exhibit an affinity for and interest in in science, I articulate in this section some of the ways in which I would make sense of and think about student learning through contributions to the well-being of the individual, society, and environment (WISE).

By its very nature, STEPWISE focuses on science and technology education as a starting point (location, origin, position) for thinking about the education of students (Bencze & Carter, 2011). In the STEPWISE approach, students are to "apply their science and technology education, including their primary and secondary research findings to take socio-political action" (STEPWISE, n.d.). The students are to focus on STSE issues; and it is here that the particular is used to frame what students are to do and to constitute the context for what and how to learn. As a teacher and citizen concerned with the well-being of my students, my own predilection is to think science education from a different standpoint—to deinstitutionalize it much in the way Italian society was one of the leading countries to rethink its approach to psychiatry. Rather than institutionalizing patients, which therefore where disconnected from the very societal relations that constitute the foundation of higher-order psychological functions and personality (Leont'ev, 1983; Vygotskij, 2005), psychiatry was deinstitutionalized leading to patients living in small groups in the community, where they were part of societal relations with others.

The authors "have encouraged teachers to help students to develop expertise and confidence in conducting correlational studies, as well as experiments, as some bases for their actions" (Bencze & Carter, 2011, p. 662). It turns out that this also was my own approach as a science teacher early in my work, as I thought science education from the standpoint of science education. Like the teachers in the STEPWISE approach, I selected the topics or phenomena and students collected their own data. Over time, while still teaching, I realized the shortcomings of this way of thinking. Students did not have a stake in the *What* and *How* of the phenomena under investigation. It was at that point that I created the conditions for students to design their own curriculum. For example, in a 6-week unit on electricity that was part of the Ontario curriculum, I provided students with the two-page description. Mostly in groups—though one of the 54 students in my three classes wanted to work on his own—students were then designing what to investigate and how. For example, one group decided to investigate superconductivity, which they expanded into a comparative investigation with regular conductors and semi-conductors during discussions with me. They planned the purchase of liquid nitrogen, negotiated with the others teachers to have an entire day to do their superconductor experiments, and organized the rental of a Dewar flask. Another group decided to develop curriculum materials for fifth-grade students. They were later teaching the unit to an actual class; and it was this teaching that constituted part of their evaluation. Evaluation, to a large extent, also was shifted to the students. Thus, in a discussion with the students, we settled on the following contributions to their term grade: 35% teacher evaluation, 5% self-evaluation, and 60% peer evaluation. In my approach to WISE, I would shift in a similar way. I would also shift the emphasis away from science and technology as the organizers and would make the curriculum problem oriented.

During the latter part of the 1990s, while conducting the research on activism as a curriculum context, I found out again that many students did not begin their activism with typical science investigations (e.g., Roth & Barton, 2004; Roth & Lee, 2004). Enticed into activism by newspaper articles and classroom visits of activists, a group of four girls wanted to document the state of the local creek in the ways typical of journalism: using photographs, audio-recorded descriptions produced while visiting various parts of the creek, and interviews with the mayor, adults, and aboriginal elders. They later reported their findings as part of an activist open-house event in the community, where their exhibit was next to those of the adult activists (Fig. 32.1). They were never forced to do the kind of investigations more typical of classical science education, such as measuring the speed of the creek water and its correlation with the frequency of different organisms or with the depth, width, and cross-sectional area of the creek. In and through their interactions with others—e.g., during whole-class presentations and discussions—they learned about what others were doing and how. They were free to add to their own work something they had learned from others, which some did, but were not held to do so—they were really engaging in *free-choice learning* (e.g., Falk, Heimlich, & Foutz, 2009).

As a teacher, I would also want to think curriculum from a whole-life perspective rather than from the position of an individual subject (Roth, 2014a). In life, we do not do science, technology, history, or mathematics separate from the concerns arising from life—unless we are working in a profession that explicitly is associated with one of these domains (e.g., as mathematician, physicist, or historian). In life, issues are messy and science and technology never have the sole or correct answer. Thus, for example, I researched for over a decade the struggle of one part of my municipality over access to community water that everyone else already has access to (Roth, 2008). It turns out that water is an issue from the perspective of health, municipal engineering, (environmental, community) politics, community development, law (ecojustice, environmental justice), geology, and agriculture. Accordingly, very different forms of expertise were brought into the process (e.g., Roth et al., 2004). It turns out that there was justified critique of the methods that the scientists used to determine water quality; and the valuable historical, qualitative and anecdotal knowledge about water levels were completely disregarded as the (false) scientific evidence was used in the (political) attempt to discount other forms of knowing.

In my own classes, students might learn mathematics while pursuing the question of the relationship between the shape of a car and wind friction that they chose in the context of the Ontario curriculum on motion. While they were working on the data analysis, I might introduce some mathematical or statistical modeling tool, or students see these tools used by others. In their focus on the relationships between speed and acceleration, the students would be confronted with relations that are the very basis for understanding the mathematical topics of differentiation and integration of a function. My students also discussed philosophical issues, for example, discussing texts by the environmentalist / scientist / broadcaster David Suzuki, the anthropologist / philosopher Gregory Bateson, or a book with a Wittgensteinian take on physics as language by the associate director of the Harvard Smithsonian. That is, in my "science" classes, science actually is but one of the things students draw on and are concerned with rather than constituting the exclusive goal that STEPWISE currently takes.

We can also draw inspiration for rethinking the theoretical foundations of STEPWISE by using the ways in which "disabled" come to be integrated in societal activity as an analogy. Often, those with some form of "disability," including autism or Down syndrome (trisomy 21), are relegated to special institutions or care facility, or simply left to the care by the family. However, a recent Canadian documentary shows that some companies have realized that employing "disabled" persons comes with economic benefits—not because the "disabled" are exploited but because they are abled in ways that make them do certain jobs better than the nominally "abled" persons (CBC, 2014a). That is, in the face of saying that he was hiring "disabled" persons, the employer describes them as being better at the job than "normal" persons. Of course, this is a contradiction in terms. It would be much better to say that not only we are different from (groups of) others but also that we are different from ourselves. We all differ and the appropriate approach to theorizing is to think every person in relation to the society that exists only in and through its differently abled members, whose abilities differ across the different activities that constitute the patchwork of society. There are similar projects with "autistic" individuals, which show that these individuals frequently are much better at certain jobs than normal ones, such as a high school graduate who works as data analyst for a software engineering company in a job normally done by university graduates (CBC, 2014b).

The authors of the STEPWISE approach find "that implementation of the [STEPWISE] theoretical framework … is extremely difficult" (Bencze & Carter, 2011, p. 660). They report that not a single teacher with whom they worked implemented the model as described and intended. One way of bringing about change that worked literally every time we used it is coteaching. In this approach two or more individuals teach together each person taking full part in the full responsibility for a course (e.g. Roth, Tobin, Zimmermann, Bryant, & Davis, 2002). This model evolved in the early 1990s when I assisted an elementary school in its effort to address the student-identified need for more student-centered learning. The teachers felt inadequate teaching science through design activities. I offered coteaching with them, providing opportunities to learn from each other. Over time, teachers became so confident in teaching by means of open-inquiry design that they partnered up with other teachers to allow them to learn teaching in this way (e.g. Roth, 1998b). Thus, as a researcher or head teacher, I would be coteaching with my peers so that they could develop the confidence and competencies required. Moreover, students would also become part of our debriefing sessions (e.g. Roth et al., 2002). This is so because my own predilection is the organization of education around the idea of the collective control over condition, and the fact that more important than knowing this or that science and technology (and associated skills) is the competence to engage in a collective endeavor, where the competencies of the group matter: scientific literacy is an emergent feature of human practice generally (Roth, 2003) and transcends any intention that begins participation and intentions (Roth, 2007a).

If I were to use WISE as my teaching framework, I would think about activism in terms of participation in collective control over conditions. I would want my students to develop an appreciation of how their participation in the collective control over condition in fact expands their control over individually relevant aspects of the conditions. An individual cannot change the municipal bylaws concerning the ways in which we use water, draw on creek-water resources, and contribute to its environmental health; but collectively, as one of our studies showed, by contributing to community activism and community-based scientific literacy (S. L. Lee & Roth, 2003c) and in networks that exercise collective control through community action (S. L. Lee & Roth, 2003b), we can change the bylaws.

Through their interactions with others, community activists produce "hybrids between formal scientific and local situated concerns" (S. L. Lee & Roth, 2003a, p. 120). Not all and often many of these activists have scientific background. They reflect the patchwork nature of the community, including lawyers, farmers, and concerned citizens in addition to some individuals with more traditional scientific savvy. These members of the community patchwork have motives determined by all the other collective activities in which they are involved; and their participation in the activist patchwork brings these other concerns to bear on the control over the environmental health of the watershed generally and the creek specifically. Participating is in the interest of the farmer, who, by contributing to the collective control also increases control over his conditions, which include the quantity and quality of the water available from the creek. The environmental lawyer contributes to the local patchwork, which in turn increases collective control over environmental issues connected with his larger project of extending sustainability through transdisciplinary research.

As a teacher oriented to WISE, I would emphasize a patchwork ontology and give up the hegemonic tendency underlying the attempt to push specific science skills (down the throats of students who have few options being a captive participant). I would emphasize the intricate relation of individual and collective control over conditions and the expansion of action possibilities (room to maneuver) through participation in collective control. But participation in collective control requires "skills" scientists often do not have: openness and willingness to accept the ways in which non-scientists frame problems (initially). The scientific and technological approach often is part of the problem rather than the solution. It is precisely by thinking in ways other than that inculcated through science and technology education—which includes inculcation into the blind spots of the scientific disciplines (Roth, 2001a)—that we come to act in communitarian and altruistic ways. We do not want indoctrination in science but perhaps more of a discussion of science as a language and tool that does fail to provide appropriate answers; my high school students came to this understanding and could make choices whether to draw on or reject the scientific ideology (e.g. Roth & Alexander, 1997). Thus, for example, my coauthor, a high school student at the time, insisted that "scientific and technological rationality cannot provide the necessary ethical dimension for making informed choices about dealing with issues such as abortion, euthanasia AIDS, or genetic engineering" (p. 139). He wanted to work for the "betterment of man," that is, the well-being of individual and society; and he became a pediatric physician.[4] These no longer have to be motives for what we do, because the underlying orientations already are in communitarian and altruistic in essence.

[4] My co-investigator and coauthor of this study, conducted during my last two years as a fulltime high school teacher, was a physics student. When he finished his education, which included a PhD in cell biology in addition to an MD, he became the third pediatric nephrologist (professor) in Canada.

32.4 Coda

One cannot but agree with the goal of putting the well-being of individuals, society, and environment (WISE) in the center of a society's concerns. This goal ought to be framing all of the activities that form society not only of schooling but also of manufacture, food production, or resource industries. In adopting such a goal, I would not limit myself to science and technology alone but would take the perspective of literacy in these areas as a collective *praxis*. There is no reason why persons of all ages would or should be excluded from such practice. In this, I see in STEPWISE one tiny piece of a bigger puzzle of realizing a changing of the world that Marx/Engels make the key point that should underlie all human activity: changing it for the well-being of *all* individuals, *all* societies, and the earth environment as a whole.

References

Bencze, J., & Krstovic, M. (2013, April). *Students' social studies influences on their socioscientific actions*. Paper presented at the annual conference of the American Educational Research Association, San Francisco, CA.

Bencze, L., & Carter, L. (2011). Globalizing students acting for the common good. *Journal of Research in Science Teaching, 48*, 648–669.

Brown, J. S., Collins, A., & Duguid, P. (1989). Situated cognition and the culture of learning. *Educational Researcher, 18*(1), 32–42.

Canadian Broadcasting Corporation (CBC). (2014a, February 26). *The economic benefits of hiring disabled employees*. http://www.cbc.ca/player/News/TV+Shows/The+National/ID/2439669551/. Accessed 31 July 2014.

Canadian Broadcasting Corporation (CBC). (2014b, August 5). *Woman with autism finds world of possibilities in new job*. http://www.cbc.ca/player/News/TV+Shows/The+National/ID/2483181649/. Accessed 5 Aug 2014.

Eckert, P. (1989). *Jocks and burnouts: Social categories and identity in the high school*. New York: Teachers College Press.

Emad, G., & Roth, W.-M. (2008). Contradictions in practices of training for and assessment of competency: A case study from the maritime domain. *Education and Training, 50*, 260–272.

Falk, J. H., Heimlich, J. E., & Foutz, S. (2009). *Free-choice learning and environment*. Lanham, MD: Altamira Press.

Fensham, P. J. (2002). Time to change drivers for scientific literacy. *Canadian Journal of Science, Mathematics and Technology Education, 2*, 9–24.

Holzkamp, K. (1979). Zur kritisch-psychologischen Theorie der Subjektivität II: Das Verhältnis individueller Subjekte zu gesellschaftlichen Subjekten und die frühkindliche Genese der Subjektivität. *Forum Kritische Psychologie, 5*, 7–46.

Holzkamp, K. (1993). *Lernen: Subjektwissentschaftliche Grundlegung*. Frankfurt am Main, Germany: Campus.

Holzkamp-Osterkamp, U. (1978). *Grundlagen der psychologischen Motivationsforschung 2: Die Besonderheit menschlicher Bedürfnisse—Problematik und Erkenntnisgehalt der Psychoanalyse*. Frankfurt am Main, Germany: Campus.

Lee, S., & Roth, W.-M. (2003a). Of traversals and hybrid spaces: Science in the community. *Mind, Culture, & Activity, 10*, 120–142.

Lee, S., & Roth, W.-M. (2003b). Re-mapping the landscape: Science educators within a community action network. *Perspectives in Education, 21*(2), 15–30.

Lee, S. H., & Roth, W.-M. (2003c). Science and the "good citizen": Community based scientific literacy. *Science, Technology, & Human Values, 28*, 403–424.
Lee, Y. J., & Roth, W.-M. (2005). The (unlikely) trajectory of learning in a salmon hatchery. *Journal of Workplace Learning, 17*, 243–254.
Leont'ev, A. N. (1959). *Problemj razvitija psixiki*. Moscow, Russia: Akademii Pedagogičeskix Nauk.
Leont'ev, A. N. (1983). Dejatel'nost'. Soznanie. Ličnost'. In A. N. Leont'ev (Ed.), *Izbrannye psixhologičeskie proizvedenija* (Vol. 2, pp. 94–231). Moscow, Russia: Pedagogika.
Luria, A. R. (1973). *The working brain*. Harmondsworth, UK: Penguin Books.
Marx, K., & Engels, F. (1978). *Werke Band 3*. Berlin, Germany: Dietz.
Marx, K., & Engels, F. (1983). *Werke Band 42: Grundrisse*. Berlin, Germany: Dietz.
Rorty, R. (1989). *Contingency, irony and solidarity*. Cambridge, UK: Cambridge University Press.
Roth, W.-M. (1995). *Authentic school science: Knowing and learning in open-inquiry science laboratories*. Dordrecht, The Netherlands: Kluwer Academic Publishers.
Roth, W.-M. (1998a). *Designing communities*. Dordrecht, The Netherlands: Kluwer Academic Publishers.
Roth, W.-M. (1998b). Science teaching as knowledgeability: A case study of knowing and learning during coteaching. *Science Education, 82*, 357–377.
Roth, W.-M. (2001a). "Authentic science": Enculturation into the conceptual blind spots of a discipline. *British Educational Research Journal, 27*, 5–27.
Roth, W.-M. (2001b). Learning science through technological design. *Journal of Research in Science Teaching, 38*, 768–790.
Roth, W.-M. (2003). Scientific literacy as an emergent feature of human practice. *Journal of Curriculum Studies, 35*, 9–24.
Roth, W.-M. (2007a). Toward a dialectical notion and praxis of scientific literacy. *Journal of Curriculum Studies, 39*, 377–398.
Roth, W.-M. (2007b). Toward solidarity as the ground for changing science education. *Cultural Studies in Science Education, 2*, 721–745.
Roth, W.-M. (2008). Constructing community health and safety. *Municipal Engineer, 161*, 83–92.
Roth, W.-M. (2009). Radical uncertainty in scientific discovery work. *Science, Technology & Human Values, 34*, 313–336.
Roth, W.-M. (2010). Activism: A category for theorizing learning. *Canadian Journal of Science, Mathematics, and Technology Education, 10*, 278–291.
Roth, W.-M. (2011). *Passibility: At the limits of the constructivist metaphor*. Dordrecht, The Netherlands: Springer.
Roth, W.-M. (2013). Activity, subjectification, and personality: Science education from a diversity of life perspective. In N. Mansour & R. Wegerif (Eds.), *Science education for diversity in knowledge society* (pp. 41–64). Dordrecht, The Netherlands: Springer.
Roth, W.-M. (2014a). *Graphing and uncertainty in the discovery sciences: With implications for STEM education*. Dordrecht, The Netherlands: Springer.
Roth, W.-M. (2014b). Rules of bending, bending the rules: The geometry of conduit bending in college and workplace. *Educational Studies in Mathematics, 86*, 177–192.
Roth, W.-M. (2014c). Reading activity, consciousness, personality dialectically: Cultural-historical activity theory and the centrality of society. *Mind, Culture, and Activity, 21*, 4–20.
Roth, W.-M. (2014d). Socio-cultural perspectives on learning science. In R. Gunstone (Ed.), *Encyclopedia of science education*. Berlin, Germany: Springer.
Roth, W.-M., & Alexander, T. (1997). The interaction of students' scientific and religious discourses: Two case studies. *International Journal of Science Education, 19*, 125–146.
Roth, W.-M., & Barton, A. C. (2004). *Rethinking scientific literacy*. New York: Routledge.
Roth, W.-M., & Lee, S. (2002). Scientific literacy as collective praxis. *Public Understanding of Science, 11*, 33–56.
Roth, W.-M., & Lee, S. (2004). Science education as/for participation in the community. *Science Education, 88*, 263–291.
Roth, W.-M., & McGinn, M. K. (1997). Deinstitutionalizing school science: Implications of a strong view of situated cognition. *Research in Science Education, 27*, 497–513.

Roth, W.-M., & McGinn, M. K. (1998). >unDELETE science education: /lives/work/voices. *Journal of Research in Science Teaching, 35*, 399–421.

Roth, W.-M., & Radford, L. (2011). *A cultural-historical perspective on mathematics teaching and learning*. Rotterdam, The Netherlands: Sense Publishers.

Roth, W.-M., Riecken, J., Pozzer, L. L., McMillan, R., Storr, B., Tait, D., et al. (2004). Those who get hurt aren't always being heard: Scientist-resident interactions over community water. *Science, Technology, & Human Values, 29*, 153–183.

Roth, W.-M., & Roychoudhury, A. (1992). The social construction of scientific concepts or the concept map as conscription device and tool for social thinking in high school science. *Science Education, 76*, 531–557.

Roth, W.-M., Tobin, K., Zimmermann, A., Bryant, N., & Davis, C. (2002). Lessons on/from the dihybrid cross: An activity theoretical study of learning in coteaching. *Journal of Research in Science Teaching, 39*, 253–282.

STEPWISE. (n.d.). *STEPWISE: Science and technology education promoting well-being for individuals, societies and environments*. Retrieved July 31, 2014 http://www.stepwiser.ca

Suchman, L. A. (2007). *Human–machine reconfigurations: Plans and situated actions* (2nd ed.). Cambridge, UK: Cambridge University Press.

Vygotskij, L. S. (2005). *Psyxhologija razvitija čeloveka* [Psychology of human development]. Moscow, Russia: Eksmo.

Willis, P. (1977). *Learning to labor: How working class lads get working class jobs*. New York: Columbia University Press.

Part IV
Afterword

Overview

Welcome to the last section of the STEPWISE edited book. It contains one chapter, an 'Afterword' written by me. In this final chapter, I have tried to review what I believe to be significant aspects of STEPWISE to celebrate and, as well, ways in which it may serve as a basis for further curriculum development, implementation in educational contexts and educational research and publication. My review of chapters in Part II of the book ('Documentaries'), in which teachers and graduate students describe particular situations of uses of STEPWISE to achieve certain kinds of educational goals, I note that STEPWISE seems to have significant potential for helping societies to become more vigilant about decisions made by powerful people and groups and, where they perceive problems, to develop and implement plans of action to address their concerns/interests. At the same time, it also seems clear that dominant perspectives and practices of school science—and, apparently, also in science teacher education and in at least one case of community youth education—are, to a great extent, antithetical to perspectives and practices inherent to the STEPWISE framework. In reviewing contributions of international scholars in Part III, similar conclusions seem reasonable. Some scholars provided examples of similar projects in their work and, moreover, indicated that such critical and activist curricular and pedagogical frameworks are needed now, perhaps, more than ever—in light of ever-increasing power from socio-economic elite and others. Using ideas and examples from their essays/reports, I engaged in some discussions about ways in which educators aiming to promote social and ecological justice/sustainability to gain some traction in our efforts to bring about a better world for all.

personal
monitoring
students
contribute
social
framework
discussion
fields
world
particularly
implement
decision-making
apparent
entertainment
action
perceive
technology
edited
potential
environmental
harms
youth
references
several
graduate
chapter
conglomerates
actions
Key
activist
perspectives
Assisting
research
societies
bring
banks
seem
contributions
networks
plans
develop
entities
environments
science
education
last
development
wellbeing
critical
STEPWISE

The Chapter

Chapter 33
Critical and Activist Science Education: Envisaging an Ecojust Future

Larry Bencze

33.1 Introduction

> We have bought hook, line, and sinker into the idea that education is about training and 'success,' defined monetarily, rather than learning to think critically and to challenge (p. 95) … [The] true purpose of education is to make minds, not careers (p. 107). … A culture that does not grasp the vital interplay between morality and power, which mistakes management techniques for wisdom, which fails to understand that the measure of a civilization is its compassion, not its speed or ability to consume, condemns itself to death (p. 103). (Chris Hedges, *Empire of Illusion*, 2009)

As a science education professor with tenure, which allows me to freely critique my university and the wider society, I believe I have no greater responsibility than to help provide citizens with expertise, confidence and motivation to critically interrogate sources of power in societies and, where they perceive indiscretions and weaknesses, take informed actions to try to make a better world. This mandate seems particularly necessary now, as agents of *neoliberal* socio-economic power appear to have rallied diverse and influential groups of actants—locally, nationally and transnationally—to causes emphasizing intense wealth concentration by elite, often at expense of wellbeing of many other individuals, societies and environments. Major instruments of such intense wealth concentration—and associated significant personal, social and environmental harms—appear to be fields of science and technology (and, likely, engineering and mathematics) and their educational counterparts, often with support from governments, transnational entities and, indeed, many citizens. Since 2006, however, I have been conducting action research aimed at understanding factors influencing educators' uses of teaching and learning approaches based on the 'STEPWISE' curriculum and pedagogical framework

---

L. Bencze (✉)
Ontario Institute for Studies in Education (OISE), University of Toronto, Toronto, ON, Canada
e-mail: larry.bencze@utoronto.ca

L. Bencze (ed.), *Science and Technology Education Promoting Wellbeing for Individuals, Societies and Environments*, Cultural Studies of Science Education 14,
DOI 10.1007/978-3-319-55505-8_33

(refer below) to enable citizens to understand problematic relationships among fields of science and technology and societies and environments and to develop and implement research-informed and negotiated actions to address relevant problems they perceive. Joining me over the years in using and/or discussing this framework have been many graduate students, teachers and colleagues—many of whom have contributed chapters to this book. I am very grateful to them for their dedication to causes of 'social justice' and 'environmental sustainability' (however complex and uncertain are these goals)—as well as for their generous contributions to this collective work.

In this last chapter of this edited collection, I provide summaries of some insights I garnered from contributors to this volume—in terms of practical and theoretical merits of STEPWISE and, related to those, possible ways forward in continuing efforts to contribute to wellbeing of individuals, societies and environments through science and technology (and other fields) education.

33.2 Needs for Activist Societies

33.2.1 Problematic Global Neoliberal Hegemony

It seems very clear that societies need cultural transformations that feature citizens willing and able to be ever-vigilant in critically analyzing and evaluating decisions made by powerful people and groups, including close scrutiny of governments in so-called 'democratic' societies—which evidence strongly suggests often are in dialectical relationships with networks of local, global and transnational self-serving capitalist entities. In agreement with many authors in this volume (admittedly, invited for such views), it is apparent that the world is virtually blanketed with influences from capitalists and other elite (Hardt & Negri, 2009; Harvey, 2005; McMurtry, 2013; Reich, 2007) who appear to have very-successfully organized networks of entities for their benefit. Their influences seem to be particularly powerful in the *neoliberal* era—which, while its origins and meaning are debatable, appears to involve considerable *intervention* in governments (vs. former *laissez faire*capitalism) and influences on networks of various other entities, collectively promoting conditions conducive to capitalist accumulation (Springer, Birch, & MacLeavy, 2016).

Although *capitalism*, an economic system prioritizing private ownership of resources and means of production for generating profit, has existed for centuries, and although it has experienced periodic crises—the most recent being the global financial crisis of 2008—over time, it appears to have become increasingly powerful. Keys to its strength and resilience appear to be its *transnational* and *discursive* characteristics. Despite signs of nationalist fervour, such as international athletic competitions like Olympic Games and curricular pronouncements about preparing citizens for competition in international markets, it is apparent that much power exists *between* nation states as a 'global capitalist network' (GCN)—apparently involving supranational (separate from nations) bodies, like the World Trade Organization (WTO), World Bank (WB), Organisation for Economic Co-operation and Development (OECD) and the International Monetary Fund (IMF), transnational corporations, think tanks, transnational advocacy groups and philanthrocapitalists (e.g., Bill Gates) that, like a giant three-dimensional spider web, also have ties with actants within nation states like pro-capitalist banks, governments, schools and universities (Ball, 2012; Hardt & Negri, 2009; McMurtry, 2013). In exercising its power, the GCN does not so much turn to military might, although it certainly can and has done so. Especially in 'developed' countries, it mainly appears to use *semiotic* influences. Through control of news and entertainment media, for instance, it has significant influence over public discourse around the world. In a similar vein, it is apparent that capitalists often exert their influences in subliminal—rather than overt, possibly physical—ways. This relates to Foucault's (2008) concept of *biopolitics*, a range of strategies and structures that enable powerful entities to influence populations. Through increased public surveillance (e.g., using video cameras and, more recently, as Matthew Weinstein points out in Chap. 28 here, through uses of drones), citizens can, in a sense, be trained to behave in ways they presume would be safe for them. At the same time, increased reporting of police actions suggesting dangers and proper behaviour can reinforce performances by people that align with priorities of the rich and powerful (Lemke, 2011).

33.2.2 Fields of Science and Technology as Instruments of Power and Destruction

Although we must think of neoliberal power as existing within networks, it seems that fields of science and technology (and, likely, engineering & mathematics) are essential instruments in this regard. On the one hand, it is apparent that such fields have benefited individuals, societies and environments in many ways. Average life spans of humans, for example, have been extended largely because of developments in life sciences and medicine and in fields of agricultural science and technology. Nevertheless, neoliberal influences over such fields (as part of larger networks) seems highly problematic. Many products and services they help generate and distribute for for-profit consumption seem harmful in many and varied ways. As discussed in numerous chapters in this book, considerable harms are linked to promotion of enthusiastic and unquestioning consumerism (Barber, 2007). Not to belabour this point, but humanity faces many potential and realized problems associated with consumerism facilitated by fields of science and technology—ranging, for example, from personal health threats linked to industrial food systems (Weber, 2009), including those stemming from increasing global obesity rates (NCD-RisC, 2016), through to massively-disruptive climate change (e.g., Klein, 2014). Although there are, undoubtedly, many ways to understand/explain culpability of fields of science and technology with such problems, analyses of reciprocal relationships between phenomena of the world and representations of them (Phenomena ←→ Representation(s)) seems helpful (Bencze & Carter, 2015). Very briefly, it is apparent that, ironically, relatively few people with minimal needs (e.g., in technologically 'advanced' societies) are convinced to develop *desires* leading them to repeatedly purchase and discard large quantities of products (e.g., cosmetics, electronics, clothing, etc.) and services (e.g., financial, entertainment, cosmetic surgery, etc.) largely on bases of various changing semiotic representations (e.g., 'sexy,' 'powerful,' 'cute,' etc.) of such commercial phenomena. Meanwhile, because companies are legally-allowed to minimize their costs in order to maximize profits through, for example, less expensive materials, energy and labour, as well as lack of costs of addressing harms linked to their products and services (e.g., cancer and heart disease), many for-profit commodities must be considered potentially, at least, harmful to individuals, societies and/or environments. Annie Leonard (2010) suggests, for example, that many manufactured household cleaning and hygiene products contain numerous untested potentially-hazardous chemicals. Similarly, it is apparent that professionals associated with pharmaceutical industries often compromise product quality—such as through testing them with young subjects, who are less likely to experience negative side-effects (Angell, 2004). To add to such consumerist concerns, some companies have been known to encourage engineers (in the private and public sector) to design products to quickly fail—perhaps leading consumers to purchase the latest innovations, sending many not-so-old products to landfills (Leonard, 2010).

While people are enthusiastically and unquestioningly consuming and discarding massive quantities of products and services, very small fractions of societies have been gaining increasing shares of wealth and wellbeing. According to a recent report for Oxfam (Hardoon, 2017), for instance, the richest 8 people in the world—such as Bill Gates, Warren Buffett and Carlos Slim—own more wealth than the roughly 3.6 *billion* people who make up the poorest half of the world's population. Moreover, this situation seems destined to get worse, according to Thomas Piketty's (2014) analyses of patterns in capitalism over at least the last century. Apart from hardships arising from lack of significant wages, many citizens of the world struggle through government policies of reduced taxation for rich citizens and corresponding reductions in social spending (Wolff, 2012). Exacerbating this problem, apparently, have been financial and legal arrangements allowing wealthy individuals and companies to avoid taxation by 'hiding' much of their wealth (at least $7.6 trillion, worldwide) in 'shell companies' in countries like British Virgin Islands, Luxembourg, Panama and Switzerland—leading to losses to governments in the order of 200 billion dollars (US) (Zucman, 2015). With such realities, it seems that a/the root of many harms to individuals, societies and environments can be attributed to the relatively few people capturing—with assistance from fields of science and technology (and others)—wealth for themselves.

33.3 Promotion of Critical and Activist Citizenship Through Science Education

In light of potential and realized harms linked to influences on fields of science and technology (and others) from governments, financiers, companies, transnational organizations, banks and others associated with intense profit concentration, many scholars, educators and others have recommended that—among a range of necessary responses to this situation—school science help prepare students to become (more) critical and activist citizens on matters of personal, social and environmental concern (e.g., Hodson, 2011; Dos Santos, 2009). As discussed in several chapters in this book, although curriculum mandates and educational research and development regarding 'STSE' (relationships amongst fields of science and technology and societies and environments) and socioscientific (relating fields of science to societies) issues have had relatively long and deep histories (Pedretti & Nazir, 2011; Sadler, 2011), it seems that many such practices stop short of encouraging students to engage in highly politicized analyses of STSE relationships and, moreover, where problems are discerned by them, to develop and implement plans of action to bring about what they perceive to be a better world (Hodson, 2011; Levinson, 2010, 2013).

This book, of course, provides readers with theory and practice surrounding 'STEPWISE' (Science & Technology Education Promoting Wellbeing for Individuals, Societies & Environments) frameworks—which greatly prioritize encouragement of teachers to help students to develop expertise, confidence and motivation for critically analyzing power relationships amongst fields of science and technology (and other, related, fields, like engineering & mathematics) and societies and environments and, where they perceive problems, to develop and implement plans of action to address them in ways they see fit. As described in Chap. 2 of this volume, teachers and others soon judged the original tetrahedral version of STEPWISE to be too impractical in contexts of formal schooling, preferring a more linear, 'deficit' (implying needs to help students overcome deficits in knowledge, skills, etc.), framework. With brief notes about each, these two versions of STEPWISE are shown in Fig. 33.1. Most authors in the Documentaries section of this book have provided examples of teaching/learning strategies based, to varying degrees, on the 'linear' version (upper right in Fig. 33.1) of STEPWISE, along with examples of student learning outcomes and rationale for their claims. Generally, such approaches have, indeed, enabled students to develop and enact research-informed and negotiated actions (RiNA) to address STSE problems/issues concerning/interesting them. As an example of such student-generated RiNA projects, a student of Mirjan Krstovic (Chap. 6, this volume) recently (2014-15 school year) chose to explore potential problems linked to a pharmaceutical company's apparent indiscretions—perhaps supported by government de-regulation—regarding quality control measures for generic (unpatented) drugs. As illustrated in Fig. 33.2, after reading about the controversy in a local newspaper (reading clockwise from the news article), she conducted some secondary research and developed an actor-network map to summarize it, then conducted studies among peers regarding their knowledge and understanding of drug company regulation. Afterwards, she used her findings to develop a video documentary (uploaded to YouTube™) that featured a pharmaceutical industry executive (a clip from which is shown in the lower-left of Fig. 33.2) answering questions about the controversy. Through this

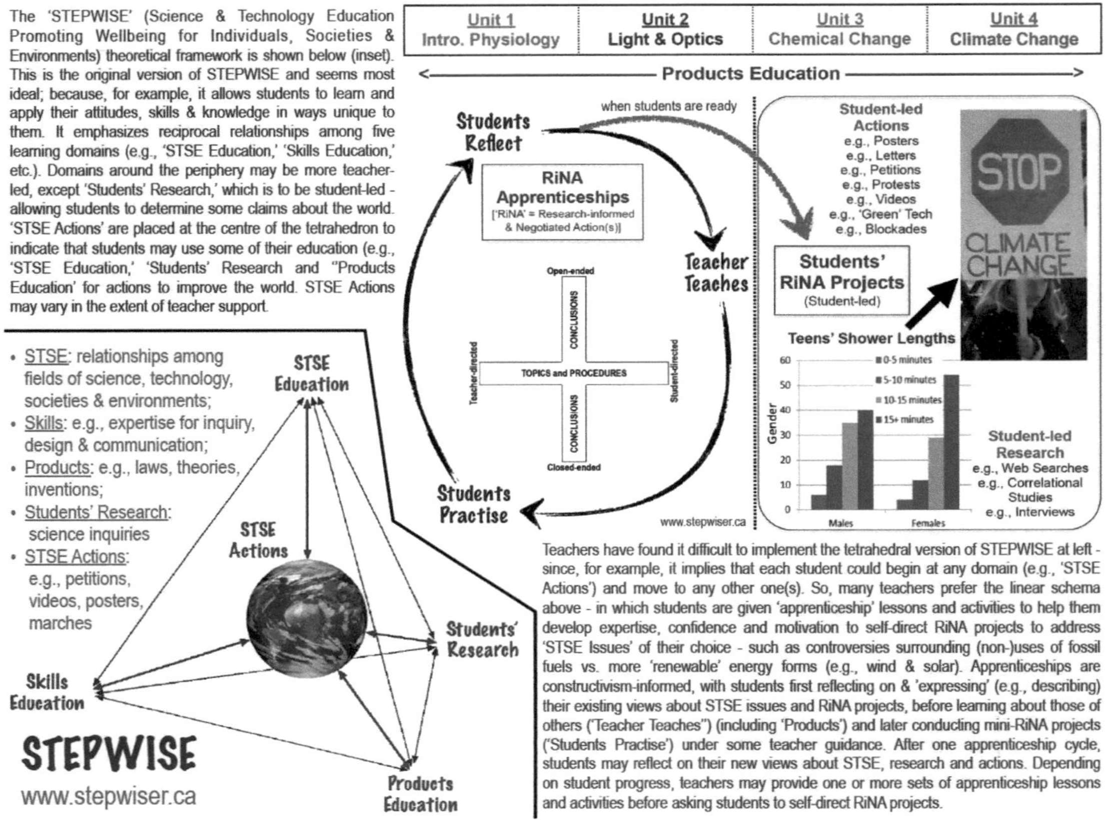

Fig. 33.1 STEPWISE theoretical and pedagogical frameworks

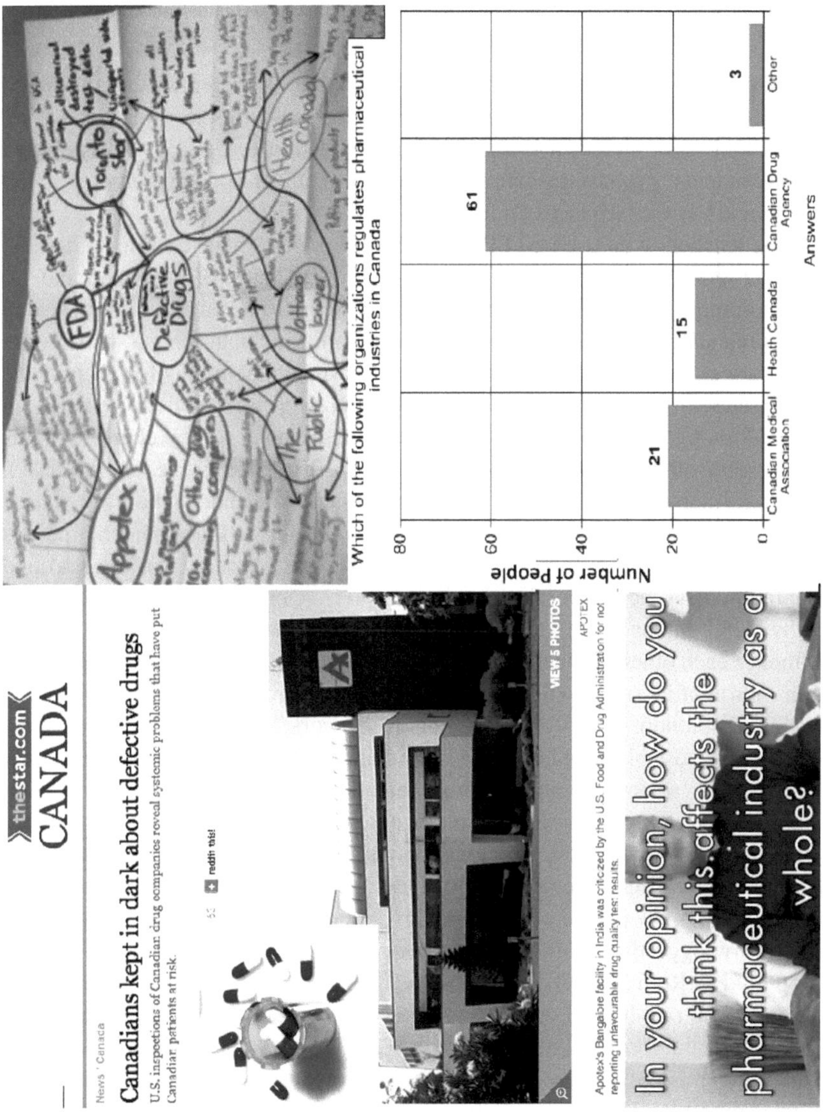

Fig. 33.2 Student RiNA project about drug testing

project, it seems clear this student had a relatively broad and deep understanding of apparent problems associated with the drug company in question:

> Apotex is able to change test results. First they produce the drugs overseas with poor working conditions, under paid employees, and minimal Health Canada inspections. Next, the drug is shipped to Canada and the records and/or ingredients are misplaced, or left behind. Once the drug arrives in Canada, new records are written that pass Health Canada's policies (Student RiNA Report, Jan. 19, 2015, p. 4).

Although perhaps using different approaches than that of apprenticeships depicted in Fig. 33.1, meanwhile, several chapters in the Commentaries section of this book provide examples of lessons and activities that, while not necessarily always based partly on students' primary research, culminated in student personal and/or social actions to address their concerns/interests.

Among teachers who used STEPWISE frameworks in their teaching, I had the longest and most direct contact with Mirjan Krstovic (Chaps. 6, 7, 8, 9, 10 and 11, 13, this volume). I worked with him, starting in Sept. 2011, as an action research facilitator regarding his various attempts at developing and field-testing with classes of secondary school students diverse lessons and student activities for each of the three apprenticeship phases (and student-ledRiNA projects) of the schema in Fig. 33.1. Although details of our collaborations are found in several chapters in this book and some other books (e.g., Bencze & Alsop, 2014; Bencze & Carter, 2015), the summary depicted in Fig. 33.3 may assist readers in understanding potential teacher pedagogies, student achievements and relevant action research findings (e.g., factors contributing to student achievements). This summary is framed around an adaptation of a schema for relationships between 'science' and 'technology' provided by Wolff-Michael Roth (2001). In this analysis, because 'World' and 'Sign' co-influence each other, distinctions between 'science' and 'technology' may not be so clear—prompting some to refer to an amalgamated process, perhaps called *technoscience* (Sismondo, 2008). Adapting this framework to the study of RiNA projects to address possibly-problematic STSE relationships, we suggest that *research* is analogous to 'science' (remembering dialectical processes), or World → Sign translations, and *action(s)* would be analogous to 'technology,' or Sign → World translations. We suspect both processes would involve negotiations (the 'N' in RiNA), even *within* individuals—acknowledging networked ontologies and social epistemologies. A key part of our thinking around using this schema to depict RiNA projects is my suggestion about existence of *ideological* gaps (along with *ontological* gaps reported by Roth (2001)). Such gaps are inconsistencies in translations between World and Sign, ontological ones being related to compositional differences (and, therefore, potential for translation) between World and Sign, while ideological gaps are *intentional* mistranslations. We suspect that both types of gaps would exist in the work of technoscientists, student activists, educators, educational researchers and others. However, as part of the *Teacher Teaches* phase of STEPWISE apprenticeships (Fig. 33.1), teachers might, for instance, enlighten students about possibilities of technoscientists representing phenomena (Signs) like climate change in ways suiting their interests and/or that of their financial sponsors (Kein, 2014; Oreskes & Conway, 2010). These might, in turn, motivate students to develop actions to address such gaps—such as the video for YouTube™ developed by a student to address

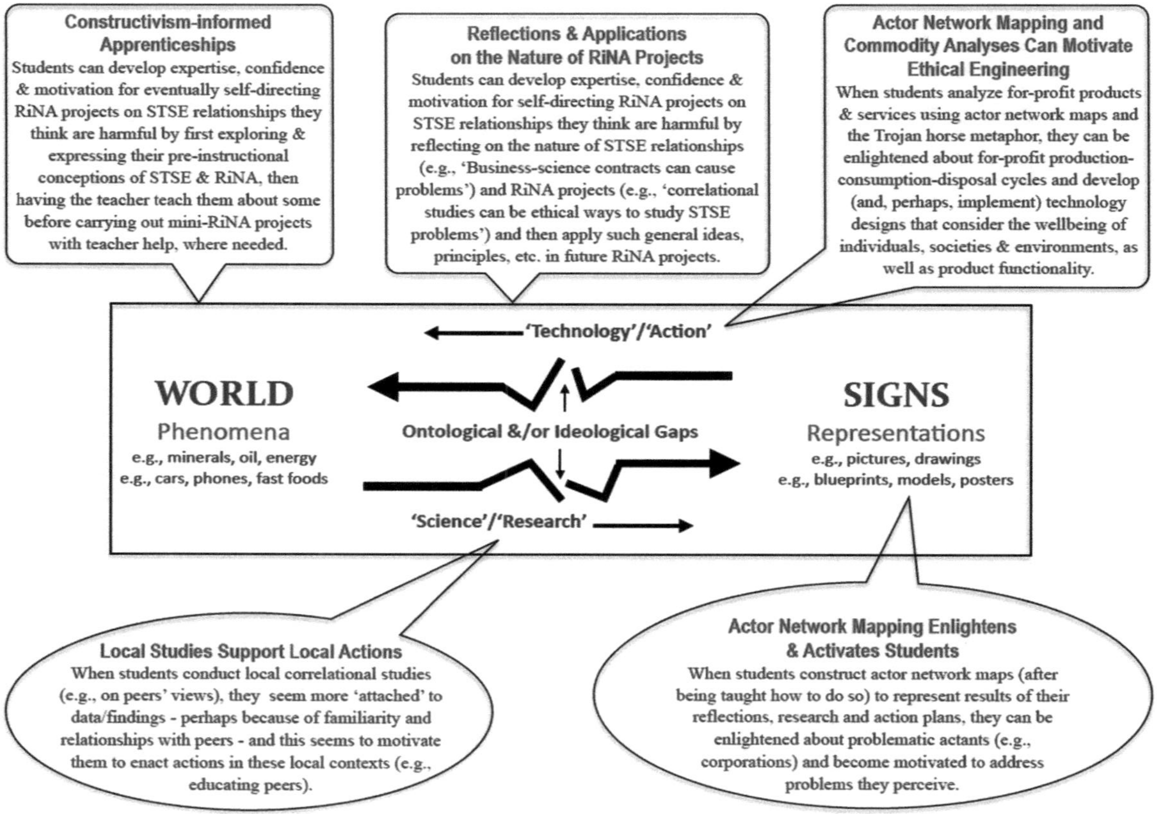

Fig. 33.3 Action research findings regarding RiNA projects

concerns surrounding questions about drug testing (World → Sign), as depicted in Fig. 33.2. In using the World ← → Sign dialectic as a basis for planning action research relating to STSE issues/problems and RiNA projects, however, it seems that we proceeded through various research and development foci in ways reminiscent of Jerome Bruner's (1960) concept of the *spiral curriculum*. In other words, we seemed to revisit perspectives and practices relating to the apprenticeship in Fig. 33.1 (upper right) at progressively more advanced levels over the six semesters (3 years) of our collaboration. We began 'simply,' exploring effectiveness of the 3-phase constructivism-informed apprenticeship—which we eventually felt should be repeated, at 'basic' (e.g., without reference to actor-network theory [ANT]) and 'advanced' (with reference to ANT) levels (see Krstovic, Chap. 6, this volume). Each semester, we explored influences of different aspects of these apprenticeships—starting with an investigation into promotion of student-led local correlational studies, which we found appeared to provide them with significant motivation for developing and implementing actions in local contexts (Bencze & Krstovic, Chap. 7, this volume). Perhaps the most significant findings relate, however, to helping students to understand and use actor-network theory (ANT) (Latour, 2005) as a basis for *depunctualizing* (Callon, 1991) commodities; that is, to help them gain awareness of many actants connected to the commodity that often are highly problematic, but unseen (punctualized). Using ANT, the student whose work was featured in Fig. 33.2 acknowledged, for example, roles for numerous actants, including: the 'FDA' (US Federal Drug Administration), 'Health Canada' (Canadian government health ministry), 'the public,' and 'destroyed test data.' Mirjan appeared to experience considerable success teaching students about ANT and to create actor-network maps using, for example, the metaphor of the Trojan horse (positive actants on the outside, harmful ones hidden inside) in association with discussions surrounding *The Story of Stuff* videos (storyofstuff.org/movies/)—which seem to be great aids to the punctualization-depunctualization construct.

33.4 Marginalization of STEPWISE

Although we have seen many wonderful examples of youth's research-informed and negotiated action projects to address problems/issues they perceive in relationships among fields of science and technology and societies and environments, including many that address diverse powerful and problematic actants, most science teachers, principals, science consultants and, to a great extent, science teacher educators we have contacted prefer not to use STEPWISE—or, even, many of its core principles—in their teaching. Generally, our experiences working with teachers, student-teachers and others suggest that many of them are very reluctant to include at least the following aspects of STEPWISE in their teaching:

- spending time exploring, reflecting upon and expressing their existing conceptions about STSE relationships and RiNA projects;
- using actor-network theory to analyze and depict STSE relationships;

- learning about power-related problems—particularly those dealing with capitalists' influences on scientists, engineers, etc., government officials and others—in STSE relationships;
- designing and self-directing research (generally);
- designing and conducting correlational and qualitative studies (vs. quantitative experiments);
- deriving their own (usually socially-negotiated with peers) conclusions from investigations;
- using data and conclusions from their research to inform planning of social actions;
- implementing plans of action to address STSE problems;
- using actor-network theory to plan or implement networked actions;
- taking actions to affect larger social change, such as through actions on powerful people and groups;
- presenting and receiving feedback about their RiNA projects in public fora.

It should also be noted that even Mirjan found that, perhaps largely due to his need to report student achievement based on teaching/learning 'expectations' defined by the government that placed STSE education in the last (from first) position in its assessment chart for the three curriculum goals (MoE, 2008, pp. 26–27), it was next-to-impossible to allow *fully* student-directed and open-ended RiNA projects (as recommended in the STEPWISE framework [Fig. 33.1]). Consequently, not unlike my experiences years ago in promoting student-led research (not consciously to be used for actions) (refer to Chap. 2, this volume), it is apparent that some sort of *invisible hand* (Smith, 1776/1976) is inhibiting the sorts of educational experiences listed above. Indeed, we arrived at a similar conclusion in one of our early studies—as we explored potential uses of the tetrahedral version of STEPWISE and realized that a more linear, perhaps less theoretically-sound, version of the framework (Fig. 33.1) was most feasible for many teachers. As described in Chap. 3 (Hoeg, Williamson & Bencze) in this volume, it appears there often are numerous structural barriers in school systems limiting the extent to which teachers can promote activities like those in the bulleted list above. In his Chap. (28) in this volume, Matthew Weinstein draws upon work of others to suggest that school science systems' reluctance to engage in complex and uncertain activities like those listed above seems related to so-called *grammar of schooling* that prioritizes instruction in discrete (reduced) packets of knowledge and skills that can be relatively-easily measured (e.g., through tests).

'Grammar of schooling' is a complex and uncertain concept and, indeed, after Jacques Derrida (1998), cannot be assumed to mean precisely any one thing. Such a claim also can be drawn from actor-network theory (Latour, 2005), which assumes any one 'thing' to be affected by and influencing dynamic networks of other entities ('actants'). Nevertheless, in light of arguments above, and supported by several scholars contributing to this book and elsewhere, much of education seems under significant influences of the global capitalist network. Peter McLaren (2000), for instance, 'dramatically' claimed that "the major *purpose* of education is to make the world safe for global capitalism" (p. 196; emphases added). Although all educational fields may be affected, it seems logical to assume science education may be crucial in contributing to capitalist causes—given importance of fields of science and technology in generating for-profit knowledge and inventions/innovations (Mirowski, 2011). Indeed, it is apparent that science education largely aims to identify and educate a few students who can serve as knowledge producers (e.g., scientists, engineers, etc.) and masses of students who, essentially, may function as knowledge consumers (Bencze & Carter, 2011). Students who may serve as knowledge producers tend to be able to quickly comprehend abstract knowledge that is rapidly-presented to them in decontextualized ways, abilities that can qualify them to provide companies, governments and others with *immaterial* labour (Reich, 2007); that is, expertise for analysing/manipulating symbols, including words, concepts, numbers and graphics, to develop and manage formulations (e.g., manufacturing & marketing) regarding commodities. At the same time, because of competitive environments to which students often are exposed, they are, essentially, being trained to take on *entrepreneurial* characteristics; that is, under pressure to be perpetually renewing their identities and competing for limited resources (Means, 2013). Meanwhile, large fractions of student populations also are prepared to serve as knowledge consumers—as compliant followers of labour instructions (e.g., from knowledge producers) and enthusiastic and unquestioning purchasers of commodities (Giroux & Giroux, 2006)—through, for example, being presented with relatively idealized conceptions of the nature of science (Hodson, 2008) that frequently involve omission of mention of problematic business-science contracts (Carter, 2005), and through teachers' intervention in students' decisions in empirical activities (Bencze & Alsop, 2009). In recent years, moreover, it is apparent that fast-emerging and widely-disseminated 'STEM' (science, technology, engineering & mathematics) education initiatives have, generally, prioritized identification and education of potential STEM workers—again often at the expense of students' critical understanding of problematic relationships among fields of science and technology (and engineering & mathematics) and societies and environments (Bencze, Reiss, Sharma, & Weinstein, in press; Gough, 2015). Overall, given various pressures on school science/STEM education, it seems that both social justice and environmental sustainability are being threatened. Such concerns have been captured in *ecojustice* education initiatives, many of which problematize fundamentals of current capitalist systems, including foci on continuous growth, competitiveness, individualized entrepreneurialism and perhaps less concern for matters of social justice and environmental sustainability (Martusewicz, Edmundson, & Lupinacci, 2015). With such threats, it seems it may be even more difficult than in the past to implement programmes like STEPWISE.

33.5 Towards Ecojust Dispositifs

There are, undoubtedly, numerous approaches that educators and others might take to encourage and enable more young people to develop expertise, confidence and motivation for critically evaluating practices and products of STEM fields (and others) and, where they identify concerns, take personal and social actions to address them. However, in light of neoliberal capitalists' successes in rallying many and varied actants cooperating in their causes, it may be that emulating such tactics—broadly—may be effective. There is, indeed, theoretical justification for this approach—in terms of *dispositif* concepts (Foucault, 2008)— discussed in chapters of this book. Derived from the French verb *disposer* (to arrange), a dispositif is an aggregate of actants to form a relatively stable apparatus that supports a particular purpose (Lazzarato, 2006). We have seen examples of this within and beyond formal schooling. In examining data from Chantal Pouliot's (Chap. 17, this volume) study of citizen actions to convince municipal authorities to address what citizens perceived to be airborne dispersal across their neighbourhoods of a toxic mixture of heavy metals (e.g., lead, arsenic, nickel and cobalt), it became apparent to us that there were at least two competing dispositifs relating to the dust situation in their city; that is, an 'activist' dispositif, consisting of some citizens, science data, appeals at city council meetings, two class-action law suits, an interactive website, a protest march, the dust, etc., and a 'development' dispositif, membership that appeared to include the mayor, some citizens, port authority members, street cleaning trucks, dust monitoring stations, the shipping company, the mining company (not local), etc. At the time of writing this chapter, this conflict continues—with the activist dispositif not satisfied with rectifying actions (e.g., sprayed dust piles at the port and increased cleaning of city streets, but continued refusal to cover the dust piles) of the development dispositif. It seems that the latter dispositif is much more powerful than the activist network. Having said that, results of the class-action suits are not yet known and, moreover, there appear to be precedents elsewhere for successes of other citizen movements—in light, for example, of successes of collectives taking over businesses that, under neoliberalism, had laid-off employees to maximize profits (Magnani, 2009).

Not only are there precedents for successes of oppositional dispositifs like those noted above, there is evidence of such phenomena in research relating to field-testing of STEPWISE approaches. Although many or most of the teachers featured in this book managed to rally many diverse actants enabling them to implement STEPWISE-informed pedagogical practices (e.g., as listed in bullets above) that seem antithetical to mainstream—apparently largely neoliberalism-influenced—practices, I have collected data relating to Mirjan Krstovic's case (Chap. 11, this volume) that seems aligned with methods of *institutional ethnography*, which often involves studies of individuals and data that may reveal hidden assumptions perhaps strongly-regulating their ways of thinking and acting (Smith, 2005). This seems to have allowed me to gain insights into ways in which Mirjan was, to a great extent, able to *teach against the grain* (Cochran-Smith, 1991). In this line of thinking, it seems that Mirjan's successes can largely be attributed to his participation in a dispositif that, together, appeared to be supportive of the kinds of practices listed in bullet-form above (under Marginalization of STEPWISE). Actants in this aggregate seemed to include, at least, the following: government curriculum sanctioning (MoE, 2008) of 'STSE' (relationships among fields of science and technology and societies and environments) education; Mirjan, as a high-energy teacher who was driven by desires for pedagogical exploration and equitable student achievement, who had a recent Masters of Education degree and whose views about science (*Naturalist-Antirealist* positions on Loving's (1991) *Scientific Theory Profile* [STP]) seemed congruent with more promotion of critical conceptions of STSE relationships; support from an educational researcher/facilitator (me); and, support from his school's principal and science department head for innovation in teaching and learning. In light of methods of institutional ethnography, however, it also seems that Mirjan was partly able to function as a relatively revolutionary teacher—for reasons beyond supportive factors itemized just above—due to his awareness of and possible shifts in his *ontological* perspectives. Ontology is any study of the nature of existence. Actor-network theory (ANT) is considered an ontological conception, positing reciprocal relationships among living, nonliving and semiotic entities (Latour, 2005). Perhaps partly through his graduate studies, although this is hard to determine precisely, Mirjan appeared to hold ontological conceptions of, for instance, his status as a teacher in relation to other teachers:

> I was known [in his previous school] for some of my radical approaches and perhaps some anti-conformist ideas (e.g., challenging the traditional 'concepts' only approach and test-teach-test method) in aspiration of a more 'balanced' science curriculum [i.e., addressing all 3 learning goals [MoE, 2008] (Feb. 26, 2013).

Such views extended, apparently, to his conceptions of societies more generally:

> You can probably argue that the level of control of the capitalist system does not exist, or [at least] not very much. ... We live in a capitalist system controlled by a very few. We are subjects of that system. ... You call it a democracy, but I cannot protest [as a teacher, under a current law] any more (Interview, Jan. 14, 2013).

Finally, his ontological conceptions seemed relatively congruent with his views about human knowledge generation (epistemology):

> I believe that construction of knowledge depends more on socio-cultural contexts rather than strictly adhering to Merton's institutional imperatives and pure logic. I support the idea that one's psychological make up would interact with one's logical reasoning, thus affecting one's judgment. In addition, I hold the position that reaching the 'truth' about knowledge is a matter of consensus amongst professionals and that strict scientific methods may not necessarily lead us to the truth about laws and theories governing physical and natural phenomena (Written Statement About Science, Dec. 12, 2012).

In a sense, we could say that, through various means, Mirjan was able to depunctualize (refer above) (Callon, 1991) many aspects of his work that often may be punctualized for other educators—perhaps serving as 'ruling relations' governing (to some extent) their professional perspectives and practices. Consequently, he was, apparently, able to contribute to formation of a dispositif supporting many (although not all) STEPWISE-informed perspectives and practices, despite opposition from its antithesis.

In light of arguments and examples above, it seems logical to suggest that mobilization of STEPWISE-informed perspectives and practices (perhaps as *ecojustice* goals) across more educational contexts may be achievable through concerted efforts to form dispositifs that cooperate to challenge neoliberal dispositifs that appear to dominate many societies. The exact composition of such an ecojust dispositif cannot, of course, be predicted or, even, recommended. As suggested by Gilles Deleuze and Félix Guattari (1987), paths of decisions are generally and perhaps should be more *rhizomatic*, acknowledging very dynamic characteristics of interactions among diverse actants. On the one hand, in line with discussions above such as that involving needs to depunctualize often-hidden possibly-problematic actants/networks, it seems clear there must be some/significant *conscientization*; while, on the other hand, given that all claims are biased and uncertain, more democratic decision-making would involve significant *praxis* (McLaren, 2000). In engaging in such critical reflective practice, perhaps contributing to conscientization, one can imagine several relevant 'stakeholders,' including: teachers; student-teachers; teacher educators; educational researchers; school and school district administrators (e.g., superintendents and principals); government education officials; business members; STEM workers (e.g., miners, technicians and engineers regarding cell phones); politicians; media personnel; etc. In thinking about organizing and facilitating negotiations among such diverse stakeholders, it seems *pragmatic*—if not as rhizomatic as desired—to imagine some more structured arrangement of them. As illustrated in Fig. 33.4, one such configuration is the *tetrahedron* that, of course, is the form of the original version of STEPWISE (Chap. 2, this volume). Again, while there are many possible arrangements one can imagine for facilitating critical reflective practices among diverse science/STEM education stakeholders, the schema in Fig. 33.4 places wellbeing for individuals, societies and environments (WISE), which may be considered related to ecojustice goals, at its geometric centre. All entities (e.g., action research teams, schools, etc.) would be in dynamic, reciprocal, tension with all the others. However, placing wellbeing in the 'centre' (acknowledging the dynamic nature of relationships) may help to prioritize ecojustice goals in minds of participants—who might be encouraged to regularly check various indicators, such as the *Canadian Index of Wellbeing* (uwaterloo.ca/canadian-index-wellbeing/). In applying this framework, we can imagine educational researchers organizing formation of action research teams, which would initiate various reciprocal relationships among other entities like those noted in Fig. 33.4. Other entities in the framework could, instead (or perhaps in parallel), initiate such a dynamic network. Of course, progress of the dispositif that emerges from such eclectic interactions cannot be predicted. Having said that, it may be that new aggregates evolve to co-support ecojustice goals if initiators make that a priority from the outset.

Authors of the Commentaries section of this book and some in the Documentaries section have provided theoretical lenses and examples that could be infused into an emerging dispositif that might start with a more structured framework like that in Fig. 33.4. Most, if not all, authors lent support to goals and, to some extent, approaches of STEPWISE. Clayton Pierce (Chap. 20), for instance, discussed similarities between STEPWISE and an after-school project in which science students investigated matters of food justice through experiences with a community garden and seemed to lead to educational actions regarding fair and sustainable food practices. His chapter also supports uses of actor-network mapping, an emphasis I also

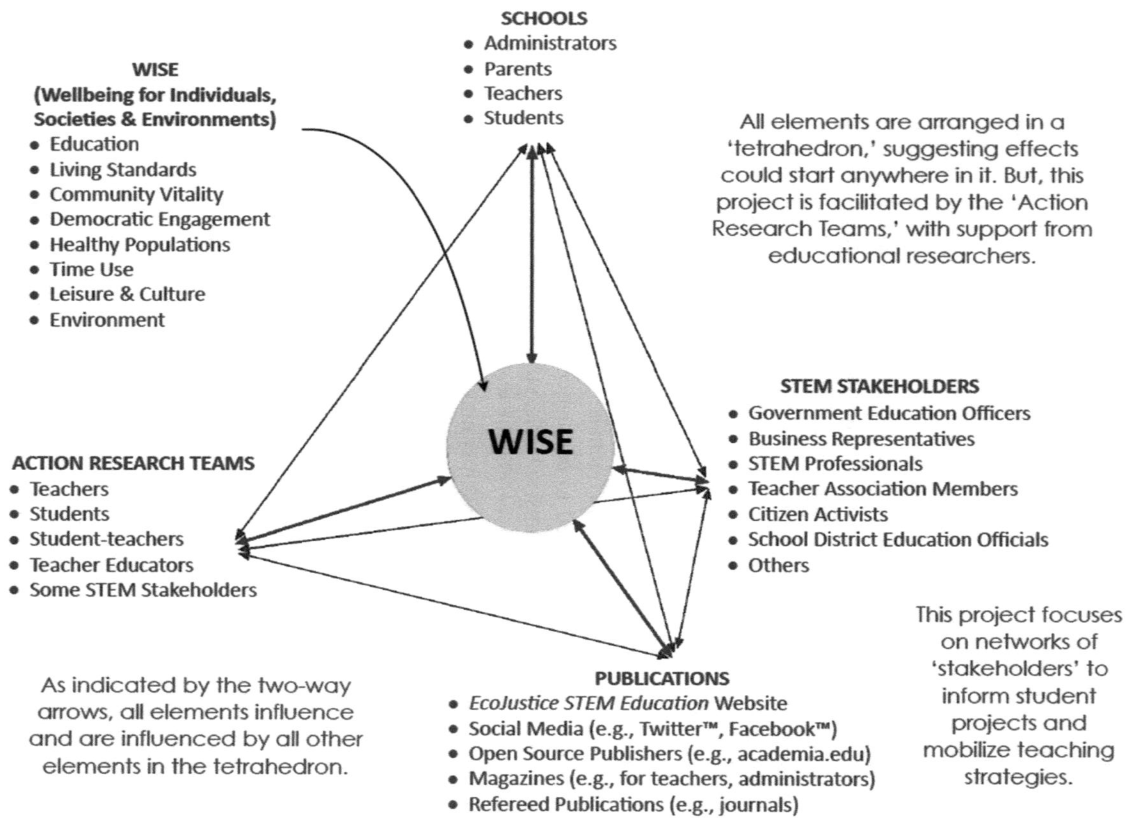

Fig. 33.4 Action research for mobilizing EcoJust STEM actions

gleaned from his book, *Education in the Age of Biocapitalism* (Pierce, 2013). Having said that, many of us supporting actor-network theory have much to consider in reading Neil Ramjewan's (with Brandon Zoras) critical analyses of this theory (Chap. 12). Cassie F. Quigley (Chap. 21), meanwhile, while writing about environmental (rather than science) education, describes motivation students gained through self-directed research that seemed to inspire more-sustainable actions regarding a local controversy surrounding pesticides uses. In these and other chapters, although not always specifically-mentioning it, authors also seemed to support the concept of de-punctualization of phenomena (Callon, 1991)—enlightening learners to various often-hidden problematic actants associated with common phenomena (e.g., food and insects). Such a process is, actually, mentioned in the Chap. (26) written by Audrey Groleau and Chantal Pouliot—in which they describe how the game, *Decide*, provides players with insights into a number of often-unseen problematic entities relating to various socioscientific issues. In support of their analyses of the Decide game, and perhaps in response to Mellita Jones' (Chap. 23) reluctance to appear to be "preaching" to her students about possibly-problematic actants, an argument in favour of such pro-active stances by educators is that some critical de-punctualization may be necessary to counter highly-pervasive advertising to children—at, apparently, increasingly younger ages—by private sector agents (Bakan, 2011).

Although authors have provided examples of ways in which STEPWISE-type programmes have been implemented, and authors like Clayton Pierce (Chap. 20) and others stress needs (perhaps now more than ever) for programmes like STEPWISE as alternatives to mainstream neoliberalism-influenced science/STEM education perspectives and practices, there are many suggestions among chapters in this book indicating struggles to do so. Melitta Jones (Chap. 23), for instance, outlines how she gradually accommodated many ideas and approaches inherent to STEPWISE. As a perhaps unintentional response to such struggles, meanwhile, Ralph Levinson (Chap. 22) describes successes that participants in the European project, *Socio-Scientific Inquiry Based Learning*, seemed to have had by taking a 'stepwise' approach to programmes like STEPWISE—in the sense of linking their perspectives and practices to a very common movement (in Europe and elsewhere); namely, inquiry-based learning (IBL). This is a movement about which I have significant reservations (Bencze & Alsop, 2009); yet, meeting, in a sense, educators 'where they're at' may, indeed, be pragmatic.

In implementing programmes like STEPWISE, several authors provided some significant theory-informed cautions. It is difficult to assemble their ideas into a coherent story, not the least because many of them draw from Gilles Deleuze and Félix Guattari (1987)—reminding us that trajectories of any programme are likely to and, moreover, *should* be unpredictable. In that regard, indeed, Jesse Bazzul and Shakhnoza Kayumova (Chap. 30) provide an interesting set of arguments around relative merits of programmes like STEPWISE that purposely aim to create *politicized subjects*—a goal that may not well-align with poststructural perspectives. Along such lines, Laura Colucci-Gray (Chap. 25) urges us to use another of Deleuze and Guattari's (1976) metaphors; that is, that of the *nomad*, being and becoming through eclectic interactions with various actants. In such journeys, these and other authors in this book allude to various complexities participants may encounter. Isabel Martins (Chap. 29) discusses various—not often certain—ways that we might consider the concept of wellbeing as a curricular and societal goal. Similarly, several authors advise educators and others to be careful when thinking of 'activism.' Matthew Weinstein (Chap. 28), for instance, suggests that the common STEPWISE practice of focusing students' attentions on critical examinations of commodities they might purchase may miss attending to some perhaps less immediate and obvious 'targets' of actions, such as uses of drones as means for surveillance and control by powerful individuals and groups. Along similar lines, Ajay Sharma (Chap. 31) suggests that, while *educational* actions developed and implemented by students directly involved in STEPWISE-informed pedagogical practices are positive, more efforts may be needed to encourage them to more-directly challenge larger systems of power. In taking actions, Matthew Weinstein also wisely recommends that students might benefit from being linked to community activists—who often have deep histories in various issues/problems. Laurence and Jean Simonneaux (Chap. 27), meanwhile, caution—among their many, rich, theoretical contributions—that actions need to strike some balances between rational and emotional justifications. Finally, again, while perhaps not intending to address such cautions, some authors here place considerable emphases on the inevitable situated nature of decision-making in such work. Lyn Carter and Jenny Martin (Chap. 24), for instance, draw on discursive psychology for this purpose. Wolff-Michael Roth (Chap. 32), meanwhile, stresses that many of the concepts and practices inherent to STEPWISE—including its apparent relatively-narrow focus on science and technology education and its communitarian goals—need to be continually treated as part of dynamic and unpredictable large-scale systems, which he effectively analyzes in terms of cultural-historical activity theory. Overall, it seems clear to see that authors of this book have provided readers with many theoretically and experientially strong perspectives and practices to consider in initiating programmes of ecojustice education.

References

Angell, M. (2004). *The truth about the drug companies: How they deceive us and what to do about it*. New York: Random House.

Bakan, J. (2011). *Childhood under siege: How big business targets children*. Toronto, ON: Allen Lane.

Ball, S. J. (2012). *Global Education Inc.: New policy networks and the neo-liberal imaginary*. Abingdon, UK: Routledge.

Barber, B. R. (2007). *Consumed: How markets corrupt children, infantilize adults, and swallow citizens whole*. New York: Norton.

Bencze, J. L., & Alsop, S. (2009). A critical and creative inquiry into school science inquiry. In W.-M. Roth & K. Tobin (Eds.), *The world of science education: North America* (pp. 27–47). Rotterdam, The Netherlands: Sense.

Bencze, J. L., & Alsop, S. (Eds.). (2014). *Activist science & technology education*. Dordrecht, The Netherlands: Springer.
Bencze, J. L., & Carter, L. (2015). Capitalists' profitable virtual worlds: Roles for science & technology education. In P. P. Trifonas (Ed.), *International handbook of semiotics, vol. 1 & 2* (pp. 1197–1212). Dordrecht, The Netherlands: Springer.
Bencze, L., & Carter, L. (2011). Globalizing students acting for the common good. *Journal of Research in Science Teaching, 48*(6), 648–669.
Bencze, L., Reiss, M., Sharma, A., & Weinstein, M. (in press). STEM education as 'Trojan horse': Deconstructed and reinvented for all. In L. Bryan & K. Tobin (Eds.), *Thirteen questions in science education* (pp. xx–xx). New York: Peter Lang.
Bruner, J. S. (1960). *The process of education*. Cambridge, MA: Harvard University Press.
Callon, M. (1991). Techno-economic networks and irreversibility. In J. Law (Ed.), *A sociology of monsters: Essays on power, technology and domination* (pp. 132–161). London: Routledge.
Carter, L. (2005). Globalisation and science education: Rethinking science education reforms. *Journal of Research in Science Teaching, 42*(5), 561–580.
Cochran-Smith, M. (1991). Learning to teach against the grain. *Harvard Educational Review, 61*(3), 279–310.
Deleuze, G., & Guattari, F. (1987). *A thousand plateaus: Capitalism and schizophrenia* (B. Massumi, Trans.). Minneapolis, MN: University of Minnesota Press.
Deleuze, G., & Guattari, F. (1976). *Rhizome: Introduction*. Paris, France: Editions de Minuit.
Derrida, J. (1998). *Of grammatology*. Baltimore: John Hopkins University Press.
Dos Santos, W. L. P. (2009). Scientific literacy: A Freirean perspective as a radical view of humanistic science education. *Science Education, 93*(2), 361–382.
Foucault, M. (2008). *The birth of biopolitics: Lectures at the Collége de France, 1978–1979* (M. Senellart, Ed.). New York: Palgrave Macmillan.
Giroux, H. A., & Giroux, S. S. (2006). Challenging neoliberalism's new world order: The promise of critical pedagogy. *Cultural Studies ↔ Critical Methodologies, 6*(1), 21–32.
Gough, A. (2015). STEM policy and science education: Scientistic curriculum and sociopolitical silences. *Cultural Studies of Science Education, 10*(2), 445–458.
Hardt, M., & Negri, A. (2009). *Commonwealth*. Cambridge, MA: Belknap Press of Harvard University Press.
Hardoon, D. (2017). *An economy for the 99%: It's time to build a human economy that benefits everyone, not just the privileged few*. Oxford, UK: Oxfam International.
Harvey, D. (2005). *A brief history of neoliberalism*. Oxford, UK: Oxford University Press.
Hedges, C. (2009). *Empire of illusion: The end of literacy and the triumph of spectacle*. New York: Nation Books.
Hodson, D. (2008). *Towards scientific literacy: A teachers' guide to the history, philosophy and sociology of science*. Rotterdam, The Netherlands: Sense.
Hodson, D. (2011). *Looking to the future: Building a curriculum for social activism*. Rotterdam, The Netherlands: Sense.
Klein, N. (2014). *This changes everything: Capitalism and the climate*. Toronto, ON: Simon & Schuster.
Latour, B. (2005). *Reassembling the social: An introduction to actor-network-theory*. Oxford, UK: Oxford University Press.
Lazzarato, M. (2006). From biopower to biopolitics. *Tailoring Biotechnologies, 2*(2), 9–19.
Lemke, T. (2011). *Biopolitics: An advanced introduction* (E.F. Trump, Trans.). New York: New York University Press.
Leonard, A. (2010). *The Story of Stuff: How our obsession with stuff is trashing the planet, our communities, and our health – And a vision for change*. New York: Free Press.
Levinson, R. (2010). Science education and democratic participation: An uneasy congruence? *Studies in Science Education, 46*(1), 69–119.
Levinson, R. (2013). Practice and theory of socio-scientific issues: An authentic model? *Studies in Science Education, 49*(10), 99–116.

Loving, C. C. (1991). The scientific theory profile: A philosophy of science model for science teachers. *Journal of Research in Science Teaching, 28*(9), 823–838.
Magnani, E. (2009). *The silent change: Recovered businesses in Argentina*. Buenos Aires, Argentina: Teseo.
Martusewicz, R., Edmundson, J., & Lupinacci, J. (2015). *Ecojustice education: Toward diverse, democratic, and sustainable communities* (2nd ed.). New York: Routledge.
McLaren, P. (2000). *Che Guevara, Paulo Freire, and the Pedagogy of the Revolution*. Lanham, MD: Rowman & Littlefield.
McMurtry, J. (2013). *The cancer stage of capitalism: From crisis to cure*. London: Pluto.
Means, A. (2013). Creativity and the biopolitical commons in secondary and higher education. *Policy Futures in Education, 11*(1), 47–58.
Ministry of Education [MoE]. (2008). *The Ontario curriculum, grades 9 and 10: Science*. Toronto, ON: Queen's Printer for Ontario.
Mirowski, P. (2011). *Science-mart: Privatizing American science*. Cambridge, MA: Harvard University Press.
NCD Risk Factor Collaboration (NCD-RisC). (2016). Trends in adult body-mass index in 200 countries from 1975 to 2014: A pooled analysis of 1698 population-based measurement studies with 19.2 million participants. *The Lancet, 387*(10026), 1377–1396.
Oreskes, N., & Conway, E. (2010). *Merchants of doubt*. London: Bloomsbury Press.
Pedretti, E., & Nazir, J. (2011). Currents in STSE education: Mapping a complex field, 40 years on. *Science Education, 95*(4), 601–626.
Pierce, C. (2013). *Education in the age of biocapitalism: Optimizing educational life for a flat world*. New York: Palgrave Macmillan.
Piketty, T. (2014). *Capital in the twenty-first century*. Cambridge, MA: Cambridge University Press.
Reich, R. B. (2007). *Supercapitalism: The transformation of business, democracy, and everyday life*. New York: Knopf.
Roth, W.-M. (2001). Learning science through technological design. *Journal of Research in Science Teaching, 38*(7), 768–790.
Sadler, T. (Ed.). (2011). *Socio-scientific issues in the classroom: Teaching, learning and trends*. Dordrecht, The Netherlands: Springer.
Sismondo, S. (2008). Science and technology studies and an engaged program. In E. J. Hackett, O. Amsterdamska, M. Lynch, & J. Wajcman (Eds.), *The handbook of science and technology studies* (3rd ed., pp. 13–31). Cambridge, MA: The MIT Press.
Smith, A. (1776). *An inquiry into the nature and causes of the wealth of nations* (p. 1976). Oxford, UK: Oxford University Press.
Smith, D. E. (2005). *Institutional ethnography: A sociology for people*. Toronto, ON: AltaMira Press.
Springer, S., Birch, K., & MacLeavy, J. (Eds.). (2016). *The handbook of neoliberalism*. New York: Routledge.
Weber, K. (Ed.). (2009). *Food Inc.: How industrial food is making us sicker, fatter, and poorer – And what you can do about it, A participant media guide*. New York: Public Affairs.
Wolff, R. D. (2012). *Democracy at work: A cure for capitalism*. Chicago: Haymarket Books.
Zucman, G. (2015). *The hidden wealth of nations: The Scourge of tax havens* (T.L. Fagan, Trans.). Chicago: University of Chicago Press.

Index

L. Bencze (ed.), *Science and Technology Education Promoting Wellbeing for Individuals, Societies and Environments*, Cultural Studies of Science Education 14,
DOI 10.1007/978-3-319-55505-8

F

G

H

T

MIX
Papier aus verantwortungsvollen Quellen
Paper from responsible sources
FSC® C105338

FSC
www.fsc.org

If you have any concerns about our products,
you can contact us on
ProductSafety@springernature.com

In case Publisher is established outside the EU,
the EU authorized representative is:
Springer Nature Customer Service Center GmbH
Europaplatz 3, 69115 Heidelberg, Germany

Printed by Libri Plureos GmbH
in Hamburg, Germany